P9-DWT-290

THE DISTRIBUTION AND ABUNDANCE *of* ANIMALS

About the authors . . .

H. G. ANDREWARTHA is reader in zoölogy at the University of Adelaide, Australia. He has carried out extensive research in many different areas on the Australian continent and has worked in English research laboratories. L. C. BIRCH, reader in zoölogy at the University of Sydney, Australia, has an impressive background of research experience in English, American, South American, and Australian laboratories.

Looking back, I think it was more difficult to see what the problems were than to solve them.

DARWIN

THE DISTRIBUTION AND ABUNDANCE *of* ANIMALS

By

H. G. ANDREWARTHA
Department of Zoology, University of Adelaide

and

L. C. BIRCH
Department of Zoology, University of Sydney

THE UNIVERSITY OF CHICAGO PRESS · CHICAGO · ILLINOIS

THE UNIVERSITY OF CHICAGO PRESS, CHICAGO & LONDON

 Published 1954. Second Impression 1961. Composed and printed by THE KINGSPORT PRESS, INC., *Kingsport, Tennessee, U.S.A.*

54-13016

To

JAMES DAVIDSON

In 1926 when James Davidson left Rothamsted Experimental Station to take up his appointment to the newly-formed Waite Agricultural Research Institute in the University of Adelaide he was expert in insect morphology and he was an acknowledged authority on the taxonomy of the Aphidae. On arriving in Australia he was quick to appreciate the almost unique opportunity which this country offers for the study of ecology. He gave up active work in morphology and taxonomy and devoted himself wholeheartedly to ecology. He worked in this new field with vision and with great integrity until his death in 1945. Andrewartha joined him in 1932; and Birch worked with him from 1940 to 1945. In dedicating this book to the memory of the late Professor James Davidson we gladly acknowledge him as a pioneer in the field in which we have worked.

Acknowledgments

WHEN this book was first being planned, we discussed our ideas with a number of our colleagues, to whom we are grateful for the privilege of sharpening our wits on theirs; especially we have to thank Dr. T. O. Browning, of the Waite Agricultural Research Institute, University of Adelaide, and Mr. J. Le Gay Brereton, C.S.I.R.O., Section of Wildlife Management, Canberra, for help received at this stage. Parts of the manuscript were read and criticized by a number of our colleagues. We mention especially Professor J. A. Moore, of Columbia University, New York, who read chapters 2, 6, 7, 10, and 16; Professor Th. Dobzhansky, of Columbia University, New York, who read chapters 15 and 16; Dr. J. M. Rendel, of the C.S.I.R.O., Genetics Section, Sydney, who read chapter 15. To research students and colleagues in our own laboratories we are grateful for valuable discussions and suggestions, particularly to Mr. L. Barton-Browne, Mr. M. A. Bateman, Miss M. A. Besly, and Mr. P. E. Madge. We are especially grateful for the generous encouragement and helpful advice given us from time to time by Professor Thomas Park, of the Zoölogy Department, University of Chicago. It is a pleasure to acknowledge this assistance from our colleagues, but none of them is to be held responsible for anything in the book, especially as we have not always followed their advice on particular issues.

The whole of the text was read and corrected by Miss H. M. Brookes. All the original line drawings were made by Mrs. P. E. Madge. Miss C. M. Hill helped to prepare the indexes and to correct proofs. Mrs. H. G. Andrewartha helped in these matters also, and with criticism at all stages in the preparation of the manuscript.

We are grateful to the editor of *Biological Reviews* for permission to reuse substantial parts of an article on diapause by H. G. Andrewartha, which was published in *Biological Reviews*, Vol. **27** (1952). To Methuen and Company of London we are indebted for permission to use the quotation from *King Solomon's Ring* by K. Z. Lorenz at the head of chapter 12. We are indebted to the editor of the *Journal of Animal Ecology* for permission to reuse Figures 13.03, 13.04, 13.05, and 13.06. Dr. K. H. L. Key, of the C.S.I.R.O., Division of Entomology, Canberra, kindly lent us the original map for Figure 12.07. We also acknowledge a number of quotations from *Animal Ecology* (London: Sidgwick & Jackson, 1927), by Charles Elton, and from *Butterflies* (London: Collins, 1945), by E. B. Ford. We have quoted the words of a number of other writers, and many of our illustrations have been adapted from those published by other authors. Full acknowledgment of all this material has been made at the appropriate places in the text.

Contents

PART I. INTRODUCTION

PART II. PHYSIOLOGICAL ASPECTS OF ECOLOGY

PART III. ANALYSIS OF ENVIRONMENT

PART V. GENETIC ASPECTS OF ECOLOGY

BIBLIOGRAPHY AND INDEXES

PART I

Introduction

CHAPTER 1

The Scope of Ecology

In solving ecological problems we are concerned with what animals do *in their capacity as whole, living animals, not as dead animals or as a series of parts of animals. We have next to study the circumstances under which they do these things, and, most important of all, the limiting factors which prevent them from doing certain other things. By solving these questions it is possible to discover the reasons for the* distribution and numbers of animals in nature.

ELTON (1927, p. 34)

1.0 THE TWO SORTS OF ECOLOGY

IN ITS full context the quotation which heads this chapter emphasized the difference between the older disciplines of anatomy, histology, taxonomy, physiology, etc., and the newer one of ecology, in which, as Elton said, we have to study the living animal in the circumstances in which it is found in nature. Our purpose, in doing this, is to explain why a certain kind of animal is found in certain areas but not in others; why they are numerous in one place but scarce in another; why they are more numerous this year than they were last; and so on. This problem has been tackled in two very different ways.

It was appreciated from the very first that the same, or a similar, group of species is likely to be found in the same sort of "habitat" (see sec. 2.2 for the meaning of "habitat"). So it became popular to study these communities of animals; and to many it has seemed as if this were the very essence of ecology. In order to help with the study of communities, Elton (1927, p. 63) used the term "niche."

The best way to indicate briefly the scope of community ecology is to explain the meaning of "niche." In the paragraph in which he used this term Elton said that all communities of plants and animals have a similar ground-plan; they all have their herbivores, carnivores, and scavengers. In a wood there may be certain caterpillars which eat the leaves of trees, foxes which hunt rabbits and mice, beetles which catch springtails, and so on. "It is convenient to have some term to describe the status of an animal in its community, to indicate what it is *doing* and not merely what it looks like, and the term used is 'niche.' . . . The 'niche' of an animal means its place in the biotic environment, *its relations to food and enemies*." Thus the caterpillar and the mouse broadly occupy similar niches because they eat plants; the fox and the ladybird also fit broadly into similar niches because they both eat other animals in the community.

There are many papers in technical journals which report studies of communities of animals without stressing, or even mentioning, the niches which the different sorts occupy. At their worst, these may be mere descriptive lists; but the better ones usually report more or less quantitatively the relative numbers of the different sorts of animals. If the methods of the plant sociologists are followed—and they often are—the different species may be classified according to their "dominance" in the community (Kontkanen, 1948, 1949).

There is a difference of opinion about the meaning of "community"; some would stress the "ecological relationship" between species, whereas others would agree that any assemblage of species which is usually found in the same "habitat" is a community (Kontkanen, 1950). The "habitat" is a subjective concept, because the boundaries must be arbitrarily fixed by the student (sec. 2.2). The disagreement about the constitution of a community is academic and not important in practice, because the enormous task of unraveling the ecological relationships in even a simple community has usually proved impracticable. Instead, we are given descriptive accounts of all the animals that may be found in a certain "habitat." Or, if this prove to be too ambitious, the study may be restricted to a taxonomic group, perhaps a family or an order. Characteristic titles for papers in this field are: "Analysis of the Animal Community in a Beech Forest"; "An Ecological Study of the Saltatoria of Point Pelee"; or "The Tree-Hole Habitat with Emphasis on the Pselaphid Beetle Fauna."

These studies are justified by the hope that collectively they will discover the laws governing niches, that is, the relationships between the members of communities. Progress has been summarized from time to time, and the reader will know books by Pearse (1926), Elton (1927), Clements and Shelford (1939), Allee *et al.* (1949), Dice (1952), and others. All will agree that there is much more to be learned in this field. Nevertheless, these studies have been going on long enough now for us to be fairly sure that, no matter how mature the science of "community ecology" may become, it is not likely to give rise to a satisfactory, or even to any, general theory about "the distribution and numbers of animals in nature." There are two reasons for this. When too much emphasis is put on the community, too little attention is paid to the species whose distribution and abundance have to be explained. The distribution and abundance of a species cannot be explained by studying only its relations with the plants and animals in its "community." There are certain other important components of environment which also require to be considered.

But there are other zoölogists who have a practical and urgent interest in the distribution and abundance of animals. These are the ones who are concerned with the insects which may harm crops or livestock, or the vertebrates which are valued for their flesh or fur or even for the fun of shooting them. Papers written by these men have such a wide range of titles that it is not possible to select characteristic ones. Only occasionally does one meet the word "ecology." When it occurs, it is used in a different context. "Ecological Studies of *Eutettix tenellus*" and "Ecology and Management of *Zenaidura macroura*" are two examples. Note that these titles imply that the ecology of a certain species has been investigated, not the "ecology" of a forest or a lake.

The two meanings are so distinct that the time may have come to give them different names. We cannot use "synecology" and "autecology" in this context. The meanings of these words have become attenuated (Chapman, 1931, p. 5; Allee *et al.*, 1949, pp. 48, 227); but in none of their meanings do they discriminate between that sort of ecology which leads to an explanation of the distribution and abundance of animals and that sort which describes the relationships of members of communities. In French a distinction is made between "la biocénotique" and "l'écologie"; and Kontkanen (1950, p. 9) suggested that "biocoenotics" might be used in all languages to cover the study of communities. If a new name is needed, this one seems to recognize the real cleavage which has developed in methods and knowledge, and it is etymologically satisfactory. But perhaps it would be best merely to speak of the "ecology" of a certain species or the "ecology" of a certain community.

The literature dealing with species of harmful insects and useful fishes, birds, and mammals is enormous. It contains, if one is prepared to search deeply, ample material from which to build a wide and satisfying general theory of ecology as we use this word to refer to the distribution and abundance of animals in nature. This has not hitherto been done; that is why we decided to write this book.

1.1 THE REASONS WHY DISTRIBUTION AND ABUNDANCE SHOULD BE REGARDED AS DIFFERENT ASPECTS OF THE SAME PROBLEM

In studying the ecology of an animal, we seek to answer the questions: Why does this animal inhabit so much and no more of the earth? Why is it abundant in some parts of its distribution and rare in others? Why is it sometimes abundant and sometimes rare? These are all problems of distribution and abundance.

The concept of *distribution* is well understood by naturalists. The distribution of a species coincides with the broad geographic limits inside which the species may be found more or less permanently established. It has become customary to separate distribution from abundance, although Elton (1927) was careful to avoid this error. However necessary this abstraction may be as a methodological device, the separation should never be allowed to persist in the final synthesis, for distribution and abundance are but the obverse and reverse aspects of the same problem. This becomes quite clear when we observe closely the way animals are actually distributed in nature. Inside the distribution there may be favorable zones where a high level of abundance is maintained; but near the limits of the distribution there may be a marginal zone which is sometimes inhabited and sometimes not and which, in general, is a zone characterized by low numbers. Outside the distribution there are none. This interaction between distribution and abundance may be illustrated by the following example, which is based on the ecology of a certain grasshopper in Australia (see sec. 13.12). We did not have quite enough empirically determined facts to complete the example, so we filled in the gaps with imaginary values. These values are possible ones, and the example may be regarded as being essentially realistic.

The distribution of these grasshoppers is bounded on one side (coastward) by a humid zone, where arable farming is practiced; it is bounded on the other side by desert. The area where the grasshoppers are found is just too arid for arable farming, but hardy grasses and shrubs provide pasturage for sheep. The weather is unkind. The summer is hot and extremely arid; evaporation from an exposed surface of water may exceed 40 inches, and this may be tenfold the amount of rain that falls during the same period. The only plants that remain green during summer are certain xerophytic shrubs and trees which the grasshoppers do not eat. The winter is the wet season, and usually there are grasses and other herbs for the grasshoppers to eat. But the rainfall is unreliable; sometimes so little rain falls that there is no food for the grasshoppers, and most of them die from starvation without laying any eggs; at other times there may be so much rain that diseases spread widely among the grasshoppers. Notwithstanding these risks, the grasshoppers are numerous throughout much of their distribution. Occasionally, when the weather is unusually kind, they may become extremely abundant and remain so for several years. They are not found in the desert to the north or in the humid zone to the south.

The chief causes of high death-rates among the grasshoppers are those which we have listed in Table 1.01. By methods which will be described fully in section 13.124, we estimated the probabilities that are shown in the first three rows in the column headed "A" in Table 1.01. These probabilities express the chance that any one generation will escape catastrophe from the cause named in the first column. It is also the chance that this catastrophe will not occur in any one year, because this species completes one generation every year. We lacked empirical evidence about the frequency of outbreaks of disease, so we filled in the fourth row with an imaginary figure. All the figures in the last two columns are imaginary. The figures under "A" refer to a certain district which is near the center of the distribution of the grasshoppers; it is a place where we know the grasshoppers are usually numerous. Locality "B" is arbitrarily supposed to be near the southern (humid) boundary, and "C" near or outside the northern (arid) boundary of the distribution of the grasshoppers.

Having set up this table, we are now in a position to develop our example. We propose to discuss the numbers of grasshoppers that are likely to be found in each locality, A, B, and C. We shall suppose for the sake of simplicity that the localities differ only in the ways set out in the table. That is to say, we shall take it that the three localities differ from one another only with respect to the frequency with which catastrophe is likely to overtake the grasshoppers living there. We shall assume that catastrophes, when they come, are of equal severity in each locality and that, during the intervals between catastrophes, the grasshoppers grow just as quickly and lay just as many eggs in each locality.

TABLE 1.01

PROBABILITIES OF "HIGH" DEATH-RATES AMONG POPULATIONS OF GRASSHOPPERS LIVING IN LOCALITIES A, B, OR C

STAGE OF LIFE-CYCLE	CAUSE OF "HIGH" DEATH-RATE	PROBABILITY* OF ESCAPING "HIGH" DEATH-RATE IN ANY ONE YEAR (OR GENERATION)		
		A	B	C
Egg in diapause	Drought during summer	0.90	0.96	0.72
Nymphs about to emerge	Drought at time of hatching	.99	1.00	.98
Active stages (nymphs and adults)	Drought during winter and spring, resulting in starvation	.90	0.96	.86
Active stages (nymphs and adults)	Excessive humidity, causing disease	.96	0.80	.99
All stages	All the above causes	0.77	0.74	0.60

* The probabilities are based on the frequencies of certain extremes of weather at the three localities.

The chief result of taking account of these additional causes of variation would be to magnify the differences between the localities. We can afford to leave them out, because our example is already sufficiently striking. We are also assuming that deaths from causes other than adverse weather may be neglected. This will be justified with respect to this species in section 13.124.

In locality A the chance that any one year (or any one generation) will be free from catastrophe from all the causes mentioned in Table 1.01 is 0.77. By expanding the expression $(0.23 - 0.77)^n$, we may calculate the probabilities that in any group of n years there will be 1, 2, 3, . . . , n years free from all catastrophes. For example, in A the probability of getting 10 years free from catastrophe is 0.77^{10}, which is 0.073. If we call the particular level of abundance which is attained after 10 years without a catastrophe a "plague," then we may say that plagues of grasshoppers are likely to occur at A about 7 times in 100 years; at B about 5 times in 100 years; and at C about 6 times in 1,000 years (Table 1.02). If we choose to consider some lesser level of abundance, say, the numbers attained after 5 years without any catastrophe, then these numbers would be present at A about 27 times in 100 years; at B about 22 times; and at C about 8 times during the same period.

We have assumed, so far, that the three areas are alike in everything except in frequency of catastrophe. But a climate which is severe in one way is likely to be severe in others as well. Unkind weather which retards the rate of increase between catastrophes is more likely to happen at B or C than at A. Also the catastrophes at B and C are likely to be more severe than those at A.

TABLE 1.02
PROBABILITIES OF GETTING GROUP OF 5 OR 10 YEARS WITHOUT "CATASTROPHE" IN THREE LOCALITIES, A, B, AND C OF TABLE 1.01

LOCALITY	PROBABILITY	
	5 Years	10 Years
A	0.270	0.073
B	0.220	0.049
C	0.078	0.006

For these reasons, the numbers attained after 2, 5, or 10 years without catastrophe are likely to be fewer at B or C than at A. When this extension is made to the original model, it becomes easy to conceive of C as a place where a grasshopper's chance to survive and multiply is so small that the species may not persist there. This locality may be "outside the distribution" of the species,

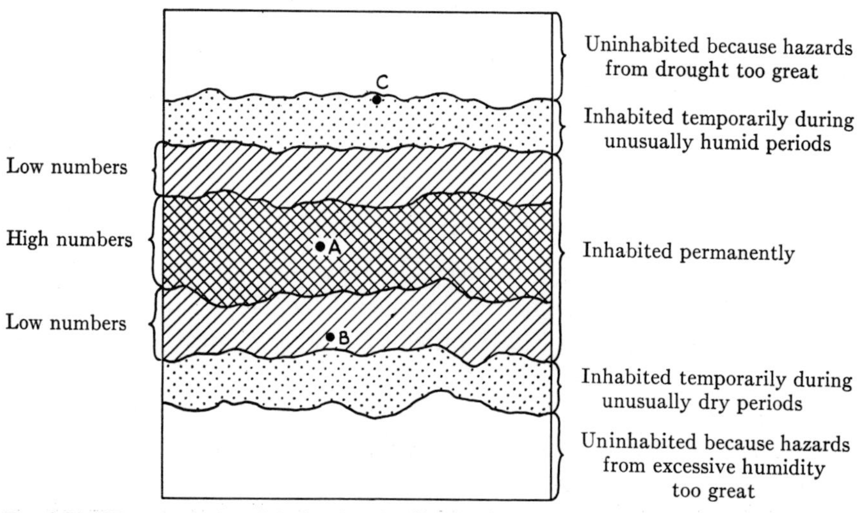

FIG. 1.01—Hypothetical model, showing the distribution of a grasshopper. The model is based on the ecology of *Austroicetes cruciata* in South Australia; the distribution is bounded on the north by desert and on the south by an area of high rainfall. For the real distribution of *A. cruciata* see Fig. 13.07. For further explanation see text; *A*, *B*, and *C* are the localities mentioned in Tables 1.01 and 1.02 and in Fig. 1.02.

or we might say that this locality may be one where the numbers are likely to be zero. It is, as we have said before, all a matter of emphasis.

Of course, boundaries of distributions do not stay fixed in the one place; they fluctuate with weather and other components of environment. This is illustrated in Figures 1.01 and 1.02. The letters *A*, *B*, and *C* refer to the localities in Table 1.01. Not only are the populations at A likely to be larger than those at B or C, but they are also likely to be more secure against extinction.

Instead of building our example around weather, we might have based it on other components of the environment which influence the numbers of animals (see chap. 2). This may have resulted in a more complicated example but would not have altered the principle. For the numbers of an animal present in an area are determined by the animal's chance to survive and multiply. This is

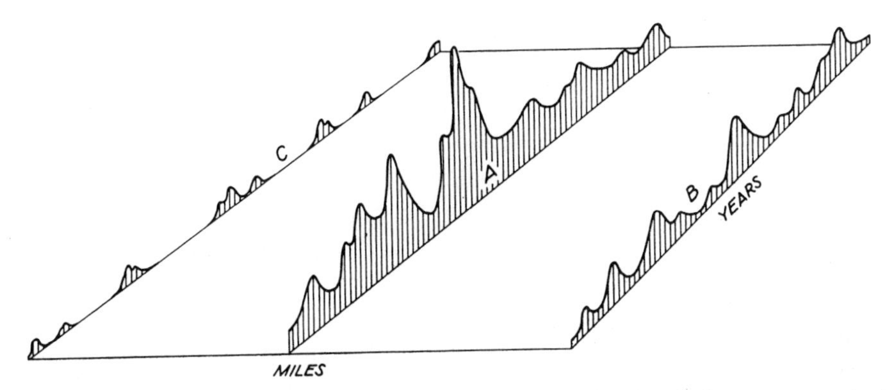

Fig. 1.02.—Hypothetical model, showing the numbers of grasshoppers during many years in the localities marked *A*, *B*, and *C* in Fig. 1.01.

limited by the environment. As the severity of the several components of environment which impose this limitation increases, the numbers decrease, eventually becoming zero as the limits of the distribution are reached. No doubt the principle is obvious enough to many readers; yet, as we stated earlier, many biologists consider distribution and abundance as distinct concepts and proceed to build up a separate set of laws to explain each. A familiar statement might be that climate may limit distribution but may not "control" abundance, because it lacks the essential quality of "competition" which is postulated—we believed erroneously—as the only mechanism whereby numbers may be "regulated." A sentence quoted from Nicholson (1947, p. 11) summarizes this viewpoint: "It [climate] can produce either tolerable or intolerable conditions; if these are intolerable, the species cannot exist; if tolerable, the species will continue to multiply indefinitely, for climate does not become any more severe, however numerous the animals may become." This statement overlooks several important characteristics of the environment of an animal, as we shall demonstrate in chapters 2, 13, and elsewhere.

It is not intended to elaborate these points in this chapter, but one example will indicate our meaning. Toward the northern limit of their distribution, grasshoppers are rare in most years; occasionally they may become temporarily abundant. This is partly because, in these parts, the spring is so unkind and so short that many grasshoppers die of starvation before they have laid any or many eggs. When the weather is kinder than usual, births may exceed deaths, and this may continue for several years. But quite often drought will reduce the numbers to a low level. The weather is "tolerable" even during the driest years; otherwise, the grasshoppers would die out; but the favorable periods are so short that there is no chance of "multiplying indefinitely." A succession of good years might result in large numbers for a period; but the probability of such a succession of good years near the edges of the distribution is so small that it rarely happens.

1.2 THE WAY WE STUDY ECOLOGY

If one wishes to explain the distribution and numbers of a certain species of animal, there is one way of going about the job which we have found especially profitable. It is best to describe the method as if it were done in three stages, although, in practice, it is usually best to have the three stages going forward at the same time.

a) The physiology and the behavior of the animal must be investigated. But, as our mentor, the late Professor James Davidson, used to say: "Knowledge is infinite." Therefore, it is necessary to exercise insight and wisdom in choosing the particular aspects of physiology and behavior which are to be studied. Although this is often largely work for the laboratory, the insight to decide which questions most need to be answered can be gained only by studying the animal in its native haunts. Sometimes the experiments will be best done in the field (e.g., sec. 5.11). The ecologist must, first of all, be a naturalist who loves to work in the open air and study his animals as he finds them in nature.

b) The physiography, climate, soil, and vegetation in the area must be investigated. The other organisms in the area must also be investigated if these seem to be important in the ecology of the species that is being studied. These observations may also provoke some questions which can be answered only by experiments in the laboratory or carefully designed experiments in the field.

c) The numbers of individuals in the population that has been selected for study must be measured as accurately as practicable. Sampling methods nearly always present a thorny problem. We shall have something to say, in passing, about this in chapter 13, but the best advice is to keep an open mind and try to solve each problem on its own merits. Such is the variability of nature that the records need to be kept during a long period; in some of the best studies, routine counts of the numbers in the population have been kept for 10 or 15 years.

The ecologist should have a sound working knowledge of statistical methods. Ultimately, his results must be expressed in numbers, and the conclusions to be drawn from them are likely to be both sounder and more profound if they are based on the proper statistical tests. Moreover, with a knowledge of statistics, the student is better able to design good experiments; without such a knowledge, it is easy to miss the essential points in the design of an experiment and to misinterpret the results.

The ecologist should be careful to avoid the *misuse* of mathematics. Ecologists, and especially mathematicians with a slight knowledge of biology, seem to be prone to the mistake of building a model with symbols which, they pretend, represent certain qualities of animals. The symbols are then manipulated according to the rules of mathematics and, finally, the conclusions are translated into words which purport to describe some law of biology. This is a tricky game. Animals are too complex to be represented by symbols. The conclusions derived from such models can, at best, have a very limited scope; at their worst they may be grossly misleading. Sometimes a discrepancy slight enough to be overlooked in the original definitions, and perhaps in the conclusions, may be revealed during an intermediate step as a contradiction (sec. 10.11). But some models may be more plausible than this, and their shortcomings may be concealed more deeply. The student should be advised to avoid this method unless he is prepared and able to think deeply and critically about the particular model that he contemplates.

It seems that many ecologists find it difficult to describe the results of their work in a style which is easy to read, yet leaves the reader with no doubts about what is meant. At least some of the obscurity may be traced to the belief that an altogether different style is needed to describe the animals in a wood from that which is used to describe the actors in a play. But the same rules for simple and lucid writing govern both cases. Even the vocabulary need not be so different as some scientists would make it. For useful guides in these matters the reader might refer to Leeper (1941, 1952) and Orwell (1946).

1.3 THE PATTERN OF THIS BOOK AND A GUIDE TO ITS USE

We shall build this book according to the pattern which we outlined in section 1.2. The three chapters in Part II will deal with particular aspects of physiology or behavior. Of course, we would not try to write, even if we could, a comprehensive discussion of either physiology or behavior. Part II is just a sample of the sort of information which an ecologist is likely to find helpful.

The "innate capacity for increase in numbers" (r_m) is a statistic which summarizes a great deal of worth-while information about an animal. Ultimately, the explanation of distribution and abundance must depend on an animal's chance to survive and multiply in the particular place where it is living. Although r_m will hardly ever be calculated precisely, a knowledge of the principles in this chapter is useful background for an ecologist. "Diapause" is included, because it occurs often and with many different aspects in the life-cycle of insects and other groups of animals. It is a complex subject which has often been misunderstood. An animal's chance to survive and multiply may depend quite largely on how well it seeks food, especially if food is sparsely distributed in the area where it lives. Dispersal is important in this regard; it may also help to determine the animal's chance of eluding capture by a predator.

Although chapter 2 contains a certain amount of general introductory material and is therefore placed in Part I, it may also serve as an introduction to Part III. In this chapter we show that the environment of any animal is divisible into four primary components. We state, without proving it at that time, that the animal's chance to survive and multiply, and hence the species' distribution and abundance, may be determined by all four components, but sometimes by one more than another. In the chapters of Part III we take the components of environment one at a time and show how each may influence the animal's chance to survive and multiply.

In Part IV we deal with the numbers of animals in nature. Chapter 13 is strictly empirical. Relatively few examples are used: it is better to discuss few thoroughly than many too briefly. In this way we can explain the important principles better. We can also describe the methods that each worker has used; for one cannot properly appreciate the results of an investigation unless he knows how the investigator got his results. Scientists sometimes neglect this educational principle. Whitehead (1947, p. 189) said that "the main structure of successful education is formed out of the accurate accomplishment of a succession of detailed tasks." Further on in the same essay he gave a timely reminder that the details of facts are exciting to the imagination in so far as they illuminate some scheme of thought. We have summarized our "scheme" or general theory in chapter 14. The detailed analyses which appear in previous chapters, especially in chapter 13, have been directed to the general end of formulating the principles of chapter 14.

We have learned from genetics that a species is more variable and plastic than used to be implied by the old-fashioned taxonomist. The newer ideas about species can be helpful to an ecologist. Also some ideas from ecology may help the geneticist to enlarge his ideas about the variation within species or the origin of new species. These are our reasons for including Part V.

The general reader and the student who is being introduced to the subject of animal ecology may prefer to skip the more difficult sections at first reading. So from chapter 1 he might proceed through the subsequent chapters in order, omitting sections 3.13, 3.2, and 3.3 in chapter 3, section 9.23 in chapter 9, and sections 10.1 and 10.223 in chapter 10. But, whatever sections are chosen for special interest, chapters 1 and 2 should be read first.

CHAPTER 2

Components of Environment

He who would be truly initiated should pass from the concrete to the abstract, from the individual to the universal.

PLATO, *The Symposium*

2.0 INTRODUCTION

ERNST HAECKEL in 1870 defined *ecology* as "the total relations of the animal both to its inorganic and its organic environment." In speaking of the animal and its environment Haeckel gave a lead which has unfortunately been ignored in many more recent attempts to define ecology. Many textbook definitions incorporate the phrase "populations of animals and their environment" or some such equivalent expression. But this leads to methodological difficulties, since every individual in a population of animals is part of the environment of other individuals. Other animals of the same kind which are present in the vicinity are of great importance to the animal: if there are too few of them, it may not be able to find a mate at the right time (sec. 9.13); if too many are present, the animal may suffer from the harmful influence of crowding (sec. 9.2). The difficulty of thinking of the "environment of a population" is that it leaves out half the picture. The phrase itself represents a false abstraction. It is an example of what Whitehead (1926, p. 75) called "the fallacy of misplaced concreteness." It tends to confuse rather than to clarify. For this reason we speak of the "environment" of the individual, regarding the population as part of the environment rather than as itself having an environment.

It is necessary to emphasize this; ecology is so complex and subtle that it is easy to fall into the error of false abstraction. We are compelled to think in abstractions, but at the same time we should discard those which experience shows to be inadequate and replace them with others which are more adequate. It is not sufficient to say with Solomon (1949, p. 10): "But because such an approach [the environment of a population] is demanded by long-established traditions of thought and because it is so familiar, we tend to overlook the serious difficulties it raises," and to leave it at that. By overlooking such errors we merely contribute to the confusion of thought and terms for which ecology is already notorious. These difficulties must be faced and surmounted.

In practice, we are always interested in populations or, more strictly, in samples from populations from which we may estimate a reliable mean or mode. But we still have to consider the influence of the environment (or some particular component of it) on the individual and then take the mean of an adequately large sample of individuals to represent the population. For example, the average fecundity of a sample of individuals gives the birth-rate among the population from which the sample was taken; the mean duration of life provides a measure of death-rate, and so on (see chap. 3). This approach to the analysis of environment is not at variance with the conception of the population as a unit in evolution. The two concepts are used together in chapter 15.

One solution which has found some favor with botanists is to disregard the distinction between the organism and its environment and to consider, instead, the organism *and* its environment as a unit which is called the "ecosystem" (Tansley, 1935). This may be logical enough, but we fail to see how it helps. The ecosystem is so all-embracing and subtle that it seems to defy analysis and is therefore hardly likely to further our chief purpose, which is to explain the distribution and numbers of animals in nature. A way which offers opportunity for clarity of thought is to consider the environment of an individual. The population is then understood as a group of individuals, each having an environment which resembles those of its neighbors but differs from theirs if only because the environment of an individual includes its neighbors but not itself.

2.1 ANALYSIS OF ENVIRONMENT: HISTORICAL

It is customary in a book of this sort to give a general definition of environment before proceeding to analyze it. We choose not to do this because of the great difficulty of pinning down anything so subtle as the ecologist's conception of environment into a rigid and formal definition. Allee *et al.* (1949, p. 1) stated, for example, that "the *environment* of any organism consists, in final analysis, of everything in the universe external to that particular organism. Those parts of the total environment that are evidently of direct importance to the organism are regarded as constituting the *effective environment*." The definition has obviously been worded broadly with the intention of making it completely general; but, even so, it is clear that the word "external" cannot have its usual meaning, for the definition would then exclude the internal "parasite" from the environment of its "host." With many insects this represents an important aspect of predation which must not be left out of any definition of environment. The definition might be restated thus: the effective environment is everything else in the universe which is of importance to the organism. But definitions as broad as this are not very useful. This one has the same disadvantages as the definition of "ecosystem" which we have already mentioned. It leaves one with the feeling of not having got any further forward.

Rather than attempt a formal definition, we prefer to classify and describe all the material things (like trees and logs and other animals, etc.) and all the qualities (like temperature, moisture, radiation, and so on) which we can think of as influencing, either separately or in interaction one with another, the animal's chance to survive and multiply. When this has been done, our conception of environment will become clear. This approach is not altogether original, but it is nevertheless unconventional, for it has been customary ever since Haeckel's time to group the components of the environment together and classify them according to what they are rather than what they do. For example, a classification which starts off by dividing environment into organic and inorganic components or biotic and physical factors is clearly based on what the components of environment are and for this reason might be called "taxonomic." We have not emphasized these sorts of affinities. Instead, we have tried to bring together components that are alike in what they do to influence the animal's chance to survive and multiply. When this approach is followed, environment may be broken down into four major components, which can be separately defined and logically analyzed (sec. 2.2). It will be shown that the environment of any animal may be fully described in terms of these components. But first it is necessary to explain why we discard certain suggestions which have been put forward and used by other ecologists.

2.11 *The Distinction between "Physical" and "Biotic" Factors*

In their definition of ecology Allee *et al.* (1949, p. 1) said that environment includes "both the physical and the biotic environments." This is the usual practice in ecological texts, and it is commonplace for authors contributing to ecological journals to accept this division either explicitly or implicitly in their writings. But these terms have not always been used with the same meaning, and sometimes there is doubt as to just what is meant. Allee *et al.* (1949, p. 227) defined biotic factors as including living organisms and nonliving organic matter. Chapman (1931, p. 155) definitely included living organisms among "biotic factors" but left the reader the choice of deciding whether food is a "biotic" or a "physical factor." He added that the choice is purely arbitrary in this case. But Clements and Shelford (1939) included only living organisms in their discussion of the "biotic complex." This is the usage which seems most common. Although the terms have not always been defined precisely, it is usual for authors to include "living organisms" and to exclude "nonliving organic matter" from their discussions of "biotic factors" (Smith, 1935; Nicholson, 1947; Solomon, 1949). Whatever procedure is followed, this division is unrealistic and has the serious deficiency that it does not help us to understand the way in which environment influences the animal's chance to survive and multiply. The difficulty is especially apparent when we consider "food" and "shelter." In Clements and Shelford's usage, "shelter" and, in the usage of Allee *et al.*, both "food" and "shelter" may be provided by either the "biotic" or the "physical factors" in the environment; yet apart from its contribution to food, the nonliving organic matter has more in common with the inorganic matter than it has with the living organisms. The nonliving matter has little in common with temperature, moisture, light, and so on with which it is usually grouped under the heading "physical factors." So, even as a "taxonomic" classification, the distinction between "physical" and "biotic factors" is unsatisfactory. As a functional classification, it has even less to recommend it. Some ecologists have supposed that the influence of "physical factors" upon the organism are "density-independent," while the "biotic factors" are "density-dependent." We shall show in the section which follows that this is not so.

2.12 *The Distinction between "Density-dependent" and "Density-independent" Factors*

The idea of analyzing environment into two sets of "factors," which differ in that the influence of one on the rate of increase (or decrease) of the population is quite independent of its density (the number of animals per unit space) and that of the other dependent on density, seems to have been first put forward by Howard and Fiske (1911). Discussing natural populations of insects, they suggested that there were certain "factors" in the environment which destroyed a constant proportion of the population, irrespective of its density: they had in mind certain components of weather, such as frosts or drought, and these they called "catastrophic factors." Certain other components of environment, such as predators or disease, they called "facultative factors," because, they said, these destroyed a proportion which varied with the density of the population. The same idea was put forward by Thompson (1928), who used "individualized" and "general" instead of "facultative" and "catastrophic."

In the meantime, Chapman (1931) quite independently drew an analogy between the rate of increase of a population of animals and the rate of flow of an electrical current through a conductor. By this analogy the animal was supposed to have a theoretical potential rate of increase which was called "biotic potential." Just as an electrical potential may be quite independent of the resistance of the conductor through which it will flow, so the "biotic potential" was supposed to be quite independent of the environment in which the animal was growing. But the environment was supposed to offer a "resistance" to the increase of the population, and this Chapman called the "environmental resistance." Depending upon the level of "environmental resistance," the animal thus realized some proportion of its biotic potential. Of course, biological systems have little in common with electrical circuits, and the unreality of this analogy is manifest when it is considered that the analogy requires that the animal's capacity for increase shall be independent of the environment, whereas, in fact, the animal's innate capacity for increase is immensely influenced by almost every component of the environment (see chap. 3).

Nevertheless, Chapman's conception of "environmental resistance" was taken up by Gause (1934) in a very special sense: he restricted it to the special case of the influence of crowding on the rate of increase of a population growing in accordance with the theory of the logistic curve (sec. 9.241). It was also taken up by Smith (1935), who used "environmental resistance" in the broad sense originally intended by Chapman. Referring back to Howard and Fiske, he suggested that "environmental resistance" could be analyzed into two components or sets of "factors," which they had called "facultative" and "catastrophic" and which he renamed "density-dependent" and "density-independent" mortality factors. The latter were supposed to destroy a constant proportion of the population, irrespective of its density. He also constructed an arithmetical model which was based partly on an extrapolation of the logistic theory as expounded by Volterra (1926) and Gause (1934) and partly on a tacit assumption that "density-independent factors" really means what the name implies. Smith's arithmetical model served to partition environmental resistance precisely into its two components, using the equation for linear regression $m = aD + b$, where m is death-rate, D is density, b is the density-independent component of "environmental resistance," and a is the density-dependent component. This equation was derived from the logistic equation, which assumes a linear relationship between death-rate and birth-rate, on the one hand, and density, on the other, and an immediate response in these rates to changes in density. We shall show in section 9.23 that these assumptions cannot be verified and must be considered to be probably untrue even for experimental populations—more so for natural populations. Moreover, the logistic theory has a very restricted application and, indeed, lacks the generality which Smith attributed to it in order to construct his model.

The additional idea that Smith combined with Chapman's "environmental resistance" was the assumption that the "density-dependent" component of "environmental resistance" influences mortality in direct proportion to the density of the population. The objections to this are twofold: it is unlikely that "density-dependent" components do act in this way (see sec. 9.234), and, in the second place, the division of environment into "density-dependent" and "density-independent factors" is misleading because all the evidence indicates that there is no component of environment such that its influence is likely to be independent of the density of the population.

An example will make this clear. Suppose that in a certain natural area there is living a population of insects of a certain sort which are exposed to frost of certain severity such that it kills some of the insects. Since some live and some die, it is obvious that the individuals vary with respect to their chance of surviving the frost. This variability has two components: (*a*) differences between individuals with respect to their innate cold-hardiness and (*b*) differences between places where the insects may live with respect to the degree of protection from frost which they afford the insects.

Cold-hardiness in the population will be distributed normally or approximately so: a large population has a greater chance of containing some very cold-hardy individuals than a small one. If a number of small populations were exposed to a frost of a certain severity, the proportion of deaths in each would vary; and the smaller the populations, the greater the variability. In a number of large populations the variability would be less, and in very large populations the death-rates would be nearly constant. Now consider two populations, one large and the other small (in equal areas), which are exposed to a frost of a certain severity. The death-rate in the large population could be predicted within narrow limits; the death-rate in the small one might vary between wide limits. Therefore, it would not be true to say that the frost was likely to kill the same proportion of insects in the two populations. Frost is not a "density-independent factor," even after the variability associated with the places where the animals may live has been arbitrarily excluded from consideration.

When we come to consider the second component of variability, we find further strong evidence against the likelihood of a "density-independent factor." It is difficult to imagine an area so uniform that all the places where animals may live provide equal protection from the elements, and it is certain that the proportion of animals living in more favorable places would vary with the density of the population. A smaller number is likely to be better protected than a larger number in the same area. This important aspect of the non-uniformity of the areas where natural populations live is dealt with toward the end of section 2.2 and in sections 12.0 and 13.02. We shall pursue it no further at this stage, except to quote from an article by H. S. Smith, in which he asserts the distinction between "density-dependent" and "density-independent" factors: "But climate so obviously limits geographic distribution and determines the average number of so many species that, even in the absence of proof, we must admit that under certain conditions it is capable of acting as a *density-dependent* factor. It seems most probable that this takes place through the existence of protective niches in the environment which are more or less limited in number" (Smith, 1935, p. 894). We would add to this that we do not know of any experimental or observational evidence which would indicate that any component of the environment characteristically destroys a constant proportion of the population, irrespective of its density.

If, then, no "factor" is "density-independent," why single out some in particular to call them "density-dependent"? It is an unfortunate name; nevertheless, it is a phrase that has been widely used, and we must find out the meaning that is intended for it and use it in that sense. The meaning of "density-dependent factor" is made clear in the following quotation from Elton (1949) (and also in several quotations from Nicholson and from Varley which we give in sec. 2.121): "It is becoming increasingly understood by population ecologists that the control of populations, i.e. the ultimate upper and lower limits set to increase, is brought about by density-dependent factors, either within the species or between species. The chief density-dependent factors are intra-specific competition for resources, space or prestige; and inter-specific competition, predators or parasites."

It may be true, as Elton says, that increasing numbers of ecologists are coming to accept this explanation. But when we search the literature for the evidence on which this "understanding" is based, we find that it is not really a *conclusion* based on scientific experiment but rather that it has more of the status of a *dogma*. It has come to be accepted because it has been asserted strongly by a number of authors (Nicholson, 1933, 1947; Smith, 1935; Varley, 1947). It is true that elaborate mathematical models have been built up on this dogma (Nicholson, 1933; Nicholson and Bailey, 1935; Smith, 1935), but these merely tell what would happen if the dogma were true: they do not, in any way, demonstrate its truth. Varley (1947) used the equations derived from the models of Nicholson and Bailey to partition empirical data gathered during a study of a natural population of the gallfly, *Urophora jaceana*. The conclusions reached by the analysis of these data depended on the dogma but could not in any way be used to demonstrate its truth (see sec. 10.223 for a fuller discussion of these experiments). De Bach and Smith (1941) used the same equations to describe the results of an experiment with the housefly *Musca* and its parasite, *Mormoniella*. The design of this experiment presupposed the truth of Nicholson's models; consequently, its results could not be used to verify the model (see sec. 10.223). The classical experiments which have been done in relation to the logistic theory (see chaps. 9 and 10) have sometimes been quoted in support of the dogma of "density-dependent" factors (Smith, 1935). We shall show in chapter 9 that very special precautions have to be taken in the design of these experiments if the results are to conform to the logistic theory; and, in any case, the conditions of these experiments are far too artificial for them to form the basis of a general theory about the numbers of animals in natural populations.

We conclude that "density-independent factors" do not exist; and there is no need to attach any special importance to "density-dependent factors" (in the narrow sense explained above) when discussing the way environment may influence the distribution and abundance of animals in nature. We shall document this statement fully in subsequent chapters and especially in chapters 13 and 14. This makes it necessary to look closely at what has been written about "competition," for, as Elton implied in the passage quoted above, much of the importance attributed to competition depends upon the importance that may be attributed to density-dependent factors.

2.121 COMPETITION

Those who accept the proposition that the density of populations can be "controlled" or "regulated" only by density-dependent factors frequently use the terms "competition" and "density-dependent factors" synonymously. This leads to the general proposition that the density of a population is determined or controlled by "competition." This point of view is best illustrated by several quotations from the writings of A. J. Nicholson and G. C. Varley. Other authors could be quoted in the same vein but these will suffice:

> The observed fact that there is a relation between the population densities of animals and environmental conditions can be explained only in terms of balance, just as the relation between the weight carried and the height reached by a balloon can be explained only in this way. Without balance the population densities of animals would be indeterminate, and so could not bear a relation to anything. . . . For the production of balance, it is essential that a controlling factor should act more severely against an average individual when the density of the animals is high, and less severely when the density is low. In other words, *the action of the controlling factor must be governed by the density of the population controlled.* Clearly no variation in the density of a population of animals can modify the intensity of the sun, or the severity of frost, or of any other climatic factor. . . . A moment's reflection will show that any factor having the necessary property for the control of populations must be some form of competition. If the severity of its action against an average individual increases as the density of animals increases, the decreased chance of survival, or of producing offspring, is clearly brought about by the presence of more individuals of the same species in the vicinity. This can only mean that the decreased chance of survival is due to increased competition of some kind [Nicholson, 1933, pp. 133 and 135].

> When considering why this relation should exist, the important point to grasp is that climate is quite unaffected by the densities of the animals it influences, and so it cannot limit these densities. It can produce either intolerable or tolerable conditions; if these are intolerable, the species cannot exist; if tolerable, the individuals will continue to multiply indefinitely, for the climate does not become any more severe however numerous the animals become. Clearly this multiplication cannot continue indefinitely. Before long competition effects will first slow down, and later arrest, further increase; or they may cause a decrease [Nicholson, 1947].

Varley (1947, p. 182) wrote, as if this must be accepted as an axiom: "The controlling factors which keep a population in balance must be affected in their severity of action by the population density on which they act."

In the paper from which the first quotation was taken, Nicholson was discussing hypothetical populations or mathematical models, although the introduction from which this quotation comes seems to be meant to be taken quite generally and to apply to natural populations. The second quotation refers quite unequivocably to natural populations. It is also quite clear that the quotation from Varley is meant to refer to natural populations.

It is not easy to understand what precisely is meant by the word "balance" in these quotations. We shall defer the discussion of this to chapter 14, because at the present we are concerned only with what these quotations have to tell us about the division of the environment into density-dependent and density-independent factors and the stress which these authors place on the importance of density-dependent factors.

The theme which dominates these quotations—it is repeated several times even in these brief extracts—is that, in a favorable "environment," numbers would go on increasing without limit (until they "filled the universe") unless they were prevented from doing this by some density-dependent factor. Alternatively, in an unfavorable "environment," numbers would go on decreasing indefinitely until the population became extinct unless the "environment" were ameliorated by a "density-dependent factor." When stated in this way, the fallacy of this view is apparent. In nature the conditions of life do not remain continuously favorable or continuously unfavorable. Circumstances are perpetually changing.

The weather is subject to diurnal fluctuations between night and day; erratic fluctuations are associated with movements of air-masses over the earth's surface; more rhythmical changes are associated with the procession of the seasons; and so on. The food and shelter in any substantial area are also subject to both short-term and long-term changes. The short-term changes may be either erratic or rhythmical and are associated with the weather. Long-term changes may be associated with ecological succession. Natural populations neither increase nor decrease continuously. While circumstances remain favorable, the numbers increase; when they become unfavorable, the numbers decrease. It is precisely because change is the general rule in nature that there is no need to invoke "density-dependent factors" to explain the mean numbers in a natural population or the fluctuations that occur from time to time. This point is fully elaborated in chapters 13 and 14.

The last part of the first quotation from Nicholson makes it very clear that in his view the universal importance of competition depends on the universal importance of density-dependent factors; the one is deduced from the other. In fact, the two terms are used synonymously. It needs to be pointed out, in order to avoid confusion, that in this usage "competition" includes a most unusual meaning. This is clear in Nicholson's statement which was quoted previously: "A moment's reflection will show that any factor having the necessary property for the control of populations must be some form of competition. If the severity of its action against an average individual increases as the density of the animals increases, the decreased chance of survival, or of producing offspring, is clearly brought about by the presence of more individuals of the same species in the vicinity. This can only mean that the decreased chance of survival is due to increased competition of some kind." If we were to suppose that the population under discussion was a species of scale insect and the controlling factor a predator, the statement would then read: "A moment's reflection will show that any factor having the necessary property for the control of the scale insects must be some form of competition. If the severity of the predator's action against an average scale insect increases as the density of the scale insects increases, the decreased chance of the scale insect's surviving, or of producing offspring, is clearly brought about by the presence of more scale insects in the vicinity. This can only mean that the decreased chance of the scale insect's surviving is due to increased competition." This implies that the scale insect actually "competes" with its brethren for the opportunity of being eaten by the predator. We do not suppose that Nicholson believes this, though it quite logically follows from the way in which he uses the word "competition" to embrace the action of all "density-dependent factors."

The literal meaning of "compete" is "together seek." We shall therefore say that competition occurs whenever a valuable or necessary resource is sought together by a number of animals (of the same kind or of different kinds) when that resource is in short supply; or if the resource is not in short supply, competition occurs when the animals seeking that resource nevertheless harm one another in the process. Starting from this broad base, we shall inquire into the importance of "competition" (in the strict sense) in nature. And although we cannot accept the argument by which Nicholson concludes that "competition," in the sense in which he uses it, is all-important in nature, it does not necessarily follow that "competition," in the strict sense in which we have defined it, is not important. There may be other grounds for considering competition important.

We shall inquire into two questions: (*a*) How often in nature do we find a population that has used up or is likely to use up all its stocks of food and other resources? (*b*) When a resource is in short supply and competition occurs, does a study of competition as such help us to explain the observed densities of the populations? To answer the first question, we must take account of food, shelter, and other requirements and consider how often the supply of these may be insufficient to meet the needs of the animals present. The first fact to notice in this study is the prevailing scarcity or rareness of most species in relation to the amount of food that is available for them (secs. 14.1 and 14.4). With herbivorous insects, for example, this is self-evident. Any field, meadow, or orchard abounds in species which feed upon the crop plants but pass quite unheeded by the farmer because they consume relatively so little of the crop. The same is equally true of more "natural" situations, such as woods, forests, or heath-lands. The same is true of carnivorous insects, though not so immediately obvious. Yet any practical entomologist who has worked in the field of "biological control" could testify to the high proportion of "useless" predators that occur in nature. Hence many of them never become numerous, even when living among a relatively dense population of prey. The fact that a relatively low density is characteristic of most animal species in nature has not escaped the notice of naturalists. Some quotations from Darwin and other prominent naturalists will illustrate this point.

"Rarity," wrote Darwin in the eleventh chapter of *The Origin of Species*, "is the attribute of a vast number of species of all classes, in all countries. If we ask ourselves why this or that species is rare, we answer that something is unfavourable in its conditions of life; but what that something is we can hardly ever tell."

The idea of low numbers (or "rareness") as a prevailing attribute of most species in nature is implicit in the following quotation from Smith (1935, p. 880), which is taken from his discussion of "rare" and "common" species: "In general, attempts to determine reasons for abundance or scarcity have been based upon studies of species which are of economic importance only. The fact that the number of species which become sufficiently abundant to damage crops is relatively small, and that such species form only an insignificant fraction of the total number of phytophagous insects is ignored. It is essential to study species which are rare in individuals as well as species which are abundant, if reasons for abundance and scarcity are to be understood."

It is common experience that the prevailing low numbers of animals in nature is not often due to shortage of food; Bodenheimer (1930), thinking of insects, says: "Any orchard, meadow or field will prove this sufficiently." The same experience is borne out in some remarks of Cockerell (1934): "I recall some observations on Coccidae [scale insects] made in New Mexico many years ago. Certain species occur on the mesquite and other shrubs which exist in great abundance over many thousands of square miles of country. Yet the coccids are only found in isolated patches here and there. They are destroyed by their natural enemies, but the young larvae can be blown by the wind or carried on the feet of birds, and so start new colonies which flourish until discovered by predators and parasites. This game of hide-and-seek doubtless results in frequent local extermination, but the species are sufficiently widespread to survive in parts of their range, and so continue indefinitely."

But there are also situations in which food, shelter, or some other resource is in short supply, and the animals that are present may be said to compete for it. Competition undoubtedly occurs, even though it may not be so widespread as sometimes imagined. But does a consideration of competition help to explain the observed density of the population? We can outline our way of answering this question by taking a few examples; an imaginary one is a useful beginning, for we can keep it simple and concentrate on the main principles. Consider a population of solitary bees of the genus *Megachile* living in an area in which there are several fencing posts. In the posts there are 500 auger-holes, each one of which will serve as a nest for one bee and one bee only, and there are no other places in this area which are suitable for nests. The bees are skilful searchers, and no auger-hole remains unused while there is a bee still requiring a nesting site. Each bee needs to make only one nest during her lifetime. Now on one occasion we shall suppose there were 500 bees present; they filled up the 500 auger-holes, and no competition occurred. On a second occasion we shall suppose that there were 1,000 bees; they also filled up the 500 auger-holes, but only after a certain amount of competition, which resulted in 50 per cent of the population being deprived of the chance of contributing progeny to the next generation. On a third occasion we shall suppose that 5,000 bees were present; they also occupied the 500 auger-holes, but only after intense competition, resulting in 90 per cent of the population failing to contribute progeny to the next generation.

To the student of evolution the most important feature in this situation, and one which he must try to measure, is the intensity of the competition between the bees, for this determines the rate of change in the genetic constitution of the population. This is the aspect of competition which has been familiar to biologists ever since the publication of *The Origin of Species*. But the competition which was observed among the bees in this example is of little immediate importance or interest to the student of ecology, who is primarily concerned with explaining the density of the present population and not its genetic composition. The study of competition in this example tells him nothing about the density of the population. On the other hand, he can predict with certainty the number of bees which will be present in the area by the simple operation of counting the auger-holes. Having counted 500 auger-holes, he would know that the area could support the number of bees that would emerge from 500 nests. This would be the same irrespective of the intensity of competition which might have gone on during the previous generation when the parents were searching for nesting sites. In other words, in this situation, the consideration of competition was quite irrelevant to the study of the ecology of the bees. What was relevant was the number of nesting sites, or, in more general terms, the quantity of the particular resource that was limiting. In this example it was nesting sites. In others it might be food, shelter, and so on. This is an unnaturally simple example, but it is sufficiently close to nature to illustrate quite nicely the principle that when the supply of some necessary resource is inadequate for the number of animals present, it is the quantity of the limited resource which determines the density of the population, not the intensity of competition (see also Fig. 14.01).

There is one exception to this rule which can be illustrated by extending the above example. The bees, as we suppose in the previous example, did not do one another any bodily harm during the competition for nests. This, incidentally, is likely to be true of the bees of the genus *Megachile* in nature, but not of *Bombus*. Let us now imagine them to be different in one particular, namely, that during the search for nests they fight with one another and some are killed or maimed so that they cannot build a nest. If there were only a small excess of bees, say 550 competing for 500 nests, the competition would not be intense, and in the end there might still be 500 sound bees which would build nests in all the 500 auger-holes. But if there were a great excess of bees at the beginning, competition might be so intense that in the end there might remain fewer sound bees than there were nesting sites. In this case some nesting sites would remain unoccupied: and competition may be justly said to be the direct cause of the smaller numbers in the next generation. Again this is an unnaturally simple example, but we believe it typifies the only sort of situation in nature in which the density of the population is determined by the intensity of competition. The full discussion of this involves an elaborate analysis of situations in which food is a limiting resource and one or more species of animals compete for it. The argument, which is a little more involved in these situations, is explained in sections 10.02 and 11.22.

We have rejected the particular usage of the word "competition" which embraces the predator-prey relationship. There is no competition between the predator and its prey; but there may, of course, be competition between the predators themselves if the prey are in short supply. But this is "intraspecific" competition. The common resource in short supply which the predators together seek is the prey which are their food. We shall see in chapters 10 and 11 that there is no need to make any basic distinction between carnivores and herbivores in considering the ecological aspects of their resources of food. When food is in short supply, whether it is animals or plants which are being eaten, competition will occur; but this does not necessarily determine population density.

Our definition of competition embraces two distinct meanings and it is necessary to consider each meaning separately in answering the question: When competition occurs, does a study of competition as such help us to explain the observed densities of the populations? Now competition may imply merely that a number of animals seek to share a common stock of food or some other resource which is insufficient for the needs of all of them. We have shown that in this meaning the concept of competition is not relevant to the ecological problem; the more realistic approach is to seek an explanation in terms of the quantity of the resource that is in short supply (see especially sec. 11.22). Or, competition may imply that the animals may harm each other as they seek to share in some common resource. In the latter case there may be no logical objection to introducing the concept of competition into the ecological problem; but it is doubtful whether this will often help toward a clear understanding of the facts. For example, we have not found it necessary to introduce the concept of competition either into chapter 13, where we discuss a number of natural populations, or into chapter 14, where we develop a general theory of population ecology (see especially sec. 13.22).

2.2 THE FOUR COMPONENTS OF ENVIRONMENT

In this chapter we are seeking a concept of *environment* which will help us to understand and explain the observed distribution and abundance of animals in nature. The meaning of "environment" is so complex that it must be broken down into simpler components which can be considered separately. That is to say, their boundaries must not overlap. But their distinctness does not mean that each is considered to exert its influence independently of the others. The study of such interactions between components of the environment will form quite an important part of the whole study. This may increase the practical difficulties of any particular study, but it raises no theoretical difficulties. The reader who has a working knowledge of statistical methods (and no ecologist should be without it) will readily appreciate what we mean when we say that this method of approach is analogous to the statistical method of partial regression. In quantitative biological studies the method of partial regression is used to measure the independent influence of each of a number of possibly related variables on some other quantity independent of their influence on one another. This analogy can be of great utility in considering environment.

The components of the environment must be defined in such a way that they can be studied individually by observation and experiment, especially their respective influences on the longevity, speed of development, and fecundity of the animal. These determine the rate of increase or decrease of the population, and this, in the long run, is at the root of all studies of distribution and abundance (see chap. 3). We agree with Allee *et al.* (1949, p. 331) that "the problem, on the one hand, is to avoid a classification so general that it is meaningless, and, on the other hand, to avoid one so specific that it is inflexible."

With these requirements in mind, we choose to divide the environment of an animal into four components: (i) weather, (ii) food, (iii) other animals, and organisms causing disease, and (iv) a place in which to live. This analysis of environment fulfils the requirements set out above. Together, the four components completely describe the environment of any animal. They are distinct, for their boundaries are clear-cut and easily perceived. It will be shown in Part III and in chapter 13 that the division of environment into these four components opens the way for a deep and logical analysis of the way in which environment influences the animal's chance to survive and multiply; and this leads to a general theory of distribution and abundance which is, we think, more adequate than others which have hitherto been proposed (chap. 14).

These components may require further subdivision. For example, the most important components of weather may be temperature, moisture, and perhaps light, particularly in the way in which plants and animals respond to the length of day. Food has several aspects, some of which are more physiological than ecological, while others are strictly ecological. These will be fully discussed in chapter 11. The influence of other organisms presents a number of subtleties; this component is readily divisible into other animals of the same kind (chap. 9) and other animals of different kinds (chap. 10); it also includes viruses, bacteria, and fungi which cause disease (sec. 10.323). There are some subtle interactions between "other animals" and "food" which are especially instructive when treated in this way (sec. 11.22). By "other animals of the same kind" and "other animals of different kinds" we mean other animals of the same species and of different species, respectively. The individuals within a species are not, of course, all of the same genotype, and in some analyses it is certainly necessary to distinguish between different genotypes within the species, as we do in chapter 16. But no distinction is made between genotypes within the species in the data of chapters 9 and 10, and so in these chapters "species" is synonymous with "kind."

A place to live is an obvious component of any environment. It is of value, from a methodological aspect, because it helps in understanding the diverse interactions between the several components (chap. 12). In nature, populations inhabit areas. In exceptional circumstances we may be interested in a population occupying the whole "distribution" of the species. More often we shall be interested in a specific part of the distribution; a particular wood, meadow, or orchard; or an acre, a square mile, a county; or a climatic or edaphic zone; and so on. Whatever the area, it will be made up of a large number of diverse local situations. Some may provide good refuge or shelter for the animal; others may afford only poor shelter or none at all. Food may be present in some situations but absent from others. Predators and other animals may be distributed in an uneven or "patchy" way over the whole area. In other words, animals are rarely so uniform that they do not manifest a wide variability with respect to the sorts of situations in which they can live. No substantial area inhabited by a population is so uniform that it does not contain a great diversity of situations which animals may inhabit. Certainly, the influence of weather (and other components of environment) will vary from one local situation to another. What appears, superficially, to be the same sort of place as another may, in fact, be quite different in its influence on the animal's chance to survive and multiply. A certain sort of place—for example, a log of a certain diameter or a burrow of a certain depth—may provide adequate shelter from frost at latitude 35° but not at latitude 45°. Another sort of place may provide a good refuge in an area which lacks an especially aggressive predator, but not if the predator be present. A place which affords one animal a good chance to survive and multiply may not be equally good for ten or twenty. This means that the place where an animal may live often owes its importance to interactions with other components of environment (sec. 12.2). This book is concerned chiefly with terrestrial animals which may be associated with certain sorts of soil, vegetation, and so on. With aquatic animals pH, concentration of certain ions, and concentration of oxygen and other gases may determine the places where they live.

There is good reason for calling our fourth component of environment "a place in which to live" instead of "habitat." The proper meaning for "habitat" is made clear in Elton's (1949) essay on population interspersion. A habitat is defined as an area which seems to possess a certain uniformity with respect to physiography, vegetation, or some other quality which the student decides is important. The limits of the habitat are decided arbitrarily by the student as *the first step* toward his study of the community. Of course, "habitat" has been used loosely and in diverse contexts (such is the fate of all ecological terms), but its proper and strong meaning is as in Elton's essay: it is an appropriate term when used in relation to communities.

According to our method, the first step in an ecological study is to choose a species of animal to work with. Then we set out to discover all the sorts of places where it may live and to evaluate them with respect to their influence on the animal's chance to survive and multiply. Our studies will also be confined within a certain area (the distribution or part of the distribution of the species which we choose to study), but our methods of arriving at the limits to our area are almost exactly the opposite of those used by the other school of ecologists in defining their habitats.

PART II

Physiological Aspects of Ecology

CHAPTER 3

The Innate Capacity for Increase in Numbers

In looking at nature, it is most necessary to keep the foregoing considerations always in mind, never to forget that every single organic being may be said to be striving to the utmost to increase in number, that each lives by a struggle at some period of its life, that heavy destruction inevitably falls either on the young or the old, during each generation or at recurrent intervals. Lighten any check, mitigate the destruction ever so little and the number of the species will almost simultaneously increase to any amount.

DARWIN, *The Origin of Species*

3.0 INTRODUCTION

ANY animal living in a particular environment may be expected to grow at a certain rate, to live for a certain period, and to produce a certain number of offspring, usually spread over a certain span of its life. For any one species, each individual in the population will have its own particular speed of development, longevity, and fecundity at different ages in its life. It is usually more useful, however, to speak of the mean values for the population. There will be a mean rate of growth of individuals in the population; a mean longevity, which is more usefully considered as a distribution of ages at which different individuals die; and a mean fecundity, which is more usefully considered as mean birth-rates at different ages of the mothers. The values of these means are determined in part by the environment and in part by a certain innate quality of the animal itself. This quality of an animal we may call its *innate capacity for increase*. We shall devote a full chapter to the discussion of this concept, because it is important in relation to the distribution and abundance of animals in nature and because, despite its importance, it has hitherto received but little attention. Animal ecologists, in stressing the environment, have tended to overlook the innate qualities of the animal or else, recognizing the need to consider the animal, have insisted on a sharp line of division between the "biotic potential" of the animal and the "resistance" of the environment. This was a mistake, for the animal's innate capacity for increase cannot be considered apart from its environment.

Nevertheless, the innate capacity for increase is a quality which is just as characteristic of the species as is, for example, its size. It is, however, a character that is more difficult to measure and define, since it may vary widely in different environments. The analogy with size may be carried a little further, for it is well known that for some animals the size may vary with such components of the environment as temperature, moisture, food, and so on. Nevertheless, as a rule, ordinary variations in the environment may not make much difference to the size of the animal, and it is often sufficient to define size without any special reference to the environment from which the animal has come. On the other hand, relatively small changes in one or another component of the environment may result in enormous differences in the animal's innate capacity to increase; so when this character is being considered, it is always necessary to define very carefully the particular environment in which the animal is living.

Environments in nature rarely, if ever, remain constantly favorable or constantly unfavorable but fluctuate irregularly between the two extremes. The animal's innate capacity for increase fluctuates correspondingly, being sometimes positive and sometimes negative. While conditions remain favorable and the innate capacity for increase remains positive, the numbers increase. If it remained so indefinitely, the species would continue to multiply until eventually it covered the earth. The elephant probably has the lowest innate capacity for increase of any animal. Even so, calculations based on the assumption of continuous increase lead to the conclusion that the progeny of one pair of elephants could populate the earth within several hundred years. Darwin estimated that a single pair of elephants could give rise to 19,000,000 elephants after a period of 750 years. But elephants do not increase like this. Adverse weather, food shortages, and a limitation of the number of suitable places in which elephants can live impose checks to increase; from time to time the rate of increase becomes zero or negative. Once the checks are lifted, the population will increase again. The rate of multiplication will be determined by the innate capacity for increase.

The concept of environment which we use in this book is described in chapter 2. Environment is considered to apply to individuals in a population but not to a population as a whole. The latter usage, though common, is so limited as to lead to endless confusion. The environment of individuals in a population is partly the other animals in the population. There is no clear-cut division between the population and its environment. But the distinction between the *individual* and its environment is clear-cut and can be defined with precision. This idea is very important in relation to the present chapter. It is desirable for the sake of simplicity to speak of the birth-rate or the death-rate within a population. This need not be confusing if it is remembered that the death-rate of a population may be found by integrating the expectation of life of all the individuals that comprise it. It is also made clear in chapter 2 that any environment may be analyzed into four general sets of components, namely, weather, food, other animals, and a place in which to live. Each one of these will be discussed in detail in Part III. In nature, one or several components may predominate to determine the *actual* rate of increase, which we shall call r. But in an experiment it is possible to exclude predators, diseases, and all other organisms of different kinds; and food, space, and other animals of the same kind can be artificially kept at optimal levels. Temperature, moisture, and the other components of weather and the quality of food cannot be excluded from the experiment in the same way, but, in an experiment, they can be kept artificially constant. We define r_m, the innate capacity for increase, as the maximal rate of increase attained at any particular combination of temperature, moisture, quality of food, and so on, when the quantity of food, space, and other animals of the same kind are kept at an optimum and other organisms of different kinds are excluded from the experiment. This is an approximate definition, because it does not mention the distribution of ages in the population. The precise definition is given in section 3.132.

3.1 THE INNATE CAPACITY FOR INCREASE

Biologists have sometimes referred to "the reproductive rate" as the characteristic defining the capacity of a species to increase in numbers. But, without the additional information of the death-rate (or survival-rate) of the offspring, this tells little about the innate capacity of the species to increase. Darwin pointed this out in *The Origin of Species.* There is no relation between the numbers of eggs laid by animals and the abundance of these animals in nature. The number in nature depends upon a balance struck between birth-rate and death-rate. It has been a characteristic feature of quite a number of ecological studies, particularly those with insects, to study one or another characteristic of the species, such as fecundity, speed of development, longevity, and so on. From such information, inferences have been made about distribution and abundance. It is quite possible that a study of one particular function alone may lead to different conclusions from those which might be drawn from the study of some other function (e.g., Holdaway, 1932). It is necessary rather to consider the sum-total effect of all those functions which make for increase in numbers and which make one species differ from another, that is to say, all those functions which influence birth-rate and survival-rate.

An animal's innate capacity for increase depends upon its fecundity, longevity, and speed of development. With a population, these are measured by the birth-rate and the survival-rate (or its inverse, the death-rate). When the birth-rate exceeds the death-rate, the population increases in numbers at a rate dependent upon the difference between them. When they are the same, the population numbers remain stable. When death-rate exceeds birth-rate, the population declines at a rate dependent upon the difference between them. This argument is simple enough. But complexity is introduced as soon as we seek to estimate quantitatively the rate at which the population increases or decreases. *The difficulties are encountered because both the number of births and the probability of death vary with the age of the animal.*

The students of human populations were the first to appreciate this principle, and Lotka (1925) derived a function which he called "the intrinsic rate of natural increase" to take into account the changes in birth-rates and death-rates with age. Demographers have coined certain expressions which we shall also use. The table which sets out the detailed distribution of birth-rate with age is called the "age-schedule of births," and the detailed distribution of deaths with age is called the "age-schedule of deaths." The particular birth-rate and death-rate which are characteristic of a particular age-group are called the "age-specific birth-rate" and the "age-specific death-rate." The "intrinsic rate of natural increase" is sometimes called the "Malthusian parameter," because of the emphasis given by Malthus to the geometric rate with which populations could theoretically increase in numbers. We shall not use the expression "Malthusian parameter" because it is misleading. There is a precise meaning in the science of biometry for the word "parameter." The symbol r_m is quite clearly a "statistic" and not a "parameter" (see R. A. Fisher, 1946, *Statistical Methods for Research Workers*, p. 7). If we knew the exact value of the "intrinsic rate of natural increase" of an animal, we would call that a "parameter" and specify it by a Greek letter. We cannot, in fact, know the parameter exactly, but we can make an estimate of its value. This estimate is termed a "statistic." There is a sense in which every reproductive pair of individuals in the population has its "intrinsic rate of natural increase," but we are always interested in the estimate of a mean value for the population. This is the statistic r_m, which we call the "innate capacity for increase."

Although r_m was originally devised for the study of human populations (Lotka, 1925), its relevance to ecology was at least implicit in A. J. Lotka's book, *Elements of Physical Biology*, published in 1925, and in R. A. Fisher's *The Genetical Theory of Natural Selection*, published in 1930. To Leslie and Ranson, of the Bureau of Animal Population in Oxford, however, must be attributed the credit of having adapted the statistic r_m for a more general use in animal ecology in their experimental determination of r_m for the vole *Microtus agrestis* (Leslie and Ranson, 1940). Since then its use has been extended to insect populations by Birch (1948, 1953*a*, *b*), Leslie and Park (1949), Evans and Smith (1952), and Howe (1953*a*, *b*). And we may expect that it will come to be more widely used by ecologists.

Before considering the way in which the appropriate tables of birth-rates and death-rates may be constructed, it is best to consider the relationship between birth-rates, death-rates, and the innate capacity for increase, r_m.

3.11 *The Relationship between Birth-Rate, Death-Rate, and the Innate Capacity for Increase*

The rate of increase of a population which has an assumed constant age-schedule of births and deaths and which is increasing in numbers in an unlimited space is given by.

$$\frac{\delta N}{\delta t} = bN - dN$$
$$= (b - d)N,$$

where t denotes time and b and d are constants representing the instantaneous birth-rate and death-rate. Now $b - d$ is the infinitesimal rate of increase which is the innate capacity for increase, r_m. Hence

$$\frac{\delta N}{\delta t} = r_m N.$$

This is the differential form of the equation describing the curve of geometric increase of an infinitely expanding population.

In the integrated form:

$$N_t = N_0 e^{r_m t}$$

where

N_0 = number of animals in time zero,
N_t = number of animals in time t,
r_m = innate capacity for increase,
e = base of Naperian logs.

Since we are dealing with geometric increase, it follows that the relationship between $\log_e N$ and t is linear. The slope of this line is the value r_m. This is made clear by the following:

$$N_t = N_0 e^{r_m t};$$

thence

$$\log_e N_t = \log_e N_0 + r_m t \quad \text{(since Naperian log of } e = 1)$$
$$= a + r_m t \quad \text{(where } a \text{ is a constant).}$$

This is the equation for the straight line with co-ordinates $\log_e N$ and t. The infinitesimal rate of increase r_m should not be confused with the finite rate of increase, i.e., the number of individuals added to the population per head per week. An example will make the distinction clear. Consider a population which multiplies ten times in every 2 weeks. The infinitesimal rate of increase may be shown to be 1.1513 by the following calculations:

$$\text{Since} \quad N_t = N_0 e^{r_m t}, \quad \text{then} \quad \frac{N_t}{N_0} = e^{r_m t},$$

$$\text{and} \quad \therefore r_m = \frac{\log_e(N_t/N_0)}{t} = \frac{\log_e 10}{2} = 1.1513 \text{ per head per week.}$$

The finite rate of increase of the same population is 3.16 per head per week, as shown by the following. Let the finite number of individuals arising from one female in one week be λ. Then, by definition,

$$\lambda = \frac{N_t}{N_0} \quad \text{when} \quad t = 1;$$

$$\text{but} \quad \frac{N_t}{N_0} = e^{r_m} \quad (\text{when } t = 1),$$

$$\text{i.e.,} \quad \lambda = e^{r_m}$$

$$= \text{antilog}_e \; r_m.$$

In this example $\lambda = \text{antilog}_e\ 1.1513 = 3.16$ per head per week. Hence a population which multiplies ten times in every 2 weeks has an infinitesimal rate of increase of 1.1513 and a finite rate of 3.16 ($= \sqrt{10}$). That is to say, the population multiplies 3.16 times per female per week. It follows that, by the end of the second week, one individual will have given rise to $(3.16)^2 = 10$ individuals, and by the end of the third to $(3.16)^3$, and so on.

3.12 *The Construction of Age-Schedules of Births and Deaths*

The generalized relationship between birth-rate, death-rate, and r_m was discussed in section 3.11 without any consideration of the complete expression of both birth-rates and death-rates in terms of age-groups in the population. The actual calculation of r_m, however, involves a knowledge of these details.

The birth-rate of a population is best expressed as an "age-schedule of births." This is a table which gives the number of offspring (or eggs in the case of insects) produced in unit time by a female aged x. This is usually designated m_x, where the suffix x denotes the age-group.[1] In these calculations we usually count only the females. If equal numbers of males and females are born, then m_x is the total number of eggs or offspring for each age-interval, divided by 2. The nature of this statistic will be clear from an examination of Tables 3.01 and 3.02 and Figures 3.01 and 3.02, which show, among other things, the age-schedule of births for two very different sorts of populations, a vole and an insect (Leslie and Ranson, 1940; Birch, 1948).

The m_x values for the vole in Table 3.01 give the mean number of live daughters born per 8 weeks per female of ages shown in the x column. The female vole begins to breed when it is about 3 weeks old. In Table 3.02 the m_x values for the weevil *Calandra oryzae* are given in terms of the total eggs laid per female per week, divided by 2, since the sex-ratio is unity. In each case

[1] **A glossary of terms used in this chapter is given at the end of the chapter.**

the values of x refer to the mid-point of each age-group. The intervals for the age-groups may be chosen quite arbitrarily and depend partly on the method by which the data were obtained. Birth is taken as zero age in the case of the vole, and the time the egg was laid is taken as zero age in the case of the

TABLE 3.01*

LIFE-TABLE, AGE-SPECIFIC FECUNDITY-RATES, AND NET REPRODUCTION-RATE (R_0) OF THE VOLE *Microtus agrestis* WHEN REARED IN THE LABORATORY

Pivotal Age in Weeks (x)	l_x	m_x	$l_x m_x$
8	0.83349	0.6504	0.54210
16	.73132	2.3939	1.75071
24	.58809	2.9727	1.74821
32	.43343	2.4662	1.06892
40	.29277	1.7043	0.49897
48	.18126	1.0815	0.19603
56	.10285	0.6683	0.06873
64	.05348	0.4286	0.02292
72	0.02549	0.3000	0.00765
		$R_0 = 5.90424$	

* After Leslie and Ranson (1940).

TABLE 3.02*

LIFE-TABLE, AGE-SPECIFIC FECUNDITY-RATES, AND NET REPRODUCTION-RATE (R_0) OF RICE WEEVIL *Calandra oryzae* AT 29° C. IN WHEAT OF 14 PER CENT MOISTURE CONTENT

Pivotal Age in Weeks (x)	l_x	m_x	$l_x m_x$	
0.5, 1.5, 2.5, 3.5	0.90	....	....	Immature stages: egg, larva, and pupa
4.5.......	.87	20.0	17.400	Adults
5.5.......	.83	23.0	19.090	
6.5.......	.81	15.0	12.150	
7.5.......	.80	12.5	10.000	
8.5.......	.79	12.5	9.875	
9.5.......	.77	14.0	10.780	
10.5.......	.74	12.5	9.250	
11.5.......	.66	14.5	9.570	
12.5.......	.59	11.0	6.490	
13.5.......	.52	9.5	4.940	
14.5.......	.45	2.5	1.125	
15.5.......	.36	2.5	0.900	
16.5.......	.29	2.5	0.725	
17.5.......	.25	4.0	1.000	
18.5.......	0.19	1.0	0.190	
		$R_0 = 113.485$		

* After Birch (1948).

insect. The fecundity-table for *C. orzyae* shown in Table 3.02 and Figure 3.02 is quite typical of many insects. In contrast to the vole, the sexually immature stages of the insect (egg, larva, pupa, and early adult) occupy a greater proportion of the total life-span. In both the case of mammals such as the vole and insects such as the weevil, the data can be obtained experimentally by observing a number (or "cohort") of individuals from the time they are born until they die and recording the number of offspring they produce as they grow older. Similar information for human populations cannot be obtained in this manner, as individuals do not always remain in the same place for such long-term observations. The age-schedule of fecundity of a

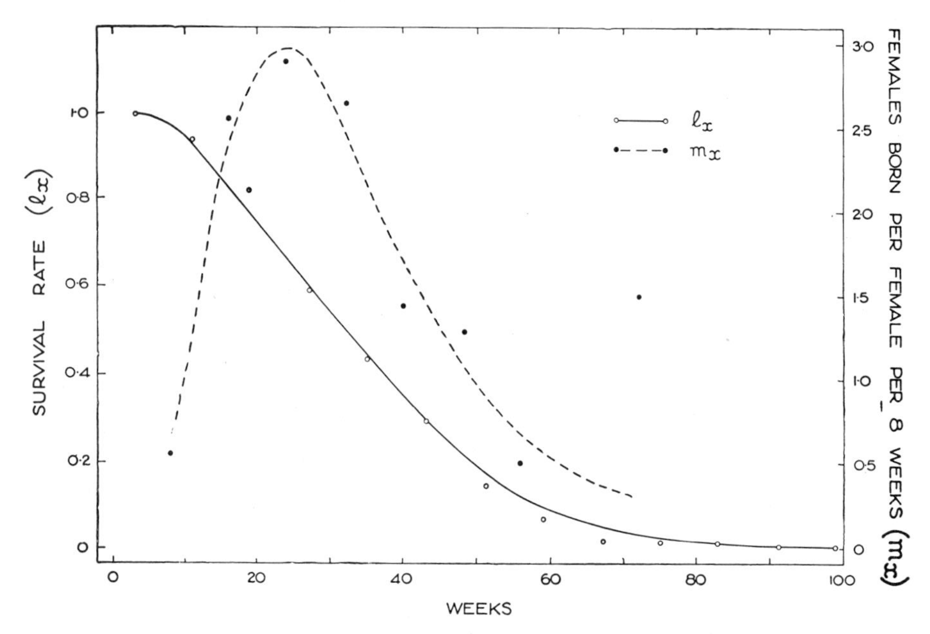

FIG. 3.01.—The life-table and age-specific fecundity curves for the vole *Microtus agrestis* reared in the laboratory. (After Leslie and Ranson, 1940.)

human population can, however, be compiled from census figures, which give details of the age at which women in the community give birth to offspring (Pearl, 1924).

The offspring from the parent-generation die at different ages. The table which gives this information is known as the "life-table." Again we usually count only the females; but life-tables can, of course, be compiled for males as well. The life-table gives the probability at birth of being alive at age x. This probability is usually designated l_x. At zero age ($x = 0$) it is designated l_0 and is taken as having a value of unity ($l_0 = 1$). The second columns of Tables 3.01 and 3.02 and Figures 3.01 and 3.02 show the life-tables for the vole *Microtus agrestis* and the weevil *C. oryzae*. These life-tables are not complete but cover the reproductive period of life, which is all that is necessary to consider in studying rates of increase. Referring to Table 3.02, the figures $x =$ 4.5 and $l_x = 0.87$ mean that, from a sample of 100 eggs of zero age, 87 per cent survive to the mid-point between the fourth and fifth week. Between the eighteenth and nineteenth weeks, only 19 per cent are still alive. In the case of both the vole and the weevil the life-table was obtained by following through the survival of a sample of individuals from birth until the last member of the population died.

Life-tables were first used by students of human populations. It is only in recent years that ecologists have begun to realize their importance. There are,

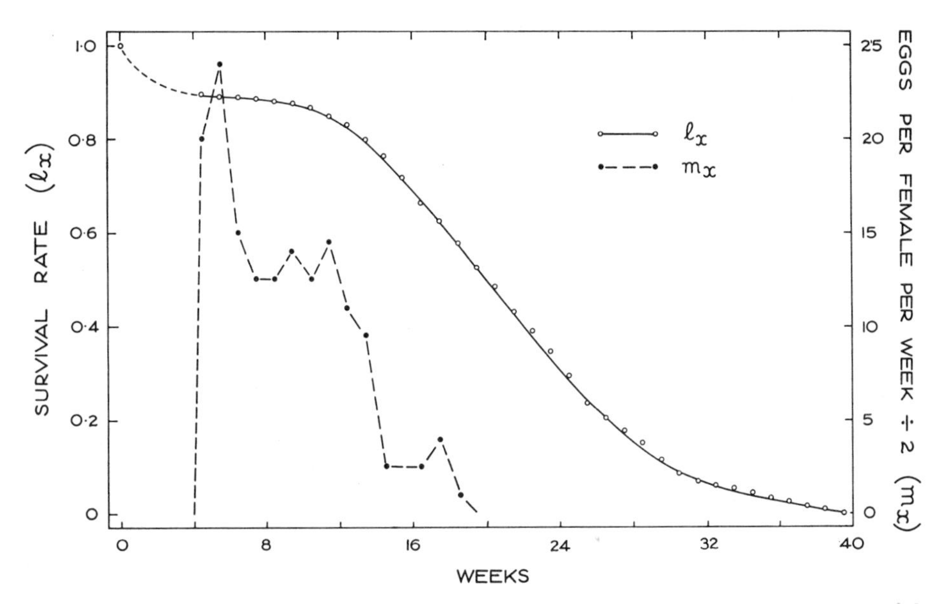

FIG. 3.02.—The life-table and age-specific fecundity curves for the weevil *Calandra oryzae* at 29° C. in wheat of 14 per cent moisture content. The curves provide all the information necessary for the estimation of the innate capacity for increase. (After Birch, 1953*a*.)

consequently, not many complete life-tables for other species. The information which is available has been largely determined since 1935, when Pearl and Miner (1935) compared the life-tables of the few organisms which had then been studied with any degree of completeness (Deevey, 1947). From the standpoint of this chapter we are primarily interested in life-tables of animals reared artificially in the laboratory. The life-table of populations in nature may be quite different, for in nature the population is exposed to numerous hazards which will destroy members of the population in different age-groups, depending upon what are the chief causes of deaths. These life-tables have been called "ecological life-tables." Such information is generally much more difficult to determine than experimental life-tables. Deevey (1947) has adequately reviewed the available information on ecological life-tables.

It is sometimes convenient to express the life-table in forms other than a survivorship or l_x table. Although alternative modes of expressing the same information do not concern us in the calculation of the innate capacity for increase, brief mention will be made of two alternatives of particular use in certain population studies. These are the d_x and the L_x curves. The curve giving number of deaths within age-intervals is know as the d_x curve. The L_x curve, also known as the life-table age-distribution or life-table age-structure, gives the proportion of individuals alive which are between ages x and $(x + 1)$ (Dublin, Lotka, and Spiegelman, 1949):

$$L_x = \int_x^{x+1} l_x \delta x.$$

In practice it is usually adequate to ignore the curvature between age-groups; hence

$$L_x = \frac{l_x + l_{x+1}}{2}.$$

The life-table age-distribution assumes special importance in the consideration of age-structure of populations, as will be shown later in this chapter.

3.13 *Calculation of the Innate Capacity for Increase*

The innate capacity for increase was first worked out in the study of human populations. Because of the relatively large part of the human life-span occupied by pre-reproductive development and the relatively low birth-rate, it is sufficient to use an approximate arithmetical procedure in calculating r_m. With small mammals and insects, in which the birth-rate and length of life are likely to be quite different from those for human populations, a more precise solution is needed. We shall follow the historical approach and give the approximate solution first.

3.131 APPROXIMATE CALCULATION OF r_m

We have seen how the geometric increase of a population is given by the equation

$$N_t = N_0 e^{r_m t},$$

where N_0 is the number of reproducing individuals at time t_0, N_t is the number of reproducing individuals at time t, and r_m is the innate capacity for increase. This relationship provides us with a means of calculating the innate capacity for increase from the age-schedule of births and the age-schedule of survival-rates (life-table), i.e., the m_x and l_x columns of Tables 3.01 and 3.02. In calculating r_m from these data, it is convenient to deal in time-intervals of generations. Now the number of individuals at the end of a generation will be

$$N_T = N_0 e^{r_m T},$$

where T is the mean length of a generation. Hence

$$\frac{N_T}{N_0} = e^{r_m T}.$$

Now N_T/N_0 is simply, by definition, the ratio of total female births in two successive generations. It is the multiplication per generation or *net reproduction-rate*. It is usually designated R_0. Thus

$$R_0 = e^{r_m T}$$

and

$$r_m = \frac{\log_e R_0}{T}.$$

The intrinsic rate of natural increase can thus be determined if the net reproduction-rate and the mean length of a generation are known. Both these statistics can be readily estimated from the tables giving the age-schedule of births and survival-rates. The calculations involve nothing more complicated than simple arithmetic.

The method of estimating R_0 is set out in Tables 3.01 and 3.02. The $l_x m_x$ products for each age-group are summed, the total being the value of R_0, i.e., $R_0 = \Sigma l_x m_x$. The net reproduction-rate of the vole, for example, is 5.904. In other words, a population of voles has the capacity to multiply 5.904 times in each generation.

The estimation of T is also straightforward. The mean duration of a generation is defined as the mean period elapsing from birth of parents to birth of offspring. Clearly, this is only an approximate definition, since offspring are not born at any one time but over a period, e.g., in the vole from about the fourth to the seventy-second week of life. Both the mean age of the mother at which offspring are born and the extremes of age at which they are born vary for each individual in the population. But we may consider the births for each generation as concentrated at one moment, with successive generations spaced T units apart, where T is the mean duration of a generation (Dublin and Lotka, 1925). It may be defined approximately as follows:

$$T = \frac{\Sigma l_x m_x x}{\Sigma l_x m_x}.$$

The figures for the product $l_x m_x$ in the last column of Tables 3.01 and 3.02 may be regarded as a frequency-distribution. The mean of the distribution is the approximate value of T. In these particular examples:

$$M.\ agrestis: \quad T = \frac{143.75}{5.904} = 24.4 \text{ weeks},$$

$$C.\ oryzae: \quad T = \frac{941.85}{113.49} = 8.3 \text{ weeks}.$$

Knowing R_0 and T, we can determine the innate capacity for increase. In the example of the weevil *C. oryzae* (Table 3.02) the estimate of r_m is made as follows:

$$R_0 = 113.49, \quad T = 8.30 \text{ weeks},$$

$$r_m = \frac{\log_e 113.49}{8.30} = 0.56.$$

Now this estimate of r_m is approximate only, since the procedure outlined for estimating the mean length of a generation (T) was approximate only. The true estimate of T necessitates a preliminary calculation of r_m, as can be seen from the following relationship:

$$R_0 = e^{r_m T}.$$

Hence

$$T = \frac{\log_e R_0}{r_m}.$$

In some cases the first approximation to r_m obtained by the above method may be accurate enough. It is sufficiently accurate for most estimates of rates of increase of human populations. With animals which have a greater rate of increase, this method may be too crude. This is the case in the example of the weevil *C. oryzae*. We described the direct method first, because it illustrates quite simply the relationship between birth-rates, survival-rates, mean length of a generation, net reproduction-rate, and the innate capacity for increase.

A glance at Figures 3.01 and 3.02 will show the importance of the nature of the curves for birth-rate and survivorship in determining the innate capacity for increase of a population. The maximal rate of increase would be given when the peaks of both curves occur at the same point on the same time-axis. If the birth-rate curve were moved farther to the right along the time-scale, the capacity of the population to increase in numbers would be lower. It is, in fact, possible to calculate exactly the outcome of moving one curve along the time-axis with respect to the other and so have a precise measure of the way in which the two components, birth-rate and survival-rate, determine the innate capacity of a population to increase in numbers.

3.132 PRECISE CALCULATION OF r_m

Lotka (1925) showed mathematically that the distribution of ages in a population in which the birth-rates and death-rates for each age-group remain constant and which is increasing in unlimited space would approach a certain distribution which he called "the stable age-distribution" because it would not vary with time. Lotka also showed that, as such a population approached its "stable age-distribution," its rate of increase also approached a certain constant, which he called the "the intrinsic rate of increase." This is the quantity which we have called the "innate capacity for increase," r_m. It will now be seen that the definition of r_m given in section 3.0 was approximate, for it did not specifiy that the population should have a stable age-distribution. The precise value of r_m may be obtained by solving the following equation:

$$\int_0^\infty e^{-r_m x} l_x m_x \, \delta x = 1,$$

where 0 to ∞ is the life-span of the reproductive stages and l_x and m_x are as defined on pages 36, 38. The accurate solution of this equation involves some tedious calculation. Various approximations are usually permissible which reduce the calculations to simple arithmetical procedures. The product $l_x m_x$ is known for each age-group (see Tables 3.01 and 3.02). Various values of r_m can be substituted in the equation, and the true value obtained by graphical interpolation (Leslie and Ranson, 1940; Birch, 1948). When the calculation was made for the weevil *C. oryzae*, the value of r_m was found to be 0.76, to be compared with 0.56 obtained by the approximate method. Further simplifications in the procedure of estimating r_m for insects have been introduced by Howe (1953*a*); these modifications enable estimates to be made with greater speed and with an accuracy which is sufficient for most purposes. The usual method of calculating r_m for human populations may be found in Dublin and Lotka (1925, appendix) or Lotka (1939, p. 68).

3.2 THE STABLE AGE-DISTRIBUTION

By definition, the innate capacity for increase, r_m, is the actual rate of increase of a population with stable age-distribution. The actual rate of increase depends upon the distribution of ages in the population. Since the distribution of ages changes with time in all populations other than those in which the distribution is the stable age-distribution, the actual rate of increase will also change with time. This will be true even though the values for l_x and m_x (the birth-rates and death-rates in the different age-groups) remain constant. For these reasons, r_m is the only statistic which adequately summarizes the physiological qualities of an animal which are related to its capacity for increasing. This is the reason for keeping the word "innate" in the definition of r_m. For the same reasons, r_m is the only statistic which serves to compare different species with respect to these qualities. The actual rates of increase which ignore the distributions of ages in the populations are no good at all for these purposes.

The stable or Malthusian age-distribution may be calculated from the life-table and the innate capacity for increase. Thus if C_x is the proportion of the population of stable age-distribution aged between x and $x + \delta x$ and b is the instantaneous birth-rate,

$$C_x = b e^{-r_m x} l_x.$$

The instantaneous birth-rate can be calculated from the following equation:

$$\frac{1}{b} = \int_0^\infty e^{-r_m x} l_x \, \delta x.$$

For the usual methods of computation, reference should be made to Dublin and Lotka (1925) and Leslie and Ranson (1940). Leslie has, however, pointed out another method of calculation which saves much of the numerical integration involved in the more usual methods (Birch, 1948). If at time t we consider a stable population consisting of N_t individuals and if during the interval of time t to $t + 1$ there are B_t female births, we may define a birth-rate as follows:

$$\beta = \frac{B_t}{N_t}.$$

Then, if we define for the given life-table (l_x) the series of values L_x by the relationship

$$L_x = \int_x^{x+1} l_x \, \delta_x$$

(the stationary or life-table age-distribution of the actuary), the proportion (P_x) of individuals aged between x and $x + 1$ in the stable population is given by

$$P_x = \beta L_x e^{-r_m(x+1)},$$

$$\frac{1}{\beta} = \sum_{x=0}^{m} L_x e^{-r_m(x+1)},$$

where $x = m$ to $m + 1$ is the last age-group considered in the complete life-table age-distribution. It will be noticed that the life-table (l_x) values for the complete age-span of the species are required for the computation of P_x and β. But where r_m is high, it will be found that, for the older age-groups, the terms $L_x e^{-r_m(x+1)}$ are so small and contribute so little to the value of β that they may be neglected.

The calculations involved are quite simple and are illustrated in the following example for *C. oryzae* at 29° C. (Table 3.03). In the present example, instead of calculating the values of L_x, the values of l_x were taken at the mid-points of each age-group. This was considered sufficiently accurate in the present instance. It should also be pointed out that, whereas only the total deaths among immature stages were required in the calculation of r_m, the age-specific mortality of the immature stages is needed for the calculation of the stable age-distribution. In this example 10 per cent of the immature stages died—and 98 per cent of these deaths occurred during the first week of larval life (Birch, 1945*a*). Hence the approximate value of L_x for the mid-point of the first week will be 0.95 and thereafter 0.90 for successive weeks of the larval and pupal period (second column, Table 3.03). The stable age-distribution is shown in the fifth column of Table 3.03. This column simply expresses the fourth column of figures as percentages. It is of particular interest to note the high proportion of immature stages (95.5 per cent) in this theoretical population. This is associated with the high value of the innate capacity for increase. It emphasizes a point of practical importance in estimating the abundance of insects such as *C. oryzae* and other pests of stored products. The number of adults found in a sample of wheat may be quite a misleading representation of the true size of the whole insect population. Methods of sampling are required which will take account of the immature stages hidden inside the grains, such, for example, as the "carbon dioxide index" developed by Howe and Oxley (1944). The

TABLE 3.03*

CALCULATION OF THE STABLE AGE-DISTRIBUTION OF *Calandra oryzae* AT 29° C. WHEN $r_m = 0.76$

Age-Group (x)	L_x	$e^{-r_m(x+1)}$	$L_x e^{-r_m(x+1)}$	Percentage Distribution $100\beta L_x e^{-r_m(x+1)}$	
0	0.95	0.4677	0.4443150	54.740	95.5 per cent total immature stages
1	.90	.2187	.1968300	24.249	
2	.90	.10228	.0920520	11.341	
3	.90	.04783	.0430470	5.304	
4	.87	.02237	.0194619	2.398	4.5 per cent total adults
5	.83	.01046	.0086818	1.070	
6	.81	.00489	.0039609	0.488	
7	.80	.002243	.0017944	0.221	
8	.79	.001070	.0008453	0.104	
9	.77	.000500	.0003850	0.047	
10	.74	.000239	.0001769	0.022	
11	.66	.000110	.0000726	0.009	
12	.59	.000051	.0000301	0.004	
13	.52	.000024	.0000125	0.002	
14	0.45	0.000011	0.0000050	0.001	
			$1/\beta = 0.8116704$	100.000	

* After Birch (1948).

nature of this stable age-distribution has a bearing on another practical problem. It adds further evidence to that developed from a practical approach (Birch, 1946*a*), as to how it is possible for *C. oryzae* to cause heating in vast bulks of wheat, when only a small density of adult insects is observed. It is a reasonable supposition that the initial rate of increase of insects in bulks of wheat may approach the innate capacity for increase and therefore that the age-distribution may approach the stable form. Little, however, is known about the actual age-distribution in nature at this stage of an infestation. Howe (1953*a*) made an estimate of the age-distribution of *C. oryzae* in a bin of wheat at the peak of an infestation and found that 11 per cent of the population were adults. This is fairly close to the theoretical estimate of 4.5 per cent of adults in a stable age-distribution at 29° C., which is shown in Table 3.03.

The stable age-distribution of the vole is shown for comparison, together with its life-table or stationary age-distribution in Table 3.04. Conditions under which the life-table distribution might be expected will be discussed in chapter 9. It will be noticed that the older age-groups are not represented in Tables 3.03 and 3.04. This does not mean that there would be none of these old animals living in those populations but that the dilution of the population with young is so great that in a random sample of a thousand or so the probability of including a vole older than 56 weeks or a weevil older than 9 weeks is small. Tables 3.03 and 3.04 indicate the sort of age-distribution which is approached by a rapidly growing population when the density of the population in terms of numbers per unit of space is relatively low.

TABLE 3.04*
STABLE AGE-DISTRIBUTION AND LIFE-TABLE AGE-DISTRIBUTION OF THE VOLE *Microtus agrestis*.

Age-Group	Stable Age-Distribution	Life-Table Age-Distribution
0	57.7	23.5
8	25.5	21.2
16	10.7	17.8
24	4.1	13.8
32	1.4	9.7
40	0.5	6.3
48	0.1	3.8
56	....	2.1
64	....	1.0
72	....	0.5
80	....	0.2
88	....	0.1
	100.0	100.0

* After Leslie and Ranson (1940).

3.3 THE INSTANTANEOUS BIRTH-RATE AND DEATH-RATE

We have seen that the rate of increase of a population at any time may be given by the difference between its crude birth-rate and its crude death-rate. We have also shown that this difference does not give the innate capacity of the population to increase in numbers, because the age-distribution of the population has not been taken into account. But the capacity of the population to increase in numbers is adequately measured by the rate of increase that the population would have if it had a stable age-distribution. This also gives an adequate basis of comparison for different populations. A population with a stable age-distribution has, of course, a birth-rate, b, and a death-rate, d. It is the difference between these two quantities which we have designated the innate capacity for increase, r_m:

$$r_m = b - d.$$

In the process of estimating the innate capacity for increase, neither the birth-rate nor the death-rate is directly estimated. They may, however, be estimated. once r_m is known.

We have already defined a birth-rate by the expression

$$\frac{1}{\beta} = \sum_{x=0}^{m} L_x e^{-r_m(x+1)}.$$

This is not, however, the same as the instantaneous birth-rate, b. The relationship between the two is given by

$$b = \frac{r_m \beta}{e^{r_m} - 1}.$$

Thus, in the example for *C. oryzae*, we have $1/\beta = 0.81167$ (Table 3.03), $r_m = 0.76$, and thus $b = 0.82$; and the difference between r_m and b is the instantaneous death-rate, $d = 0.06$.

3.4 THE INFLUENCE OF ENVIRONMENT ON THE INNATE CAPACITY FOR INCREASE

It is now clear why the innate capacity for increase cannot be expressed quantitatively except for a particular environment. In the case of poikilothermic animals such as insects, temperature and moisture are two of the important components which influence longevity and fecundity and therefore the innate capacity for increase. Temperature may be less important for homoiotherms, though for any particular homoiotherm it is hardly possible at this stage of our knowledge to say just how temperature might influence its capacity to increase in numbers. Any component of the environment (other than than those which have been excluded by definition—see sec. 3.0) may influence the value of the innate capacity for increase. We shall give three simple illustrations of this principle.

TABLE 3.05

INFLUENCE OF TEMPERATURE AND MOISTURE ON FINITE RATE OF INCREASE (λ)* OF *Calandra oryzae* (SMALL "STRAIN") AND *Rhizopertha dominica* IN WHEAT AT CERTAIN COMBINATIONS OF TEMPERATURE AND MOISTURE CONTENT

C. oryzae						*R. dominica*					
TEMP. (°C.)	Per Cent Moisture Content of Wheat					TEMP. (°C.)	Per Cent Moisture Content of Wheat				
	14	12	11	10.5	10		14	11	10	9	8
13.0......	0	0	0	0	0	18.3......	0	0	...	...	0
15.2......	1.007	0	0	0	0	22.0......	...	1.11	0	...	0
18.2......	1.15	...	1.10	0	0	26.0......	...	...	1.28	...	0
23.0......	1.54	1.35	...	...	.	32.3......	1.99	...	...	...	0
25.5......	1.83	...	1.36	1.07	0	34.0......	2.14	...	1.47	1.38	0
29.1......	2.15	1.50	1.42	0.96	0	36.0......	...	...	1.30	0	0
32.3......	1.65	...	...	0	0	38.2......	1.34	1.19	0	0	0
33.5......	1.13	1.00	0	0	0	38.6......	0	0	0	0	0
34.0......	...	0	0	0	0						

* $\lambda = e^{r_m}$ = rate of increase per female per week. After Birch (1953*a*).

Temperature and moisture both influence the innate capacity for increase of these two grain beetles, *C. oryzae* and *R. dominica*. In many ways these two beetles are suitable for laboratory studies; it is possible to determine their age-specific fecundity and survival-rates at various combinations of temperature and moisture. From this information the innate capacity for increase can be calculated for various combinations of temperature and moisture within the range in which the species survive. There is a zone of temperature and moisture within which the capacity for increase is greatest. There is a zone beyond which no increase can occur at all. This is shown in Table 3.05 and is illustrated in Figure 3.03. The rate of increase has been plotted in the figure as a finite rate of increase (λ). This is the multiplication of the population per female per week.

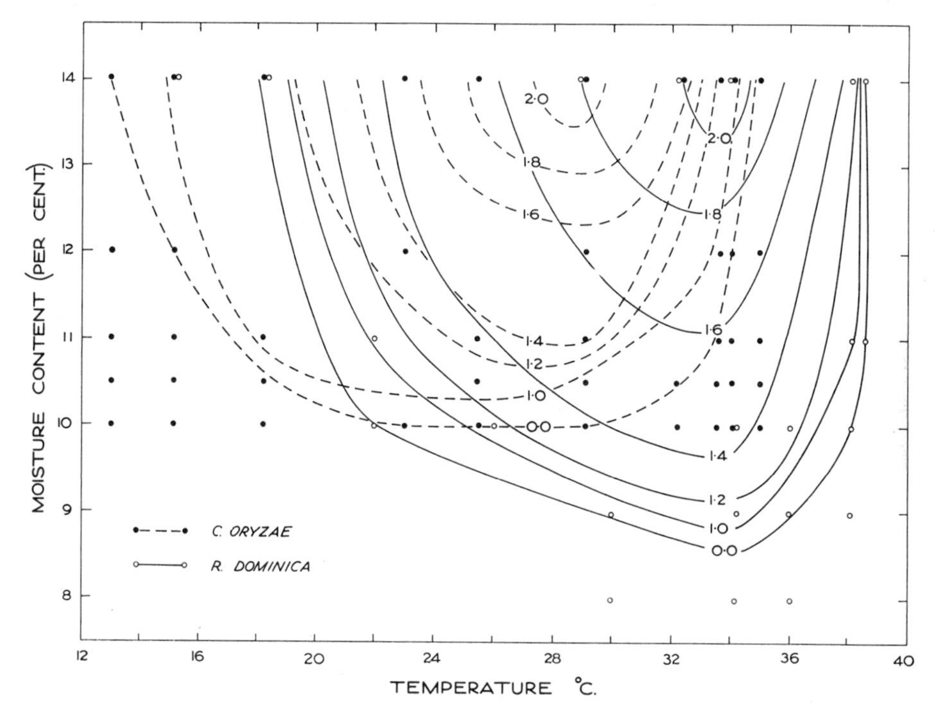

FIG. 3.03.—The finite rate of increase (rate of increase per female per week λ) of *Calandra oryzae* (small "strain") and *Rhizopertha dominica* living in wheat of different moisture contents and at different temperatures. The points on the graph show combinations of temperature and moisture at which experiments were done; the values of λ at these combinations are given in Table 3.05. The lines have been drawn through equal values of λ. (After Birch, 1953*a*.)

It is a rather more familiar concept than the infinitesimal rate of increase, r_m. These values have been plotted on a graph with temperature and moisture as co-ordinates, and lines have been drawn through points of equal value of λ. The isopleth for unity defines the zone of temperature and moisture in which populations could maintain their numbers but could not increase ($r_m = 0$). When λ is less than unity (r_m is negative), the death-rate exceeds the birth-rate, and the population eventually dies out. The time it takes to become extinct will depend upon the value of λ. The zero isopleth ($r_m = -\infty$) shows the zone in which no insects of reproductive age are added to the population; if any eggs are laid, the death-rate is 100 per cent in the immature stages. For purposes of our present discussion, we can confine our attention to finite rates of increase. For other purposes, it is more convenient to consider the infinitesimal rate of increase, r_m. The relationship between r_m and λ over the range of values of λ shown in Figure 3.03 are summarized in Figure 3.05. This figure permits one expression to be readily changed to the other.

From Figure 3.03 we deduce that *R. dominica* might be expected to thrive better than *C. oryzae* in wheat that is hot and dry. But *C. oryzae* might be able

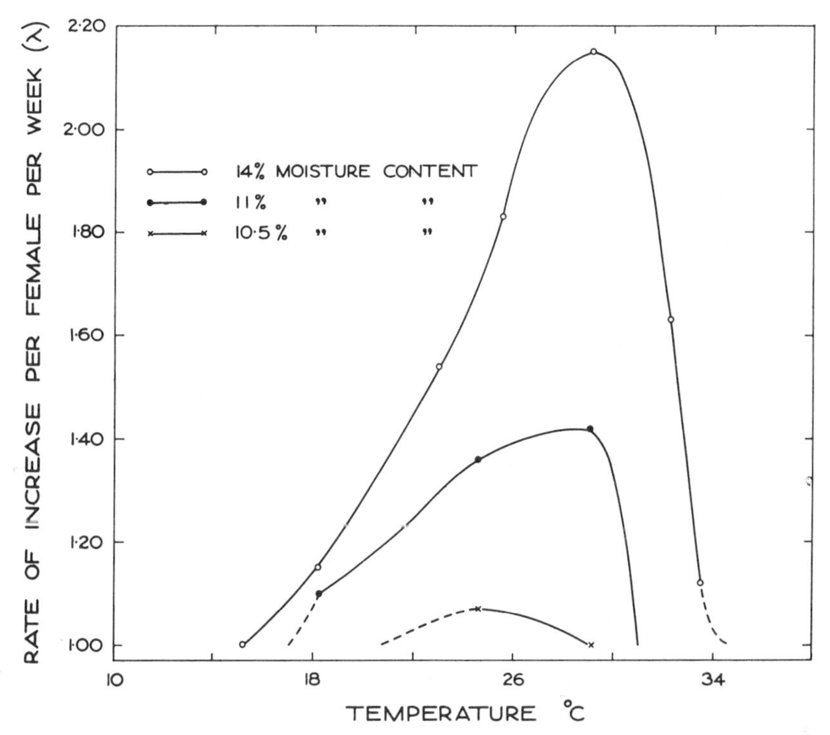

FIG. 3.04.—The finite rate of increase (λ) of *Calandra oryzae* (small "strain") living in wheat of different moisture contents and at different temperatures. (After Birch, 1953*a*.)

to multiply in wheat so cold that *R. dominica* could not maintain a population there. We might also expect the geographic distributions of the species to be related to climate in a way that is indicated by the position of the isopleths for zero-increase in Figure 3.03. The observed distributions of the two species in Australia agree with this hypothesis: *C. oryzae* is a serious pest wherever wheat is stored; *R. dominica* is also a pest in the warmer parts of the mainland but is virtually absent from Tasmania, which is colder.

If we were to consider only the information given in Figure 3.03, we might expect that populations would be able to persist in any place where the temperature and the moisture content of the wheat exceeded the values indicated by the isopleths for zero on the graph. But we would expect the rate of increase to be greater at certain combinations of temperature and moisture content than at others. The rate of increase is conveniently measured by λ. Values of λ for *C. oryzae* calculated for a number of combinations of temperature and moisture are shown in Figure 3.04. Reference back to Figure 3.03 shows how the two species differ in these respects. For example, at a temperature of 33° C. and in wheat of 12.5 per cent. moisture content, *C. oryzae* would take 16 weeks to reach a population of the size which *R. dominica* could reach in 10 weeks. The range

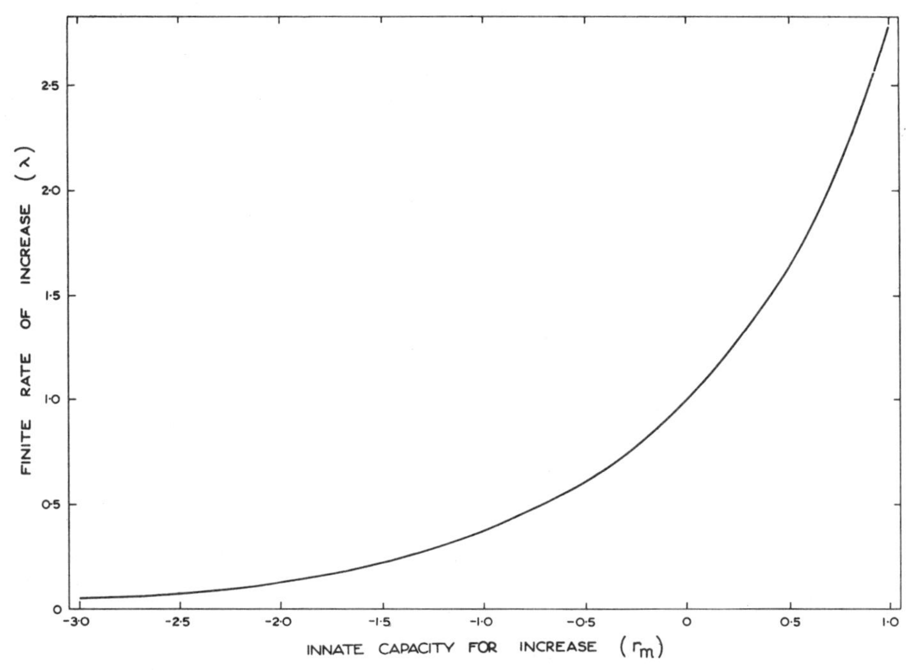

FIG. 3.05.—The relationship between the innate capacity for increase (r_m) and the finite rate of increase (λ).

of λ which is positive for the two species extends from λ = 1 to λ = 2.15. For the beetle *Ptinus tectus* the range is less, extending from λ = 1 (at about 13° C.) to a maximum of λ = 1.331 at 27° C. (Howe, 1953*a*). At first glance, neither of these ranges may seem great. However, quite small differences in the value of λ will cause big differences in the numbers which would be reached after several weeks. This can be readily appreciated by looking at Figure 3.06, which shows the multiplication per female after different numbers of weeks for different values of λ.

Another instructive feature of a graph such as Figure 3.03 is that it enables us to see at a glance the combinations of temperature and moisture at which two species have the same capacities for increase. These are shown by the intersections of isopleths of the two species. This information is tabulated in Table 3.06.

When we come to consider natural populations, it is necessary to take into account certain checks to increase which were artificially excluded from the experiments upon which Figure 3.03 was built. With these graminivorous species, predators are unimportant; they are not as a rule, influenced by the

TABLE 3.06
COMBINATIONS OF TEMPERATURE AND MOISTURE AT WHICH *Rhizopertha dominica* and *Calandra oryzae* HAVE SAME FINITE RATE OF INCREASE ($\lambda = e^{r_m}$)

λ	Temp. (°C.)	Moisture Content of Wheat (Per Cent)
1.0	23.5	10.3
1.2	21.5	12.0
1.4	24.5	11.7
1.6	28.3	12.8
1.8	30.0	13.0

distribution of their stocks of food because they can usually rely upon men to carry them to where their food is. But a colony breeding freely in a bin of wheat may be destroyed when the bin is emptied; or a period may be set to their multiplication by a change in the weather. If catastrophes come with

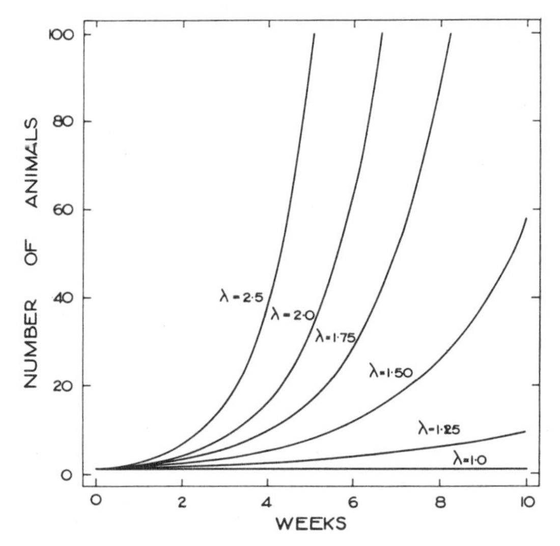

FIG. 3.06.—The relationship between the finite rate of increase (rate of increase per female per week λ) and the size of the population resulting from one female.

equal frequency in two places, then, other things being equal, the animals will be more numerous in the place where their innate capacity for increase is greater. Alternatively, in a place where λ is low, a much longer interval between catastrophes may be required if the animals are to become so numerous that the species may be considered "common" in that area.

Extending this idea to cover comparisons between different species whose requirements and behavior are essentially similar but which differ with respect to their innate capacities for increase, we might say that the common species will be those which have a sufficiently high value of λ to enable them to reach large numbers in the time available between catastrophes. The rare species will be those in which the innate capacity for increase is low relative to the intervals which are allowed them for multiplication between catastrophes. This hypothesis is supported by the values of r_m estimated by Howe (1953*a*) for 9 species of Ptinidae. One of these beetles, *Ptinus tectus*, is regarded as a serious pest of stored wheat in Britain. Some of the others are widely distributed, but none is numerous enough to be considered a serious pest. A glance at Table 3.07 shows that r_m was greatest for *P. tectus*, which is also the most numerous species in nature. The four species which follow *P. tectus* in Table 3.07 gave

TABLE 3.07
INNATE CAPACITY FOR INCREASE, r_m, AND FINITE RATE OF INCREASE (λ)* AT 25° C. FOR 9 SPECIES OF PTINIDAE

Species	r_m	λ
Ptinus tectus	0.395	1.462
Gibbium psylloides	.235	1.265
Trigonogenius globulus	.227	1.255
Stethomezium squamosum	.178	1.195
Mezium affine	.160	1.173
P. fur	.094	1.099
Eurostus hilleri	.072	1.075
P. sexpunctatus	.044	1.045
Niptus hololeucus	0.043	1.044

* $\lambda = e^{r_m}$ = rate of increase per female per week. After Howe (1953*a*).

values for λ which were about 18 per cent less than that for *P. tectus*. A value for λ of 1.25 indicates a tenfold increase in just over 10 weeks. But none of these species is considered a serious pest. A colony of one of these species, left undisturbed in a large bin of warm grain, would undoubtedly become numerous in due course. The fact that, in nature, none of these species is ever found in large numbers suggests that a period is set to their increase by the weather, by the emptying of the bin, or by some other "catastrophe" which is likely to be repeated before the colony has had time to become numerous. The risk for *Eurostus hilleri* may be even greater, for its innate capacity for increase is smaller still. During the 13 years which have elapsed since it was first discovered in Britain, it has spread widely over the country, but nowhere has it become abundant.

Birch (1953*a*) compared the innate capacity for increase of the small and the large "strains" of *C. oryzae* in wheat and in maize. Table 3.08 shows that the small "strain" had a higher value of λ than the large "strain" in wheat, but the reverse was true in maize. In grain storages in Australia the small "strain" is common in wheat and the large "strain" is common in maize. But the small "strain" is so rare in maize and the large "strain" so rare in wheat that they have not yet been recorded as breeding in these in grain stores. Again there is a correlation between the value of λ at a favorable temperature and the abundance of the species.

A parallel might be drawn between the beetles which we have discussed in this section and the grasshoppers which were discussed in section 1.1. We explained how periods of drought or excessive humidity might reduce a large population of grasshoppers to a small remnant, which would multiply again when the weather became more kindly. With the insects which live in stored grain, the catastrophe is more likely to come when the bin is emptied, so that the grain may be used; or during periods of prolonged storage the grain may become intolerably hot (Birch, 1946*a*). In each case a period is set to increase by recurrent catastrophes which may wipe out the whole population or reduce it to a remnant. This is characteristic of many natural populations (chap. 13).

The concept of innate capacity for increase which we have discussed in this chapter is an abstraction from nature. In nature we do not find populations

TABLE 3.08
FINITE RATE OF INCREASE* FOR THE SMALL AND THE LARGE "STRAINS" OF *Calandra oryzae* IN WHEAT AND MAIZE AT 29.1° C. AND 70 PER CENT RELATIVE HUMIDITY

"Strain"	Wheat	Maize
Small	2.15	1.52
Large	1.76	1.55

* Rate of increase per female per week, $\lambda = e^{r_m}$. After Birch (1953*a*).

with stable age-distributions: weather fluctuates; food may be scarce or sparsely distributed; and an animal's chance to survive and multiply usually depends on how many and what sorts of other animals are associated with it. For these reasons the actual rate of increase r, which may be observed in a natural population and which we discuss at length in subsequent chapters, is much more complex than the theoretical r_m of this chapter. In chapter 9 we show how r departs from r_m under the influence of too few or too many other animals of the same kind. In chapter 10 we discuss other organisms of different kinds, chiefly with respect to their influence on one component of r, namely, the death-rate. The theory of innate capacity for increase is basic to all these discussions, although, of course, the statistic r_m, as we have defined it, does not have any real meaning in relation to a natural population.

Our definition of r_m is arbitrary. We might have defined it broadly so that it was influenced by all the components of environment. This would amount to saying that r_m is the actual rate of increase which would be observed in a natural population if only the ages in the population were distributed as in the stable age-distribution. There would be no profit in such a broad definition because there would be no way in which we could measure r_m experimentally.

Nor could we study it in natural populations, because natural populations probably do not have stable age-distributions.

It is better to limit the meaning of the innate capacity for increase arbitrarily as we have done, because then r_m can be measured precisely by experiment. So long as its limitations are remembered, it may then form a powerful aid to further thought about the distribution and numbers of animals in nature.

GLOSSARY OF TERMS

THE m_x TABLE—This gives the age-schedule of female births (or eggs destined to become females). For any particular parental age-group of pivotal age x, m_x is the number of female births.

THE l_x OR LIFE-TABLE—This gives the age-schedule of survival. For any particular age-group of pivotal age x, l_x is the proportion of individuals alive at the beginning of the age-interval.

THE d_x TABLE—This gives the age-schedule of mortality. It is another way of expressing the life-table. For any particular age-group of pivotal age x, d_x is the proportion of individuals dying within this age-interval.

THE NET REPRODUCTION RATE, R_0—This is the multiplication per generation. It is expressed as the ratio of total female births in two successive generations.

THE MEAN DURATION OF A GENERATION, T—This is the mean time from birth of parents to birth of offspring.

THE INNATE CAPACITY FOR INCREASE, r_m—This is the infinitesimal rate of increase of a population of stable age-distribution (sec. 3.0).

THE FINITE RATE OF NATURAL INCREASE, λ—This is the multiplication per female in unit time of a population of stable age-distribution. This is best defined by the equation $\lambda = e^{r_m}$.

THE STABLE AGE-DISTRIBUTION—This is the age-distribution which would be approached by a population of stable age-schedule of birth-rate and death-rate (i.e., m_x and l_x constant) when growing in unlimited space.

THE L_x TABLE OR LIFE-TABLE AGE-DISTRIBUTION—This gives the proportion of individuals alive between the ages x and $x + 1$ in the life-table. It is the age-distribution of a population which is stationary, the total number of births being constant and equal to the number of deaths over the same period and having an unchanging age-schedule of death-rates.

CHAPTER 4

Diapause

4.0 INTRODUCTION

DIAPAUSE occurs as an important adaptation in many species of insects, enabling them to persist in regions from which they might otherwise be killed out by extremes of climate, or to maintain high numbers in an area which might otherwise support only a few. Diapause usually occurs in that stage of the life-cycle which is highly adapted to resist the rigors of the climate. In the grasshopper *Austroicetes cruciata*, diapause occurs in the egg stage. The diapausing egg, by virtue of a remarkably effective mechanism for preventing evaporation and yet providing for the rapid absorption of water as opportunity offers, is able to withstand prolonged exposure to dryness and thus survive the arid summer (Birch and Andrewartha, 1942). In other species, which live in the cool, temperate zones of the Northern Hemisphere, diapause may occur in a stage in the life-cycle which is able to withstand extreme cold. Or diapause may be of value to the species because it synchronizes the life-cycle with the weather and thereby insures that the active stages of the life-cycle shall be present when there is an abundance of food and the weather favors rapid development and a high survival-rate (secs. 4.6 and 8.12).

The "hibernation" found in certain vertebrates serves to protect them from the extremes of weather during the unfavorable season and is to this extent analogous to diapause in insects. The "anabiosis" found in certain Protozoa, Rotifera, and Nematoda (sec. 7.0) serves the same purpose. Diapause occurs also in certain Crustacea, mites, and snails; but it has been studied less thoroughly in these groups than in insects, so we shall draw most of our examples for this chapter from the insects.

The first experimental demonstration of the phenomenon which we now call "diapause" was made by Duclaux (1869). He observed that eggs of the silkworm *Bombyx mori* failed to hatch when they were kept indoors in a warm room during winter. So he kept some in an icebox for 40 days, and then brought them into the warmth; most of them hatched. Subsequently, Duclaux took some eggs which had been in the warmth for 50 days and placed half of them at −8° C. and half at 0° C. for 2 months. When they were returned to the warmth, 94 per cent of the latter and 13 per cent of the former hatched promptly. Among those which had been at −8° C., many which failed to develop remained alive, because 55 per cent of them eventually hatched in the spring after having spent the winter out-of-doors. Duclaux concluded that the eggs of *Bombyx* required to be exposed to cold before they could develop and that the most suitable temperature for this was probably a little above 0° C.

These experiments passed unnoticed, and no name was given at that time to the phenomenon which Duclaux had discovered. Wheeler (1893), in an embryological paper, coined the term "diapause" to describe a particular stage in the embryonic development of the grasshopper *Xiphidium*. This term has never been accepted or used by embryologists. Henneguy (1904) lifted it from its embryological setting and used it to refer to the physiological state of dormancy or arrested development which he had observed in a number of insects; in this usage "diapause" gained wide currency. For a time it was used to cover almost any state of arrested development. Shelford (1929) suggested that the term "diapause" be restricted to cases in which development is arrested "spontaneously" and does not respond immediately to any ordinary amelioration of the environment. He suggested the term "quiescence" for the simpler case in which development is temporarily inhibited by an unfavorable environment and may be resumed as soon as the hindrance is removed. There may be cases in which the first cause of diapause may lie in the environment, but the final cause may nevertheless be "spontaneous," that is, internal to the animal.

It is not always easy to be sure whether a particular case of arrested development should be classed as diapause. Sometimes the eggs, but more often the young larvae, of "parasitic" Hymenoptera will undergo a period of arrested development within the body of the prey, particularly of those prey which are themselves in diapause. This is, of course, a valuable adaptation which insures that the life-cycles of the predators and prey will synchronize, and it has often been described as diapause (Flanders, 1944; Birch, 1945*b*). It may be that these cases include examples of veritable diapause, but it may be difficult to obtain critical evidence that the predator is not merely reacting to the unfavorable environment created by the prey in diapause. The bug *Rhodnius prolixus* passes through five nymphal instars; and in each instar it normally takes only one large meal of blood. If the meal is large enough (in nature it usually is) to distend the abdomen, this will cause a nervous stimulus to be carried to certain neurosecretory cells situated at the base of the brain. These cells then secrete a hormone, and in due course the bug molts, passing into the next instar. Development in *Rhodnius* may be arrested in any one of three ways. Bugs may survive unfed for many weeks, and a number of bugs which received occasional small meals remained alive without developing for many months. Bugs which were decapitated shortly after they had engorged remained alive without developing for more than a year. And, third, when the bugs were reared in sterile surroundings so that they contained no symbionts (*Actinomyces rhodnii*), development was arrested in the fourth or fifth instar (Wigglesworth, 1934, 1948*c;* Brecher and Wigglesworth, 1944). In all three cases the bugs resumed development and molted when a developing bug was grafted to the dormant one in such a way that the body fluids of the two could intermingle through living tissue. It is now quite clear that arrested development is due to the absence of a hormone: in the first case because the requisite nervous stimulus was lacking; in the second case because the endocrine gland had been removed when the bug was decapitated; and in the third case (Wigglesworth suggests) because the hormone may not be produced in the absence of symbionts.

Wigglesworth (1948*c*) pointed to the analogy between the arrested development of *Rhodnius* and the general case of diapause in other insects, suggesting that the immediate cause of diapause may be a failure in the secretion of hormones necessary for growth. This seems quite likely, but we must remember that, even so, this discovery reveals only one of the later links in a chain of events; the physiologist may well be equally interested in all the links, but the ecologist must needs be more interested in the earlier ones. Thus with *Rhodnius* the cause of the failure in secretion of the hormone is relatively simple—the absence of a large blood meal—whereas a similar failure in a diapausing pupa of *Platysamia* or a diapausing larva of *Cydia* may be related to the environment by a longer and more complex chain of causes. The early links in this chain (which are of first importance in ecology) have been discovered for many species (sec. 4.4), but the intermediate ones remain mysterious (see sec. 4.5). For further discussion of the distinction between diapause and quiescence the reader might refer to Andrewartha (1952, pp. 51–52).

4.1 VARIATIONS IN INTENSITY OF DIAPAUSE AND VOLTINISM

With some species, every individual in every generation enters diapause; typically, there is one generation each year. The life-cycle is called "uni-voltine," and the diapause "obligate." In these species diapause occurs so consistently that it seems to be quite independent of any stimulus from outside the animal. On the other hand, there are many species in which one or several generations in which few or no individuals enter diapause may be followed by a generation in which most or all individuals enter diapause; typically, there will be several generations each year. This sort of life-cycle is called "multi-voltine," and the diapause "facultative." Facultative diapause of this sort invariably depends upon an appropriate stimulus from outside the animal (see sec. 4.4). These extremes are linked by a graded series of intermediate sorts of life-cycle (see Table 4.01).

The grasshopper *Austroicetes* has a uni-voltine life-cycle in which an intense diapause occurs in the egg stage of every individual in every generation. The

TABLE 4.01

LIST OF SPECIES ILLUSTRATING TYPES OF LIFE-CYCLE AND STAGES IN WHICH DIAPAUSE OCCURS

SPECIES	ORDER	LIFE-CYCLE		REFERENCE
		Stage*	Type†	
Acantholyda erythrocephala	Hymenoptera	PP	I	Schwerdtfeger (1944)
Acronicta rumicis	Lepidoptera	P	?	Kozhanchikov (1938)
Acrydium arenorum	Orthoptera	A	II	Sabrosky *et al.* (1933)
Aedes geniculatus	Diptera	L1	III	Roubaud (1935)
A. triseriatus	Diptera	E3	III	Baker (1935)
Aeropus sibiricus	Orthoptera	E2	I	Bei-Benko (1928)
Agriotes sp.	Coleoptera	L	?	Evans (1944)
Alsophila pometaria	Lepidoptera	E	I	Flemion and Hartzell (1936)
Anopheles barberi	Diptera	L1	III	Baker (1935)
A. bifurcatus	Diptera	L1	III	Kennedy (in lit.)
A. plumbeus	Diptera	L1	III	Roubaud *et al.* (1933)
Antherea pernyi	Lepidoptera	P	III	Zolotarev (1938)
Archips cerasivorana	Lepidoptera	E1	I	Baird (1918)
Austroicetes cruciata	Orthoptera	E2	I	Andrewartha (1943*b*)
Biston sp.	Lepidoptera	P	?	Van Emden (1933)
Bombyx mori	Lepidoptera	E1	I, III	Kogure (1933)
Cacaecia rosana	Lepidoptera	L2	III	Marchal (1936)
Calocasia coryli	Lepidoptera	P	?	Kozhanchikov (1938)
Camnula pellucida	Orthoptera	E2	I	Moore (1948)
Cephus cinctus	Hymenoptera	PP	I	Salt (1947)
Ceratophyllus fasciatus	Aphaniptera	L2	IV	Bacot (1914)
Chloealtus conspersus	Orthoptera	E	I	Carothers (1923)
Choristoneura fumiferana	Lepidoptera	L1	I	Baird (1918)
Chorthippus parallelus	Orthoptera	E	I	Sansome *et al.* (1935)
Circotettix verruculatus	Orthoptera	E	I	Carothers (1923)
Cleonus punctivatus	Coleoptera	A	?	Kamenskii *et al.* (1939)
Contarinia tritici	Diptera	L2	II	Barnes (1943)
Croesus septentrionalis	Hymenoptera	PP	?	Kozhanchikov (1938)
Cryptus inornatus	Hymenoptera	L2	III	Simmonds (1948)
Cydia pomonella	Lepidoptera	L2	III	Theron (1943)
Daseochaeta alpium	Lepidoptera	P	?	Kozhanchikov (1938)
Deilephila sp.	Lepidoptera	P	III	Heller (1926)
Dermacentor variabilis	Acarina	L, A	?	Smith and Cole (1941)
Diataraxia oleracea	Lepidoptera	P	III	Way *et al.* (1949)
Diatraea lineolata	Lepidoptera	L2	III	Kevan (1944)
Dichromorpha viridis	Orthoptera	E	I	Carothers (1923)
Diparopsis castanea	Lepidoptera	P	III	Pearson *et al.* (1945)
Drosophila melanogaster	Diptera	L1	IV	Alpatov (1929)
D. nitens	Diptera	A	IV	Bertani (1947)
Ephestia elutella	Lepidoptera	L2	II	Waloff (1949)
Epiblema strenuana	Lepidoptera	L2	III	Rice (1937)
Epilachna corrupta	Coleoptera	A	III	Douglass (1928)
Euproctis phaeorrhaea	Lepidoptera	L1	I	Grison (1947)
Eurygaster integriceps	Hemiptera	A	I	Fedetov (1946*b*)
Exeristes roborator	Hymenoptera	L2	III	Baker and Jones (1934)
Gilpinia polytoma	Hymenoptera	PP	I, III	Smith (1941)
Grapholita nigricans	Lepidoptera	L2	I	Langenbuck (1941)
Gryllulus commodus	Orthoptera	E1	IV	Browning (1952*a*)
Habrobracon brevicornis	Hymenoptera	A	IV	Skoblo (1941)
Haltica ampelophaga	Coleoptera	A	III	Picard (1926)
Heliothis armigera	Lepidoptera	P	IV	Ditman *et al.* (1940)
Hesperotettix pratensis	Orthoptera	E	I	Carothers (1923)
H. viridis	Orthoptera	E	I	Carothers (1923)
Laspeyresia molesta	Lepidoptera	L2	III	Dickson (1949)
Leptinotarsa decemlineata	Coleoptera	A	III	Breitenbrecher (1918)
Lipara lucens	Diptera	L2	II	Varley and Butler (1933)
Listroderes obliquus	Coleoptera	A	I	Dickson (1949)
Loxostege sticticalis	Lepidoptera	L2	III	Pepper (1937)
Lucilia sericata	Diptera	L2	IV	Mellanby (1938)
Lymantria dispar	Lepidoptera	E3	I	Tuleschkov (1935)
L. monacha	Lepidoptera	E3	I	Tuleschkov (1935)

Species	Order	Life-Cycle		Reference
		Stage*	Type†	
Malacosoma disstria	Lepidoptera	E3	I	Hodson *et al.* (1945)
M. neustria	Lepidoptera	E3	I	Tuleschkov (1935)
Mantis religiosa	Orthoptera	E	I	Salt and James (1947)
Melanoplus bivittatus	Orthoptera	E3	I	Salt (1949)
M. differentialis	Orthoptera	E2	I	Bodine (1929)
M. femur-rubrum	Orthoptera	E3	I	Salt (1949)
M. mexicanus	Orthoptera	E3	I	Parker (1930)
Melissopus latifereanus	Lepidoptera	L2	III	Dohanian (1942)
Melittobia chalybii	Hymenoptera	L2	III	Schmieder (1933)
Melolontha sp.	Coleoptera	L	?	Berezina (1940)
Mydaea platyptera	Diptera	L2	III	Roubaud (1922)
Orgyia antiqua	Lepidoptera	E2	I	Christensen (1937)
Orthopodomyia pulchripalpis	Diptera	L1	III	Tate (1932)
Otiorrhynchus cribricollis	Coleoptera	A	I	Andrewartha (1933)
Oxya chinensis	Orthoptera	E	IV	Pemberton (1933)
Papaipema nebris	Lepidoptera	E1	I	Decker (1931)
Phlebotomus papatasii	Diptera	L2	III	Theodor (1934)
Phytodietus fumiferanae	Hymenoptera	L2	I	Wilkes (1946)
Pieris brassicae	Lepidoptera	P	III	Richards (1940)
Platyedra gossypiella	Lepidoptera	L2	III	Squire (1940)
Platysamia cecropia	Lepidoptera	P	I	Williams (1946)
Popillia japonica	Coleoptera	L1, L2	IV	Ludwig (1932)
Pyrausta nubilalis	Lepidoptera	L2	I, III	Arbuthnott (1944)
Reduvius personatus	Hemiptera	nymphs	I	Readio (1931)
Rhagoletis completa	Diptera	P	I	Boyce (1931)
Sarcophaga falculata	Diptera	P	?	Roubaud (1922)
Scleriphron caementarium	Hymenoptera	L2	?	Bodine and Evans (1932)
Sitodiplosis moselana	Diptera	L2	I	Barnes (1943)
Spalangia drosophilae	Hymenoptera	L2	III	Simmonds (1948)
Sphecophaga burra	Hymenoptera	L2	III	Schmieder (1939)
Telea polyphemus	Lepidoptera	P	II	Dawson (1931)
Timarcha tenebricosa	Coleoptera	E3	?	Abeloos (1935)
Trichogramma cacaecia	Hymenoptera	L2	III	Marchal (1936)

* Stage E1, embryo in an early stage, usually before segmentation is complete; stage E2, embryo partly developed, usually near the close of anatrepsis; stage E3, embryo fully grown and nearly ready to hatch. L1, larva at the close of an early or intermediate instar; L2, larva at the close of the final instar. PP, prepupa; P, pupa; A, adult.

† Type I, strictly uni-voltine life-cycle due to the presence of an obligate firm diapause in every individual of every generation. Type II, virtually uni-voltine life-cycle, with most individuals of every generation entering diapause. Type III, multi-voltine life-cycle, with facultative diapause usually occurring in some individuals of most generations and most individuals of some generations. Type IV, uni-voltine or multi-voltine life-cycle, with diapause occurring in a few individuals of some generations. After Andrewartha (1952).

intensity of the diapause may be illustrated by the following observations. Many thousands of eggs have been examined, but none has ever been found in the autumn that contained other than an early-stage embryo (Fig. 4.01, *A* to *D*). When eggs were collected in the field shortly after they were laid and held at 30° C. for 9 months, none developed beyond the stage shown in Figure 4.01, *D*, though nearly all remained alive (Andrewartha, 1943*b*). The intensity of the diapause is further emphasized by the slowness with which it disappears. Eggs were collected from the field at intervals of 2 weeks during the winter and incubated at temperatures between 19° and 32° C. Early in the winter more eggs hatched at the lower temperatures (see Table 4.02), and the rate of development increased relatively less with increasing temperature. But later in the winter nearly all the eggs hatched at all temperatures, and the speed of development increased more rapidly with increasing temperatures, up to 32° C. (Birch, 1942).

Diapause occurs in the egg stage of many Acrididae, particularly nonmigratory plague grasshoppers (Andrewartha, 1945). But the intensity of diapause varies in the different species. In contrast to the diapause found in eggs of *Austroicetes*, that in the eggs of the grasshopper *Melanoplus mexicanus* is "weak." Eggs of this species were collected in the field in Minnesota and incubated at 27°, 32°, and 37° C. The proportions of nymphs to emerge at

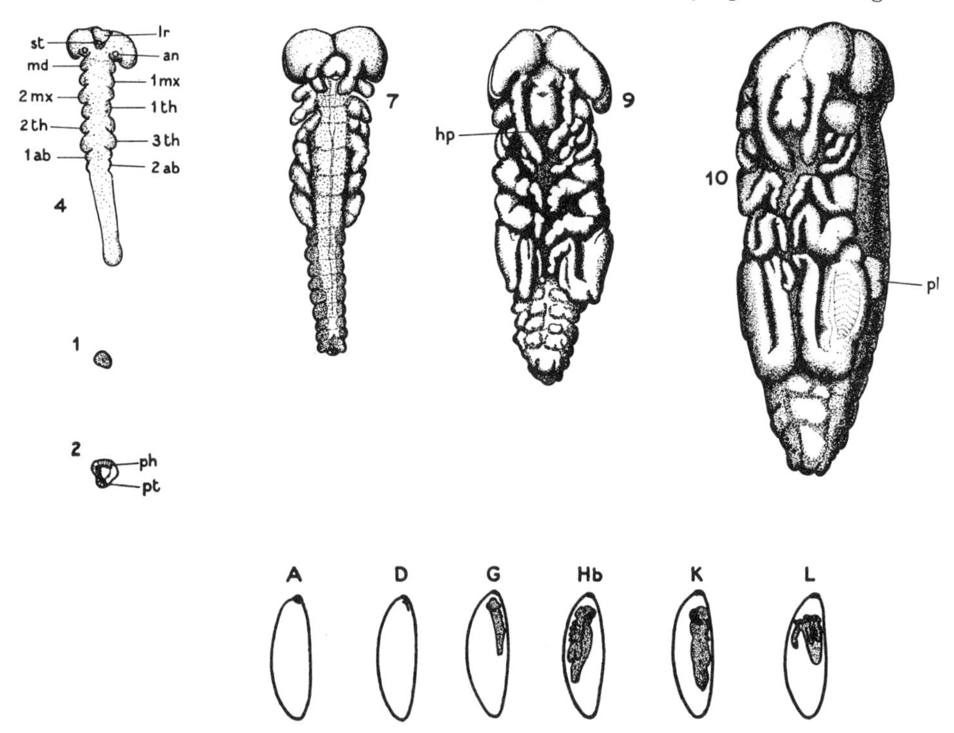

FIG. 4.01.—Stages in the morphogenesis of the embryo of *Austroicetes cruciata* numbered according to the stages of Steele (1941). The increase in size and the changes in position of the embryo in the egg may be followed in the series *A* to *L*, the development of the body form in series *1* to *10*. *A* corresponds to *1* and *2*, *Hb* to *10*. Diapause sets in at stage *A*. Diapause-development may proceed actively while the embryo remains in stage *Hb* or *K*, and, with diapause-development nearly completed, katatrepsis begins (stage *L*). *ph*, protocephalic region; *pt*, protocormic region; *lr*, labrum; *st*, stomodeal opening; *an*, antenna; *md*, mandible; *hp*, hypopharynx; *1 th.*, *2 th.*, *3 th.*, thoracic legs; *1 mx.*, *2 mx.*, first and second maxillae; *1 ab.*, *2 ab.*, first and second abdominal segments; *pl*, pleuropodium. (After Steele, 1941.)

each temperature were 45, 40, and 37 per cent, and the mean incubation periods were, respectively, 26, 32, and 46 days (Parker, 1930). This response closely resembles that given by the eggs of *Austroicetes* which were collected late in the winter (see Table 4.02), and it may be concluded that the "intensity" of the diapause in *M. mexicanus* is much less than that in *Austroicetes*. In fact, the original diapause in *M. mexicanus* may be about equivalent to the remnant left in *Austroicetes* as late as May 29 (Table 4.02). Nevertheless, *M. mexicanus* is strictly uni-voltine, at least in the northern parts of its distribution.

Diapause in the egg stage of *Gryllulus* is even less intense than that in *M. mexicanus*, but the species is nevertheless uni-voltine throughout its distribution in South Australia (Swan and Browning, 1949). Figure 4.02 shows that in *Gryllulus* about 22 per cent of the eggs had no diapause (at least none that

TABLE 4.02*
NUMBERS OF EGGS (AS PER CENT OF TOTAL) OF *Austroicetes* WHICH HATCHED WHEN INCUBATED AT DIFFERENT CONSTANT TEMPERATURES

Date of Collection	19° C.	22.5° C.	30° C.	32° C.
May 16..........	16	5	0	0
May 29..........	95	68	67	19
June 16..........	91	91	96	91

* The eggs were collected from the field at fortnightly intervals during winter. After Birch (1942).

could be demonstrated by incubating them at 27° C.); and, of the rest, 70 per cent entered a diapause which was quite "weak," disappearing slowly at 27° C. (Browning, 1952*a*).

The codlin moth *Cydia* is a good example of a multi-voltine life-cycle associated with facultative diapause. In the northern parts of its distribution one

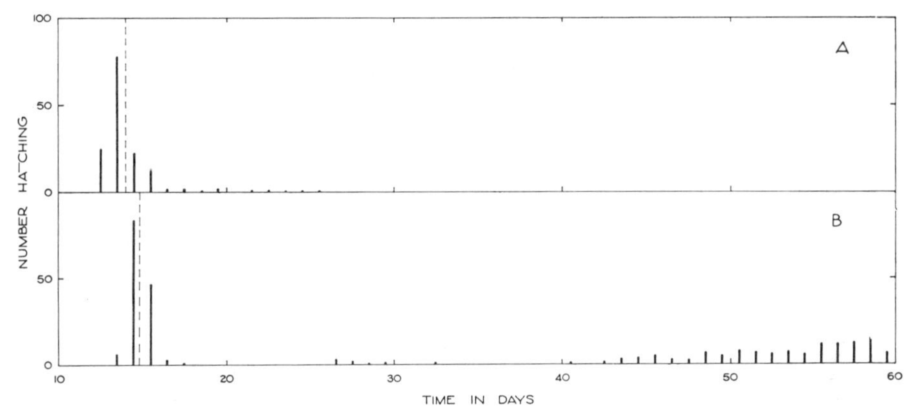

FIG. 4.02.—The eggs of *Gryllulus commodus* were incubated at 27° C. *A*, after having been stored at 13° C. for 30 days from the time that they were laid; *B*, eggs were placed at 27° C. as soon as they were laid. These frequency diagrams illustrate the marked differences in the distributions of the incubation periods of the two groups of eggs. The abscissae refer to the duration of the incubation period and the ordinates to the frequencies with which incubation periods of particular durations were observed. The broken line in *A* indicates the mean duration of the incubation period for all eggs and in *B*, the mean duration of the incubation period for all eggs which hatched in less than 20 days. (After Browning, 1952*a*.)

generation each year is usual, but there are generally a few individuals which complete two generations. Farther south, two or even three generations are usual. There is rarely a simple all-or-none difference between generations. Garlick (1948), during a 5-year study of *Cydia* in a two-generation district in Ontario, found that the percentage of first-generation larvae which transformed without diapause varied from 27 to 57 per cent, with a mean of about 48 per cent. In a three-generation area it would be usual to find a few of the first generation entering diapause, more of the second, and all of the third. In view of the known occurrence of geographic races of insects with different voltinism, it is possible that there is a genetic explanation for the differences in voltinism in *Cydia* from different regions; but this has not been investigated, and for the present it would seem that differences in the environment are more important. Similar variations in the life-cycle of *Loxostege* are clearly phenotypical, since large differences in voltinism may occur in the same area in different years. In Montana, for example, the percentage of first-generation larvae to enter diapause may vary from less than half of 1 per cent to over 60 per cent (Pepper, 1938).

4.2 THE OCCURRENCE OF DIAPAUSE IN VARIOUS STAGES OF THE LIFE-CYCLE

Diapause is commonplace in the embryonic, larval, and pupal stages of insects; it also occurs in nymphs and adults. Diapause may occur in diverse stages in the development of the embryo of different species, but the stage of development is usually characteristic of each species. Thus with the grasshopper *Austroicetes*, diapause sets in a few days after the eggs have been laid (Andrewartha, 1943*b*); at this time the embryo resembles a small plate (Fig. 4.01, *A*). With *Melanoplus differentialis* the embryo develops to the end of anatrepsis (about the stage shown in Fig. 4.01, *D*) before diapause sets in. But in the related *M. mexicanus* and *M. femur-rubrum*, development is not interrupted until dorsal closure is completed and the embryo appears to be nearly ready to hatch (Parker, 1930; Slifer, 1932; Moore, 1948). In the moths *Malacosoma*, *Lymantria*, and *Porthetria*, diapause occurs at a stage when the embryo appears to be nearly ready to hatch (Tuleschkov, 1935; Hodson and Weinman, 1945).

Diapause occurs in the first larval instar of the tortricid *Choristoneura fumiferana*. In this species the eggs develop without interruption and hatch about 14 days after they are laid. But the first-instar larva retains, undigested in its gut, the "plug" of yolk which was inclosed when the embryo completed dorsal closure. It does not feed during its first summer; instead, it spins a loose cocoon in which it remains, in diapause, until after the winter (Graham and Orr, 1940). There are not many examples of diapause in the intermediate larval instars of holometabolous insects, but Ludwig (1932) has described an interesting example in *Popillia*. When 80 larvae were reared at 20° C., 32 entered diapause at the end of the second instar and 48 at the end of the third instar. The mean duration of the period from hatching to pupation was 189.5 days for the first group and 190.7 for the second. This indicated that, once diapause had been completed in the second instar, it did not recur during the third, but if the second instar were completed without diapause, then development was arrested at the close of the third instar. On the other hand, diapause is commonplace in the final larval instar of Lepidoptera and other Holometabola (see Table 4.01 for examples).

Diapause is well known in the pupae of certain moths, notably the saturniids *Platysamia*, *Samia*, *Cecropia*, and the noctuid *Diparopsis* (Pearson and Mitchell, 1945; Williams, 1946). There are indeed many species which hibernate or aestivate as pupae, but the critical information is lacking which is required to distinguish diapause from a "quiescence." The same is true of many species which hibernate or aestivate as adults. This habit is commonplace among the Coleoptera and Hemiptera. A number of examples of authentic diapause in the adult stage are listed in Table 4.01.

4.3 THE INFLUENCE OF ENVIRONMENT ON THE DURATION OF DIAPAUSE

In considering diapause it is helpful to think of development in terms of its morphological aspect, *morphogenesis*, and its physiological aspect, which, by analogy, may be called *physiogenesis*. Diapause may then be considered (at least for ecological purposes) as a stage in physiogenesis which must be completed as a prerequisite for the resumption of morphogenesis. In the more complicated or intense diapause, a whole chain of stages in physiogenesis may need to be completed before morphogenesis may be resumed. We shall therefore speak of the *diapause-stage*, meaning that stage in the life-cycle during which morphogenesis is more or less at a standstill; and we shall use *diapause-development* to mean the physiological development, or physiogenesis, which goes on during the diapause-stage in preparation for the active resumption of morphogenesis. The resumption of active morphological growth and differentiation may be taken to indicate the completion of the diapause-stage. This criterion does not usually raise any practical difficulties, although in nature clear-cut distinctions between diapause- and nondiapause-stages in the life-cycle are unusual. It is generally possible to demonstrate at least slight morphological changes during diapause; and it seems unlikely that the actual physiological processes which constitute diapause-development are restricted to the diapause-stage.

4.31 *Temperature*

Since diapause may often be an adaptation which preserves the species in a resistant stage during the rigors of winter, it is not surprising that exposure to adequate low temperature should be one of the surest ways of bringing the diapause-stage to an end. But the role of low temperatures in promoting diapause-development may be far from simple, as may be seen from the following account of diapause in the egg stage of the grasshopper *Austroicetes*. Diapause sets in at an early stage in the morphogenesis of the embryo; it is intense and prolonged, and eggs are unusually uniform in the intensity of the diapause that is present. The influence of temperature is intricately related not only to the stage reached by the embryo but also to the history of the egg; and the reader who requires to follow the complexities of the relationships should consult the original literature (Steele, 1941; Birch, 1942; Andrewartha, 1943*b*).

Throughout the early embryonic development of *Austroicetes*, there are a number of stages nicely balanced between morphogenesis and diapause, each of which must be completed as a prerequisite before the next may proceed. When the egg is laid, it has the inherent capacity to develop without delay to the stage where a blastoderm has been laid down and an early-stage embryo (Fig. 4.01, *A*) has been differentiated. But from that stage it may not develop further unless it is exposed (preferably alternately) to low temperatures within the range which will permit diapause-development to proceed rapidly (5°–13° C.), and high temperatures (17°–35° C.) which permit morphogenesis to go on. Some progress may be made at intermediate temperatures between 15° and 20° C., because the temperatures favoring diapause-development and growth overlap in this range. Ideally, development at alternating temperature should proceed until the embryo reaches the end of anatrepsis (Fig. 4.01, *D*). At this stage the response when the egg is exposed to constant low temperature is maximal. If morphogenesis stops far short of this stage, no amount of exposure to low temperatures will be adequate; if it stops only a little short of this stage, then a more prolonged exposure to low temperatures will be required; and if exposure to alternating temperatures be prolonged after the stage which is normally reached at the end of anatrepsis, then monsters may be produced which cannot develop normally, no matter how long they may subsequently be exposed to low temperatures. In nature the egg enters the winter with the embryo at or near the end of anatrepsis, having been brought to this stage by the diurnal temperatures of autumn. Indeed, the more the complexities of diapause in the egg of *Austroicetes* are unraveled, the more it is seen as a beautiful example of an adaptation which is delicately attuned to a harsh environment (see sec. 4.6).

None of the nymphs which emerged in the laboratory from eggs which had been incubated as described in the preceding paragraph would take food when it was offered to them. But nymphs which emerged from eggs which had completed all, or nearly all, their development in the naturally fluctuating temperatures of the field fed quite freely when food was offered them in the laboratory. This matter was not pursued further, but there is a suggestion here that ability to hatch in an apparently healthy manner may not be the ultimate criterion for the completion of diapause-development. Perhaps this observation may serve to link diapause in *Austroicetes* to the type found in

Malacosoma and *Choristoneura*. In the former the fully developed embryo enters diapause inside the egg, with its gut fully distended by an undigested "yolk-plug" which is liquefied and absorbed only after an adequate exposure to low temperatures. In the latter the larva hatches with its gut still distended by the "yolk-plug." The larva promptly spins a hibernaculum, in which it spends the winter. Again the yolk is liquefied and absorbed only after an adequate exposure to low temperatures.

It is likely that complexities of the sort described for *Austroicetes* occur in the embryonic diapauses of other species. For example, the eggs of *Melanoplus differentialis* at the time they are laid have the inherent capacity to develop without delay at constant high temperatures to the stage when the embryo has almost completed anatrepsis. Morphogenesis is then interrupted by a diapause that will disappear during exposure to low temperatures; but when eggs were exposed to low temperatures within a day or so of being laid, no response was obtained, and those eggs which still remained alive entered diapause at the usual stage (Slifer, 1932; Burdick, 1937). When eggs of *M. bivittatus* were incubated at a constant high temperature in the laboratory, development proceeded without interruption until the embryo had reached an advanced stage of development, with dorsal closure complete and eyespots near the anterior pole of the egg. At this stage the egg entered a diapause which responded in the usual way to exposure to low temperatures. In nature many eggs enter the winter in this stage, and these hatch normally and promptly in the spring; but others which may have been laid late in the summer enter the winter in a much earlier stage. These eggs resume development in the spring, but a proportion of them fail to hatch promptly; instead they enter a diapause at the same stage as those which have been exposed to constant high temperatures from the time they were laid. An interesting sidelight on the complexity of diapause in this species comes from the observation that in the former case not only does the diapause involve the arrest of morphogenesis, but also the heart-beat is stopped and the embryo does not pulsate. In the latter case diapause involves only the cessation of morphogenesis; the heart continues to beat and the embryo pulsates more or less normally (Salt, personal communication). This latter type is more like the diapause occurring in the postembryonic stages of the life-cycle of many species. For example, larvae of *Cydia* may wander actively during the diapause-stage and may even spin several hibernacula, if necessary, without becoming competent to resume development.

Diapause in the postembryonic stages usually (though not invariably) responds to low temperature also. The eggs of the lymantriid *Euproctis* hatch during July, and the larvae reach the second or third instar by autumn. At this stage they spin a characteristic "nest," in which they overwinter. Diapause-development proceeded during winter, and by the end of February the larvae were competent to resume development if they were placed at 25° C. in an atmosphere saturated with water vapor (Sanderson, 1908; Grison, 1947). The influence of cold on the diapausing larva of the codlin moth *Cydia pomonella* is well known. The fully fed larva spins a dense hibernaculum in which it overwinters. When diapause-development has been completed, the larva will, if placed in the warmth, cut its way out of the hibernaculum, spin a flimsy cocoon, and pupate. This may occur in a few individuals after prolonged exposure to 25° C., but most larvae will eventually die without completing diapause if

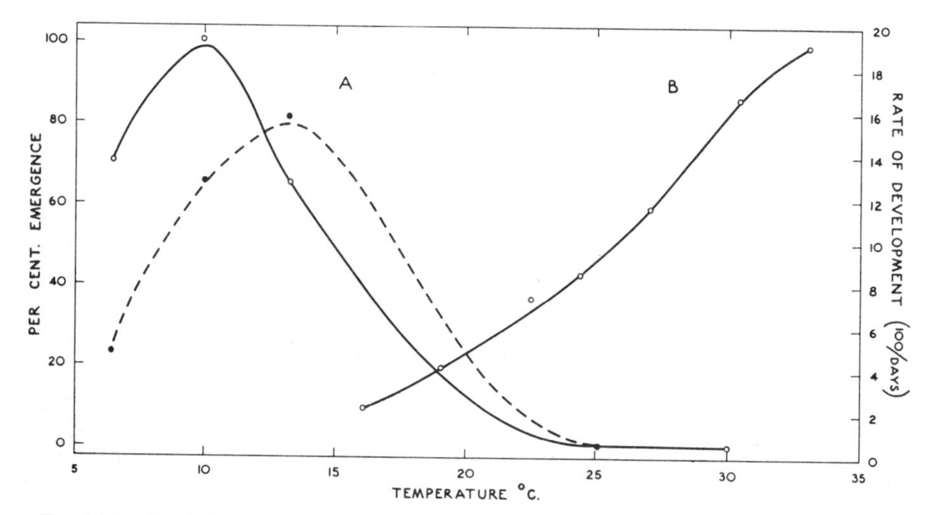

FIG. 4.03.—The influence of temperature on diapause-development and morphogenesis in the embryo of *Austroicetes cruciata*. *A*, the ordinate shows the proportion of eggs to complete diapause-development during 60 days at the specified temperature. *B*, the ordinate shows the proportion of post-diapause development completed each day at the specified temperature. Note that the temperature ranges for the two processes overlap at intermediate temperatures between about 15° and 20° C. *Complete line:* eggs from South Australia; *broken line:* eggs from Western Australia. (After Andrewartha, 1952.)

they are held indefinitely at this temperature. For healthy completion of diapause, the larvae require to experience several months at low temperatures. Thus Putnam (1949) found that larvae collected in the autumn and placed at 2°–4° C. required to be held at this temperature at least until the end of February; when larvae were removed to the warmth before this, pupation was erratic and egg-production poor.

Prolonged exposure to a constant low temperature promoted diapause-development in the characteristic prepupal diapause in the sawflies *Cephus* (Salt, 1947) and *Gilpinia* (Prebble, 1941) and in the pupae of other insects characterized by a diapause in this stage, for example, the fly *Sarcophaga* (Roubaud, 1922) and the moth *Platysamia* (Williams, 1946). The adults of *Drosophila nitens* are rather unusual. When reared in captivity, they entered a seasonal diapause during the winter, even when the temperature and illumination were kept constant. After an exposure of 10 days to 0° C., the flies were competent to resume normal reproductive activities (Bertani, 1947).

It is now known that the particular range of low temperatures which favors diapause-development varies with different species. Many species inhabit cold, temperate regions, where they may be at times exposed to temperatures well below 0° C.; and a number of workers have been misled by supposing that diapause would be influenced more by lower temperatures than by higher ones. They have been influenced in this by the assumption that diapause is due to an "inhibiting" substance which may be destroyed at low temperatures; and they have argued that the lower the temperature, the greater the destruction of the inhibiting substance. Since this is the least tenable of the hypotheses advanced to explain diapause (sec. 4.5), it is not surprising that these efforts should most often meet with failure. With many species, diapause-development proceeds most rapidly at some temperature above 0° C. In the case of *Austroicetes*, which occurs in a climatic zone where the winter is mild and freezing temperatures are rare, it is not surprising that diapause-development should proceed best at temperatures above 0° C. Figure 4.03 shows that in this species the speed of diapause-development increased from 6.5° to 10° C.; it decreased but was still appreciable at 13.3° C.; and by 25° C. had decreased to zero. Temperatures around 10° C. were the most effective for the larvae of *Cephus* (Salt, 1947) and *Alsophila* (Flemion and Hartzell, 1936). A temperature somewhere between 5° and 10° C. was the most effective for the eggs of *Malacosoma* (Flemion and Hartzell, 1936).

Although much more work would have to be done before a generalization could be made, it seems likely that the typical relationship between low temperature and the rate of completion of diapause may be expressed as a sigmoid curve, which, for most species, leaves the axis somewhere near 0° C., reaches a maximum at a temperature not far below the lowest temperature for morphological development. The curve declines to zero at a temperature somewhere within the lower reaches of the temperature-range in which morphological development is possible (see Fig. 4.03).

The disappearance of diapause with low temperatures is not completely general, and certain exceptions are of special interest in ecology. The moth *Diparopsis* has a facultative diapause in the pupal stage. The larvae feed on cotton in the river districts of Nyassaland. Diapause could not be detected at low or at high temperatures but only at intermediate temperatures (Pearson and Mitchell, 1945). If the eggs of the locust *Locustana* are kept in warm, dry soil from the time they are laid, all the eggs will develop until the embryo approaches the completion of anatrepsis. At this stage, development is inhibited by the lack of moisture; but if the eggs are then transferred to moist soil, a variable number, usually of the order of 90 per cent, will resume development and hatch without interruption. The remainder do eventually develop and hatch, but only irregularly and after 2 or 3 months. If, however, the eggs are not moistened at this stage but are left dry and warm for several months, then, on moistening, all the eggs resume development and hatch promptly (Faure, 1932; Matthée, 1951). In this case diapause-development proceeds within the temperature-range appropriate for morphogenesis; moreover, it continues while morphogenesis is being inhibited by dryness. This doubtless is to be regarded as a special adaptation related to the arid environment which *Locustana* normally inhabits.

4.32 *The Influence of Humidity on the Duration of Diapause*

The survival-value of diapause as an adaptation may be related to unfavorable aridity during summer, as well as to unfavorable low temperatures during winter. This will be discussed in more detail in section 4.6; it suffices for the present that the relationship is undoubted. By analogy with the influence of low temperatures on the completion of diapause, it might be expected that moisture (or lack of it) in the environment might be causally related to diapause-development. This is a complex subject and one in which it is difficult to discriminate between cause and effect. It is undoubted that in nature the absorption of water is often associated with the resumption of morphogenesis after diapause. And experiments have been done in which the eggs of a grasshopper which had been in diapause were made to resume development by treating them artificially so that they would absorb water. But there are also species which may lose and absorb water repeatedly during the diapause-stage without this causing diapause to disappear or diminish in any way. Andrewartha (1952) reviewed the available evidence and concluded: "The examples given above illustrate the intricate requirements of the organism for water at different stages of its development and the difficulty of unravelling the true causal relationship between water and diapause. But the evidence indicates that the absorption of water during aestivation or hibernation, however necessary it may be for healthy post-diapause development, is rarely if ever the primary stimulus which initiates diapause-development."

4.33 *The Influence of Light on the Duration of Diapause*

The mosquito *Aedes triseriatus* develops through several successive generations during summer; but in the autumn the mature embryo, apparently ready to hatch, enters diapause, notwithstanding the fact that at this time the temperature may be higher than in spring, when development is resumed. Baker (1935) collected eggs containing fully developed embryos during September and soaked them several times: none hatched. He then kept the eggs dry at 80 per cent relative humidity, exposing half of them to 16 hours and the other half to 10 hours of light each day for 5 weeks. Both lots were then soaked. Two larvae emerged from the lot which had been exposed to light for 10 hours each day, but the water containing the other group swarmed with larvae. The mosquito *Anopheles claviger* enters diapause in the autumn as a larva in the second or third instar. Kennedy (personal communication) collected 120 second-stage larvae from the field in the autumn and divided them into three batches, which received 14, 8, and 0 hours of light each day. Each lot was at the same temperature, and otherwise all were treated in the same way. Nearly all the larvae molted once to reach the third instar, but after that the larvae in the group experiencing 14 hours of light each day were virtually the only ones to develop any further. After 3 weeks, 39 of 40 larvae in this group had pupated; there were 18 pupae from the group in total darkness, and 6 from those experiencing 8 hours of light daily. Both these experiments showed that exposure to long days allowed diapause-development to be completed. In nature these species resume activity at the end of the winter.

4.34 *The Termination of Diapause after Wounding and Other Treatments*

There are a number of species in which diapause comes to an abrupt end when some sort of wound or shock is imposed upon the animal. In nature this may happen when a predator pierces the integument and deposits an egg inside the body of the diapausing animal. The artificial methods that have been used include pricking with a hot or cold needle, electric shock, dipping in acid or xylol, or exposure to dilute vapor of xylol, chloroform, or some similar substance. Several examples were described by Andrewartha (1952). Although there are one or two exceptions, it is generally true that these methods have been successful only when the diapause was "weak" or indefinite.

4.4 STIMULI WHICH DETERMINE THE INCEPTION OF DIAPAUSE

Diapause may function as an adaptation which preserves the species in a resistant stage during the unfavorable season and at the same time keeps the life-cycle in step with the rhythm of the season. This follows more or less inevitably with life-cycles of types I and II (Table 4.01), but it also happens more often than not that, in multi-voltine life-cycles, diapause sets in about the same time each year.

The influence of the season is well illustrated by the uni-voltine life-cycle of the bug *Reduvius personatus*. The winter is spent in the third, fourth, or fifth nymphal instar. The bugs become adult in the spring, and eggs are laid more or less continuously from June to September. These hatch after about 20 days, so that toward the end of summer the population may include representatives of all five nymphal instars; but, irrespective of the stage of development which may have been reached, all will have entered diapause by the middle of November (Readio, 1931).

In a 5-year study of *Cydia pomonella* in Ontario (a two-generation area for

Cydia), Garlick (1948) found that the proportion of the larvae of the first generation, which transformed without diapause, varied from 27 per cent in 1945 to 57 per cent in 1942, with a mode for the 5 years very close to 50 per cent. Whether a particular larva would enter diapause or not seemed to be independent of both genotype and generation but was determined quite rigorously by the time of the year; the date on which larvae ceased pupating and, instead, entered diapause was remarkably constant, not differing by more than ± 3 days from August 24 for the 5 years. This is well shown by the data in Table 4.03, which is taken from Garlick (1948).

TABLE 4.03*

PROPORTION OF *Cydia* LARVAE ENTERING DIAPAUSE IN RELATION TO DATE AT WHICH MATURITY IS REACHED

Date of Maturity	Per Cent in Diapause	Date of Maturity	Per Cent in Diapause
July 20	10	August 9	70
July 25	10	August 13	90
July 31	20	August 15	95
August 3	30	August 25	99
August 7	60	August 27	100

* After Garlick (1948).

More experiments could be quoted, and, although there are exceptions, it is generally true that in multi-voltine life-cycles the incidence of diapause is usually closely related to the season of the year. Temperature, humidity, duration of daylight, and maturity of food all vary with the seasons, and each one may, in a particular case, be the stimulus that determines diapause.

Another generalization which holds almost without exception is that individuals which are going to enter diapause grow more slowly and often become heavier than those which are not going to enter diapause. It has been suggested that the slow rate of development (no matter how it may have been induced) may be the direct cause of diapause (Cousin, 1932; Knocke, 1933; Tuleschkov, 1935). When diapause happens to be induced by temperature or food, the same stimulus will also cause the animal to grow more slowly, and it is therefore impossible to discover the causal relationship between slow growth and diapause. The same difficulty persists even when diapause is genetically determined. For example, Arbuthnott (1944) selected two races from wild populations of the moth *Pyrausta nubilalis;* one was strictly uni-voltine, being homozygous for a firm obligate diapause; the other was multi-voltine, being homozygous for a facultative diapause. When the two races were reared side by side in circumstances which inhibited diapause in the multi-voltine race, the individuals of the multi-voltine race developed more rapidly than did those of the uni-voltine race (see Table 4.04).

But a number of cases in which diapause may be induced by exposure, during an early stage of the life-cycle, to a certain photoperiod have been thoroughly studied (sec. 8.12); in these cases diapause would seem to be the cause of slow growth rather than vice versa (Andrewartha, 1952).

4.41 *The Influence of Temperature on the Inception of Diapause*

With the bug *Reduvius personatus*, diapause may occur in any of the last three nymphal instars and usually appears in the autumn, when the temperature is falling. The lower temperature at this time is not the only, nor yet the chief, cause of the onset of diapause; for at this time of the year the temperature may be higher than in the spring, when morphogenesis is proceeding quite

TABLE 4.04*

NUMBER (EXPRESSED AS PER CENT OF TOTAL) OF LARVAE OF UNI-VOLTINE AND MULTI-VOLTINE RACES OF *Pyrausta* WHICH HAD PASSED SPECIFIED INSTAR 15 OR 20 DAYS AFTER HATCHING FROM EGG

INSTAR	AFTER 15 DAYS		AFTER 20 DAYS	
	Uni-voltine Race	Multi-voltine Race	Uni-voltine Race	Multi-voltine Race
Third	86.4	99.4	100.0	100.0
Fourth	26.8	90.7	97.2	99.2
Fifth	6.0	75.6	60.7	88.0

* After Arbuthnott (1944).

actively. But the temperature at which bugs are reared may nevertheless modify the tendency toward diapause. When bugs were reared at 22° C., they invariably entered a prolonged diapause at the close of the third or fourth instar, and all those which had experienced diapause in the third instar entered diapause again at the close of the fifth instar; the only ones to escape this second period of rest were those which had just experienced a prolonged diapause in the fourth instar. An interesting feature of the development at 22° C. was that a certain amount of diapause-development seemed to be necessary before the fourth ecdysis could take place, but it did not seem to matter whether this occurred in the third or the fourth instar. On the other hand, when the bugs were reared at 32° C., growth proceeded without interruption throughout the third and fourth instars, diapause being postponed until the close of the fifth instar (Readio, 1931). The larva of the beetle *Popillia* is influenced by temperature in a similar way. When larvae were reared at 20° C., most of them experienced a prolonged diapause in the second instar; the few which escaped it in the second instar had a prolonged resting period in the third instar. But at 25° C. all the larvae developed beyond the second molt without prolonged delay, and diapause was postponed until the end of the third instar. A few larvae were a little tardy in the second instar, and these usually spent a correspondingly shorter time in diapause in the third instar (Ludwig, 1932). Like *Reduvius*, *Popillia* seems to require at any particular temperature a fixed amount of diapause-development, but it seems immaterial whether this takes place in the second instar or the third or partly in each (see Table 4.05). Also as in *Reduvius*, rearing the larvae at a lower temperature not only increased the time spent in diapause but also caused it to set in at an earlier stage in the life-cycle. In these two species the tendency toward diapause is strengthened when the insects grow slowly at a lower temperature; but there is no indication as to whether tardy growth or low temperature is the more directly influential in bringing about this result.

TABLE 4.05*
DURATION OF LARVAL INSTARS OF *Popillia* REARED AT 20° AND 25° C.

Group†	No. Completing Second Instar		Duration of Instars in Days							
			20° C.				25° C.			
	20° C.	25° C.	1st	2d	3d	Total	1st	2d	3d	Total
I........	32	94	30.4	33.3	125.8	189.5	15.6	16.9	102.5	135.0
II........	48	12	29.1	70.0	91.6	190.7	21.5	19.6	91.4	132.5

* After Ludwig (1932).
† Groups arbitrarily decided on basis of duration of second instar, group I having a short and group II a long second instar.

A "weak" diapause was induced in some 10–50 per cent of the pupae of *Heliothis armigera* when the larvae were reared at 19° C., but diapause never appeared in the pupae if the larvae were reared and allowed to pupate at 25° C. or some higher temperature. The temperature seemed to exert its greatest influence during the period when the larvae were actively feeding; once pupation had occurred, the temperature had no further influence on the inception of diapause (Ditman *et al.*, 1940).

In the adult stage, diapause may take the form of a failure to ripen eggs or sperm and may be manifest by an extended pre-oviposition period. This is the case with *Habrobracon brevicornis*, whose larvae live inside the larvae of *Heliothis*, devouring the body contents of their prey (Skoblo, 1941). When *Habrobracon* was reared through its larval stages at temperatures between 25° and 30° C., the ensuing adults required a pre-oviposition period of 2 days at 24°, 3 days at 18° and 8 days at 15° C. But when the larvae were reared at 18° C., the adults entered diapause, and the pre-oviposition period was extended to 100 days at 18° or 33 days at 24° C. Since a proportion of the prey (*Heliothis*) also entered diapause when reared at 18° C., it is not clear whether the response observed in *Habrobracon* was due primarily to temperature or to food. This is characteristic of the difficulties that are encountered when one tries to discover the causes of diapause.

4.42 *The Influence of Food on the Inception of Diapause*

The food of most phytophagous species varies in quality and moisture content with the season; and similar rhythmical seasonal changes may occur in food of carnivorous species, particularly if their prey exhibits a seasonal diapause. For this reason it is often difficult to decide whether an observed relationship between diapause and food is independent of temperature, light, and other components of the environment which also display a concomitant seasonal rhythm. There are, however, a number of experiments available in which these other influences have been adequately excluded and from which it seems clear that the quality of the food may be the chief stimulus determining diapause.

The larva of *Trichogramma cacaeciae* lives inside and feeds on the contents of the eggs of the tortricid *Cacaecia rosana;* it may be artificially constrained to breed in the eggs of *Mamestra brassicae* (Marchal, 1936). The prey, *Cacaecia,* is strictly uni-voltine; egg-masses containing from 30 to 60 eggs are deposited usually on quince trees toward the end of June and during July. The eggs begin to develop immediately, reaching after a few days a stage in which the embryo is a simple germ-band imbedded in the yolk. Then diapause sets in, and the embryo remains in this stage for the rest of the summer and throughout the winter; the eggs resume development in the spring and hatch about the middle of April. The parasite *Trichogramma* is bi-voltine and polymorphic. During July macropterous females lay eggs in the newly laid eggs of *Cacaecia,* choosing, as a rule, only the central 8 or 10 eggs of the egg-mass. The eggs of the predator hatch promptly and develop without interruption into fully grown maggots, which occupy the entire space within the eggs which are their prey. At this stage the larva of *Trichogramma* enters diapause. Its gastric sac can be seen fully distended with the erstwhile yolk of the *Cacaecia* egg, and in this condition it spends the remainder of the summer and all the winter. It pupates early in the spring and emerges about March as an adult micropterous female. The 8 or 10 micropterous females of *Trichogramma* which emerge at this time from each egg-mass proceed to lay eggs in the remaining eggs of the same egg-mass, which by this time are no longer in diapause. The eggs laid by micropterous *Trichogramma* into post-diapause eggs of *Cacaecia* develop into larvae which do not enter diapause but, instead, pupate a few days after the larva becomes fully grown. During late June and July these pupae give rise to macropterous females, which fly off in search of newly laid eggs of *Cacaecia.*

Up to this point these observations do not establish the independent influence of food on the inception of diapause in *Trichogramma.* For the one type of food, namely, tissues of prey in diapause, is present only during summer, and the other type, namely, tissues of prey from which diapause has disappeared, is available only in spring; and the diapause which is observed in the summer brood of *Trichogramma* might equally well have resulted from some quality of temperature or light which is appropriate to July but lacking in April, when the nondiapause generation develops. But the matter was put beyond doubt by a further series of experiments, in which *T. cacaecia* was constrained to breed in the eggs of *Mamestra brassicae,* a species in which diapause does not occur. Starting in March with brachypterous females which have overwintered as diapausing larvae in the eggs of *Cacaecia,* Marchal bred one nondiapause generation in *Cacaecia* and then, between July and November, bred, in the eggs of *Mamestra*, five successive macropterous generations without a sign of diapause. The sixth generation in *Mamestra* entered the winter. The survival-rate during the winter was low; relatively few of the parasites hibernated successfully, and most of those which survived to emerge in the spring were brachypterous.

These experiments show that whatever influence meteorological events may have on the inception of diapause in *Trichogramma*, it is overshadowed by the influence exerted by the condition of its food, namely, the tissues of the egg in which the predatory larva is living. The critical difference is between diapausing and nondiapausing prey; and it seems difficult to escape the conclusion that this is due to the direct influence of the quality of the food on the inception of diapause.

The quality of the food available to the larva appears to be causally related not only to the inception of diapause but also to the development of polymorphic forms in the eulophid *Melittobia chalybii*, which lives and feeds inside many species of bees and wasps. As many as 800 predators may be reared on one prey; and in this case the critical differences in the quality of the food seem to be due to selective feeding on the part of the predators; the first few to hatch consume the blood and body fluids, leaving the late-comers to eat the remaining less easily assimilated tissues (Schmieder, 1933, 1939). Macropterous females usually leave the larva on which they were reared and seek out a fresh prey. They may live as active adults for about 70 days, laying eggs at the rate of 4 or 5 each day throughout this period. The first 15 or so eggs laid in the fresh prey emerge after about 14 days as brachypterous adults (mostly female), each of which during a relatively short life (2–30 days) may lay about 50 eggs in the same larva from which they have emerged. These eggs, together with the remaining eggs laid by the original macropterous female, give rise to larvae, all of which enter a diapause which may endure for about 60 days. It is possible to interfere artificially with this rhythm, and the results given in Table 4.06 indicate that whether or not any individual becomes a diapausing macropterous type or a nondiapausing brachypterous type depends entirely on the condition of the prey in which it develops. In other words, the occurrence of diapause in *Melittobia* is independent of the polymorphic form of the mother and her age but is determined completely by the quality of the food available to the larva.

"Fresh" prey had not had any larvae previously reared in them, and "spent" prey had already served to nourish from 12 to 20 larvae of *Melittobia*. Later eggs in fresh prey were procured by transferring females which had been ovipositing for some time to fresh prey; and in the converse experiments newly emerged females were placed on spent prey and thus constrained to lay their first few eggs in prey in which about 20 larvae had already been reared. In the case of the brachypterous females laying their first few eggs in fresh prey, this was repeated for seven successive generations with consistent results, although in nature this type of female invariably gives rise to diapausing macropterous progeny.

The species which have been discussed so far are all carnivorous; the same phenomenon may be observed in herbivorous species. The caterpillars of *Euproctis* (Lymantriidae) feed upon the leaves of apple trees. Eggs are laid in June and hatch in July. Diapause occurs in the larvae usually at the end of the

TABLE 4.06
INFLUENCE OF FOOD DURING LARVAL STAGE ON DIAPAUSE AND POLYMORPHISM IN *Melittobia*

CONDITION OF PREY	PARENT		OFFSPRING	
	Form	Age	Form	Development
Fresh	Brachypterous	First few eggs	Brachypterous	Nondiapause
Fresh	Brachypterous	Later eggs	Brachypterous	Nondiapause
Spent	Brachypterous	First few eggs	Macropterous	Diapause
Spent	Brachypterous	Later eggs	Macropterous	Diapause
Fresh	Macropterous	First few eggs	Brachypterous	Nondiapause
Fresh	Macropterous	Later eggs	Brachypterous	Nondiapause
Spent	Macropterous	First few eggs	Macropterous	Diapause
Spent	Macropterous	Later eggs	Macropterous	Diapause

* Summarized from data given by Schmieder (1933).

second or third instar. But when the newly emerged larvae were fed the youngest foliage it was possible to select, they completed development without diapause (Grison, 1947). This experiment shows that the strict uni-voltinism of *Euproctis* observed in nature is impressed upon the population by the rhythmical seasonal changes in the quality of the food provided by the leaves of its perennial deciduous host.

The changing composition of the food, particularly changes in the relative proportions of fat and water, seems to be the chief cause of diapause in the larvae of *Platyedra* and *Diatraea*, two subtropical species which feed on annual plants. Both species are multi-voltine, with diapause likely to occur at any season of the year and apparently independently of the weather (Squire, 1937; Kevan, 1944). The composition of the food of the larvae of *Platyedra*, living inside the cotton boll, may vary between wide extremes, depending on whether seeds are immature or approaching ripeness. Squire has estimated that the average fat content of the food ingested by larvae which entered the boll when it was 30 days old is 16 per cent, compared with 27 per cent for larvae which entered the boll when it was 50 days old (Squire, 1939, 1940). And the food of larvae which enter the bolls 2 days before they split will average 30 per cent of water, compared with 75 per cent for larvae which enter the bolls 14 days before they split. As the crop ripens and the population of *Platyedra* increases, the larvae are forced more and more onto the riper bolls, and many larvae may be found in the bolls which have opened; the changing composition of the food is reflected in the increasing numbers of the larvae entering diapause. Between January and May the water content of samples of seeds taken from the field changed from 80 to 20 per cent, and the proportion of larvae entering diapause increased from 5 to 62 per cent. The relationship between the water content of the food and the proportion of larvae entering diapause is shown by the following figures which are quoted from Squire (1939):

Percentage water content of seeds......	0–20	30–40	50–60	70–80
Percentage of larvae in diapause.......	62	60	27	5

The increased fat and decreased water content of the food is also reflected in the composition of the food reserves stored by the larvae. The fat content (expressed as a percentage of the total dry matter) of diapausing larvae was found to be 43 per cent, compared with 28 per cent for nondiapausing larvae; and the water content of the former was 10 per cent below that of the latter. The strength of these correlations is apparent; and in this case it seems safe to accept the obvious relationship as being also the causal one, provided that a note of warning is sounded. Kozhanchikov (1938) found the fat content of diapausing larvae of *Pyrausta* to be 10 per cent above that for nondiapausing larvae; and the same general observation has been made for *Cydia* (Baumberger, 1914). Now it is known for these species that diapause is determined at an early stage by influences other than food (Arbuthnott, 1944; Dickson, 1949), and therefore for these species the correct inference is that a predisposition toward diapause causes the larva to accumulate an increased proportion of fat, and not the reverse. Nevertheless, the complexities of diapause are such that it would be poor logic to argue that the causal relations established for one species are likely to hold *in detail* for another. So on present evidence it seems most likely that diapause in *Platyedra* is indeed caused by an excessive accumulation of fat as a result of the ingestion of food which is rich in fat and poor in water. Diapause in *Diatraea* appears to be like that described for *Platyedra.* The larvae of *Diatraea* tunnel in the stems of maize. More larvae enter diapause when they feed on plants which are approaching maturity. The tendency toward diapause is independent of the season, occurring readily in both the wet and the dry season whenever the food consists of old and woody plants (Kevan, 1944).

The wild strain of *Ephestia elutella* found in warehouses in England is virtually uni-voltine, and the voltinism is not markedly influenced by food; but a strain has been selected from this wild material which is multi-voltine when reared within a narrow range of temperature and in highly "favorable" food (Waloff, 1948). Small departures in "favorableness" of food result in an increasing amount of diapause; and on the less favorable foods this strain of *Ephestia* becomes virtually uni-voltine: diapause, once induced by the unfavorable food, may endure for many months. This is illustrated by the data given in Table 4.07, which are condensed from the data by Waloff (1948). A larva uses about 48 grains of Manitoba wheat when the grain is plentiful, because it eats only the embryos; when offered only 10 grains, the larva is forced to feed on the endosperm, with the consequence that all the survivors enter diapause. The results with soya-bean flour conform to those reported above for *Platyedra*. It would seem that for this strain of *E. elutella* the quality of food influences the amount of diapause and, in general, the foods which favor rapid growth and high survival-rate result in less diapause than those on which growth is slow and survival-rate low.

TABLE 4.07*

INFLUENCE OF FOOD ON INCEPTION OF DIAPAUSE IN *Ephestia elutella* REARED AT 25° C. AND 75 PER CENT RELATIVE HUMIDITY

Food	Survival-Rate (Per Cent)	Duration of Larval and Pupal stages for Nondiapausing Larvae (Days)	Proportion in Diapause as Per Cent Survivors
Manitoba No. 1 wheat	87	50	0
Wheat germ	100	42	0
Soya-bean flour, 1.3 per cent fat	40	61	0
Peas	21	82	14
Soya-bean flour, 7 per cent fat	31	67	32
English wheat	47	..	100
10 grains Manitoba wheat	24	..	100

* After Waloff (1948).

4.43 *The Influence of "Maternal Physiology" on the Inception of Diapause*

There is a wealth of information available, arising mostly from Japanese work, to show that in *Bombyx* the diapause-pattern which characterizes one generation may be due to the physiological constitution of the females of the preceding generation. Kogure's work on the influence of light and temperature on the inception of diapause, which is described in section 8.12, is a striking example. Light and temperature acting on the developing embryo so impress a particular pattern on its physiogenesis that this pattern persists throughout the larval stages and ultimately becomes manifest in the type of egg laid by the adult female (Kogure, 1933). The type of egg with respect to diapause and color (which are closely linked) is determined entirely by the phenotype of the mother. The eggs of the spring brood of bi-voltine or multi-voltine races are invariably pale (except for a few colored ones, which always remain in diapause for the remainder of the summer); but the eggs of the autumn brood are colored. Toyama (1913) selected a number of races differing in the color of the egg membrane and made an extensive series of crosses. With one exception (crimson), the color of the egg and, without exception, the voltinism in the F_1 generation were determined by the constitution of the mother, irrespective of how she had been mated. This was all the more striking, since further crossing indicated that the genes for egg color and voltinism contributed by the male were dominant. Similar results were obtained by Uda (1923). In the races with which he worked, the brown color of the diapausing egg was recessive to gray. Brown was inherited as an ordinary recessive when the results were measured by the proportion of moths in the F_1 generations laying brown or gray eggs, but not when the results were measured by the color of the eggs in the F_1 generations; without exception, in the F_1 generations the eggs exhibited the maternal coloration (and voltinism), irrespective of whether this was dominant or recessive. For example, the backcross brown female crossed with F_1 male produced all brown eggs instead of the expected 1:1 ratio of brown to gray; but when the F_1 generation became adult and were classified according to the proportion of females laying brown or gray eggs, the expected frequencies were observed. Similar results were reported by McCracken (1909), who studied the inheritance of uni-voltinism and bi-voltinism in *Bombyx.* A different approach by Umeya (1926) confirmed the results given by these genetical experiments. When ovaries were transplanted during the larval stage from uni-voltine to bi-voltine individuals and in the reverse direction, the incidence of diapause in the resulting eggs was always characteristic of the host's race and never of the donor's.

Diapause sets in at the end of anatrepsis in the embryonic development of *Melanoplus differentialis*. Relative to *Bombyx*, the diapause in *Melanoplus* is "weak" and variable. It is completed slowly in the majority of eggs when they are kept at 28° C. continuously from the time of laying. It is known that at 28° C. about 32 days of active development are required to complete morphogenesis (that is, excluding the time spent in diapause). The time spent in diapause at 28° C. may be used as a measure of the intensity of diapause, and Burdick (1937) found that this time decreased steadily from the beginning to the end of the season. Eggs laid at the beginning of the season required about 78 days at 28° C. to complete diapause-development, compared with 28 days for those laid by the same group of females late in the season. This may have been due to any combination of causes, such as temperature, photoperiod, food, or the age of the female. But, whatever the cause, diapause must have been determined before the egg was laid; for, once they had been laid, all the eggs were treated alike. In *Phlebotomus papatasii* diapause occurs at the close of the final larval instar; yet in this species, too, it may be determined before the egg is laid. Roubaud (1927, 1928) collected wild *Phlebotomus* during October and allowed them to have a meal of blood. The eggs were subsequently separated into two groups: those which had been laid by females which laid their first eggs within 5 days of feeding and those from females which waited 6–16 days after feeding before laying any eggs. The eggs from the former group gave rise to larvae which completed their development without diapause, but 93 per cent of those from the second group entered diapause at the close of the fourth instar.

Delay in the manifestation of a facultative diapause is quite usual: the stimulus which evokes diapause may operate during an early stage of the life-cycle, but diapause may not be manifest until a much later stage; this is especially obvious when light (photoperiod) is the stimulus which evokes diapause (sec. 8.12). The special feature about the examples discussed in this section is that the delay extends through one generation into the next. With *Bombyx*, diapause in the embryo is determined by the condition of the cytoplasm in the egg. And the experiments of Fukuda (1940, 1951*a*, *b*, 1952) and Hasegawa (1952) show that this is dependent upon secretions from the subesophageal ganglion.

When the subesophageal ganglion was removed surgically from a late-stage caterpillar or pupa, the resulting moth laid only nondiapausing eggs. When the subesophageal ganglion from another caterpillar or pupa was implanted into the body of an individual which had been determined to lay nondiapausing eggs, its condition was changed so that it had a strong tendency to lay eggs which entered diapause; this happened independently of whether the implant came from individuals which had or had not been determined to lay diapause eggs. When the subesophageal ganglion and the brain, still united by the para-esophageal connectives, were implanted into the body of an individual which had been determined to lay nondiapausing eggs, the result depended on whether the implant came from an individual which had or had not been determined to lay diapause eggs. Fukuda inferred from these experiments that the subesophageal ganglion produces a hormone which influences the moth to lay eggs which enter diapause; and this may happen even after the subesophageal ganglion has lost its nervous connection with the brain. The brain may inhibit the production of the diapause-determining hormone from the subesophageal ganglion, provided that nervous connection between the two is maintained. Whether the brain does so inhibit the subesophageal ganglion depends on the photoperiod and temperature experienced by the moth while it was an embryo.

There would seem to be three ways in which the secretion from the subesophageal ganglion might influence the constitution of the cytoplasmic parts of the egg. (*a*) It might organize the inclusion of a "growth-inhibiting factor" in the eggs. All the evidence is against this. (*b*) It might so organize the metabolism of the fat body into egg-yolk that the diapause egg comes to have a different (intractable) sort of yolk from that of the nondiapause egg. (*c*) Its influence might be to deprive the egg of an essential enzyme system. If it were possible to find out what goes on during the egg's exposure to low temperatures (i.e., what constitutes diapause-development), it might provide the answer to this problem. There are three hypotheses which might be tested if the appropriate physiological experiments could be devised. (*a*) In the egg in diapause, the yolk may be different from that in the nondiapausing egg. The embryo may be unable to utilize it as food because one of the earlier stages in the process may require an enzyme system which is active only at low temperatures. (*b*) The yolk is not different. It cannot be used to promote growth because an essential part of the enzyme systems is lacking; this may be active at normal temperatures but requires to be synthesized at low temperatures. (*c*) The yolk is different, and an essential part of the enzyme systems is lacking as well; the yolk remains intractable until such time as the appropriate enzymes have been synthesized at low temperatures; then the next stages may proceed at normal temperatures. From this point it is appropriate to pass to a consideration of a general theory of diapause.

4.5 THEORY OF DIAPAUSE

According to one hypothesis, diapause results from an accumulation of an inhibiting substance and disappears when this has been destroyed. The inhibitor has been postulated as accumulated waste products of metabolism, causing a sort of autointoxication or "asthenobiosis" (Roubaud, 1922), or as *Latenzstoff* (Goldschmidt, 1927) or as "X" or "diapause-factor" (Bodine, 1932); and Salt (1947) postulated a second substance which he called the "Y"-factor, to explain his observations on diapause in *Cephus*. No substance corresponding to these hypothetical inhibitors has yet been demonstrated empirically. According to another hypothesis, diapause is to be attributed to a deficiency of water in the tissues (Slifer, 1946) or, what amounts to the same thing, to an excess of fatty and other dry material (Baumberger, 1914, 1917; Squire, 1940). It is, indeed, commonly recorded that diapausing individuals tend to be heavier and to have a lower water content than those which do not enter diapause. Also the water content characteristically tends to increase during post-diapause development. But Andrewartha (1952) showed that the evidence mostly indicates that the reduced water content of the tissues, so characteristic of insects in diapause, is likely to be an effect rather than a cause. A third hypothesis attributes diapause to the absence of a particular growth-promoting hormone (Wigglesworth, 1936, 1948*c*; Williams, 1946).

The evidence for the hormone theory of diapause was summarized by Andrewartha (1952), and there can be little doubt that this theory adequately explains the final stages in the diapause occurring in all postembryonic preimaginal stages of the life-cycle. The evidence is less complete for the other stages, but at least there is nothing from them to contradict the theory. Briefly, this theory states that whenever diapause occurs in a postembryonic preimaginal stage of an insect, it always occurs at the close of a stadium, never interrupting a stage of active growth. So the completion of diapause is always followed by ecdysis. The hormonal control of molting is so well documented now that it is an obvious step to postulate that the *immediate* cause of diapause is the failure of the neurosecretory cells of the brain to secrete the molting hormone. This hypothesis has been amply verified by experiment (Williams, 1946, 1947, 1948).

But the hormone theory accounts for only the last links in what might be a long chain of processes. In order to account for some of the earlier links, Andrewartha (1952) proposed the "food-mobilization hypothesis." For a full discussion of the evidence on which this theory is based, the reader might consult Andrewartha (1952, pp. 85–95). Its essential features are: (*a*) There is no evidence that the neurosecretory center functions differently in the individual which has emerged from diapause as compared with the one which has never been in diapause. Therefore, it is reasonable to assume that the stimulus which "triggers" the neurosecretory center is the same in post-diapause as in nondiapause individuals. (*b*) Wigglesworth (1934) showed that *Rhodnius* molted only after a large meal (sec. 4.0). The experiments of Bounhiol (1938) with the larvae of *Bombyx*, of Piepho (1941) with the larvae of *Galleria*, of Titschak (1926) with the larvae of *Tineola*, and other experiments with other species in which diapause does not occur indicate that the timing of ecdysis is related to the ingestion and metabolism of food: with *Bombyx* and *Galleria*, ecdysis occurred, provided that a certain minimum amount of feeding had taken place during that stadium; with *Tineola*, several extra ecdyses were intercalated into the life-cycle when the larvae fed on very dry food and had therefore to "work" harder to get an adequate store of water from their food. (*c*) If one may assume that the stimulus comes from the breaking down of the food reserves stored in the fat body and elsewhere, rather than from the mere accumulation of these reserves, then it is reasonable to assume that diapause occurs because the food reserves which have been accumulating during the actively developing stage fail, for some reason or another, to break down at the critical stage when diapause occurs.

The hypothesis does not specify (because the literature provides no clues to this) whether the failure on the part of the diapausing individual to "mobilize" its food reserves at this critical stage is due to the intractable nature of the food itself or to the absence of some essential enzyme system. This is the same problem which was posed when we were considering the particular case of diapause in the egg of *Bombyx* (sec. 4.43), but it is now clear that this problem is quite general. It is indeed the most fundamental problem still outstanding in the study of diapause.

4.6 THE ADAPTIVE VALUE OF DIAPAUSE

The firm obligate diapause which occurs in the egg stage of the grasshopper *Austroicetes* confers upon the species three "advantages" which are nicely related to the climate of the area where it lives: (*a*) it insures that the active stages shall be present only in that season of the year which is most favorable for their development and survival; (*b*) it insures that the grasshopper shall spend the hot, dry summer as a diapausing egg; (*c*) the egg in diapause, because of its remarkable resistance to water loss (sec. 7.233) and its low level of respiration, is the only stage in the life-cycle which could resist desiccation during exposure to the extremes of heat and dryness experienced during summer in this area. With species from the Northern Hemisphere the diapause-stage may be especially resistant to cold. The principle is the same, and these three "advantages" are characteristic of diapause as an adaptation in other species. Since all three occur together in *Austroicetes*, this may be usefully described as an example.

The ecology of *Austroicetes* has been described in a series of papers by Andrewartha (1939; 1943*a*, *b;* 1944*a*, *b*, *c*), Birch and Andrewartha (1941, 1942, 1944), Steele (1941), Birch (1942), and Andrewartha and Birch (1948). The climate in the area in Australia where *Austroicetes* occurs is like that of the Mediterranean region of Europe (Davidson, 1936*c*). The summer is hot and arid, the the winter cool and moist. The soil may gain moisture from rainfall in excess of that which it loses by evaporation for about 5 months during the winter, and this constitutes the growing season for most of the vegetation in this area; for the remainder of the year the soil remains parched, and the temperature at 1 inch below the surface may be as high as 50° C. at times. Only the most xerophytic shrubs and trees remain green, for the evaporation from a free-water surface during the summer may be as much as 4 or 5 feet. In this severe climate *Austroicetes* is able to maintain a large population, and at irregular intervals the grasshoppers become so numerous that they constitute a serious menace to agricultural crops. The chief hazards which the grasshopper encounters have been set out and evaluated by Andrewartha and Birch (1948). They are desiccation of the egg during summer, starvation of the active stages during spring, and epizoötics of fungal disease while the active stages (nymphs and adults) are present. In each instance the presence of diapause reduces the hazard.

The eggs are laid in the soil during early summer (November–December); the soil is usually air-dry by this time. The chorion, though more impermeable than that of many species from more humid climates, affords quite inadequate protection from desiccation in this climate. But the eggs are packed into an oötheca and surrounded by a moist secretion from the accessory glands which affords temporary protection. The eggs develop without delay to the stage where the embryo has been separated from the primitive blastoderm and has begun to sink into the yolk; it is still quite minute. At the same time, the cuticle is laid down with its wax layer, and the hydropyle is differentiated; the egg then enters a firm diapause in which it remains throughout the summer. Eggs collected from the field at this stage and held at 30° C. may remain alive without developing for upward of a year. But the diapause is such that the egg is competent at any time to respond quantitatively to exposure to low temperatures; and a limited amount of morphogenesis becomes possible at high temperatures for any particular short period of exposure to cold. This enables the egg to respond to the diurnal temperatures of the autumn; and morphogenesis slowly advances to about the stage represented by the end of anatrepsis (Fig. 4.01 *K*). The egg enters the winter in this stage. If the egg in this stage continues to be exposed alternately to cold and warmth, it becomes unhealthy and dies, chiefly because the embryo becomes gross and is unable to complete katatrepsis and the egg fails to accumulate a store of liquified yolk, which is normally the next step in physiogenesis. But in nature this does not happen, for the low temperature of winter does not favor morphogenesis. During winter, diapause-development proceeds: it is made evident by the accumulation of a store of liquefied yolk particularly in the posterior third of the egg; this enables the embryo to complete katatrepsis and grow in a healthy manner when it is again exposed to warmth. The nymphs hatch about September (southern spring); they mature rapidly, and the adults lay eggs as summer approaches (November–December). Usually the vegetation dries up, and the adults die from starvation before they have laid all the eggs of which they are potentially capable. Thus the egg stage occupies 9 months of the year.

Artificially, by suitably manipulating the temperature of the environment, the duration of the egg stage may be shortened to a minimum of about 3 months. But in nature no ordinary variation in the weather will materially shorten the duration of the egg stage. In this climate there can never be sufficient accumulation of "cold" during the summer to allow morphogenesis to proceed beyond the end of anatrepsis. During the winter, the temperature is always adequate for diapause-development and at the same time too low to permit more than a limited amount of morphogenesis. The result is that the eggs usually hatch during the first week in September, and the date of hatching rarely departs from this by more than a week or so. During the next few months the temperature is favorable for the development of the nymphs, and there is usually plenty of food for them. Occasionally during an unusually severe drought the nymphs may die in large numbers from starvation (Birch and Andrewartha, 1941); or an unusually humid period may cause an outbreak of fungal disease. But the danger from drought is less at this time of the year than later, and the danger from disease is less than earlier in the season. If the eggs were to hatch during autumn and the nymphs be present during winter, they might find plenty of food, but the chance of widespread disease would be greater; if they were to hatch when they are first laid during summer, there would be no food for the nymphs. The presence of diapause in the egg stage insures that the active stages shall be present only at that season when their survival-rate is likely to be high.

The egg of *Austroicetes*, once the cuticle and its associated wax has been laid down, becomes highly resistant to water loss and remains so while the egg continues in diapause. While the egg remains in diapause, the yolk is granular and opaque throughout. As diapause-development proceeds, the posterior third of the egg becomes filled with a nongranular translucent fluid, and at the same time the egg's capacity to retain its moisture decreases (Birch and Andrewartha, 1942). During the summer and autumn no fluid is apparent, for the small amount of liquefaction which may result from chilling during a cool night is rapidly taken up by a correspondingly small amount of growth on the part of the embryo, and the egg remains firmly in diapause. The chief hazard at this season is drought; the responses of the early-stage egg are such as to insure that it will remain firmly in diapause during summer and so remain highly resistant to desiccation. The degree of protection conferred upon the egg by the presence of diapause may be inferred from the figures in Table 4.08.

TABLE 4.08*

RATE AT WHICH EGGS OF *Austroicetes* IN DIFFERENT STAGES OF DEVELOPMENT LOST WATER AT 20° C. AND 22 PER CENT RELATIVE HUMIDITY

Date Eggs Were Collected from Field	Stage	Mean Loss of Weight (Mg. per Egg per Day)
November 11	Pre-diapause	0.44
December 23	Diapause	.02
June 19	Post-diapause	0.18

* From Birch and Andrewartha (1942).

It is clear that in *Austroicetes* diapause contributes to the species' ability to maintain a high population in the particular climatic zone in which it occurs. Not only the presence of diapause but also its nature influence the limits of the distribution of the species. Coastward from the zone occupied by *Austroicetes*, the winter becomes more humid and lasts longer; inland it becomes shorter and more arid; the limits of the distribution of *Austroicetes* in both directions are well defined (Andrewartha, 1944*b*). If the eggs of the grasshopper were to hatch later, the risk of starvation for the active stages would be greater, and presumably the limits of its distribution on the arid side would be moved toward the coast; conversely, if they hatched earlier, the risk of widespread disease would be greater, and the limits of the distribution on the humid side would move farther inland. It is easy to conceive of slight changes in the nature of diapause in *Austroicetes* that might result in a shrinking of the area which the species would be able to inhabit. Perhaps in the absence of diapause the species would disappear. With *Austroicetes* this cannot be put to the test, for the diapause is universal and obligate.

From the accounts given in section 15.112 of the introduction into North America of *Pyrausta* and *Gilpinia*, it may be inferred that the diapausing races in these species have successfully colonized areas where the nondiapausing races have been unable to survive. The multi-voltine race of *Pyrausta* has been established in the northeastern coastal regions since before 1917. Yet there is no evidence that this race has spread westward. On the contrary, it has continued to be restricted to this zone; and the populations now present in the areas of the Great Lakes and to the southwest of them have almost certainly been derived from the uni-voltine race which originally colonized this area. A similar inference may be made with respect to *Gilpinia*. In Europe the 12-chromosome nondiapausing race makes up 90 per cent of most populations, and it is likely that individuals of this race have been introduced into North America together with those of the 14-chromosome diapausing race. Yet, if they have been introduced, they have failed to survive, for it is the latter which have alone become established in North America.

The first of the three functions of diapause discussed in relation to *Austroicetes* is quite general. In most species in which diapause occurs, it may be recognized as an adaptation which serves to synchronize the life-cycle of the animal with the rhythm of the environment and to insure that the active stages shall be present during the favorable season. The case of the predator whose life-cycle is synchronized with that of its prey is a particular case of this general rule. Less generally, the presence of diapause in a particular stage may confer upon it a resistance which it might otherwise lack, and this stage may be more resistant than any other in the life-cycle. Or all functions may combine in the one species, as in *Austroicetes*. It is not necessary to discuss the details for particular species. The general pattern persists, but the variation in detail is unending.

CHAPTER 5

Dispersal

The idea with which we have to start is, therefore, that animal dispersal is on the whole a rather quiet, humdrum process, and that it is taking place all the time as a result of the normal life of the animals.

ELTON (1927, p. 148)

5.0 INTRODUCTION

IF ALL the places where the individuals of a species are living were mapped and a boundary drawn around the area, this would be a picture of the distribution of the species at this particular moment. The animals are usually distributed unevenly (sec. 13.02), and this pattern of distribution (as distinct from the whole area that is occupied in a general way) is sometimes referred to as the "dispersion" of the species. In this chapter we are not concerned with "dispersion" as used in this sense but with *dispersal* in the sense that Elton (1927) used this word. He made a distinction between *distribution*, which relates to the places that happen to be occupied at any particular moment, and *dispersal*, which refers to movement away from a populated place, resulting in the scattering of at least some of the original population. Dispersal may be on the grand scale, bridging ecological barriers and leading to an extension of the distribution of the species. The spread of harmful species of insects by commerce from the Old World to the New and in the opposite direction provides some striking examples of the consequences of the artificial dispersal of these species across ecological barriers. Darlington (1938) showed that the fauna of the Greater Antilles (Cuba, Hispaniola, etc.) was closely related to that of Central America, which is several hundreds of miles away over the ocean. Cyclonic tropical storms (hurricanes), arising in Central America, sometimes pass over these islands. The chance that any one storm or the storms of any one year might transport an animal as large as a land snail from Central America to a suitable breeding place in the Greater Antilles may be slight; but when this is multiplied by the many thousands of years available, then Darlington's suggestions that these storms may account for the remarkable similarity of the two faunas becomes quite credible. Considerations of this sort are important for students of animal geography, but we do not pursue them in this book. We are more interested in dispersal over shorter distances, the sort of scattering that results not in the extension of the distribution beyond ecological barriers but merely in the reshuffling of the individuals within the area in which the animals are distributed. For this is the aspect of dispersal which is important for the student who wants to understand the abundance of animals and the limits imposed by environments on distributions.

In looking at populations in nature, one is impressed by the relative scarcity of most sorts of animals. The collector may know, from long experience, just the kind of place to look for a particular species, but he may count himself lucky if he finds a specimen in the first half-dozen places that he searches. This emptiness of so many of the places that are suitable for occupation is a striking and characteristic feature of nature and superimposes still another pattern on the distribution of animals, on top of the basal design determined in the first place by the distribution of food and the places that are suitable for occupation. It also adds to the rate at which the pattern changes. Even in an area where the distributions of food and places to live remain much the same, spots that are occupied today may become vacant tomorrow and re-occupied next week or next year. But usually the distributions of food and places to live are constantly changing because of the ordinary processes of growth and senescence and ecological succession.

In these circumstances the motility of the species, its readiness to seek a new place to live, and the "skill" with which it searches are likely to be important in determining its chance to survive and multiply and therefore its over-all abundance. The powers of dispersal possessed by a species may also be important when it has to live with a predator or predators which may annihilate the colony in a local situation. A species with high powers of dispersal may be constantly colonizing new situations, keeping ahead of the predator and thereby maintaining itself in greater abundance than one with poorer powers of dispersal. As a corollary, the predator with high powers of dispersal may tend to reduce the abundance of the prey more than one that is more sluggish (secs. 5.3, 10.321, and 14.2).

Two quotations may serve to illustrate and emphasize the general principles outlined in the preceding paragraphs. The first from Nicholson (1947) relates to prickly pear (*Opuntia* spp.) in Queensland and the moth (*Cactoblastis*) which feeds during its larval stage upon the cactus:

> After the introduction of this moth into Australia it increased in abundance at a terrific rate, quickly destroying the prickly pear over large areas around the points of liberation. After some time the point was reached at which very little prickly pear remained, whereas there were in the country countless millions of moths which had bred upon the prickly pear destroyed during the previous generation. Most of the caterpillars coming from the eggs of these moths died for simple lack of food, and most of the remaining prickly pear was destroyed at the same time. But this did not produce complete eradication of the prickly pear. By chance, or because of some favourable circumstance, here and there prickly pear plants survived. Similarly, even at this time some few of the moths succeeded in finding odd plants on which to lay their eggs, although, because of the great reduction in the numbers of *Cactoblastis* in the previous generation, many plants were not found. The end result,

which still persists, is that prickly pear is scattered in small isolated groups, with wide intervals between them. In a few of these *Cactoblastis* is still to be found, and these are generally doomed to complete destruction because *Cactoblastis* is able to increase rapidly in numbers on them. In the other groups of prickly pear, which have not so far been found by *Cactoblastis,* the pear tends to spread; but sooner or later is found by the moth, and the destruction of these groups is achieved shortly afterwards. In the meantime seed is scattered in new places, so maintaining the existence of prickly pear. Consequently, within Queensland as a whole, there is a low density of prickly pear, and a low density of *Cactoblastis,* which vary little from year to year; but at the same time there is continual fluctuation in space.

The second quotation is taken from Errington's (1943) analysis of mink predation on the muskrat (*Ondatra zibethicus*). He referred to the higher fecundity of the northern subspecies of the muskrat, and then wrote:

It appears more reasonable that a higher biotic potential on the part of *O. z. zibethicus* would be of far less advantage in enabling it to cope with enemies than in permitting this sub-species quickly to fill up depopulated habitats . . . and the characteristic instability of water levels of so many prairie marshes of northern United States may offer a greater test of survival value than is experienced by either the Maryland or the Louisiana subspecies. Other regions of the continent admittedly are not free from water fluctuations, but, to get an idea of marsh vicissitudes of the northern plains and prairies, the reader should take into account not only the Great Drouth of the Thirties but also the lesser drouths of "normal" years that affect one marsh and miss its neighbour, the irregular and unpredictable "dry spells" that may expose a given marsh bottom half a dozen times in as many months or that may depopulate a habitat one season and leave it with good water levels and lush vegetation—and barren of muskrats—the next.

Unlike predation, which operates chiefly as a secondary or an incidental factor in muskrat mortality, drouths of the north-central prairies typically cut deeply into muskrat populations and may bring about the ". . . important case" referred to by Wright (1940, p. 243) "where local populations are liable to frequent extinction, with restoration from the progeny of a few stray immigrants."

The powers of dispersal possessed by the species are thus seen to be chiefly important in relation to the patchiness of the distributions not only of food and places to live but also of the other animals in the area. These qualities of the environment are discussed elsewhere (particularly in chaps. 2, 11, 13, and 14).

This chapter is concerned with the dynamic phenomenon of dispersal: it is essentially a study of animal behavior, of how animals move about from place to place, and especially their *readiness* to move. The literature abounds in descriptive references to this quality of animals; for, indeed, its importance is usually clear to anyone who observes natural populations, but there are relatively few precise and detailed studies.

5.01 *The Innate Tendency toward Dispersal*

The diverse adaptations in plants for the dispersal of seeds are commonplace objects for wonder and admiration. Thistledown features in story and song and winged or barbed seeds of other plants are also well known. With a plant, winged seeds and the like are obviously adaptations for dispersal. Paradoxically, because motility is a primary quality of animals, adaptations among them for dispersal are less obvious. To demonstrate adaptations for dispersal in animals, it is necessary to distinguish movement associated with the ordinary activities of searching for food, a mate, and other requirements from movement which leads to finding a new place to live or for progeny to live. The individuals of some species may be observed to move actively, or even to travel great distances, in search of food or in response to some other stimulus. There may be cases which seem, at first sight, to be obvious examples of dispersal; yet if the movement is chiefly devoted to a search for food and does not lead to the occupation of new places and the formation of new colonies, it can hardly be counted as such.

The case of the swift *Micropus apus* in Europe may be quoted as a warning against the too facile acceptance of the obvious. There are many records, going back over many years, of migration-like movements of the swift during summer, which is the breeding season. These have been variously interpreted as a late spring migration, an early autumn migration, or an after-nesting dispersal. Koskimies (1947) has shown that they are, in fact, none of these things but merely flights undertaken in direct response to bad weather or some phenomenon associated with the weather. The swifts move away from the area while the weather is dominated by the warm front of a cyclonic depression. Flights of this sort may continue for as long as a week, and, since the swift may fly as far as 1,000 kilometers in a day, the distances covered may be great. The young nonbreeding birds may not return to where they started from. To this extent the flights may be counted as dispersive. But the breeding birds probably do return to the same place because they have nests or nestlings to look after.

The correlation between the flights and the weather is well established by Koskimies' observations, but the direct stimulus may only be inferred. The swift feeds only on insects taken from the air during flight. During this sort of weather, which is characterized by downward currents and warm rain, there are virtually no insects left in the air. Experiments have shown that the adult swift may not survive more than 2 days without food, and it seems most likely that these flights, despite their magnitude, merely represent a search for food. Experiments have shown that the nestlings may survive 7 days' starvation. This is most unusual: the nestlings of most insectivorous birds succumb after 2 days without food. The swift is highly specialized with respect to its food and beautifully adapted to this kind of diet by virtue of the extraordinary powers of flight in the adult and the altogether unusual capacity in the young to survive prolonged starvation.

The swift doubtless engages in other movements, perhaps at other stages of the life-cycle or at other seasons of the year, which are chiefly dispersive. Indeed, with most species, much of the movement that can be observed is to be attributed to the innate tendency toward dispersal. This may sometimes be observed directly; and, in the absence of opportunity for direct observation, it may be inferred by the appropriate statistical analysis of samples taken from natural populations which may or may not have had marked individuals added to them. Or behavior relating to dispersal may be studied in the laboratory.

Students of evolution, particularly the school which emphasizes the importance of geographic isolation for speciation, have often recorded examples of dispersal and used them as an argument against the likelihood of sympatric speciation. Mayr (1947) summarized this point of view: "Dispersal is conspicuously neglected in all schemes of sympatric speciation even though it is one of the basic properties of organic nature. There is a dispersal phase in the life-cycle of every species. It is the adult stage in most insects with wingless larvae. . . . It is known that certain birds are extremely sedentary and live throughout the breeding season or even their adult life within a restricted territory. However, even in these species, there is a certain amount of dispersal before the juvenile takes up its first territory. . . . Sessile marine organisms always seem to have a larval stage in which they may be dispersed by ocean currents for hundreds of miles and during which populations within this area are thoroughly mixed up."

This innate tendency toward dispersal has been observed also in species which may be constrained to stay within a local breeding area by the most trivial of ecological barriers. A relatively narrow strip of arable land would seem to provide a complete ecological barrier to the moth *Panaxia dominula*, which is to be found in marshy woodlands where reeds grow. Two isolated small populations of *Panaxia* which live in two such areas near Oxford have been studied by Sheppard (1951*b*). Each area is quite small; they are surrounded by farmland and are about 1 mile apart. Three varieties of this moth are known: two of them, *medionigra* and *bimacula*, are, respectively, the heterozygote and the homozygote of the gene *medionigra*, which shows no dominance (Fisher and Ford, 1947). In one colony this gene has fluctuated between 2.9 and 11.1 per cent for 12 years; in the other colony it has not been found during 3 years' observations. The gene *medionigra* appears to be absent from the second colony, and this may be due to the failure of the moths to cross the mile of farmland which separates the two colonies, especially as some hundreds of moths have been captured, marked, and released, but none bearing the distinctive mark of one colony has been found in the other. Yet even this species, which is so easily restrained by a seemingly trivial ecological barrier, is imbued with the usual instincts for dispersal within the confines of its small breeding area.

The adults usually emerged in the morning, and they usually copulated in the afternoon of the same day. In a series of experiments, females were marked in the field either during or shortly after copulation. In no case was a marked moth found in the same place the next day. As the females rarely laid many eggs the day they emerged and the number laid during the first day after copulation was less than 20 per cent of their total output this meant that the moths, notwithstanding the presence of plenty of food in the vicinity, always dispersed to some other part of the breeding area before laying the greater proportion of their eggs (Sheppard, 1951*b*).

At the other extreme there are the migrants which regularly traverse seemingly enormous barriers. Among the migratory butterflies, *Danaus plexippus* occasionally crosses the Atlantic Ocean, and *Vanessa cardui* regularly flies from Africa across the mountains to northern Europe (Williams *et al.*, 1942). The difficulty here is to decide to what extent these great flights contribute to dispersal and to what extent they resemble the seasonal migrations of those birds which may fly over half the circumference of the earth, merely to return to the same nest in the same territory the next year (sec. 5.4).

The innate tendency toward dispersal, which seems to be present to a greater or smaller degree in all animals, may be accentuated by crowding, hunger, warmth, wind, and so on. As we measure dispersal in nature (sec. 5.1) or study dispersive behavior in the laboratory (sec. 5.2), we often demonstrate, at the same time, the influence of certain stimuli on the dispersal of the animals.

5.1 THE MEASUREMENT OF DISPERSAL IN NATURAL POPULATIONS

Davidson and Andrewartha (1948*a*, *b*) kept daily records for 14 years of the number of *Thrips imaginis* in roses. Every morning at 9 o'clock, 20 roses of a particular kind (Cecil Brunner) were picked from the same hedge in a large garden. The thrips breed in a great diversity of flowers, including roses; but they do not breed, or hardly at all, in the Cecil Brunner roses because these lack fully developed stamens. But they are strongly attracted to these flowers, which therefore serve admirably as a "trap" for the active winged adults. The numbers found in the roses depend upon (*a*) the density of the population in the garden and (*b*) the activity of the thrips in seeking out the flowers. And, indeed, the data can be analyzed into these two components.

Thrips are quite small insects, and at times the number found in one rose exceeded 500, though the majority of records showed less than 100 thrips per rose, and 275 out of a total of 1,773 records showed less than 10 thrips per rose. The data for 8 years (1932–39) during October–November (southern spring) were used to measure the activity of the thrips in seeking out the flowers. At this time of the year the density of the population was usually increasing steadily, though often rapidly, and it was possible to eliminate the variance associated with the increase in density with time by a polynomial expression of the third degree (see Fig. 6.08).

The data were transformed to logarithms, because, for a study of activity, we were interested not in the position or slope of the trend line itself but in the departures of the individual daily records from the smooth curve. These measured the daily movements of the thrips into and away from the roses. The proper basis for comparisons from day to day was the proportion of the population that was moving into the flowers. This was given by the logarithmic scale.

Figure 6.08 shows that, when the influence of temperature and rainfall are taken into account (sec. 6.22), the departures of the individual daily records from the trend line were of the same order when the population was sparse as when it was dense. That is, the activity of the thrips in seeking new places to live was independent of the density of the population and, by inference, equally independent of any external stimuli associated with shortage of food or shelter or so on. In fact, the only components of the environment that could be shown to be associated with activity of the thrips were temperature and rainfall (sec. 6.22).

These results were confirmed by other records kept during the course of this study. A hot and arid summer is characteristic of the Adelaide district. Every year as the summer approaches, the ephemeral plants which clothe the meadows, parks, and roadsides die, leaving only a dry stubble. The thrips which had been living in the flowers disappeared: the population, which had been widespread and largely continuous throughout the countryside, contracted and broke up into small scattered units. The only places where the thrips were able to live through the summer were especially favored areas; most of them, like the garden from which our records were taken, contained a wide variety of plants, both perennial and annual, because they were tended and watered artificially during the dry period. In such situations there were always a few flowers in which the thrips could live and breed, though for diverse reasons (secs. 11.12 and 13.116) they never became numerous even in these places except in the spring.

With the return of the rainy season in the autumn, the seeds of the ephemeral plants germinate; during the winter they grow and in the spring they flower. During the 14 years of this study it was our practice to collect samples of these flowers as soon as they appeared, and invariably we found that they quite early became colonized by *Thrips imaginis*. At this time of the year, our routine counts would often show less than 10 thrips per rose, and it is quite clear that the thrips in the garden were not suffering any of the effects of crowding or overpopulation. Yet they were leaving the garden quite freely and seeking new places to live, apparently in response to an innate instinct for dispersal which is independent of any stimulus associated with shortage of food or space or any other resource.

5.11 *The Use of Marked Individuals*

Dobzhansky and Wright (1943, 1947) in each of four experiments, liberated from 3,000 to 5,000 marked adults of *Drosophila pseudoobscura* in an area where the density of the wild population of this species was estimated to be about 6

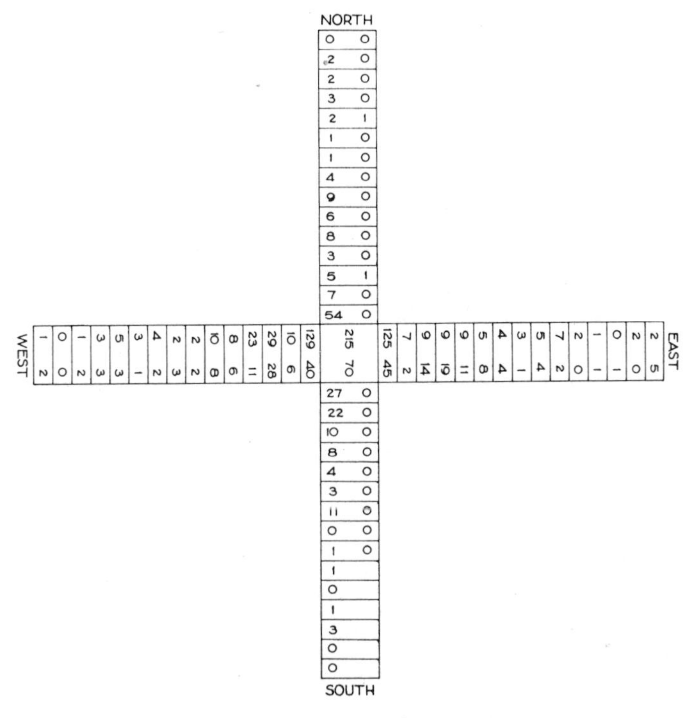

FIG. 5.01.—The design of the trapping experiment used by Dobzhansky and Wright in their experiments on the dispersal of *Drosophila pseudoobscura*. The traps were arranged in a rectangular cross. The marked flies were liberated at the center of the cross. The column of figures on the left indicate the numbers of marked flies caught on the first day, and those on the right the numbers caught on the second day. (After Dobzhansky and Wright, 1943.)

per 100 square meters. The marked flies were orange-eyed mutants which were easily distinguishable from the wild flies. The area contained 61 traps arranged in the form of a rectangular cross, with the traps spaced along the four arms of the cross at intervals of 20 meters (Fig. 5.01). The flies were liberated near the center of the cross. The traps were examined once each day at the same time, and the numbers of marked and wild flies recorded; they were then released again near the trap from which they had been taken. In this way the distribution of the marked flies at the end of each day was discovered, and the dispersal of the population was measured by daily changes in the shape and variance of the distribution. In an experiment of this sort, the rate of increase of the variance provides a measure of the rate of dispersal of the population, and the shape of the distribution provides additional information about the behavior of the flies in relation to dispersal. For example, if the population is homogeneous with respect to rate of dispersal and if dispersal is random with respect to direction and to other members of the population, then the distribution in one direction (e.g., along a line of traps) should conform to the familiar normal curve of the statistician, with the mode at the place where the flies were released. If, however, the dispersal is random in these two respects but the population is heterogeneous with respect to the rate of dispersal, the curve, while remaining symmetrical, will be either flatter or steeper than the normal curve (Fig. 6.06). The best way to measure this is to calculate the ratio

$$Ku = \frac{n\Sigma r^4 f}{\Sigma r^2 f} \quad \text{(Fisher, 1948, p. 52).}$$

In this expression Σ has its usual meaning as the symbol of summation, f is the number of flies caught in a trap, r is the distance between the trap and the place where the flies were released, and n is the total number of flies caught in all the traps. For the normal curve the ratio $Ku = 3$; for a flat curve it is greater, and for a steep curve less, than 3.

Table 5.01 shows the value of Ku on each day for the four experiments. Values of Ku in excess of 3 for all but the last day or two of each experiment indicate a pronounced flatness in the curves, particularly in the earlier stages of each experiment. The low values for Ku toward the end of each experiment may be an artifact, if by then some of the far-ranging individuals had flown

TABLE 5.01*

VALUES FOR KURTOSIS ($Ku = n\Sigma r^4 f/\Sigma r^2 f$) AND STANDARD DEVIATION (S.D.) OF DISTRIBUTIONS IN METERS, DAY BY DAY, FOR EACH OF DOBZHANSKY AND WRIGHT'S FOUR EXPERIMENTS WITH *Drosophila*

DAY	EXPERIMENT							
	I		II		III		IV	
	S.D.	*Ku*	S.D.	*Ku*	S.D.	*Ku*	S.D.	*Ku*
1	39	9.8	59	7.6	58	10.4	68	8.3
2	57	5.7	92	5.0	94	4.4	95	5.9
3	74	4.2	102	4.4	131	2.8	136	4.2
4	72	4.3	117	3.6	129	3.0	177	4.0
5	64	4.5	122	4.0	133	2.7	171	3.0
6	84	3.5	159	3.6	171	1.8	...	..
7	93	3.1	161	2.7	190	1.9	...	..
8	114	2.4	...	..	...	..	...	..
9	97	3.9	...	..	...	..	...	..

* After Dobzhansky and Wright (1943).

beyond the last traps in the lines. The high values for Ku indicate heterogeneity in the population. While most flies moved relatively short distances each day, there were always a few far-ranging ones which flew much farther. Similar results were found with another experiment with the same species in a different district in California (Dobzhansky and Wright, 1947) and in Brazil with *D. willistoni* (Burla *et al.*, 1950).

Because of the high values for *Ku* in these experiments, the familiar methods appropriate for the normal distribution overestimate the variance; and Dobzhansky and Wright (1947, p. 312) said that a more satisfactory estimate is given by the expression

$$s^2 = \frac{\pi \Sigma r^3 \bar{f}}{\Sigma r \bar{f} + c}.$$

In this expression, Σ has its usual meaning as the symbol for summation, r is the distance of the trap from the center of the cross, $\bar{f}$ is the mean number of flies found in traps at this distance from the center, and c is the number of flies found in the central trap.

The rate at which variance increased indicated the speed with which the flies dispersed. If s_1^2 was the variance at the end of the first day and the flies continued to disperse at an even rate, then the variance should have increased by the same amount each day, giving by the tth day $s_t^2 = ts_1^2$. If, however, the rate of dispersal fell off with time (perhaps as the population became sparser), then the daily increments to the variance should have become progressively smaller. The day-by-day values for s^2 for the four experiments are given in Table 5.02. The daily increments in variance are erratic, owing largely to the

TABLE 5.02*

DAILY TEMPERATURE ($T°$ C.) VARIANCES (s^2 IN SQUARE KILOMETERS × 1,000), AND INCREMENTS IN VARIANCE (i), FOR DOBZHANSKY AND WRIGHT'S FOUR EXPERIMENTS WITH *Drosophila*

DAY	EXPERIMENT											
	I			II			III			IV		
	T	s^2	i	T	s^2	i	T	s^2	i	T	s^2	i
1	13	3.2	..	21	7.3	...	21	8.5	..	22	10.6	...
2	19	5.0	1.8	22	12.8	5.5	22	12.8	4.3	22	15.2	4.6
3	19	7.4	2.4	21	13.6	0.8	25	17.6	4.8	23	23.9	8.7
4	15	7.2	−0.2	22	16.0	2.4	22	17.7	0.1	26	39.4	15.5
5	13	6.1	−1.1	20	19.0	3.0	23	17.3	−0.4	20	28.6	−10.8
6	18	8.0	1.9	23	29.9	10.9	24	21.5	4.2	..	...	...
7	17	8.7	0.7	17	25.4	−4.5	23	28.5	7.0	..	...	...
8	17	11.9	3.2	..	...	...	..	...	..	..	...	...
9	15	10.5	−1.4	..	...	...	..	...	..	..	...	...

* After Dobzhansky and Wright (1947).

influence of temperature and perhaps other meteorological events on the activity of the flies (sec. 6.22), but they give no indication of becoming smaller with time. Since the density of the population undoubtedly decreased with time (as the liberated flies became distributed over an increasing area), these experiments indicated in *Drosophila*, as in *Thrips*, an instinctive urge to disperse, which is independent of the density of the population. Dobzhansky and Wright (1947) estimate that in the most favorable circumstances found in these experiments it would require about 5 months for the standard deviation in one direction to reach 1 kilometer. In a normal distribution this means that about one-third of the flies would have traveled more than 1 kilometer from the point of release by the end of 5 months.

The results of these experiments as summarized in Tables 5.01 and 5.02 have provided a neat demonstration of two qualities in *Drosophila*: (*a*) popula-

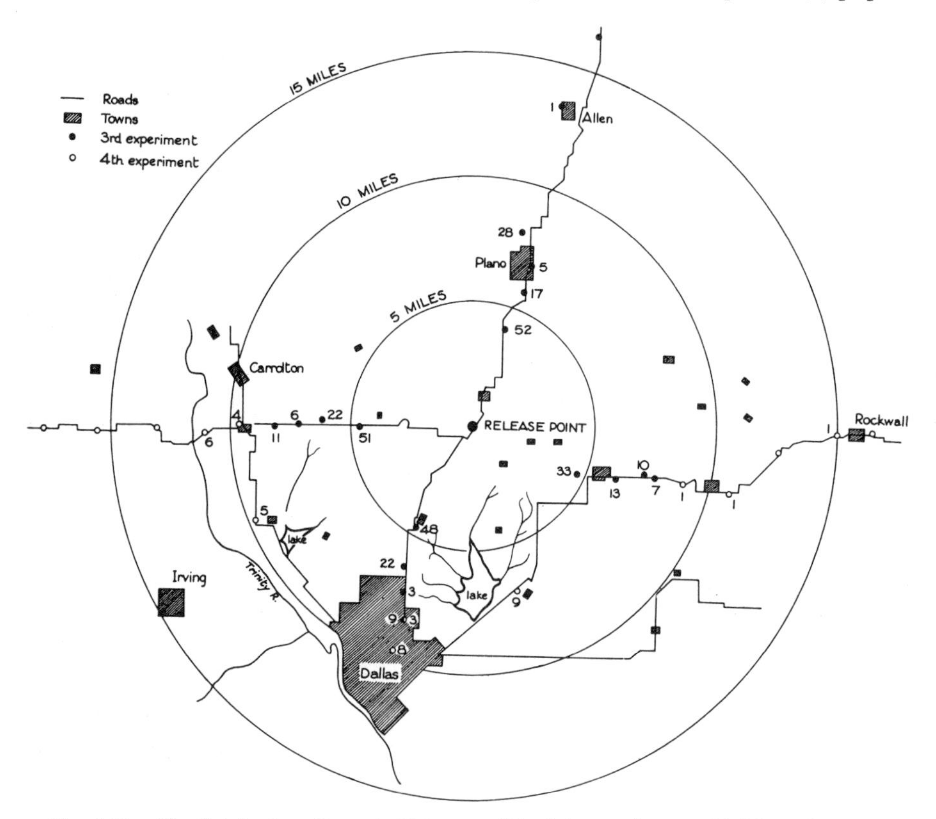

FIG. 5.02.—The distribution of traps in Bishopp and Laake's experiments with *Musca domestica* and *Chrysomyia macellaria*. The traps were placed in an irregular rectangular cross. Regularity was sacrificed, in order to have the traps in suitable local situations. This turned out to be an important part of the design of the experiment. The figures indicate the number of flies caught in each trap. (After Bishop and Laake, 1921.)

tions tend to disperse at a speed that is independent of the density of the population; and (*b*) the flies in a population are heterogeneous with respect to this instinct. In other words, most individuals possess an instinct toward dispersal, but there are a few in which it is more strongly developed than usual. Ecologists, experienced in the observation of natural populations, may feel that these nice experiments of Dobzhansky and Wright have given precise expression to a phenomenon which is quite commonplace not only among insects but with other animals also, including vertebrates.

Bishopp and Laake (1921) did an experiment with marked flies (mostly *Musca domestica* and *Chrysomyia macellaria*); the experiment was repeated four times between June and October. They arranged 16 traps in the form of a rectangular cross. There was no trap in the center, where the arms intersected, which was the place where the marked flies were liberated (Fig. 5.02).

An important practical detail in the design of the experiments concerned the precise placings of the traps. These were arranged, as far as practicable, at uniform intervals along the lines running away from the center; but, where necessary, complete consistency in this regard was sacrificed in favor of selecting a situation which was expected to attract the flies from a distance, that is, a place which, experience had taught, would already be supporting a relatively dense population of wild flies (secs. 12.0, 12.3, and 13.02). This turned out to be a fortunate arrangement, because the traps in the different situations were found to be uneven with respect to the relative numbers of the different species caught in them. For example, if *Musca domestica* from the wild (unmarked) population was the most abundant species in a trap, then it was generally found that the marked flies in this trap would usually contain a relative excess of *M. domestica* also. This indicated that the flies were first attracted to the environs of the trap and only secondarily by the trap. If the precaution of carefully selecting the place for the trap had been neglected, the numbers of flies recaptured (considering the few traps that it was practicable to handle and the large distances involved) might have been too few to justify any reliable conclusions. Instead, it turned out to be a most successful experiment. The flies were caught in the vicinity of packing houses, where they were abundant, dusted with a red chalk, and then released at the appropriate place. The details of numbers and species liberated and the numbers recaptured are given in Table 5.03. The results indicate a pronounced tendency toward dispersal. In all but the fourth experiment, where the most distant trap was 17 miles away, some individuals of both *M. domestica* and *Chrysomyia macellaria* were found in the outermost traps, and doubtless others passed beyond them. At least a few individuals covered these distances surprisingly rapidly. Some *M. domestica* were recaptured within 24 hours, over 6 miles from where they had been liberated; *Phormia regina* about 11 miles, 48 hours later; and *C. macellaria* 10 miles, after 48 hours.

The authors analyzed the direction of the flight with respect to the wind. Relatively few flies were found upwind from the place where they had been liberated, but rather more had moved across the wind than with it, indicating an active dispersal rather than a passive drifting with the wind. As with other experiments of this sort, temperature and humidity exercised a big influence on the activity of the insects. This, together with the shorter distances to the traps, accounted for the relatively greater numbers recaptured in the first experiment. Bishopp and Laake said that many of the flies must have passed by (probably after a brief visit) several farmhouses, small towns, and other favorable places, in order to reach the more distant traps. This shows that *M. domestica*, *C. macellaria*, and *P. regina* were not merely seeking the nearest food or breeding place, but they also displayed an instinct for dispersal which was independent of these direct external stimuli.

The experiments of Gilmour *et al.* (1946) with the sheep blowfly were designed primarily to measure the density of a natural population of flies in a particular area, but they also provided information about the dispersal of the

TABLE 5.03*
DETAILS OF RELEASES AND RECAPTURES IN BISHOPP AND LAAKE'S FOUR EXPERIMENTS WITH MIXED SPECIES OF FLIES

Experiment No.	I	II	III	IV
Total no. flies of all species liberated	18,000	33,000	60,000	60,000
No. different species as per cent of total:				
Musca domestica	70	55	42	66
Chrysomyia macellaria	25	36	55	29
Phormia regina	1	8	1	3
Other species	4	1	2	2
No. marked flies (all species) recaptured in traps specified distances from place of liberation:†				
0.5 miles	3,416 (4)	.....	.....	.....
1 mile	534 (4)	.....	.....	.....
2 miles	636 (6)	207 (5)	.....	.....
3–4 miles	66 (2)	239 (9)	57 (1)	.....
5–7 miles	.....	13 (2)	233 (11)	9 (1)
8–10 miles	.....	.....	55 (4)	21 (5)
Over 10 miles	.....	.....	.....	9 (10)
Distance (miles) of farthest trap from place of release	4	5	8	17
Greatest distance traveled (miles):				
Musca domestica	4	5	8	13
Chrysomyia macellaria	4	5	8	15
Phormia regina	4	5	.....	11

* After Bishopp and Laake (1921).
† Numbers in parentheses indicate numbers of traps.

flies. Three separate experiments were done in December, January, and March (southern summer), and in each the same procedure was followed. Altogether, 102 traps were uniformly spaced, about ¾ mile apart, to cover a circular area having a diameter of 8 miles (Fig. 13.01). About 40,000 flies were reared in the laboratory, stained with neutral red, and then released in the center of the circle. The traps were examined each day for the next 2 or 3 days. The details of the numbers released and recaptured, together with the numbers of wild *Lucilia cuprina* estimated to be in the area, are given in Table 5.04. The results were analyzed graphically and statistically. Figure 5.03 indicates that the flies dispersed individually and at random with respect to the points of the compass, despite the fact that in this experiment (III) a wind with an average velocity of 20 miles per hour was blowing across the area during the second day.

Making these assumptions (that the flies dispersed individually and in random directions) and the additional one that the population was homogeneous with respect to dispersive activity, Gilmour *et al.* derived a mathematical expression to describe the expected distribution of the marked flies at the end of a particular time. The observed distribution, as revealed by the traps, was

TABLE 5.04*

DETAILS OF NUMBERS OF MARKED FLIES LIBERATED AND RECAPTURED AND ESTIMATED DENSITIES OF POPULATIONS OF WILD FLIES IN THREE EXPERIMENTS WITH *Lucilia cuprina*

EXPERIMENT NUMBER	NUMBER OF MARKED FLIES			ESTIMATED DENSITY OF WILD POPULATIONS (FLIES PER ACRE)
	Liberated	Recaptured		
		1st Day	2d Day	
I (December)	36,300	782	452	1.7
II (January)	38,700	1,456	386	0.4
III (March)	37,950	725	432	1.4

* After Gilmour *et al.* (1946).

compared with the expected one (Fig. 5.04). For the observed data the ratio of marked to wild flies for each trap was used instead of the number of marked flies caught. It had been demonstrated quite clearly (sec. 13.02) that the natural population was distributed unevenly over the area; the distribution of

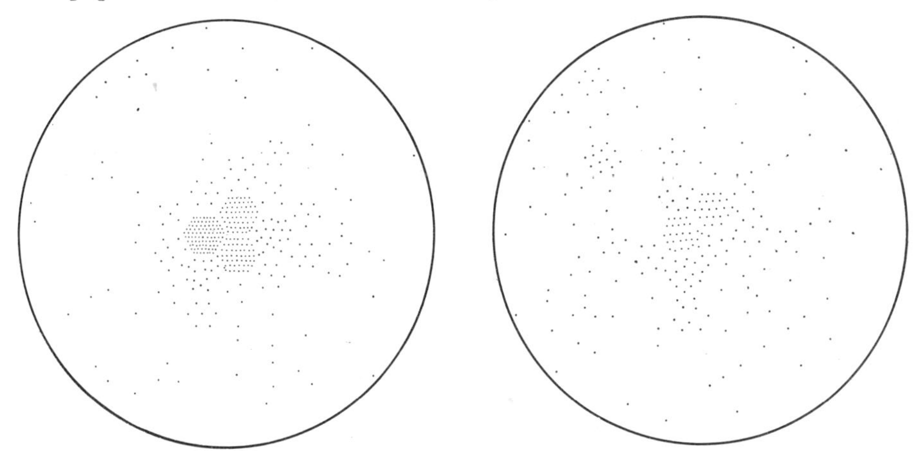

FIG. 5.03.—Scatter diagrams to illustrate the distribution of marked flies on two successive days in the third experiment of Gilmour *et al.* (1946). Each dot represents two flies, and the dots representing the catch of each trap have been spread evenly over a hexagon, with the trap in the center. (After Gilmour *et al.*, 1946.)

the marked flies tended to assume the same pattern, and correcting the raw data in this way made an allowance for this pattern. The agreement between observed and expected distributions (Fig. 5.04) is reasonably close and gives no cause to doubt that the flies move at random to one another and to the points of the compass. The discrepancies relate chiefly to a deficit in the relative numbers of marked flies found in the traps nearest to the place where they were liberated. This is best explained by assuming heterogeneity in the population: the more sluggish ones, which remained closer to the place where they were liberated, were less readily trapped than the more active ones, which had moved farther afield. That all the samples liberated included some very active individuals is confirmed by the observation that in each experiment at least a few flies were found in the most distant traps (4 miles) within 24 hours of being liberated. This is like the heterogeneity found by Dobzhansky and Wright in

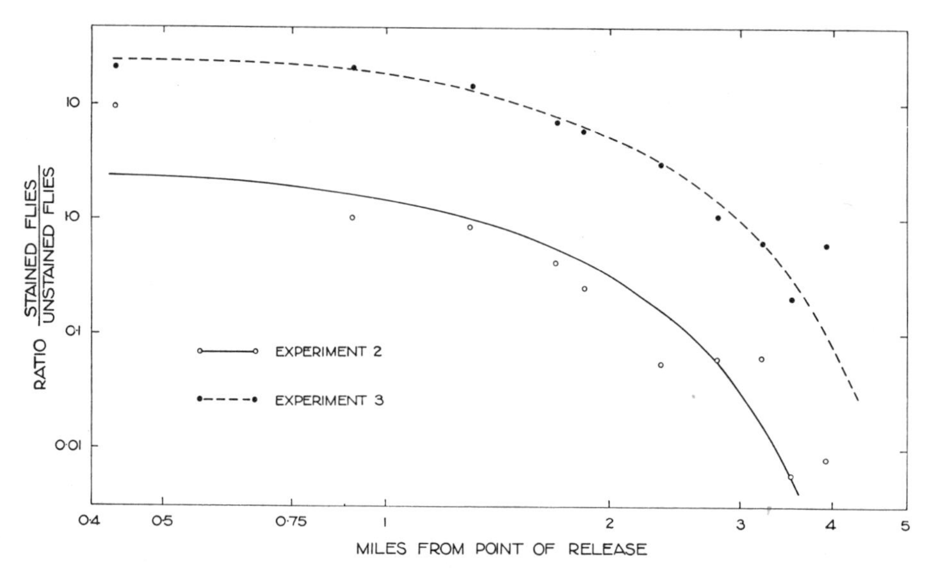

FIG. 5.04.—The curves show the numbers of marked flies expected in traps relative to the distance from the place where they were released. The circles represent the observed data. The heterogeneity of the experimental area was partly corrected by relating the number of marked flies in each trap to the density of the population of wild flies around the same trap. (After Gilmour *et al.*, 1946.)

Drosophila. The speed of dispersal was measured by the radius of the circle beyond which half the flies had passed by the end of the second day (Table 5.05).

TABLE 5.05

SPEED OF DISPERSAL OF *Lucilia cuprina* IN RELATION TO TEMPERATURE, RELATIVE HUMIDITY AND WIND

Experiment	Speed of Dispersal* (Miles)	Temperature† (° C.)	Relative Humidity† (Per Cent)	Wind† (Miles per Day)
I (December)	<0.43	17.3	63	208
II (January)	1.70	25.0	27	173
III (March)	1.40	18.7	25	149

* The speed of dispersal was measured by the radius of the circle (in miles) beyond which half the flies had passed by the end of the second day (data from Gilmour *et al.*, 1946).

† Mean for 3 days: the day that flies were liberated and the 2 succeeding days.

As usual with insects, activity was influenced chiefly by temperature, but atmospheric humidity and wind seemed also to be influential. Dispersal tended to be more rapid with low relative humidity and in the absence of strong wind.

Steiner (1940) measured the dispersal of marked adults of the codlin moth (*Cydia pomonella*) in apple orchards. During the course of experiments extending over 4 years, he marked and liberated some 10,000 moths and recaptured 744 in traps scattered around the orchard. As in other experiments of this sort, most of those recaptured were found in traps relatively close to the place where the moths were liberated; but each year some longer flights were recorded, the maximum varying from 663 to 2,079 feet. The results indicated that some moths had made long flights, passing by both traps that were attractively baited and apple trees providing suitable places for breeding; and it was inferred that in all the samples liberated there were at least a few moths in which the urge to disperse was stronger than these counter attractions.

In the experiments decribed in this section, insects were marked with paint, colored chalk, or by the use of a distinctive mutant. Radioactive materials have also been used successfully for tagging purposes. Jenkins and Hassett (1950) tagged 3,000,000 larvae of the mosquito *Aedes communis* with radioactive phosphorus, near Churchill in northern Canada. Recoveries of tagged adults were made for 1 month at distances up to 1 mile from the place where they were released. In experiments with cherry fruit flies, *Rhagoletis singulatus*, which were fed on sucrose solutions containing radioactive phosphoric acid, detectable amounts of radioactivity were found 1.5 months after treatment (Lindquist, 1952). Such methods may prove to be useful adjuncts to the more traditional methods of marking.

5.12 *Dispersal to and from Winter Quarters and Other Specially Restricted Areas*

With many species, not necessarily migratory in the sense of section 5.4, the distribution may undergo marked cyclical fluctuations associated with the seasons or with a particular stage in the life-cycle. Many species in cool, temperate climates can survive the winter only in places that offer extra protection from the weather. These locations may be discrete and quite small relative to the broad areas occupied during summer. Since the cyclical movement to and from winter quarters usually takes place into territory that is unpopulated, there is an opportunity to study this special aspect of dispersal quantitatively without any special technique, such as marking. In warm, temperate climates characterized by an arid summer, many insects and other invertebrates aestivate, and the same phenomenon of dispersal may be observed in the autumn. With certain species, for example, among the midges of the genus *Culicoides*, the larvae live in highly specialized places, often very small in area, and the adults disperse from these (Kettle, 1951). Similar situations may be produced artificially, for example, during the course of ordinary farming operations which depopulate one area while leaving a natural population living near by.

Frampton *et al.* (1942) traced the dispersal of the leafhopper *Macrosteles divisus* from an old meadow into a field of potatoes (Fig. 5.05, *A*). The meadow harbored a natural population of the insect, and the area occupied by the potatoes had been depopulated by plowing and otherwise preparing the ground for planting the crop. The insects were not counted directly, but their relative numbers were inferred from the distribution of a virus disease which they alone transmit to the potatoes. The pea weevil *Bruchus pisorum* hibernates as an

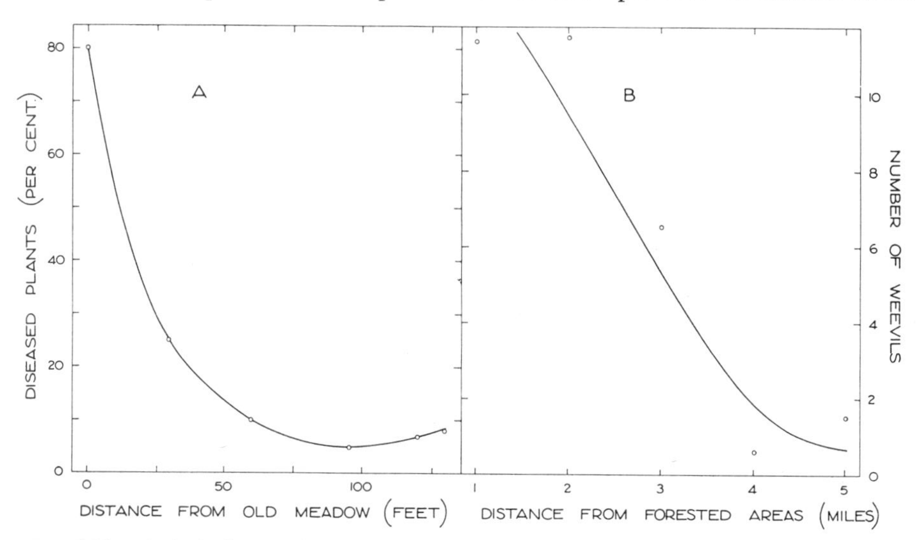

FIG. 5.05.—*A*, the incidence of virus disease in a field of potatoes relative to the distance from the old meadow which carried a natural population of the vector (*Macrostelus divisus*) (after Frampton *et al.*, 1942). *B*, The numbers of pea weevil *Bruchus pisorum* found in pea fields relative to their distance from the forests where *Bruchus* had spent the winter. (After Wakeland, 1934.)

adult in timbered areas in parts of North America, flying in the spring to adjacent fields of peas, where they breed during the summer. Wakeland (1934) estimated the density of populations of *Bruchus* in fields of peas and related this to their distances from the forests where they had spent the winter (Fig. 5.05, *B*). The distances in Figure 5.05, *A*, are expressed in feet; in Figure 5.05, *B*, in miles; Fulton and Romney (1940) studied the northward dispersal of the beet leafhopper *Eutettix tenellus* from its winter breeding grounds in Nevada and Utah; a similarly shaped curve was obtained, but with distances expressed in hundreds of miles.

Wolfenbarger (1946; see also Wadley and Wolfenbarger, 1944) has summarized similar data for a large number of small organisms, and this paper should be consulted in the original for further details. He fitted every set of data to one or the other of the two arbitrary expressions:

$$Y = a + b \log x, \quad \text{or} \quad Y = a + b \log x + \frac{c}{x}.$$

In these expressions Y is the density of the population at distance x from the source, and a and b are constants. The first expression implies that the density of the population decreases by a constant amount for equal multiples of the distance from the source; and the second implies that the decrease in density is greater than this by an amount proportional to the distance. The theoretical basis for these expressions is obscure. The empirical justification for using them is that some of the sets of data collected by Wolfenbarger followed the calculated curves quite closely; but there are others that do not; Wolfenbarger did not apply statistical tests to any of the curves that he calculated.

The bloodsucking adults of the midge *Culicoides impunctatus* occur fairly extensively in moorland and woodland in certain parts of Scotland. The larvae live in damp soil and have such special requirements that they tend to be restricted to quite small areas. Kettle (1951) explored one such breeding ground and found that it was restricted to an area about 90 by 40 yards. This and one other small breeding area seemed to supply all the adults found on one 5-acre experimental plot. The dispersal of the adult midges from these larval breeding grounds was studied by recording the weekly total of midges caught on 22 sticky traps evenly spaced over the 5 acres. The numbers decreased as the distance from the breeding ground increased; and Kettle found that the expression

$$\log Y = \log a + bx$$

fitted the observed data within the limits of experimental error (Fig. 5.06). In this expression Y is the density of the population at distance x from the source, a is the density of the population at the source, and b is the average rate at which the logarithm of the density decreases with distance. The expression implies that the density of the population decreases by a constant proportion for unit increase from the source; this seems a more realistic theory than that implied by Wolfenbarger's equations.

The decrease in density with increasing distance from the source may be divided into two components: one which may be called "dilution," due to the increasing area over which the population is spread, and the other which may be called "mortality" because some of the midges die or disappear from the area. If the component due to "mortality" be assumed to be zero, then a value for b in the above equation may be calculated geometrically; the difference between this value and the empirical one provides a means of estimating the "mortality"-rate in the population relative to the distance from the source. Kettle estimated for the midge *C. impunctatus* in the experimental area which he studied that half the females in the population died within 76.7 yards of the breeding ground; for the males the comparable figure was 73.6 yards. By similar calculations he estimated that the average distance (away from the breeding grounds) flown by the females was 81.4 yards and by males 79.0 yards.

This aspect of dispersal may be important in particular circumstances; but it is the dispersal of local groups through an area that is already populated by the species that the ecologist is more likely to meet and need to measure. We have discussed in the preceding sections several studies which are of general interest in this regard. In the next section we shall discuss the influence of

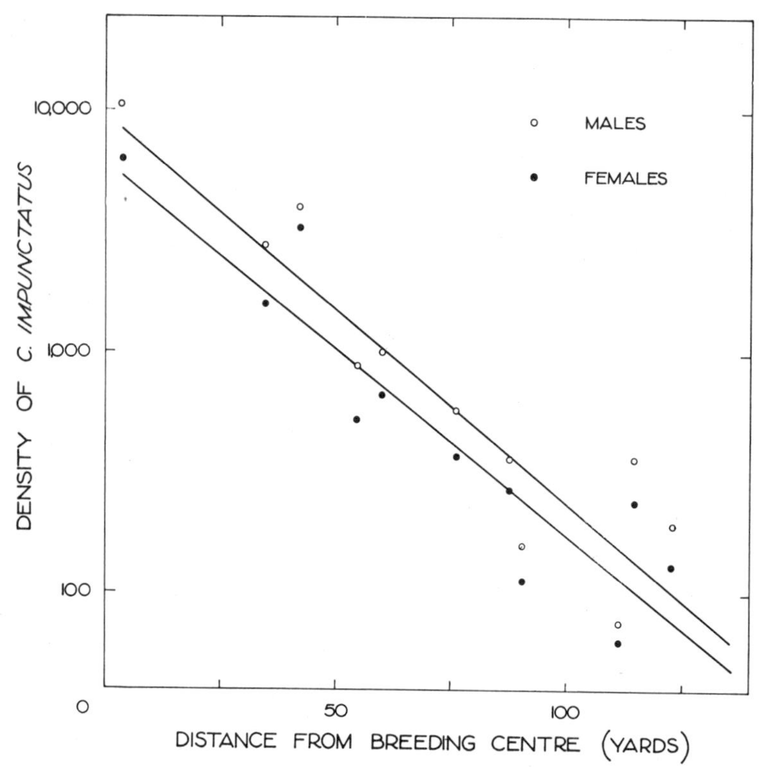

FIG. 5.06.—The relationship between the number of adult *Culicoides* caught on sticky traps and the distance of the trap from the larval breeding ground of the midge. The theoretical curve is calculated from the expression $\log Y = \log a + bx$. (After Kettle, 1951.)

wind on the distribution of small terrestrial animals and ways in which this may be studied.

5.13 *Dispersal by Wind*

5.131 DISPERSAL BY WINDS NEAR THE GROUND

Wingless insects, mites, and other animals, provided that they are small enough, may float in the air for distances that are quite comparable with those flown, crawled, or walked by the larger kinds. The small larvae of the gipsy moth *Lymantria dispar* were shown by Collins (1915) to have been carried at least 13 miles in this way; and Stabler (1913) showed that the mite *Bryobia praetiosa* could be collected in numbers from the air at least 650 feet from the nearest center of population.

Quayle (1916) demonstrated the dispersal of the first-instar nymphs of the scale insect *Saissetia oleae*, by counting the number of young ones on a group of citrus trees from which the previous generation of *Saissetia* had been exterminated by fumigation with cyanide. The experimental area was a rectangle carrying 400 trees; it was surrounded on three sides by trees which had not been fumigated and on which there was a dense population of *Saissetia*. The numbers of immigrants on the central plot of fumigated trees were counted during August, because by this time the insects had all settled down on the leaves and twigs and dispersive activity for that generation had ceased. The density of the population of young *Saissetia* on the outside rows of trees (those nearest the source of invasion) varied from 53 to 2,020 on 100 leaves; nearer the center of the plot there were fewer, the numbers varying from 10 to 300 on 100 leaves. Converted to numbers per tree or per acre, these figures represent a stupendous dispersal over relatively short distances, ranging up to 200 feet. The only important means of transport would be by air.

Quayle also extended these observations by collecting the young insects from the air on sheets of sticky "tanglefoot" paper. He exposed 21 sheets at heights varying from 2 to 6 feet above the ground and at distances varying from 10 to 450 feet from the nearest scale-infested tree; the average distance was 70 feet. The sheets were exposed for no more than 3 days; they were put out on June 28, when the dispersive activity was judged to be near its maximum. The total number of young *Saissetia* trapped on 21 sheets was 7,262, and the range was 31–1,056 per sheet. The greatest number was on a sheet which was equidistant from four heavily infested trees, each 10 feet away, and the lowest number was on the most distant sheet (450 feet); but the four sheets ranging from 100 to 450 feet trapped a total of 508 *Saissetia*.

The size of these sticky sheets is not given, but by any computation the number of first-instar *Saissetia* in the air in the vicinity of these citrus trees must have been enormous. Experience suggests that these figures are quite usual. Certainly they are not to be attributed to unusually high winds. These small creatures can, if they will, cling so tightly that no wind would unloose them. Their presence in such relative abundance in the air indicates an active participation on their part: they have an instinct which makes them launch themselves into the air. This is not surprising, since dispersal to new uninhabited places is almost the only defense which these otherwise sessile animals have against predators, which often feature importantly in their environments (secs. 5.3, 10.321, 14.2).

Some small insects, though winged, also depend largely on air currents for their dispersal and are therefore appropriately mentioned in this section. The work of Broadbent (1948), Kennedy (1950*a*, *b*), Johnson (1950), and others on the dispersal of winged aphids may be taken as an example. Because winged aphids cannot make headway against even a gentle wind if this exceeds about

2 miles per hour, their dispersal has often been called "passive." However true this may be, there is no longer any doubt that the dispersive "flights" are started by the insects' actively launching themselves into the air and are terminated by an equally definite action by the aphid. Kennedy (1950*b*) watched the behavior of *Myzus persicae* in the vicinity of two small trees (about 4 feet high), one peach and one spindle. The former is the normal winter host for this aphid, the latter is not a host. He described his observations as follows: "Although random as between 'right' and 'wrong' hosts, alightments were not passive. Flying aphids are not 'combed out' of the air by plants as they are by sticky traps. Alightments took place only when the air was still enough for the insects to be in control of their own movements, and they clearly directed themselves to their landing place at least over the last few centimetres. Since *M. persicae* made such directed alightments equally on peach and spindle it seems most improbable that the smell of the host directs aphids to it, although smell possibly plays its part after alightment. The alighting response appeared to be a non-specific *visual* one, evoked by objects looming up along the path of the flying insect."

Kennedy (1950*b*) also counted the numbers of *M. persicae* alighting on and taking off from the peach and the spindle. Aphids were extremely abundant in the air at this time, and he estimated that on some days more than 10,000 aphids alighted temporarily on the two trees. The counts showed no significant difference between the number alighting on or leaving the two trees. There must have been a slight difference in favor of the peach, since in the end this retained a fairly dense population of aphids and the spindle had none; but this was too small to be detected by the method of counting. In other words, after a brief stay, to rest and perhaps to feed awhile, all the aphids that had alighted on the spindle left it again, and nearly all those from the peach did likewise. Their instinct was toward dispersal—not to settle on the first favorable food plant that was found. This urge toward dispersal was also emphasized by Kennedy (1950*a*) in the following observation: "Now *Myzus persicae* seems to be an outstandingly restless aphid compared with, for example, *Aphis rhamni* Fonsc. and *Aphis fabae* Scop., judging by the relative frequency with which their nymphs are found abandoned by the winged mothers."

The common observation that there are as many aphids to be seen in the air on calm days as on windy ones is further evidence that the insects actively launch themselves on these "flights." They are not readily blown from their resting places even during a high wind, but they launch themselves freely into relatively calm air (Davies, 1939; Greenslade, 1941; Broadbent, 1946). If the wind freshens, those that are near enough to the ground seek shelter again; but those high enough to lose sight of the ground, being unaware of their relative movement, continue to drift with the wind, often flying against it.

Broadbent, in a series of trapping experiments, found more aphids in air 6 feet above the ground than at 3 feet or 1 foot.

In these experiments the traps were cylinders 5 inches in diameter and 12 inches long, coated with adhesive grease. Twelve such traps were exposed over a field of potatoes, and the daily records of the number of aphids found adhering to them gave an indication of the numbers taking part in these dispersive flights (Table 5.06). Broadbent found that from 30 to 50 per cent of all the

TABLE 5.06*

MEAN NUMBER OF APHIDS CAUGHT PER TRAP (CYLINDER 12 × 5 INCHES IN DIAMETER) BETWEEN JUNE 19 AND AUGUST 8

Height above Ground (Inches)	Per Cent of Total	Number of Aphids per Trap	
		Total, June 19–August 8	Mean Daily
8	15	359	7
36	37	906	18
68	48	1,164	23

* Traps exposed over a field of potatoes; means based on daily counts from 4 traps at each level. Data from Broadbent (1948).

aphids on the traps were on the lee side (i.e., they had alighted on the trap against the wind), and significantly more of those on the lee side had alighted on yellow traps than on black. These observations confirmed those of Kennedy (quoted above) that, provided that the wind is not excessive, the alighting is a positive action made by the aphid. These small winged insects clearly have powerful instincts toward dispersal. The importance, for our purpose, of the experiments and observations which have been done with aphids is that they provide a nice example of the way in which dispersive behavior may be demonstrated in the field.

5.132 THE PRESENCE OF SMALL ANIMALS IN THE UPPER AIR

The early workers in this field realized that small animals, like newly hatched caterpillars, mites, and aphids, which habitually launched themselves into the air, as well as being transported horizontally by winds near the surface, might also be carried to great heights by ascending currents. Collins (1915) placed traps on the tops of the highest towers he could find on high ground and recorded the capture of newly emerged larvae of the gipsy moth. Felt (1925) studied the movements of balloons, inflated to a minimal buoyancy, and showed that these might drift for hundreds of miles in the currents in the upper air; he suggested that gipsy-moth larvae and other small insects might also be expected to be dispersed in this way.

With the improvement of aircraft, it was possible to test these hypotheses, and several elaborate studies were made, particularly in America. Coad (1931) estimated from the results of some 100 flights at different heights, that the air between 50 feet and 14,000 feet above 1 square mile of territory near Tallulah, Louisiana, normally contained about 25,000,000 insects and other small animals, many of which were still alive when they were captured. The greatest numbers were found in the first 1,000 feet, but some were found at all levels up to 14,000 feet. In general, the larger and heavier ones and the stronger fliers were found at the lower levels, and the smaller ones, the weaker fliers and wingless ones, constituted most of the "catch" at the higher altitudes. Glick (1939) collected a total of 24,784 insects, spiders, and mites during some 1,000 hours' trapping at altitudes between 200 and 5,000 feet. This investigation lasted for 5 years, during which routine sampling of the upper air was practiced not only during the day but at night also. The numbers were greater at the lower altitudes (Table 5.07). Other studies in Europe and elsewhere (Hardy and Milne,

TABLE 5.07*

MEAN DENSITY OF AERIAL POPULATION OF INSECTS, SPIDERS, AND MITES AT DIFFERENT ALTITUDES

Altitude (Feet)	Cubic Feet of Air per Animal	Mean No. Animals Trapped during 10 Minutes' Flying
200	6,800	13
1,000	16,000	5
2,000	31,000	..
3,000	55,000	..
5,000	118,000	1

* Data from Glick (1939).

1937, 1938*a*, *b;* Glick, 1942; Freeman, 1945; and others) confirmed these results, so that Johnson (1951), in reviewing the subject, was able to write: ". . . on most warm days, the air over a few square miles of country up to several thousand feet contains a population of millions of insects of a great variety of species which, as a common and regular feature of their lives, are blown by the wind in unsuspected quantities over great distances."

In addition to establishing the presence in the upper air of these animals, most of the workers quoted above also sought to correlate the numbers of animals in the air at a particular time with the present or very recent weather. The results were rather contradictory and inconclusive. Glick (1939) wrote: "At times when the weather seemed to be ideal for insects, only a few or none would be taken. Again, when weather conditions were 'bad,' insects were taken in considerable numbers."

This is not surprising, for all these attempts to explain the fluctuations in aerial populations were based on hypotheses that were far too simple (Johnson, 1951). No matter how "favorable" the weather may be for dispersal, the numbers in the air must depend upon the density of the breeding populations on the ground and the behavior of the individuals in launching themselves into the air. These two phenomena must form the basis for any study of aerial populations; and it is rather likely that progress will be greater if attention is concentrated on one species at a time.

Johnson (1951) kept continuous records of the number of the bean aphid (*Aphis fabae*) in the air over a bean crop for 11 consecutive days. During this period the weather was relatively unchanging, but the density of the population of aphids on the beans was increasing rapidly. The number of aphids caught in the suction trap per hour (the equivalent of the number present in 19,000 cubic feet of air) rose fairly steadily from about 30 at the beginning to nearly 1,000 toward the end of the period, despite nearly uniform weather throughout. There was also a marked diurnal rhythm in the numbers caught in a suction net at 2,000 feet, and this reflected the activity of the aphids near the ground. During the night they cling to the plants, but shortly after sunrise they begin to launch themselves into the air; this activity reaches a peak before noon and again late in the afternoon. Many of these air-borne aphids are picked up by vertical currents, and some were recaptured in the suction net at 2,000 feet. Very few were captured during the night or up to 8:30 A.M.; the numbers increased during the day and reached a maximum toward evening, declining again at night. The aphids may have disappeared from this level (2,000 feet) at night, because they continued to be dispersed upward or horizontally, without being replenished from the ground; or, if downward currents had developed, they may have been returned to the ground.

What may happen to animals that have been carried to the upper air seems not yet to have been studied very closely. Some at least of those that are captured are alive, but many may died before returning to the ground. Even with those which do return while still alive, except with fertilized or parthenogenetic females, the considerations discussed in section 9.1 may well make this means of dispersal of little importance in the ecology of most of the species in which it occurs. It may be that the important thing is the instinct which makes these animals launch themselves into the air. This results in many of them being blown by winds near the surface, and, of these, a sufficient proportion comes to rest in places where they may survive and multiply. On the other hand, those which happen to be caught by ascending currents and carried high may have a negligible chance of finding a suitable place to live. In other words, these may have to be regarded as casualties which the species can afford to lose so long as some of its members, taking the more direct surface route, become successfully dispersed. These questions still await study, for we cannot yet answer them with precision with respect to any species.

With the gipsy moth it would seem that those which are dispersed through the upper air have little chance of contributing to the next generation. Collins (1915) records that this species was introducted into Medford, Massachusetts (northeastern United States) in 1869. Every effort was made to prevent its spread, and for this reason a full record of its dispersal was kept during the next 40 years. It spread outward from Medford at an average speed of 2.5 miles per annum (i.e., per generation), the speed being greatest (5 miles per annum) toward the northeast and least (1.5 miles per annum) toward the west and southwest. This can all be accounted for by surface winds; apparently none among all those which were carried into the upper air survived to establish new colonies elsewhere. On the other hand, Elton (1925) found numerous individuals of the aphid *Dilachnus piceae* accompanied by the syrphid *Syrphus ribesii* on Spitzbergen, and the circumstances of their arrival there indicated that they had been blown from the Kola peninsula some 800 miles away. Barber (1939) found the fly *Euxesta stigmatias* breeding in an isolated field of sweet corn in Florida in circumstances which made it seem most likely that they had been blown in from the West Indies. Dorst and Davis (1937) traced the dispersal of the beet leafhopper *Eutettix tenellus* from winter breeding grounds across 200 miles of desert (where there was no food for the leafhopper except the experimental plots used for trapping them) to the beet-growing areas of Utah. Felt (1925) pointed out that the moth *Heliothis armigera* rarely, if ever, overwinters north of southern Pennsylvania and the Ohio River, but it sometimes becomes very abundant on sweet corn in southern Canada. Felt also listed a number of other species from the tropics or subtropics which were unable to survive a northern winter, yet from time to time were found breeding in Canada during the summer. With the exception of the aphids recorded by Elton on Spitzbergen and the beet leafhopper *Eutettix*, the others mentioned in these examples are of the larger, stronger-flying sorts which are rarely found at the greater heights. Interesting though these and many other similar observations may be as evidence of dispersal over hundreds of miles of species that are not truly migratory, it is the very nature of this problem that we cannot tell merely from observation whether they had been blown chiefly by surface winds or by currents in the upper air.

So it must be concluded that, whereas studies of the animals present in the upper air have clearly established the regular occurrence of surprisingly large numbers of many species of small animals there, they have not yet established beyond reasonable doubt that this form of dispersal is important in the ecology of any of these species. We shall therefore not pursue this subject any further in this book. It is to be hoped that the many interesting problems that have been opened up will continue to be investigated actively, so that it may become possible in the future to evaluate the probable importance of dispersal through the upper air in the ecology of these very small animals.

5.2 LABORATORY EXPERIMENTS TO DEMONSTRATE DISPERSIVE ACTIVITIES

The dispersive activities which are such a striking feature of the behavior of locusts, not only in their gregarious swarms but also in the solitary phase (sec.

5.4), may be observed also in the nymphs of these species. The "marching" of gregarious bands of nymphal locusts is such a conspicuous and characteristic feature of their behavior that it has repeatedly been described from observations in the field (see Uvarov, 1928, for a full summary and discussion). But nymphs in the solitary phase are also very restless, and Kennedy (1939) described how nymphs of *Schistocerca gregaria* wandered, during the course of a day, from 100 to 400 meters from the place where they had started. The distances covered by the wingless nymphs are relatively short compared with the more spectacular movements of the winged adults and are relatively unimportant for the dispersal of the population. Fundamentally, the behavior of the two stages is similar, but it may not be practicable to study the dispersal of the adults within the confines of a laboratory experiment, whereas the experiments of Ellis (1951) show that the nymphs may very profitably be studied in this way. We shall describe these experiments to show what may be done in the laboratory to elucidate problems arising from observations made in the field.

When a number of nymphs of the African migratory locust, *Locusta migratoria migratorioides*, were placed in a cage with a floor about 1 foot square and with a discrete source of light (a 40-watt incandescent globe) in the ceiling, they would, in certain circumstances, proceed by crawling and jumping to travel more or less continuously and consistently in one direction around and around the cage. This behavior was called "marching." It was measured by placing a number (usually about 5) of marked nymphs in a cage with a number of others and recording, hour by hour, the proportion of time spent marching and the average speed of the marching. By both criteria it was found that marching depended not only upon certain components of the environment but also upon the condition of the nymph.

Figure 5.07 shows that the proportion of time spent marching increased during the course of an experiment lasting 8 hours and was greater by about double in the cage where there was no food. There is a common explanation for these two phenomena: the nymphs marched very little for the first hour or so after a meal, while the gut was still distended with food; both the proportion of time spent marching and the speed of the marching reached an asymptote 3 or 4 hours after a meal. Marching is a group phenomenon: it is not likely to occur when a nymph is placed in a cage by itself. Both the proportion of time spent marching and the speed of marching were influenced by the number of other nymphs in the cage. This is illustrated in Figure 5.07 where the curve for 5 numphs rose more slowly and reached a lower maximum than did the curve for 50 nymphs. Marching also depended upon the sort of lighting. It occurred when a single incandescent globe was placed in the ceiling of the cage, but not when the cage was illuminated by diffused light. Both temperature and wind influenced the amount of marching and its speed. Figure 5.08 shows that the time spent marching reached a maximum within the range 30° –33° C., which is also the range that is optimal for the development of this species.

Figure 5.08 also shows that the proportion of time spent marching was greater when a wind of 30 meters per minute was blowing through the cage than in still air. Ellis did a number of subsidiary experiments to try to elucidate the cause for this phenomenon and reached the conclusion that it may be a direct response to the stimulus of moving air.

The condition of the nymph influences the amount and speed of marching. Those which had been reared in a crowd and had the morphological characters

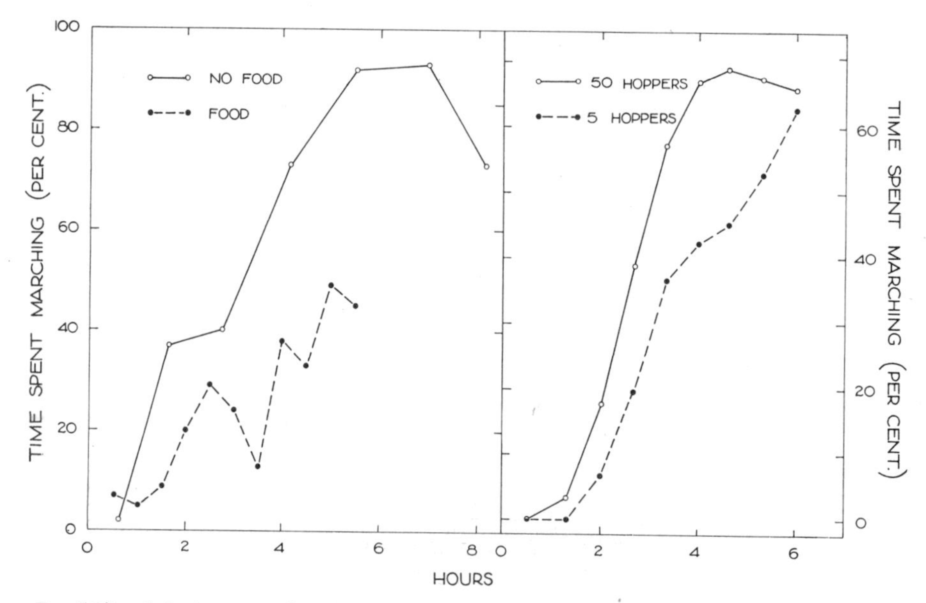

FIG. 5.07.—*Left*, the proportion of time spent marching by fourth-instar nymphs of *Locusta migratoria* in relation to the period since the last meal. *Right*, the proportion of time spent marching in relation to the number of other nymphs in the cage. (After Ellis, 1951.)

of the phase *gregaria* marched more, and more rapidly, than those which had been reared in isolation and belonged to the phase *solitaria* (Fig. 5.09). Within the life of any individual the tendency to march is at a minimum immediately before and after each molt and reaches a maximum in each instar between molts (Fig. 5.09).

The results of these experiments show that marching is not, as has sometimes been thought, a response to unfavorable circumstances. On the contrary, the most vigorous marching occurred at temperatures which are optimal for the normal life and development of the species; even with fresh food in the cage, the nymphs still marched, although less vigorously than in cages where there was no food. It is clear that marching is a spontaneous activity occurring as part of the normal behavior of nymphs of *Locusta* and contributing to their natural dispersal. Similar behavior on the part of winged adults is well known to result in their dispersal over enormous distances.

With other species, such as small, weak-flying, or wingless insects or spiders, etc., whose dispersal depends more on their drifting with air currents (secs. 5.131 and 5.132), it may be important to study how they become air-borne. For example, Johnson and Southwood (1949) found during a year's experiments with nets flown at heights varying from 50 to 4,000 feet that one species, *Lygus pratense*, outnumbered all other species of bugs caught in the nets: of a total of 36 Heteroptera, 21 were *L. pratense*, and the remainder were distributed among

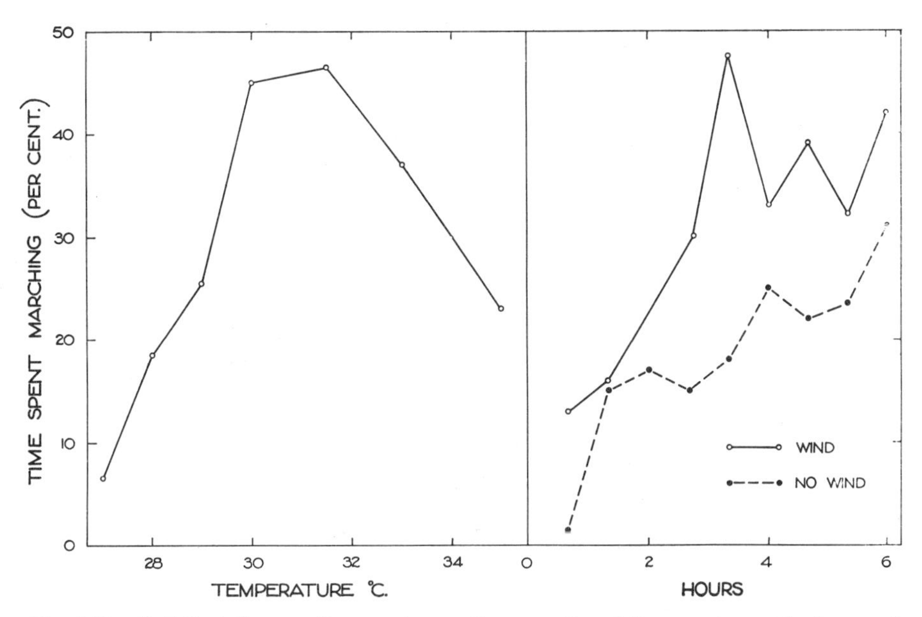

Fig. 5.08.—*Left*, the influence of temperature on the proportion of time spent marching by nymphs of *Locusta* in the fourth instar. *Right*, the influence of wind on the proportion of time spent marching by nymphs of *Locusta* in the fourth instar. (After Ellis, 1951.)

9 other species. Now *L. pratense* is a common bug in England but not relatively so common as to outnumber all other species in this proportion. The inference is that *L. pratense* differs from most other species of the same group in some detail of its behavior, which results in its being present in the air, drifting, in relatively great numbers. This is the sort of observation that might profitably be followed up in the laboratory.

Dispersal by wind is an important feature in the ecology of the spruce budworm *Choristoneura fumiferana* (Wellington and Henson, 1947). The second-instar larvae in the spring wander to the tips of the branches, whence they launch themselves into the air on silken threads, which add to their buoyancy. Wellington (1950) placed larvae in an apparatus which maintained a uniform temperature and a gradient in humidity. He measured the rate at which the larvae crawled and found that this reached a maximum at a particular humidity. He measured moisture with a specially designed microevaporimeter. The maximal activity for second-instar larvae occurred when the evaporimeter registered between 0.10 and 0.13 cu. mm. per minute (Fig. 7.09). Wellington has calculated that the weather which would favor the transportation of the larvae, once they are in the air, would also be likely to produce, at the surface of the branch, a humidity which would stimulate the larvae to crawl vigorously. He also pointed out that such increased activity is likely to result in a large proportion of the larvae being exposed on the tips of the branches at a time

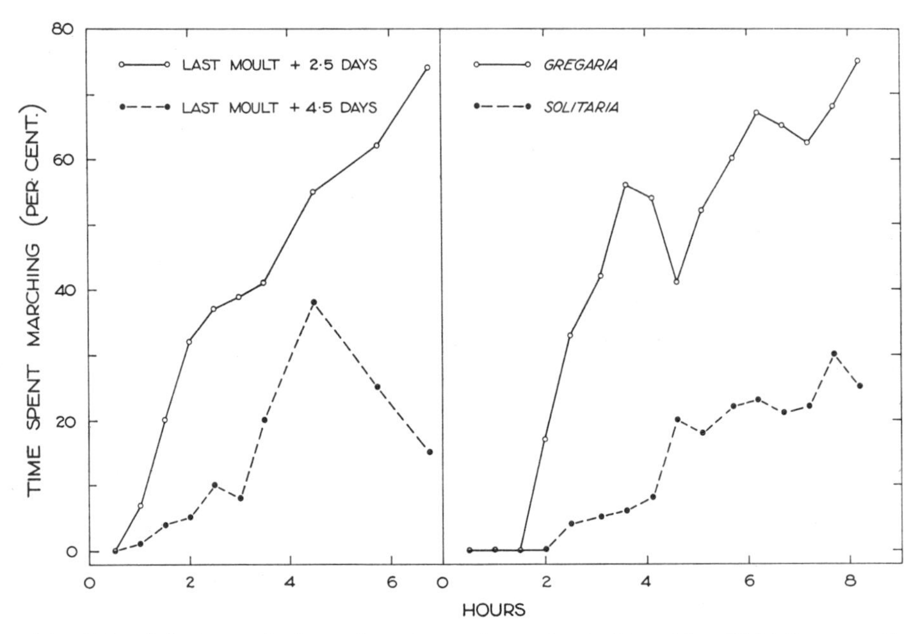

FIG. 5.09.—*Left:* changes in the tendency to march associated with molting. *Right:* the proportion of time spent marching by fourth-instar nymphs of *Locusta migratoria* in phase *gregaria* compared with phase *solitaria*. (After Ellis, 1951.)

when they are most likely to be successfully transported by aerial currents.

Dispersal by wind is important in the ecology of aphids. Once they are in the air, they may be transported involuntarily by the wind; but first the aphids must leave the plant on which they are resting by voluntarily taking flight. They do this much more readily in some circumstances than others, and this is reflected in the varying numbers found in the air at different times of the day and in different weather (secs. 5.131 and 5.132). So Broadbent (1949) placed adults of two species, *Brevicoryne brassicae* and *Myzus persicae*, in an apparatus in which temperature, light, humidity, and pressure could be controlled, and he recorded the proportion undertaking flight in the circumstances of each experiment. He found that *Brevicoryne* flew more readily than *Myzus* and that young adults of both species flew more readily than older ones. With both species, periods of activity alternated with periods of inactivity. Starving increased activity for the first hour or two. Light made little difference above a threshold of 100 lumens per square foot; below this, activity was reduced; and darkness inhibited it completely. At 24° C. the aphids flew readily at all humidities tested between 50 and 100 per cent relative humidity; at 32° C. flight was sometimes inhibited at the higher humidities. The aphids flew more readily when the humidity was changed from high to low, and less readily when the change was in the opposite direction. They also flew more readily when the barometric pressure was fluctuating than when it was steady. Broadbent concluded that, in nature, the aphids would probably be stimulated to fly by the fluctuations in humidity and pressure which might occur near the surface of the plant during a bright day of calm or moderate wind.

5.3 THE SPECIAL IMPORTANCE OF DISPERSAL IN A PREDATOR

We show in section 10.32 that there are broadly two schools of thought about predation. Students of wildlife who accept Errington's theory argue that the amount and quality of "cover" (i.e., suitable places to live) in a particular territory determine the density of the population and that the numbers and kinds of predators have relatively little influence on the "carrying capacity" of the area. If this be so, then no special interest is attached to the study of dispersal (or other special qualities) of predators as such.

But the success of many ventures in the biological control of insect pests makes it quite clear that, with many insects at least, predators may be of major importance in determining the abundance of the prey. The difference would seem to arise because there are many species of insects which, unlike the small fur-bearing mammals and game birds which are the chief subjects for study by students of wildlife, live in places that afford no protection (in the physical sense) from predators. There may be no place on a tree where a scale insect may be out of reach of its predator, the ladybird; and for this reason the arrival on the tree of ladybirds may presage the extinction of that particular colony of scale insects. But before the colony has become extinguished, some individuals may have floated away and come to rest on another tree, where there may have been neither scale insects nor ladybirds. They may flourish here. How abundant the new colony becomes and how long it lasts may depend on the period that elapses before it is found by ladybirds. Although in this case the prey may have no secure place to hide, yet the continued existence of the species may depend upon this perpetual game of hide and seek which it plays with its predators. The abundance of the prey may depend to quite a large extent on just how good the predator is at seeking.

The importance of dispersal in predaceous insects was emphasized in an article by Flanders (1947), in which he analyzed the qualities of hymenopterous "parasites" in relation to their effectiveness as agents of biological control.

"Effectiveness" is used in the sense of capacity to reduce and maintain the density of the population of the species preyed upon to a level such that it no longer does much harm to crops. "Parasite" is used in the sense peculiar to entomology: i.e., the special sort of predator which lives, during its larval stage, inside the body of the prey and destroys it. Flanders first pointed out:

In many cases it is known that the population level of a host species is determined by the host-searching capacity of the free-living gravid females of the parasitic insect. The actual level attained is determined by those attributes of the female parasite that contribute to its host-searching capacity.

He described the circumstances in which the searching goes on:

When a host is under biological control, it has a characteristic "colonial" or "spotted" distribution (Smith, 1939). The host population consists of groups of individuals, which fluctuate in density independently of one another. One group may be exterminated by a parasite, but in the meantime some host individuals will have migrated and established new groups. The parasite must therefore follow its host, and so must equal it in powers of locomotion. . . . The power of locomotion appears to be correlated with the number of eggs laid per day. Females that deposit their full complement of eggs at one insertion of the ovipositor (*Schizaspidia tenuicornis*) or within a few days (*Coccophagus trifasciatus* Compère) tend to have a much lower power of locomotion than those of a parasite species such as *Metaphycus helvolus* (Compère) which, at 80° F. oviposit at a rate of about 9 eggs per day for 1 or 2 months. The female of *M. helvolus* is very active and apparently travels considerable distances between ovipositions. Observations indicate that populations of *M. helvolus* may disperse a distance of 25 miles in one year. In the case of *Comperiella bifasciata* Howard parasitic on *Aonidiella* spp. with an oviposition rate of 2 or 3 eggs per day for about a month, the writer observed that the dispersion from the point of liberation, when hosts were abundant, was considerably less than one mile annually.

Species such as *Coccophagus gurneyi* Compère and *Dibrachys cavus* Walker which are kept out of artificial enclosures with difficulty, having the ability to find and pass through small crevices, obviously have a high power of locomotion.

The importance of dispersal in predators is discussed further in section 10.321

5.4 MIGRATION

The migrations of certain birds and fishes are too well known to need recapitulation here, and we mention them merely to point out that this is a phenomenon which is only slightly related to dispersal as we have been discussing it in this chapter. Birds may circumnavigate continents or fly halfway across the earth; eels cross the Atlantic; salmon migrate from mountain stream to deep ocean. All these migrations are mass movements of the breeding population as a whole; they do not essentially or normally involve any scattering of the population or result in the occupation of new territory. It is not the main migratory flight but rather the scattering which may occur at the end of it or the aberrations of individuals which fail to follow the path of the flock which represent the dispersive activities of these species. We shall not follow this matter further in this book.

The migrations of butterflies and locusts resemble those of birds and fishes, in that whole populations or substantial parts of populations may move en masse from one place to another; but they differ in certain important respects. The insects do not live long enough to make the return journey: if this is undertaken, it must be by members of a later generation; *Danaus plexippus* may be an exception to this rule. Whether, and to what extent, a return journey is made remains quite unknown for the great majority of the species of insects that are recorded as migratory. In a few species, for example, *D. plexippus*, *Vanessa atalanta*, and *Plusia gamma*, the evidence leaves little room for doubt that northward flights predominate in the spring and that in the autumn most of the movement is toward the south. For a number of other species (about a dozen, according to Williams *et al.*, 1942), the evidence, though not completely convincing, points in the same direction. There are hundreds of other species, well known to be migratory, for which there is still no evidence that a return journey is attempted. This poses a problem: unless an adequate proportion of the progeny of those which migrate north in the spring return in the autumn at least far enough south to find a suitable place for breeding during the winter, the persistence of the habit of migration in these species seems at variance with the accepted theory of evolution and natural selection. It seems probable, therefore, that most species which are incapable of surviving the winter in the north do undertake at least some part of the return journey; but this has been missed because it is not conspicuous. Certainly, none of them follows the ancestral paths back and forth twice a year with the fidelity shown by the migratory birds. In other words, the migratory flights of insects, being less compact, may result in a considerable measure of scattering of the original population and are therefore relatively more important in relation to studies of dispersal. For a full discussion of the migrations of insects, the student is referred to Williams (1930) and Williams *et al.* (1942). The two examples given below illustrate the nature of the problem and the methods which may be used in studying it.

Beall (1941*a*, *b*) described the autumnal exodus of *Danaus plexippus* from Ontario, based on 7 years' observations. During the spring and summer in Ontario the population included many old and sexually mature individuals; but toward autumn these all died, and their places were taken by young and sexually immature butterflies, which migrated southward in this condition. The migration was the outcome of voluntary flight mostly near the ground, rarely more than 20 feet up, and often against the wind. Usually in a wind that was too strong for them, the butterflies sought shelter, but sometimes, for example over water, they might be blown backward while still heading south. It is not known where these migrants went or where they bred; but Beall (1946) has produced some interesting evidence which suggests that the butterflies which migrated back into Ontario in the spring were related to those which had left there in the autumn rather than to the populations which occur in other regions of North America which he had studied.

Beall and Williams (1945) studied the size of *Danaus plexippus* occurring in different regions of North America. They measured the length of the forewing and found that, by this criterion, the butterflies in eastern Canada were larger than those in a number of other regions (Table 5.08). A statistical analysis

TABLE 5.08*

AVERAGE LENGTH OF FOREWING (IN MILLIMETERS) OF *Danaus plexippus* COLLECTED IN DIFFERENT REGIONS OF NORTH AMERICA, BASED ON 1,553 MEASUREMENTS

REGION	FEMALES		MALES	
	No.	Mean Length of Forewing (mm.)	No.	Mean Length of Forewing (mm.)
Mexico	15	49.3	17	50.6
California	98	50.3	198	50.3
Louisiana	34	50.2	71	50.6
Florida	8	50.9	11	51.4
Minnesota	13	48.3	26	50.3
Ontario	537	51.9	525	52.3

* Data from Beall and Williams (1945).

showed that the variance between groups was about 20 times the variance within groups, indicating a high level of significance. The difference was due primarily to the large size of the butterflies from Ontario relative to all the others, and the significance of the difference was due mainly to the homogeneity within most groups. The chief interest in these results, for our present purpose, is that they indicate a possible method of recognizing the presence of immigrants among a resident population, and perhaps of tracing the origin of the immigrants. The method might, in fact, serve as a substitute for the technique of marking individuals which has been so successfully practiced in studying the migrations of birds, less so with fishes, and scarcely successful at all with migratory insects. Beall (1946) put the method to the test with *Danaus*, with some very interesting results.

He first analyzed the 1,062 records (Table 5.08) for autumnal emigrants from Ontario year by year and found no significant differences between the samples for the different years or any evidence of heterogeneity within these samples. In other words, it seemed certain that a single homogeneous population had been breeding in Ontario summer after summer, despite the fact that each autumn the area was depopulated by emigration to the south and repopulated in the spring by immigrants. The next step was to measure the immigrants living in Ontario during a particular summer and compare them with those which had left the same area during the previous autumn. This was done during the summer of 1941, and Beall found a small but significant difference between the two groups. He concluded that the immigrants were not the same individuals which had emigrated the previous autumn but were the progeny of these or very similar butterflies which had been bred elsewhere in the meantime.

The same method was used to investigate the problem of where the emigrants go after leaving Ontario. It was known that a similar southerly migration of *Danaus plexippus* may occur in Louisiana and that this usually takes place about a month after the height of migratory flights from Ontario. Beall tested the hypothesis that these were the same individuals, by comparing the size of those caught in Louisiana during late October and early November with those

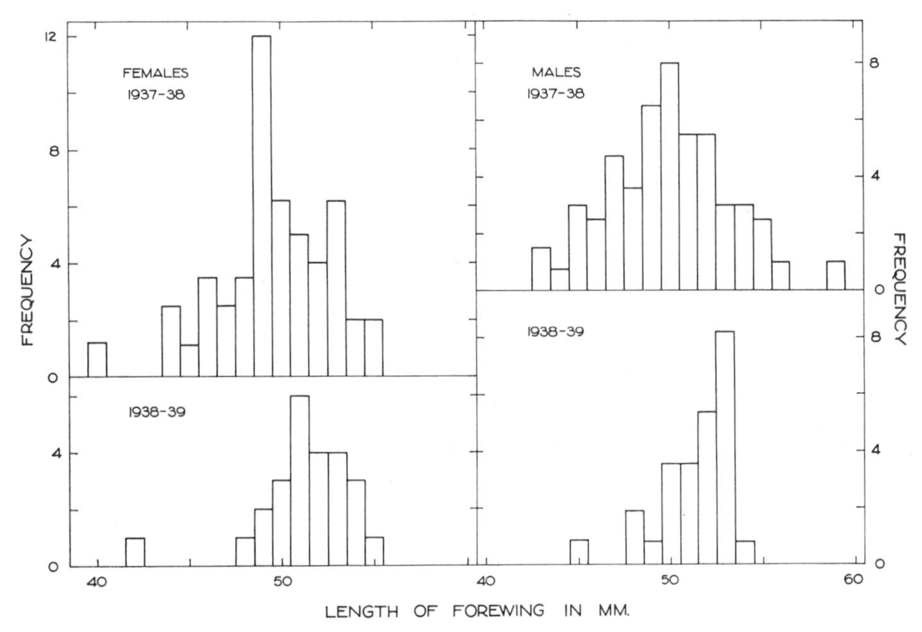

FIG. 5.10.—Frequency histograms showing the distributions of the size of *Danaus* captured in California during winter. Note the heterogeneity of the populations with respect to size. (After Beall, 1946.)

which had left Ontario in September. He found that the *Danaus* from Louisiana were much smaller and yet just as homogeneous as the larger ones from Ontario, and therefore he concluded that they could not have originated in Ontario.

It was also known that many *Danaus* normally pass the winter in California. Beall measured samples collected in California at intervals during several winters and found that they differed in two important respects from all other populations that he had studied. The average size varied considerably from year to year and from month to month (Table 5.09), and the frequency-distributions of measurements revealed a heterogeneity greater than in any of the other populations studied (Fig. 5.10). Both these observations could be explained by postulating (*a*) that the population of *Danaus* resident in California during the winter is not derived from the same source but is a mixture of populations which differ with respect to the size of the individuals comprising them, and (*b*) that the composition of the mixture changed from time to time, depending upon the chance arrival of butterflies from different regions.

TABLE 5.09
AVERAGE LENGTH OF FOREWING IN *Danaus plexippus* COLLECTED IN CALIFORNIA DURING THE WINTER

Date		Females		Males	
		No.	Average Length of Forewing (mm.)	No.	Average Length of Forewing (mm.)
1937	October	7	48.9	20	48.4
	December	20	49.2	25	49.8
1938	February	15	50.9	8	50.4
	March	7	49.4	18	50.6
	October	10	49.7	10	51.1
1939	January	7	52.3	7	52.3
	February	8	52.2	11	50.9
1942	February	21	50.5	87	50.5

* Data from Beall (1946).

The largest were comparable with those from Ontario, and the smallest were smaller than those from Louisiana; the population as a whole differed from those from both these places. These results are highly suggestive but inconclusive, at least with respect to detail. They are especially interesting for their method, which may well be equally fruitful when applied to other species, particularly those in which migratory movements occur chiefly at night or are for some other reason difficult to observe directly.

The moth *Plusia gamma* is well known as a migratory species which migrates every year, sometimes in large numbers, into Britain from Continental Europe. This species is unable to survive the winter in Britain, and, with very rare exceptions, none is found there from November to April. Usually during May the first immigrants appear, and others continue to arrive during the summer. Meanwhile, breeding occurs, and it is usually possible for one or two generations to be completed during summer, so that by August or September the species is usually more abundant than it was earlier in the summer. Fisher (1938) has described the movement of *P. gamma* in England, based on records made during five years. She had an enormous number of different observations to analyze, because in England an "Insect Immigration Committee" has, for a number of years, kept an organized watch on the movements of migratory insects. Diagrams were prepared in which arrows indicated the direction of flight. One such diagram (Fig. 5.11) shows clearly that there is no correlation between the direction of flight and the direction of the wind. On the other hand, the direction of flight is closely related to the season of the year: most of the moths were flying north in the spring and south in the autumn. This indicates a true migration in both directions; but there is no evidence for this species of how much scattering occurs or to what extent the same population occupies the same areas in succeeding years. Nor is there any indication of what stimuli prompt the migrations or govern their directions.

The migrations of locusts are governed to a large extent by wind. Unlike *Danaus* and *Plusia*, locusts often fly at a height of several thousand feet. The circumstances in which they take to the wing most successfully are also those which are likely to produce powerfully ascending currents of air which may carry the locusts upward involuntarily (Rainey and Waloff, 1951; see also

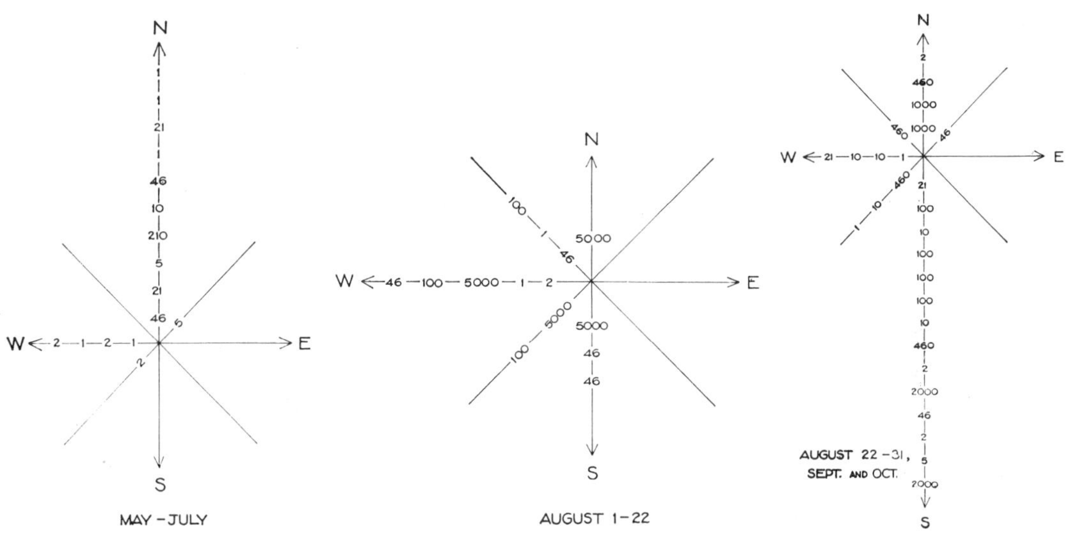

Fig. 5.11.—Analysis of the direction of flight of *Plusia gamma* observed in England at different seasons of the year. The length of the arrow is proportional to the number of the flights observed; the figures indicate the number of moths seen in each flight. (After Fisher, 1938.)

sec. 6.22). The direction of flight of a locust flying near the ground seems to be independent of the direction of the wind. When the wind exceeds the speed at which the locust can make progress against it, the insect may either change direction and fly downwind or else seek shelter on the ground (Waloff and Rainey, 1951). Once a locust is out of sight of the ground, it no longer perceives changes in the direction or velocity of the wind, and it may remain aloft for a long time. In a strong wind it may be blown for great distances.

Waloff (1946*a*) traced the migration of one swarm of *Schistocerca gregaria* from Morocco to Portugal. Waloff (1946*b*) analyzed a great many records of the migrations of swarms of the same species in eastern Africa during six years: and Rainey and Waloff (1948) did the same for the area around the Gulf of Aden. The results were consistent. Although swarms, particularly when they were near the ground, were often observed to fly against or across the wind, yet the major movements of the swarms were clearly the outcome of their having, for the most part, been blown downwind with the prevailing winds, especially those at 2,000–3,000 feet. The study in eastern Africa, in particular, showed that the movements and therefore the distribution of swarms were largely governed by temperature and wind—temperature sufficient to establish upward convection and to get the swarms on the wing and wind to blow them for big distances. There was no evidence that the migrations led, except by chance, to favorable places for breeding; but when, in the course of their wanderings with the wind, the locusts arrived at a place that was moist enough, they would develop to sexual maturity and lay eggs. Notwithstanding that the migrations of locusts are largely involuntary relative to those of birds or butterflies, they nevertheless display a certain regularity because most species occur in zones where seasonal winds, trade-winds or monsoons, prevail. Thus in eastern Africa the main trend for swarms of *S. gregaria* from October to March (the period of the northeast monsoon) was southward and southwestward, whereas during April to June (the period of the southeast trade) the major trend was northward and northeastward (Waloff, 1946*b*).

For many years it was held that the solitary phase of locusts did not migrate; but more recently the careful observations of a number of workers, especially Rao (1942), leave little doubt that migration is a regular feature of the behavior of phase *solitaria* also. In northwestern India *Schistocerca gregaria* regularly migrate during spring from the coastal "reks" to the inland valleys; this movement seems to occur whether the population is made up of gregarious swarms or solitary individuals. During summer, with the prevailing southwesterly winds, the locusts migrate into Sind and the southwest Punjab; during autumn, with the withdrawal of the monsoon, these areas become very dry; easterly or northeasterly winds prevail; and the locusts return with the winds to the coastal areas. Most of these movements, whether by gregarious swarms or solitary individuals, involve all or most of the population; but the relatively haphazard character of the migrations inevitably results in some scattering of the populations, and to this extent they are of interest in relation to dispersal.

Although *migration* is essentially a different phenomenon from *dispersal* in the rather narrow sense in which we are discussing it in this chapter, it nevertheless is an adaptation which has a high survival value, and for very much the same reasons. The Acrididae, like all other groups of animals, include many species that are rare and a few that are abundant. Those that are abundant enough to be regarded as pests may be broadly classified into two groups, the locusts and the plague grasshoppers. The plague grasshoppers characteristically are found in subhumid or semiarid zones, where rainfall is seasonal and one season or another may be severe either from drought or cold. The locusts, on the other hand, characteristically live in zones of uncertain rainfall, often in quite arid areas. With only very minor exceptions, all the plague grasshoppers have, and none of the locusts has, a diapause-stage in the life-cycle (Andrewartha, 1945). Diapause is an adaptation which enables the animal to survive during an unfavorable season and doubtless contributes largely to the success and abundance of those species possessing it (sec. 4.6). The locusts, lacking diapause, nevertheless maintain themselves in abundance; and one is tempted to suggest that with them, in the climatic zones that they inhabit, migration, instead of diapause, has been developed as an adaptation enabling a high rate of survival and making for greater abundance.

An interesting parallel has been reported by Wiltshire (1946), who, as a result of 10 years' observations of Lepidoptera in the Middle East, was able to divide the 10 common species that occurred in one area into two groups. Eight species, lacking diapause in any stage of the life-cycle, were strongly migratory, breeding on the lowlands during winter and the uplands during summer. The two species which lacked any tendency to migrate spent the unfavorable season in diapause.

5.5 DISPERSAL IN RESPONSE TO CROWDING

There are two aspects to the subject of dispersal in response to crowding. The spectacular "migrations" of lemmings and other small rodents and of locusts, army worms, and other invertebrates are well known. All these are animals whose numbers are likely to fluctuate between wide extremes (Elton, 1927; Uvarov, 1928; MacLulich, 1937; Moran, 1949, 1950). Lemmings are the best known; but similar gregarious migrations are recorded for mice, voles, hares, rabbits, and other rodents. Elton's (1927, p. 132) description of the Norwegian lemming is striking:

"The Norwegian lemming lives normally on the mountains of Southern Norway and Sweden, and on the arctic tundras at sea-level farther north. Every few years it migrates down into the lowland in immense numbers. The lemmings march chiefly at night, and may traverse more than a hundred miles before they reach the sea, into which they plunge unhesitatingly, and continue to swim on until they die. Even then they float, so that their dead bodies form drifts on the shore. This migration, a very remarkable performance for an animal the size of a small rat . . . is caused primarily by over-population in their mountain home, and the migrations are a symptom of the maximum in numbers which is always terminated by a severe epidemic; and this reduces the population to a very few individuals. After such a 'lemming year' the mountains are almost empty of lemmings."

The great exodus of the majority from the breeding grounds allows the species to persist, but it is difficult to understand how this behavior may have arisen. For those that migrate perish, or, if they do not, their progeny must live in less favorable places, and there is no evidence of a return to the breeding grounds, while those that stay behind thrive and multiply in the original breeding ground, which is the place best suited to the species. This phenomenon seems not to contribute anything to the building-up of new centers of population or to an extension of the distribution, except very temporarily, and therefore we shall discuss it no further. Spectacular emigrations following an increase in numbers are well known in grouse, the snowy owl, the arctic fox, and some squirrels. As with the lemming, the emigration seems to be associated with high numbers and to be relatively independent of the amount of food. Had there been no emigration, the whole population might have starved, but the emigrants usually leave before there is any noticable shortage of food (Dymond, 1947).

In a less spectacular fashion, crowding may merely accentuate the normal instincts toward dispersal. When this occurs locally, it may increase the rate at which new colonies are being founded and thus lead to an increased density of the population of the area as a whole. This phenomenon is relevant to our discussion, though we may doubt whether it is as important as it is sometimes considered to be. Kluijver (1951) in his comparative study of the biology of the great tit, *Parus major*, in a breeding area of about 129 hectares was able to show that relatively more birds emigrated from the area when the population was dense. And Mayr (1952), in his review of this article, added: "Similar correlations between high population density and emigration have been established by many authors for various species of vertebrates and invertebrates."

This may be so, and those who accept the "density-dependent" dogma may have little difficulty in accepting this statement; but it must be remembered that instincts and adaptations for dispersal, as we have shown and as Elton recognized in the quotation which precedes this chapter, are universal among animals, and dispersal is going on all the time from populations both dense and sparse. To demonstrate that it is going on relatively more actively in a dense population than in a sparse one may require very critical experiments; and with nonterritorial animals it may be even more difficult than with Kluijver's great tit. Even in this species Kluijver observed that, for the most part, fluctuations in numbers synchronized in 5 localities for the 12 years that he observed them. The excessive emigration associated with denser populations may well have led, as with the lemming, into uninhabitable regions, dooming the participants to destruction. There is also the further consideration that only a few species are common, relative to the many that are rare.

5.6 DISPERSAL IN RELATION TO DISTRIBUTION AND ABUNDANCE

Dispersal is a broad subject with many sides: the student of animal geography or paleontology will see it in different perspective from the student who is chiefly interested in the numbers of a species in a particular part of its distribution or in the limits imposed by environment on distribution. We have seen (sec. 1.1) that distribution and abundance are merely different aspects of the same phenomenon, and it is from this point of view that we have discussed dispersal in this chapter.

It is the heterogeneity of nature, of the areas that populations may inhabit, of the places where animals may live and the consequent colonial or "spotty" distribution of the populations of the animals themselves that makes dispersal so important. Heterogeneity in space may, on occasion, be studied and explained in terms of geology, pedology, topography, climatology, or mathematical probability; for example, in a tropical forest with its enormous diversity of species, chance must play a large part in determining the pattern of distribution of the plants and hence of the food and places where animals may live. Heterogeneity in time may be related to weather, the ordinary processes of growth, senescence, and decay, and ecological succession. Temporal changes bring spatial changes in their wake, and the species which is not adapted to provide for the future may be expected not to persist. As we have seen, it is a characteristic of most species that they devote a large part of their collective energy to the founding of new colonies. Small species with poor power of locomotion may rely on wind or other agencies. With them, adaptation may take the form of behavior patterns which enable them to become air-borne and an enormous fecundity that offsets the great destruction of life inevitable with this method of dispersal. With larger species, better equipped with organs of locomotion, special instincts may be evolved, conferring on the individual a high capacity for searching. This makes for less wastage, but, even so, the urge for dispersal may be stronger than the instincts for the preservation of the individual.

Fundamentally, there is no difference between herbivores and carnivores; but we have an additional interest in the carnivore as a predator, and therefore we study its dispersal from a different aspect. That this is relative and not very fundamental is clear from the passage about *Cactoblastis* quoted in section 5.0.

It has been said (for example, see Elton, 1927, p. 155) that dispersal does not lend itself to deliberate or quantitative study and that we shall always need to rely on chance observations for most of our information about this subject. It is true that dispersal is very difficult to measure and that there is an enormous amount to be learned by keen observation. But we have shown in this chapter that there are ways of making quantitative studies of dispersal, and in the future, as the great importance of this subject becomes increasingly appreciated, these are sure to be extended and improved.

PART III

Analysis of Environment

CHAPTER 6

Weather: Temperature

Climate plays an important part in determining the average *numbers of species, and periodical seasons of extreme drought or cold, I believe to be the most effective of all checks.*

DARWIN, *The Origin of Species*

6.0 INTRODUCTION

DARWIN considered that in "the struggle for existence" the indirect influence of weather operating through the food supply was more important than its direct influence on the animals themselves. There can be no doubt that weather may, in this indirect way, greatly influence the animal's chance to survive and multiply. We shall have something to say about this elsewhere, particularly in chapters 13 and 14. For the present we are concerned with the direct influence that weather may have on the speed of development, fecundity, and longevity of animals; these are the three components of the innate capacity for increase, r_m (sec. 3.1). Weather may also influence the dispersal of animals and other aspects of their behavior (secs. 5.1 and 6.22). It is convenient to consider, first, the independent influence of the separate components of weather, temperature, humidity, light, and so on and then later to try to describe how, in nature, the influence of one may depend on its interaction with the others. Temperature and humidity have been widely studied. Light, air movements, and atmospheric pressure have also been shown to influence development or behavior in particular cases. The volume of the literature in this field is now enormous. Some of it has been summarized and listed by Allee *et al.* (1949). Our selection of examples is restricted to the few which are sufficient to illustrate our central theme, namely, the principles which govern the distribution and numbers of animals in nature and the kinds of studies which bring these out most clearly.

6.1 THE RANGE OF TEMPERATURE WITHIN WHICH AN ANIMAL MAY THRIVE

It is well known that animals of one sort or another thrive high on mountains or at high latitudes and in the heart of equatorial continental deserts. The extremes of temperature from the hottest to the coldest places where animals live are great. It has been claimed that the larvae of certain Diptera thrive at temperatures as high as 55° C. or even higher; but we shall probably be near the mark if we accept the opinion of Brues (1939) that 52° C. is about the highest temperature at which any animal is known to thrive indefinitely. At the other extreme, there is the beetle *Astagobius angustatus*, which is known to carry on its life-cycle in ice grottos, where the temperature range is between

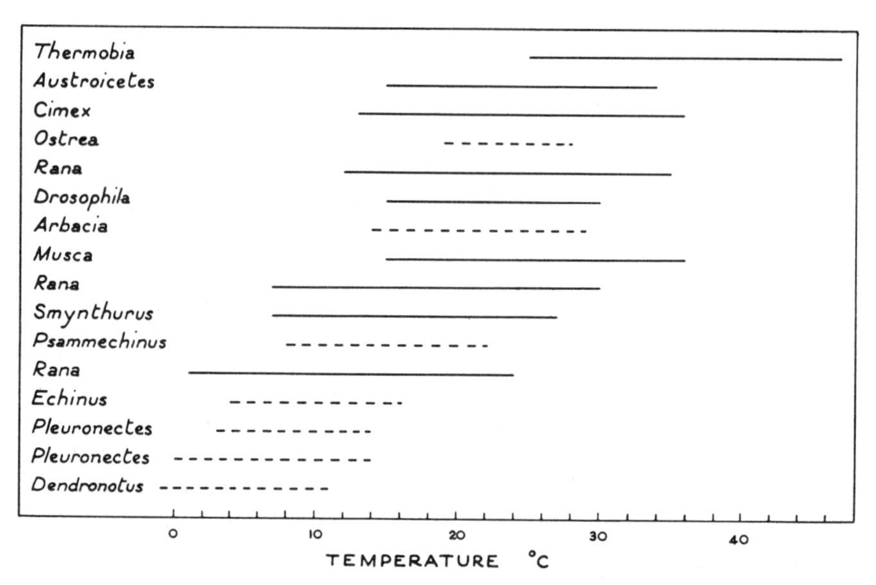

FIG. 6.01.—Showing the range of temperature favorable for development of a variety of aquatic (*broken lines*) and terrestrial or amphibious animals (*solid lines*). Note (*a*) that the ranges for the aquatic animals are mostly much shorter than for the terrestrial ones and (*b*) the wide variation between species in the limits of the favorable range; for example, the upper limits of the favorable range for *Dendronotus*, which lives in the ocean at high latitudes, is below the lower limit for *Thermobia*, which inhabits bakehouses. The three lines for *Rana* refer to three different species. (Data taken largely from Moore, 1940*a*.)

− 1.7° and 1.0° C. (Allee *et al.*, 1949). But no individual species is known which can thrive over such a wide range as from 0° to 50° C. The range of temperature favorable to any particular species is related to the prevailing temperatures in the places where the animal usually lives. Not only the average temperature but also its variability will be reflected in the physiology of the particular animal. Those that live in cold places (or are active during the cold season in warm temperate zones) have a favorable range lower than that for animals from warmer zones. For example, the favorable range for the development of the eggs of *Smynthurus viridis* (Collembola) extends from 7° to 27° C. (Davidson, 1931); but with *Austroicetes cruciata* (Orthoptera) the favorable range for the development of eggs extends from 15° to 34° C. (Birch, 1942). The former develop during the winter in the warm, temperate climate of southern Australia, the latter develop during the spring in a more northerly (somewhat warmer) part of the same general climatic zone. This principle is further illustrated in Figure 6.01, in which the species are arranged in order of the median temperature of their favorable ranges.

It will be noticed in Figure 6.01 that all the animals with narrow ranges (16° C. or less) were aquatic, living in places where the difference between minimal and maximal temperature is small, and that most of those with wide ranges (20° C. or more) were terrestrial or amphibious, living in places where the temperature is more variable. Moore (1940*b*) tabulated the favorable range for 41 aquatic animals from among Crustacea, Echinodermata, Mollusca, Tunicata, and Pisces and found that for 37 of them (91 per cent) the range was 16° C. or less; for 17 (42 per cent) the range was less than 14°; and 2 fishes were restricted to the narrow range of 8°. Six amphibians and 2 terrestrial insects listed by Moore had ranges varying from 22° to 25° C.; and he cited numerous other terrestrial arthropods and amphibians for which the range, though not determined precisely, was known to exceed 20° C.

If an animal is exposed to a low, or a high, temperature which is outside the limits of the favorable range, it may be killed directly, or it may continue to live for an indefinite period, yet fail to grow or produce any young. The result for the population is much the same in either case. These two aspects of "unfavorable" temperatures will be discussed separately: the former in sections 6.3 ff., and the latter in sections 6.2 ff. in relation to the influence of temperature, within the favorable range, on behavior and "activity," speed of development, and fecundity.

6.2 THE INFLUENCE OF TEMPERATURE WITHIN THE FAVORABLE RANGE

6.21 *The Preferred Temperature, or "Temperature-Preferendum"*

When animals are allowed to move freely along a temperature gradient, they usually congregate between quite narrow limits of temperature. This narrow band of temperature has been called the "preferred temperature" or the "temperature-preferendum." For example, when Doudoroff (1938) placed a number of the fish *Gisella nigricans* in a temperature gradient extending from 20° to 30° C., they arranged themselves at the different temperatures with the frequencies shown in Figure 6.02. Most of the fish congregated at a temperature near 26°; ordinates from the abscissa at 25° and 28° inclose about 75 per cent of the total area under the curve, indicating that about three-quarters of the fish came to rest within this 3° band of temperature; over 60 per cent stayed between the limits 25° and 27° C. The preferred temperature varied, up to a certain ceiling, directly with the temperature at which the fish had been living. Beyond this ceiling the preferred temperature became independent of the temperature to which the fish had been acclimatized. This is illustrated by Figure

6.03, which is redrawn after Fry (1947), who called this upper limit above which the preferred temperature cannot be raised by further increasing the temperature at which the fish had been living the *final preferendum*. He considered this to be a useful characteristic to measure for ecological studies.

Natural populations may not be homogeneous with respect to the "preferred temperature." For example, Wilkes (1942) found a trimodal distribution for this character in a population of the chalcid parasite *Microplectron*. The

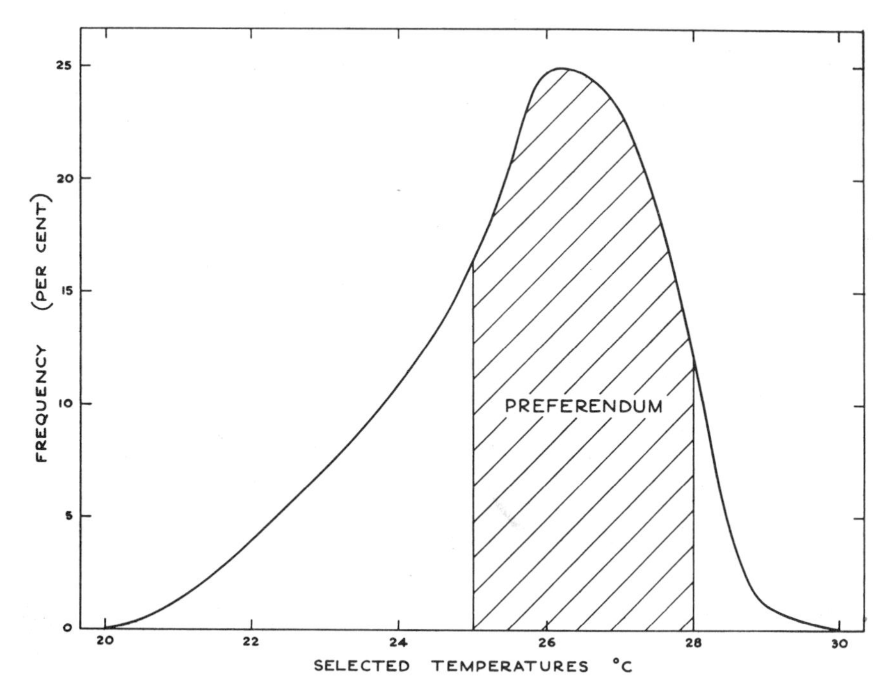

FIG. 6.02.—Showing the "preferendum" for the fish *Gisella nigricans*. The height of the graph indicates the proportion of the fish choosing each temperature. About 75 per cent of the fish congregated in the temperature indicated by the shaded area. (Modified after Doudoroff, 1938.)

greatest frequency was at about 25° C., with two minor ones at about 15° and 8°. By selecting and inbreeding the wasps which congregated at the lower temperature, he produced a distinct race in which the distribution was virtually unimodal: this race had a "preferendum" at about 9° C. (sec. 15.111). These results are decisive because of the large numbers of animals tested and the large differences obtained. Very often experiments with preferred temperatures with terrestrial animals are more difficult to interpret than those done with aquatic animals. The reason for this is that whenever there is a gradient in temperature, there are also, inevitably, gradients in the relative humidity or the saturation deficit (or both) of the atmosphere (sec. 7.234).

Indeed, Wellington (1949*b*), after a thorough investigation of the zones of temperature and humidity preferred by larvae of the moth *Choristoneura* *fumiferana*, came to the conclusion that it was not influenced by temperature to any extent that could be detected. Instead, the situations in which they congregated were determined entirely by the evaporative power of the atmosphere in these zones (sec. 7.12; Fig. 7.02). Wellington then examined the (quite extensive) literature dealing with the experimental aspect of this subject and concluded that, with few exceptions, the results attributed to temperature were obtained without the influence of humidity being adequately eliminated

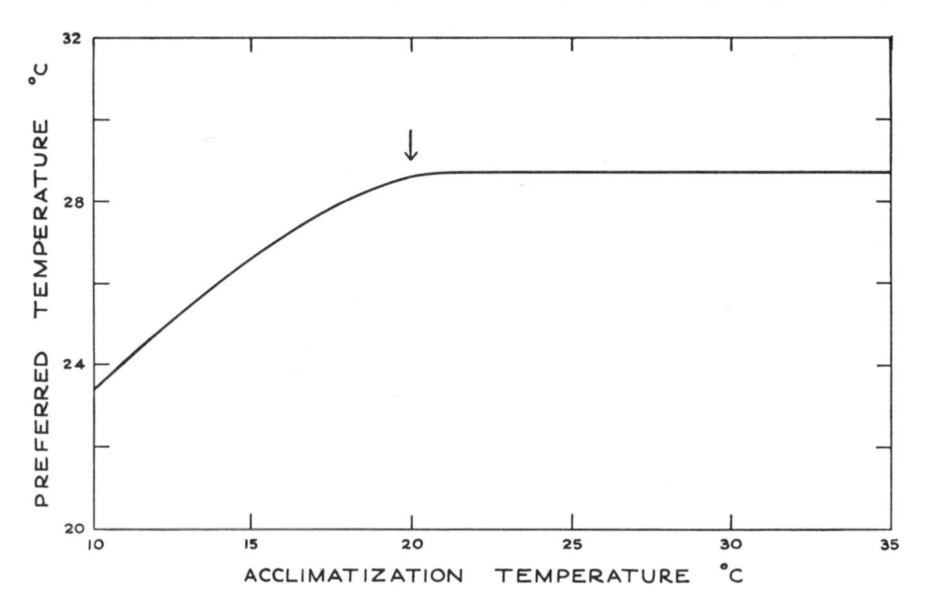

Fig. 6.03.—Showing the "final preferendum" temperature for the fish *Gisella nigricans*. For explanation see text. Note how the "final preferendum" increases with the temperature of acclimatization up to a limit and then becomes independent of the temperature of acclimatization. (Data from Doudoroff, 1938.)

in the experimental design. This may be so, but we find it hard to accept the implication that temperature is usually so unimportant as it seems to be with *C. fumiferana*, however difficult it may be to demonstrate an independent response to temperature in cases where the influence of humidity is dominant. In addition to the more recent papers by Wellington (1949*a*, *b*, *c*), the subject has been reviewed by Deal (1941) and Fraenkel and Gunn (1940).

The question as to whether temperature or humidity exercises the greater independent influence on the animal seeking a "preferred" zone in which to rest is not easily resolved by ordinary field observations on behavior. Because changes in absolute humidity are slow and small relative to changes in temperature, fluctuations in the evaporative power of the atmosphere, whether measured directly, as "evaporation" or indirectly as "saturation deficit" (sec. 7.234), follow fluctuations in temperature closely (Fig. 6.04). When observational data, such, for example, as the behavior of a swarm of locusts, is found to be correlated with temperature, the correlations with evaporation and saturation deficit will be of the same order. This follows from the simple rule in statistics which states that if two (or more) variables x and z are closely correlated positively and if a third variable y is correlated with x, then the correlation between y and z will be of the same order and sign as that between y and x. This does not preclude the possibility of a relationship between y and x which is independent of that between y and z and vice versa; but more compli-

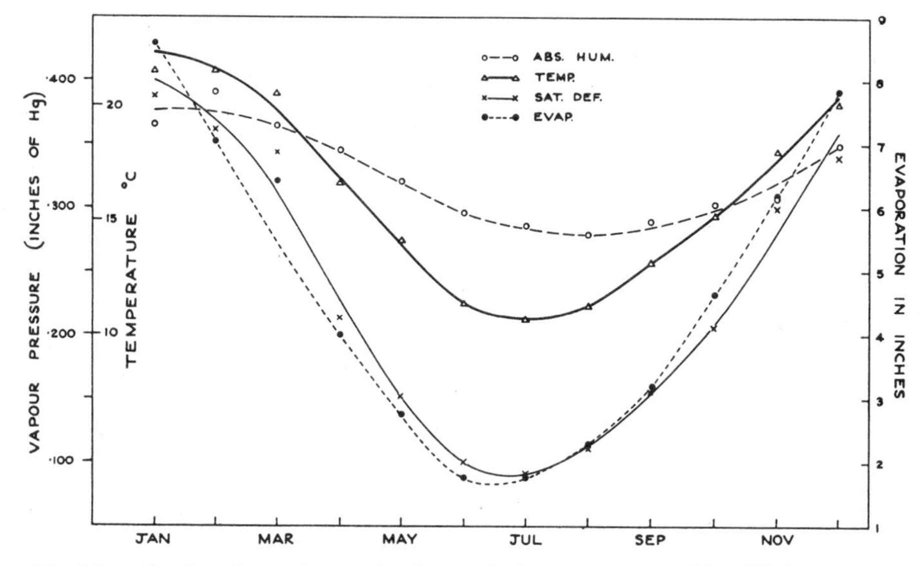

FIG. 6.04.—Showing the secular trends of atmospheric temperature and humidity at the Waite Institute, Adelaide. The curves are based on mean daily records for 23 years. Note that changes in absolute humidity are slight and gradual compared to the more abrupt changes in saturation deficiency and evaporation, which are closely related to temperature as well as to absolute humidity.

cated techniques are needed to evaluate this. The student who wishes to assimilate the "philosophy" behind this argument should read the section on *partial regression* in a good modern textbook on statistical methods. The immediate practical outcome is that causal relationships may be overlooked by concentrating on an obvious simple correlation; but the descriptive value of such calculations may remain unimpaired. Field observations are often related to temperature because it is more easily measured: Clark (1947*a*) got an adequate estimate of the temperature associated with the behavior of locusts by simply placing a thermometer with a blackened bulb among the insects in the situations that they occupied. And Wellington (1949*a*) painted the bimetallic strip of a thermograph black and exposed it in the foliage of spruce trees. Such calculations are valuable for descriptive purposes, provided that we remember that in this method temperature is being allowed to "speak for" humidity and any other variable in the environment which is correlated with temperature and that therefore the causal relationships may be quite other than what they seem to be at first sight.

The behavior of locust nymphs and adults in the field has been the subject of much study because it is generally accepted that high numbers in the outbreak area are not by themselves sufficient to produce the gregarious and migratory behavior which is such an essential feature of the destructiveness of these insects. Given high numbers, it is still necessary to have a stimulus which will bring about "aggregation" or crowding while the insects are in an active or excitable condition. Provided that the terrain is suitable, temperature may act in just that way, for the insects tend to crowd into local situations where the temperature approaches their "preferendum." With *Chortoicetes terminifera* this is about 42° C. (Clark, 1947*a*, 1949).

Clark studied the behavior of early-stage nymphs of this locust in an outbreak center in New South Wales. The first requirement of large numbers was satisfied in this case; for at the time when the eggs had just hatched, the first-instar nymphs were estimated to occur in the vicinity of the egg beds in numbers up to 2,000 per square yard in some situations. Notwithstanding the fact that their parents had been gregarious, the first-instar nymphs were neither markedly gregarious nor excitable. They soon distributed themselves over the area in relation to the occurrence of green plants suitable for food. Most of them spent the night sheltering in the vegetation or in other places. At daybreak the temperature in these situations was about 2° C., and the hoppers remained there motionless. They became more active as the temperature increased, until at 20° C. "disability" due to cold could no longer be detected and their movements were "normal." Between 20° and 45° C. the hoppers sought out and crowded into the warmest possible situations at ground level, taking up what the author calls the "basking formation." They crowded closely together, often with their bodies touching and broadside to the sun. On cool days basking groups persisted until late in the afternoon, usually breaking up as the shadows lengthened and the temperature fell. On hot days the basking groups would disperse as the temperature approached 45° C., and the hoppers would seek more shady places; ordinarily the groups would form again later in the afternoon. This behavior was governed by the existence of a "preferendum" at about 42° C. Outbursts of jumping would occur in the basking groups at irregular intervals, and as time went on the innate gregariousness and excitability of the hoppers increased. The first occurrence of gregarious mass migration was observed in the second instar in situations where the numbers were highest; elsewhere in situations where the numbers were lower, it was not observed until the fourth or fifth instar. The author summarizes his observations in the following words:

"The change from individualistic to gregarious behaviour is a result of increasing responsiveness of hoppers to the presence and movements of others.

It is effected by a period of crowding during which mutual contact, both visual and mechanical, becomes probably the most common experience of hoppers. External influences, especially temperature, play an important causal role in crowding. A period of crowding is apparently essential for the progeny of swarms of this species to *develop* gregarious behaviour, i.e., the capacity for mass migration. This is probably true for other locusts. The length of the necessary period of crowding varies in relation to population density and, probably, temperature conditions." Kennedy's (1939) account of the behavior of the desert locust *Schistocerca gregaria* in an outbreak center in the Sudan was in general agreement with the results described here for *Chortoicetes*. Kennedy concluded that "aggregation" of *Schistocerca* was largely the result of a "diurnal regime resembling in many ways that of phase *gregaria*, determined largely by temperature differences and changes, and involving prolonged basking on the small patches of bare sheltered and sunlit ground among the vegetation. As the patches are few in relation to number of locusts present, aggregation occurs."

Bogert (1952) observed similar behavior in certain lizards which live in Florida and Arizona. The lizards moved between sunny and shady places according to the temperature of their bodies. They were sensitive to small differences in temperature and usually were able, by this method, to keep the temperature of their bodies within a range of about 3° C. In certain sorts of weather the lizards would remain abroad all day, but on a clear summer's day, with the sun near its zenith, nearly all the lizards would be found sheltering in shady places. Bogert pointed out that lizards are able to carry on their "normal" activities only within a relatively narrow range of temperature. He suggested that this may explain why the distributions of large reptiles are restricted to the tropics. The large body, with its relatively small surface area, would take a long time to warm up. Bogert also pointed out that amphibians do not behave like this, nor do they seem to have "temperature preferenda."

We leave this subject now and pass on to another aspect of the influence of temperature on the activity of animals which, although it is of great importance in ecology, has been less studied than the matter of "preferenda."

6.22 *The Influence of Temperature on the Rate of Dispersal*

In chapter 5 we discussed the tendency that animals have to move about, leading to the dispersal of the population. In chapter 14 we give reasons for considering this sort of activity one of the most important characteristics of an animal, helping to determine the limits of its distribution and the level of its abundance, particularly when predators are important or food is sparsely distributed in the area where it lives. It is therefore important to understand how this activity may be influenced by the animal's environment. The several examples given in this section show how temperature may influence the dispersal-rate and indicate some of the ways in which this influence may be measured.

Perhaps the simplest approach to this problem is provided by the laboratory experiment in which temperature is carefully controlled. The speed of movement of the experimental animals is then observed directly, as the temperature in the place where they are living is varied at will. Such experiments do not, of course, supply any information about the behavior of animals living naturally, but they may provide valuable background for the interpretation of the more complex observations that may be made in nature.

The speed at which goldfish (*Carassius*) could swim at different temperatures was measured by Fry and Hart (1948). As was to be expected, the fish could swim more rapidly at high temperature than at low within the favorable range. But an interesting and important result was that the speed at which the fish could swim was determined not entirely by the present temperature, but it depended also on the temperature at which they had recently been living. In other words, the phenomenon of acclimatization, which is important in relation to the limits of the favorable zone (sec. 6.322), can also be observed in relation to activity within the favorable zone.

In these experiments a uniform group of young fish was divided into 7 lots and kept at 7 different temperatures between 5° and 35° C. until they had become fully acclimatized. They were then placed, one at a time, in a special rotating tank which stimulated them to keep swimming and in which it was possible to measure the speed at which they swam at any temperature. The maximal speed which the fish could maintain continuously for 2 minutes was measured and called the "cruising speed." When the cruising speed was measured at the same temperature as had been used for acclimatization, the maximal cruising speed was attained at about 28° C. But when the fish had been acclimatized at some temperature other than that at which the cruising speed was subsequently measured, it was found that the maximal cruising speed was always attained by fish which had been living near the temperature at which the test was made. The cruising speed was markedly less for fish that had been living at temperatures either much above or much below the temperature at which the test was made (Fig. 6.05).

It is well known that after the weevil *Calandra oryzae* has been breeding for a number of generations in a large bin of wheat, the insects will all be concentrated within a foot or two of the surface, irrespective of whether the colony started near the surface or in the depths of the bin. Birch (1946*b*) showed that this was due to the influence of temperature on the dispersal of the adult weevils. He placed small colonies of weevils on the surface and at various depths in several experimental bins of wheat. As the colonies developed in the depths, they caused the temperature of the wheat immediately around them to rise.

As the temperature reached and passed 32° C. (due to the heat of metabolism of the immobile young stages in the grain), the adults moved away to cooler places. Eventually, the layer of wheat near the surface was the only place which was cool enough to support the colony. These experiments showed that *C. oryzae* dispersed scarcely at all until the place where they were living became uncomfortably hot.

Another example of the direct approach to this problem is given by the observations of Gunn *et al.* (1948) on the behavior of migrating swarms of the

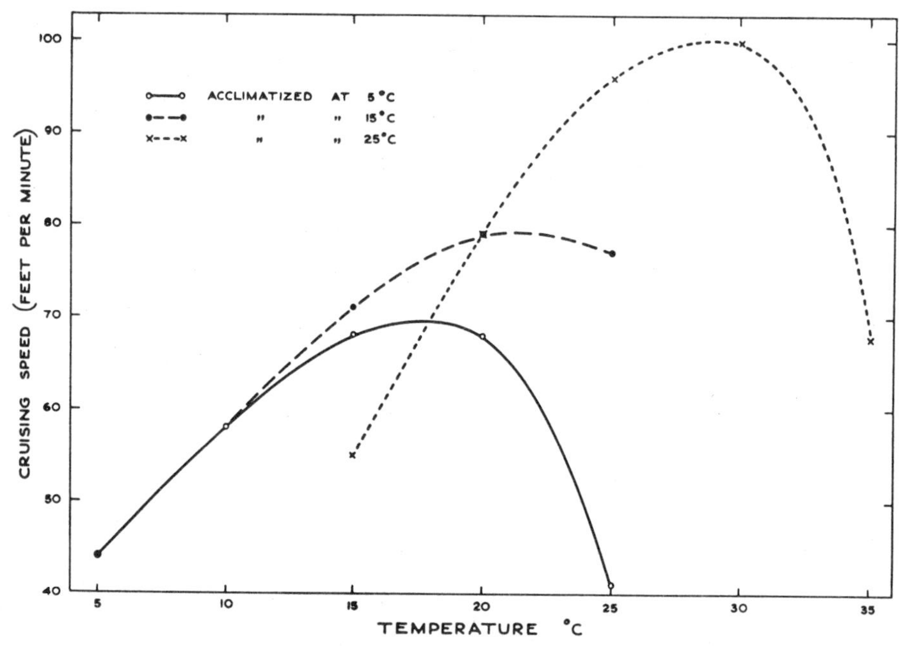

FIG. 6.05.—Showing the influence of acclimatization on the "cruising" speed of goldfish at various temperatures. Note how the temperature at which maximal "cruising" speed was attained was influenced by the temperature of acclimatization. (Data from Fry and Hart, 1948.)

locust *Schistocerca gregaria* in Kenya. Swarms of this locust often spend the night "roosting" in trees or shrubs. In the morning they descend to the ground. After spending about an hour on the ground basking in the sun, they usually take off on a gregarious migratory flight. A consistent routine is followed. At first, the insects merely bask in the sun; as the temperature rises, they begin to make short low flights, but at first there are very few in the air at any one time; later the numbers making flights increase, and each one stays aloft longer, so that, as time passes, more and more are seen in the air at once, flying over their comrades on the ground. Eventually enough are moving to stimulate the rest to take off more or less en masse. According to Gunn *et al.* (1948), the moment at which this mass flight begins is related to temperature but not to humidity or any other component of the environment which they measured. Swarms were occasionally observed to begin migrating when the temperature was as low as 14°–16° C.; once a swarm waited until the temperature had risen to 23°. But, as a rule, migration began when the temperature was between 17° and 22° C. Migration began at a lower temperature when the maximal temperature for the previous day had been low, and vice versa. In other words, the influence of temperature on the beginning of migration was modified by some stimulus which was correlated with the maximal temperature of the previous day. The explanation advanced by Rainey and Waloff (1951) is as follows. It is much easier for locusts to become "air-borne" and therefore to begin migrating when there are upward thermal currents near the surface of the ground. Therefore, migration usually does not begin until these currents develop or, in their absence, begins only at a relatively high temperature, when the locusts are more vigorous. Upward thermal movements usually develop with rising temperature in the morning, but the temperature at which they appear depends on the temperature of the air some distance above the ground. This is related to the maximal temperature of the previous day. In this way the correlation between the temperature at which migration begins and the maximal temperature of the previous day is explained. Provided that the temperature of the air exceeds a certain threshold (about 16° C.), the swarm will begin migrating just as soon as the thermal movements of the air are sufficient to enable them to become "air-borne." In the absence of favorable air currents, the locusts may still become "air-borne," but this requires much more energy on their part and does not happen until the temperature approaches 22°–23° C. Confirmatory evidence for this explanation of the data is given by the complementary observation that in the afternoon migratory swarms usually cease flying about the time that the temperature falls to 19°–23° C. It is of interest to recall that Kennedy (1939) reported that the solitary phase of this locust was stimulated by light to take off on its dispersal flight, provided that the temperature exceeded a certain threshold (sec. 5.2).

With other species it may be impracticable to make direct observations of this sort. But with suitable experiments, using the appropriate statistical methods, it is still possible to make precise inferences about the influence of temperature on the dispersive behavior of these species. A method which has the merit of being direct and simple is exemplified by the experiments of Dobzhansky and Wright (1943). Between 3,000 and 5,000 marked flies (*Drosophila pseudoobscura*) were liberated in an area where a natural population of the same species was already living at an average density of from 4 to 8 flies to each 100 square meters of territory. In each of four experiments the marked flies were liberated all together at the same place. Traps were then set out, spaced 20 meters apart and arranged in the form of a square cross with the center of the cross at the place where the flies had been liberated. During the next 5–10 days the flies in each trap were recorded daily and then set free again at the same place where they had been caught. The results for any one day gave the distribution of the flies on that day, and a comparison of the distributions on successive days gave information about the behavior of the flies—their rate of dispersal and so on. Figure 6.06 illustrates the distribution of the marked flies on two successive days in one experiment. One characteristic of the distribution is its variance, $\Sigma fr^2/\Sigma f$, where Σ has its usual meaning as the symbol of

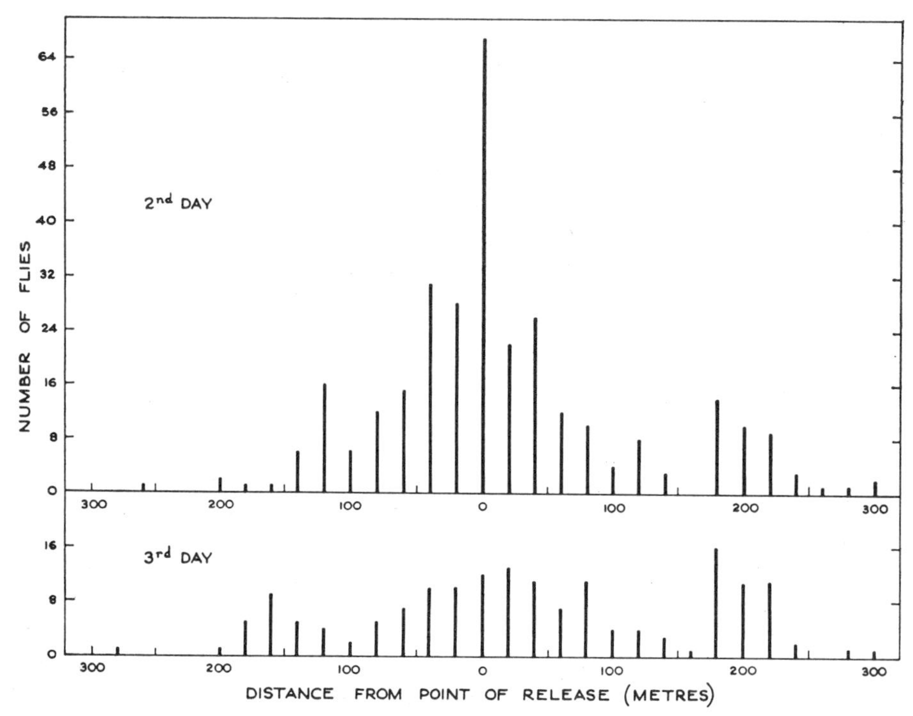

FIG. 6.06.—Showing the distribution of marked flies (*Drosophila*) on two successive days in one of Dobzhansky's experiments. See text for explanation. The unit for the abscissae in this diagram is 20 meters—the distance between traps; the ordinates show the number of flies in each trap. (After Dobzhansky and Wright, 1943.)

summation, f is the frequency (i.e., the number of flies in the trap), and r is the distance of the trap from the origin of the distribution.

If the variance at the end of the first day was s^2 and then, if the flies continue to disperse at the same rate, the variance at the end of the second day will be $2s^2$ and at the end of the tth day ts^2. On the other hand, if the rate of dispersal falls off with time (as the flies become less crowded, for example), then the daily increment in variance will be less than s^2, and the variance of the distribution on the tth day will be less than ts^2. In these experiments (as can be seen from Table 6.01) the rate of dispersal did not fall off with time; on the last (tth) day of each experiment the variance was about t times greater than it had been on the first day. This means that the variance increased, on the average, by about s^2 each day; but Table 6.01 also shows that the increase was far from uniform from day to day. A large part of this variability could be attributed to the influence of temperature on the activity of the flies.

When the daily increments in variance were correlated with daily temperature, the coefficient for linear regression of variance on temperature was found to be 2.31 ± 0.50. This means that, on the average, the daily increase in variance increased by 2.31 units (400 square meters) for each increase of 1° F. The regression was not truly linear, for the rate of dispersal was relatively slow below 20° C. and much faster above this temperature. The general conclusions to be drawn from these experiments, which are relevant to this section, are:

TABLE 6.01*

VARIANCES OF DISTRIBUTIONS OF MARKED FLIES ON SUCCESSIVE DAYS AFTER LIBERATION

Day	I	II	III	II¹	III¹	IV¹
1 (s_1^2)	3.8 (279)	8.7 (584)	8.4 (674)	7.6 (337)	9.0 (295)	11.3 (635)
2	8.1 (609)	21.0 (354)	22.2 (532)	22.5 (220)	30.5 (236)	22.8 (306)
3	13.6 (369)	25.8 (238)	42.6 (228)	22.7 (128)	52.2 (133)	46.5 (102)
4	13.0 (171)	34.5 (276)	41.6 (166)	35.0 (168)	40.3 (69)	78.4 (73)
5	10.4 (94)	37.5 (145)	......	35.7 (90)	44.3 (123)	73.1 (51)
6	17.8 (215)	......	......	63.2 (78)	73.0 (58)	......
7	21.4 (81)	......	......	64.5 (79)	90.4 (39)	......
8	32.4 (69)	......	......	......	......	......
ts_1^2	30.4	43.5	33.6	53.2	63.0	56.5

* Experiments I, II, and III based on traps arranged along 4 radii at 90°. Experiments II¹, III¹, and IV¹ based on traps along 2 radii at 180°. This arm of the original cross was extended by adding traps to each end after the flies had reached the perimeter of the original area. The figures in parenthesis give the number of flies caught that day (Σf). After Dobzhansky and Wright (1943).

(*a*) apart from the first day, when some 3,000–5,000 flies were liberated at one place, the density of the population had no appreciable influence on the rate of dispersal; (*b*) the temperature exerted a significant influence on the rate of dispersal, and, in particular, this increased with temperature when the temperature exceeded 20° C. Certain other inferences about the behavior of *Drosophila* in a natural population may be drawn from the results of these most interesting experiments, and these have been discussed in the appropriate places elsewhere (sec. 5.11).

When the raw data to be analyzed are derived, not from an experiment of the sort that was done with *Drosophila*, but rather from observations made on an undisturbed natural population, then other methods of analysis may be needed. The statistical device of partial regression is often useful. As an example we may describe the quantitative study of the movements of the flower-inhabiting thrips (*Thrips imaginis*) made by Davidson and Andrewartha (1948*a*, *b*). This species is strongly attracted to roses, and these flowers were used as traps. The insects were breeding in a variety of other flowers in and around the garden, and these were the reservoir from which the thrips came to the freshly opened roses. Daily records of the numbers of thrips in roses were kept for the months August–December (southern spring) for 14 years. A characteristic series of records is shown in Figure 6.07, which represents the daily numbers of thrips in roses during 1932.

The seasonal trend and the enormous daily fluctuations about the trend shown for this year are quite typical for all the other years for which records were kept. The first step in analyzing the data is to transform the numbers to logarithms: this is done chiefly because proportional changes in numbers are more instructive than additive changes. It is then possible to account for the

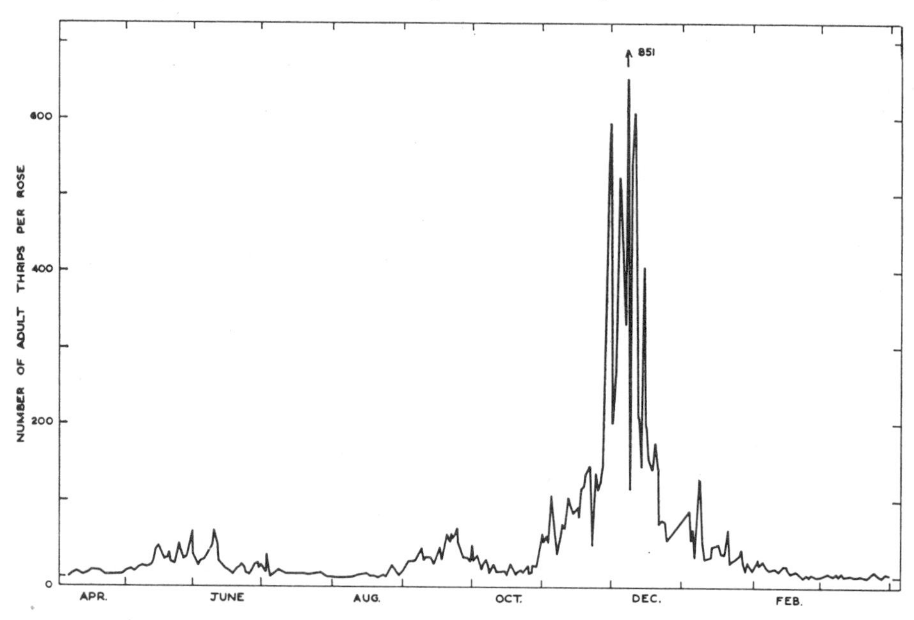

FIG. 6.07.—Showing the daily mean numbers of thrips per rôse for 1932–33. Note the seasonal trend and the marked daily fluctuations about the trend. (After Davidson and Andrewartha, 1948*a*.)

systematic trend of numbers with time by fitting empirically a curvilinear regression of the form

$$Y = a + bx + cx^2 + dx^3 ,$$

where Y represents the logarithm of the number of thrips observed and x represents days. A typical curve is illustrated in Figure 6.08. The systematic trend which this curve describes is due chiefly to the natural increase of the population with time, but it may include other components associated with systematic trends in the weather. The important point to note is that the curve has accounted for the trend with time, whatever may be its cause, so that the residual variance (represented by the departures of the observed points from the trend line in Fig. 6.08) is independent of time. These departures are partly (probably largely) due to daily variations in the degree of activity displayed by the thrips in moving into and out of the "trap" flowers. On a warm day more thrips will move out of the flowers in the surrounding fields, where they have been breeding, and find their way into the newly opened roses, which are sampled daily. It is possible to evaluate precisely the degree of association between the "activity" of the thrips (as represented by the residual departures from the trend line) and the various components of environment which, experience suggests,

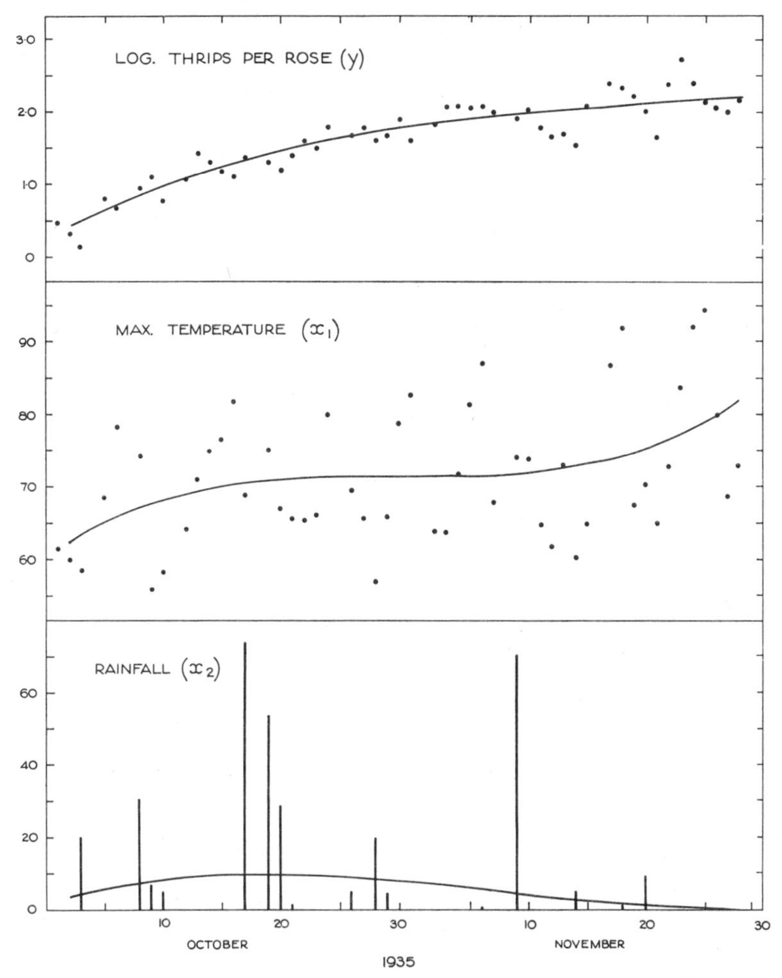

FIG. 6.08.—Showing (*a*) daily records of the number of thrips per rose during the spring of 1935 (the numbers have been expressed as logarithms); (*b*) the daily maximal temperature for the same period in degrees Fahrenheit; and (*c*) the daily rainfall. The unit is one-hundredth of an inch. The trend was in each case obtained by calculating the appropriate polynomial of the form $Y = a + bx + cx^2 + dx^3$. (After Davidson and Andrewartha, 1948*b*.)

may be important. The final step in the analysis is to solve the equation for partial regression, which may be written,

$$Y = b_1x_1 + b_2x_2 + b_3x_3 + \ldots,$$

where Y represents the departure of the observed data from the trend line in Figure 6.08; $x_1, x_2, x_3, \ldots$, represent temperature, humidity, and other components of environment, also expressed as departures from the trend line with time; and b_1, b_2, b_3, . . . , are the partial regression coefficients such that b_i measures the association of y with x_i, independent of any relationships which might exist between these and any other variable in the equation.

In the present example it was found that the independent influence of temperature was highly significant and that, on the average, the numbers of thrips in the flowers on any one day increased by 25 per cent for each increase of 5° C. in the daily maximal atmospheric temperature. The only other component of environment which was found to be significantly related to the activity of the thrips was daily rainfall. But this was less important than temperature. In the particular circumstances of this investigation the measurable activity of the thrips which could be attributed to temperature was about three times that associated with rainfall.

A striking and most important feature of the data illustrated in Figure 6.08 is that, for any given fluctuation in temperature, the relative increase or decrease in the numbers of thrips (i.e., the magnitude of the departures from the trend line) is about the same, whether the absolute numbers of thrips are low or high. This indicates that the movements of the thrips to and from the flowers is not dependent on density; it is not the outcome of jostling or any other manifestation of numbers. But it could be the outcome of an innate "wanderlust," possessed to an equal degree by thrips living in solitude and those living in a crowd. This tendency may be modified by several components of the environment, of which temperature has been shown to be an important one.

Williams (1939, 1940) was the first to apply the method of partial regression to get a precise measure of the independent influence of different components of the environment on the dispersal of insects in nature. He counted the numbers of insects of all kinds caught each night in a light trap that ran continuously for 4 years at Rothamsted. He related these numbers to the daily records of temperature, humidity, moonlight, and so on. The results indicate that relatively more insects came into the trap in warm weather than in cold and that none of the other components of the environment that were measured influenced the counts as much as temperature. These two papers are especially valuable for their full discussion of method. We have not used them for our main example in this section because, with so many species being considered together, the interpretation is not so clear as in the simpler case with *Thrips imaginis*.

6.23 *The Influence of Temperature on the Speed of Development*

It has long been known that poikilothermic animals complete their development more rapidly in warm weather than in cool. The speed of development at different constant temperatures has been measured for many species, and several mathematical expressions have been proposed which purport to describe the relationship of speed of development to temperature. For technical reasons it is inevitable that constant temperatures are used in such experiments.

6.231 CONSTANT TEMPERATURE

Figure 6.09 illustrates two alternative ways that are often used to represent the speed of development at different temperatures. The temperature is plotted as the abscissae, and the ordinates may be either the duration of the particular

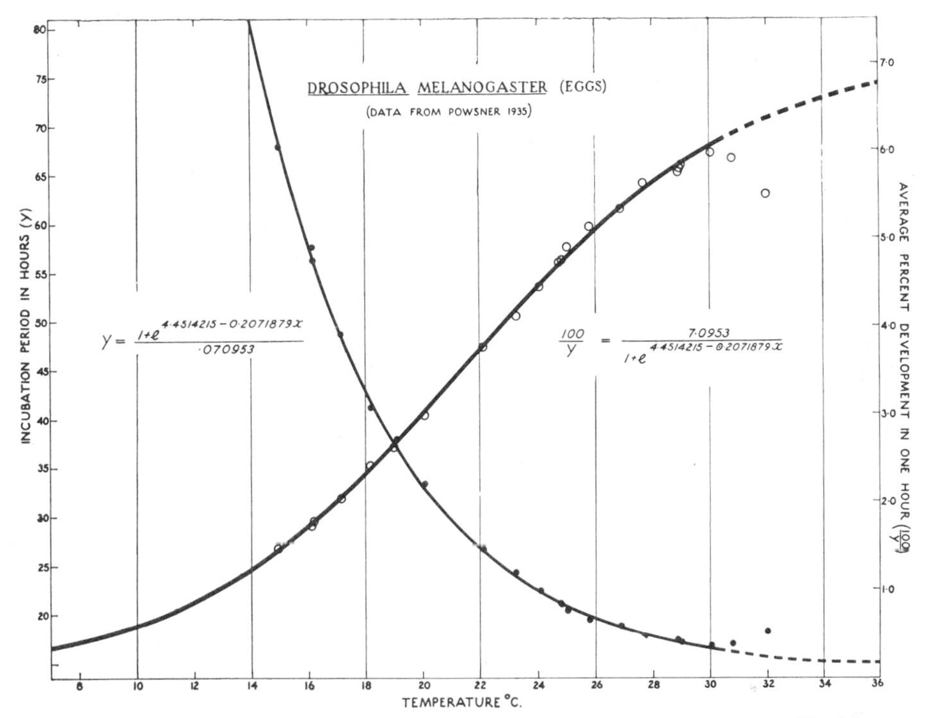

FIG. 6.09.—Showing the speed of development of eggs of *Drosophila melanogaster*. The hours required to complete the egg stage are plotted against temperature (*descending curve*), and the reciprocal of this time is also plotted against temperature (*ascending curve*). The points plotted on the diagram are observed values in each case. The ascending logistic curve was fitted to the reciprocals of the observed times. The calculated values for the descending curve were the reciprocals of the calculated values for the logistic curve. (After Davidson, 1944.)

stage that is being studied or, more often, the reciprocal of this measurement. Commonly, the latter is multiplied by 100, so that it becomes per cent development per unit time (day or hour). The transformed data are the usual starting point for generalized mathematical expressions or curves. But note that the reciprocal is a highly specialized conception of speed of development and has a number of peculiar properties. For example, the slope of the curve is a function of the duration of the particular stage under consideration. This becomes immediately apparent by a simple analogy. Two vehicles travel at identical constant speed along a straight road. One has twice as far to go as the other. If the "speed" of the cars be estimated as a percentage of the journey completed in an hour, one will be judged to have twice the "speed" of the other. This peculiar property of the reciprocal scale makes it valueless for comparisons either between the same stage in different animals or between different stages in the same animal.

Fry (1947) examined a number of the mathematical expressions which have been proposed to describe the relationship of temperature to speed of development. He classified these into "physiological" or "biological," according to the purpose for which they were intended. He designated the expressions of vant'Hoff, Arrhenius, and Bělehrádek as "physiological" and that of Janisch and the well-known method of "thermal sums" as "biological." Davidson (1944) called the equations of vant'Hoff and Arrhenius "theoretical" expressions, and the others, including the logistic curve which he was himself proposing, "empirical." We think this is the better classification.

The Arrhenius equation is usually written as follows:

$$\frac{Y_2}{Y_1} = e^{\frac{1}{2}\mu\left(\frac{1}{x_1} - \frac{1}{x_2}\right)},$$

where Y represents speed of development at temperature x measured in degrees absolute, and μ is a constant. The vant'Hoff equation is usually written:

$$Q_{10} = \left(\frac{Y_1}{Y_2}\right)^{\left(\frac{10}{x_1 - x_2}\right)},$$

where Y represents the speed of development and x represents temperature on the centigrade scale, and Q_{10} is a coefficient. These expressions may be written:

$$\mu = 4.6(\log Y_2 - \log Y_1) \times \frac{1}{(1/x_1) - (1/x_2)}$$

and

$$\log Q_{10} = 10(\log Y_1 - \log Y_2) \times \frac{1}{x_1 - x_2}.$$

By hypothesis, μ and Q_{10} are constant. When the equations are written in this form, it becomes clear that a constant value for μ implies a linear relationship between the logarithm of the speed of development and the reciprocal of temperature on the absolute scale; and a constant value for Q_{10} implies a linear relationship between the logarithm of the speed of development and temperature on the centigrade scale. But Bělehrádek (1935) pointed out that the reciprocal of the absolute temperature is practically a linear function of temperature on the centigrade scale between the limits 0° and 40° C. So that the expressions of Arrhenius and vant'Hoff are virtually equivalent; both imply that the proportional increase in speed of development produced by a given difference in temperature is constant throughout the temperature range at which an animal may develop. If μ fits any particular set of empirical data, then Q_{10} should fit equally well, and vice versa. In practice, it has been shown repeatedly that Q_{10} and μ are far from constant and that they vary in a systematic way with temperature. Bělehrádek (1935) quoted numerous authors who had found either μ or Q_{10} to be inconstant. We shall illustrate the inconstancy of μ with just one example taken from the work of Bliss (1926). Working with the prepupal stage of *Drosophila*, he found the following values for μ:

Between 12° and 16° C. μ = 33,210
Between 16° and 25° C. μ = 16,800
Between 25° and 30° C. μ = 7,100

This is but a quantitative example of the principle which Barcroft (in Fry, 1947) recognized when he pointed out that over a range of temperature for which the animal is best suited the temperature-metabolism curve is relatively flat. For example, between 6° and 20° C., the standard metabolism of the trout is doubled but that of the goldfish is increased fivefold. This is about the temperature range that best suits the trout, but the goldfish is adapted to much higher temperatures (Fry, 1947). When we consider that Q_{10} and μ were originally designed to describe the relationship of temperature to a single chemical reaction, it is hardly surprising that they should be inadequate when they are related to the immensely complex chain of reactions which proceed during the morphogenesis of an animal. These coefficients have also been applied, seemingly indiscriminately, to measurements of rate of activity, such as heartbeat, rate of nervous conduction, velocity of locomotion (*Paramecium*), and so on. These are much simpler processes than development, and it may be that a simpler expression will describe them. But we are not immediately concerned with this usage, except to point out that there is no good reason for expecting measurements of activity to be equivalent to measurements of metabolism. In view of the chemical complexities of even the simplest morphogenesis, it seems unlikely that a simple (or, indeed, any) theoretical expression will be found that adequately relates temperature to the speed of development. In the meantime, the purposes of ecology are best served by seeking a satisfactory empirical equation to describe this relationship.

The best known of the empirical equations is the hyperbola:

$$y(x - a) = K.$$

In this equation y is the duration of development, usually expressed in days or hours, at temperature x, which is usually expressed in degrees centigrade.

When y is converted to its reciprocal, which is the conventional measure for speed of development, this equation becomes:

$$\frac{1}{y} = k + bx,$$

where y and x have the same values as before and k is a new constant equal to $-ab$. This is the equation of a straight line, and a can be shown to be the value of x where the value of y becomes zero, i.e., the temperature at which the "speed of development" becomes zero. It has been called the "threshold of development." The supposed linear relationship between temperature and the speed of development has been widely accepted in the past (sec. 6.234). However, it is now well known, and any well-conducted experiment will demonstrate, that the temperature-time curve approximates to the hyperbola for only a very short range of temperature, departing widely from it at either end of the favorable range. Moreover, it is never possible to get empirical points near the theoretical "threshold," so that the value of a is determined by extrapolation—a statistical procedure which in this instance has virtually no biological meaning.

Realizing that the hyperbola is inadequate, Bělehrádek (1935 and earlier papers) proposed the equation

$$Y = \frac{a}{x^b},$$

where Y is time required to complete development at temperature x, which is usually expressed in degrees centigrade, and a and b are constants. This is the equation of an exponential expression which becomes linear on the logarithmic scale, taking the form

$$\log Y = \log a - b \log x,$$

and b is seen to be the coefficient of linear regression of $\log y$ on $\log x$. In this abstract sense b has the meaning of a temperature coefficient; but with both axes of the graph on the logarithmic scale, it is difficult to attribute any real meaning to b. However, since this is an empirical equation, the proper criterion for its usefulness is how well it follows the empirical data. It will not be sufficient just to rely on visual inspection, for it is well known that the double-log scale greatly minimizes the apparent discrepancies from the curve, though these remain to be demonstrated by the appropriate statistical methods. But even by visual inspection the empirical equation of Bělehrádek can be shown to be inferior to the logistic expression which we describe below (Fig. 6.10).

The logistic equation was derived by Verhulst, and rediscovered by Pearl (Pearl and Reed, 1920), to describe the trend of increase in a population growing where the space was limited (sec. 9.21). In this context the equation has a theoretical basis; but Davidson, having observed the consistent recurrence of a sigmoid trend in temperature-development data, fitted the Pearl-Verhulst logistic curve as a purely empirical description of this trend. He searched the literature for experiments that appeared to have been carried out with the necessary precision. He found that in every case the logistic curve followed the observed data more closely than did any of the older expressions. He published

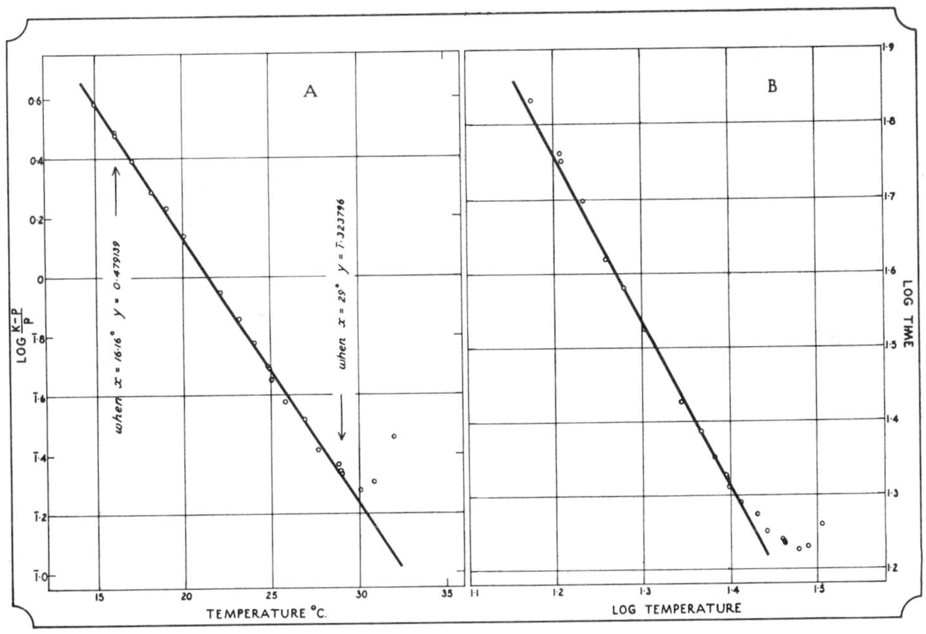

FIG. 6.10.—Showing *A*, the linear transformation of the logistic curve; *B*, the straight line derived from the Bělehrádek equation for the same data for *Drosophila*. Note that the former gives a closer fit over a greater proportion of the temperature range at which development may proceed. (After Davidson, 1944.)

his findings in two brief papers (Davidson, 1942, 1944), which should be consulted in the original, for they constitute a completely adequate exposition of the use of the logistic equation to relate temperature to speed of development of insects. The logistic curve is realistic, giving an easily comprehended picture of the trend of speed of development at different constant temperatures. It is easily calculated directly from the empirical data, and, in addition, it has the merit of being the most adequate empirical description of this relationship that has so far been suggested. From now on, it should be used in preference to all the older expressions.

The logistic equation may be written

$$\frac{1}{Y} = \frac{K}{1 + e^{a-bx}},$$

where Y represents the time required to complete development at temperature x, which is usually measured in degrees centigrade, and a, b, and K are constants. K defines the upper asymptote toward which the curve is trending; b defines the slope of the curve; and a relative to b fixes its position along the x axis. This may be the more easily comprehended by considering that the logistic curve is a bi-symmetrical sigmoid curve having a point of inflection (the steepest point on the curve) whose co-ordinates are $1/Y = K/2$ and $x = a/b$. The ordinate $1/Y$, being the reciprocal of the time required to complete a particular stage, measures the proportion of the total development completed in unit time; it thus serves as a conventional measure of speed of development. In practice, it is convenient to multiply this by 100 and express time in days (or hours), so that the right-hand side of the equation expresses per cent development per day (or per hour).

The constants K, a, and b are readily calculated from the data, making use of the transformation

$$\log_e \frac{K - y}{y} = a - bx.$$

This is the equation of a straight line. Given the value of K and any two values of y corresponding to particular values of x, the constants a and b are readily computed. The usual method is to make a first approximation for K and two values of y from a freehand graph, then, using the linear transformation given above, to improve these values by several successive approximations. The whole process of fitting may be carried out speedily and with quite adequate accuracy by using graphical methods throughout (Pearl, 1930, p. 420; Davidson, 1944). Alternatively, once K has been fixed by interpolation, the constants a and b may be calculated by the usual method of linear regression (Davidson, 1944).

Applying this method to Powsner's (1935) data for the eggs of *Drosophila*, Davidson (1944) obtained the equation

$$\frac{100}{Y} = \frac{7.0953}{1 + e^{4.451422 - 0.207188x}}.$$

The resulting curve is shown in Figure 6.09. It will be seen that the observed points cluster quite closely about the calculated curve between 15° and 30° C. Above 30° the harmful influence of high temperature is manifest from the pronounced decrease in the speed of development above that temperature. The reciprocal of the logistic is also shown in Figure 6.09. This is calculated directly from the logistic by taking the reciprocals of the calculated values. It transforms the data back into their original form, namely, time required to complete development. It seems less instructive than the simple logistic curve, but it may be useful in certain *ad hoc* calculations. Figure 6.10, *A*, shows the straight line obtained by plotting log $(K - y)/y$ against temperature; and Figure 6.10, *B*, is included so that the logistic may be compared with Bělehrádek's equation. The superiority of the former is evident.

So far, the logistic equation has been applied to data for the speed of development of insects (Davidson, 1942, 1943*a*, *b*, 1944; Birch, 1944*a*, 1945*c*;

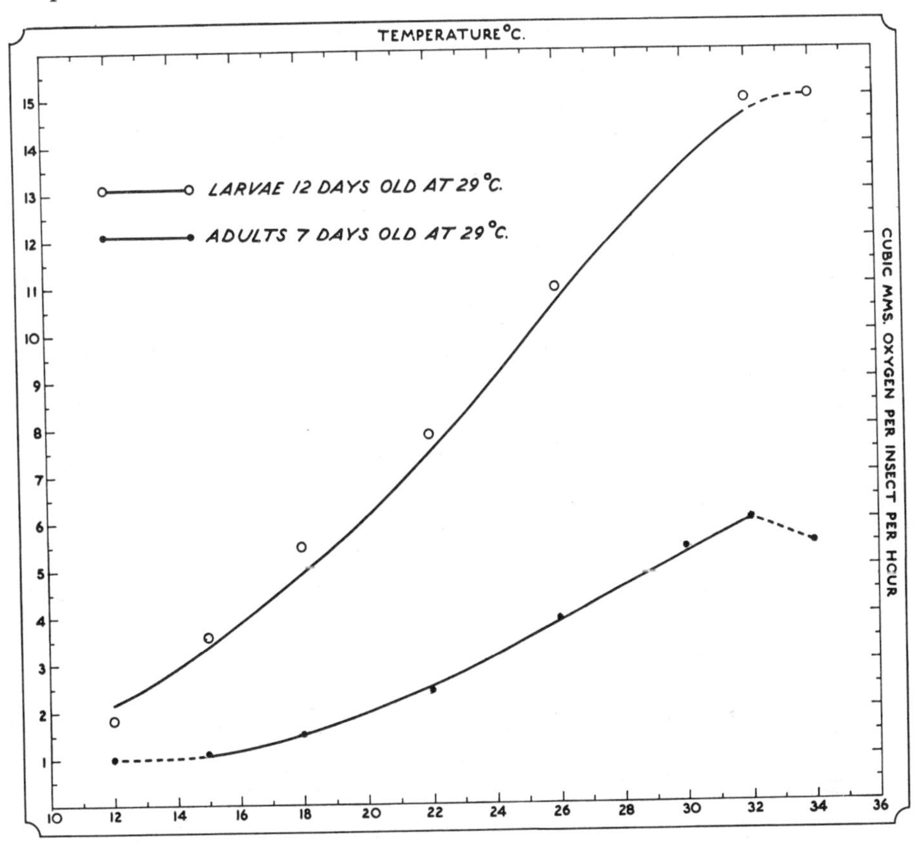

FIG. 6.11.—Showing the relationship between temperature and the rate of oxygen consumption by larval and adult stages of *Calandra oryzae*. (After Birch, 1947.)

Birch and Snowball, 1945; Browning, 1952*b*). But Birch (1947) found that it also was appropriate for data relating the rate of oxygen uptake to temperature in developing larvae of *Rhizopertha dominica* and *Calandra oryzae* (Fig. 6.11).

It was pointed out above that the logistic equation gives a curve which has a point of inflection at the co-ordinates $1/Y = K/2$, $x = a/b$; that is, the steepness of the curve increases up to this point and then gradually decreases. This means that for any given increment in temperature the increase in the speed of development is greatest for median temperatures and gradually falls off at higher temperatures. The fact that the acceleration fell off after the point of inflection led Davidson (1944) to suggest that the unfavorable influence of high temperature may be beginning to make itself increasingly evident from this point onward. But this is a theoretical consideration which has little bearing on the usefulness of the logistic equation as an empirical description of the relationship between temperature and the speed of development of insects. There are a number of considerations which make it desirable to refrain from imputing a theoretical foundation for the logistic equation.

It has been shown that the χ^2 test may be used to test the goodness of fit of the logistic curve (Birch, 1944*a;* Browning, 1952*b*), provided that the final approximations of the constants K, a, and b are found by solving the appropriate equations for maximal likelihood. The χ^2 test essentially measures the magnitude of the departures of the observed values from the hypothetical ones relative to the intrinsic variability of the material being observed. The precision of the test is enhanced by increasing the number of replicates at each temperature.

Browning (1952*b*) applied the χ^2 test to his own data for eggs of *Gryllulus*, to those of Birch (1944) for eggs of *Calandra*, and to those of Powsner (1935) for eggs of *Drosophila*. The last are those used by Davidson (1944) and reproduced in Figure 6.09. These three sets of data were chosen because in each case they had been derived from adequately large samples (total exceeded 500 in each case) and the experimental work had been done precisely. The values for χ^2 for these three sets of data are set out in Table 6.02. The relevant curves are

TABLE 6.02*
GOODNESS OF FIT OF CALCULATED CURVES SHOWN IN FIGURES 6.09, 6.12, AND 6.13

Source of Data	χ^2	Degrees of Freedom	P
Eggs of *Drosophila* (Powsner)........	1,053.6	17	<0.0000
Eggs of *Calandra* (Birch)............	29.6	2	<0.0000
Eggs of *Gryllulus* (Browning).........	82.6	3	<0.0000

* After Browning (1952*b*).

shown in Figures 6.09, 6.12, and 6.13; from these figures it may be seen that in each case the departures of the observed points from the hypothetical curve are small indeed. Yet the final column in Table 6.02 indicates that in each case discrepancies as large as these would not have occurred by chance once in many thousands of trials. This means that there is a significant proportion of the variance due to differences in temperature left over after the logistic curve has accounted for its portion. Or, in more colloquial terms, we might say that in each of the three cases we have been able to show that there are odds of very much better than 10,000 to 1 against the logistic curve's being the "true" representation of the relationship between temperature and the speed of development.

There are sound theoretical reasons for expecting just this result. Fortunately, Powsner (1935) published not only the data for the egg stage of *Drosophila* which we have been discussing but also equally precise data for the larval and pupal stages. Browning (1952*b*) re analyzed these data first for each stage separately and then by adding together the time required for egg and

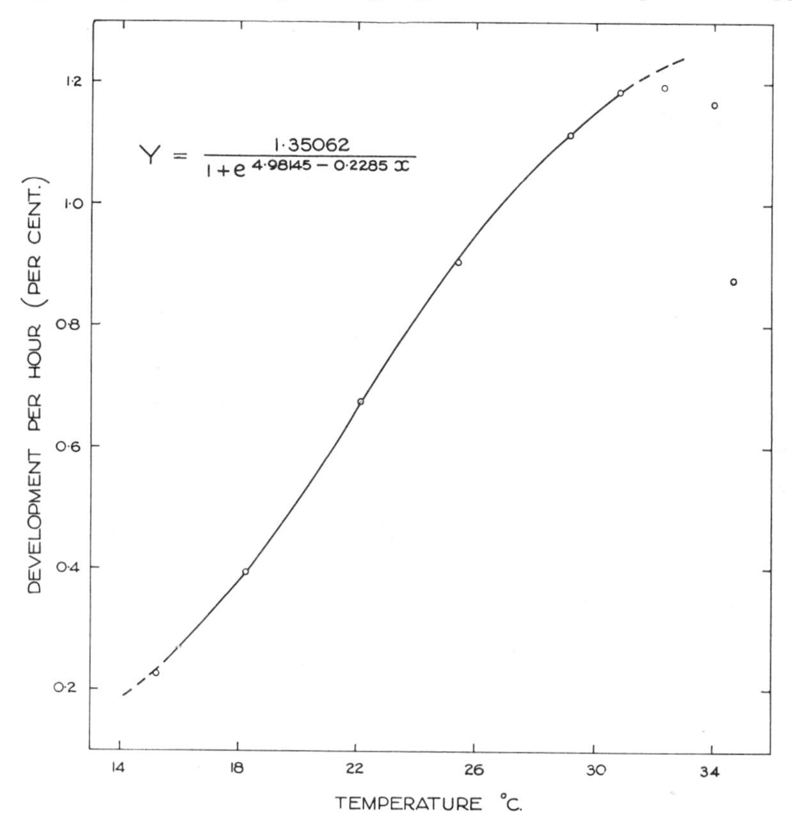

FIG. 6.12.—Showing the relationship between temperature and the speed of development of the egg stage of *Calandra oryzae*. (After Birch, 1944*a*.)

larval stages, and, finally, for egg, larval, and pupal stages taken together (Fig. 6.14; Table 6.03). Two striking results stand out. First, the curves

TABLE 6.03*
GOODNESS OF FIT OF CALCULATED CURVES SHOWN IN FIGURE 6.14

Stage in Life-Cycle of *Drosophila*	χ^2	Degrees of Freedom
Egg	1,053	17
Egg + larva	4,952	18
Egg + larva + pupa	66,581	10

* After Browning (1952*b*).

for the different individual stages differ not only in the magnitude of the constants *a* and *b* but also in the range of temperature they occupy. Second, the best fit was for the egg stage, the poorest for the combined egg, larval, and pupal stages. The second of these results could have been predicted from the first. For even if the relationship between temperature and speed of development for the egg stage were truly expressed by the logistic equation and simi-

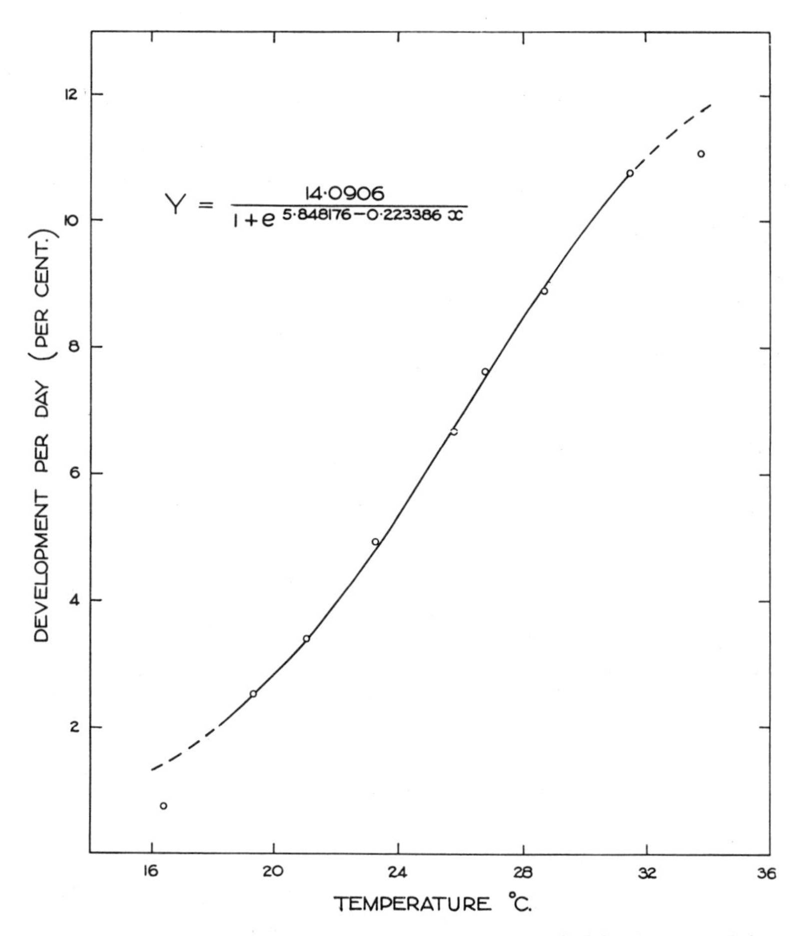

FIG. 6.13.—Showing the relationship of temperature to the speed of development of the egg stage of *Gryllulus commodus*. (After Browning, 1952*b*.)

larly for the larval stage, still the two stages added together could conform to the logistic equation only if the constants a and b were the same in each of the two primary curves. In this case a and b are demonstrably different for the egg and larval stages (Fig. 6.14). A moment's reflection will indicate that even a seemingly uniform stage like embryogenesis is hardly likely to consist either of one uniform process or even of a series of successive processes so similar that the curves describing them will have the same constants a and b. It is therefore unlikely that any but the shortest and simplest stage of morphogenesis will be truly represented by the logistic equation; and even this still remains to be demonstrated.

These considerations, important as they may be to the theory of the subject, need not detract from the usefulness of the logistic curve as an empirical description of the relationship between temperature and the rate of development of poikilothermic animals. Figures 6.09, 6.12, and 6.13 show that the

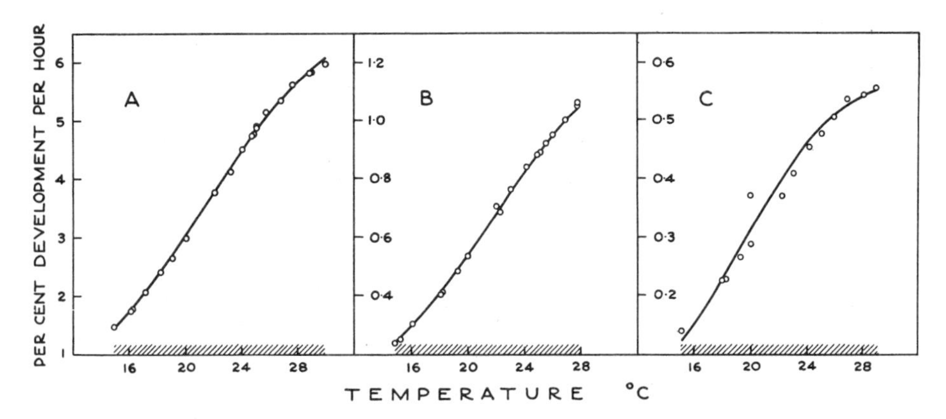

FIG. 6.14.—Showing the relationship between temperature and the speed of development of *A*, the egg stage; *B*, the egg and larval stages taken together; and *C*, the egg, larval, and pupal stages taken together, of *Drosophila melanogaster*. Note that the deviations from the trend line become progressively greater from *A* to *C*. (After Browning, 1952*b*.)

departures of the observed points from the trend lines are small in each case; and the calculated trend can be taken as adequate for most ecological purposes.

6.232 FLUCTUATING TEMPERATURES

In nature, animals live in places where the temperature fluctuates; so it is necessary to see how far the conclusions reached in the preceding section, which were based on experiments done at constant temperatures, may be applied to animals living in nature. There is no simple answer to this question. The matter is complicated by the following considerations: (*a*) The life-cycle may include a diapause-stage which needs to be completed at a lower temperature than that which favors the more obvious processes of morphogenesis (chap. 4). (*b*) The fluctuations may include extreme temperatures which are harmful, and even short exposures to these may impair the animal's competence to develop healthily at a favorable temperature. (*c*) Different stages in the life-cycle may have different limits to the favorable range and may respond differently to temperature within the favorable range. (*d*) Healthy development may proceed during short or intermittent exposures to extreme temperatures which would be harmful or even lethal if experienced continuously. (*e*) The relationship between temperature and the speed of development is not linear. We shall examine each of these five phenomena in the order in which they have been listed, beginning with diapause.

a) One of the most characteristic features of diapause is that it disappears during adequate exposure to "low" temperature (chap. 4). Diapause-development (that is, the physiological processes which culminate in the disappearance of diapause) proceeds most rapidly, as a rule, over a range of temperature that is lower than the favorable range for morphogenesis, although the two ranges

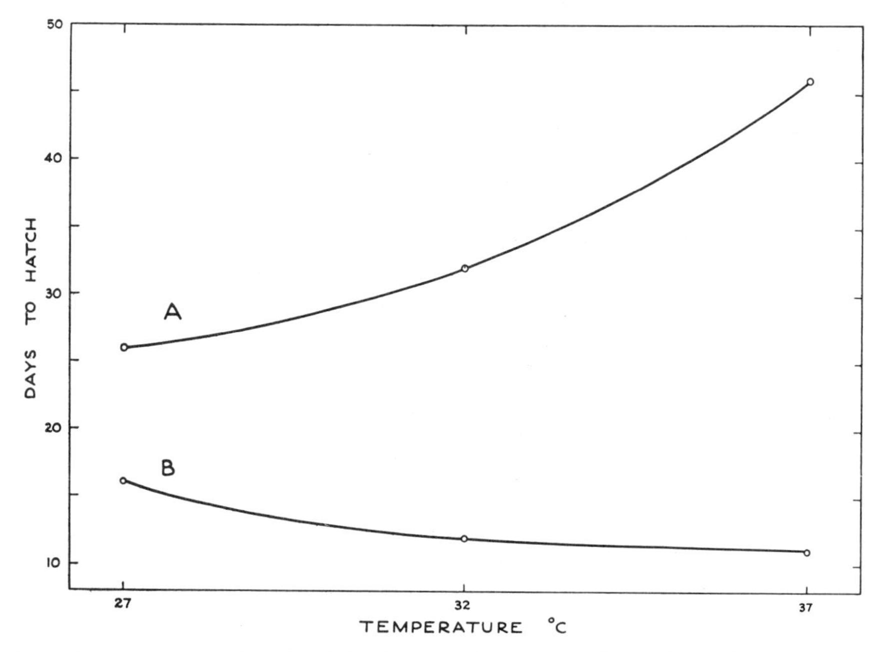

Fig. 6.15.—Showing the duration of development of the eggs of *Melanoplus mexicanus* at constant temperature. *A*, without any exposure to cold; and *B*, after 240 days at 0° C. Note that the slope of curve *B* is characteristic of "normal" nondiapause insects; *A* slopes in the opposite direction, because the weak diapause is somewhat "firmer" at 37° than at 27° C. (Data from Parker, 1930.)

usually overlap (Fig. 4.03). In cases where a clear-cut and firm diapause is present, no difficulties of interpretation need arise, and investigation will usually reveal that a specific, and often prolonged, exposure to low temperature is required before morphogenesis may proceed healthily or at all. But it was made clear in section 4.3 that diapause is not a clear-cut, all-or-none type of phenomenon. There are innumerable instances in nature of weak or incipient diapause and still others, like *Austroicetes*, which may be found at certain times of the year with the remnants of what was originally a firm diapause (sec. 4.1). In these cases the animals (or in marginal cases, where the diapause is more pronounced, a proportion of the animals) are competent to complete development at constant temperatures within the favorable range, without first experiencing cold, and the presence of diapause is thus not completely self-evident. But development is markedly more healthy in the lower part of the temperature range, and characteristically the increase in speed of development with increasing temperature is relatively small or may even be negative (Figs. 6.15, 6.16). Exposure to fluctuating temperature with the lower temperature in the range favorable for diapause-development or, alternatively, a single prolonged exposure to such a temperature usually results in a pronounced acceleration in the speed of development at higher temperatures (Fig. 6.16, *B*).

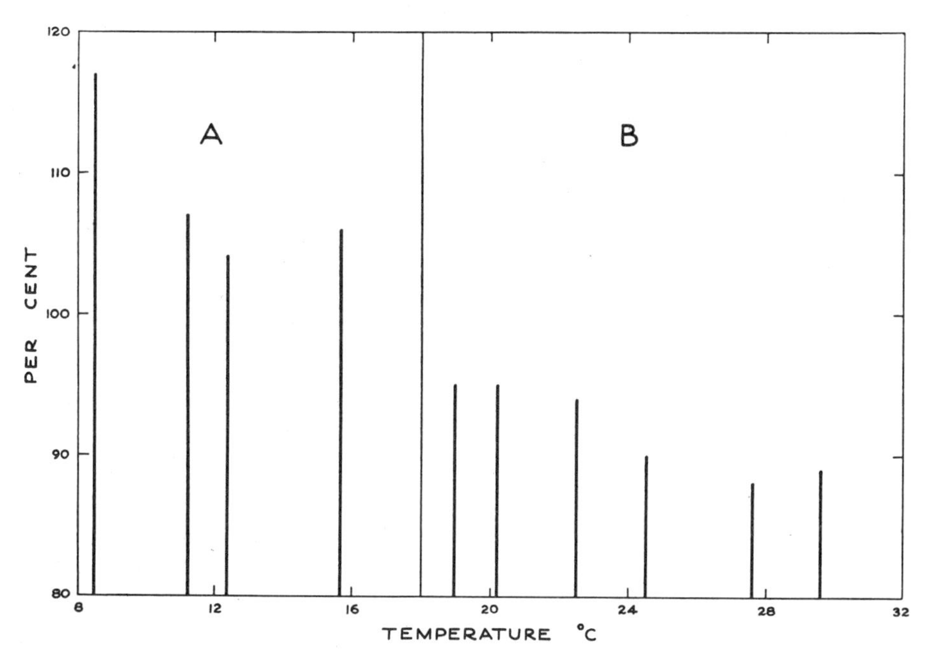

Fig. 6.16.—Showing the relative time required to complete development by "control" and "experimental" eggs of *Rana pipiens*. The ordinates show the time required for the "experimental" eggs to complete development, expressed as a percentage of the time required by the "control" eggs at the same temperature. *A:* All eggs completed the first part of their development at some temperature above 18° C. *B:* All eggs completed the early part of their development at some temperature below 18° C. Note that a weak or incipient diapause retarded the development of eggs represented in *A*, whereas the disappearance of this diapause from eggs represented in *B* caused the "experimental" eggs to develop more rapidly than the "controls." (Data from Ryan, 1941.)

But it is not correct to attribute this acceleration either to the "stimulating" influence of fluctuating temperatures or to acclimatization. It is more instructive to recognize it for what it is, namely, a manifestation of diapause. The eggs of *Melanoplus mexicanus* will serve as an example in which diapause is relatively well developed (Parker, 1930). The pupa of *Calliphora erythrocephala*, the eggs of *Locusta migratoria*, and the eggs of *Rana pipiens* are species in which diapause, as it is usually understood, does not occur. But with carefully planned experiments, making proper use of the appropriate statistical techniques, it is possible to demonstrate a small response to low temperature which is best interpreted as a "diapause-effect," i.e., it is a manifestation of a very weak or incipient diapause (Ahmad, 1936; Ryan, 1941). These four examples will now be discussed briefly.

The well-known work of Parker (1930) with eggs of *Melanoplus mexicanus* has often been quoted in textbooks and elsewhere as an example of the "stimulating" influence of low temperature. This is unfortunate, for it is indubitably an example of a weak diapause. A proportion of the eggs hatched without any exposure to low temperature; but the response to increasing temperature was quite the reverse of that usually found, the incubation period being longer, the higher the temperature. This response was reversed after a prolonged exposure to low temperature; not only was the usual increase in speed of development with increasing temperature observed, but the actual time at each temperature was also shortened (Fig. 6.15).

The pupae of *Calliphora erythrocephala* are apparently quite normal in their response to temperature in the favorable range: the data in Table 6.04, second column, would give no reason to suspect the presence of any sort of diapause. Yet when pupae were exposed to 5° C. for 8 days, the time required to complete development was reduced by about 10 per cent at each temperature except 30° (Table 6.04, third col.). The duration of the pupal period at 18° and 23° was also reduced by about the equivalent amount when pupae were exposed on alternate days to 5° and 18° or 23° (Ahmad, 1936). Ahmad considered

TABLE 6.04*

DURATION OF PUPAL PERIOD OF *Calliphora* IN DAYS AT DIFFERENT CONSTANT TEMPERATURES AND REDUCTION IN DURATION OF PUPAL PERIOD DUE TO PRELIMINARY EXPOSURE OF 8 DAYS AT 5° C.

Temperature (° C.)	Duration of Pupal Period (No Exposure to Cold) (Days)	Reduction in Pupal Period Due to Exposure of 8 Days at 5° C. (Days)
14.8	22.1	2.1
18.4	14.6	1.5
23.0	10.9	0.9
27.0	8.8	0.5
30.0	7.8	0.1

* After Ahmad (1936).

whether these results might be due to morphogenesis going on at 5° C., but rejected this possibility. There seems little doubt that here we are concerned with the same phenomenon as in the egg of *Melanoplus*, the difference being chiefly one of degree. In *Melanoplus* the diapause, though weak, is prominent enough to be recognized as a true diapause; in *Calliphora* a true diapause as such does not occur, yet the beginnings of an incipient diapause must be there, for the responses to constant temperature differ significantly from those to fluctuating temperature and in a way that is different from *Melanoplus* only in degree. In the eggs of *Locusta migratoria* the diapause-effect is even weaker than in pupae of *C. erythrocephala*, for maximal response was obtained after 1 day's exposure to 5° C.; in some circumstances, for example, with eggs that had completed 25 per cent of their development, an exposure of 4 days at 5° C. also resulted in a small but significant decrease in the time required to complete development. But continuous exposure for longer periods to 5° C. was harmful and apparently impaired the eggs' competence to develop healthily when they were returned to a favorable temperature (Ahmad, 1936).

Ryan (1941) incubated eggs of the frog *Rana pipiens* at a series of constant temperatures. "Control" batches remained at the same temperature throughout the experiment, but "experimental" batches came to these same temperatures for the latter part of their development after having completed the earlier part at either a higher or a lower temperature. The period required by the "experimental" eggs to complete a particular stage in their embryogenesis was then compared with the period required by the "control" eggs to complete the same stage. In Figure 6.16 the period required by the "experimental" eggs is expressed as a percentage of the period taken by the "control" eggs for the same stage at the same temperature (indicated on the abscissae). The results were typical of a "weak" or incipient diapause. Eggs which had completed their early development at a high temperature and were then placed at a "low" temperature developed more slowly than did the controls, which had been at the low temperature all the time. This could be interpreted as a disappearance of "diapause" from the "controls" at low temperature. When the transfer was made in the opposite direction, Ryan found that "experimental" eggs which had completed the first part of their development at "low" temperature and were then transferred to "high" temperature developed more rapidly than "controls" which had been at "high" temperature all along. Again this is just what would have been expected with a weak diapause. The "diapause" had disappeared from the "experimentals" while they had been at "low" temperature, but not from the "controls" which had been at "high" temperature all along. Also thoroughly characteristic of a weak diapause is the gradient with temperature which is indicated by Figure 6.16, *B*. The higher the temperature at which the "experimentals" were placed, the more their advantage over the "controls." This is because some diapause-development would be possible at the lower temperatures in the "high" range but not at the higher ones (sec. 4.31).

b) The harmful influence of extreme temperature is discussed in section 6.3. At present we wish to illustrate the principle that when temperature fluctuations include temperatures outside the favorable range, it is possible for temporary exposure to the extremes to harm the animal so that its normal speed of development is not maintained when it is returned to a favorable temperature. A continuous exposure of 16 days at 5° C. was sufficient to kill about half the eggs of *Locusta migratoria;* all the eggs were dead after an exposure of 32 days at 5° C. It is clear that 5° C. is outside the limits of the tolerable range for eggs of *Locusta*. Moderately long (sublethal) exposures to 5° C. result in a slowing-down of the speed of development when the eggs are returned to a favorable temperature. The degree of retardation depends upon the duration of exposure to 5° C., the stage of development of the embryo, and the particular favorable high temperature at which development subsequently proceeds. This is indicated by the data in Table 6.05, which is condensed from the more extensive

TABLE 6.05*
INFLUENCE OF EXPOSURE TO 5° C. IN RETARDING SPEED OF DEVELOPMENT OF EGGS OF *Locusta* IN RELATION TO DURATION OF EXPOSURE, STAGE OF DEVELOPMENT OF EMBRYO, AND TEMPERATURE OF INCUBATION

Incubation-Temperature.....	27° C.		30° C.		33° C.		37° C.	
Days at 5° C...............	8	16	8	16	4	8	4	8
Per cent development:								
0..................	−0.07	...	0.63†	...	0.46†	0.69†	0.57†	...
25..................	.02	0.61†	− .12	0.66†	− .22†	− .06	.17†	0.35†
50..................	0.75†	1.48†	0.70†	1.48†	0.40†	0.86†	0.76†	0.97†

* Figures in the body of the table are differences in incubation periods (in days) obtained by subtracting controls from treatments. A dagger indicates significance at $P = 0.02$. After Ahmad (1936).

data given by Ahmad (1936). The retardation in the speed of development indicated in the table must be due to the harmful influence of continuous exposure to low temperature. The temperature (5° C.) and the duration of exposure (in some instances no more than 4 days) are both moderate, judged by the standards of temperate climates; but *Locusta* is an animal of tropical and subtropical regions, where temperatures of this order are unlikely except as part of the diurnal fluctuations, that is, they are not likely to be experienced for more than a few hours at a time. High temperatures outside the favorable range may have a similar harmful influence. For example, Ludwig and Cable (1933) found that when pupae of *Drosophila melanogaster* were exposed for 1 day to 33° C., they were unable to maintain their normal speed of development when returned to 25° C. The stages most affected were male pupae which were 1, 2, or 3 days old and female pupae 1 or 2 days old.

c) With many species, particularly among the holometabolous insects, different stages of the life-cycle live at different times of the year or in different sorts of places and experience quite different ranges of temperature. It would be surprising if these stages responded similarly to temperature; the differences are often so obvious that experimental demonstration is not necessary. But with species in which all stages of the life-cycle live together in a seemingly uniform place, it is necessary to look into the matter a little more closely. For example, with the rice weevil *Calandra*, all stages may be found all the year round living together in the same handful of grain. The egg stage of *Calandra* required, on the average, 10.4 days to complete its development at 18.2° C. and 3.6 days at 29.1° C. This indicated a proportional increase of 2.84 in the duration of the egg stage as the temperature decreased from 29.1° C to 18.2°. For the same pair of temperatures the proportional increases in the duration of the larval, prepupal, and pupal stages were, respectively, 4.78, 2.06, and 3.77. These are large and significant differences (Birch, 1945*c*). The same principle may be illustrated by measuring the rate of oxygen consumption of the different stages. With larvae of *Calandra* (12 days old at 29° C.) the oxygen consumption per insect per hour increased from 3.59 cu. mm. at 15° C. to 14.95 cu. mm. at 32° C., giving a proportional increase of 4.17. For adults (7 days old at 29° C.) the relative increase for the same pair of temperatures was 5.55. With larvae of *Rhizopertha*, another inhabitant of stored grain, the oxygen consumption per insect per hour was 1.41 cu. mm. at 22° C. and 4.84 cu. mm. at 38°, indicating a proportional increase of 3.43. With adults, the proportional increase for the same pair of temperatures was 2.95. With the *Drosophila melanogaster* the duration of the egg stage at 27.8° C. was 17.8 hours, and at 18.2° C., 41.4 hours, indicating a proportional increase of 2.34. The proportional increase in the duration of the larval stage for the same pair of temperatures was 2.62 (Powsner, 1935). These examples were quoted because they all come from experiments that were done with unusual care and precision. Many others could be given which point to the same conclusion.

Eclosion from the egg and transformation to pupa or adult are but easily perceived end-points in a process which consists of many more stages than these. It may be more difficult to demonstrate that the latter also respond differently to temperature, but there is no reason to suppose that they would differ in this respect from the longer, better-defined stages that have been studied. In section 6.231, in the discussion of the logistic curve, further evidence was advanced toward this conclusion.

If, then, it is true that the successive stages in development are likely to respond differently to temperature, it must be recognized that this introduces an error which cannot be avoided. But experience suggests that this error may be small, particularly if the stages that are considered together are kept as few as practicable (sec. 6.231, especially Figs. 6.09 and 6.14 and Table 6.03).

d) We shall now consider the possibility that healthy development may proceed during short or intermittent exposures to extreme temperatures which would be harmful or perhaps lethal if experienced continuously. This is the reverse of what was discussed in paragraph *b* above. It has been investigated by Birch (1942) for eggs of *Austroicetes cruciata* and by Ludwig and Cable (1933) for pupae of *Drosophila melanogaster*. Post-diapause eggs of *Austroicetes* will develop and hatch at constant temperatures between 16° and 33° C. But at 33° the harmful influence of continuous high temperature is manifest from the flattening of the temperature-development curve about this point. When eggs were incubated on alternate days at 33° and 19° C., it was found that the rate of development at 33° C. was greater than when the eggs were exposed to this temperature continuously. The pupae of *Drosophila* developed and the adults emerged at constant temperatures between 15° and 33° C. At 10° and at 34° C. no flies emerged, although at 10° C. many puparia contained fully formed flies which had failed to emerge; a few were also found in the puparia at 34° C. When pupae were placed on alternate days at 10° and 20° C., it was estimated that about 2.4 per cent of the total development had proceeded at 10° C. Similarly, it was shown that a little development was possible during intermittent exposures to 8° but not at 7° C. The design of these experiments does not completely preclude the possibility of these results being due to a diapause-effect, as in *Calliphora* and *Locusta*, but the alternative explanation seems more likely in this case.

e) If the relationship between temperature and speed of development were linear, then the speed of development of an animal in an environment where temperature is fluctuating within the limits of the favorable range could be accurately determined by linear interpolation on the temperature-development curve, using the arithmetic mean of the fluctuating temperatures. This is, of course, the procedure followed in the classical method of "temperature-summation" (Shelford, 1927). But we have already shown in section 6.231 that the relationship between temperature and speed of development is distinctly curvilinear (Fig. 6.09). It follows that the estimates got by linear interpolation lack precision and should not be used when precise estimates of the speed of development are required.

But it does not follow that fluctuating temperature within the favorable range is not equivalent to the corresponding constant temperature. On the contrary, the little experimental work that has been done in this field indicates that fluctuating temperature within the favorable range is closely equivalent to the corresponding constant temperature; and a precise estimate of speed of development can be made by a modified temperature-summation method which takes into account the curvature of the graph (sec. 6.234). For example, in one experiment the eggs of *Austroicetes* were exposed on alternate days to the following pairs of temperatures, 19° and 31° C., 23° and 27° C., and 23° and 31° C. In every case the speed of development was the same as that found at the corresponding constant temperature (Birch, 1942). Pupae of *Drosophila* were exposed on alternate days to the following pairs of temperature, all within the favorable range, 21° and 29° C., 23° and 28° C., 15° and 25° C. In no case was there any difference from results at the corresponding constant temperature (Ludwig and Cable, 1933).

These results are typical of others that might be quoted, and it seems that, provided that the diapause-effect is not influencing the results, short-term (e.g., daily) fluctuations in temperature within the favorable range may safely be considered to be equivalent to constant temperatures in this range. This is not to say that the average speed of development is the same as that at a constant temperature equal to the mean of the fluctuating temperatures. This would imply that the relationship between temperature and speed of development is linear, which it is not (sec. 6.231). When the fluctuations include extremes outside the favorable range, it is necessary to consider the following possibilities: (*a*) that there is a diapause-effect, (*b*) that short exposures to extreme temperature may impair the animal's competence to develop at a favorable temperature, and (*c*) that healthy development may be possible during short exposures to extreme temperatures that would be harmful if the exposure were prolonged. Provided that these restrictions on the scope of data from experiments at constant temperature are recognized, there is no reason why this should not continue to be the chief method used in the laboratory to study the influence of temperature on the speed of development.

6.233 SPEED OF DEVELOPMENT AS AN ADAPTATION

We saw in section 6.1 and Figure 6.01 that the limits of the favorable range of temperature may be correlated with the range of temperature characteristic of the places where the species usually lives. A similar adaptation may be recognized in relation to the influence of temperature on the speed of development. For example, Ide (1935) studied the distribution of mayflies (Ephemeroptera) in mountain streams in Canada. Quantitative records were kept at intervals throughout the year of the mayflies living in six different "stations" along one such stream. A "station" was a situation where the mayflies usually breed, selected for its similarity to all the other "stations" in all important

TABLE 6.06*

SEGREGATION OF SPECIES OF MAYFLIES IN A MOUNTAIN STREAM IN RELATION TO TEMPERATURE

"Station"†	Temperature (° C.)	No. of Species	No. of Species Not Found Higher Up	No. of Species Not Found Lower Down
1	9.0	7	..	..
2	16.3	15	8	1
3	19.5	16	2	..
4	21.5	22	8	1
5	20.5	21	2	1
6	24.0	29	6	..

* After Ide (1935).
† Station 1 (the coolest) is near the source, and station 6 (the warmest) is farthest from the source of the stream.

respects except temperature. The temperature was lowest near the source, and the water became warmer toward the lower reaches of the stream. For example, on one typical summer day the maximal temperature of the water at the six "stations" varied from 9° for the one nearest the source to 24° C. for the one farthest from the source (Table 6.06). There were more species living in the lower (warmer) situations than in the higher (colder) ones. For example, of the 15 species normally found at station No. 2, there were 8 species which were never found in station No. 1. Similarly, 8 of the 22 species from station No. 4 were never found in any situation colder than this. There was no evidence that cold of the order experienced at any of these six situations was directly harmful to any of the species.

Hence it must be some other aspect of low temperature that is limiting the distribution of the mayflies. Now, mayflies complete one full generation each year. The adults emerge between June and August, depending on the species. They live for only a few days. The rest of the year is spent in the egg or nymphal stages in the water. Ide's researches make it quite clear that the chief adaptation enabling a species to inhabit a colder situation is its ability to develop more rapidly at low temperatures and thus to complete its life-cycle within the year in these colder situations near the source of the stream. The same

TABLE 6.07*

RELATIONSHIP BETWEEN SPEED OF DEVELOPMENT OF EGGS OF FOUR SPECIES OF *Rana* AND TEMPERATURE OF WATER IN WHICH THEY USUALLY BREED

SPECIES OF *Rana*	NORTHERN LIMIT OF DISTRIBUTION	BREEDING SEASON	USUAL TEMPERATURE OF WATER AT BREEDING TIME (° C.)	HOURS REQUIRED TO DEVELOP FROM STAGE 3 TO STAGE 20			RATIO OF COLUMN *b* TO COLUMN *c*
				12° C. (*a*)	16° C. (*b*)	22° C. (*c*)	
sylvatica	67°	Late March	10°	205	115	59	1.95
pipiens	60°	Early April	12°	320	155	72	2.15
palustris	51°–55°	Mid-April	15°		170	80	2.12
clamitans	50°	June	25°		200	87	2.29

* Moore (1942).

principle may be important in relation to the northern distribution of species, because the species living at high altitudes near the source of the stream were largely the same as those found near sea-level in high latitudes.

The same principle has been demonstrated in Amphibia (Moore, 1939, 1940*a*, 1942). The morphogenesis of the amphibian embryo may be followed through a series of well-defined stages. These were described for the frog *Rana sylvatica* by Pollister and Moore (1937). Using the time required to develop from stage 3 to stage 20 as a criterion of the speed of development, Moore compared four species of the genus *Rana* (Table 6.07). He found a close relationship between the temperature of the water in which the species normally breeds and the speed of development of the eggs at constant temperature. The eggs of species from the colder places (e.g., farther north, or earlier in the season) developed more rapidly at low temperatures than did those of species from warmer places (farther south, or later in the season). On the other hand, at moderate temperatures the increase in speed of development with rising temperature was less for the northern species. The most southern species, *R. clamitans*, was able to develop from stage 3 to stage 20 in 45 hours at 33° C.—a temperature which was lethal to the other three species.

6.234 THE SPEED OF DEVELOPMENT IN NATURE IN RELATION TO TEMPERATURES RECORDED IN THE FIELD

This matter has, for a long time, been of interest to "applied" biologists, particularly entomologists. It would be useful to be able, by means of meteorological records, to predict the date of the emergence of the spring brood of an insect pest or the duration of a generation. But the matter is not so simple as it may seem at first. The raw data from which we usually have to work consist of (*a*) reasonably precise information, derived from laboratory experiments, about the time required to complete the different stages of the life-cycle at a number of constant temperatures: (*b*) information, derived from observations in nature, about the behavior of the animal, particularly the sorts of places where it lives and the seasons of the year when the various stages of the life-cycle are present: and (*c*) records of temperature and other components of weather. For a limited investigation it may be possible to measure the temperature in the precise situations where the animal lives. But often this will be impracticable. For example, if it is desired to refer back a number of years, the standard records kept by the meteorological services will, almost certainly, be the only ones that are available.

The first step is to arrive at a satisfactory law relating temperature to the speed of development (sec. 6.231). Nearly all the early attempts to solve this problem centered around the hyperbola and its reciprocal, the straight line. Reibisch (1902) plotted the reciprocal of the time required for the development of fish eggs against temperature, drew a straight line through the observed points, and defined the point where this line cut the abscissae as the "threshold of development." Sanderson and Peairs used the same method to determine the "developmental zero" for 12 species of insects. Krogh (1914) showed that the "developmental zero" so determined was not necessarily a true threshold, since quite appreciable development might occur at temperatures well below this point. He also showed that the speed of development at higher temperatures near the upper limits of the favorable range was appreciably slower than might be expected from an extrapolation of the straight line.

But the hyperbola may be accepted as descriptive of the relationship between temperature and the time required for development within a limited range of temperature, without necessarily attributing any biological significance to a (see below). Simpson (1903) was probably one of the first to develop the concept of the "thermal constant" expressed in units of "day-degrees." This follows from the equation of the hyperbola, $K = y\,(x - a)$, where K is the thermal constant; y is time required to complete development; and, since a is the "developmental zero," $x - a$ is the "effective temperature." Since the reciprocal of the hyperbola is a straight line which cuts the abscissae at a, it also follows that when x is variable (as, for example, temperature in nature), K may still be calculated by summing $d\ (x - a)$, so long as the appropriate temperature x may be assigned to each period d. As a crude approximation, a day may be taken as the unit for d, and the mean of the daily maximal and minimal temperature may be substituted for x. This is the basis for the familiar practice of "temperature summation." This method was used in the well-known papers of Glenn (1922, 1931) on the codlin moth. He introduced a correction for the falling-off in speed of development at high temperature which had been demonstrated by Krogh, but he ignored the existence of a true threshold somewhere below the theoretical one, a. Shelford (1927), in his monumental work on the same insect, made allowance for departures from linearity at both ends of the medial range but not in the medial range itself. That is to say, the linear transformation of the hyperbola remained the chief basis of his extremely complicated computation of "developmental units." The "developmental unit" was introduced to replace the "day-degree." It made allowance for humidity as well as temperature. But the chief refinement associated with this method was due to the substitution of the hour as the unit for which temperature was measured, rather than the day. We shall show below that the use of the mean daily temperature may introduce substantial errors.

The use of the logistic equation in place of the hyperbola allows for the curvature of the temperature-development relationship throughout the whole range of temperature. The logistic equation formed the basis of a quantitative study of the rate of development of the eggs of the grasshopper *Austroicetes* in the field (Andrewartha, 1944*a*). This is an instructive example, because in it one meets a number of the difficulties that usually arise in this sort of study. (*a*) The eggs of the grasshopper are laid in the soil at a depth of about 1 inch. So they experience temperatures which are quite different from those registered in the meteorological screen. But it was necessary to use standard meteorological records because it was desired to analyze the records for 50 years back. (*b*) The only records available for the zone where the grasshoppers occurred were daily maximal and minimal air temperatures. (*c*) On numerous occasions the minimal daily temperature was lower than the threshold below which no appreciable amount of development occurs. On these occasions development was proceeding for only part of the day. So it was unrealistic and inaccurate to use the mean temperature for the full day to estimate the amount of development. (*d*) And, finally, there were no records of soil temperature in the area where the grasshopper occurred. But there was a station about 150 miles away for which records of temperature in the soil were available. So, first of all, the data for this station were analyzed as if the grasshopper eggs had been located in the soil alongside the thermometer, and then the results were interpreted in terms of the air temperatures recorded in the area where the grasshoppers occurred. The analysis is best described in four steps.

a) *Measurement of "effective" temperature.*—No appreciable development goes on in the eggs of *Austroicetes cruciata* at temperatures below about 12° C. But the daily minimal temperature of the soil at a depth of 1 inch was frequently below this. So a horizontal line was drawn on the thermograph chart, cutting off that part of the graph exceeding this temperature each day. Development was considered to have been proceeding only during that part of the day which was so delimited. The best estimate of the mean temperature for this part of the day was got by measuring the area with a planimeter and divid-

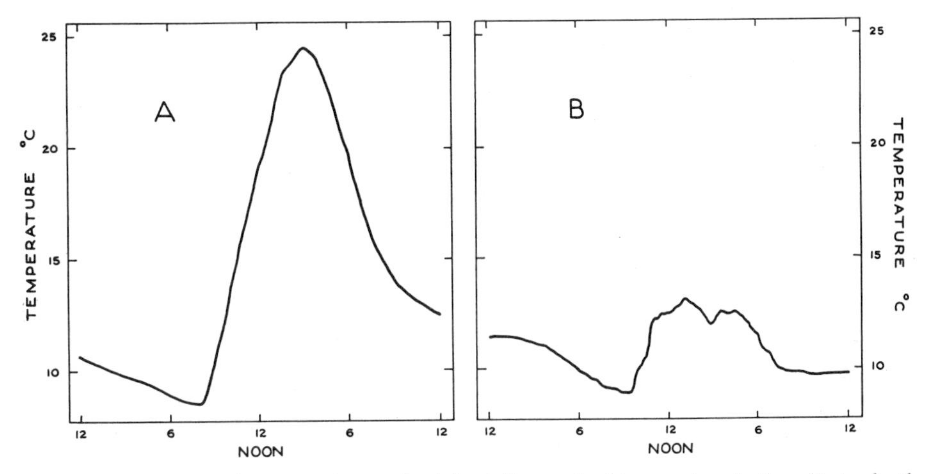

FIG. 6.17.—Showing thermograph records of the daily march of temperature in the soil at a depth of 1 inch at the Waite Institute, Adelaide. *A*, characteristic record for a bright sunny day in spring (September), and *B*, characteristic record for a dull cloudy day in winter (July).

ing this area by the length of the base line. Alternative estimates of the mean "effective" temperature were made (1) by taking the arithmetic mean of the maximum plus 12° C. and (2) by reading from the thermograph chart the maximal and minimal temperature for each 2-hourly period each day during the period when development was considered to be proceeding and taking the arithmetic mean for all these readings. The three different estimates are compared in Table 6.08. Both short-cut methods underestimated the mean "effective" temperature compared with the "true" mean temperature measured with the planimeter. But the second estimate (col. C of Table 6.08) came closer to the "true" mean temperature estimated with the planimeter than did the first. This is because the temperature did not change uniformly with time (Fig. 6.17). The first method takes much less account of the curvature and irregularities in the thermograph record.

b) Estimates of the amount of development.—The logistic curve was used to express the relationship between temperature and the speed of development (expressed as per cent of total development completed per day). The amount of development completed in a month was estimated by three different methods, and then, for purposes of comparison, this was expressed, in Table 6.09, as the mean daily speed of development for each month. (A) the unit of time was taken as 2 hours. The mean temperature for each 2-hourly period was calculated from the thermograph record. The speed of development appropriate to each 2-hourly period was read off directly from the temperature-development curve. These quantities were then averaged to give the mean speed of development for the day. (B) That part of the day during which development was considered to be proceeding was taken as the unit. The mean temperature was calculated as the maximum plus 12° C. divided by 2. The appropriate speed of development was read off the temperature-development curve. And then, since this

TABLE 6.08*
"Effective" Daily Temperature Estimated by Three Methods

Month	"Effective" Daily Temperature (° C.) 1 Inch under Soil				
	Planimeter (A)	$\frac{\text{Max.} + 12}{2}$ Daily (B)	$\frac{\text{Max.} + \text{Min.}}{2}$ 2-Hourly (C)	Difference A − B (D)	Difference A − C (E)
1938 June	14.42	14.09	14.24	+0.33	+0.18
July	15.15	14.60	15.00	+ .55	+ .15
Aug.	15.86	15.31	15.88	+ .55	− .02
Sept.	18.32	17.79	18.14	+ .53	+ .18
1939 June	15.42	14.89	15.19	+ .51	+ .21
July	14.59	14.21	14.50	+ .38	+ .09
Aug.	14.81	14.51	14.68	+ .30	+ .13
Sept.	17.69	17.02	17.52	+ .62	+ .17
Mean	....	....	....	+0.48	+0.14

* After Andrewartha (1944*a*).

rate might apply to only part of the day, the appropriate correction was applied to express the result as the daily speed of development. (C) The day was the unit. The mean temperature was calculated by taking the mean of the maximum and the minimum. The mean daily speed of development was then read off directly from the temperature-development curve.

The results of the three methods are compared in Table 6.09. Method A, which, being based on 2-hourly intervals, makes full allowance for the threshold for development and also allows for most of the curvature in both the thermograph record and the temperature-development curve, leaves little room for error and may be used as a standard by which the other two methods may be judged. Method B, which allows for the threshold but ignores the curvature both in the thermograph record and in the temperature-development curve, underestimates the amount of development by amounts ranging from 5.7 to 24.0 per cent. Method C, which ignores the threshold and the curvature of both the temperature record and the logistic curve, was, as might have been expected, grossly misleading. It underestimated the amount of development by as much as 68 per cent in one month. Neither method seems sufficiently precise.

(*c*) *The correlation between maximal temperature and amount of development per day.*—Since the mean daily temperature, even when corrected for the threshold of development, was inadequate, it was necessary to carry the analysis one step further. The "true" daily speed of development (i.e., the daily figures from which the monthly means in col. A of Table 6.09 were derived) was correlated with the daily maximal temperature of the soil at a depth of 1 inch. A close curvilinear relationship was demonstrated (Fig. 6.18). This relationship would be quite consistent if the thermograph curve were the same shape each day. The variability was associated with cloud, wind, and other meteorological events which modified the shape of the thermograph curve. The variability was, however, not great, and it was possible to use the daily

TABLE 6.09*

ESTIMATED MEAN DAILY RATE OF DEVELOPMENT OF EGGS OF *A. cruciata*.

MONTH	MEAN DAILY SPEED OF DEVELOPMENT AS PER CENT DEVELOPMENT PER DAY				
	$\frac{\text{Max.} + \text{Min.}}{2}$ 2-Hourly (A)	$\frac{\text{Max.} + 12}{2}$ Daily (B)	$\frac{\text{Max.} + \text{Min.}}{2}$ Daily (C)	Difference as Per Cent of Column A	
				B − A	C − A
1938 June	0.35	0.33	0.09	5.7	68.5
July	0.52	0.44	0.18	15.4	65.4
Aug.	0.82	0.68	0.54	17.1	34.2
Sept.	2.22	1.97	2.14	11.3	3.6
1939 June	0.60	0.54	0.51	10.0	15.0
July	0.44	0.39	0.18	11.4	59.1
Aug.	0.50	0.38	0.26	24.0	48.0
Sept.	1.80	1.47	1.65	12.8	8.4

* Column A is based on mean temperature for 2-hourly period; column B on mean "effective" temperature; and column C on mean daily temperature. After Andrewartha (1944*a*).

maximal temperature of the soil at a depth of 1 inch to get a reasonably precise estimate of the daily speed of development.

d) The relationship between maximal temperatures of soil and air.—The calculations up to this point have been based on thermograph records of temperature in the soil at a station some 150 miles away from the zone where the grasshoppers occur. In this zone no records were kept of temperature in the soil. It was necessary to relate the temperature of the air to the temperature of the soil. This was done by comparing the two at the station where they were both kept, with the hope that the same differences would recur elsewhere in the same general climatic region. This was the step where there was the greatest room for error. It would not have been necessary, had there been adequate records of the temperature in the soil available for the area where the work was to be done. But this is the sort of difficulty which is all too familiar to ecologists.

The analysis which has just been described formed part of an ecological investigation aimed at understanding the reasons for the known distribution and abundance of the grasshopper *Austroicetes*. In this particular case the estimated date of hatching of the nymphs in the spring was calculated for each year for 50 years back. In the few years in which the "expected" date of hatching could be compared with an observed date of hatching, the two estimates agreed within a few days.

The purpose of Shelford's (1927) study of the codlin moth was different. He wanted to provide a method whereby farmers might be able to anticipate the date of the emergence of the codlin moth each year, so that they might more

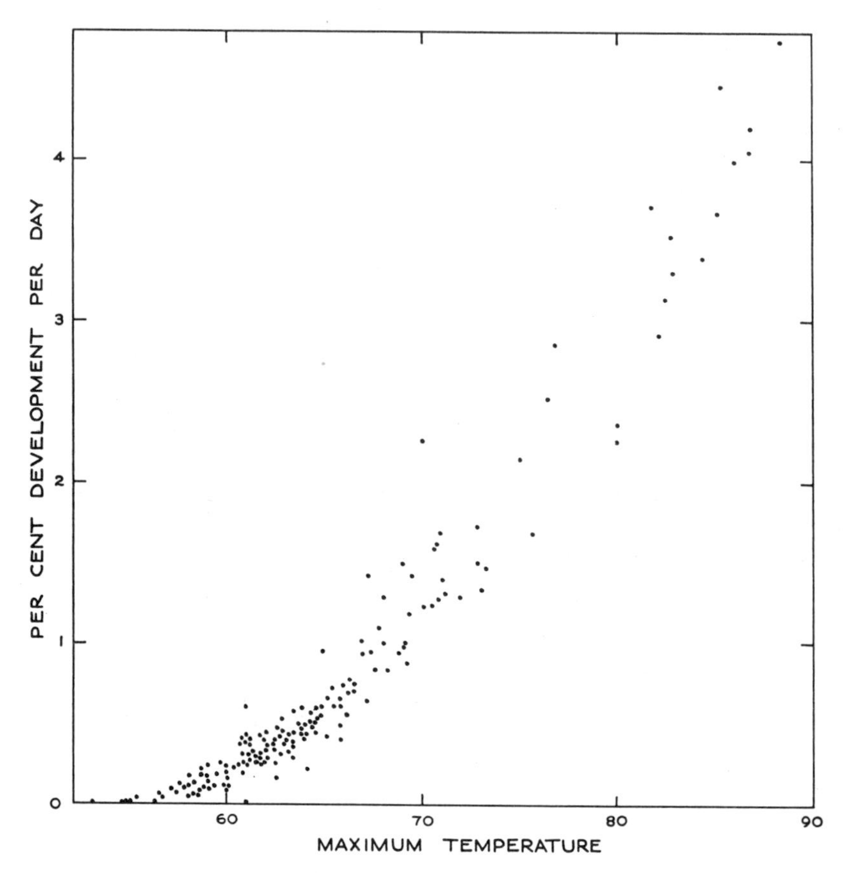

FIG. 6.18.—Showing the curvilinear relationship between maximal daily soil temperature at a depth of 1 inch and the mean daily speed of development of eggs of *Austroicetes*. The maximal temperature was read from thermograph records. The speed of development was computed for each 2-hour period throughout the day and then averaged. (After Andrewartha, 1944*a*.)

accurately time their spraying program. This is an entirely different matter. To be successful, the computations require to be highly precise. Sufficient has been said in this section to indicate just how complicated and laborious the measurements and calculations may need to be to get precise results from this sort of analysis. It may well not be worth while as a routine practical measure: in the case of the codlin moth, experience has taught that it is much better to hang a few baited traps in the orchard and time the spraying program with reference to the direct evidence provided by the appearance of moths in the traps. But for a research project, as in the case of the work with *Austroicetes*, which has just been described, we may be more interested in the past than the future, and the labor of getting the most precise estimates possible may be well worth while. It is in this connection that the principles discussed in this section have their chief interest.

6.24 *Influence of Temperature on Fecundity and the Rate of Egg-Production*

The limits of the favorable range of temperature for egg-production and oviposition are often of the same order as those for the development of the immature stages, but need not necessarily be similar. Thus *Calandra oryzae*, living in wheat of 14 per cent moisture content, is able to complete its life-cycle at temperatures between 15° and 34° C. Oviposition also occurs over this range, but so few eggs are laid at 15° and 34° that it would be more realistic to consider this process to be limited to a rather narrower range, say 17°–33° C. (Birch, 1945*c, d*). With *Microbracon hebetor*, Harries (1937) found that the development of the immature stages could be completed between 12° and 32° C. but that oviposition was virtually restricted to the range 16°–36° C., although a few eggs were laid at 14° C.

The number of eggs laid during an animal's lifetime (i.e., its fecundity) may be strongly influenced by components in the environment other than temperature, notably moisture, food, and the number of other animals present in the environment. The daily number of eggs (i.e., rate of egg-production) depends upon these things, too, and, in addition, is influenced by the age of the animal. For example, Dick (1937) studied oviposition in a number of beetles and found that the fecundity and rate of egg-production were influenced differently by age for different species; in some species copulation caused an increase in both the fecundity and the rate of egg-production; food, humidity, and in some species the frequency of suitable situations for oviposition were also important. These considerations complicate the gathering of experimental data on the influence of temperature on fecundity and rate of egg-production and also the interpretation of field observations on this subject. Birch (1945*d*) studied the oviposition of two beetles commonly found in stored grain, *Calandra oryzae* and *Rhizopertha dominica*, in a carefully controlled environment and was able to measure not only the direct influence of temperature but also its interaction with the moisture content of the food, the number of other animals in the environment, and the age of the ovipositing females (secs. 3.12, 3.4, and 9.222). The influence of temperature in relation to the number of other animals of the same sort may be seen in the information set out in Table 6.10. The number of eggs laid reached a maximum at a median temperature of 26° C. for the sparser population and 29° for the denser one. At temperatures above and below this, fewer eggs were produced. With *Thrips imaginis*, very few eggs were laid at 8°, but within the range 13°–23° C. there was a tendency for more eggs to be laid, the higher the temperature. Owing to the variability of the data, these differences were nonsignificant, but it is likely that with larger samples the trend could be established more precisely (Andrewartha, 1935). Similarly with the European cornborer *Pyrausta nubilalis*, Vance (1949) found that the moths laid more eggs at 29° than at 21° or 32° C. (Table 6.11).

TABLE 6.10*

TOTAL NUMBER OF EGGS LAID BY *Calandra oryzae* IN WHEAT OF 14 PER CENT MOISTURE CONTENT AT DIFFERENT TEMPERATURES

Temperature (° C.)	1 Insect per 10 Grains	1 Insect per 50 Grains
23	264	266
26	265	384
29	296	344
32	197	...
35	5	...

* After Birch (1945*d*).

The aphid *Toxoptera graminum* is ovoviviparous. Young are produced throughout adult life, but rather more rapidly by the young adult than by the older one. The total number produced during the lifetime of one female depended on the temperature at which it was living (Wadley, 1931). Figure 6.19 illustrates this relationship. With both the winged and the wingless form of

TABLE 6.11*

INFLUENCE OF TEMPERATURE ON NUMBER OF EGGS LAID BY *Pyrausta* KEPT AT 96 PER CENT RELATIVE HUMIDITY AND OFFERED WATER TO DRINK

Temperature (° C.)	No. Fertile Eggs per Moth
21	708
25	758
29	823
32	533

* After Vance (1949).

Toxoptera, the shape of the curve relating temperature to fecundity is the same, but that for the winged form is distributed over a lower range of temperature than that for the wingless form.

These four examples are typical of others that could be quoted. They indicate that the fecundity of insects tends toward a maximum at a moderately high temperature, which may be relatively near the upper limit of the favorable range. As the upper or lower limit of the favorable range is approached, the fecundity declines abruptly (Fig. 6.19).

The influence of temperature on the rate of egg-production by insects (as distinct from the total eggs produced during a lifetime) has been reviewed by Harries (1939). The seven examples which he used consistently indicated the same trend. Below a certain well-defined lower limit the rate of egg-production was virtually zero. As temperature increased, the rate of egg-production increased up to a maximum and then decreased at still higher temperatures, until the upper limiting temperature was reached, when the rate again became zero. For the temperature range within which the rate of egg-production was increasing, the trend with temperature followed a sigmoid course (Fig. 6.20). Norris (1933) found that most females of *Ephestia kühniella*, which had been mated with one male only, produced fertile eggs at 27° and at 30° C., provided that the prepupae and pupae of both males and females had been reared at a

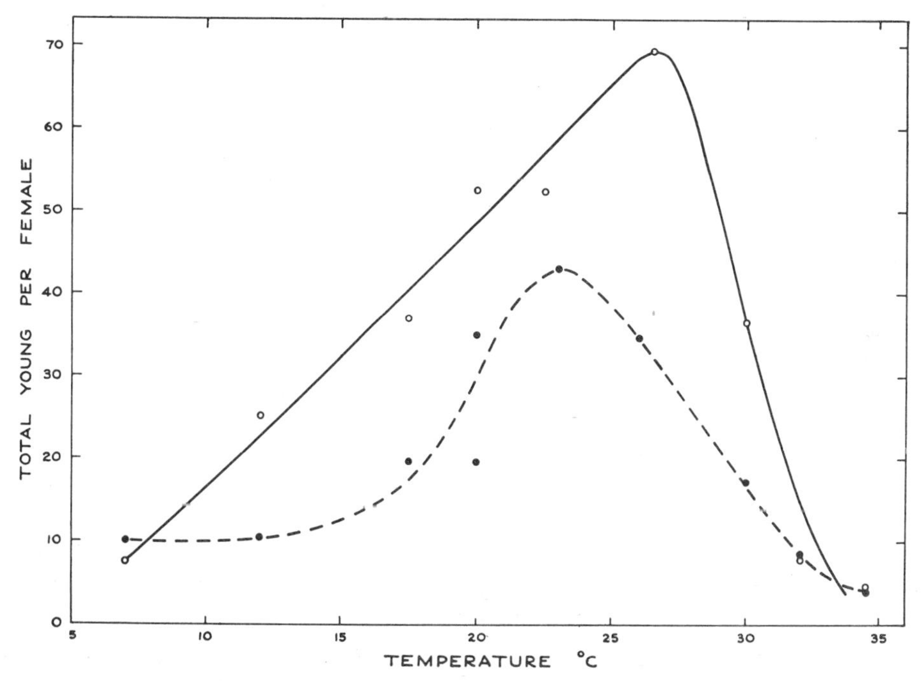

FIG. 6.19.—Showing the influence of temperature on the fecundity of the aphid *Toxoptera graminum*. *Complete line*, the winged form; *broken line*, the wingless form. (Data from Wadley, 1931.)

favorable temperature lower than 27° C. As the temperature at which the prepupae and pupae were reared increased from 27° to 30° C., the proportion of females which laid no fertile egg increased from 10–59 per cent to 96–100 per cent. There was a certain variation between "cultures," but these experiments showed that exposure to unfavorably high temperature during the prepupal and pupal stages made the adults sterile.

The results from experiments with constant temperatures may not necessarily be equivalent to what happens with fluctuating temperatures in nature. There is evidence that, at least in some circumstances, a short period at low temperature may result in an increased rate of egg-production on return to a higher temperature (Dick, 1937). Figure 6.21 illustrates this for *Tribolium confusum*. After 100 days at a constant temperature of 27° C., the rate of egg-production had fallen from an initial rate of more than 8 per day to about 3 per day. On the 101st day the insects were placed at 18° C. and kept at this temperature for 8 days. On being returned to 27° C. on the 109th day, the insects began producing eggs more rapidly, quickly reaching a peak of about 5 eggs per day, which was nearly double the rate for the period just before they had been placed at low temperature. At a constant temperature of 27° C. the rate began to decline again but required about 70 days to decrease to the origi-

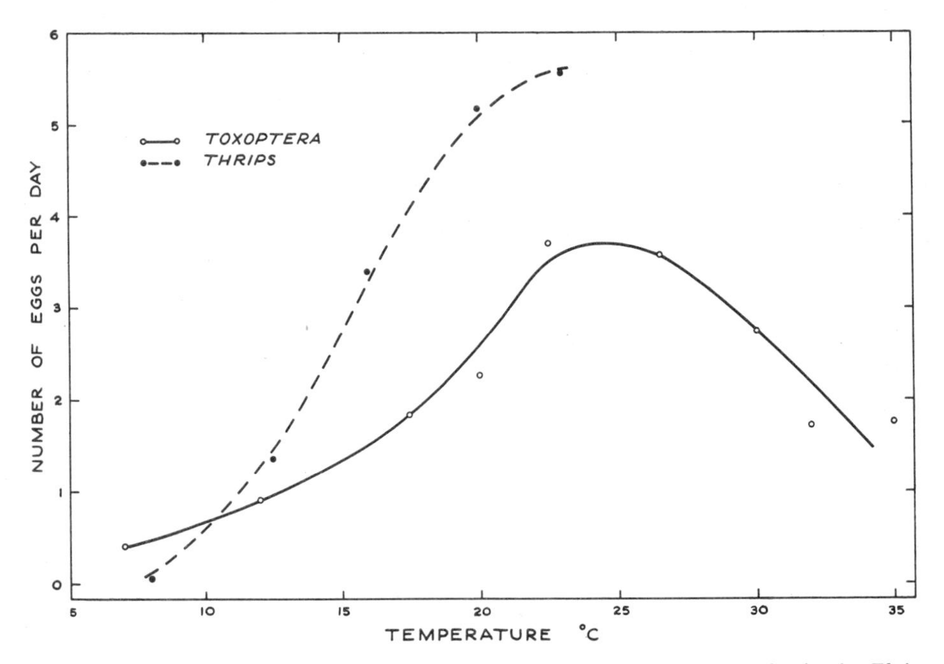

FIG. 6.20—Showing the relationship between temperature and speed of egg-production for *Thrips imaginis* and *Toxoptera graminum*. (Data from Wadley, 1931; and Andrewartha, 1935.)

nal figure of 3 eggs per day. There can be no doubt that in this case a short exposure to 18° after the insects had been living at 27° for 100 days caused a marked increase in the rate of egg-production at 27° and that this increase was sustained for at least 60 days.

During a 10-year study of the European cornborer *Pyrausta nubilalis*, in the field in Ontario, Stirret (1938) accumulated a great volume of quantitative data on the flight activity, fecundity, and oviposition-rate of this species. The results of this study of a natural population confirmed the results of the laboratory experiments discussed in this section. The fecundity was estimated by counting the numbers of eggs laid each night on a sample plot of 100 plants of corn, and at the same time the number of moths flying in the plot was estimated independently. The average number of eggs per moth varied from 2.5 in 1928 to 15.2 in 1930. There was an apparent relationship with temperature, which suggested that most eggs were laid in those years when the mean daily air temperature during the flight period was about 23° C. and the atmospheric humidity was equivalent to a saturation deficiency of about 7 mm. of mercury.

The males of some mammals become sterile when they are exposed continuously to high temperature. Gunn, Sanders, and Granger (1942) found that in certain hot districts in western New South Wales rams were sterile during the hot season. The semen began to degenerate about 4 weeks after the date

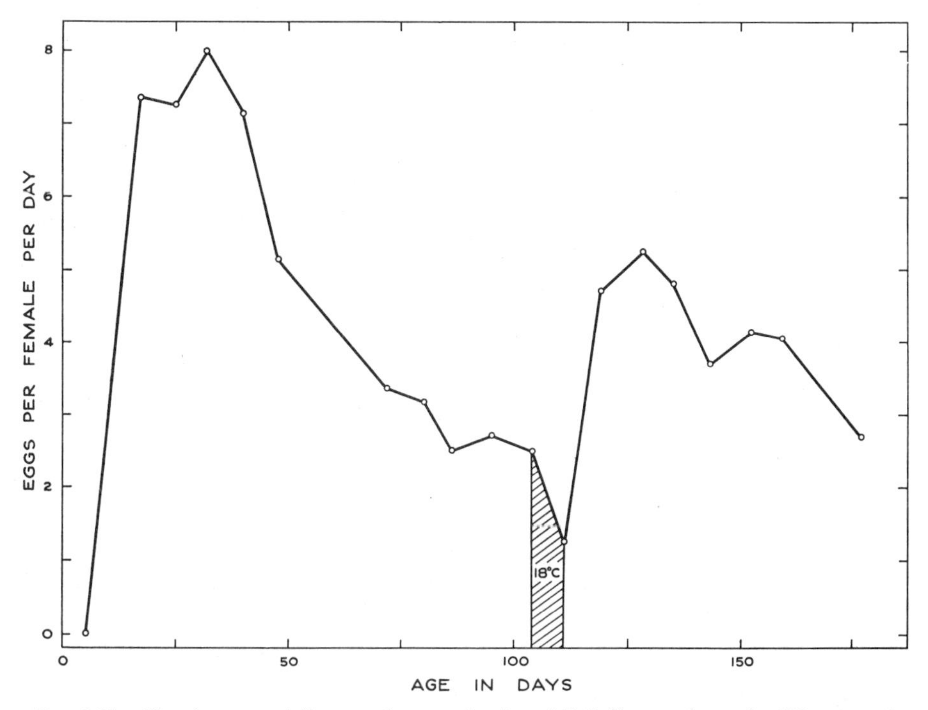

FIG. 6.21.—Showing mean daily rate of egg-production of *Tribolium confusum* for 180 consecutive days. The insects were at 27° C. for all the time except for 8 days from the 101st to the 108th day. Note sharp and sustained increase in rate of egg-production after 109th day. (Data from Dick, 1937.)

when the daily maximal temperature began consistently to exceed 32° C. Testes which had degenerated became normally active again about 2 months after the hot season had ended. By keeping rams in hot rooms and by insulating the testes during the cool season, Gunn, Sanders, and Granger showed that temperature was the chief cause of the sterile condition. Moore and Quick (1924) and Cowles (1945) got similar results with other species of mammals by heating the testes artificially or by transplanting them into the body. Cowles (1945) suggested that the habit of mating in the spring, which is commonly found in birds and other warm-blooded animals which live in the temperate zone, might be an adaptation which is associated with the tendency for the males to become sterile during hot weather.

6.3 THE LETHAL INFLUENCE OF TEMPERATURE OUTSIDE THE FAVORABLE RANGE

In the sense in which we have so far been using it, the favorable zone is delimited by the temperatures above and below which growth and multiplication cease to be possible, even though the animal may live for a long time at these temperatures. But the favorable zone may be considered in another way, in which the emphasis is placed on the survival of the *individuals* rather than on the continuance of the *population*. In sections 6.31 and 6.32 we discuss certain experiments in which animals were exposed to lethal low or lethal high temperatures.

6.31 *The Lethal Influence of Low Temperature*

The lethal influence of low temperature has been the subject of much study, especially by entomologists in Europe and North America; for the survival-rate during winter, together with the multiplication during summer, may largely determine the numbers of a certain pest in a certain area. Species from temperate climates often have a stage in the life-cycle which is especially adapted to survive during exposure to extreme cold: and these are said to be "cold-hardy." On the other hand, with species from tropical or subtropical climates, all stages in the life-cycle and, with species from temperate climates, the stages of the life-cycle that are usually present during summer may lack the capacity for becoming dormant when the temperature falls below that which favors active development. These forms are, as a rule, readily killed by exposure to moderate temperature, often well above 0° C.

6.311 THE THEORY OF "QUANTITY FACTOR" AND "INTENSITY FACTOR"

Because the animal may have to be exposed to moderately low temperature for a considerable period before it dies, ecologists sometimes speak of the "quantity factor" as distinct from the "intensity factor," which refers to the instantaneous killing of a cold-hardy individual by the freezing of its tissues (Payne, 1926*a*, *b*, 1927*a*). Payne coined these terms after studying the lethal influence of cold on representative insects from three groups which in nature are exposed to very different levels of cold. The oakborers were taken to represent those which are exposed during winter to temperatures many degrees below 0° C.; certain aquatic forms were taken to represent those that might experience temperatures as low as 0° C. but not below; and certain insects, usually considered to be of subtropical origin, which commonly infest stored grain, served as examples of the group which is used to living in situations where the temperature is usually well above 0° C. The "quantity factor" refers chiefly to the lethal influence of "low" temperatures (usually above 0° C.) on this third group: the term implies that the exposure to "cold" requires to be maintained for a considerable period—days or weeks. It is well known that representatives of the first two groups may survive unharmed for many months at these temperatures. On the other hand, most of them die instantly when their tissues freeze, as, of course, do the members of the third group also. The temperature required to bring about this freezing of the tissues is the measure of the "intensity factor" (see below, next section).

This theory seems to have dominated the outlook of most students of this subject up to the present time. See, for example, the summaries published by Uvarov (1931), Salt (1936), Mellanby (1939), Wigglesworth (1939, p. 364), and Luyet and Gehenio (1940). But, as Salt (1936, 1950) pointed out, the division of insects into those which respond to the "quantity factor" and those which respond to the "intensity factor," which is implicit not only in the original theory but also in most of the later work based on it, is unreal. Representatives of the "nonresistant" group (e.g., *Ephestia*) die after very short (virtually instantaneous, e.g., 10-minute) exposures to nonfreezing temperatures below 0° C. On the other hand, representatives of the "resistant" group may also succumb to exposure to temperatures that do not result in the freezing of their tissues—provided that the temperature is low enough and the exposure long enough. Moreover, the temperature required to produce freezing depends upon the duration of exposure; it is not, as has hitherto been widely assumed, independent of time. As a consequence, it seems that the conventional "undercooling point," got by rapidly cooling the animal until a "rebound" is observed (see below), indicates only the temperature required to bring about freezing after an instantaneous exposure; it gives little information about the time-temperature responses of the animal to cold. It is clear that the distinction between "quantity factor" and "intensity factor" is not so simple as it has seemed.

The consequences for ecology of the long predominance of this too simple hypothesis is that the study of the lethal influence of nonfreezing temperatures has unfortunately been largely restricted to temperatures above 0° C. And the information gathered about the lethal influence of freezing temperatures has been unnecessarily abstract, dealing with the temperature required to produce instantaneous freezing, but ignoring for the most part the interaction between temperature and time which is an important feature of nature. It is necessary to consider the following principles: (*a*) Some sorts of animals may continue to live even after their tissues have been frozen; nevertheless, they succumb to adequately prolonged exposure to adequately low temperature. (*b*) All other sorts die instantly when their tissues freeze, the time and temperature required to produce freezing varying widely. (*c*) Many, perhaps most, sorts can be killed by adequate exposure to "low" temperatures which are not low enough to freeze the tissues; but the low temperature above which the animal may live indefinitely varies enormously; it may be well above 0° C. for some and well below 0° for others. (*d*) In all three categories the death-rate depends upon both time and temperature. In the sections which follow we give a few selected examples to illustrate these principles.

6.312 THE LETHAL INFLUENCE OF NONFREEZING TEMPERATURES

It may reasonably be supposed that for any particular individual there is a precise limit to the time during which it may survive exposure to a particular lethal temperature. If we could observe this limit directly for a sufficiently large number of individuals drawn at random from the population which we choose to study, we would be in a position to calculate directly, in the ordinary way, the mean duration of the fatal exposure and ascribe a variance to it. These two statistics would adequately describe the lethal influence of temperature on the population under study. But unfortunately, especially with high temperature and nonfreezing lethal low temperature, we usually have no way of telling by direct observation the precise moment when death has been irrevocably determined. It may be necessary to wait for a day or two, sometimes much longer, after the exposure to lethal temperature has been terminated and the animals returned to a favorable temperature, in order to see which ones will die (sec. 6.323). In this case the statistical method known as "probit analysis" is appropriate.

If the problem is to measure the duration of exposure to a certain temperature which is required to kill any proportion of the population, we begin by exposing a number, say six samples of 50 or more animals each, to this temperature for varying periods. The periods are chosen with the expectation that a small proportion of the animals will die as a result of the shortest exposure and that the longest exposure will kill nearly all those experiencing it. Alternatively, the problem may be to measure the particular temperature required to kill any proportion of the animals when they are exposed to it for a certain period. Then the procedure is varied to the extent that all the samples are exposed for the same period but to a range of temperatures chosen with the expectation that few will die as a result of exposure to the highest temperature and that exposure to the lowest temperature will kill most of them. In either case the animals are then removed to a favorable temperature. After a suitable interval, say several days or longer, the number of deaths in each sample is recorded.

If these numbers, expressed as proportions, are plotted against the treatment (or dosage), they will fall along a sigmoid curve, which is the familiar "dosage-mortality" curve of the toxicologist. This curve is the one that can also be obtained by progressively summing the ordinates of the ordinary frequency polygon. Provided that the frequency-distribution of deaths against "dosage" is normal or nearly so, the sigmoid "dosage-mortality" curve becomes a straight line when the proportions are transformed to probits (Finney, 1947).

If the distribution is not normal, it may usually be made so by transforming the dosage to logarithms or some other appropriate scale.

Once the straight line has been formed, it is easy to read off, by linear interpolation, the dosage (temperature or duration of exposure) which would be expected to kill any proportion of the population. Usually the most instructive figure is the particular dosage required to kill 50 per cent of the population. Strictly, this measures the median of the distribution, but it is also a good estimate of the mean. This is equivalent to the mean we would calculate if we could write down by direct inspection of each individual the duration of exposure just required to kill it. We may sometimes wish to know the dosage required to kill some other proportion of the population. There is no difficulty in this, but it must be remembered that, unless the size of the samples is increased, the estimates will be less precise. To estimate the dosage required to kill 95 per cent the samples must be 3 times, and for 99 per cent 10 times, as large as those required to estimate, with the same degree of accuracy, the dosage required to kill 50 per cent.

The "probit" analysis has become standard practice in "applied" laboratories where insecticides or drugs are being tested. Unfortunately, in the field of "fundamental" ecology, where it would often be equally appropriate, the method has hardly been used, presumably because far too many biologists have been unfamiliar with the statistical procedures which enable precise methods to be used in this subject. The methodological mistakes which recur most frequently in the literature are: (*a*) the animals have been kept at the unfavorable temperature until they are visibly "dead," notwithstanding the fact that death must have been determined long before it could be detected in this way; (*b*) we are told the duration of the exposure required to kill all the animals in the sample. This depends on the size of the sample and is not a statistic that can be measured precisely by any method of sampling short of including the whole population in the sample. The probit method enables the investigator to extract most information from his results. Failure to design experiments so that the method can be applied leaves the reader groping for the little bit of information that can be got. It is very little extra trouble to design suitable experiments, and the subsequent analysis is not difficult. It is fully set out in Finney (1947).

The bedbug *Cimex* may be taken as an example of a species which is not capable of becoming dormant at temperatures below the favorable range for development. The influence of nonfreezing temperatures down to 1° C. on the survival-rate of eggs of *Cimex* has been discussed in detail by Johnson (1940). This is a paper which should be studied in the original by the student who is interested in methodology. Using the probit transformation, Johnson was able to make exact comparisons between the survival-rates of eggs of different ages and histories, exposed to various low temperatures at several different levels of humidity.

In one series of experiments, eggs of a uniform age (0–24 hours at 23° C.) were exposed to constant temperatures between 1° and 12° C. for periods ranging from 7 to 50 days. At the conclusion of each exposure the eggs were removed from the low temperature and incubated at 23° C. Failure to hatch was taken as the criterion for death. A typical experiment is set out in Table 6.12, which shows the results when eggs were exposed in a moist atmosphere (saturation deficit about 1 mm. of Hg) at 4.2°, 9.8°, and 11.7° C. The mortality is expressed both as a per cent and as a probit. The former is transformed to the latter simply by reference to the appropriate table (in Finney, 1947) or Fisher and Yates (1948). Table 6.12 also includes the estimated values of *b*, L.D. 50, and L.D. 99.99. The first is the coefficient for linear regression of probit on exposure time, i.e., it measures the slope of the probit line. The second and third are the estimated durations of exposure corresponding to death-rates of 50 and 99.99 per cent, respectively. The variances for *b* and L.D. 50 are also included because these quantities are necessary in the calculation of the statistical significance of the results.

TABLE 6.12*
DEATH-RATE OF EGGS OF *Cimex* EXPOSED FOR VARIOUS PERIODS TO 4.2°, 9.8°, AND 11.7° C. IN MOIST AIR

4.2° C.			9.8° C.			11.7° C.		
Exposure (Days)	Death-Rate		Exposure (Days)	Death-Rate		Exposure (Days)	Death-Rate	
	Per Cent	Probit		Per Cent	Probit		Per Cent	Probit
9	4.2	3.27	7	10.1	3.72	7	9.4	3.7
16	16.9	4.04	14	7.5	3.56	14	9.7	3.7
23	52.5	5.06	21	19.9	4.15	21	36.7	4.7
30	86.3	6.09	28	65.8	5.41	29	59.2	5.2
35	94.5	6.60	35	86.3	6.09	35	69.4	5.5
40	91.8	6.39	42	100.0	8.72	42	90.7	6.3
44	94.5	6.60				46	94.0	6.6
50	100.0	8.72				49	96.2	6.8
b	0.1139 (0.000151)†			0.1273 (0.000241)			0.0800 (0.000058)	
L.D. 50	22.9 (1.120)			25.8 (0.774)			26.7 (1.409)	
L.D. 99.99	55.6			55.0			73.3	

* After Johnson (1940).
† Figures in parentheses are variances.

Figure 6.22, *A*, shows the death-rate, in per cent, plotted against duration of exposure, and Figure 6.22, *B*, shows the same data plotted as probits. The advantages of transforming the data to the scale in which the regression becomes linear are that it enables a precise estimate of the exposure time required to kill any specified proportion of the eggs; and it makes it possible to attribute a significance to the observed differences. Figure 6.22, *B*, indicates that the probit lines for 4.2° and 9.8° C. are nearly parallel and that both slope more steeply than the probit line for 11.7° C. On the other hand, if we compare the co-ordinates on the abscissae for probit 5 (ie., the value for L.D. 50), we find that 9.8° and 11.7° C. come close together, whereas 4.2° is well away to the left. These two statistics, namely, L.D. 50 and b, are the two most instructive ones to extract from the probit line. The former is the most reliable

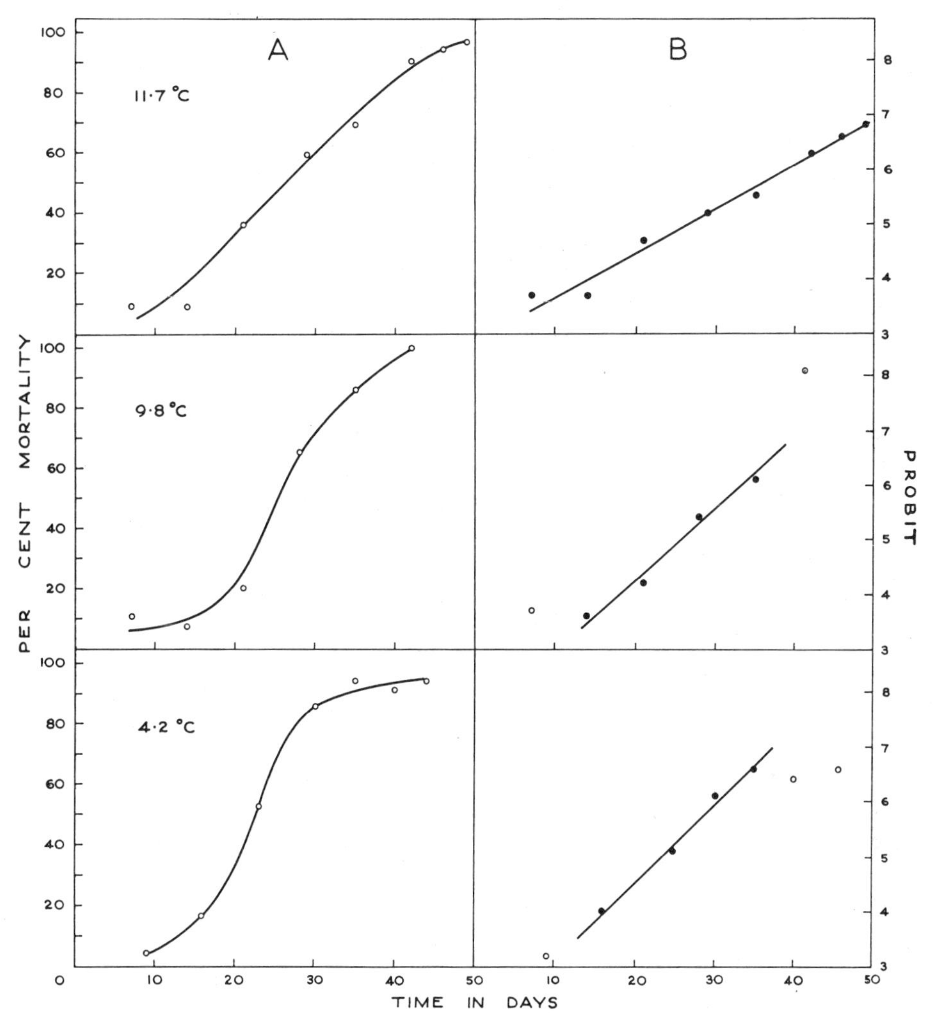

FIG. 6.22.—Showing the death-rate for eggs of *Cimex* exposed to 4.2°, 9.8°, or 11.7° C. for a varying number of days. *A*, the per cent mortality is plotted against time, giving a sigmoid curve, *B*, by transforming per cent to probit, the sigmoid curve becomes a straight line. (Data from Johnson, 1940.)

measure of the animals' capacity to survive at a given temperature, and the latter is the best comprehensive measure of the relative toxicity of short and long exposures. A large value for b, i.e., a steep slope to the probit line, indicates that there is relatively little "scatter" between the exposures that kill few and many; conversely, a small value for b, i.e., a gently sloping probit line, indicates a wide margin between the exposures required to produce these extremes in the death-rate.

It will be instructive to see what measure of significance may be attributed to the differences which we have pointed out above. Table 6.13 sets out the appropriate statistics. The difference between the regression coefficients for 4.2° and 9.8° C. is less than its standard deviation; hence, clearly, we may consider these lines to be parallel. But the difference between the regression co-

TABLE 6.13
SIGNIFICANCES OF DIFFERENCES IN VALUES OF b AND L.D. 50 GIVEN IN TABLE 6.12

Comparisons	Differences	S.D.	t	P
b for 9.8° C. − b for 4.2° C....................	0.0164	0.098	. .	. . .
b for 11.7° C. − b for 4.2° C....................	0.0339	0.0145	2.3	<0.05
L.D. 50 for 11.7° C. − L.D. 50 for 4.2° C.......	3.8	1.590	2.4	$<.05$
L.D. 50 for 9.8° C. − L.D. 50 for 4.2° C........	2.9	1.376	2.1	$<0.1>0.05$

efficients for 11.7° and 4.2° C. exceeds its standard deviation 2.3 times. Such a difference would have occurred by chance in less than 5 per cent of tests; hence we may consider that these lines differ in slope. This means that the range in the extreme exposure times which, on the one hand, just allowed all the eggs to survive and, on the other hand, was just sufficient to kill them all was the same at 4.2° as at 9.8° C., but was narrower at both these temperatures than at 11.7°. The wider range at 11.7° may be associated with the fact that some development is possible at 11.7° (although the eggs cannot hatch at this temperature) and the fact that the survival-rate depends upon the stage of development of the eggs (see below).

The difference between L.D. 50 for 11.7° and 4.2° C. exceeds its standard deviation in the ratio 1:2.4; with 12 degrees of freedom, this gives a probability of something less than 5 per cent that this difference has occurred by chance, and it may reasonably be concluded that eggs die more rapidly at 4.2° than at 11.7°. The difference between L.D. 50 for 9.8° and 4.2° exceeds its standard deviation in the ratio 1:2.1; a difference of this magnitude might have occurred by chance about 7 times in a hundred trials; and if we had to rely on this evidence alone, we should be cautious about accepting this difference as proof that 9.8° C. is more lethal than 4.2° C. In fact, Figure 6.23 shows that the curve is relatively steep between these temperatures, indicating that the death-rate is changing rather rapidly with respect to temperature.

The variances for L.D. 99.99 (i.e., the estimated exposure required to kill all the eggs) are not given in Table 6.12. They would inevitably be much larger than those for L.D. 50, and it may well be that none of the values given in Table 6.12 for L.D. 99.99 would differ significantly from any other. Although the L.D. 50 (i.e., the exposure required to kill half the animals in the sample) is the statistic that can be estimated most reliably; nevertheless, the estimate of L.D. 99.99 derived from the probit line is better than any that can be obtained by attempting to measure empirically the exposure required to kill all the animals. The chief reason for this is that in the empirical determination the exposure depends not only on the temperature but also on the size of the sample.

When the temperature was reduced below 4° C., a much shorter exposure was fatal. The time required to kill 50 per cent was shortest at the lowest temperature tested (namely 1.1° C.); it increased rather steeply up to 7° C.;

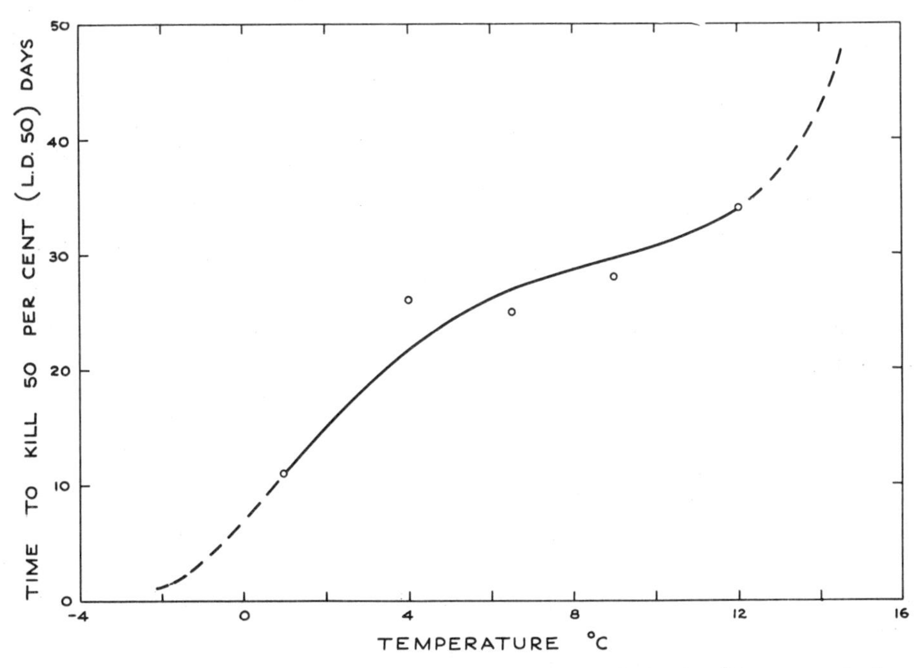

Fig. 6.23.—Showing the relationship between temperature and the time of exposure in days required to kill half the individuals in a sample of eggs of *Cimex*. The curve obviously approaches infinity at the lower incipient lethal temperature and approaches zero at some low temperature which was not measured. The broken lines are extrapolations which illustrate the limits toward which the curve is tending. (Data from Johnson, 1940.)

above 7° the curve flattened somewhat (Fig. 6.23). Since the eggs can hatch with less than 50 per cent mortality at temperatures above 15° C., this may reasonably be taken to be about the incipient lethal low temperature (i.e., the temperature above which the animal may live indefinitely without being killed by temperature); and in extrapolating the curve in Figure 6.23 we have drawn it asymptotic to 15° C. Johnson did not use temperatures lower than 1.1° C., so that the extrapolation of the curve at its lower end is largely hypothetical, but it is based on empirical data for other species.

The tissues of most insects probably have a freezing point in the range −1° to −2° C. If a batch of animals is cooled below the freezing point of the tissues, some will freeze and some will not; this is related to their capacity for "undercooling," which we discuss below. If they are of the sort which cannot withstand freezing, all those that freeze will die. But some of those which have not been frozen will have been fatally injured if the exposure to low temperature has been long enough. There are indications from the literature (e.g., Nagel and Shepard, 1934; Salt, 1936) that animals which lack the capacity to become dormant may be killed without being frozen by as little as 10–20 minutes' exposure to temperatures of the order of −10° C. For example, Salt (1936) showed that an exposure of 15 minutes to −15° C. could be fatal to adults of the beetle *Tribolium*, even though their tissues had not been frozen. About 75 per cent of the beetles died as a result of this exposure; of these, about 55 per cent had been frozen, leaving about 20 per cent that had definitely been killed without being frozen. To this figure must be added a proportion, which cannot be determined, of the 55 per cent which had accumulated a lethal dosage of cold before freezing. We do not know how many of these there might be, but it would certainly seem safe to count more than 20 per cent, perhaps 30 per cent, mortality due to the lethal influence of nonfreezing temperature during an exposure of 15 minutes to −15° C. Nagel and Shepard (1934) estimated that about half the eggs of *Tribolium* died from an exposure of 45 minutes to −6° C.; with sixth-instar larvae, half died after 30 minutes at −12° C. and 12 hours at −4° C. These figures overestimate the number of deaths due to nonfreezing temperatures, because at these temperatures some of the insects must have undercooled and frozen, and all of these may not have died without being frozen.

It seems reasonably certain that the L.D. 50 (due to nonfreezing temperature) continues to decrease with decreasing temperature and, at some low temperature, approaches zero. We have therefore extrapolated the lower part of the curve in Figure 6.23 on these grounds and have made it approach closely to zero. There need be little doubt that a generalized curve of this general shape describes the relationship between low temperature and the death-rate due to lethal effects other than freezing for the large class of animals which lack the capacity for dormancy when exposed to temperatures below the favorable range for development. The curve may vary in position and in other details for each particular sort of animal; also for different stages of the life-cycle of the same animal; or for the same stage which has been differently acclimatized or has in some other way had a different history.

Thus Johnson (1940) found that the survival-rate of the eggs of *Cimex* exposed to low temperature depended on the age of the egg. When eggs were exposed to 7.7° C. in moist air, there was little difference between eggs that were from 1 to 4 days old at 23° C. But after the fourth day they became more susceptible to cold as they became older (Fig. 6.24).

Johnson was unable to demonstrate any results from acclimatization. Eggs which had been produced and laid at 15°, 18°, and 23° C. were exposed to 10°

in moist air; the time required to kill 50 per cent was the same for all three lots. Robinson (1928) tried to acclimatize adults of *Calandra* by exposing them to a series of descending temperatures: 72 hours at 10°, 65 hours at 7.2°, 48 hours at 4.4°, and 36 hours at 1.6° C. These beetles were then exposed to −1.1° alongside controls which came to the low temperature directly from 23°. The time required to kill about 50 per cent of the controls was 90 hours, compared to 28 hours for those that had been "acclimatized." This result might have been

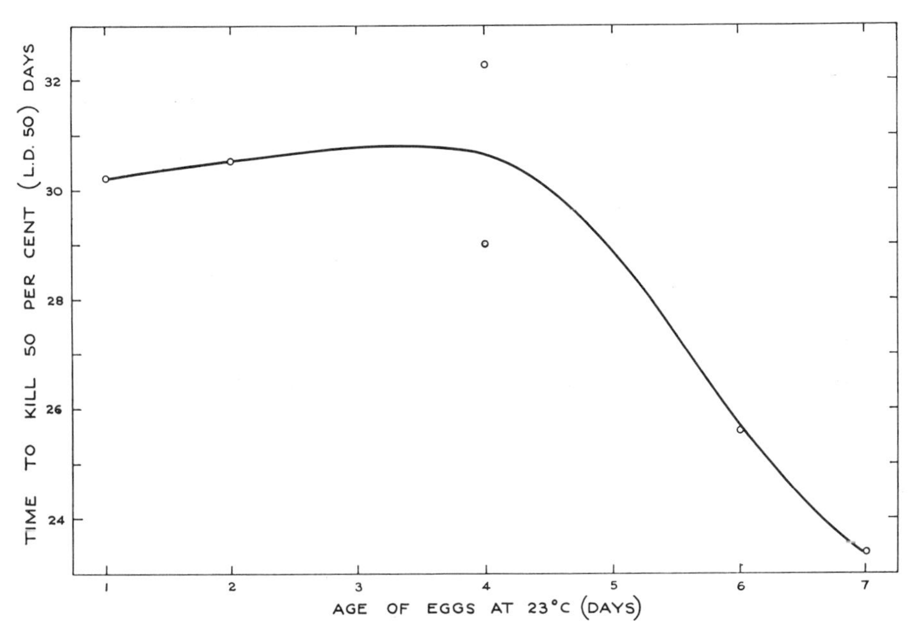

Fig. 6.24.—Showing the way in which the stage of the developing embryo influenced the survival time for eggs of *Cimex* held at 7.7° C. and 90 per cent relative humidity. (Data from Johnson, 1940.)

anticipated, for the "acclimatization" temperatures were below the incipient lethal low temperature for *Calandra*. During their so-called "acclimatization" the beetles were actually experiencing lethal temperatures and were accumulating a sublethal dose of cold; they died in a shorter time at −1.1° C. because they had already been "part-killed." This experiment is often quoted as proof that *Calandra* (and perhaps other similar animals) cannot be acclimatized. The temperatures at which it would be reasonable to acclimatize *Calandra* and other animals lacking the capacity for dormancy are those which lie just inside the favorable zone for development. So far as we know, this experiment has not been done, and we can not yet say whether this sort of animal does or does not possess the capacity to become acclimatized.

Most of the common insect pests of stored products are usually considered to have originated in tropical or subtropical zones. All those that have been studied appear to lack the capacity to become dormant in any stage of the life-cycle; they are therefore unable to survive exposure to temperatures below the range that is favorable for development. For example, Nagel and Shepard (1934) estimated the time required to kill 50 per cent of several stages of *Tribolium*. Their results are summarized in Table 6.14.

Robinson (1928) found that about half the sample of adult beetles of *Calandra oryzae* died after an exposure of 8 days at 7.2°, 4 days at 1.6°, or 2 days at −1.1° C. With adult beetles of *Oryzaephilus surinamensis*, 30 days at 10° C. or 4½ days at 2° were required to kill about half of those in the samples (Thomas and Shepard, 1940). Similar results have been reported for species that normally live in warm climates or warm situations and therefore are not usually exposed to the hazards of a cold winter. The fruit fly *Ceratitis capitata* is usually considered to have originated in tropical Africa. It has since spread to a number of warm, temperate areas, including South Africa, Australia, and the Mediterranean region of Europe; but it is not known from any cool, temperate region. Nel (1936) found that neither eggs, larvae, nor pupae could

TABLE 6.14*
ESTIMATED L.D. 50, IN HOURS, FOR DIFFERENT STAGES OF *Tribolium*

Temperature (° C.)	Egg			Larva		Pupa
	1–24 Hours	48–72 Hours	120–32 Hours	3d Instar	6th Instar	
12	118	258	157	...	...	...
7	43	214	110	134	149	258
0	5	....	...	...	...	...
−4	1.5	3.5	...	9	12	...

* Ages expressed in terms of hours at 27° C. From Nagel and Shepard (1934).

survive exposure to temperatures in the range 0°–3° C. for more than 1–3 weeks. A significant proportion of the eggs of *Locusta migratoria* and pupae of *Muscina stabulans* were killed during exposures of 8–16 days at 5° C., but the same exposure did not reduce the survival-rate of pupae of *Calliphora erythrocephala*. The first is a subtropical species, the second ranges into cool, temperate zones, but the pupae are usually found in the warmer situations; the last is widely distributed in cool, temperate climates, and the pupae may occur in relatively exposed situations.

Many of the forms that live in temperate climates differ markedly in their resistance to cold at different stages of the life-cycle. The stage that is normally present during the summer may resemble the subtropical form that we have been discussing, in that dormancy is not possible and the incipient lethal low temperature corresponds more or less to the lower limit of the range favorable for development. The stage that is normally present during winter may be much harder to kill with cold, because it is able to become dormant when the temperature falls below that at which development is possible. For example, the Japanese beetle *Popillia japonica* overwinters in the second or third larval instar. In these stages it is capable of becoming dormant and may survive moderately low temperatures almost indefinitely. But both the adult and the first-instar larva lack this capacity. The newly emerged first-instar larva cannot survive 15 days at 10° C. (Ludwig, 1928). The small parasitic wasp *Diadromus collaris* usually overwinters as an adult, in which stage the females, in particular, are able to withstand several months at 4.5° C. But the larvae and pupae, which are the stages normally present during summer, die after 3 or 4 weeks at this temperature (Given, 1944). The red scale of citrus, *Aonidiella aurantii*, may overwinter in all stages in the milder parts of its distribution, but nearer its northern limits the adult tends to be the only stage to survive the winter. Munger (1948) placed all stages of the scale, including the very early "white-cap" stage, the "first moult," "second moult," gray adult, and mature adult in an outdoor "lath house" in California at the beginning of winter. All stages had been reared at 25° C. During the winter the temperature of the air in the lath house ranged from 0.5° to 7° C. By the end of winter, deaths among the adults amounted to 12 per cent; the deaths among other stages exceeded 50 per cent; from 73 to 96 per cent of the youngest ("white-cap") stage had died.

With insects, the capacity for dormancy seems to be most highly developed among those that are also able to enter the state of diapause (chap. 4). Diapause is often associated with the stage of the life-cycle that is present during winter. In these cases diapause seems usually to be associated with a high degree of "cold-hardiness." But cold-hardiness is not restricted to diapausing individuals. It may be developed by others that are quiescent without being in diapause. These forms seem capable of living almost indefinitely at moderately low temperatures; nor would it seem that they are readily killed by moderate exposures to extremes of low temperature, provided that they are not frozen. But this aspect of the subject has scarcely been studied, and there is little that can be said about it. It seems best to discuss the "cold resistance" of these dormant and cold-hardy forms in relation to the severity of the exposure required to freeze their tissues. This is a subject which has received much more attention.

6.313 THE LETHAL INFLUENCE OF FREEZING TEMPERATURES

It is characteristic of the water in the body fluids of animals that it tends to remain unfrozen, in a supercooled state, when the animal is chilled below 0° C. The strength of this tendency varies enormously in different animals, as we shall see below. The supercooling occurs because the water is present in a finely divided condition in association with colloids. For this reason and also because one organ or tissue may be quite different from another in this respect, the first outcome of chilling is almost certainly the formation of small isolated ice crystals here and there throughout the animal; doubtless there will be more in some tissues than in others. These, being insulated from one another and from the rest of the unfrozen water by the presence of colloids (which tend to become dehydrated in the vicinity of the ice crystals and thus enhance the insulation), grow slowly and tend not to "inoculate" the unfrozen supercooled water. Sooner or later, however, a threshold is passed or an "accident" occurs; the insulation breaks down somewhere; inoculation takes place and sets off a wave of freezing which sweeps through most parts of the animal's body (Salt, 1950). This is illustrated in Figure 6.25 (*solid line*). The curve illustrates the way the

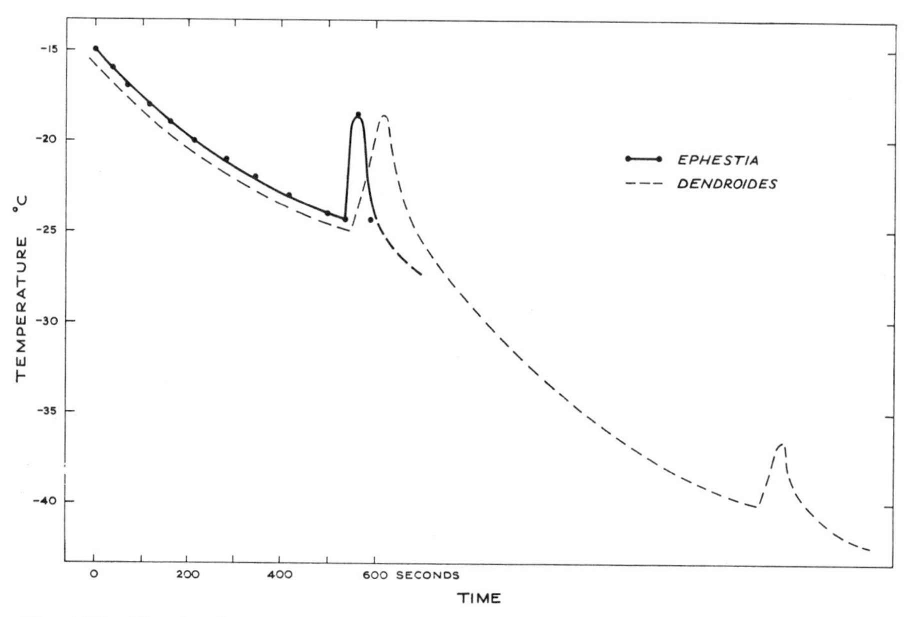

FIG. 6.25.—Showing characteristic curves for the temperature of an insect which has been abruptly transferred from room temperature to some very low temperature (e.g., −40 to −50° C.). Complete line for one that dies after the first rebound (based on data for *Ephestia* from Salt, 1936): Broken line is a hypothetical curve for an insect that survives the first rebound, as, for example, *Dendroides*.

temperature of an insect (*Ephestia*) changed when it was transferred abruptly from room temperature to a low temperature. It will be noticed that the temperature fell steeply and fairly smoothly to −24.3° C., when it abruptly increased to −18.6° and then proceeded to fall again. Supercooling ceased at −24.3° C.; the wave of crystallization which followed liberated enough heat to raise the temperature by 5.7°; then, as this heat was absorbed, the temperature began to fall again. The point at which supercooling came to an end is known as the "undercooling point"; the rise in temperature is the "rebound"; and the temperature to which the rebound rises has been called the "freezing point."

Usually the freezing that occurs during the rebound is sufficiently sudden and thorough to kill the animal; but exceptions to this rule also occur. It is not known what proportion of the total water in the tissues is frozen at the end of the rebound, for this is a matter that has received scarcely any attention. Almost certainly it will vary widely. Using the dilatometer method of Bouyoucos and working with fluids expressed from the bodies of a number of different insects, Sacharov (1930) measured the proportion of the total water in the body fluids that was frozen at different temperatures (see Table 6.15). These

TABLE 6.15*

PROPORTION OF BODY WEIGHT MADE UP BY WATER AND PROPORTION OF BODY WATER FROZEN AT CERTAIN TEMPERATURES

Species	Water as Per Cent Live Wt.	Ice as Per Cent Total Water				
		−3.9° C.	−5.8° C.	−7.8° C.	−11.1° C.	−17.4° C.
Euproctis larvae in diapause	71.8	..	...	...	5.1	15.2
Euproctis larvae feeding	82.9	..	5.0	44.9	...	...
Euxoa larvae hibernating (autumn)	71.4	..	27.0	...	...	...
Euxoa larvae hibernating (spring)	75.9	..	28.4	...	...	...
Euxoa larvae active (summer)	84.7	..	67.4	...	...	...
Scoliopteryx adult in hibernation	48.7	..	0.0	...	28.9	53.4
Plagionotus larva in hibernation	54.1	..	...	...	0.0	3.5
Locusta nymph active	86.2	0.0	20.9	...	...	...
Apis adult active	74.0	..	37.0	...	73.9	...

* Data from Sacharov (1930).

figures may be quite different from what would obtain in the entire insect at the same temperature, but they indicate wide differences between species and stages. Evidence from experiments done in vivo is scarce, but a second undercooling point followed by a second rebound at a lower temperature has been observed on several occasions, from which it may be inferred that some water remained unfrozen after the first rebound. For example, Payne (1926*b*), working with a group of hardy beetle larvae (her oakborer group), observed the usual undercooling points and consequent rebound. The temperature at which the rebound occurred was highly variable but was above −20° C. for all the species. The insects apparently survived this wave of freezing, for all those that were thawed were found to be still alive. But if the insects were chilled still further, a second rebound was found somewhere below −40° C. This was always fatal (Fig. 6.25, *broken line*). Duval and Portier (1922) noticed that larvae of *Cossus* (Lepidoptera) had an undercooling point of about −12° C., but most larvae survived short exposures to −15° C. The larvae were chilled further, and in a few instances a second rebound was noticed at about −20° C. All larvae died after a short exposure to −21° C. Kozhanchikov (1938) observed that diapausing prepupae of *Croesus* (Lepidoptera) were "frozen" at −6° C.; yet he claimed to demonstrate a small amount of "thermostable" respiration down as low as −20° C. Since it is a priori unlikely that respiration would occur in the absence of liquid water, respiration at this temperature may be taken as evidence for the presence of unfrozen water at −20° C. in an insect that had appeared to be frozen at −6° C.

But the species that can be revived after they have experienced the first undercooling point and rebound are exceptional. With most species the freezing that accompanies the rebound is sufficient to kill them, and hence interest has centered around the first undercooling point. Most of the work on this subject has been based on the classical experiments of Bachmetjew (1907). He pushed a fine thermocouple junction into the tissues, usually the thorax of the insects with which he worked, and recorded its temperature as it was chilled. Bachmetjew identified the undercooling point as the lethal temperature, but he somewhat illogically considered that death was determined only when this temperature was reached for the second time, i.e., on the way down after the rebound. He also considered the temperature reached at the top of the rebound to be the "true freezing point" of the animal's tissues. This misconception has been perpetuated and has had a very wide currency, but a moment's reflection will show that it is largely determined by events which are quite unrelated to the nature of the tissues, i.e., to the temperature at the time the rebound starts, the amount of supercooled water that freezes during the rebound, and the weight, specific heat, and conductivity of the objects in the immediate vicinity of the animal when the experiment is being done. For a complex animal like an insect, the "freezing point" can hardly have a rational meaning; but if we like, we may attribute an arbitrary meaning to it and consider it the highest temperature at which a substantial amount of ice will form in the tissues when every effort has been made to reduce supercooling to a minimum. Whenever this has been done, the "freezing point" has been found to be very close to 0° C.—certainly not below −2° C. (Salt, 1950).

The degree of supercooling that is obtained with a certain individual depends on a number of things. When its temperature is falling continuously, the spontaneous termination of the supercooled state becomes increasingly likely, the lower the temperature, down to about −50° C.; beyond this limit the probability decreases with decreasing temperature (Luyet and Gehenio, 1940). But experience shows that the undercooling point is relatively independent of the speed of chilling, provided that this is relatively fast, i.e., fast enough for the undercooling point to be reached in minutes rather than hours (Salt, 1950). On the other hand, at a steady temperature the incidence of the spontaneous termination of the supercooled state becomes a function of time; the longer the animals are held at a constant (low) temperature, the more of them will experience the rebound. When a sample of 18 diapausing prepupae of the sawfly *Cephus* were held at −20° C. for 120 days, they became frozen at the rate indicated in Table 6.16.

Since exposure to low temperature in nature has to be measured in hours or days rather than in minutes, the undercooling point which eliminates the influence of time is clearly a highly arbitrary concept, having only limited descriptive value (Salt, 1950). It follows that for an adequate description of the responses of animals to freezing temperatures, it is necessary to use the same methods that were discussed in the preceding sections in relation to non-freezing temperatures. That is to say, the lethal influence of a particular (freezing) temperature can be adequately described only in terms of the duration of the exposure required to kill the animal—and, for technical reasons, the best quantity to measure is the duration of exposure required to kill half of a random sample from the population. In this case it is possible to make the measurement directly on each individual, if it can reasonably be assumed that the moment the undercooling point occurs is always the moment at which death is inevitably determined.

TABLE 6.16*

TIME REQUIRED FOR "REBOUND" TO OCCUR IN DIAPAUSING PREPUPAE OF *Cephus* HELD FOR 120 DAYS AT −20° C.

Day	1	2	3	4	5	8	13	15	16	28	33	42	66	71
No. of larvae freezing	0	0	2	1	1	1	1	1	1	1	1	2	1	1

* On the 120th day, 4 were still unfrozen. Data from Salt (1950).

These are the fundamental limitations to the concept of the "undercooling point." They are reinforced by certain technical considerations: for example, the degree of undercooling may be greatly modified by prodding or jarring the animal or allowing it to wriggle, by the presence or absence of water in contact with the animal, and by wounding. Even the mild wound caused by inserting a fine thermocouple junction into the tissues may greatly raise the temperature of the undercooling point, so that results got by earlier workers using this method may not be compared with those of later workers with improved methods. Nevertheless, for special purposes these limitations do not matter. For example, the undercooling point, provided that it is measured in a standard way, may be a useful guide to the relative "cold-hardiness" of an individual.

6.314 COLD-HARDINESS

Kozhanchikov (1938) recognized that insects could be grouped into three classes according to their response to low temperatures. The ones in the first group cannot survive for any considerable period if the temperature falls below the lower limit of the range favorable for normal development. In other words, this sort cannot become dormant at low temperatures; they must either develop or die. This group is made up of (*a*) species that live in, or originated from, tropical or subtropical climates, e.g., *Calandra*, *Cimex*, and *Locusta* (see sec. 6.312); there are many of these species in which no stage in the life-cycle is capable of becoming dormant at temperatures below the favourable range; (*b*) species from temperate climates in which there is a certain stage in the life-cycle adapted to survive the winter but all other stages (i.e., those that are normally active during summer) resemble the subtropical and tropical forms in lacking the capacity to become dormant at low temperatures. Most members of this first group reach an undercooling point and rebound at a temperature only a few degrees below 0° C. There are exceptions, notably among those which are adapted to live in dry places or on dry food, e.g., *Calandra*, *Ephestia*. These may have undercooling points that are much lower. But this is not of much importance. All the members of this group, irrespective of their undercooling points, die when exposed to nonfreezing temperature; they are not "cold-hardy" in any meaning of this term.

The second group includes the forms that are able to become "quiescent" in the sense that Shelford (1927) used this term. That is, while remaining competent to develop at any time if they should be placed in the warmth, they may nevertheless survive, inactive but healthy, during prolonged exposure to cold. This group, popular belief to the contrary notwithstanding, is rare, and in searching the literature it is difficult to find many authentic representatives of it. Kozhanchikov (1938) placed the prepupal stage of *Agrotis segetum* (Lepidoptera) and the larval stage of *Lasiocampa quercus* in this group. The pupa of *Heliothis armigera* probably comes in here too.

The third group contains the diapause-stage of species from temperate climates. Many of the species that are adapted to live out-of-doors in temperate climates hibernate in a particular stage of the life-cycle; and frequently this stage and no other is capable of entering a state of diapause (sec. 4.2). The chief difference between members of the second and third classes is that the former become dormant directly in response to low temperature, but, with the latter, dormancy has usually arisen in response to some quite different stimulus (sec. 4.4). This may be primarily genetic and largely independent of environment; or, if environmental, the stimulus may have been operative at a much earlier stage in the life-cycle. The members of the two classes have in common a capacity to survive unharmed through quite prolonged exposures to nonfreezing low temperatures; usually they are killed only if their tissues are suddenly or extensively frozen, as, for example, during the rebound. For this reason, and in contrast with the first group, the members of the second and third groups might be said to be "cold-hardy"; but this term is usually reserved more specifically for the more cold-hardy members of these groups: those whose undercooling points are well below 0° C. The most cold-hardy of all may even survive the freezing that accompanies the first rebound and die only after a second wave of freezing, probably associated with a second undercooling point and rebound.

The degree of cold-hardiness, as measured by the undercooling point, differs widely between species. It may change in the one individual and usually does in rhythm with the season; or it may be modified artificially by desiccating the animal. There is a close correlation between the undercooling point and the amount of water in the tissues. This is brought out by the figures in Table 6.17, which has been compiled from data from Ditman *et al.* (1943), Payne (1926*b*,

1927*b*), Salt (1936), and Kozhanchikov (1938). There are exceptions to this rule. The water content of larvae of *Ephestia kühniella* changed not at all as it passed from the actively feeding stage to the prepupal stage, but the undercooling point changed from −6° to −21° C. (Salt, 1936). Salt also found several other exceptions; but, as a rule, it is true that the lower the proportion of water in the body, the greater the cold-hardiness.

The adult beetle of *Leptinotarsa*, preparing for diapause, becomes dehydrated even in a moist, cool environment in the presence of plenty of lush food. This is

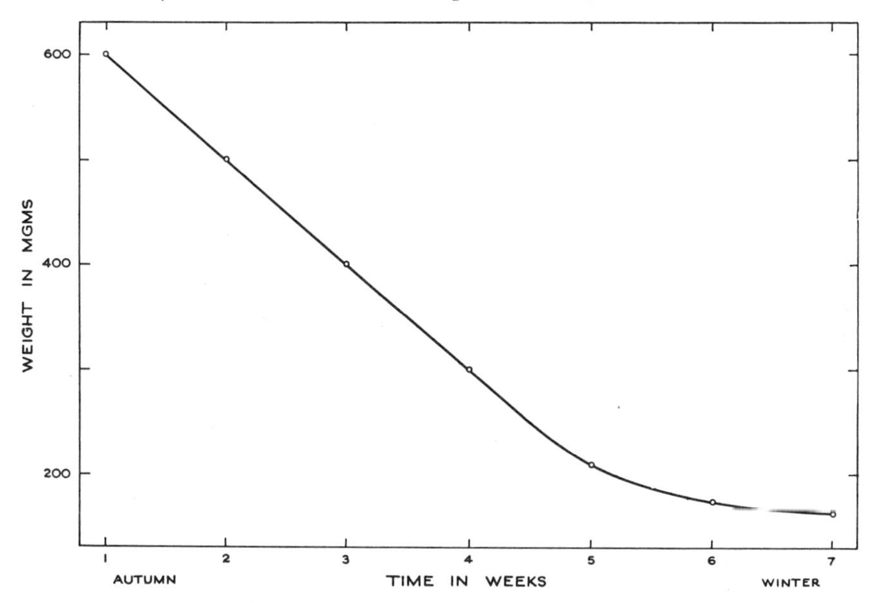

FIG. 6.26.—Showing the way in which the larva of *Diacrisia virginica* becomes dehydrated as it enters diapause at the close of the summer. (Data from Payne, 1927*b*.)

characteristic of diapause, whether it is a preparation for hibernation or for aestivation. In the former case dehydration is a valuable adaptation because thereby cold-hardiness is enhanced. The caterpillars of *Diacrisia virginica* at the close of summer weighed 600 mg. They entered diapause in this stage, and during the next 5 weeks lost weight (chiefly water) at the rate of about 80 mg. per week. As the live weight of the caterpillars approached 200 mg., the rate of loss slowed down greatly, and eventually the weight remained fairly steady at about 180 mg. (Fig. 6.26). The undercooling point became lower as the water-content was reduced, but the freezing associated with the first rebound point was sufficient to kill the caterpillars up to the point where the weight-loss curve began to flatten out. But once the dehydration had proceeded to the point where the live weight of the caterpillar was about 200 mg., the first rebound ceased to kill them; a much lower temperature (about −40° C.) was required.

The diapausing larva of the oakborer *Dendroides canadensis* becomes dehydrated in the same way, but not to the same extent, as *Diacrisia*. Figure 6.27 shows the water content during autumn and the seasonal trends in cold-hardiness in *Dendroides*. The "survival temperature" is the same as the undercooling point during September–October and May–June, but during winter it is very much lower; for *Dendroides* resembles *Diacrisia*, in that, once it has dried

TABLE 6.17*
RELATIONSHIP BETWEEN PROPORTION OF LIVE WEIGHT MADE UP BY WATER AND UNDERCOOLING POINT, FOR 11 SPECIES OF INSECTS

Species	Stage	Water as Per Cent Live Wt.	Undercooling Point (° C.)
Pyrausta	5th-instar larva, feeding	67–69	−11 to −15
	5th-instar prepupa, hibernating	56–57	−21 to −25
Anasa	Adult, feeding (summer)	63–68	−12 to −15
	Adult, hibernating (winter)	57–61	−17 to −23
Synchroa	Larva, summer normal	.....	− 6 to −13†
	Larva, summer dehydrated (24 hr. over $CaCl_2$)	.....	−12 to −26†
Popillia	Larva, hibernating normal	.....	− 5 to − 7†
	Larva, hibernating, dehydrated to 50% of original water content	.....	*ca* − 28†
Musca	Puparia, normal	.....	−12 ± 3.2
	Puparia, dehydrated to lose 10% of original weight	.....	−22 ± 1.8
Agrotis	Larva, feeding	88	− 2 to − 3
	Prepupa, hibernating	73	−10 to −11
Loxostege	Larva, 4th-instar, feeding	85	− 2 to − 3
	Prepupa, 5th-instar, hibernating	60	Lower than −20
Croesus	Prepupa in diapause	61	Lower than −20
Acronicta	Pupa in diapause	63	Lower than −20
Lymantria	Eggs in diapause	62	Lower than −20
Cydia	Prepupa in diapause	59	Lower than −20

* Data from Ditman *et al.* (1943), Payne (1926*b*, 1927*b*), Salt (1936), Kozhanchikov (1938).
† Results probably got by piercing tissues with thermocouple junction.

out beyond a certain point, it is able to survive the freezing associated with the first rebound and dies only after a second undercooling point has occurred; this is usually about −40° (Payne, 1926*a*, 1927*b*). In northeastern United States the corn earworm, *Heliothis armigera*, hibernates as a pupa. Barber and Dicke (1939) exposed a number of pupae out-of-doors; some were in dry soil, and some were in moist soil. The death-rates in the two groups are shown in Table 6.18. Since the moist soil was a better conductor than the dry, these

TABLE 6.18*
DEATH-RATES IN TWO GROUPS OF PUPAE OF *Heliothis armigera* EXPOSED OUT-OF-DOORS DURING WINTER

EXPERIMENT	DEATH-RATE AS PER CENT	
	Moist Soil	Dry Soil
First	27.4	0.0
Second	52.0	1.3

* The experiment was repeated during a second winter. After Barber and Dicke (1939).

pupae were probably colder than those in the dry soil. But the difference between the two groups may have been due to the failure of the pupae that were kept moist to become cold-hardy. Hodson (1937) found that the bug *Leptocoris trivittatus* did not become more cold-hardy when it was desiccated; and Sweetman (1929) found the same with the beetle *Epilachna corrupta*. These may be regarded as exceptions to the general rule.

One outcome of the seasonal fluctuation in cold-hardiness, which is illustrated in Figure 6.27, is that when some constant test (for example, ex-

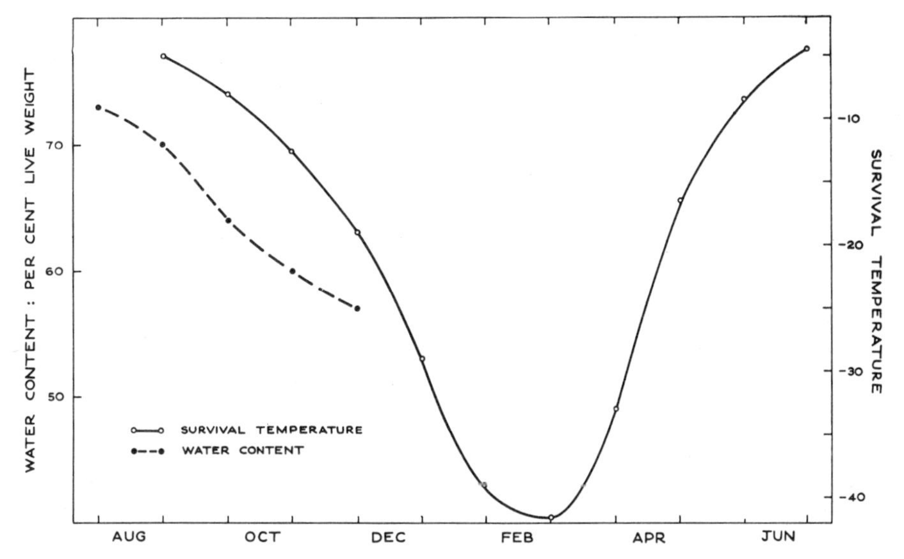

FIG. 6.27.—Showing the falling water content (*broken line*) of larval *Dendroides canadensis* as it enters diapause in the autumn, and the seasonal trend in cold-hardiness (*complete line*) of larval *Dendroides*. (Data from Payne, 1927*b*.)

posure to $-20°$ C. for 3 hours) is applied to samples of the population at different seasons of the year, different proportions of the animals in the samples will become frozen at different seasons. For example, if this test were applied to, say, 100 *Dendroides* in September, most of them might experience an undercooling point and rebound and become frozen; but if the same test were repeated in December, perhaps the proportion freezing would be less than half of the total number in the sample. An individual which happened not to freeze in September would contain about the same low amount of ice in its tissues as one that did not freeze in December; and one that froze in December would contain about the same large amount of ice as one that froze in September. But because many more individuals in the sample of 100 froze in September than in December, there would be, in the aggregate, more ice present in the September sample. This is all very elementary, and it must surely be clear that no useful purpose is served by coining the misleading term "bound water" to describe that portion of the supercooled water which happens to be located in those individuals which happen not to have experienced an undercooling point and rebound. Yet this is just the experiment that Robinson (1928) did with *Callosamia* and *Phyllophaga*, which has been so widely and uncritically quoted as evidence that cold-hardiness is due to the presence of "bound water." Now that Kistler (in Salt, 1936) has clearly shown that "bound water" in Robinson's experiments was no more than supercooled water—and, indeed, a moment's thought along the lines indicated in this present paragraph will make this quite clear—there seems to be no reason for allowing this misleading conception to persist.

6.315 SURVIVAL-RATE IN NATURE IN RELATION TO FREEZING TEMPERATURES

The reduction in cold-hardiness in the spring (Fig. 6.27) is associated with the completion of diapause-development and the consequent increase in water content which is characteristic of the post-diapause-stage. This is a reason why a population which has withstood great extremes of cold during winter may suffer high mortality during a late spring frost, when the temperature is much less extreme than it was during winter. As an example we refer to the observations of Payne (1926*b*), who studied a population of oakborers which during winter may withstand temperatures as low as −40° C. In the fall the undercooling point usually goes down ahead of the temperature, acting as a factor of safety. In the spring when the insects have begun to lose their hardiness, a sudden cold snap or a blizzard is fatal to many of them. During 1924 there was a sudden cold wave during the first part of the month, with a blizzard on April 1. After this, about 43 per cent of the larvae were found to be dead. If hazards of this sort were severe and frequent, the individuals that were tardy in completing diapause would be favored. But it so often happens that the favorable period for growth and reproduction is short during spring and early summer, so that selection pressure may also be operating in the other direction.

The northern limit of distribution of a species may often be determined by low temperatures. A striking example, which also illustrates the point made in the first chapter that distribution and abundance are but different aspects of the same phenomenon, is provided by the moth *Heliothis armigera* in eastern North America. The species is a serious pest of corn and cotton in the southern states, where it is well established and maintains relatively high numbers; in the absence of any important natural checks, the numbers are kept down only by routine destruction of the caterpillars with insecticides. Farther north in the region of New York, most overwintering pupae are killed by cold. The survivors have their numbers augmented during summer by immigrants from farther south; but, on the whole, the species is less abundant than it is in the southern states (Barber and Dicke, 1939). Farther north still, in Canada, it is said that every pupa is killed every year and that the summer population is derived entirely by immigration from farther south. The numbers fluctuate sporadically and are never very high (Pond, 1948). Sacharov (1930) considered that the distribution of the cutworm *Euxoa segetum* as a pest in the region of the lower Volga may be largely explained in terms of the influence of freezing temperatures on overwintering larvae. The species is permanently established throughout the region, but it multiplies to become a pest quite infrequently, and then only in certain areas which are characterized by a thicker and more reliable covering of snow. Elsewhere, where the cover of snow is thin or lacking, temperatures in the soil at the depth where the larvae overwinter are generally so low that most larvae are destroyed most years and the population never gets a chance to increase to large numbers. Sacharov observed the rise and decline of one outbreak. Unusually heavy falls of snow during several winters prior to 1924 permitted the numbers to increase in several areas in the region of the lower Volga. The outbreak came to an end during the winter of 1925–26, when the cover of snow was unusually thin everywhere and temperatures of the order of −12° C. were registered in soil in places where the overwintering larvae were numerous. These temperatures were fatal. This is just another instance of the way in which distribution and abundance are under the control of the same sorts of influences.

With a species that is well adapted to withstand cold, low temperature may not be an important hazard over most of its distribution. This would seem to be the case with *Cephus* in the Canadian prairie; for Salt (1950) reported that, of the many thousands that he had collected from the field over a period of years, he had not found one that had been killed by the cold of winter. Similarly, with the cornborer *Pyrausta*, Stirrett (1931, 1938) showed that the death-rate during winter in Ontario was negligible, rarely exceeding 10 per cent. This was because the temperature during winter in the corn-growing areas of Ontario was rarely, if ever, low enough to kill the cold-hardy diapausing fifth-instar larva of *Pyrausta*. As we have seen, a cold-hardy stage is characteristic of many species of temperate climates. Nevertheless, extreme fluctuations in weather may occasionally cause many deaths, even in species that are usually secure.

The beetle *Dendroctonus brevicornis* lives in the bark of several sorts of pines in northwestern North America, in areas where freezing temperatures occur in the winter. The insect is adapted to withstand cold, and the death-rate during most winters is small. But during the winter of 1924–25 unusually low air temperatures of the order of −27° C. were recorded, and on this occasion from 25 to 80 per cent of the *Dendroctonus* in the bark were killed by cold. An outbreak of the related *D. frontalis* in West Virginia came to an end after the severe freeze of 1892–93 (Miller, 1931).

A high death-rate during winter may be the rule rather than the exception. The bug *Perillus bioculatus* preys upon the Colorado potato beetle, *Leptinotarsa decemlineata*, and in certain circumstances may be important in reducing its numbers. Knight (1922) during 7 years' observations of *Perillus* in New York and Minnesota observed a high death-rate during winter each year.

> Under natural conditions of hibernation the mortality of the bugs must be very high during most years. It has been the writer's experience during six or seven years' observation in the potato fields in New York and Minnesota that very few bugs appear in the fields the following spring where in the preceding fall they were abundant. Under natural conditions it appears that probably not more than five per cent. of the bugs that go into hibernation come out safely in the spring. In New York and Minnesota this may well be due to the fact that many of the bugs seek hibernation in situations where their fatal minimum temperature is reached during the cold winters. In the spring of 1921 the largest number of bugs appeared in the field that the writer has ever observed. The fall of 1920 was very favorable for the bugs that went into hibernation, and following this the winter was unusually mild. The mild winter probably allowed many bugs to hibernate safely in situations where during ordinary cold winters their fatal temperature would have been reached. Such conditions would seem to account for the greater abundance of the bugs in the spring of 1921

Knight did not add, though he might have, that during an unusually cold winter the converse would hold and only those very few bugs which had happened to find especially safe situations would survive. This is an example of the interaction between the animal's "place to live" and other components of environment (in this case low temperature) which we discuss in section 12.2. This principle is of great importance in understanding the distribution and abundance of animals in nature. Salt (1950) summarized this principle neatly. Insects that cannot survive the formation of ice in their tissues "depend for their survival on the insulating protection of their hibernacula (soil, plant debris, snow, ice, etc.) and on their ability to undercool. The combination of these two protective factors may be sufficient that mortality as a result of freezing is a rarity in some species. More often the protection is incomplete, with the result that those individuals in the more exposed hibernacula and with the least ability to undercool perish. In severe winters this fraction may be temporarily large and may seriously deplete the population. If a species has been expanding its range into colder regions, one severe cold period can eradicate it in such areas, restricting it once again to its normal range."

6.32 *The Lethal Influence of High Temperatures*

There has been much less work done on the lethal influence of high temperatures than of low temperatures. With terrestrial animals it may be difficult to devise an experiment which adequately measures the influence of lethal high temperatures independently of the influence of moisture. Moreover, it is often more interesting to study the influence of moisture because, in nature, harmful high temperatures are most likely to occur in deserts or in warm, temperate zones where the summer is arid. In such places the danger from evaporation is likely to be more pressing than the danger from heat. With aquatic animals, especially with certain species of fishes which live in shallow lakes or rivers in temperate zones, high temperatures may sometimes cause a lot of deaths. Two aspects of the lethal influence of temperature on fishes have received special attention, namely, the limits of the tolerable zone and the influence of acclimatization on the "heat resistance" of the animal.

6.321 THE LIMITS OF THE TOLERABLE RANGE

If an animal is exposed to an extreme high or an extreme low temperature, it will quickly die; and no practical difficulties are raised when we say that death was caused by the exposure to unfavorable temperature. As the temperature becomes somewhat less extreme, death may approach more slowly, but it may still be clear that exposure to unfavorable temperature was the chief cause of death. Moving farther and farther away from the extremes, we come to a zone of moderate temperature where it is not reasonable to attribute any lethal influence to temperature. Fry (1947) called the two limits to this median tolerable zone the "incipient lethal low temperature" and the "incipient lethal high temperature"; and he went on to say that above and below the median tolerable zone are the "upper lethal zone" and the "lower lethal zone," respectively (Fig. 6.28). Can we discover the limits of the median tolerable zone by experiment?

In order to discover these limits, we need to know the least extreme temperature which would be fatal. Since the time required for a fatal exposure increases as the temperature becomes less extreme, this means that we need to know the least extreme temperature at which the time required for a fatal exposure becomes indefinitely long. When we come to do the necessary experiment, we find that, as the limit is approached, the duration of exposure becomes so long that eventually we may be unable to decide whether the animal died of old age or from some harmful influence of temperature. Thus there is no empirical way to measure these limits precisely. In practice, it is necessary to substitute arbitrarily some definite period for the "indefinitely long" exposure required by the theory. For example, in their experiments with young speckled trout (*Salvelinus*) Fry *et al.* (1946) used periods that varied from 1,200 to 5,000 minutes, depending on the circumstances. But in each case they chose a period that was sufficiently long to justify the assumption that any fish which was still alive at the end of this period would have been able to survive an indefinitely long exposure to the particular temperature of the experiment. In this way an approximate result can be obtained.

6.322 ACCLIMATIZATION

The limits of the tolerable zone are not "fixed" but depend to some extent upon the condition of the animal. For example, it has been shown, especially with fish, that the temperature at which an animal has been living may have quite a profound influence on the limits of the tolerable zone. This effect is usually called "acclimatization." Figure 6.28, which has been constructed from the data of Fry *et al* (1942) for the goldfish, shows that the tolerable range moves up the temperature scale and becomes narrower, the higher the temperature at which the fish had been living before the experiment began. There is an upper and a lower limit to the temperatures at which acclimatization goes on, and there is a corresponding limit imposed on the variation in the

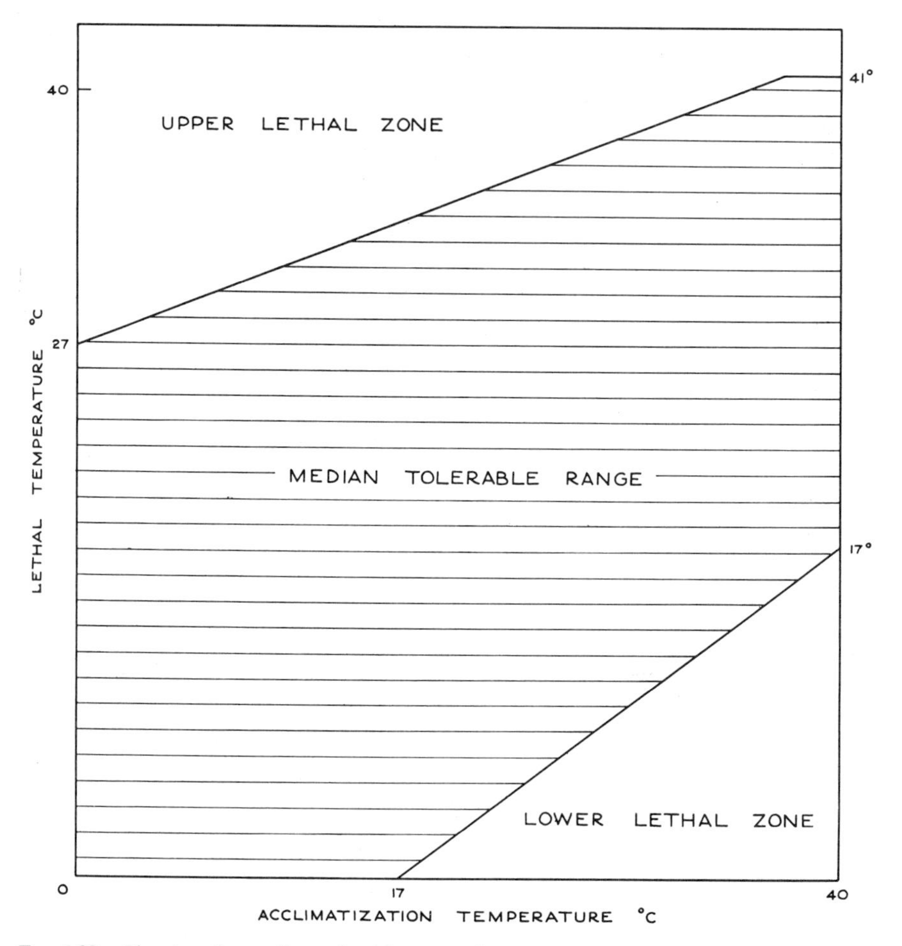

FIG. 6.28.—Showing the median tolerable range of temperature for goldfish and how the lower and upper lethal limits depend upon the temperature to which the fish had previously been acclimatized. (Modified after Fry, Brett, and Clausen, 1942.)

incipient lethal temperatures which can be brought about by acclimatization. Fry (1947) considered these limits to be characteristic of the species. Thus with goldfish the acclimatization effect is not accentuated by exposures above 41° C., and the limit above which the incipient lethal low temperature cannot be raised by acclimatization at high temperature is 17° C. (Fig. 6.28). Taking into account the full potential influence of acclimatization, the lowest incipient lethal low temperature for the goldfish may be below 0° C., and the highest incipient lethal high temperature may be near 40° C.

Acclimatization influences not only the upper and lower incipient lethal temperatures but also the duration of exposure to a lethal temperature which an animal may withstand. Figure 6.29, *A*, illustrates this. It has been constructed from Fry *et al.*'s (1946) data for yearling speckled trout. The diagram indicates clearly that the duration of exposure to lethal high temperature re-

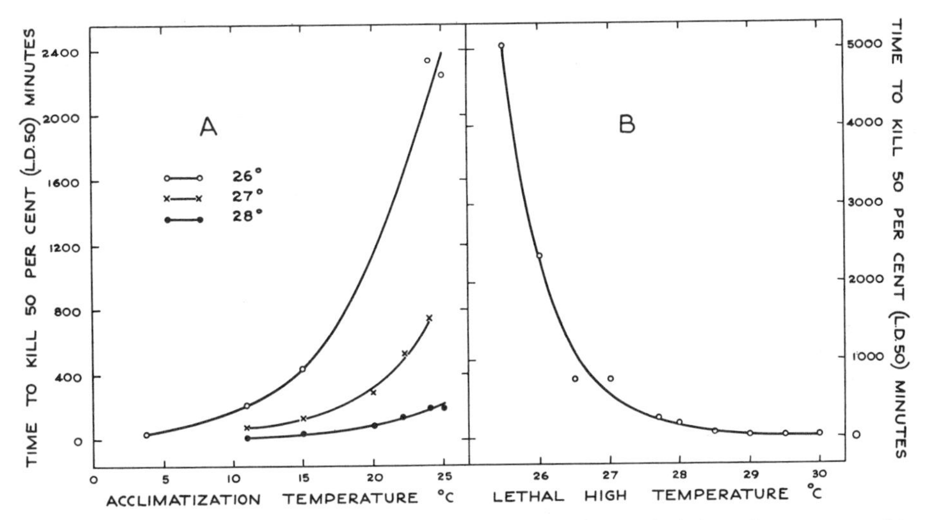

FIG. 6.29.—*A*, showing the influence of acclimatization on the survival time of trout, exposed to three lethal temperatures, 26°, 27°, and 28° C. The influence of acclimatization was most pronounced when the lethal temperature was 26° C. But at all three temperatures fish which had been acclimatized at a high temperature lived longer than those which had been living at a lower temperature. *B*, showing the survival time for trout exposed to lethal high temperatures ranging from 25.5° to 30° C. The fish had all been living at 24° C. before the experiment began. (Data from Fry, Hart, and Walker, 1946.)

quired to kill half the fish was less for those that had been living at a low temperature than for those that came from the higher temperature.

In nature, acclimatization occurs from one season to the next, as was well shown by Brett's (1944) data for bullhead (*Ameiurus*) living in Lake Opeongo, Ontario. He collected fish from the lake at intervals from July, 1940, to September, 1941, and measured the high temperature required to kill half the sample during an exposure of 12 hours. Figure 6.30 shows that the seasonal trends are most striking. The capacity to become acclimatized to changing levels of temperature is a useful adaptation for fish which live in lakes and rivers in a temperate climate. In Ontario the extremes of temperature are such that fish inhabiting these situations occasionally suffer severe mortality from the direct influence of high temperatures (Huntsman, 1946). If they lacked this adaptation, the death-rate would doubtless be higher, and the occasions on which many died would happen more frequently, with the consequence that the average numbers in the area might be lower. One might expect similar phenotypical differences to occur between populations from different climatic zones, but this may be complicated by the existence of genetic differences in such populations. Hart (1952) compared the lethal temperatures of fresh-water fish from different localities in North America. He studied 14 species. Geographic differences in the upper lethal temperature were found only in three species—*Notropis cornutus*, *Gambusia affinis*, and *Micropterus salmoides*, each

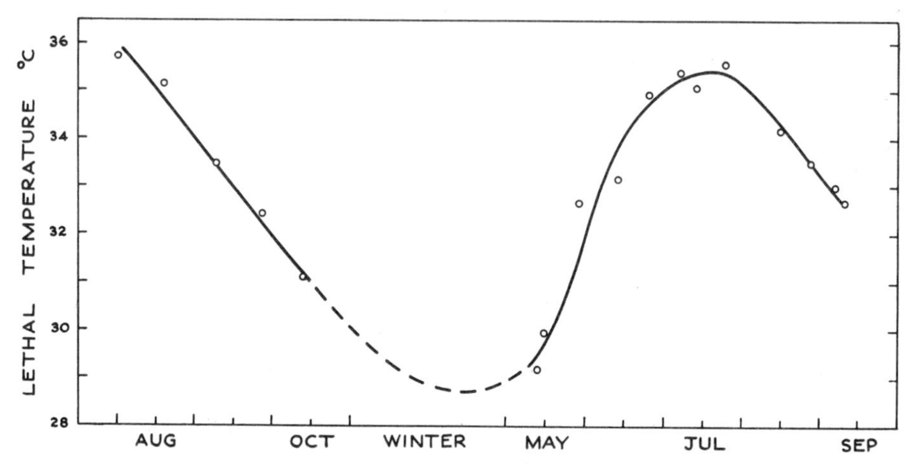

FIG. 6.30.—Illustrating the seasonal trend in the "heat resistance" of the bullhead *Ameiurus*. The ordinate gives the lowest temperature at which just half the sample of fish were dead after an exposure of 12 hours. The fish have clearly become acclimatized to the prevailing temperature of each season. (After Brett, 1944.)

from southern Canada and Tennessee. In each case he was able to find morphological differences between the populations from the different regions, and he recognized them as subspecies. Hart suggested that the similarity in the upper lethal temperatures for the other 11 species may have been due to their great capacities for acclimatization; or perhaps it was due to the relatively low temperatures prevailing in the places where these fishes lived. Since none of them ran any risk of being killed by high temperature, there was little likelihood of any of them being selected for this character.

6.323 DELAY IN THE MANIFESTATION OF THE LETHAL INFLUENCE OF HIGH TEMPERATURE

We mentioned in section 6.31 that it is often not possible to tell merely by looking at an animal the precise moment when it has had a fatal "dose" of high or low temperature. Sometimes the harm that has been done cannot be detected until much later in the life-cycle. This is well illustrated by Darby and Kapp's (1933) experiments with maggots of the fruit fly *Anastrepha*. In nature the maggots of *Anastrepha* live surrounded by the juicy contents of the fruit, so it is possible to measure the influence of high temperature on survival-rate, independent of the influence of moisture. Darby and Kapp exposed batches of larvae in their third instar to a series of temperatures between 36° and 46° C. for 4 hours. The deaths from heat were negligble up to about 40°; an exposure of 4 hours to 42° killed about half of the larvae; and at 46° all were dead at the end of 4 hours. In another experiment, batches of larvae were exposed to 40.5° C. for various periods. About 11 hours at 40.5° were required to kill about half the larvae; and all were dead after 15 hours.

Darby and Kapp not only recorded the number of deaths at the end of the 4 hours' exposure to the various high temperatures but also gave the survivors the opportunity to continue their development and pupate at a favorable temperature. Most of them did complete their larval development, but relatively few of them were able to pupate. Many died at this stage. The percentage of surviving larvae which also survived to become pupae is shown in Table 6.19. It is clear that exposure to high temperatures did a certain amount of harm which could not be measured by counting those which died immediately. A large proportion of those which apparently recovered were nevertheless unable to complete their development, dying at the next "critical" stage in their life-cycle. This "delayed-action" effect of high temperature is well known to students of this subject. Often the survivors appear quite healthy at first and may feed and grow apparently quite normally until some critical stage, e.g., ecdysis or metamorphosis, is reached. Then the harmful influence of the previous exposure to high temperature becomes manifest in the unusually high death-rate at this stage (Larsen, 1943).

TABLE 6.19*

DELAYED HARMFUL INFLUENCE OF HIGH TEMPERATURE ON LARVAE OF *Anastrepha*

Temperature (° C.)	Per Cent Pupae
40.8	53
41.4	49
41.8	41
41.9	13

* The figures show the proportion of larvae which, having survived exposure to high temperature for 4 hours, were subsequently able to pupate. After Darby and Kapp (1933).

6.324 THE "HEAT RESISTANCE" OF ANIMALS THAT LIVE IN WARM PLACES

In section 6.233 (and see Fig. 6.01) we showed how animals that normally inhabit warm places are adapted to develop at higher temperatures than those from cold places. Similarly, as might be expected, the capacity to survive exposure to extreme temperatures is related to the temperatures which prevail in the places where the animal lives. For example, the nymphs of the firebrat *Thermobia domestica*, which normally inhabits such hot places as the hobs of bakers' ovens, may live indefinitely at temperatures as high as, or higher than,

42° C. But the nymphs of the related silverfish, *Lepisma saccharina*, which, being a common household pest in temperate climates, inhabits much cooler places, are unable to withstand temperatures above about 36° C. (Sweetman, 1938, 1939). Buxton (1924) found certain species of Orthoptera living in the desert in Palestine on bare earth with a temperature of 60° C.; but it is unlikely that even these desert forms could survive for very long at such a high temperature. Chapman *et al.* (1926) measured the temperature in a number of situations around a sand dune in Minnesota. On a bright summer's day the temperature on the surface of the sand was 51° C., which is comparable with that found in the desert. But at the same time the temperature of the air about 12 inches above the surface of the sand was 27° C., and that of the sand 12 inches below the surface was 38° C. On a cloudy day with rain falling, the temperature of the air 12 inches above the surface was 18° C., and that of the sand 12 inches below the surface was 22° C. The fauna of the dune included the sand wasp *Bembex* and its parasite *Dasymutilla*. The former constructs deep burrows in the sand and stocks them with flies, which serve as food for the larvae. The latter seek out these burrows and lay their eggs on the larval *Bembex*. The adult *Bembex* was inactive at temperatures below about 25° C., so that its activity was largely restricted to bright days. Yet it succumbed rapidly to high temperatures above about 42° C. Nevertheless, it spent a substantial part of its time on bright, sunny days burrowing in the sand. It contrived to get through the surface layers, where the temperature of the sand was so high that exposures of more than a few minutes would be fatal to the wasp, by short bursts of digging interspersed with frequent visits to the layers of air above the dune where the temperature was tolerable. But its parasite, *Dasymutilla*, being wingless, might not evade the temperature of the surface sand except by entering the burrows of *Bembex* or by climbing the sparse vegetation on the dune. Neither device affords adequate protection, as the insect must spend a lot of time crawling on the surface of the sand. It is able to survive on the dune because it is adapted to withstand high temperature. Chapman *et al.* (1926) found that when a batch of *Dasymutilla* females were gradually warmed during 160 minutes from 0° to 56° C. the first sign of heat stroke appeared at 52°, and all the insects were knocked over by the heat when the temperature reached 55°; in a comparable experiment with adult *Bembex*, the corresponding temperatures were 42° and 44° C. These authors also observed: "In the normal course of the day all the insects leave the surface of the sand when its temperature nears 50° C. Some climb grasses and some enter their burrows, while others fly about some distance above the sand making hurried landings to enter their burrows. The female mutillids were consistently the last to retreat when the temperature rose and the first to return to the open sand when the temperature fell."

6.4 SUMMARY

We have seen in the preceding sections how temperature may influence behavior, speed of development, and longevity and that acclimatization may be important with respect to all these aspects. Leaving behavior aside for the moment, it is possible to epitomize the remainder in a single hypothetical diagram. This has been done in Figure 6.31. The zone of lethal low temperature has been divided into two subzones; A_{I} is the zone of freezing temperatures, which is chiefly important in relation to forms that can become dormant, and, for the most part, this means diapausing individuals: A_{II} is the zone of nonfreezing lethal low temperature, which is chiefly important with respect to nondormant forms. In zones A_{I} and A_{II} and in zone C the influence of acclimatization is indicated by drawing the curves as a band rather than a line. In both zones A and C, interest centers on the lethal influence of temperature, and this is expressed as the duration of exposure required to kill half the sample of animals (L.D. 50). This approaches infinity at the incipient lethal temperatures, i.e., at the limits to the zone of favorable temperature, and approaches zero at extremes of low temperature A or high temperature C. The central zone is the zone of favorable temperature, and here interest centers on the speed of development which is represented in the diagram by portion of a logistic curve.

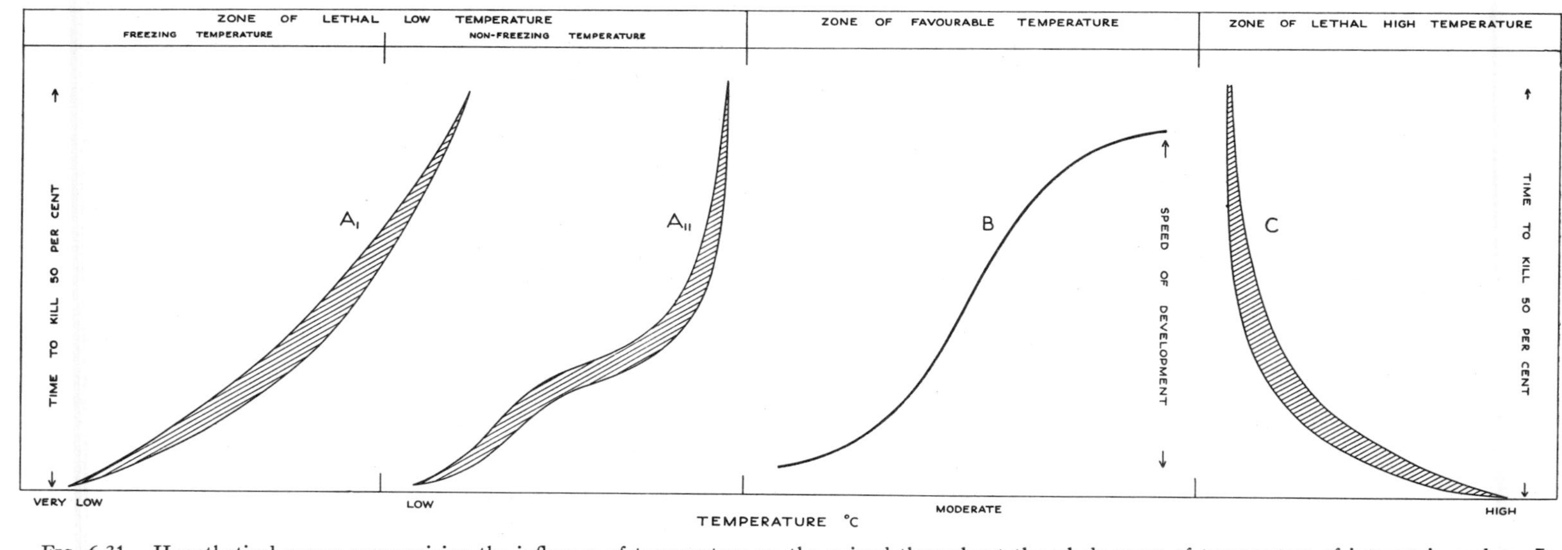

FIG. 6.31.—Hypothetical curves summarizing the influence of temperature on the animal throughout the whole range of temperature of interest in ecology. In zones *A* and *C*, interest centers on the lethal influence of temperature and is expressed as the duration of exposure that is lethal. In *A*, separate curves are drawn for hardy and nonhardy forms. The curves are drawn as bands rather than lines, to indicate the influence of acclimatization. In zone *B*, interest centers on speed of development, and this has been expressed as part of a logistic curve. For further explanation see text.

CHAPTER 7

Weather: Moisture

7.0 INTRODUCTION

Most animals require to keep the proportion of water in their bodies constant within rather narrow limits. This is done by maintaining a balance between the water taken in through the mouth or through the integument and the water lost by excretion, by transpiration through the integument, and as a waste product of respiration. The mechanisms involved are many and varied, for some animals live in places so wet that they have to do work to prevent the accumulation of excessive amounts of water in their tissues, while others live in places so dry that they can survive there only because they possess remarkably effective mechanisms for reducing the amount of water lost from the body. The inhabitants of fresh water represent the one extreme, those of the desert sand dunes the other. All the intermediate levels are occupied as well, and animals of one sort or another may be found living in situations where the moistness ranges from absolutely wet to very nearly absolutely dry. Of course, most individual species are restricted to a relatively narrow part of this range: it is a matter of common experience that a species from the desert cannot, as a rule, survive in a humid climate and vice versa.

With an aquatic animal, the dryness of the place where it is living is best expressed in terms of the osmotic pressure of the fluid surrounding it. The sea has an osmotic pressure which exceeds that of the body fluids of teleost fishes. In this sense the sea is "dry," and the teleost fish is able to live in the sea only by virtue of its capacity to resist the tendency to lose water. A river has an osmotic pressure less than that of the body fluids of the animals that live in it, and they are adapted to prevent the accumulation of excessive water in their tissues. In estuaries where sea and fresh water mix, the osmotic pressure may fluctuate from place to place and time to time, and the animals that live in the estuary require to be able to withstand these fluctuations. But species that live exclusively in the sea or exclusively in fresh water need not be so adapted, for the moisture in the places where they live remains virtually constant, in marked contrast with the pronounced fluctuations in humidity found in most places where terrestrial animals live.

With terrestrial animals, moisture is best considered in terms of the humidity of the atmosphere in the place where the animal is living, though it is, of course, necessary to take into account the presence of liquid water for drinking or in the food and as a physical hazard, e.g., the flooding or water-logging of soil. There is also the chance that excessive moisture, either as free water or as water vapor may favor the spread of epizoötics of fungal or bacterial disease. Among terrestrial animals, the active stages of most of the Mollusca, Turbellaria, Nematoda, Onycophora, Amphipoda, Isopoda, and a few of the Insecta and Acarina, as well as most of the Amphibia, characteristically require to live in situations where the atmosphere is saturated with water vapor, or nearly so, for most of the time. Consequently, they are to be found associated with damp soil or vegetation or with pools and intertidal regions. In section 7.31 we give several examples of representatives of these groups which have become adapted to survive the loss of a relatively large proportion of the water from their bodies. These species have been able to colonize relatively more arid situations than have closely related members of the same group which have not become so adapted. But since even the most hardy of these animals continues to lose water rapidly in a dry atmosphere, there are quite strict limitations to the effectiveness of this device. None of these animals is found in really dry situations except those which have also developed a specially resistant resting stage in the life-cycle.

Certain species of terrestrial animals are able to survive in an area where the humidity is likely to fluctuate widely because the different stages of the life-cycle require different levels of moisture in the environment. The best-understood examples of this are among the insects which enter diapause in one stage of the life-cycle. At the onset of diapause the water content of the tissues becomes greatly reduced, and in this condition the insect may, while remaining quite inactive, withstand extremes of dryness which would be fatal to the active stages. Several examples of this phenomenon are discussed in section 4.6. The egg of the grasshopper *Austroicetes cruciata* is able to maintain a balance of water in its tissues against the strains of both excessive wetness and excessive dryness. During summer it remains alive in parched, sun-baked soil without relief from any rain for weeks or months at a time (Andrewartha and Birch, 1948); and during winter it is in contact with wet soil in which the osmotic pressure is less than that of the tissues of the egg. A remarkable group of cells called the "hydropyle" is responsible for regulating the movement of water into and out of the egg (see sec. 7.233).

Among the tardigrades, rotifers, and nematodes, there are species in which the active stages may be virtually aquatic, living in the films of water, in damp soil, or on moist vegetation. They cannot survive except where there is free water or the atmosphere is saturated with water vapor. But these animals also have in the life-cycle a resting stage, a cyst which may be able to survive for a long time in very dry situations. Leeuwenhoek in 1701 was probably the first to observe that when the water in which they had been living dried up, certain tardigrades and rotifers entered a resting stage which was able to survive prolonged exposure to extreme dryness (Hall, 1922). In 1764 Baker revived some nematodes after they had been continuously dry for 27 years. The famous Italian biologist of the eighteenth century, Abbot Spallanzani, confirmed these early experiments on tardigrades and rotifers (Schmidt, 1918). In 1860 the Société de Biologie in Paris set up a special commission under Broca to study, in carefully controlled experiments, the revivification of rotifers and tardigrades which had been severely desiccated. This remarkable phenomenon was given the name *anabiosis* by Preyer (1891). It is not known in what way, if at all, it is related to diapause in insects: at least, it produces the same result by allowing these species to live in places where the amount of moisture may fluctuate extremely.

The vertebrates and insects are often quoted in textbooks as being the two groups that have been most successful in colonizing dry land; and, indeed, it is these two groups that are characteristically inhabitants of the more arid situations. The large size of the vertebrate reduces the relative loss of water by transpiration and also enables it to travel considerable distances for a drink. The insect can survive in a dry place despite its small size because it possesses a remarkably impermeable cuticle. In addition, there are other physiological attributes which are important, as are behavior patterns which cause the animal to avoid unfavorable extremes of wetness or dryness.

Therefore, this chapter is divided into three main parts. Section 7.1 deals with the behavior of animals in response to gradients of moisture; section 7.2 deals with the more physiological aspects of water balance in terrestrial animals; and the remaining sections are devoted to the discussion of the more strictly ecological aspects of moisture as a component in the environments of terrestrial animals.

We do not pay much attention to aquatic animals in this chapter, because, as we explained above, their environments tend to be very constant with respect to moisture. With aquatic animals, water is more important in relation to the subject matter of chapter 12.

7.1 THE INFLUENCE OF MOISTURE ON THE BEHAVIOR OF TERRESTRIAL ANIMALS

The tendency, which can be readily observed in some animals, to move in a definite direction along a gradient of humidity until they come to rest in some particular part of it may sometimes be recognized as an adaptation which helps the animal to avoid the extremes of wetness or dryness and bring it toward the places where moisture is more favorable. The danger from excessive dryness is merely desiccation; the dangers from excessive wetness are twofold: there is the obvious danger from drowning when it is free water that is in excess, as well as the more subtle one that high atmospheric humidity may favor the development and spread of epizoötics of fungal or bacterial disease.

In studying the behavior of animals in gradients of moisture, whether in nature or in artificial surroundings in the laboratory, it is well to remember that the animal may, at the same time, be influenced by other stimuli from, for example, gradients in temperature, light, air currents, and so on, and that there may be quite strong interactions in the responses to the different stimuli. We shall consider first, in section 7.11, several experiments in which moisture was the only variable and then proceed in subsequent sections to consider more complex experiments in which certain other variables have been introduced.

7.11 *Moisture-Preferences of Animals at Constant Temperatures*

In the experiments described in this section, one of two methods has been used. Animals have been placed in a series of chambers each one of which provided a moisture gradient over a relatively narrow range of humidity. All the chambers were kept at a constant temperature. Alternatively, the animals were placed in one chamber which provided a humidity gradient covering a wide range from very dry to very humid. With experiments of these types, the responses of the animals in the gradients may be recorded in a number of ways. It is sometimes sufficient merely to count the numbers of animals which gather at the wet end and the dry end after a certain time has elapsed. Alternatively, the time spent in any part of the gradient may be recorded. This involves a complete record of the paths of each individual animal. It has the advantage of providing additional information on the mechanism of orientation and movement in the gradient.

When the wood louse *Porcellio scaber* was offered a range of humidities, it tended to collect in the moist end of the gradient (Gunn, 1937). In dry air the animals were incessantly active, while in moist air they hardly moved at all, with the result that few remained at the dry end and in due course the majority were to be found at the wet end (see Fig. 7.01). Larvae of the wireworm *Agriotes* tended to avoid dry air and moved toward the moist end of the gradient, provided that the relative humidity exceeded 70 per cent; in air that was drier than this, they seemed to be unresponsive to gradients in the humidity (Lees, 1943*a*). They behaved in the same way in a moisture gradient in soil (Lees, 1943*b*). In moist air (i.e., with relative humidity in excess of 70 per cent) intense reactions were observed; differences in relative humidity of 7 per cent (at 17° C.) were sufficient to produce a reaction in almost every individual in the sample. Less pronounced reactions, which were nevertheless statistically significant, were observed when the differences in humidity were much smaller than this. When the gradient extended from 92 per cent relative humidity at one end to 99.5 per cent at the other, 190 larvae, out of a total of 200, were found at the moist end; the remaining 10 were in the middle of the gradient.

Pronounced responses to small differences in relative humidity have also been demonstrated for certain other species. For example, Waloff (1941) showed that in certain circumstances *P. scaber* showed a strong preference for the moist end of a gradient which extended from 86 to 100 per cent relative humidity; and adults of *Tribolium castaneum* strongly preferred 95 per cent relative humidity to 100 per cent (Willis and Roth, 1950).

Some species in certain circumstances choose the dry end of the gradient. Pielou and Gunn (1940) found that the adults of *Tenebrio molitor* moved rapidly toward the dry end of a gradient when they were placed at the moist end,

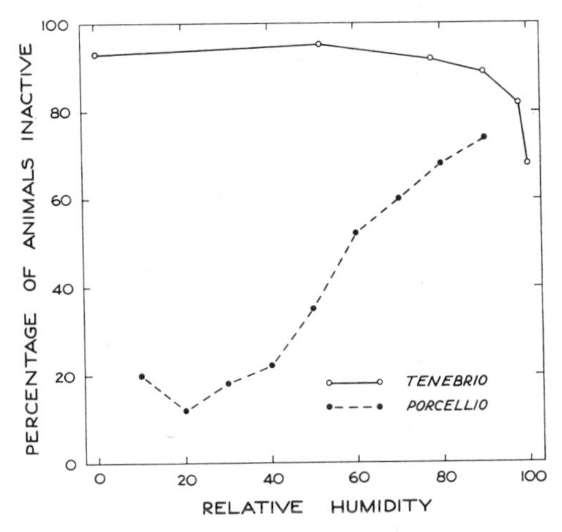

FIG. 7.01.—Activity of the adult meal worm *Tenebrio molitor* and the wood louse *Porcellio scaber* in uniform humidities. Each point shows the proportion of animals inactive. (After Gunn, 1937; and Pielou and Gunn, 1940.)

provided that the relative humidity exceeded 70 per cent. In air that was drier than this, the beetles were relatively inactive and showed little tendency to change their position in the gradient (see Fig. 7.01). The body louse *Pediculus humanus corporis* showed a tendency to move from the moister to the drier air within the range 100–75 per cent relative humidity; in air that was drier than this, they showed no tendency to change their position in the gradient (Wigglesworth, 1941). Adults of the beetle *Ptinus tectus* seem to be unusual. According to Bentley (1944), they responded to gradients in humidity in dry air and in moist air, but not in the intermediate range between 70 and 90 per cent relative humidity. Below 70 and above 90 per cent the beetles tended to move toward the drier part of the gradient.

Vertebrates also may show responses to humidity gradients. The American toad *Bufo americanus* spent less time at the dry end of a humidity gradient than at the wet end. The horned lizard, on the other hand, spent more of its time in the dry end of the same gradient, though its responses were less distinct than those of the toad (Shelford, 1918). Shelford and Martin (1946) found that pheasant chicks tended to congregate at a relative humidity of 79 per cent in a gradient ranging from 40 to 92 per cent. Domestic chicks congregated in the range bounded by 77 and 92 per cent relative humidity.

7.12 *The Behavior of Animals in a Moisture Gradient when Temperature or Some Other Component Is Varied as Well*

There are two ways of determining the moisture-preferences of animals at different temperatures. The reactions of the animals may be studied in a humidity gradient at a number of constant temperatures. Alternatively, the animals may be placed in a temperature gradient. Provided that the air is neither completely dry nor completely saturated, a moisture gradient will be established in the same chamber, since the evaporation will be different at different temperatures. The gradient in evaporation can be regulated as desired, just as it is in a humidity gradient at constant temperatures (Wellington, 1949*b*).

An example of the first type of experiment is provided by Savory's (1930) experiments on the spider *Zilla x-notata*. He found that this spider went to the moist end of a gradient when the temperature was 5° C. but to the dry end when the temperature was 20° C. or higher. This is evidently a clear-cut case of reversal in reaction to humidity as a result of changed temperature.

Wellington (1949*c*) studied the reactions of the larvae of the spruce budworm *Choristoneura fumiferana* in an apparatus where a gradient in temperature was associated with a gradient in evaporation. Wellington measured evaporation directly in his gradients and found this a more useful index than either relative humidity or saturation deficit. The behavior of the larvae was studied in four different sorts of gradient, which are described in the legend of Figure 7.02. The histograms show the final distributions of second-instar larvae in these four gradients. The larvae always congregated within the one range of evaporation-rate, namely, 0.13–0.14 cu mm/minute, irrespective of temperature. The same data have been rearranged in Figure 7.03, with temperature forming the abscissae of the graph. There is an apparent preference for different temperatures in each experiment; but this is not a response to temperature at all, but a preference for the same rate of evaporation at different temperatures. An interesting feature of these experiments, particularly from the point of view of methodology, was that the response, so clearly brought out when moisture was measured as rate of evaporation, was not apparent when the units were relative humidity or saturation deficit. The rate of evaporation is a function of the saturation deficit and certain other factors. In the present instance we must conclude that these "other factors" were exerting an important influence on the behavior of the larvae. Wellington confirmed these results with many other experiments and reached the general conclusion that the larvae of the spruce budworm always tend to congregate in a particular zone of evaporation-rate, wherever that zone may be in the temperature gradient. Also, he observed that the different instars tended to prefer different zones in the gradient of evaporation-rate (see Fig. 7.04). After the larvae had been in the apparatus for 3 hours, their strong preference seemed to weaken, and they tended to wander more widely (Wellington, 1949*c*).

The wood louse *Oniscus asellus* in a moisture gradient in complete darkness

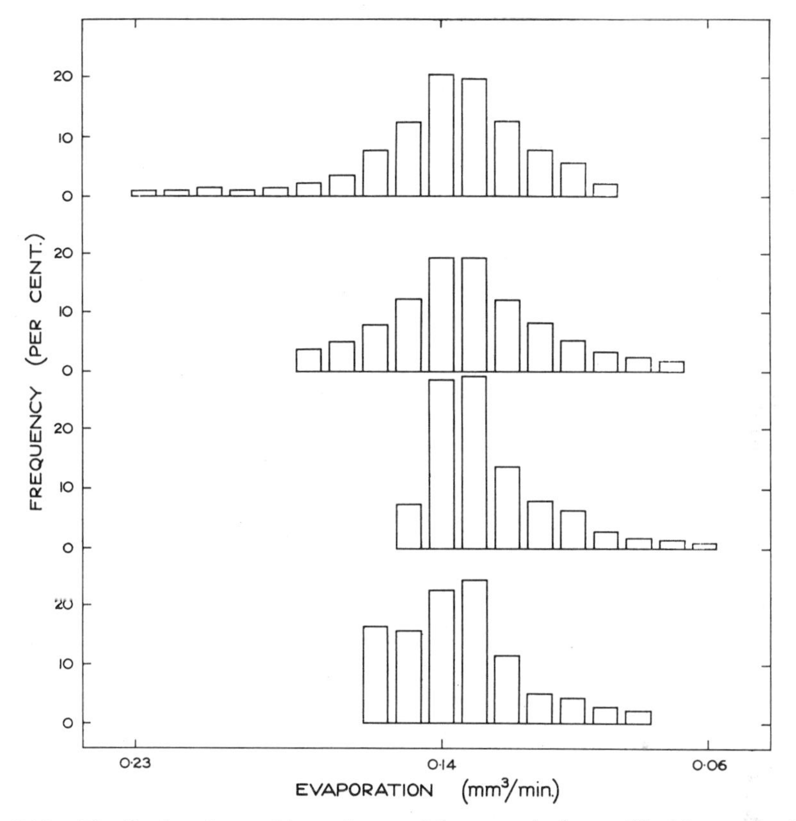

FIG. 7.02.—Distribution of second-instar larvae of the spruce budworm *Choristoneura fumiferana* in response to evaporation in four sorts of gradient: *A* (top), in a temperature gradient in a dry atmosphere. The 36° C. isotherm lay above the zone where the rate of evaporation was 0.23 mm³/minute. *B*, in a temperature gradient in a moist atmosphere. The 36° C. isotherm lay above the zone where the rate of evaporation was 0.18 mm³/minute. *C*, in a temperature gradient in a very moist atmosphere, with the 36° C. isotherm above the zone where the rate of evaporation was 0.15 mm³/minute. *D* (bottom), in a gradient of evaporation at a constant temperature of 20.6° C. The maximal rate of evaporation did not exceed 0.16 mm³/minute. (After Wellington, 1949*b*.)

tended to wander toward the moister end of the gradient and stay there. Waloff (1941) placed a number of *Oniscus* in an apparatus of which one end was moist and bright and the other dry and shaded from the light. The animals at first moved away from the light toward the drier end, thus reversing their normal response to moisture in the darkness. After staying awhile at the dry but dark end, they moved toward the moist end, notwithstanding that it was light. Perhaps the animals became desiccated while at the dry end, and in this condition they were attracted more strongly by the moisture than they were repelled by the light. An alternative, though less attractive, explanation is that desiccation caused a reversal in the reaction to light.

7.13 *The Condition of the Animal in Relation to Its Behavior in a Moisture Gradient*

There are a number of experiments which show that the behavior of an animal when it is placed in a moisture gradient depends not only on the circum-

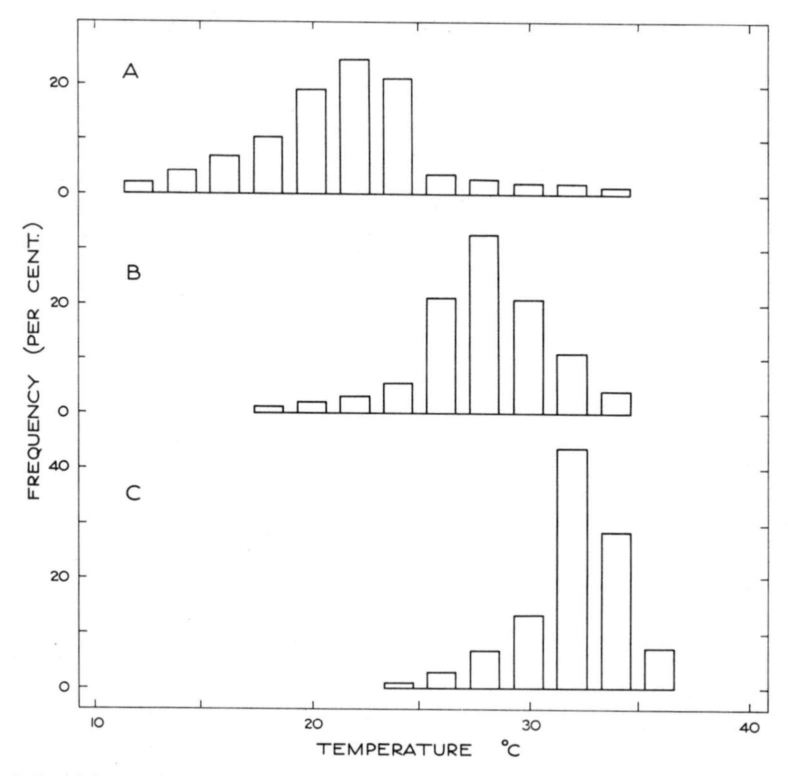

FIG. 7.03.—Distributions of second-instar larvae of the spruce budworm *Choristoneura fumiferana* in a gradient of temperature at three levels of humidity. The same data which were used to construct Fig. 7.02, *A*, *B*, and *C* are here plotted against temperature instead of evaporation. *A*, in a temperature gradient in a dry atmosphere; *B*, in a temperature gradient in a moist atmosphere; *C*, in a temperature gradient in a very moist atmosphere. (After Wellington, 1949*b*.)

stances of its surroundings but also on the condition of the animal itself. Roth and Willis (1951) observed the distribution of samples of the flour beetles *Tribolium castaneum* and *T. confusum* in an olfactometer which gave the beetles the choice of moving into a stream of dry air (near 0 per cent relative humidity) or a stream of moist air, which was nearly saturated with water vapor. The beetles, which were taken from where they had been living in whole-meal flour containing 10–12 per cent of water and placed immediately in the apparatus, almost invariably came to rest in the dry air. On the other hand, beetles from the same colony, which had been kept for several days without food in air at about 0 per cent relative humidity before being tested in the olfactometer, almost invariably came to rest in the moist air. With shorter periods of starvation, i.e., about 1½ days with *T. castaneum* and 3½ days with *T. confusum*,

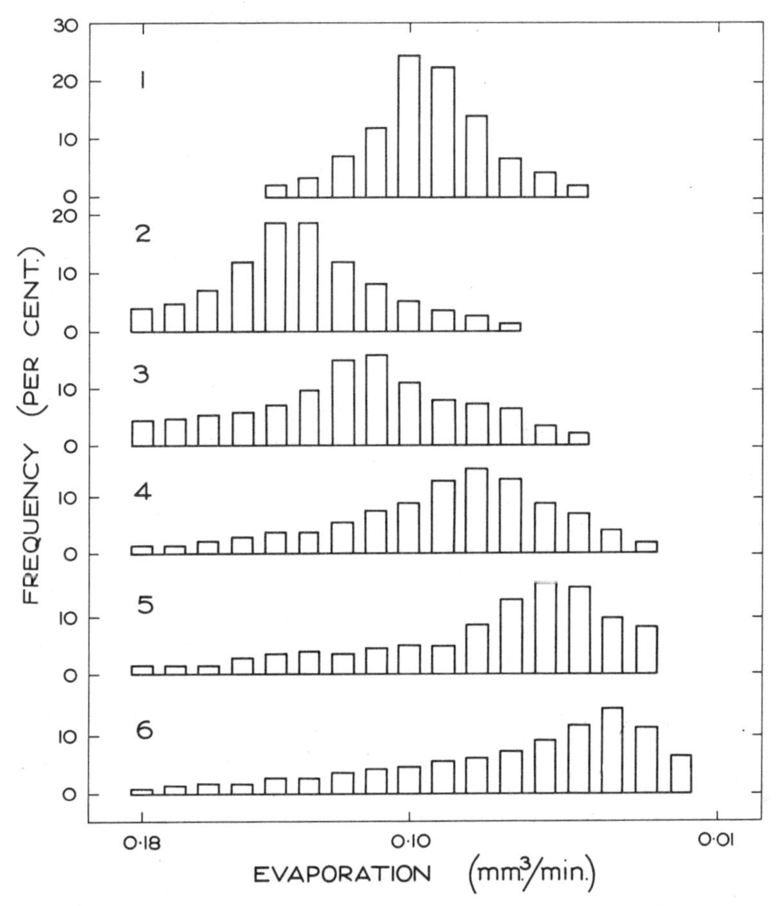

FIG. 7.04.—The evaporation-rates preferred by the six larval instars of the spruce budworm *Choristoneura fumiferana*. Note that the instars differed in their preferences. (After Wellington, 1949*b*.)

the response was intermediate; about half the beetles chose the dry air and half the moist. The results of these experiments are summarized in Figure 7.05. The graphs show (1) that these beetles responded definitely and immediately to the gradient in moisture; (2) that the normal tendency to seek out the dry air was completely reversed when the beetles had been starved for 5–6 days in dry air; and (3) that, although both species are alike in these two attributes, they differ quite markedly in hardiness, for with *T. castaneum* the normal preference for dry air was largely reversed after 2 days' starvation, while with

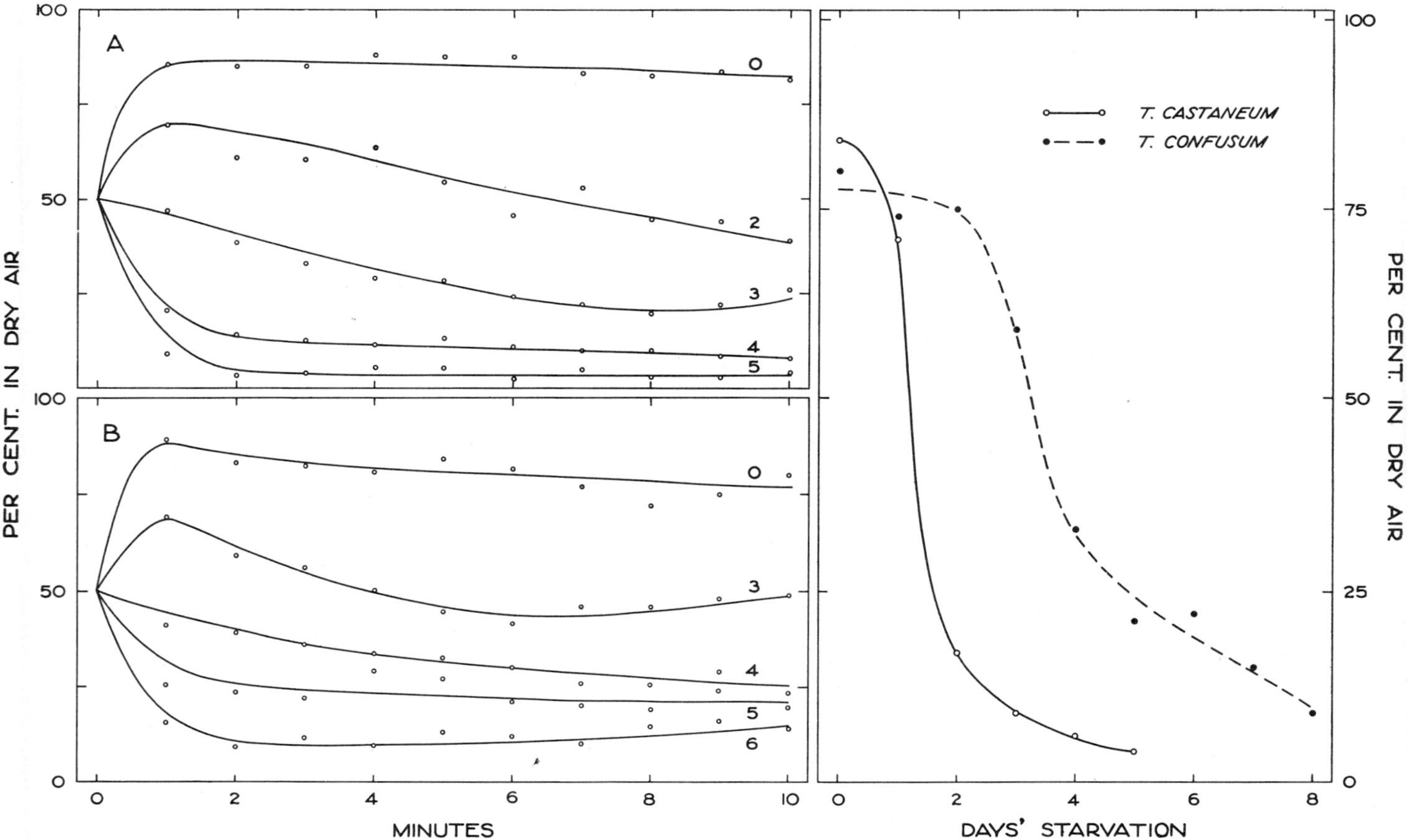

FIG. 7.05.—The preference of adult beetles, *Tribolium*, for moist or dry air and the way this may change in beetles that have been starved and desiccated. *A, Tribolium castaneum:* the proportion of beetles in the dry air stream based on readings taken once a minute for the first 10 minutes after the beetles were placed in the olfactometer. The numbers opposite each curve indicate the number of days the beetles had been starved and desiccated before being tested. *B*, the same as *A* but with *T. confusum*. Note that both species give a definite reaction during the first minute or two, which was usually maintained with little alteration throughout the experiment. *C*, showing the period of starvation and desiccation required to reverse the normal preference of the beetles for dry air. Note that *T. confusum* was hardier than *T. castaneum*. (From data of Roth and Willis, 1951.)

T. confusum at least 4 days were required to produce an equivalent result.

Roth and Willis carried these experiments one step further. Beetles (females of *T. castaneum*) that had been starved in dry air for 4 days (by which time only 7 per cent of them were in the condition to choose the dry air) were then starved for a further 60–70 hours in moist air with a relative humidity in excess of 90 per cent, during which time they continued to lose a little weight (from 69 to 67 per cent of the original weight), but the proportion of water in the body increased slightly, from 52 to 58 per cent. When these beetles were placed in the olfactometer, 42 per cent chose the dry air. The majority (58 per cent) still chose the moist air. Nevertheless, during 60–70 hours' exposure to moist air the condition of the beetles had changed sufficiently for 35 per cent of them to have recovered their normal preference for dry air. Another lot of beetles (males of *T. confusum*) were starved in dry air for 6 days. At the end of this time 13 per cent preferred dry air, and 87 moist. They were then placed in flour that had been oven-dried to a moisture content of 0.13 per cent and kept for 66 hours in this in a desiccator containing calcium chloride. After this treatment, 67 per cent of the beetles preferred the dry air. The opportunity to eat, even though the food was oven-dry flour, had so altered the condition of the beetles that 54 per cent of them had regained their normal preference for dry air.

These striking results are not to be explained in terms of water content alone, for the adults of *Tribolium*, like the larvae of *Tenebrio*, possess the remarkable attribute of retaining a relatively constant proportion of water in their bodies while being starved in dry air (Roth and Willis, 1951). We can only say that the results depend upon the condition of the insect, and leave it at that.

Certain other experiments with other species indicate the same conclusion. For example, adult beetles of *Ptinus tectus*, which had been living in dry flour and had become desiccated before being placed in an apparatus that maintained a gradient in the humidity of the air, were more active than the controls, which had been living in moist flour. At low humidities they showed less tendency to avoid the moist end of the gradient, and at high humidities their normal reaction was reversed, and they collected in the moist region (Bentley, 1944). The behavior of desiccated cockroaches *Blatta orientalis* was similar; they congregated at the moist end of the gradient, in contrast with the controls, which tended to move toward the dry end (Gunn and Cosway, 1938). Larvae of the spruce budworm which had become desiccated chose a moister part of the gradient than the controls. This is illustrated in Figure 7.06.

The reversal or partial reversal of behavior of desiccated animals has an obvious value for the animal. It is more difficult to explain the peculiar behavior of the body louse *Pediculus humanus corporis*, which tends to prefer the humidity it has been living in prior to the experiment. Those kept at 95 per cent relative humidity and then given the choice of 95 and 10 per cent tended consistently to avoid the dry air; but some of those which had been living at 10 per cent before being tested tended to avoid the moist end of the gradient (95 per cent relative humidity). After a time they began to move in the opposite direction, and then they avoided the dry end of the gradient (Wigglesworth, 1941). This complex behavior remains unexplained.

Shelford's (1918) observations of various invertebrates and vertebrates in gradients of temperature, moisture, wind velocity, and salinity (with aquatic animals) convinced him that animals became more sensitive to the stimulus

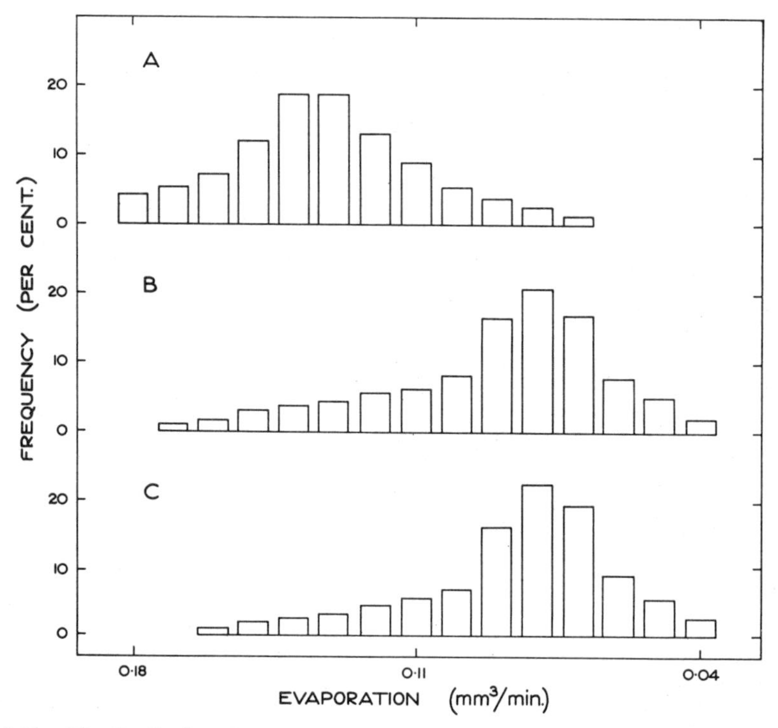

FIG. 7.06.—The distribution of normal and desiccated second-instar larvae of the spruce budworm *Choristoneura fumiferana* in an evaporation gradient. *A*, distribution obtained in the first hour; *B*, the same after 24 hours in the apparatus; *C*, a group of larvae previously dried for 1 hour over calcium chloride and then observed an hour after having been placed in the apparatus. (After Wellington, 1949*c*.)

after encountering its maximal intensity in the experiment. The increased sensitivity lasted for varying periods, depending upon the species. A reversal of behavior, however, such as is given by *Ptinus tectus*, involves more than an increased sensitivity to the same stimulus.

7.14 *Special Sensory Organs in Insects for Perceiving Differences in Atmospheric Humidity*

Moisture in air at constant temperature may be measured in units of absolute humidity, relative humidity, saturation deficit, or rate of evaporation. In certain circumstances, e.g., in a high but constant wind, the rate of evaporation from a free water surface might be very nearly proportional to the saturation deficit of the air; but the rate of evaporation from an insect may not be so simply related to the saturation deficit. It is probable that an insect responds either to the relative humidity of the air or to its evaporating power, but not to both, unless it possesses two distinct sorts of receptors. A receptor which absorbed or lost water until it came into equilibrium with the vapor pressure of the air would register relative humidity like the hair of an ordinary hair-hygrometer; but a receptor which continuously lost water into unsaturated air would register the rate of evaporation, which might in some circumstances, though not necessarily always, be simply related to the saturation deficit. Some insects seem to perceive differences in relative humidity and others in evaporation or saturation deficit.

We have already seen in section 7.12 that the behavior of *Choristoneura* could be explained in terms of evaporation but not of saturation deficit or relative humidity. Lees (1943*a*) studied the movements of the wireworm *Agriotes* in gradients of humidity at three constant temperatures. At the dry end of his apparatus the relative humidity varied from 90 to 92.5 per cent; at the moist end it was about 100 per cent. The intensity of reaction to the gradient was recorded as $100[(W - D)/(W + D)]$, where W and D are the numbers of animals in the wet and the dry half of the gradient. This index would take a value of zero if the animals were indifferent to humidity. Values found in experiments are plotted against relative humidity in Figure 7.07, *A*, and against saturation deficit in Figure 7.07, *B*. The correlation between intensity of reaction and saturation deficit is clearly better than that with relative humidity, suggesting that the receptors on *Agriotes* may be like an evaporimeter. Lees (1943*a*) was unable to identify a receptor in *Agriotes*, though he showed by amputation that the response was dependent upon the integrity of the appendages of the head. In contrast with *Agriotes*, the responses of the mosquito *Culex fatigans* (Thomson, 1938) and the beetle *Tenebrio molitor* (Pielou, 1940) were more closely correlated with relative humidity than with saturation deficit. It might be inferred that in them the receptors are like a hygrometer. Pielou and Gunn (1940) were successful in identifying certain receptors on the antennae of *Tenebrio*. As the antennal segments were progressively amputated, the animals responded less and less strongly to differences in humidity, eventually failing to respond at all. A reaction still occurred when only four of the eleven segments remained on each side. There was no reaction when only three segments were left. The receptors responsible for perception of moisture are "pit-peg" organs scattered along the length of the antennae. The animals did not respond unless the number of receptors left exceeded a certain "threshold." The removal of maxillary palps, on the other hand, had no effect. Hygroreceptors have also been found on the antennae of *Tribolium confusum*,

T. castaneum, and *T. destructor* (Roth and Willis, 1951) and on the antennae of *Blattella germanica* and *Aedes aegypti* (Roth and Willis, 1952).

7.15 *Orthokinesis, Klinokinesis, and Klinotaxis*

Observations on a variety of invertebrates indicate that their behavior in response to a gradient in humidity or to other stimuli may be classified into three types, to which the names *orthokinesis*, *klinokinesis*, and *klinotaxis* have

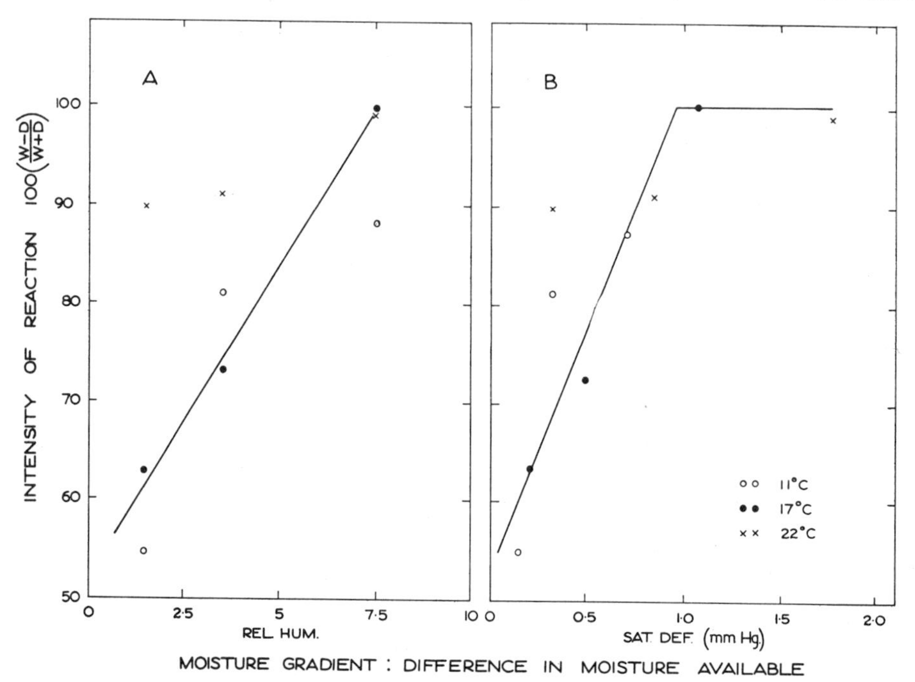

FIG. 7.07.—The relationship between intensity of reaction of *Agriotes* larvae in a moisture gradient to *A*, relative humidity, and *B*, saturation deficit. (After Lees, 1943*a*.)

been given. For example, *Porcellio scaber* tends to move at random with respect to the direction of the gradient in humidity, but it moves very much more rapidly in dry air than in moist. An individual which happens, in the course of its random movements, to find its way into the region of moist air will slow down and become motionless, with the result that, in due course, most of the animals are to be found at rest at the moist end of the gradient (Fig. 7.01). This sort of behavior is called "orthokinesis" (Fraenkel and Gunn, 1940). The adults of the meal worm *Tenebrio molitor* behave similarly, except that they are most active in moist air, so that they come in due course to congregate in the drier parts of the gradient (Gunn and Pielou, 1940). With the body louse *Pediculus humanus corporis* the frequency of random turning movements increases as it approaches an unfavorable zone. The result of this is to prevent the animal from proceeding far into the unfavorable zone if the reaction is strong and immediate. If the reaction is weak, it results in a convoluted course in the unfavorable zone (Wigglesworth, 1941). This is an example of "klinokinesis." When the larvae of *Agriotes* meet with a change of moisture, they sometimes show a directed recoil from the dry air. This is an example of "klinotaxis," because it is a directed and not a random movement. It accounts for only a small part of the movements of *Agriotes* in a moisture gradient (Lees, 1943*a*). The examples given in Table 7.01 indicate that orthokinesis and klinokinesis occur rather frequently and often in the same individual and that klinotaxis is rare.

TABLE 7.01

Classification of Behavior of Certain Insects and Crustaceans in a Humidity Gradient

Species	Type of Movement			Author
	Orthokinesis	Klinokinesis	Klinotaxis	
Porcellio scaber	+	+	−	Gunn (1937); Waloff (1941)
Oniscus asellus	+	+	−	Waloff (1941)
Armadillidium vulgare	+	+	−	Waloff (1941)
Tenebrio molitor (adults)	+	+ (slight)	−	Gunn and Pielou (1940)
Agriotes (larvae)	+	−	+ (slight)	Lees (1943*a*)
Pediculus humanus corporis	−	+	−	Wigglesworth (1941)
Locusta migratoria	+	+	−	Kennedy (1937)
Blatta orientalis	+	−	−	Gunn and Cosway (1938)
Choristoneura fumiferana				
Larvae, second instar	+	+ (strong)	−	Wellington (1950)
Larvae, sixth instar	+ (strong)	+	−	Wellington (1950)

7.16 *The Survival-Value of Response to Differences in Humidity*

The experiments described in the preceding sections show how some animals which usually live in moist places have become adapted to recognize quite small differences in humidity and to move away from places where dryness might become a danger. Certain other species which are adapted (by virtue of water-conserving devices) to live in dry places are equally sensitive to differences in humidity but tend to move in the opposite direction along the gradient. Two good examples are *Tenebrio* and *Tribolium*. This behavior can doubtless be classed as an adaptation that has good survival-value for the species. But there are some puzzling examples of animals which, when tested in laboratory experiments, moved toward the dry end of a moisture gradient, although in nature they seem to survive and multiply better in moister situations. Examples of this puzzling behavior include *Locusta migratoria* (Kennedy, 1937), *Culex fatigans* (Thomson, 1938), *Blatta orientalis* (Gunn and Cosway, 1938), and *Pediculus humanus corporis* (Wigglesworth, 1941). It is possible that in nature these animals are preserved from moving into places where the moisture is unfavorable by response to some stimulus other than humidity.

As a good example of how a complex pattern of behavior in nature can come to be understood, we mention Wellington's (1950) experiments on the spruce budworm *Choristoneura*. The larvae are in diapause when they hatch from the egg, and they immediately seek out a place in which to spin their hibernacula for the winter. The larva's chance to survive during winter and early spring depends upon the place where it hibernates being neither too wet nor too dry. Wellington observed that although the newly emerged larva had a strong innate tendency to begin spinning, it was prevented from doing so by exposure to air in which the humidity was outside the range "preferred" by that particular

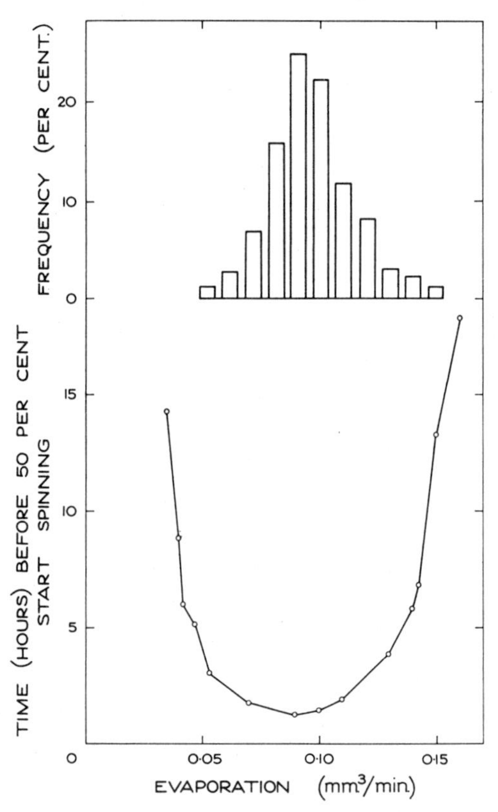

FIG. 7.08.—The relationship between rates of evaporation at which first-instar larvae of the spruce budworm *Choristoneura fumiferana* congregated in a moisture gradient and the time taken by the larvae to begin constructing their hibernacula at different rates of evaporation. *Above:* frequency-distribution of first-instar larvae in an evaporation-gradient; *below:* time required for 50 per cent of the larvae to begin spinning at constant rates of evaporation. (After Wellington, 1950.)

instar (see Fig. 7.08). The second-instar larva emerges from its hibernaculum early in spring, crawls to an outer branch on the tree, and there seeks to penetrate a needle or a bud. At this time of the year there are many situations, particularly in the area where the larvae are emerging from hibernation, where moisture may be excessive; but the tendency to turn most quickly in the zones where the moisture is favorable results in a congregation of larvae in these places (Fig. 7.09). The second "crawling-peak" shown in this figure occurred consistently, and Wellington (1950) suggested an explanation for it. Over the surface of opening buds the pressure of water vapor is higher than in the air surrounding the old foliage, where the second-instar larvae are wandering. Because the larvae are more active in air that is nearly saturated with water vapor, they may have a better chance of penetrating the bud scales than they would at lower humidities.

Sixth-instar larvae in the field inhabit natural feeding tunnels, in which the

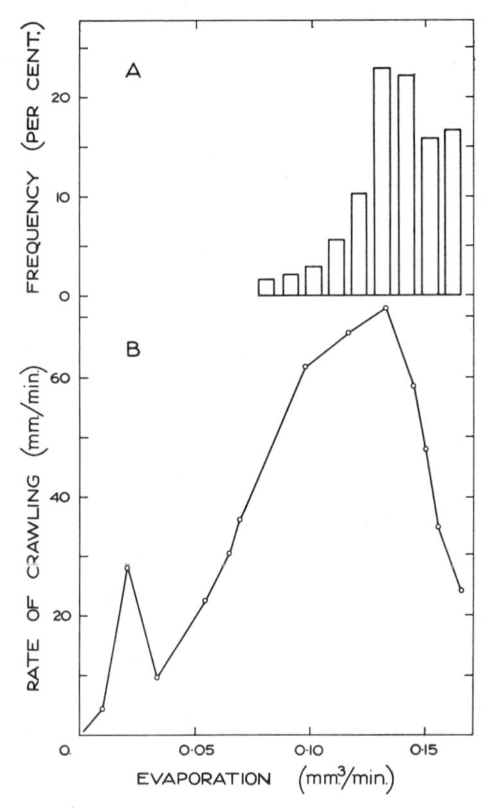

FIG. 7.09.—The relationship between the rate of crawling of the second-instar larvae of the spruce budworm *Choristoneura fumiferana* at different constant rates of evaporation and the rates of evaporation at which larvae congregated in a moisture gradient. *A*, distribution of larvae in a gradient of evaporation; *B*, speed of movement of larvae at different constant rates of evaporation. (After Wellington, 1950.)

rate of evaporation is normally almost identical with that of the preferred zone. When these tunnels are partly destroyed, as happens in older tunnels in which much feeding has gone on, air penetrates from outside, and the rate of evaporation rises. The larvae become active enough in the drier air to leave their damaged home. If tunnels are flooded suddenly, larvae are activated by rates of evaporation close to saturation and react by dropping.

These experiments in the laboratory and observations in the field show how beautifully the spruce budworm is adapted to its environment. Each larval instar is adapted to respond to differences in moisture in such a way that the larvae throughout their life are led to congregate in those places where their chance of survival at any particular stage is greatest.

Pittendrigh (1950) studied the vertical distributions of the mosquitos *Anopheles homonculus* and *A. bellator* in a rain forest in Trinidad. The mosquitos were found concentrated at different heights at different times during the day. The air was moister near the ground, but the gradient varied during the day. The mosquitos moved up and down so as to remain in the place where the humidity was the same. There was a consistent difference between the two species: *A. homonculus* was always found nearer to the ground than *A. bellator;* and the former was active during the moister parts of the day. Presumably, the individuals of each species improve their chances of surviving and multiplying by staying in a place where the humidity suits them best.

Buxton (1923) and Spencer (1928) described a number of remarkable adaptations in the behavior of animals living in deserts. The instinctive behavior of the animal in the presence of a gradient in moisture leads it to seek out and occupy local situations which are not so arid as the general surroundings in the desert. For example, the frog *Lymnodynastes ornatus* in central Australian deserts makes burrows in the beds of temporary streams, digging until it reaches moist sand. The day is spent in the burrow; but at night, when the temperature drops and the humidity rises, it comes to the surface and seeks its food, mostly nocturnal beetles. A fresh burrow is made each day. Another frog, *Helioporus pictus*, which lives in the same deserts, makes permanent burrows in which it lives during periods of drought. The survival of both frogs depends upon their finding moist soil at depths below the dried-up watercourses, and their instincts are nicely adapted for this requirement.

7.2 THE MECHANISM OF WATER BALANCE IN TERRESTRIAL ANIMALS

It is necessary for all but a few sorts of terrestrial animals to keep the proportion of water in their bodies constant within quite narrow limits. The few special exceptions mostly belong to two classes: (*a*) the resting stage in the life-cycle of some species, often associated with diapause in insects and anabiosis in tardigrades and rotifers, and (*b*) a few unusual species, mostly among mollusks and amphibians, which are able to withstand repeated temporary desiccation of the tissues without this being associated with any special resting stage (see sec. 7.31). Apart from exceptions such as these, animals cannot afford to lose, even temporarily, more water than they gain. Water may be lost by excretion, egestion, evaporation through the integument, and as a waste product of respiration. Water may be gained by eating, drinking, or absorption through the integument. If the source of supply is limited, as it often is, particularly with insects, then losses from all sources must be correspondingly reduced.

Terrestrial animals are remarkable for the diversity of efficient adaptations which have been developed in response to the urgent need to conserve and maintain the essential supply of water in the tissues. It is even claimed for some insects that they can retain in their tissues the water which is usually lost as a by-product of respiration in the oxidation of fats and carbohydrates. And it is said that they can increase the stocks of water in their bodies by oxidizing the fats and carbohydrates in excess of their ordinary requirements. These adaptations have long been of interest in relation to the evolution of terrestrial animals. From another aspect they come very much into the province of the ecologist who has to understand the distribution and abundance of terrestrial animals.

7.21 *The Intake of Water from the Surroundings*

7.211 INTAKE OF WATER THROUGH THE ALIMENTARY CANAL

Terrestrial animals obtain water in their food, and some also drink water. Bees and butterflies are commonly seen at the edges of pools, sipping water. The longevity and fecundity of some Lepidoptera are reduced when free water is not available. Muscid flies, e.g., houseflies and blowflies, must be supplied with water to drink if they are to be kept in laboratory cultures. Further examples are listed by Leclerq (1946*a*), who said he had collected 500 species of all orders of insects drowned in pools in summer, victims of their thirst. Most insects which live on dry food, such as grain-meal, dried fruit, etc., live without drinking. There is the odd exception of the adult spider beetle *Ptinus tectus*, which requires a certain amount of water in addition to that which it gets with its food. Ewer and Ewer (1941) found that both longevity and fecundity were reduced if the beetles were not supplied with water to drink. The majority of insects, however, appear to obtain their water in ways other than drinking.

The food of insects varies from liquid food, such as blood and plant sap, moist food with a high water content, such as leaves, which may contain over 90 per cent water, to dry materials, such as cereals or wood, which may contain from about 5 to 20 per cent water. Insects and other arthropods which ingest large quantities of liquid food, such as blood and plant sap, obtain far more water than they need in order to take in the necessary food constituents. Besides an excess of water, there is an excess of chlorides in the food of bloodsuckers. The excess water and chlorides have to be eliminated (see sec. 7.231).

Terrestrial gastropods drink water, though some snails can remain alive without it for months; *Helix pomatia* remained alive for 10 months without water. Snails remained alive because they could retreat into their shells and aestivate or hibernate; but slugs died in a few days when they were not provided with water (Howes and Wells, 1934*a*, 1934*b*).

Reptiles were the most primitive land vertebrates to drink water through the mouth and absorb it by the alimentary canal (Gray, 1928). Typical Amphibia do not drink. They take in water over the whole surface of their bodies. The skin of reptiles is impermeable to water. It is largely due to this that life on dry land became possible for the reptiles, but it brought with it the necessity of learning to drink.

7.212 THE INTAKE OF LIQUID WATER THROUGH THE INTEGUMENT

The impermeability of the skin or the cuticle to water enabled vertebrates and invertebrates to colonize dry land. But Amphibia and some invertebrates can absorb water through the integument. These sorts are usually found in moist places.

Frogs absorb water through the skin, but experiments with *Rana pipiens* showed that it absorbs only liquid water; no water was absorbed from an atmosphere that was saturated with water vapor (Adolph, 1933). When the frog was immersed in solutions hypotonic to the blood, it gained water; but the rate of entry was much less than would be expected from the osmotic gradient between the blood and the surrounding fluid. Normally, at 20° C., 31 per cent of the water content of the body of *R. pipiens* was replaced each day by water absorbed through the skin and excreted subsequently by the kidneys. When the skin was removed or the brain destroyed, water entered four to five times as quickly. This indicates that the skin, under the control of the nervous system or pituitary gland, is responsible for the resistance of the normal animal to osmosis. It indicates that the skin, as well as the kidney, is important in regulating the water content of the body. The way in which the water content of the body is regulated has not been fully explained, but it has been suggested that hormones are important, because extracts of the posterior lobe of the pituitary increased the rate at which water entered. It seemed to increase the permeability of the skin. The increase was greatest in species which live in dry places and least in those which live in wet places (Fig. 7.10; Steggerda, 1937).

Earthworms are terrestrial animals, but in their mode of water balance they behave like fresh-water organisms (Ramsay, 1949). When *Lumbricus terrestris* was living in a fluid medium, water entered through its skin by osmosis. The body fluids are hypertonic to those that surround the worm in the places where it usually lives, and the skin is permeable to water. Excess water is excreted by the nephridia as hypotonic urine. In all these respects the earthworm resembles the frog, but it is not known whether the earthworm has any control over the osmotic uptake of water. Adolph (1927) found that when a desiccated earthworm was allowed to recover in water, its rate of water intake increased many fold, even though the osmotic pressure of the body fluids may not have increased more than twofold. This phenomenon, however, requires to be further investigated before conclusions are drawn about osmotic control in the earthworm.

Some insects and ticks can absorb water through the cuticle. Larvae of the sawfly *Cephus cinctus* (Salt, 1946), ticks (Lees, 1947), and doubtless others absorb water in this way. Eggs of grasshoppers and locusts require to absorb water from the surrounding soil during their development; so far as is known, they require it in the form of free liquid water. The passage of water into and

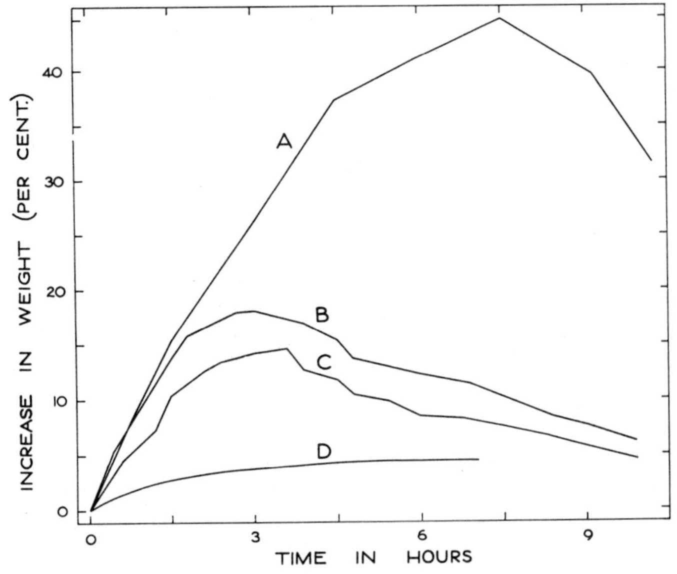

FIG. 7.10.—The increase in weight of various amphibians due to absorption of water following injection of extracts of the posterior lobe of the pituitary. *A*, toad, *Bufo americanus*, usually found in dry places, *B*, frog, *Rana pipiens*, usually found in marshes but not streams; *C*, frog, *R. clamitans*, usually spends part of its time in pools, *D*, mud puppy, *Necturus maculatus*, usually spends all its time in pools. (After Steggerda, 1937.)

out of the egg is not simply the result of osmotic forces but is controlled in a most remarkable way by the activity of "hydropyle cells." These are a small group of exceptionally large cells situated at the posterior pole of the egg, immediately under a specialized part of the cuticle known as the "hydropyle." Water can enter and leave the egg only through the hydropyle, since the rest of the cuticle is practically impermeable to water. The hydropyle was first demonstrated in the American grasshopper *Melanoplus differentialis* (Slifer, 1938). It has since been found to occur in other grasshoppers and locusts; the Australian plague grasshopper *Austroicetes cruciata* (Birch and Andrewartha, 1942), the Canadian *M. bivittatus* (Salt, 1949), and the African locust *Locustana pardalina* (Matthée, 1951). For an understanding of the mechanism of water balance in the grasshopper egg, it is necessary to give a brief account of the nature of the protective egg membranes and the specialized portion known as the hydropyle. We shall describe these structures as they occur in *L. pardalina*, which has been closely studied by Matthée (1951). When the egg is first laid, it is covered by a thin chorion, which has a thin wax layer on the inner surface. Both these layers are secreted by the ovarioles of the parent-grasshopper. They are adequate for the immediate protection of the egg when it is laid in the soil, but after a few days the serosal cells of the egg secrete a hard, tough layer of

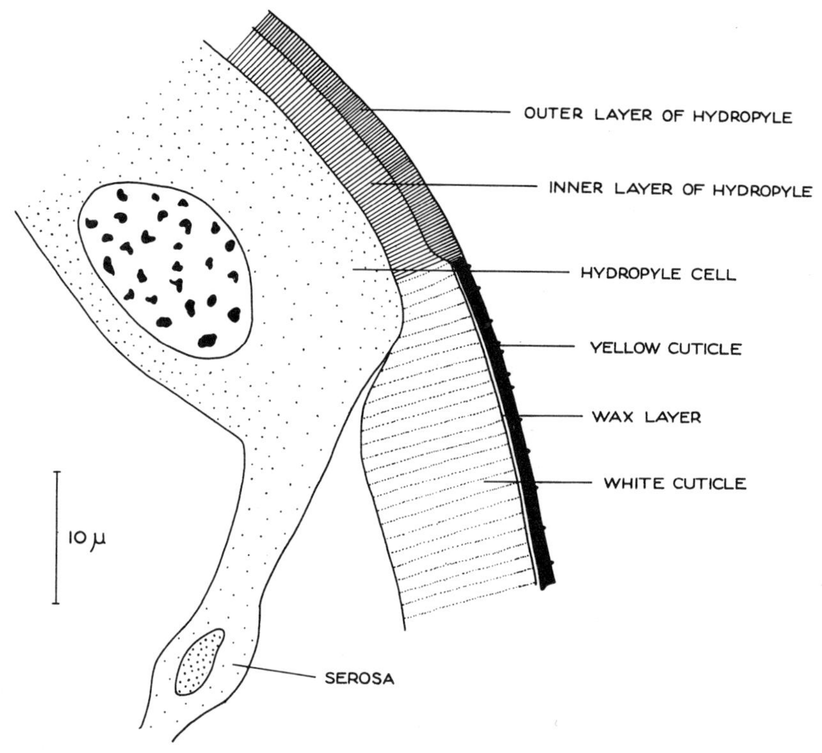

FIG. 7.11.—Section of the cuticle at the posterior end of the egg of *Locustana pardalina* with the chorion removed and showing the special water-absorbing mechanism. Lines at right angles to the surface represent pore canals. (After Matthée, 1951.)

"yellow" cuticle, to be followed by the secretion inside it of a thicker, soft layer, the "white" cuticle. Between the two there is a thin layer of hard wax. This is the layer responsible for most of the impermeability. The primary wax layer beneath the chorion breaks down after the cuticle is secreted. At the posterior end of the egg the membranes are specialized as the hydropyle. The outer layer of the hydropyle is continuous with the yellow cuticle but is several times thicker. Pore canals run at right angles to the surface (Fig. 7.11). The inner layer of the hydropyle is continuous with the white cuticle of the rest of the egg. It also contains pore canals, but they are much finer than those in the outer layer. The wax layer of the rest of the cuticle is almost certainly absent in the region of the hydropyle. Beneath the hydropyle are the large hydropyle cells which originally secreted it. They are continuous with the serosal cells, which lie under the white cuticle.

The eggs of *Locustana* absorb water from the surroundings during their development and become quite turgid (Fig. 7.12). Little water is absorbed during the first few days because of the absence of the hydropyle. Matthée (1951) has shown that water is absorbed through the hydropyle against an

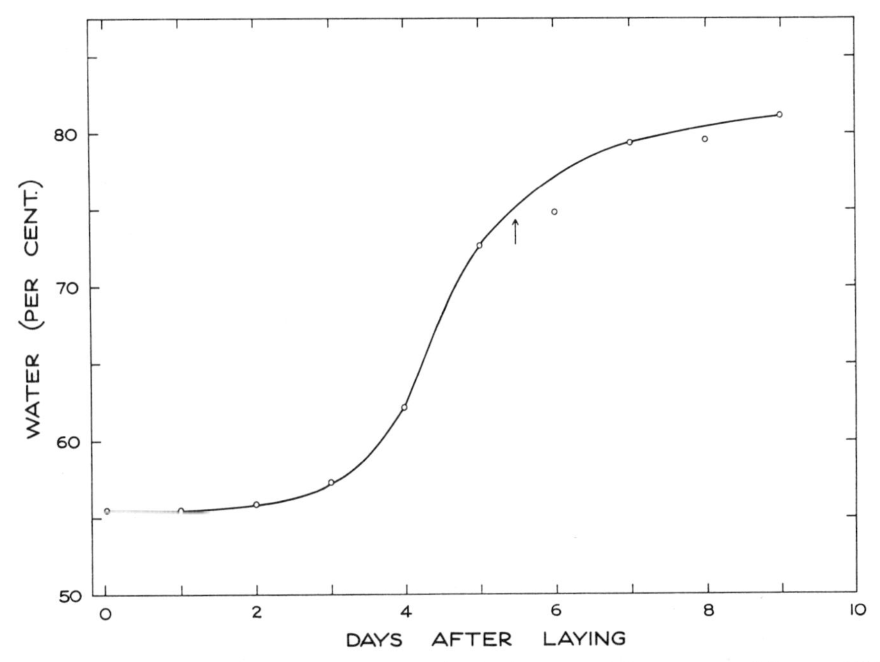

FIG. 7.12.—The increase in water content of developing eggs of *Locustana pardalina* at 35° C. The arrow indicates the beginning of katatrepsis. (After Matthée, 1951.)

osmotic gradient. Water was absorbed from glucose solutions having an osmotic pressure of 14.56 atmospheres (equivalent to a 2.1 per cent NaCl solution) when the osmotic pressure of the yolk of the eggs was only 11.92 atmospheres (equivalent to a 1.7 per cent solution of NaCl). When the eggs were deprived of oxygen in an atmosphere of nitrogen, they did not absorb water, though they remained alive. This is evidence of the necessity of an expenditure of energy in the absorption of water against an osmotic gradient. The outer layer of the egg (chorion) is quite porous, and water readily reaches the hydropyle through it. The outer surface of the hydropyle is, however, sealed off by a thin layer of protein. It is this protein layer which prevents water from being lost from the egg when it is in a dry atmosphere. But in the presence of water it dissolves, exposing the large pore canals in the outer layer of the hydropyle. Ultimately, the water reaches the cytoplasm of the small pore canals in the inner layer of the hydropyle. This contact evidently provides the stimulus for the active absorption of water by the hydropyle cells. Water is actively absorbed by them and then secreted into the posterior extremity of the egg between the serosa and cuticle, where it accumulates. This accumulation serves as a reservoir from which the embryo obtains its water. It also facilitates the revolution of the embryo during katatrepsis. When the egg cuticle dries out, in the absence of free water, the hydropyle cells secrete the protein which blocks the pore canals in the hydropyle and so prevents the loss of water. This extraordinary living valve makes it possible for the egg to take advantage of the moisture in soil when it is wet and prevents loss of water into soil which is dry. Slifer (1946) suggested that in the eggs of *Melanoplus differentialis* the hydropyle may be covered with a thin layer of wax which prevents water from entering or leaving the egg when it is dry. Her evidence for this is the increased permeability of the egg after immersion in xylol.

The activity of the secretory hydropyle cells may vary during the development of the egg, and this could account for the different rates with which water is absorbed during the development of some acridid eggs. The pattern of water absorption varies in different species. In *M. differentialis* (Bodine, 1929) and *M. bivittatus* (Salt, 1946) the increase in moisture content of the egg was interrupted during diapause; in the former species water was absorbed chiefly during katatrepsis, but in the latter it was absorbed chiefly during anatrepsis. The eggs of *Austroicetes cruciata* absorbed about 40 per cent of the water which they needed during the process of diapause-development. The remaining 60 per cent was absorbed during post-diapause-development (Birch and Andrewartha, 1942). It is probable that hydropyle cells occur in other insects besides acridids. Groups of enlarged serosal cells have been described by various authors in a number of insects. Matthée (1951) pointed out that the serosal cells immediately under the micropyle in *Anurida maritima*, described by Wheeler (1893) and called the "precephalic organ" (Imms, 1906), are analogous to hydropyle cells. The "indusium" in the egg of the mantid *Paratenodera* (Hagan, 1917) appears to be a similar structure. Nothing is known about the function of these cells in these two species. There are, however, other species which possess enlarged serosal cells at the micropylar region and which also absorb water during development, e.g., *Tachycines* (Gryllidae) (Krause, 1938), *Pteronarcys* (Plecoptera) (Miller, 1940), and *Notostira* (Hemiptera) (Johnson, 1934, 1937). Matthée pointed out that the serosal cells in the eggs of these insects quite probably play a part in the absorption of water similar to the role of the hydropyle cells in grasshopper eggs.

7.213 THE INTAKE OF WATER VAPOR THROUGH THE INTEGUMENT

Some insects and ticks are able to absorb water vapor from air which is not saturated. This is of interest, since it indicates absorption of moisture against a vapor-pressure gradient. In the adult female of the tick *Ixodes ricinus*, the haemolymph has an osmotic pressure equivalent to that of 1 per cent NaCl solution and would therefore be in equilibrium with the water vapor in air in which the relative humidity was 99.4 per cent. But the unfed tick can take water from an atmosphere in which the relative humidity is 92 per cent or more. There is thus a passage of water vapor from a pressure corresponding to 92 per cent relative humidity to one of 99.4 per cent relative humidity (Lees, 1946*a*). The movement of water molecules against a vapor-pressure gradient involves the expenditure of energy in secretory activity analogous to the secretory activity of the hydropyle cells of the grasshopper egg, which can absorb water against an osmotic gradient (sec. 7.212). This uptake of water molecules against a gradient is quite analogous with the uptake of salts from very dilute media characteristic of a number of fresh-water animals, including mosquito larvae.

In dry air ticks lose water. If the humidity is raised progressively, a point is reached, namely, at 92 per cent relative humidity, where evaporation ceases. At this relative humidity the unfed tick neither loses nor gains in water. Lees (1946*a*) called it the "equilibrium relative humidity." Water is taken up at higher relative humidities. But it is taken up only so long as the water content of the tick's haemolymph is depleted, as it would be if the tick had been previously desiccated. It is known that this absorption of water in the tick takes place through the cuticle and not through the tracheal system, since blocking of the spiracles in no way interferes with the process. It is a function of the hypodermal cells below the cuticle, in much the same way as the hydropyle cells below the egg cuticle in some insects function to absorb water against a gradient. Lees (1946*a*) closed the anus of a tick with cellulose paint and thus prevented the tick from excreting water. He found that such ticks, when exposed to humid air, ceased to increase in weight after the body had replenished its normal stocks of water. This showed that the hypodermal cells ceased absorbing water and that the normal balance of water in the tissues was maintained in this way rather than by excretion. The process of water balance in the tick resembles that of the frog in this respect, for the water content of the frog is under the control of the skin rather than the kidneys (sec. 7.212). Lees (1946*a*) found that seven other species of ticks, distributed among five genera, were all able to absorb water from atmospheres varying from 82 to 96 per cent relative humidity, depending upon the species.

Wigglesworth (1931*a*) showed that the bug *Cimex lectularius* absorbed water from unsaturated air. Breitenbrecher (1918) showed the same for the beetle *Leptinotarsa decemlineata*. The larvae of *Tenebrio molitor* absorbed water from air with a relative humidity of 88 per cent (Mellanby, 1932*a*); the grasshopper *Chortophaga viridifasciata* from air with 82 per cent relative humidity (Ludwig, 1937); and the flea *Xenopsylla cheopis* from air with relative humidity of 50 per cent (Edney, 1947). In similar experiments with terrestrial Crustacea the isopods *Armadillidium* and *Ligia* were able to absorb moisture from unsaturated air only when the relative humidity was as high as 98 per cent. No gain in water took place below this relative humidity (Edney, 1951).

In a series of experiments of a rather different sort, Govaerts and Leclerq (1946) exposed various species of insects for varying periods in air which was saturated with water vapor—but 8 per cent of this was heavy water. Samples of the water taken, after an adequate period, from the tissues of the insects were found in each case to contain about 8 per cent of heavy water, indicating that the water in the insect's body had come into equilibrium with the water in the air, at least with respect to heavy water. With the larva of *Tenebrio molitor*, equilibrium was reached after 13 days; with the adults of this species equilibrium was reached after 9 days. From the results of these experiments Govaerts and Leclerq concluded that there is a continuous gaseous exchange of water vapor through the cuticle of insects; but it is far more likely that the exchange of heavy water which they demonstrated in these experiments took place through the tracheal system during the course of respiration. Otherwise, it is difficult to account for all the examples (including the adults of *T. molitor*) which are known to be unable to gain water by absorbing it through the cuticle even from air that is nearly saturated with water vapor.

The absorption of water from unsaturated atmospheres must have considerable survival-value. All the animals considered in this section lose water in dry air, and their chance of survival must be enhanced by their ability to replenish such losses without necessarily having to find a place where the air is saturated with water vapor. For example, the tick *Ixodes ricinus* is restricted to relatively moist places, because it loses water in dry air and its life-cycle is such that it has to go for long periods without food. But it can survive in places where the air is not quite saturated, by virtue of its capacity to replenish its water supply by absorbing water vapor from air that is 92 per cent (or more) saturated. In the absence of this capacity, the distribution of *I. ricinus* would be considerably smaller than it is (Lees, 1946*a*; Milne, 1950*a*; see also sec. 12.22).

7.22 *The Conservation of "Metabolic Water"*

Berger (1907) found that larvae of the meal worm *Tenebrio molitor* maintained a constant proportion of water to dry matter in their bodies when starved in unsaturated air; and Buxton (1930) confirmed these early experiments. Even when starved for a month in an atmosphere of zero humidity, the larvae retained a constant water content in their bodies. Buxton argued that more water was lost in a dry atmosphere as compared with a moist one and that this loss was made good by "water of metabolism" formed during the oxidation of fats and carbohydrates and retained in the body. He considered that the rate of production of metabolic water was determined by the humidity of the atmosphere surrounding the insect.

Besides the meal worm, a number of other insects are known to maintain a constant proportion of water in their bodies in a dry atmosphere e.g., adults of *Lachnosterna implicata* (Sweetman, 1931), larvae of the flour moth *Ephestia kühniella*, and the beetles *Tribolium confusum* (Roth and Willis, 1951) and *Dermestes vulpinus* (Fraenkel and Blewett, 1944). These insects and others that live on dry food or in dry places must be under considerable stress to maintain an adequate amount of water in their bodies, and the experiments of Buxton and others show that they do indeed possess remarkably effective mechanisms for conserving water. Buxton's hypothesis about the conservation of metabolic water implies that the insects, by doing work, can dry out the air before it is expired from the spiracles and thereby retain in their bodies some of the water that is produced as a by-product of respiration. This is quite a credible hypothesis, especially as the absorption of water from an unsaturated air through the cuticle of a tick (Lees, 1946*a*) and the absorption of water by the grasshopper egg against an osmotic gradient (Matthée, 1951) would seem no less remarkable and may require similar mechanisms. But, so far, none of the experiments which have been done to test this hypothesis has sufficed to verify it beyond doubt.

If Buxton's hypothesis is true, it would seem to follow as a consequence that insects fasting in dry air would show a higher rate of metabolism than insects fasting in moist air. Mellanby (1932*a*, *b*, 1936) was not able to detect such differences. For example, he found with *Cimex* that bugs fasting in moist air used up food reserves at about the same rate as those fasting in dry air. Mellanby (1936) doubted this hypothesis on the grounds that an additional oxidation of food to obtain metabolic water results in an additional flow of air through the spiracles, both to let oxygen in and to let carbon dioxide escape. He suggested that this would also result in a corresponding additional loss of water.

An extension of the original theory states that fat stores in animals that live in deserts serve as a protection against desiccation. Mellanby (1942) doubted this theory also on the following grounds. According to this theory, fats are more use than carbohydrates, since every 100 gm. of fat completely oxidized gives 110 gm. of metabolic water, but every 100 gm. of carbohydrates gives only 55 gm. of water. Mellanby pointed out that fat requires more oxygen to oxidize it than do carbohydrates; 255 gm. of oxygen are required for the complete combustion of 100 gm. of fat, but only 213 gm. of oxygen are required for the combustion of the same weight of carbohydrates. The important feature, he claimed, is the amount of water produced per unit of oxygen consumed, since the more oxygen taken in, the more water is lost in respiration. Mellanby considered that these and other arguments indicated that carbohydrates are a better source of metabolic water than fats. If food reserves were used in this way, one would therefore expect them to be stores of glycogen or some other carbohydrate rather than fat. But all this theoretical argument overlooks the possibility, which is indeed implicit in Buxton's original theory, that the insect, by doing work, is able to remove water vapor from the air before it is expired. This is the critical question which still requires to be answered by a properly designed experiment.

Fraenkel and Blewett (1944) claimed to have provided conclusive evidence for the conservation of metabolic water; but, in fact, the design of their experiments was not adequate to distinguish between two alternative explanations, and therefore their experiments left this critical question unanswered. The constancy of the water content of the insects in their experiments in moist and dry air could have been due to (1) conservation of metabolic water or (2) dehydration of undigested food. The latter alternative was suggested as an explanation of the constancy of the water content of *Tenebrio molitor* larvae by Schulz (1930), but his suggestion was not substantiated by evidence. Unfortunately, Fraenkel and Blewett's experiments do not provide for the possibility that the insects in their experiments were able to dehydrate undigested food. They reared larvae of three insects, *Tribolium confusum*, *Ephestia kühniella*, and *Dermestes vulpinus* in food of different moisture content. Temperature and relative humidity were kept constant during the experiments. They measured: (*a*) dry weight of food eaten (this was taken as equivalent to the original weight of food minus the weight of the residue left; but the residue contained undigested food which had passed through the alimentary canal, as well as excrement passed out in the feces); and (*b*) dry weight and water content of pupae. From these two measurements, the ratio of dry weight of food eaten to dry weight of pupa was obtained. This gave the number of grams of food required to give 1 gm. of dry matter of insect tissue. The results of one of their experiments with *E. kühniella* are summarized in Table 7.02. The difference in absolute water content of the pupae reared from

TABLE 7.02*

RELATIONSHIP BETWEEN MOISTURE CONTENT OF FOOD AND AMOUNTS OF FOOD CONSUMED TO PRODUCE EQUIVALENT WEIGHTS OF DRY MATTER IN TISSUES OF *Ephestia kühniella* REARED IN FLOUR AT 25° C.

Moisture Content of Food (per cent)	Relative Humidity (per cent)	Wet Weight of Pupa (Mg.)	Per Cent Water Content of Pupa	Ratio Dry Weight Food to Dry Weight Pupa
14.4	70	25.5	68	6.3
6.6	20	18.7	66	9.0
1.1	0	15.8	64	12.7

* After Fraenkel and Blewett (1944).

larvae in dry air and moist air was remarkably small. But more food was eaten by those living on the drier foods. Thus 1 mg. of dry matter in the insect's tissues was produced from 6.3 gm. of dry matter in the food at 70 per cent relative humidity, but it required 12.7 gm. dry matter in the food at zero relative humidity. The results indicated that more food was eaten in dry atmospheres; but the water may have been obtained from this food either by dehydration or by oxidation of the carbohydrates and fats. Fraenkel and Blewett (1944) stated that, at 1 per cent relative humidity, less than 7.6 per cent of the water in the pupa can be derived from water ingested with the food; the rest they presume must be derived from water of metabolism. This and other comparable figures in their calculations are based on the amount of food digested. But they did not allow for the possibility of water being extracted from undigested food passed through the alimentary canal, since this was not measured. The critical experiment in which the food residue is divided into its constituents of undigested food and excretory products has not yet been done, so far as we know.

We can conclude from Fraenkel and Blewett's experiments that these three insects worked harder to produce 1 gm. of dry tissue when the food was dry than when the food was moist. But the experiments do not tell us whether the extra work was done in dehydrating undigested food or in the oxidation of fats and carbohydrates for the production of metabolic water.

The characteristic behavior of the Australian plague grasshopper *Austroicetes cruciata* prior to laying its eggs suggests that the animal may obtain the water necessary for the eggs and the frothy secretion in which the eggs are laid by oxidizing food reserves in its body. The grasshopper lays its eggs in late spring, a time of the year when the amount of green food available is quite small, evaporation is high, and water is scarce. She lays about 200 eggs in the course of several weeks. Not only does she find the necessary 600 mg. of water for the eggs themselves, but she covers each batch of 20 eggs with a frothy secretion which makes the surrounding soil quite wet. Where does this water come from? One of the most striking features about grasshoppers which are about to lay eggs is the almost incessant activity of the females in the area of the egg bed. Throughout the warm part of the day, whenever they are not feeding or ovipositing, the females are flying backward and forward about 6 feet above the surface of the ground. This activity can be proved to be a prerequisite for egg-laying, for grasshoppers which were prevented from partaking in these flights laid no eggs. If increased respiration inevitably results in increased loss of water through the spiracles, this seems an incredible sort of behavior for a species otherwise beautifully adapted for life in an arid environment. If, however, it is possible to gain water from the tissues by oxidizing fat and carbohydrate, then this otherwise aimless activity might appear to be a useful adaptation.

7.23 *The Loss of Water to the Surroundings*

Besides possessing special adaptations for obtaining water from dry surroundings, terrestrial animals have adaptations which enable them to reduce the amount of water lost to the surroundings. Water is lost in excretion, egestion, respiration, and by transpiration through the surface of the body. We shall now consider ways in which loss through these channels is reduced in certain terrestrial animals.

7.231 EXCRETION AND EGESTION

The amount of water lost with the excrement depends upon the amount of water in the food and the moistness of the place where the animal is living. Sapsuckers, like aphids, and bloodsuckers, like ticks, take in far more water than they need. They have special devices for getting rid of the excess. In the bloodsucking bug *Rhodnius prolixus* (Wigglesworth, 1931*b*) and the tick *Ixodes ricinus*, excess water is eliminated by the Malpighian tubules. In ticks of the genus *Ornithodorus* the excess water and salts are excreted by special coxal glands (Lees, 1946*b*).

In contrast to these animals which live on liquid food, there are many animals which live in dry surroundings on dry food. For them, water is a scarce commodity, and the problem is not to get rid of it but to save it. We find that such animals have highly efficient devices for reducing the amount of water lost in excretion and in the feces. There appear to be two main devices by means of which this is brought about: (*a*) by reabsorption of water in the rectum or the tubules of the excretory organ or (*b*) by excretion of nitrogen as uric acid, which is insoluble and hence nontoxic. Consequently, there is no need for a large flow of water to wash it away, and no harm is done if it accumulates as a crystalline mass. In many insects water is reabsorbed by the rectal glands from the feces before they leave the rectum, e.g., in *Tenebrio molitor* and *Pediculus humanus*. In others it is reabsorbed in the lower part of the Malpighian tubules (Wigglesworth, 1931*b*). In birds the cloaca receives the rectum and the ureters. In this common chamber, water is reabsorbed, resulting in a semifluid excrement. The kidneys of birds and mammals, for the first time in evolution, produced a concentrated urine hypertonic to the blood. Useful as this would surely be in lower vertebrates, it occurs only in birds and mammals, and that by virtue of the special loop in the kidney tubules known as the "loop of Henle," in which the absorption of water back into the bloodstream takes place. In terrestrial reptiles and frogs from deserts, the glomerulus is reduced in response to the need for the conservation of water, but there is no section of the tubule which is efficient for the reabsorption of water. The correlation of uric acid excretion with life in dry environments can be demonstrated by an examination of the excretory products of different species within groups of animals, such as gastropods, crustaceans, and reptiles, within which the transition from aquatic life to terrestrial life can be studied.

Gastropods.—Needham (1935) was the first to use this method when he studied the littorinid periwinkles, a series which ranges all the way from strictly aquatic to strictly terrestrial species. He measured the amount of uric acid accumulated in the tissues of the whole animal and compared this with the amount found in the nephridium, on the assumption that this gave a good indication of the extent to which uricotelic metabolism occurred. This assumption has since been questioned by Dresel and Moyle (1950). Nevertheless, Table 7.03 shows a steady increase in the relative amount of uric acid in the tissues of the animal as a whole, as the series proceeds from *Littorina littorea*, which is strictly aquatic, to *L. neritoides*, which is strictly terrestrial. Aquatic

TABLE 7.03*

URIC ACID CONTENT OF NEPHRIDIA AND BODY TISSUES OF SERIES OF GASTROPODS OF GENUS *Littorina* IN RELATION TO DRYNESS (HEIGHT ABOVE LOW TIDE) OF SITUATIONS IN WHICH THEY ARE USUALLY FOUND

SPECIES	DESCRIPTION OF SITUATIONS WHERE SPECIES OCCUR	URIC ACID CONTENT (MG/GM WET WT.)	
		Whole Body	Nephridium
L. littorea	Wholly marine	0.21	1.4
L. obtusata	Intermediate intertidal zone	0.37	5.1
L. rudis	Intermediate intertidal zone higher than *L. obtusata*	1.05	....
L. neritoides	Terrestrial; extends 7 feet above high-tide mark	1.83	....

* After Needham (1935).

gastropods excrete much ammonia and little uric acid, whereas terrestrial gastropods secrete little ammonia and much uric acid. The uric acid content of the nephridium of marine operculates is very low. Terrestrial snails and slugs, on the other hand, such as *Helix*, *Limax*, and *Arion*, contain much uric acid. *Helix pomatia* empties its nephridium only once every 2 to 3 weeks. Others probably excrete even more rarely. The transitional littorinids shown in Table 7.03 are intermediate in the uric acid contents of their nephridia when compared with completely marine and completely terrestrial gastropods (Needham, 1935).

Amphipods and isopods.—Certain representatives of these two groups are entirely aquatic; others are semiaquatic and live at different levels of the seashore and in moist places like leaf-litter in rain forests, e.g., amphipods of the genus *Talitris* in Australia; still others, like wood lice, are fully terrestrial. The nitrogen excretion of eleven species of isopods and amphipods from these various places was studied by Dresel and Moyle (1950). Somewhat to their surprise, they found that all the species studied were essentially ammonotelic, since more than 50 per cent of the total nonprotein nitrogen excreted was ammonia; some 5–10 per cent was uric acid in terrestrial isopods, as well as in the fresh-water *Asellus aquaticus*. Table 7.04 shows, however, that the amount of nitrogenous excretory products was less in the terrestrial species than in others. It seems that adaptation to life on land has been attended by a suppression of nitrogen metabolism rather than by the transformation of toxic ammonia to less toxic products, such as urea or uric acid. A surprising feature of these results is that terrestrial Crustacea excrete nitrogen as ammonia, even in the absence of a plentiful supply of water. This raises the question, as yet unanswered, of whether these invertebrates are less susceptible to the effects of ammonia than are the more highly evolved insects and vertebrates. Perhaps this is the way in which ammonotelism is possible among these terrestrial Crustacea.

TABLE 7.04*

MEAN QUANTITIES OF NONPROTEIN NITROGEN EXCRETED BY SERIES OF AMPHIPODS AND ISOPODS RANGING FROM AQUATIC TO TERRESTRIAL SPECIES

Species	Description of Situations where Species Occur	Mg N/10 Gm/24 Hours	
Gammarus locusta	Marine littoral zone	4.9	Aquatic (mean = 3.5)
Marinogammarus marinus	Marine littoral zone	1.1	
M. pirloti	Marine littoral zone	2.9	
Gammarus zaddachi	Estuaries	6.0	
G. pulex	Fresh water	2.3	
Asellus aquaticus	Fresh water	2.6	
Orchestia sp.	Land and water	2.0	Semiterrestrial (mean = 1.7)
Ligia oceania	Land and water	1.3	
Oniscus asellus	Dry land	0.3	Terrestrial (mean = 0.35)
Porcellio laevis	Dry land	0.3	
Armadillidium vulgare	Dry land	0.4	

* After Dresel and Moyle (1950).

Onycophora.—This group is interesting because it comes between the annelids and arthropods and uricotelic metabolism has been acquired in response to the needs of life on dry land (Manton and Heatley, 1937).

Chelonia.—Moyle (1949) showed that the turtles could be classified into three groups according to the composition of their nitrogenous excretory products, viz., one group which is essentially ureotelic, a second group in which urea predominates but some ammonia is produced as well, and a third group in which urea and ammonia are produced in approximately equal quantities. The usually accepted theory of the evolution of the Chelonia, based on morphology and paleontology, states that the tortoises and turtles have evolved from an early amphibious stock. Some of these returned to the water and became strictly aquatic, others remained on land but became associated with moist situations, while a third group became strictly terrestrial and became adapted to live in quite dry situations. If we assume that the original stock of the Chelonia was ureotelic, Moyle's biochemical studies may be considered to support the evidence from morphology and paleontology regarding the evolution of this group. The species from the desert all belong to the first group, which is ureotelic; the second group includes all the semiterrestrial species; and the aquatic species belong to the group in which urea and ammonia are excreted in about equal proportions. This suggests that the aquatic members of the Chelonia are descended from terrestrial ancestors.

The general conclusion to be drawn from comparing the nitrogen excretion of animals ranging from aquatic to terrestrial is that as the supply of water decreases (or as the drain on water increases), so the tendency to excrete nitrogen as uric acid or urea increases. This has reached its climax in terrestrial

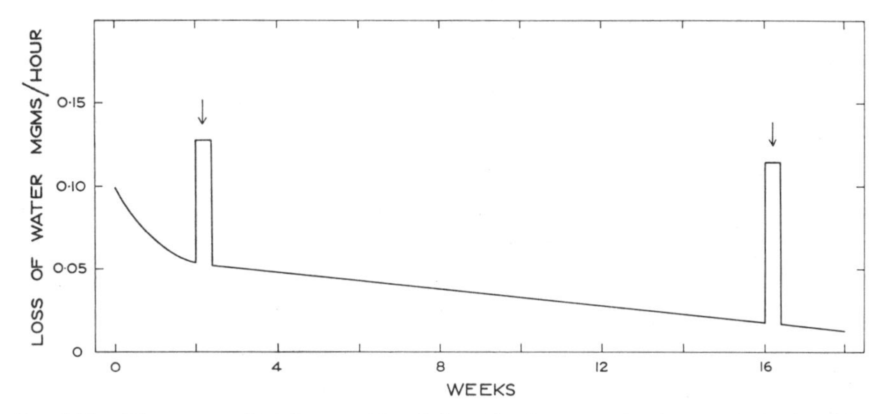

FIG. 7.13.—The rate at which larvae of *Tenebrio molitor* lost water in air at a relative humidity of 0–15 per cent, before and after 5 per cent carbon dioxide was added to the air. The arrows show when carbon dioxide was introduced, with consequent opening of the spiracles. (After Mellanby, 1934*a*.)

insects, some toads, reptiles, and birds which live in dry places. The Crustacea, as we noticed above, form an exception to this general rule.

7.232 RESPIRATION

At one time it was considered that the insect cuticle was quite impermeable to water and that water was lost only through the spiracles and with the excrement (Hazelhoff, 1928; Buxton, 1932*a*). This view was also supported by Mellanby (1934*a*), working with the larvae of *Tenebrio molitor*, *Tineola bisselliella*, and larvae and adults of *Xenopsylla cheopis*. He compared the rate at which water was lost from insects in normal air and in air of the same humidity but with 5 per cent carbon dioxide added to it. This caused the spiracles to remain permanently open. He found that in some cases water was evaporated from the insects as much as seven times more rapidly than from the controls, in which the closing mechanism for the spiracles was still functioning normally (Fig. 7.13). On the other hand, the spiracles of the larval flea (*Xenopsylla*) lack a closing device, and with this insect the addition of carbon dioxide to the air made little difference to the rate at which water was evaporated from the larva. Mellanby also found that the adult flea with its spiracles fully open lost water at about the same rate as the larva, despite the fact that the cuticle of the larva is thin and that of the adult thick. These experiments led Mellanby to the conclusion that practically all the water evaporated from an insect's body is lost through the tracheal system; and he deduced from this that insects inhabiting dry places have more need for effective means of closing the spiracles than for a thick cuticle. It is now known that the laws governing the evaporation of water from the insect are not so simple as this.

Ramsay (1935*a*) was able to show that transpiration from the cockroach *Periplaneta americana* took place through both spiracles and cuticle. Blocking the spiracles with wax did not completely prevent loss of water, and at temperatures above 30° C. this loss was considerable. It is not possible to estimate the relative amounts of water lost through the spiracles and the cuticle in these experiments, because the total loss of water through spiracles and cuticle together was not determined. Instead, Ramsay attempted to measure the loss through the spiracles directly by painting the cuticle with shellac to make it impermeable and at the same time fixing the spiracles open with wax. But, unfortunately, the shellac was not effective in completely preventing some loss of water through the cuticle.

With some species of insects, e.g., the nontracheate Collembola, which live in moist situations, the cuticle is relatively permeable to water, and the rate at which water is lost through the cuticle into dry air may be quite considerable (Davies, 1928). In others the cuticle is remarkably impermeable, and most water is lost through the tracheae, but this can be regulated by mechanisms for opening or closing the spiracles, as in the larvae of *Tenebrio molitor*. In others the cuticle is impermeable, but the rate of transpiration through the spiracles cannot be controlled. The larva of the flea *Xenopsylla cheopsis* is an example of this type, in which, as Mellanby quite rightly concluded, most of the water is lost through the spiracles. The Onycophora are like the flea larvae in this respect. They possess a cuticle but are unable to control loss of water from their spiracles. They are found only in damp situations, and this is probably associated with the tremendous number of spiracles (in some places only 80 μ apart), which are permanently open (Manton and Ramsay, 1937). For example, *Peripatopsis* lost water twice as quickly as an earthworm, forty times more rapidly than the caterpillar (*Amphipyra pyramida*), and eighty times more rapidly than the cockroach.

7.233 TRANSPIRATION THROUGH THE INTEGUMENT IN ARTHROPODA, GASTROPODA, AMPHIBIA, AND REPTILIA

The arthropod cuticle, especially that of insects, has long been the subject of study, but it is only in recent years that its structure and function have been elucidated at all fully. It is convenient to begin the story with the ecological experiments of Gunn (1933), though, of course, it started much earlier than this. Gunn measured the rate at which the cockroach *Blatta orientalis* lost water in atmospheres of different relative humidities and at different temperatures. The rate of loss of water was slow at temperatures below 30° C., but above this temperature the rate of loss suddenly increased (Fig. 7.14). Gunn thought this was due to a change in type of respiration from simple diffusion in and out of the trachea at temperatures below 30° C. to an active process of ventilation by pumping movements at 30° C. and above. This explanation was not wholly convincing, and the empirical facts of this experiment remained as a challenge to the physiologist. Ramsay (1935*a*) repeated

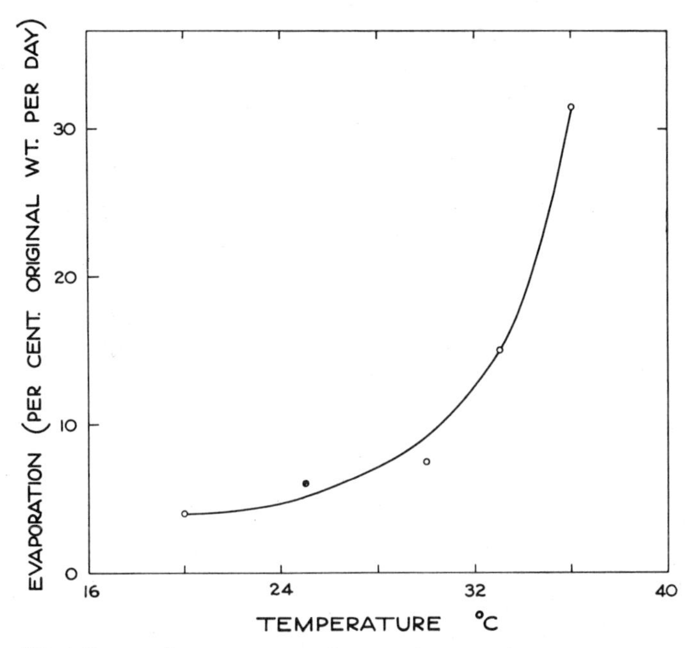

FIG. 7.14.—The influence of temperature on the rate of evaporation of water from the cockroach *Blatta orientalis* at a saturation deficit of 20 mm. mercury. (After Ramsay, 1935*a*, modified from Gunn, 1933.)

Gunn's experiments on the cockroach *Periplaneta americana*, but this time blocking the spiracles with wax so that effects due to change in the type of respiration would be eliminated. In these experiments water was lost more slowly (because the spiracles were blocked with wax), but the sudden increase in the rate at 30° C. which Gunn had noticed was still quite evident in Ramsay's experiments. Evidently, some change other than in respiration was responsible for the break at 30° C. Ramsay noticed that drops of water on the cuticle were covered with a film which showed interference patterns. Above 30° C. it underwent a change of phase. He suggested this might be a lipoid which at lower temperatures might help to make the cuticle impermeable to water. This observation interested Wigglesworth and was the starting point for his many experiments, which have led to a full understanding of the structure and function of the insecta cuticle.

Chance was prominent in the early stages of this story. The cockroach is exceptional in having a wax layer of a low melting point. Had Gunn, and later Ramsay, chosen to work with almost any other common insect, neither the abrupt change in the rate of water loss nor the occurrence of the oil film on the surface of the integument would have been noticed at the temperatures used in their experiments. These two important clues might have been missed, were it not that the cockroach is such a common and useful animal for laboratory experiments.

The early experiments of Gunn and Ramsay were largely ecological. The next stage, represented by the work of Wigglesworth and others, was strictly physiological. The outcome has been an understanding of the structure and function of the insectan cuticle which has opened up new procedures in ecological investigation and will illuminate ecological inquiries in a way unimagined a few years ago.

The experiments of Wigglesworth and others have demonstrated the existence of a very thin wax layer, often no more than 0.25 μ thick, forming the outermost layer but one of the epicuticle of insects. The waterproofing properties of the cuticle largely reside in this layer (Wigglesworth, 1948*b*). The detailed structure of the cuticle does not really concern us here. Suffice it to say that, as known in *Rhodnius prolixus* and *Tenebrio molitor* (Wigglesworth, 1947, 1948*a*), it consists of six different layers. The surface layer in *Rhodnius* is a thin cement layer of unknown composition, secreted by the dermal glands. Beneath the cement is the thin wax layer, which, together with the rest of the cuticle, is secreted by the epidermal cells. One of the most remarkable properties of these cells is that the slightest injury to the wax layer is reflected in the behavior of these living cells; they react as though they have been wounded, immediately clustering toward the injured site and secreting fresh wax to repair the damage (Wigglesworth, 1937). Wigglesworth pointed out that the abrasion of the wax layer must be a common occurrence among insects, and if they were unable to repair slight damage, they would soon cease to be impermeable to water. The superficial cement layer to some extent protects the wax from abrasion.

The function of the wax layer in preventing evaporation was first demonstrated in two ways. The wax layer was abraded with "alumina dust," and the rate at which water was lost from insects so treated was compared with that from normal insects. The rate of loss of water increased greatly in the abraded animals. The second type of experiment was rather similar to the original experiment of Gunn with the cockroach. The rate at which water was lost was measured in dry air at different temperatures. When rate of loss was plotted against temperature, curves similar to that in Figure 7.14 for the cockroach were obtained (Wigglesworth, 1945). Curves for the cockroach *Blattella germanica*, the leaf-eating larva of the sawfly *Nematus*, and the caterpillar *Pieris brassicae* all showed an abrupt increase at about 30° C. But insects which normally live in dry places and which can survive long periods without feeding, such as larvae of *Tenebrio molitor* and nymphs of *Rhodnius prolixus*, showed this abrupt increase 20°–30° C. higher. The temperature at which evaporation suddenly increases is known as the "critical temperature." Wigglesworth considered that the effectiveness of the wax layer is dependent upon orientation and close packing of the long wax molecules which are loosened at the critical temperature. Below the critical temperature, water is lost through the cuticle only very slowly. In contrast to insects which normally live in dry places, the group of insects from the soil which normally live in atmospheres saturated with moisture all show a high rate of transpiration, even at the lowest temperature. In larvae of *Bibio marci*, for example, the rate of loss approaches that from a free-water surface, and there is no visible break in the curve. In other insects from soil—for example, wireworm larvae of the genus *Agriotes* and larvae of *Tipula oleracea* and *Hepialus lupulinus*—evaporation at low temperatures is less, and there is evidence of a rather ill-defined critical temperature.

The evaporation of water from the larvae and pupae of the blowfly *Calliphora erythrocephala* are shown in Figure 7.15. Larvae in the feeding stage which were removed from meat, prepupae which had left the meat, and fully darkened puparia, all showed curves which are substantially the same. In contrast, there is a marked displacement of the curve, with a great reduction of evaporation at high temperatures, in puparia 3 days old. This increase in impermeability is due not to any change in the puparium but to the addition of the delicate cuticle of the pupa within, for when the shell of the puparium was peeled off, leaving the young pupa fully exposed, there was only a relatively slight increase in permeability at high temperatures. The pupal cuticle, though exceedingly thin and fragile, is nevertheless remarkably effective at preventing evaporation (Wigglesworth, 1945).

Lees (1947), using Wigglesworth's method, measured the critical temperature for a number of species of ticks (Fig. 7.16) and found that this was closely related not only to the systematic classification of the group but also to the rate at which water was transpired into an unsaturated atmosphere, as well as to the humidity of the places where the species usually live. The results of these observations and experiments may be summarized as follows: (*a*) For the Ixodidae the critical temperature for the wax layer ranged from 32° to 45° C., whereas all the species of Argasidae came within the range 62°–75° C. The lowest critical temperature for an argasid was 17° C. above the highest for an ixodid. (*b*) When the species were arranged in the order of their critical temperatures, this was very close to the order got by arranging them according to the rate at which they lost water at temperatures below the critical point. In other words, there was a good correlation between the hardness of the wax and the permeability of the cuticle at ordinary temperatures. (*c*) The Argasidae, in which the wax is hard and the cuticle highly impermeable, are characteristically inhabitants of hot, dry places, relatively much more arid than those where Ixodidae are found. For example, *Ornithodorus savigni* is found in the desert in Arabia and Africa, often buried in loose dry sand near the resting places of caravans. The Ixodidae are rather more

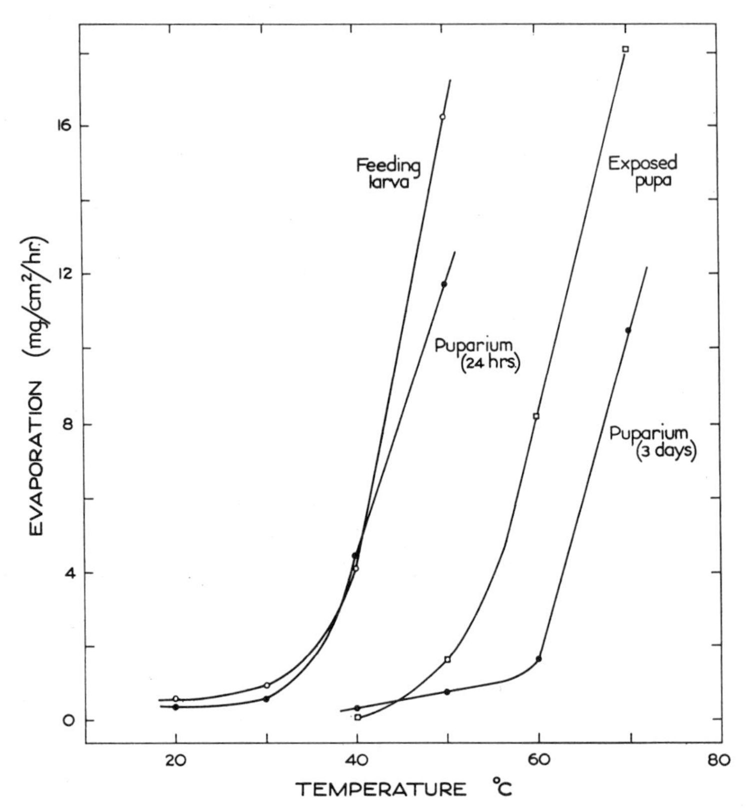

FIG. 7.15.—The rate of evaporation of water from larvae and pupae of *Calliphora erythrocephala* at different temperatures. (After Wigglesworth, 1945.)

variable with respect to the humidity of the places where they are found and include some like *Ixodes ricinus*, which are restricted to relatively moist situations. This species lives in rough hill and moorland pastures in Britain, where dense low vegetation provides shelter that remains fairly moist (Milne, 1950*a*; see also sec. 12.22). The critical temperature of the cuticular wax of this species was about the lowest of the group studied by Lees.

Davies and Edney (1952) measured the evaporation of water from five species of spiders at different temperatures. The curves for rate of evaporation at different temperatures showed well-defined critical temperatures which were different for each species. Since abrasion with inert dusts also produced an increase in the rate of evaporation, they concluded that these spiders, like insects and ticks, probably have a discrete layer of wax in the cuticle.

Transpiration through the cuticle of Crustacea and Myriapoda has been investigated by Edney (1951) and Cloudesley-Thompson (1950). There is no indication of a critical temperature in the three species of wood lice and the one millipede studied. This suggests the absence of a waxy layer in the epicuticle of these arthropods. Nevertheless, the various species differ in the

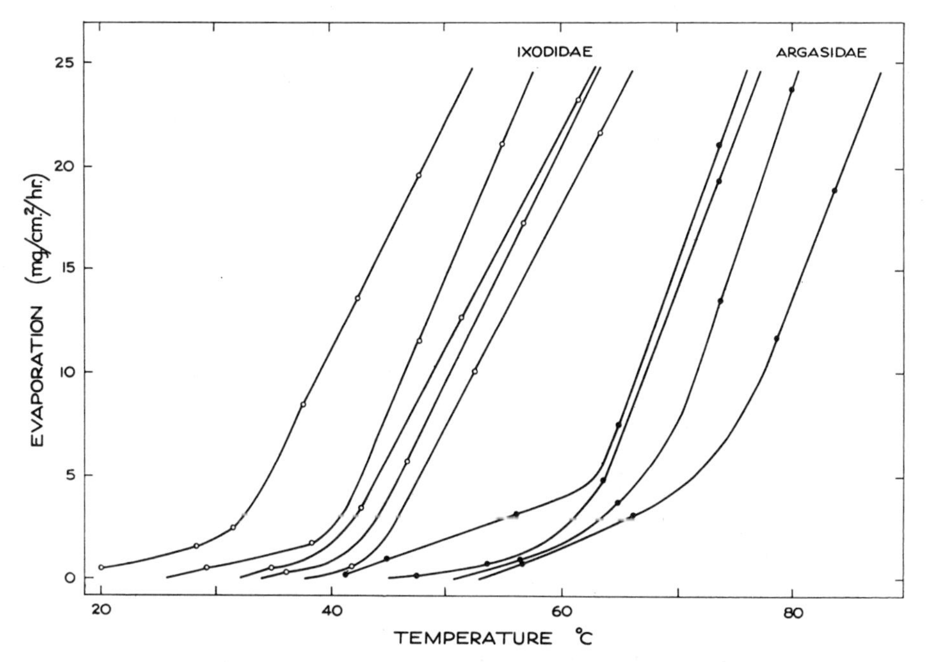

FIG. 7.16.—The rate of evaporation of water from dead engorged ticks at different temperatures. From left to right the curves refer to species *A* to *I* in the order in which they appear below. *A*, *Ixodes ricinus; B, I. hexagonus; C, Amblycmma americanum; D, Dermacentor andersoni; E, Hyalomma savignyi; F, Ornithodorus moubata; G, Argas persicus; H, Ornithodorus acinus; I, Ornithodorus savignyi.* (After Lees, 1947.)

rates at which water is lost through the cuticle (Fig. 7.17). Edney suggested that these differences may be associated with differences in the amount of tanning of the cuticle or its impregnation with lipoids. An interesting aspect of Edney's results is that the animals studied can be placed in a series from *Ligia*, which loses water most rapidly, through *Philoscia*, *Oniscus*, *Porcellio*, *Cylisticus*, and *Armadillidium nasatum* and *A. vulgare*, which is the most resistant one in the series. It is of interest to note that, of this series, *Ligia* lives on the seashore in a region where moisture is plentiful and *Armadillidium* is the common terrestrial wood louse of England and so lives in relatively dry situations. Little is known of the moisture levels in the places where the other species occur.

The eggs of insects and ticks may also be protected from desiccation by a layer of wax resembling that found on the epicuticle of the active stages, but differing from this in certain important respects. A primitive method of waterproofing eggs occurs in the ticks. Lees and Beament (1948) described how, as part of the process of oviposition, a glandular organ, the organ of Géné, is everted and holds the egg, a waxy secretion is transferred to the egg and is roughly spread over it by the movements of the egg in Géné's organ. It is

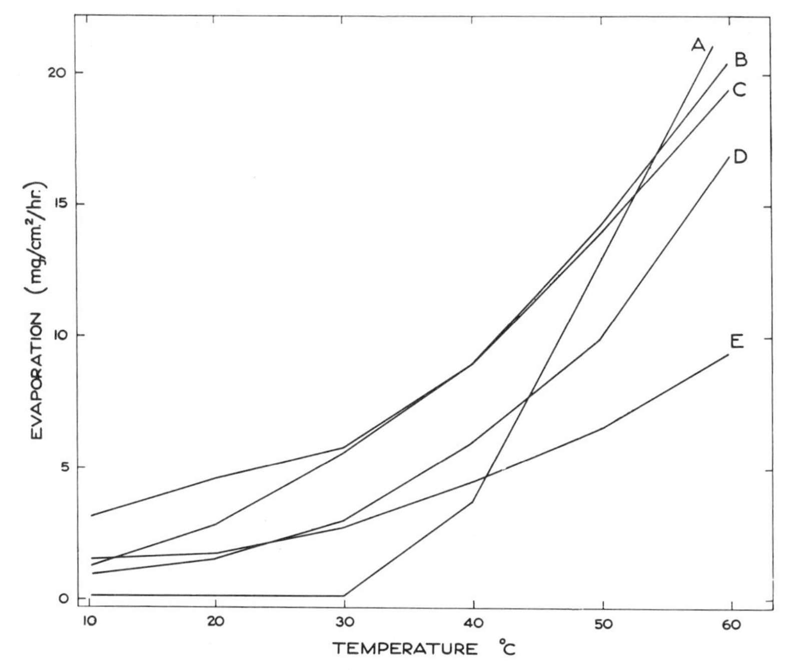

FIG. 7.17.—The rate of evaporation of water from wood lice and a millipede (*Glomeris*) compared with the cockraoch *Blattella germanica. A, Blattella germanica; B, Philoscia muscorum; C, Glomeris marginata; D, Porcellio scaber; E, Armadillidium spp.* (After Edney, 1951.)

completed by the natural spreading properties of the wax, which comes to cover the egg in a layer 0.5–2.0 μ in thickness. The wax layer waterproofs the egg: if the egg was artificially prevented from touching the organ of Géné, it quickly shriveled and died. In the tick *Ornithodorus moubata*, the wax is supplied solely by Géné's organ. In *Ixodes ricinus* waterproofing takes place in two stages. An incomplete covering of wax is first smeared over the egg during its passage down the vagina. The secretion probably comes from the accessory glands. Waterproofing is then completed in Géné's organ.

Evidence of the part played by the wax layer in the eggs of ticks is provided by experiments on the rate of transpiration of water from the egg at different temperatures. The evaporation curve exhibits a pronounced break at temperatures characteristic of the species (Fig. 7.18). They resemble the curves for the cuticles of adults (Fig. 7.16). The breaks in the curves correspond to the melting points of the cuticular waxes. The critical temperature for the eggs of ixodid ticks ranged from 35° to 44° C. Higher critical temperatures were found for argasid ticks. The susceptibility of the eggs of a given species is of the same order as that of the adults of the same species.

With the bug *Rhodnius*, and probably many other insects, a wax layer is

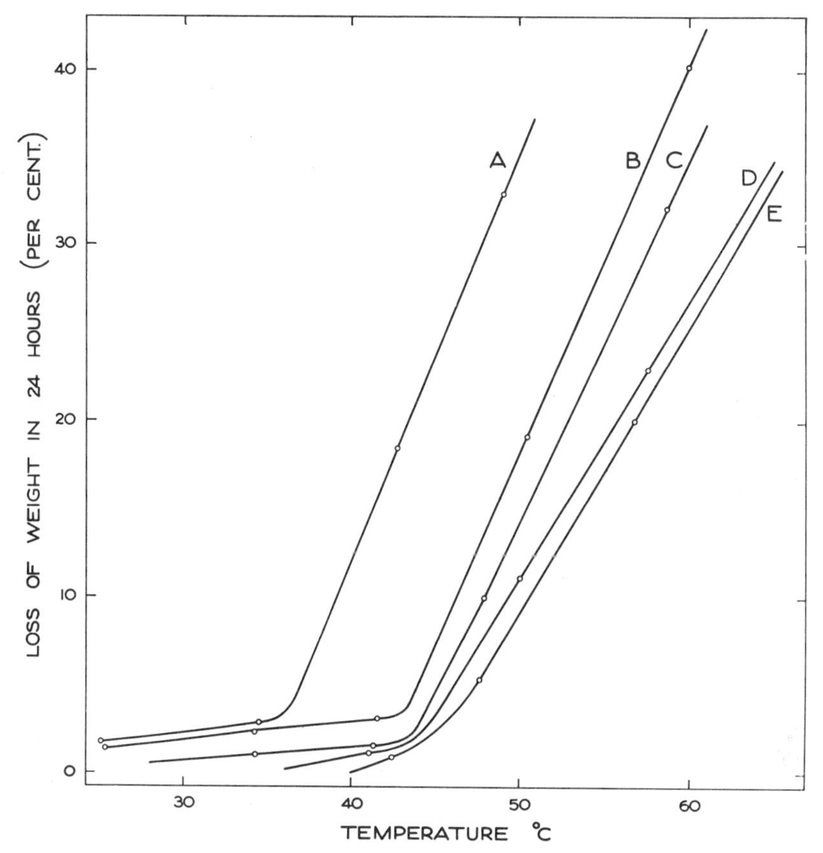

FIG. 7.18.—The rate of evaporation of water from the eggs of ticks at different temperatures. *A, Ixodes ricinus; B, I. canisuga; C, Dermacentor andersoni; D, Ornithodorus moubata; E, O. acinus.* (After Lees and Beament, 1948.)

secreted by the oöcyte about the time that it is released from its follicle in the ovary, after the chorion has been laid down. In the newly laid egg this wax layer is found just under (i.e., on the inner surface of) the chorion. In *Rhodnius* this chorion consists of seven layers of proteinaceous material, no part of which is impermeable to water. The newly laid egg is protected from desiccation solely by the extremely thin layer of wax, which is not more than $\frac{1}{2}$ μ thick. The wax is similar to what occurs in the epicuticle of the adult, having a critical temperature of about 43° C. As the egg develops, a secondary layer of wax, with a critical temperature of 68° C., is laid down (Beament, 1946*a*, *b*).

The eggs of the locust *Locustana pardalina*, at the time that they are laid, also contain a layer of wax just under the chorion; but, in addition, a secondary layer of wax is laid down quite early in the course of the egg's development. This is secreted by the serosal cells of the developing embryo after the layer of yellow cuticle has been formed but before the secretion of the white cuticle is complete. The wax layer thus lies between the yellow and white layers of embryonic cuticle and in this protected position confers upon the egg the almost incredible impermeability which it exhibits (Matthée, 1951). The eggs of other desert-inhabiting grasshoppers have a cuticle that is equally impermeable, and presumably the same mechanism is present. For example, in nature the eggs of the Australian grasshopper *Austroicetes cruciata* may spend as much as 6 months just below the surface of parched, sun-scorched soil (sec. 4.6). It has been shown (Andrewartha and Birch, 1948) that some eggs of *Austroicetes cruciata* remained alive after 80 days in air-dry soil, during which time they were exposed to evaporation equivalent to the loss of 40 inches of water from a free-water surface. During this period the egg, which might have started with 3.3 mg. of water, might have lost about 2.4 mg. and retained 0.9 mg. (Birch and Andrewartha, 1942).

Although the eggs of these desert-inhabiting grasshoppers may survive unharmed in the soil during prolonged periods of drought, they nevertheless require moisture in the soil in order to develop. During development, after the secretion of the embryonic cuticle, the eggs of *Austroicetes* increase in size to a remarkable extent, owing to the absorption of water, which, as we have seen above, is regulated by the activities of the hydropyle.

But as the egg increases in size, the egg membranes become more permeable to water, and the egg loses some of its capacity to conserve the water in its tissues. Figures 7.19 and 7.20 show the differences in the rates at which eggs of *Austroicetes cruciata* lost weight, when exposed at 20° C. in air with relative humidity 55 per cent, at different stages of their development. These eggs, which had all been laid at about the same time in November (southern spring), were collected at intervals during the following autumn and winter, as shown by the dates on Figures 7.19 and 7.20. The slopes of the curves in Figure 7.19 indicate the rate at which the eggs lost weight by evaporation of water. The data were fitted to equations of the form, $Y = A + B\xi_1 + C\xi_2$, in which B is the coefficient for linear regression and measures the average slope of the curve. By comparing the value of the coefficient B for eggs collected at different dates, it is possible to trace the changing permeability to water of the egg membranes. This has been done in Figure 7.20. The newly laid eggs (November) are so permeable because they lack the cuticle with its associated layer of wax. The extreme impermeability of the membranes which characterizes the egg throughout the summer and is still apparent at the end of April (southern autumn) is associated with diapause. The egg membranes become more permeable as diapause-development nears completion. This is not associated with any morphological change that can be detected; and Jahn (1935, 1936) suggested that the increased permeability of the membranes in *Melanoplus*

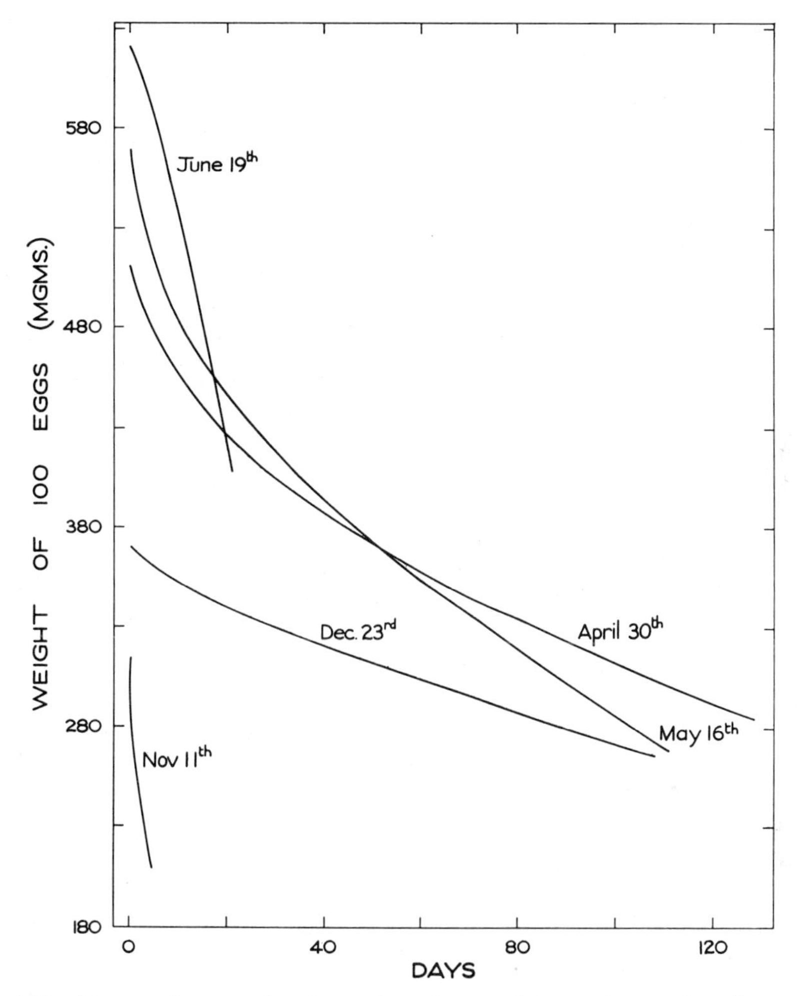

FIG. 7.19.—The rate of evaporation of water from the eggs of the grasshopper *Austroicetes cruciata* at 55 per cent relative humidity (20° C.) collected from the field in South Australia on the dates shown. (After Birch and Andrewartha, 1942.)

may be due to the fact that they become stretched as the egg absorbs water and swells during development. Similar changes have been observed in *Locustana pardalina* (Matthée, 1951) and *Melanoplus differentialis* (Thompson and Bodine, 1936).

So far in this section we have discussed certain adaptations among arthropods which serve to retard or prevent the transpiration of water through the cuticle and how these may be related to the humidity in the places where the animal usually lives. We also know quite a lot about the rate at which water is lost in relation to the humidity of the air, and we discuss this in section 7.234. This section will be concluded by several brief references to similar adaptations among gastropods, amphibians, and reptiles.

Snails and slugs have very permeable skins, through which a continuous and

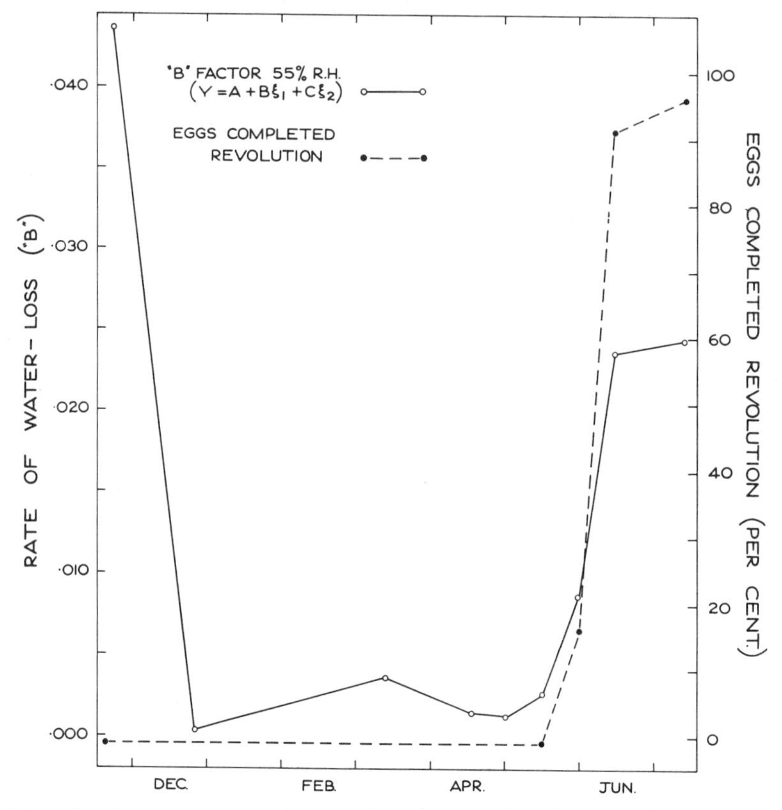

FIG. 7.20.—In the egg of the grasshopper *Austroicetes cruciata* the cuticle becomes increasingly permeable to water as the egg emerges finally from diapause and enters on the last stages of its development. The magnitude of the coefficient *B* is a measure of the rate at which eggs lost water in air at 20° C. and 55 per cent relative humidity. The proportion of the eggs which had completed katatrepsis (revolution of the embryo) is a measure of the extent to which diapause-development had been completed in each sample. (After Birch and Andrewartha, 1942.)

rapid evaporation takes place in unsaturated air. The rate of loss is probably unsurpassed by terrestrial animals of any other group (Howes and Wells, 1934*a*, *b*). An active slug in unsaturated air would have to be drinking all the time at a rate of about 2–3 per cent of its body weight per hour. It would thus never be able to venture far from a supply of water. Instead of being thus restricted, it appears that the animals have the capacity to endure great changes in their water content, undergoing alternate phases of dehydration and hydration. The period of these cycles is about 7 days for snails and a few days for slugs. In one extreme example the weight of *Helix pomatia* was observed to increase over 50 per cent in 24 hours.

Slugs which received food but no water lost weight rapidly and died in a few days. Snails, on the other hand, have the advantage of being able to retire into their shells. Their water content may become greatly reduced, and they may survive in this condition for months without water. The entrance to the shell is closed by a thin film of dried mucus. Snails can also retire in this way in the winter time, but then they close the mouth of the shell with a dense calcareous plate called the "epiphragm." Aestivation is presumably a protection against drought, while hibernation is a protection against cold.

Amphibia lose water through the skin more readily than do reptiles, birds, or mammals. Some Amphibia are able to live in relatively dry places. Thorson and Svihla (1943) found that all the Amphibia they studied, whether they normally lived in arid or humid situations, lost water rapidly in dry air. The species which normally lived in dry situations were able to survive much greater losses of water from their bodies than were those from wet ones (Table 7.07). For example, *Scaphiopus hammondii* occurs in semiarid regions and lives away from open water most of the year. It can lose 48 per cent of its weight as water without dying. But *Rana grylio*, an aquatic species, can lose only 30 per cent without dying (sec. 7.31).

Reptiles have highly impermeable skins. Gray (1928) compared the rate at which water was lost from the newt *Triton cristatus* and the lizard *Lacerta viridis*. The newt lost about 28 per cent of its body weight in water per day, whereas the lizard in the same circumstances lost only about 5 per cent of its weight per day. This rate of loss was about the same, whether the lizard was alive or dead. After removal of the skin, it lost weight at about the same relative rate as the newt. It is probable that the success of the reptiles on land is largely due to the slow rate of transpiration of water through the skin. The success of the few terrestrial amphibians is probably due not to any impermeable skin but to the ability to survive large losses of water.

7.234 THE RELATIONSHIP BETWEEN WATER CONTENT OF THE AIR AND THE RATE AT WHICH WATER IS EVAPORATED FROM THE BODY OF AN INSECT

Even with those species which can survive the loss of a considerable proportion of the water from their tissues, there is a definite limit to the amount that can be lost without the animal's dying. So with most species the capacity to endure prolonged exposure to dry air, when relatively little water is being taken in with the food, depends very largely on the rate at which water is evaporated from the body. The publication in 1924 of Bacot and Martin's paper on the longevity of the flea *Xenopsylla cheopis* stimulated discussion on the laws governing the evaporation of water from insects. These are now quite clear; but so much that has been written on this subject has been based on misconceptions that it is desirable to discuss the matter in some detail.

Bacot and Martin (1924) measured the duration of life of fleas (*Xenopsylla cheopis*) that were kept without food in air with different moisture contents. The results indicated a relationship between the longevity of the flea and the saturation deficit of the air. And they formulated the hypothesis that the rate of evaporation of water from an insect is proportional to the saturation deficit of the surrounding air. This was an interesting hypothesis, though it could not be fully tested by reference to their own data, which were not adequate for the purpose.

This hypothesis was prematurely accepted by ecologists, and in 1935 Mellanby stated the so-called "saturation-deficiency law" in these words: "If individual insects of the same species, in identical morphological and physiological states, are exposed to atmospheres with different humidities, then the rates at which the insects lose water will be proportional to the saturation deficiency, provided that the saturation deficiency is not above a maximum figure which is peculiar to the species. From this it follows that when insects are killed by desiccation, the length of life of the insects will be inversely proportional to the saturation deficiency of the air." The so-called "law" is not, of course, a law but merely a hypothesis, which we shall proceed to show does not fit the empirical facts. Indeed, it is difficult to understand how this "law" came to be so readily accepted, for a moment's consideration of the relevant physical principles will show it to be inherently unlikely.

The theoretical basis of the saturation-deficiency hypothesis as applied to animals rested on an approximation and extension of the work of Dalton on evaporation from free water surfaces. Dalton considered that the rate of evaporation from a free water surface should be proportional to the difference between the vapor pressure at the evaporating surface and the vapor pressure of the air above the water surface. The ecologists who took over this idea were mistaken in supposing that the difference in vapor pressure between the water surface and the air above was equivalent to the saturation deficit of the air. Their second error was to suppose that the surface of an insect was equivalent to a free water surface.

Now the difference in vapor pressure of water at the surface of the water and in the air above is the same as the saturation deficit of the air only if the temperature of the water surface is the same as that of the air; usually, however, it is lower. It is generally assumed that the layer of air immediately above the water surface is saturated with water vapor at the temperature of the surface. There is thus a gradient in vapor pressure from the saturated vapor pressure of air at the temperature of the surface to a vapor pressure in the air above. The difference between these two vapor pressures might be called the "vapor-pressure difference." It is this difference in concentration of water molecules and not the saturation deficit of the air which determines the rate at which water evaporates from the surface of the liquid. In the unusual circumstance of the water surface being at the same temperature as the air, this vapor-pressure difference is the same as the saturation deficit. As Leighly (1937) has pointed out, if the water surface and the air next to it have a lower temperature than the air above, the concentration of water vapor in the air above the surface, though it does not saturate the air, may exceed the concentration in the cooler, saturated air next to the water surface. Water will then be deposited from the unsaturated air on the surface of the water, as in the deposition of dew. If the gradient in the concentration of water vapor is in the opposite direction, being greatest at the water surface, then evaporation will take place, but not otherwise.

This principle can be expressed quite simply in the equation:

$$\begin{aligned} E &\propto p_0 - p_1 \\ &= k(p_0 - p_1), \end{aligned}$$

where E = rate of evaporation per unit area, p_0 = vapor pressure of air immediately above the surface of the water, p_1 = vapor pressure of the air, and k = a constant. But this expression is true only if the distance between the evaporating surface and the point where the vapor pressure is p_1 is the same for all evaporating surfaces. Since this distance, which we shall call x, is obviously not constant, it must be included in the equation for evaporation thus:

$$E = \frac{k(p_0 - p_1)}{x}.$$

Even then the expression is accurate only if the change in vapor pressure from the surface to a distance x from the surface is linear. This is not always so (Leighly, 1937). The general expression of the gradient is, therefore, the derivative $\delta p/\delta x$, and the general equation for evaporation becomes

$$E = k\,\delta p/\delta x.$$

The value of k varies with the temperature of the system, because the rate of diffusion of water-vapor molecules varies with temperature. The rate of diffusion at 30° C., for example, is 25 per cent greater than the rate at 0° C. (Ramsay, 1935*b*). We cannot then regard this as an expression which is independent of temperature.

All these arguments concern evaporation in still air. If, on the other hand, there is a constant current of air above the evaporating surface, this tends to maintain a surface of constant concentration of water vapor at a constant short distance from the evaporating surface. The gradient will be very steep, and the evaporation correspondingly more rapid. Various equations have been developed for expressing the rate of evaporation in moving air (Ramsay, 1935*b*; Leighly, 1937). Because of the great influence of air currents on evaporation, they cannot be ignored in systems in which air is moving over the evaporating surface. The bare statement, which may be found fairly often in ecological writings, that evaporation from a free water surface or a simple wet object, like a piece of wet blotting paper, is proportional to the saturation deficit of the air is misleading, because it excludes three important factors, viz., the difference between the temperatures of the evaporating surface and of the air above, the influence of temperature on the rate of diffusion, and the influence of turbulence and air movements on the rate of evaporation. This statement becomes doubly misleading when it refers to an insect, because of the complex nature of the evaporating surface. Some water may be evaporated from the surface of the cuticle, but the rate of movement of water through the cuticle is so slow that the surface can hardly be regarded as "wet." As for the water vapor that is lost through the spiracles, the evaporating surface for this must be made up of all the surfaces of the internal ramifications of the tracheal system. This may be regarded as a "wet object," but it can hardly be considered a simple one.

Even if this internal evaporating surface of an insect behaved like a simple "wet object," a piece of wet blotting paper, for example, there would still be sound reasons, in addition to those advanced above, for not expecting the rate of evaporation to be simply related to the saturation deficit of the outside air and to be independent of temperature. We have seen above that the rate of evaporation depends upon the degree of turbulence and air movement. The particular air currents that are important in this case are not so much the gross movements of air outside the insect as the flow of air along the tracheal system of the insect. This is a function of temperature, as can easily be demonstrated by measuring the rate of respiration at different temperatures. But the movement of air through the tracheal system may not be just a simple function of temperature, because many insects have closing devices associated with the spiracles, and it is unlikely that these are influenced only by temperature. Another complication is introduced by the fact that at least some species, e.g., *Tribolium*, *Tenebrio*, *Ephestia*, are able to maintain the proportion of water in their bodies constant while being starved in dry air. Whatever may be the merits of the hypothesis of the conservation of metabolic water (see sec. 7.22), it would seem likely that these species are able, by doing work, to retain in their tissues some of the water that would otherwise be lost by evaporation. If this is so, it would indeed be misleading to assume that Dalton's law can be simply applied to the living insect. It is far better to proceed cautiously, taking care that conclusion does not outstrip empirical fact, and to stop at frequent intervals to test hypothesis by well-conceived experiment.

Ramsay's (1935*a*) experiments were designed to measure the rate of evaporation through the cuticle and spiracles separately. He first covered the cuticle with shellac and left the spiracles open and then left the cuticle intact, while blocking the spiracles with wax. In neither case was the blocking completely efficient, but the results are instructive. Figure 7.21, *A* and *B* shows that the relationship between saturation deficit and rate of evaporation, both through the cuticle alone and through the spiracles alone, was approximately linear for any one temperature. The increase in the rate of transpiration

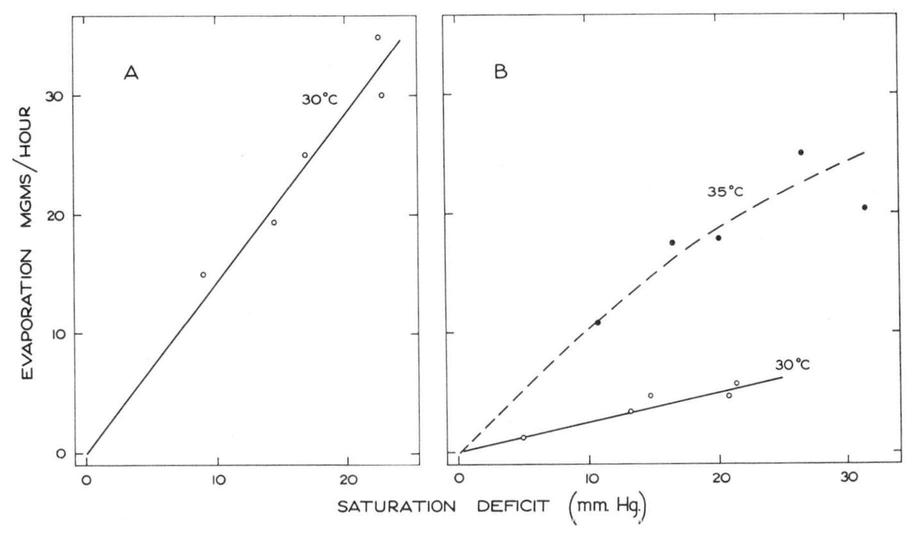

FIG. 7.21.—*A*, the relationship between rate of evaporation from the thoracic spiracles of the cockroach *Periplaneta americana* at 30° C. The animal was held transversely to a wind moving at a constant velocity of 10 meters/second. The cuticle of the animal was covered with shellac to reduce loss of water. *B*, the relationship between rate of evaporation from the body surface of the cockroach *P. americana* and saturation deficit at 30° and 35° C. The animal was held head-on to a wind moving at a constant velocity of 4 meters/second. The spiracles were blocked with wax. (After Ramsay, 1935*a*.)

through the cuticle above 30° C. was due to a change in the system, because the wax undergoes a transition at about 30° C. (see sec. 7.233 and Fig. 7.14).

In certain other experiments, the rate of evaporation of water from the insect was measured without discriminating between what was lost through the cuticle and what from the spiracles, but it may be assumed that the greater part of the water was lost from the spiracles. Experiments with larvae, prepupae, and pupae of *Popillia japonica* (Ludwig and Landsman, 1937), *Blatta orientalis* (Gunn, 1933), and larvae of *Tenebrio molitor* and *Tineola biselliella* (Mellanby, 1934*b*) consistently indicated that the relationship between the saturation deficit and the rate at which water was evaporated from the body at any one temperature could be represented by a straight line; but the lines for different temperatures were quite different both in slope and in position.

Since, in these experiments, at any one temperature the rate of evaporation from the insect was found to be a linear function of the saturation deficit, it follows, from what was said above, that (*a*) the temperature of the evaporating surface was close to that of the surrounding air and (*b*) the rate of movement of water vapor out of the insect was so slow that no appreciable gradient in water-vapor pressure was built up near the insect. All the foregoing experiments were done with insects that would remain relatively quiet and immobile during the experiment, and one might expect them to fulfil these conditions. But it might be quite otherwise with insects in active motion, for it is known that muscular activity, e.g., the fluttering of wings, may raise the body temperature as much as 9° C. above that of the surrounding air (Oostuizen, 1939; Krogh and Zeuthen, 1941; Krogerus, 1949). This is just another example of the need to be careful not to formulate generalizations which outstrip the empirical facts.

The "saturation-deficiency law" which we have been discussing states that evaporation from an insect is *proportional* to the saturation deficit of the surrounding air. We have seen that in particular circumstances evaporation is a *linear function* of the saturation deficit, but a straight-line relationship does not necessarily imply proportionality. The general statement that loss of weight (R) is a linear function of the saturation deficit (S) may be expressed in symbols:

$$R = a + KS.$$

This is the general equation for a straight line. In the particular case when $a = 0$, the equation becomes

$$R = KS.$$

The first equation describes line A in Figure 7.22, which cuts off an ordinate at some distance from the origin (when $S = 0$, $R = a$); the second equation describes line B in Figure 7.22, which passes through the origin of the graph (when $S = 0$, $R = 0$). The second equation and line B represent the special case when loss of weight is proportional to saturation deficit. Failure to appreciate this very elementary piece of mathematics has caused some biologists to misinterpret their experimental data.

The error arises as follows: (*a*) It has been assumed that all loss of weight was due to evaporation of water. (*b*) On this assumption, the data have been tested for proportionality. (*c*) When this hypothesis was not confirmed, the significance of the more general linear relationship was missed. This has been clearly pointed out by Johnson (1940) in his analysis of Mellanby's (1932*b*) data for evaporation from the bedbug *Cimex*. Figure 7.23 shows that in Mellanby's experiments the loss of weight from the bugs at any one temperature was a linear function of the saturation deficit within the range 0–90 per cent relative humidity; but the slope of the line varied with temperature. The lines in Figure 7.23 do not pass through the origin of the graph. Because a is not equal to zero, the ratio R/S is not constant, and Mellanby was therefore correct in stating that in these experiments loss of weight from *Cimex* was not proportional to saturation deficit; but he was mistaken in concluding that the relationship for *Cimex* was therefore different in principle from that

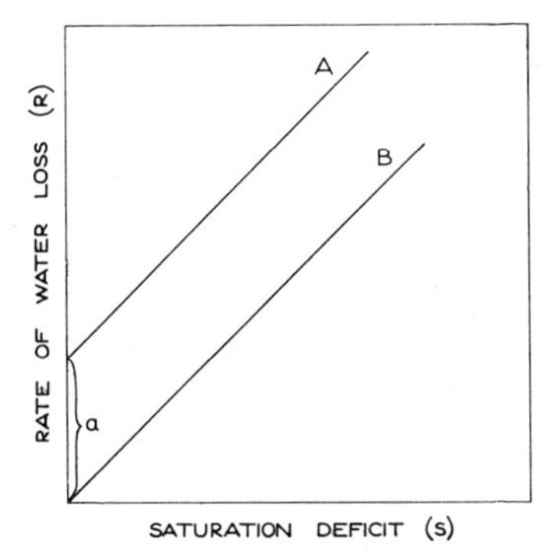

FIG. 7.22.—Graphs illustrating the hypothetical relationship between saturation deficit (S) and the rate at which a fasting insect loses weight (R). A, the general relationship when $R = a + KS$; B, the special relationship when R is directly proportional to S and therefore $R = KS$ (i.e., R/S is a constant).

for other insects, in which the ratio R/S is found to be constant. The difference is in degree, not in principle. The positive value for a in the equation

$$R = a + KS$$

merely indicates that loss of weight in these experiments could be divided into two components: a, which was independent of saturation deficit, and $(R - a)$, which was proportional to saturation deficit. If a represents weight lost by excretion or some other source than evaporation and $(R - a)$ represents evaporation (E), we still have

$$E = KS,$$

which means that evaporation was proportional to saturation deficit in Mellanby's experiments, just as it was in those other cases when a was found to be zero.

7.3 THE LETHAL INFLUENCE OF DRY AIR

The lethal influence of dry air may be expressed in terms either of the longevity of individuals or of the survival-rate of a population. The latter is a simple function of the former (as we shall see below) and is therefore just an alternative way of expressing the same thing. It is sometimes more convenient or more instructive to make a direct measurement of the survival-rate instead of the longevity. Both measurements (when taken in relation to the lethal influence of dry air) are functions of (*a*) the rate at which the animal loses water into unsaturated air and (*b*) the extent to which it can withstand desiccation of its tissues. The laws governing the rate of evaporation from the body are discussed in section 7.2; the severity of desiccation required to kill certain

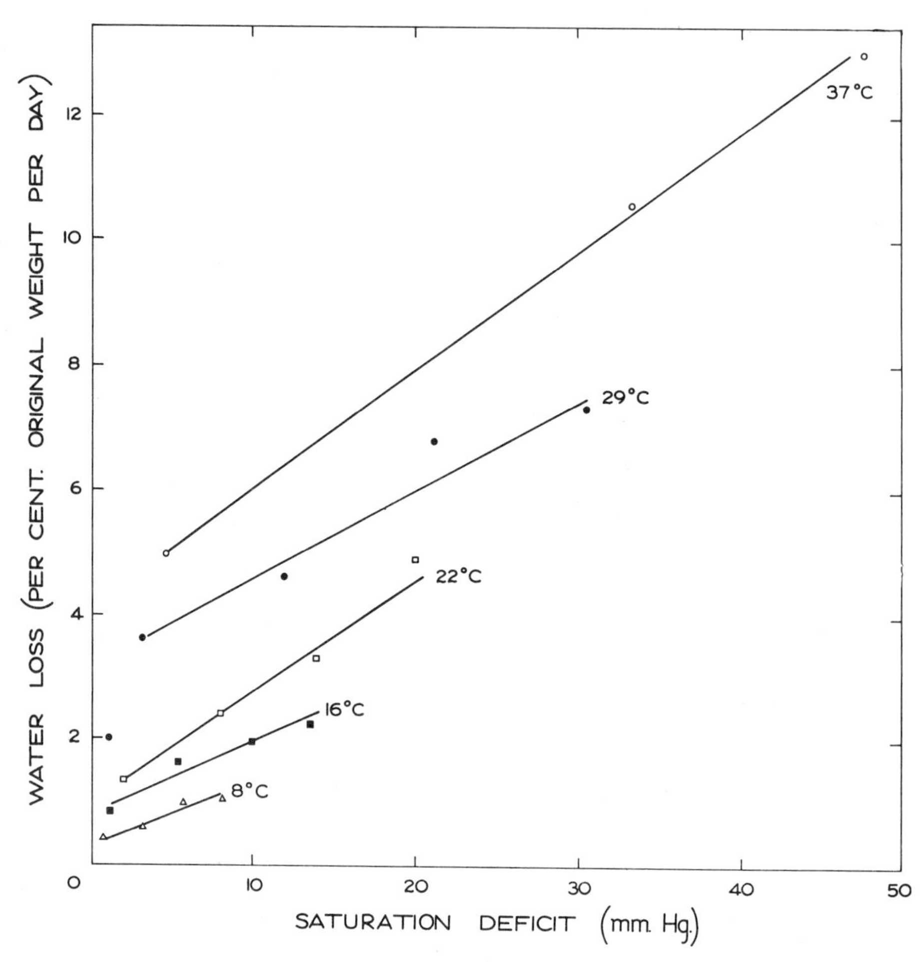

FIG. 7.23.—The relationship between rate at which fasting adults of the bedbug *Cimex lectularius* lost weight and saturation deficit at different temperatures. Regression lines have been drawn through the observed points. (After Johnson, 1940.)

animals in section 7.31; and the relationship between saturation deficit and longevity and survival-rate in sections 7.32 and 7.33.

7.31 *The Severity of Desiccation Required To Kill Certain Animals*

Different species of animals and different stages in the life-cycle of the same species vary enormously with respect to the period that they may remain alive after having lost a substantial proportion of the water from their tissues. Some species have special resting stages in the life-cycle which may be associated with the phenomena of diapause and anabiosis. These sometimes become so desiccated that they retain only a fraction of the water that was originally in their tissues, and in this condition they can survive for months or even years. This would seem to be true of the anabiotic stages of rotifers and tardigrades; though it appears that no one has measured their water content, it must be extremely low (sec. 7.0). Similarly, in the adult beetles *Otiorrynchus cribricollis* (Andrewartha, 1933), *Listroderes costirostris* (Dickson, 1949), and *Leptinotarsa decemlineata* (Breitenbrecher, 1918), the water content of the tissues is reduced, in preparation for aestivation, to the point where there seems to be virtually no fluid left. It appears that no one has measured the water content of these beetles either; but if they are examined by opening up the abdomen from above, the viscera can be seen, completely dried out, a tough flat strip adhering to the floor of the abdomen.

Few, if any, animals in an active stage of the life-cycle can survive even moderate desiccation for a prolonged period; but many animals can survive quite a substantial loss of water from their tissues for a brief period. For example, Schmidt (1918) dried out the earthworm *Allolobophora foetida* over calcium chloride and found that some worms could be revived after having lost 62 per cent of their original weight, equivalent to 73 per cent of the water originally present in their tissues. They could not, however, survive in this desiccated condition for more than a day or two and then only if they were kept at a low temperature. The desiccated worms were wrinkled and immobile, having the appearance of a "corpse" or "mummy." Individuals which had lost more than 83 per cent of their original water could not be revived, and it seemed as if desiccation to this degree would be almost instantaneously fatal. In Table 7.05 we have set out similar data for a number of other animals. Very little precise information is available in this field. The data in Table 7.05 are far from precise for the following reasons: (*a*) We are not told how long the animals could remain alive with such small proportions of water. With most of them it would probably be no more than a few hours or, at most, several days; but with the egg of *Austroicetes* or the larva of *Tenebrio* the period might be several months. (*b*) The figures given in the table apply not to the mean of a sample but more often to a few of the more hardy individuals. This, for reasons explained in section 6.312, is not very valuable information; the most instructive quantity to measure is the desiccation necessary to kill the "average" individual or, alternatively, just half the sample.

The capacity to survive brief periods of desiccation has appeared as an adaptation in some members of at least two groups of animals (Gastropoda and Amphibia) which lack an impermeable cuticle and therefore lose water relatively quickly in unsaturated air. Broekhuysen (1941) studied six species

TABLE 7.05

AMOUNT OF WATER WHICH CAN BE LOST FROM BODIES OF VARIOUS ANIMALS AND STILL LEAVE THEM ALIVE

Animal	Per Cent Loss of Body Weight	Per Cent Loss of Water	Author
Mollusks:			
Various snails and slugs	50–60	....	Howes and Wells (1934*a*, *b*)
Limax tenullus	80	....	Künkel (1916)
Crustaceans:			
Various wood lice	50	....	Edney (1951)
Annelids:			
Allolobophora foetida	61	73	Schmidt (1918)
A. chlorotisus	70	83	Hall (1922)
Lumbricus terrestris	43	....	Jackson (1926)
Insects:			
Tenebrio molitor larvae	53	100	Hall (1922)
Austroicetes cruciata (grasshopper); the most resistant of its diapause eggs	70	87	Birch and Andrewartha (1942)
Amphibians:			
Amblystoma punctatum (salamander)	47	....	Hall (1922)
Rana pipiens (for other frogs see Table 9.02)	31	45	Thorson and Svihla (1943)
Reptiles:			
Chrysemys marginata (turtle)	33	....	Hall (1922)
Scleroporus spinosus (desert lizard)	48	....	Hall (1922)
Mammals:			
Peromyscus leucopus (wood mouse)	31	....	Hall (1922)
Mus musculus (house mouse)	24	....	Hall (1922)

of gastropods which live in the intertidal region of the seashore in South Africa. As the tide recedes, these animals become exposed to the atmosphere and suffer a certain amount of desiccation. The exposure is more prolonged and therefore likely to be more severe for those species that occur high up in the littoral region and less severe for those at the lower levels. The distribution of the six species is shown in Figure 7.24. The widths of the bands are drawn in proportion to the usual abundance of the animals at the various heights above low tide. Broekhuysen first measured the rate at which each species lost water in unsaturated air. He demonstrated that there were no significant differences between the species in this regard: they all lost water at about the same rate. He then set out to test the hypothesis that those species which normally live high up and suffer the more severe exposure would be adapted to survive the loss of a greater proportion of the water from their tissue than would those from lower down. The snails were desiccated in air over calcium chloride and weighed at regular intervals. After different amounts of water had been lost, they were submerged in water to see which were alive and which were dead. The results, set out in Table 7.06, seem amply to verify the hypothesis.

The position of these gastropods as they are distributed from wet to dry parts of the intertidal zone is clearly correlated with their tolerance to dryness expressed as the minimal amount of water loss which causes death. The only

TABLE 7.06*

MINIMAL LOSS OF WATER SUFFICIENT TO CAUSE DEATH IN SIX SPECIES OF GASTROPODS

SPECIES	MINIMAL LOSS OF WATER WHICH CAUSED DEATH (PER CENT WET WT.)	
	Room Temp.	39°–40° C.
Oxystele sinensis (wet)	3	3
Cominella cincta	10	8
O. tigrina	7	4
Thais dubia	15	13
O. variegata	13	10
Littorina knysnaensis (dry)	15	14

* After Broekhuysen (1941). The species are arranged in order of occurrence in the intertidal zone: *O. sinensis* occupies the wettest situations and *L. knysnaensis* the driest.

marked exception to the sequence is *Oxystele tigrina*. Although it inhabits fairly high levels, it is most commonly confined to pools or damp places and so is probably less exposed to desiccation than a form inhabiting open rock at the same level.

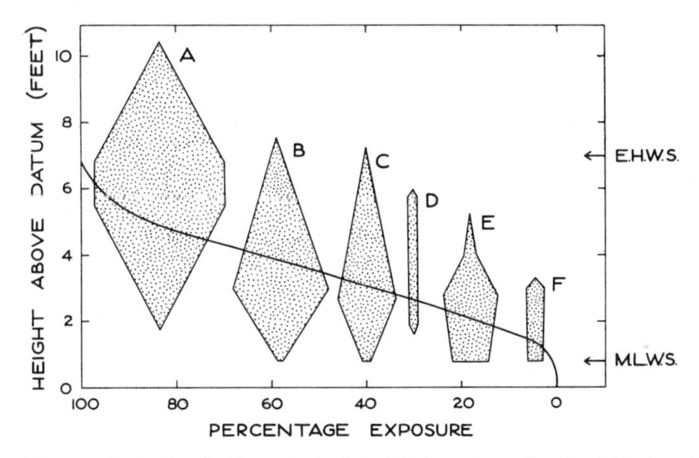

FIG. 7.24.—The vertical distribution of six intertidal gastropods; *A*, *Littorina knysnaensis; B*, *Oxystele variegata; C*, *Thais dubia; D*, *O. tigrina; E*, *Cominella cincta;* and *F*, *O. sinensis*, in relation to exposure. Vertical heights of the shaded areas represent vertical distribution, while the horizontal dimensions of these areas represent abundance at each vertical level. Superimposed on the figure is a curve showing percentage exposure at the various levels. The base line is 1 foot below "minimum low water of springs" (*M.L.W.S.*). "Extreme high water of springs" (*E.H.W.S.*) is the extreme high-water level, allowing an additional 2 feet for wash zone. (After Broekhuysen, 1941.)

Thorsen and Svihla (1943) demonstrated the same sort of adaptation among Amphibia. They measured the minimal loss of water required to kill ten species of toads and frogs ranging from species that were strictly aquatic to those which were strictly terrestrial. The results are set out in Table 7.07. The striking thing about these data is the similarity between species within the same genus: they seem to live in the same sorts of places and to be able

TABLE 7.07*
MINIMAL LOSS OF WATER SUFFICIENT TO KILL TEN SPECIES OF AMPHIBIA

Species	Av. Minimal Lethal Loss of Water Expressed as:		Description of Sort of Places Where Species Live
	Per Cent Body Weight	Per Cent Body Water	
Scaphiopus holbrookii	48	60	Terrestrial burrows
S. hammondii	48	60	Dry-land burrows
Bufo boreas	45	56	Terrestrial
B. terrestris	43	55	Terrestrial
Hyla regilla	40	50	Terrestrial but moister situations than *Bufo*
H. cinerea	39	49	Terrestro-arboreal
Rana pipiens	36	45	Terrestro–semiaquatic
R. aurora	34	43	Semiaquatic
R. sphenocephala	33	....	Semiaquatic
R. grylio	30	38	Aquatic

* The species are arranged in order of the dryness of the places in which they usually live. After Thorsen and Svihla (1943).

to lose about the same amount of water without dying. The big differences are between genera. The terrestrial genera *Scaphiopus* and *Bufo*, which live in the drier situations, were able to survive the greatest loss of water; the aquatic genus *Rana* was the most susceptible to desiccation. As with the gastropods of the previous experiment, these observations showed that the Amphibia which can live in rather drier places can do so, not by virtue of being able to retain their body water, but by being able to live for brief periods without it. In this respect the gastropods and amphibians differ from most of the larger terrestrial vertebrates and the insects. The former are able to survive in a dry atmosphere partly because their integument is less permeable but chiefly because their large size reduces the relative amount of water lost and at the same time enables them to seek out water to drink. The insects, on the other hand, have developed a remarkably impermeable integument which enables them to live in dry places despite their small size.

7.32 *The Relationship between Saturation Deficit and Longevity*

Let us first consider a hypothetical example of an insect for which the duration of life at any one temperature is dependent upon the amount of water it loses. When a certain quantity of water has been lost, the insect dies. We are, of course, not interested in single insects but in numbers of insects in a sample. The argument, however, is the same, and in the latter case we would be interested in the quantity of water which must be lost before 50 per cent of the animals died. Secondly, let us suppose that the rate of loss of water at any one temperature is directly proportional to the saturation deficit of the air, i.e., $R = kS$ (Fig. 7.25, A). From the first assumption it follows that the more rapid the loss of water, the shorter the life of the insect, because the critical level of water in the insect is reached sooner; in fact, longevity must be inversely proportional to the amount of water lost in unit time. And from the second assumption it follows that the amount of water lost in unit time is directly proportional to the saturation deficit. It follows that longevity will be inversely proportional to saturation deficit at any one temperature. So when longevity is plotted against saturation deficit, we would obtain the inverse of a linear relationship, which is a hyperbola (Fig. 7.25, *B*.)

This relationship is best explained with reference to a hypothetical example. Consider a group of identical insects which are being starved in air that is not

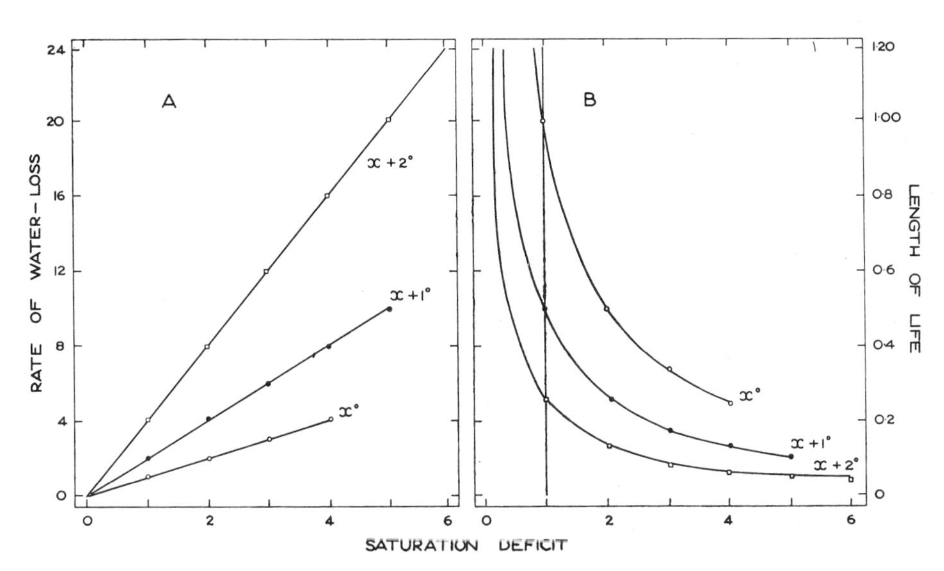

FIG. 7.25.—Graphs illustrating hypothetical influence of saturation deficit on longevity of an insect at temperatures $x°$ to $(x + 2)°$ when it is assumed that longevity is determined only by the rate of water loss. *A*, the relationship between saturation deficit and the rate at which water is lost from the insect. *B*, the relationship between saturation deficit and the longevity of the insect. Longevity is represented by the reciprocal of the rate of water loss. It is inversely proportional to the saturation deficit, as shown by the hyperbolas. The parts of the curves on the left of the vertical line have no biological significance, since insects are not immortal. (After Johnson, 1940.)

saturated with water vapor. Suppose the temperature is the same for all, but the humidity is different for each. Assume (*a*) that these insects will die when they have lost a certain constant (k_2) amount of water from their bodies, and (*b*) that the relationship between the rate at which an insect loses water and the saturation deficit may be expressed either by (1) $R = a + k_1S$ or (2) $R = k_1S$, which are the equations explained in section 7.234. Now it is clear, since a constant amount of water (k_2) has to be lost before the insect dies and since R is the rate at which water is lost, that the duration of the insect's life must be inversely proportional to R, i.e.,

$$L = k_2 \times \frac{1}{R}.$$

By substitution, this becomes either

$$L = k_2 \times \frac{1}{a + k_1 S} \tag{1}$$

or

$$L = k_2 \times \frac{1}{k_1 S} = \frac{k_2}{k_1} \times \frac{1}{S} = \frac{K}{S}. \tag{2}$$

The first is the general equation for a hyperbola, and the second is the equation for the particular hyperbola which is appropriate when the rate at which water is lost is proportional to the saturation deficit. But, in either case, provided that there is a linear relationship between the rate at which water is lost and the saturation deficit, then the data for longevity, when plotted against the saturation deficit, must fit a hyperbola. Bacot and Martin (1924) and others that followed them missed this point. They calculated the quantity LS from their data for *Xenopsylla*, and, because this was not a constant, they concluded that the relationship between longevity and saturation deficit was not hyperbolic. The correct inference to be drawn from the inconstancy of LS was that their data did not fit the particular hyperbola which is appropriate when the rate at which water is lost is proportional to the saturation deficit. Bacot and Martin should have fitted their data not to this particular equation but to the general equation for a hyperbola.

Johnson (1940) made this test with data for the longevity of first-instar nymphs of *Cimex*, starved at constant temperature and different humidities. The results are shown in Figure 7.26. The smooth lines in this figure are hyperbolas which were fitted to the experimental data. At constant temperatures of 7° and 15° C. (and, no doubt, at intermediate temperatures) the relationship between mean duration of life and saturation deficit is hyperbolic over the range of humidity studied. At temperatures about 15° C., however, the hyperbola is no longer a good fit to the data: indeed, the points seem to lie approximately along a straight line. It is reasonable to conclude that the rate of water loss alone determines longevity in the experiments conducted at 7° and 15° C. At higher temperatures some other, unknown, influence must have begun to be important. Johnson's (1942) analysis of Kirkpatrick's (1923) data on the longevity of *Oxycarenus hyalinipennis* and, as well, less complete data on other insects by other authors show that the relationship between longevity and saturation deficit is hyperbolic at any one temperature over a wide range of temperatures for a number of insects. Johnson points out that a complete confirmation of the hyperbolic relationship between longevity and saturation deficit cannot exist in nature. The arm of the curve along the ordinate cannot be asymptotic, since insects are not immortal. Thus deviations from the curve must occur at high humidities; there are other reasons, too, why deviations might be expected at high humidities, for at high humidities, instead of desiccation being the only important cause of death, other harmful influences, such as disease, may become of chief importance (see sec. 7.4).

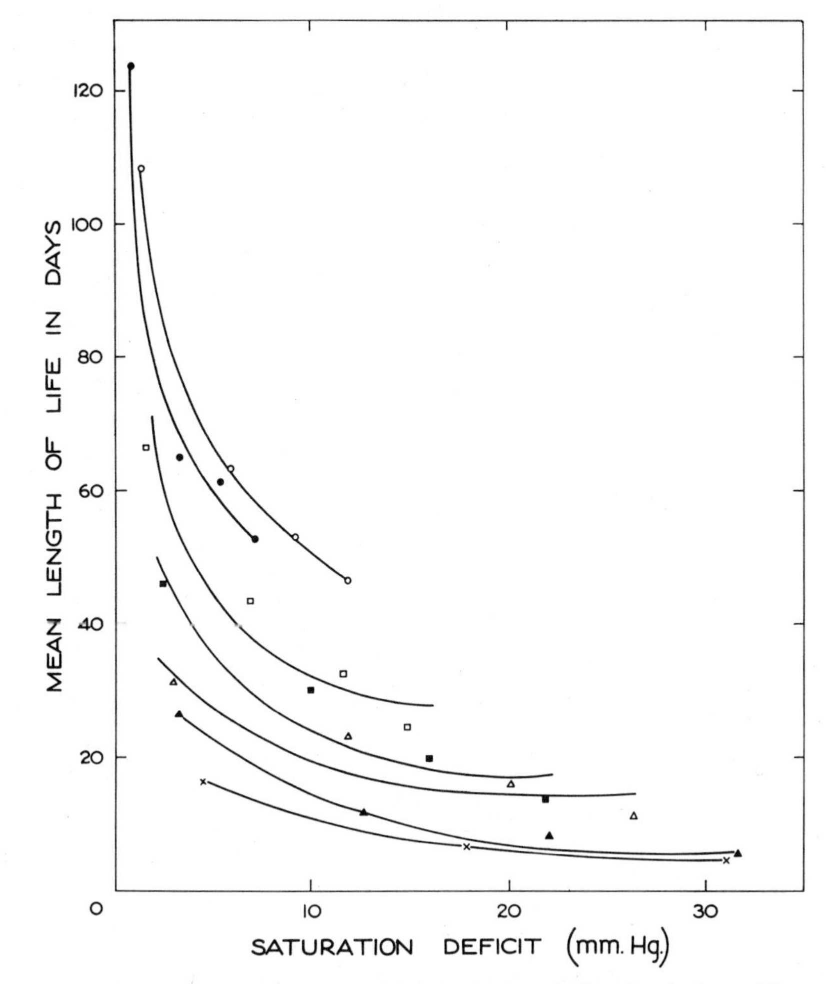

FIG. 7.26.—The mean duration of life of unfed first instars of *Cimex lectularius* at different temperatures in atmospheres of different humidities. Hyperbolas have been calculated to fit the data. The divergence of the observed points from the calculated curves indicates the extent of the divergence of the data from the theory set forth in Fig. 7.25. (After Johnson, 1940.)

7.33 *The Relationship between Saturation Deficit and Survival-Rate*

The relationship between saturation deficit and survival-rate is best explained with reference to some of the earlier attempts to formulate a general law: they were not very successful, for reasons which we shall see. For example, Maercks (1933) measured the death-rate in samples of eggs of *Habrobracon* exposed for a certain period in air at several different constant temperatures and humidities. He also analyzed similar data for the eggs of *Melanoplus*. He plotted the proportions of deaths in eggs from each treatment on a graph, of which the co-ordinates were temperature and relative humidity. He noticed that lines drawn across the graph to link points of equal death-rate coincided with lines joining points of equal saturation deficit, and he concluded from this that the relationship between death-rate and saturation deficit was linear. Of course, this does not follow; and Johnson (1942) showed quite clearly that, provided that temperature was not influencing the death-rate independently of humidity, the lines on Maercks's graph joining points of equal death-rate must coincide with lines joining points of equal saturation deficit, irrespective of what relationship there may be between death-rate and saturation deficit. In fact, when the data for *Habrobracon* are plotted, in the obvious and straightforward way, on a graph having for its co-ordinates death-rate and saturation deficit, the resultant curve is sigmoid, but the data are not adequate to determine its shape or position precisely.

Mellanby (1935) started from the assumption that the "saturation-deficiency law" is true, i.e., that the rate at which water is lost from an insect is proportional to saturation deficit and independent of temperature. From this assumption he went on to argue that the death-rate must therefore be proportional to the product of saturation deficit *times* duration of exposure. This conclusion is misleading on three counts: (*a*) The rate at which water is lost is not determined merely by the saturation deficit independent of temperature. (*b*) For any one temperature the rate at which water is lost may be a linear function of the saturation deficit, but it is not necessarily proportional to it. (*c*) A linear relationship between the rate at which water is lost and the saturation deficit does not connote a similar relationship between saturation deficit and death-rate. On the contrary, if the one is linear, the other must be sigmoid. This is best demonstrated with reference to a simple hypothetical example.

Consider a random sample of insects of the same species being starved in air that is not saturated with water vapor. The temperature and humidity remain constant, and the insects are observed at regular intervals, in order to record the number of deaths since the previous inspection. Now any particular insect will die when it has lost a certain amount of water, i.e., the amount that is fatal for that particular individual. But, like all other living things, these insects will be variable with respect both to the amount of water they can lose before dying and to the rate at which they lose it. A few will die after losing very little water and may lose this rapidly, while a few at the other extreme may survive the loss of a lot of water and may lose it relatively slowly; but most will come between the two extremes. This is indicated in Figure 7.27, *left*, which shows the number of animals in the sample which died for any particular amount of water lost. It takes the form of the normal frequency curve. An alternative way of plotting the same data would have been to plot the cumulative total of deaths against time. This has been done in Figure 7.27, *right*, which is seen to be sigmoid and is the familiar dosage-mortality curve (see sec. 6.312). This experiment could have been done in a slightly different way. A number of different samples of insects taken at random from the same population might have been exposed in unsaturated air, with temperature and period of exposure constant for all the samples but with humidity

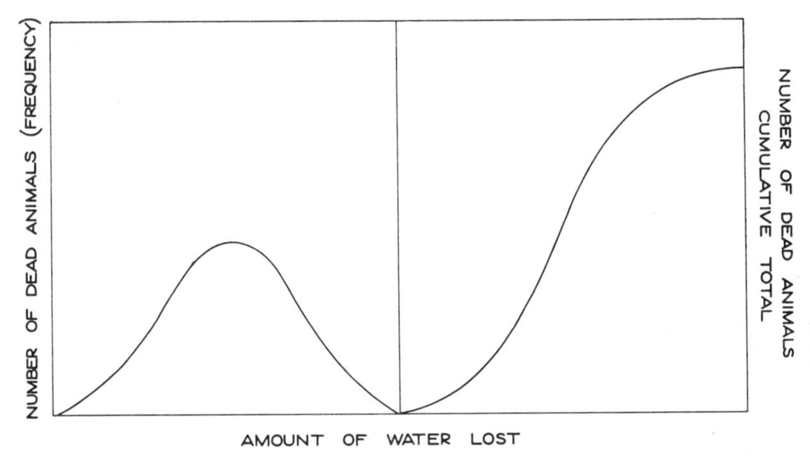

FIG. 7.27.—Hypothetical curves to illustrate the relationship between humidity of the air (dryness) and the death-rate among insects that have no means of replenishing the water lost from their bodies by evaporation. *Left:* the data expressed as a frequency-distribution: the number of deaths is plotted against the fatal "dose" of desiccation expressed as proportion of water lost; *right:* the same data expressed as a dosage-mortality curve: the cumulative total of deaths is plotted against the same values along the horizontal axis.

varied for the different samples. In this case the experimental technique may be a little more complicated, but the principle is the same. As before, the severity of the treatment (or dosage) is measured by the product of saturation deficit and duration of exposure, and the results may be expressed either as a frequency curve or a dosage-mortality curve. In both these experiments one of the factors in the treatment (or dosage) was kept constant while the other varied. A more complicated experiment could be done with both period of exposure and humidity variable, but the principle remains the same. A still more complicated experiment would introduce a third variable, temperature, but a three-dimensional graph would be required to express the results in this case (see Figs. 7.30 and 7.31).

These theoretical conclusions have been verified by the results of experiments. For example, Birch (1944*b*) determined the death-rate in eggs of *Calandra oryzae* exposed to a series of constant temperatures between 15° and 32° C. and a range of humidities betwen 20 and 90 per cent relative humidity. Within this range the lethal influence of temperature could be ignored except for the lot exposed at 15° C. When the death-rate was plotted against the product of saturation deficit and duration of exposure, the observed points followed a sigmoid trend, with a different curve for each temperature (Fig. 7.28). The smooth curves in this diagram have been fitted to the data by the method of probit analysis (sec. 6.312). At most temperatures one dosage-mortality curve did not give as good a fit as a combination of two different curves to cover different parts of the range. The probable reason for this is a

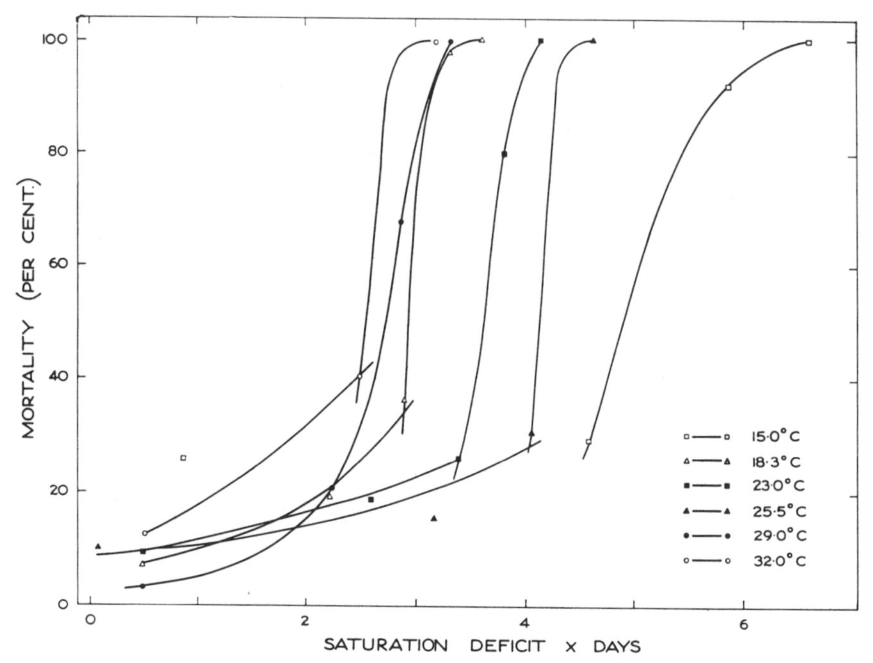

FIG. 7.28.—The relationship between saturation deficit × time and the death-rate among eggs of *Calandra oryzae* at different temperatures within the medial range of temperature. (After Birch, 1944*b*.)

skewness in the frequency curve for minimal lethal water loss. This is indicated by the flatness of the curves over the lower part of their range. This also shows that the influence of dryness is not severe until a critical degree of dryness is reached. Birch's (1944*b*) data for the death-rate of eggs of the beetle *Rhizopertha dominica* confirm this interpretation. Another example of a sigmoid dosage-mortality curve for similar data is provided by Larsen's (1943) experiments with the eggs of *Musca domestica*.

An alternative method of presenting this sort of data is shown in Figure 7.29. The products of saturation deficit × duration of exposure are plotted as the abscissae; temperatures are plotted as the ordinates; the curves show the combinations of these two components that are required to kill half the insects in the sample (this is just an alternative way of expressing the lethal treatment for the "average" individual). Since temperature is not directly lethal within this range, this diagram provides further evidence that temperature directly influences the rate at which water is lost. The graphs also indicate that the eggs of *Rhizopertha dominica* are relatively more resistant to dryness than are the eggs of *Calandra oryzae.*

The examples used so far in this section have all referred to the egg stage of insects which do not require to take in water from outside during development.

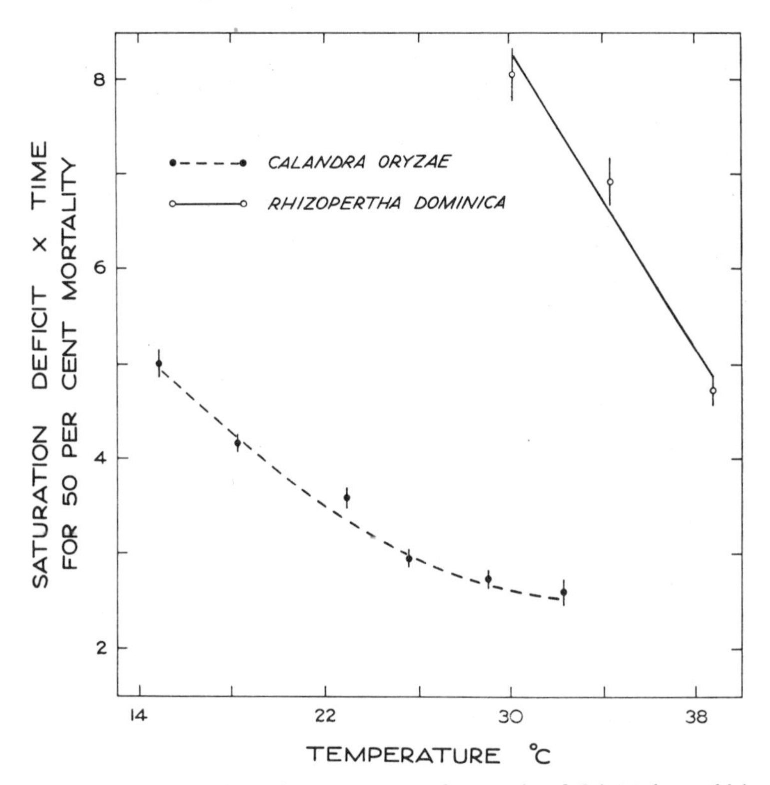

FIG. 7.29.—Various combinations of temperature and saturation deficit × time, which result in the deaths of 50 per cent of the eggs of the beetles *Calandra oryzae* and *Rhizopertha dominica.* The vertical lines show the variance of each point corresponding to odds of $P = 0.05$ (i.e., the line is drawn about the observed mean ± standard deviation × 1.96). Smooth lines are calculated polynomial curves. (After Birch, 1944*b*.)

With the active stages of the life-cycle, this sort of experiment becomes more difficult, because the animal may take in water with its food. With insects that live on "stored products," such as grain, flour, or dried fruit, we have a rather complicated situation because the moisture content of the food tends to be determined by the relative humidity of the air independently of temperature, while the influence of moisture on the insect depends on temperature and is most closely related to the saturation deficit of the air.

Figures 7.30 and 7.31 show the results of some experiments with the immature stages of two beetles that live in stored wheat. A known number of eggs were placed in wheat kept at a certain temperature and moisture content and were left there long enough for them to complete the life-cycle. The death-rate was expressed as the proportion which failed to reach the adult stage. It was noticed that most of the deaths occurred among the eggs and first-instar larvae. The subsequent stages seemed more hardy. But no account is taken of this in the graphs, which are based simply on all the deaths that occurred

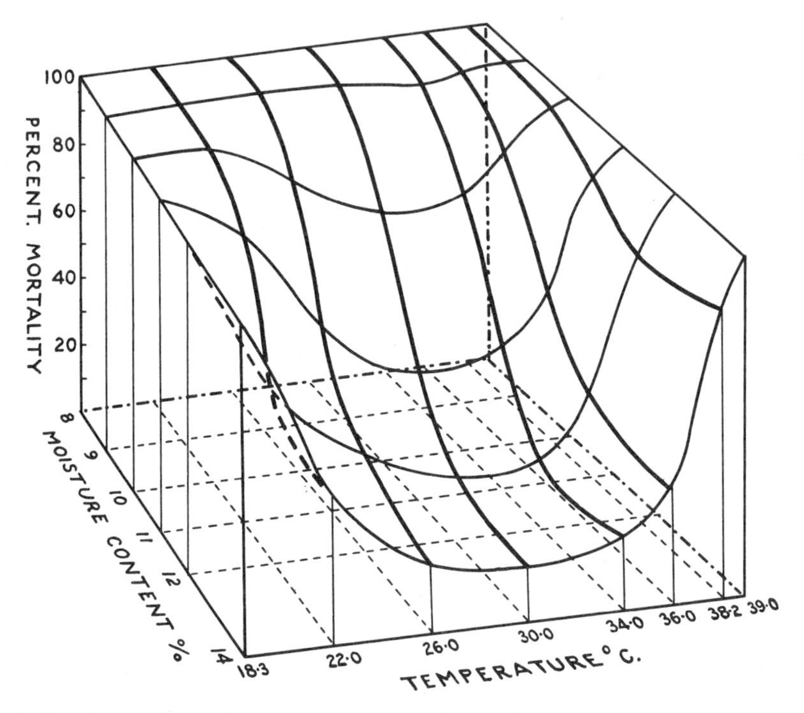

Fig. 7.30.—The death-rate in immature stages of *Rhizopertha dominica* in wheat at various combinations of temperature and moisture content. The complex interactions are well brought out by the three-dimensional graph. (After Birch, 1945*a*.)

during the development from egg to adult. If one imagines the graphs in Figures 7.30 and 7.31 as hollow cubes, inside which certain curves have been drawn, then the top surface of the cube represents the level of 100 per cent death-rate. With both *Calandra oryzae* and *Rhizopertha dominica* the individual curves which relate death-rate to moisture content of wheat, at any one temperature, are sigmoid and are indeed dosage-mortality curves, though in a complicated experiment like this it is not easy to express the "dosage" precisely in any simple units. In Figure 7.30 the three sigmoid curves for the temperatures 26°, 30°, and 34° C. are parallel or nearly so; the small departures from the parallel are nonsignificant. This means that with *R. dominica* the death-rate was independent of temperature between the limits 26°–34° C. In other words, the curves relating death-rate to temperature at each moisture content are flat between 26° and 34° C.; but they slope steeply outside this range, indicating that below 26° and above 34° C. temperature influences death-rate profoundly. And the diagram, taken as a whole, indicates a strong interaction between temperature and moisture content of wheat in their respective influences on death-rate.

With other foods, for example, growing plants, the situation is so complicated

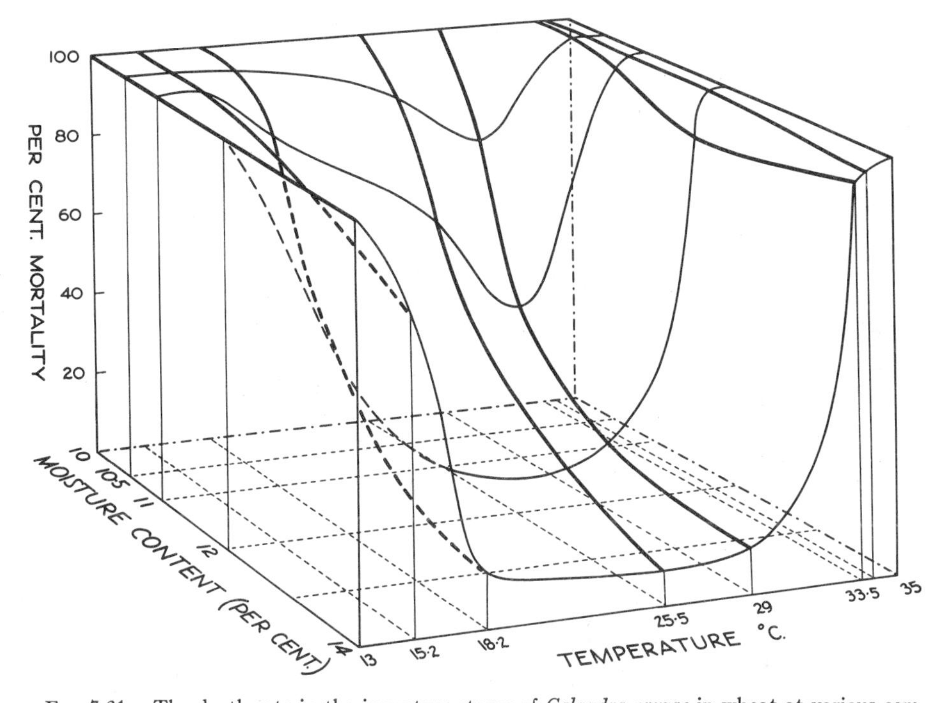

Fig. 7.31.—The death-rate in the immature stages of *Calandra oryzae* in wheat at various combinations of temperature and moisture content. Compare this with Fig. 7.30. The three-dimensional graphs illustrate nicely the complex differences between the two species. (After Birch, 1945*a*.)

that it may be difficult, if not impossible, to discover any simple general law relating the death-rate to the humidity of the air.

7.34 *Drought-Hardiness and Acclimatization*

Some animals are drought-hardy throughout their lives, while others become drought-hardy only during certain stages of their life-cycles. This is related to the level of dryness in the sorts of places where they usually live. For example, certain lizards and ticks which live in deserts are drought-hardy all their lives (Breitenbrecher, 1918; Lees, 1946*a*); this is largely due to the very slow rate at which they lose water from their bodies. This sort of adaptation is characteristic of all stages of the life-cycle for animals that live in deserts.

But in regions where drought is seasonal, the life-cycles of the animals may be so regulated that a special drought-hardy stage coincides with the dry season.

The Australian plague grasshopper *Austroicetes cruciata* is found in a region where the summer is hot and dry, and it is adapted for life in this climate by virtue of a diapause in the egg stage which insures that only the drought-hardy, diapausing eggs shall be present during summer and that the less hardy stages (post-diapause eggs, nymphs, and adults) shall be present only during late winter and spring, when the weather is not so dry (sec. 4.6). The lucerne flea *Smynthurus viridis* is unable to survive during the nymphal and adult stages for more than a few hours except when the atmosphere is nearly, though not quite, saturated with water vapor. The newly laid egg and the egg that is about to hatch are only slightly hardier. Nevertheless, *Smynthurus* is widespread and abundant in southern Australia in a region where there may be 5 months or more of continuous drought during summer (Figs. 11.01, 11.02). This is made possible by a special adaptation in the egg stage; those eggs which are laid as summer approaches complete only about half their development and then aestivate in this stage, which is extremely drought-hardy (Davidson, 1932*b*, 1933). In the weevil *Otiorrhynchus cribricollis* the drought-hardy stage is associated with diapause in the adult. After feeding for a while in the spring, the adults bury themselves in the dry soil and remain there without food through the hot, dry summer. Their bodies, when examined, appear to be quite desiccated, but the survival rate is high (Andrewartha, 1933). Weevils of the genus *Listroderes* are similarly adapted, by means of a diapause in the adult stage, to survive in regions where the summer is hot and arid. In insects, diapause, is often associated with drought-hardiness, just as it is with cold-hardiness (sec. 6.334). Indeed, the same individual is often both drought-hardy and cold-hardy. This lends support to the suggestion that there is a causal relation between diapause and hardiness with respect to both drought and cold.

On the other hand, there are species which, without entering diapause, become drought-hardy as they become quiescent in response to dryness; and this is quite a different phenomenon. As an example we quote the experiments of Breitenbrecher (1918) with the potato beetle *Leptinotarsa decemlineata*. Normal active beetles moved toward the light and away from the ground. In one series of experiments a batch of normally active beetles were divided into three lots. Each lot was buried in soil to a depth of 15 cm. One lot of soil was kept moist, the second lot was less moist, and the third was kept dry. After eight days the insects in the dry soil had lost weight, and the others weighed about the same as at the beginning of the experiment. Beetles from the moist and the less moist soils still showed the "normal" reaction to light, but those from the dry soil showed reversed reactions, they moved away from light and toward the ground. When placed on potato plants, the beetles from the moist and those from the less moist soils moved upward and began to feed on the uppermost leaves. Within a few days they began to lay eggs. But those from the dry soil moved down the stems of the plants and burrowed into the soil, where they remained until the first rain, 2 weeks later. Then they emerged and laid eggs on the potato plants. It is easy to see how this change in behavior in response to desiccation gives the animal a better chance of surviving in regions subject to drought, for it insures that the eggs are laid only when circumstances are likely to be favorable for the hatching larvae.

At the same time that these beetles were assuming the behavior appropriate to quiescence, they were also becoming drought-hardy. In a further series of experiments Breitenbrecher gradually desiccated some beetles by keeping them in cases on dry sand and offering them only food that was somewhat wilted. When desiccation had proceeded to a certain point, the behavior of these beetles was reversed, and they buried themselves in the sand. They were kept buried in dry sand for 8 months, and at the end of this time 90 per cent of them were still alive. Another lot of beetles, which were comparable in every way except that they did not experience the preliminary gradual desiccation, were also buried in dry sand alongside the first lot. Every one of this second lot died before the 8 months were up; they were not drought-hardy. Breitenbrecher did a number of other experiments which demonstrated that "drought-hardy" beetles in this experiment were "quiescent" (in the sense that Shelford used this term: see sec. 4.0) and not in diapause.

7.4 THE LETHAL INFLUENCE OF EXCESSIVE MOISTURE

The distributions of many insects are restricted by excessive wetness. We know this because some species do not spread into moist places, and within their permanent distributions they are more numerous during dry weather. It is relatively easy to give examples of this. It is much more difficult to analyze the harmful influence of excessive moisture on any one species. Cook's (1924, 1926, 1929, 1930) classic studies on the distribution and abundance of the pale western cutworm *Porosagrotis orthogonia* demonstrated this phenomenon nicely (sec. 13.13). He showed that outbreaks of *Porosagrotis* in Montana occurred after unusually dry weather. When the rainfall during spring was less than 4 inches, an increase in damage occurred the following year in an average of 8 out of 10 years. In Alberta, Seamans (1923) used the number of wet days in May and June as a basis for predicting the abundance of *Porosagrotis*. A "wet day" was one on which more than a quarter of an inch of rainfall was recorded. He showed that more than 10 such days are followed by a decrease in damage by *Porosagrotis*. Cook (1926) showed that *Porosagrotis* was able to maintain a permanent population in an area which had an average of 1 year in 10 which was dry enough for increase. Regions with a greater frequency of wet years were outside the zone of permanent distribution of this species (sec. 13.133).

Many grasshoppers and locusts are limited by excessive moisture. The nymphs of *Melanoplus mexicanus* and *Camnula pellucida* die in large numbers during periods of wet, cloudy weather when the relative humidity is close to saturation (Parker, 1930). The Australian locust *Chortoicetes terminifera* leads a precarious existence when it spreads south from its inland breeding grounds to the moister wheat belt, where the winter is cool and humid. In this area the eggs of *Chortoicetes*, lacking any protective diapause, are liable to hatch after the autumn rains. The nymphs, ill adapted for life in the wet and the cold, die at an early stage, and the locust plague comes to an end. The only circumstances in which the plague persists after reaching this zone is when the autumn rains are greatly delayed and do not come while the soil is still warm enough to hatch the eggs. In this event the eggs stay safely quiescent in the soil during winter and hatch in the spring. This is an unlikely event in this region of reliable winter rainfall, and locust plagues are correspondingly infrequent and short-lived. The southern limits of the permanent breeding grounds may be defined in terms of the monthly ratio of precipitation to evaporation (P/E) and temperature (Andrewartha, 1940). Similarly, the coastward limit to the distribution of the grasshopper *Austroicetes cruciata* in Australia may be defined in terms of moisture. In eastern Australia the limit corresponds to the isopleth (line joining equal values) for P/E equals 1.0 for September; in Western Australia, where the season is earlier, the same isopleth for August defines the coastward limit to the distribution. Coastward of these lines the weather is too wet during the spring, and the nymphs cannot survive except very sparsely (Andrewartha, 1944*b*). Within the zone of its permanent distribution the species becomes most abundant during relatively dry years. This is not entirely due to the higher survival-rate among the nymphs on these occasions; dry years also result in more extensive areas suitable for egg beds (Clark, 1947*b*). Another well-documented example of an insect that multiplies following dry years is the spruce budworm *Choristoneura fumiferana* in Canada (Wellington *et al.*, 1950). For outbreaks to occur, the summer must be dry, but dry weather during spring and autumn also seems to be favorable. There need be little doubt that excessive moisture imposes a limit on the distribution and influences the abundance of the species mentioned in these examples and others that we know. But there is not always a simple or straightforward explanation of how this happens (sec. 13.14).

Excessive moisture may simply drown the insect. In wet years in southeastern United States large numbers of the overwintering pupae of the corn earworm *Heliothis armigera* die from drowning. Barber and Dicke (1939) constructed artificial burrows in which they placed pupae. The burrows were flooded for different periods during the winter. Some pupae floated to the top of the burrows, and they were able to survive for 20 days in this condition. But those which were completely submerged died in less than 10 days. A higher proportion died at 21° than at 4.5° C. (sec. 12.22).

For the most part, the influence of excessive moisture is indirect. For example, in the presence of excessive moisture the animal may not become cold-hardly, so that it dies during winter from excessive cold (sec. 6.314). Interactions between food and moisture are, of course, commonplace. Wellington *et al.* (1950) pointed out that the production of staminate flowers was decreased in wet years, thus decreasing the amount of highly favorable food available for larvae of *Choristoneura*. But by far the most important way that excessive moisture influences the death-rates among natural populations of insects is by favoring the spread of epizoötics of virus, fungal, and bacterial diseases (sec. 10.323). Among the examples we mentioned at the beginning of this section, the association of disease with excessive moisture was probably important in relation to *Austroicetes* (Andrewartha, 1944*b*), *Chortoicetes* (Andrewartha, 1940), *Melanoplus* and *Camnula* (Parker, 1930), and *Porosagrotis* (Cook, 1926). Sometimes the egg beds of *Austroicetes cruciata* are flooded in winter, and many die as a result; but Clark (1947*b*) considered that in these cases death was more likely to be due to fungal disease than to drowning. Parasitic fungi, bacteria, and viruses usually grow and spread most rapidly during wet weather or in wet places. The spores of fungi are not formed in dry air, and even if the spores are present, they may not germinate. Rapid growth of parasitic fungi cannot take place if the air is dry (Steinhaus, 1946). As an example of entomophagous fungi which need moisture, Steinhaus quotes the fungi *Beauveria bassiana*, *Spicaria farinosa*, and *Aspergillus flavus*, each of which can infect insects only when the relative humidity is greater than 70 per cent. Infection is fastest in a saturated atmosphere.

Excessive moisture may be harmful to insects and ticks without killing them. Bataillon (quoted by Buxton, 1932*a*) found that silkworm larvae did not pupate in moist air. The fecundity of some insects and ticks is reduced in moist air (sec. 7.5), and their speed of development may be decreased (sec. 7.6). Perhaps these effects of excess moisture are due to a "drowning" of the tissues, which has its deleterious effect long before the animal itself dies. There is really no reason to suppose that death by drowning takes place only when the animal is submerged in water. At relative humidities higher than 88 per cent the tsetse fly *Glossina tachinoides* does not feed and eventually dies (Buxton and Lewis, 1934). The lives of the flies are shortest at high relative humidities and longest at a relative humidity of 44 per cent (Fig. 7.32). At relative humidities lower than 44 per cent, reduced longevity is probably due to loss of water, but this cannot be the explanation for the shorter life at high relative humidities. Buxton and Lewis did not find any difference in the total water content of flies at different relative humidities, but they suggest that at the higher humidities there may be a harmful accumulation of water in some of the insects' organs. Johnson and Lloyd found that 68 per cent of wild flies in northern Nigeria were gravid in dry seasons, but only 20 per cent in wet seasons.

MacLagan (1932*a*) noticed that the immature lucerne flea *Smynthurus*

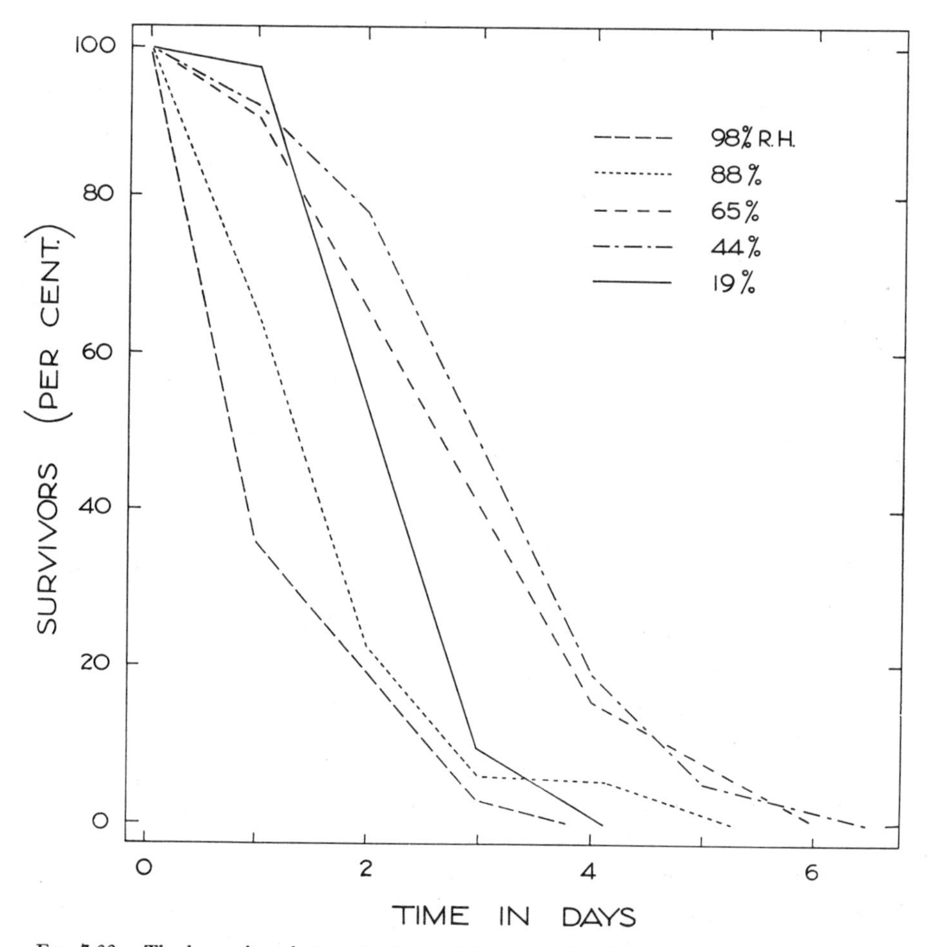

FIG. 7.32.—The longevity of starved adults of the tsetse fly *Glossina tachinoides* (males and females) at 30° C. in air at different relative humidities. (After Buxton and Lewis, 1934.)

viridis had a distended appearance in air which was saturated with water, and they lived for a shorter time at 100 per cent relative humidity than at 90 per cent. They also laid fewer eggs in soil which was saturated with water than in moist but unsaturated soil (Davidson, 1932*a*).

Certain experiments by Cunliffe on ticks which normally live in dry places also indicated that excessive moisture interfered with their normal physiology.

He kept *Ornithodorus savignyi* without food at three different humidities. One lot were kept over calcium chloride, another lot were provided with a little water, and another lot received excess water (the experiment was not described more precisely than this). The death-rate during development from egg to adult was 55 per cent for those in dry air (over calcium chloride), 73 per cent for those for which a little water was provided, and 91 per cent for those receiving most water. Similar results were obtained with *O. moubata.* The first lot completed their development most rapidly (Cunliffe, 1922). The opposite has been observed with species which usually live in moist places. For example, with *Ixodes ricinus,* which is found in moist upland pastures in Britain, none survived in air that was less than 70–80 per cent saturated with water vapor, depending on the temperature (MacLeod, 1935).

These results can be explained without reference to disease. The species which is adapted to life in a moist place (*Ixodes*) rapidly loses water through its cuticle in dry air; but it requires plenty of water and actively absorbs water from moist air to replace its losses; moreover, it is able to regulate quite precisely the quantities absorbed (see secs. 7.212 and 7.213). On the other hand, *Ornithodorus,* being adapted to life in the desert, has a remarkably impermeable cuticle and loses water very slowly, even in the driest air, but it "drowns" when surrounded by too much moisture.

7.5 THE INFLUENCE OF MOISTURE ON FECUNDITY

Hamilton (1950) kept newly emerged adults of the African locust *Locusta migratoria* in the laboratory at different temperatures and humidities and showed that they lived longer, the lower the humidity, but that they failed to produce any eggs at all below a certain threshold—about 40 per cent relative humidity. One interesting set of results relates to the period which elapsed before the locusts became "sexually mature," i.e., before the first eggs were laid (Fig. 7.33). At 70 per cent relative humidity the locusts matured sooner than in air which was drier or moister than this. The length of life was also shortest at 70 per cent relative humidity, becoming greater as relative humidity was increased or decreased from 70 per cent (Fig. 7.33). Comparison of the two curves in Figure 7.33 shows that as the time taken for sexual maturity increased (at relative humidities greater and less than 70 per cent), the length of life also increased. But this increase in length of life occurred only within the zone of humidity within which the locusts could eventually become sexually mature. At higher and lower relative humidities, the length of life became shorter again. This is shown by the two peaks in the lower curves in Figure 7.33. Thus within the zone of relative humidity, 40–80 per cent (at 32.2° C.) the length of life depended upon the rate of sexual maturity, but outside this zone it depended directly upon the relative humidity.

The desert locust *Schistocerca gregaria* responded to moisture similarly to *Locusta*, except that in the range of low humidities in which no eggs were matured the locusts continued to survive, under certain circumstances, for many months. If at any time during this period they were exposed to air containing adequate moisture, they matured eggs in about the same time as a newly emerged adult at the same humidity (Hamilton, 1950). This phenome-

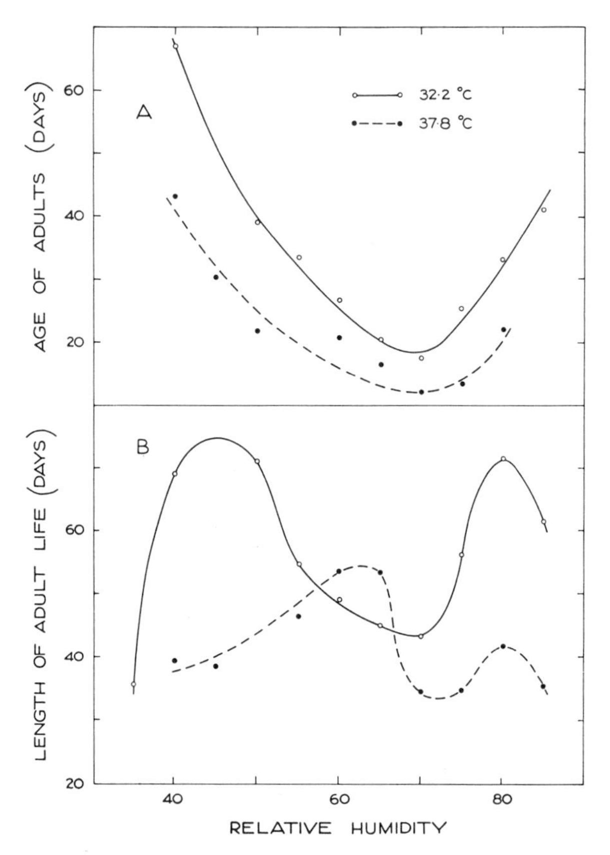

FIG. 7.33.—*A:* the time required for adults of *Locusta migratoria migratorioides* to become sexually mature at different relative humidities; *B:* the longevity of adults of *L. migratoria migratorioides* in air of different relative humidities. (After Hamilton, 1950.)

non is also well known from nature, and it is erroneously called "diapause" by nearly all students of acridology. It is, however, a clear-cut example of quiescence (see sec. 4.0).

Hamilton also recorded the total number of eggs laid during a lifetime and found that there was an optimal humidity at which fecundity was greatest. With *Locusta*, for example, this was 70 per cent relative humidity (Fig. 7.34). Quite a number of insects are like these locusts, in that fecundity is at a maximum at a definite optimal relative humidity, above and below which the number of eggs decreases. The moth *Pannola flammea* produced fewer eggs in saturated air than in air at 90 per cent relative humidity. In saturated air, only 20 per cent of the females copulated (Zwolfer, 1931). The lucerne flea *Smynthurus viridis* laid fewer eggs in soil which was saturated than in soil which was moist but not saturated (Davidson, 1932*a*).

Not all insects show a definite optimal relative humidity for egg-laying. The fecundity of the weevil *Calandra oryzae* increased with the moisture content

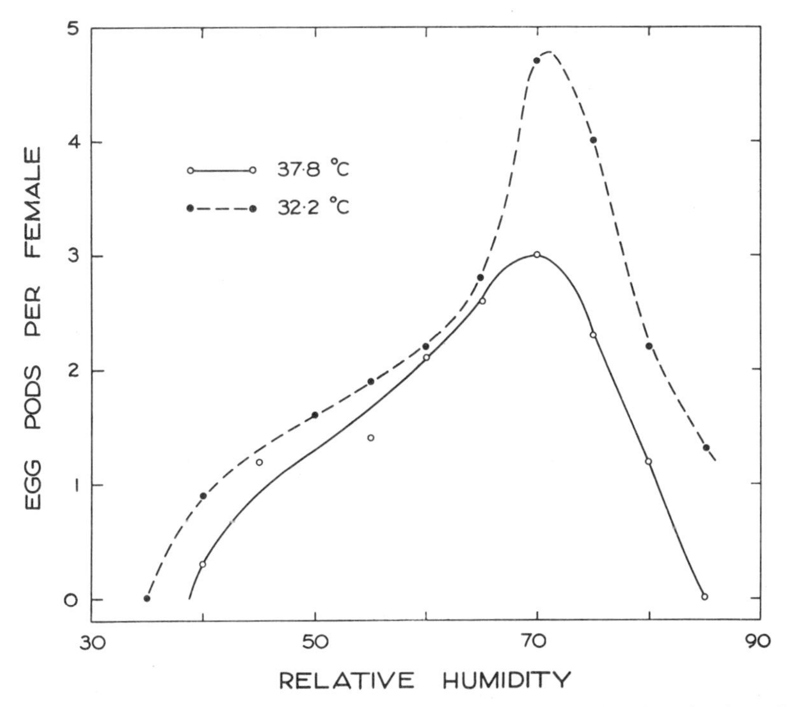

FIG. 7.34.—The number of eggs laid by *Locusta migratoria migratorioides* kept in air at different relative humidities. (After Hamilton, 1950.)

of the wheat it was feeding on from zero in wheat at 9 per cent moisture content (34 per cent relative humidity) to a maximum in wheat of 14 per cent moisture content (70 per cent relative humidity) (Fig. 7.35). But there was no decrease in fecundity in moister wheat, at least as moist as 20 per cent; it is not practicable to experiment with wheat moister than that because of the development of molds (Birch, 1945*d*). The fecundity of *Calandra granaria* showed a similar trend with relative humidity at 27.5° C., but at lower temperatures there was an optimal humidity above and below which fecundity fell off (Eastham and McCully, 1943). Thus, at 27.5° C., the reactions of *C. granaria* resembled those of *C. oryzae*, but at lower temperatures its reactions to humidity resembled those of *Locusta*. Among the many insects and ticks in which fecundity was not adversely influenced by saturated air are the potato beetle *Leptinotarsa decemlineata* (Breitenbrecher, 1918), the European corn borer *Pyrausta nubilalis* (Huber *et al.*, 1928), and the sheep tick *Ixodes ricinus* (MacLeod, 1935).

The relative influences of temperature and moisture on the fecundity of *Calandra oryzae* and *Rhizopertha dominica* is shown in Figures 7.36 and 7.37. There is a pronounced interaction between temperature and moisture. The zero lines in these graphs show the combinations of temperature and moisture beyond which these beetles did not lay eggs. Within an optimal range of tem-

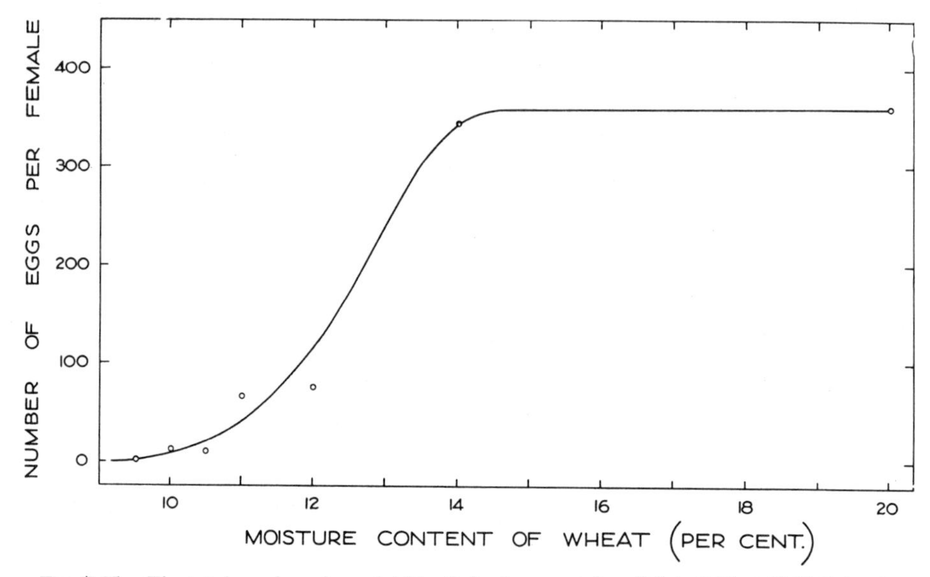

FIG. 7.35.—The total number of eggs laid by *Calandra oryzae* (small "strain") at 29.1° C. in wheat of different moisture contents. There was 1 insect per 10 grains. (After Birch, 1945*d*.)

perature they were able to lay eggs in wheat which was drier than the driest wheat in which eggs were laid at temperatures on either side of the optimal range. But at no temperature could *C. oryzae* lay eggs in wheat drier than 10 per cent moisture content; the corresponding moisture content of wheat for *R. dominica* was 9 per cent. Graphs of this sort are useful in showing the combinations of temperature and moisture which prevent eggs from being laid and which therefore prevent any increase in numbers.

Animals may be desiccated in a stage prior to the reproductive stage, and this may influence the fecundity of the adults. Ludwig (1942, 1943) kept pupae of the luna moth *Tropaea luna* and the cynthia moth *Samia walkeri* at different relative humidities. When the adults emerged, he kept them at room temperature and humidity, which was favorable for egg-laying. The numbers of eggs these moths laid are shown in Figure 7.38. The number of eggs laid by the cynthia moth was almost independent of the humidity at which the pupae had been reared within the limits of 27–100 per cent relative humidity (*middle line*, Fig. 7.38). Pupae which developed in air drier than 27 per cent relative humidity produced moths which laid fewer eggs. This was evidently due to a failure to deposit eggs, since the total number of eggs formed was not different at the different relative humidities (*top line*, Fig. 7.38). By the same criteria, pupae of the luna moth were also drought-hardy but less so than those of the cynthia moth (*bottom line*, Fig. 7.38). The pupae of these moths hang in trees and might be expected to be well adapted to retain their moisture, even

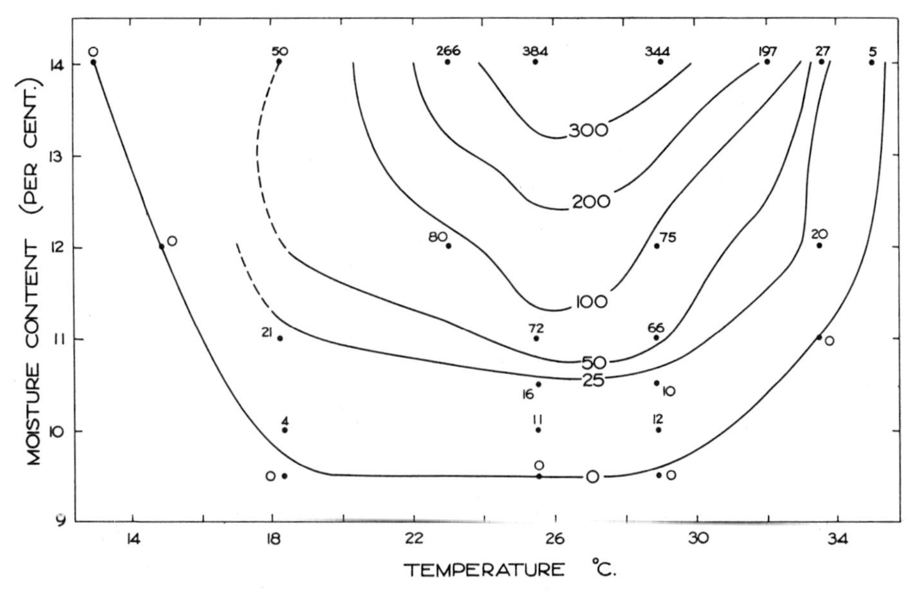

FIG. 7.36.—Lines of equal egg-production for the small "strain" of *Calandra oryzae* at different temperatures in wheat of different moisture contents. Small figures are experimental values. (After Birch, 1945*d*.)

in dry air. If similar experiments were done with pupae of species that are normally found in moister places, one would expect to find them less drought-hardy.

7.6 THE INFLUENCE OF MOISTURE ON SPEED OF DEVELOPMENT

With some insects, development ceases almost completely in dry air, although they do not die except after prolonged exposure. For example, Davidson (1936*a*) kept eggs of the locust *Chortoicetes terminifera* in an open-air insectary in dry soil for 5 months. They remained alive but quiescent; development was resumed when the soil was remoistened. This is a valuable adaptation for a species that lives in an area that is nearly a desert, where rainfall is unreliable and prolonged droughts are quite usual. If the eggs were to complete their development in dry soil, there would be a big risk that the nymphs would hatch when there was no food for them. But in this climate the rains which moisten the soil also bring forth a flush of green herbage (Andrewartha, 1937). The influence of moisture on the speed of development of the nymphs of *Locusta migratoria* and *Schistocerca gregaria* is shown in Figure 7.39, *A* and *B*. There was an optimal humidity at which nymphs developed most rapidly. This occurred at 70 per cent relative humidity for *Locusta*. The speed of development fell off in air that was drier or moister than

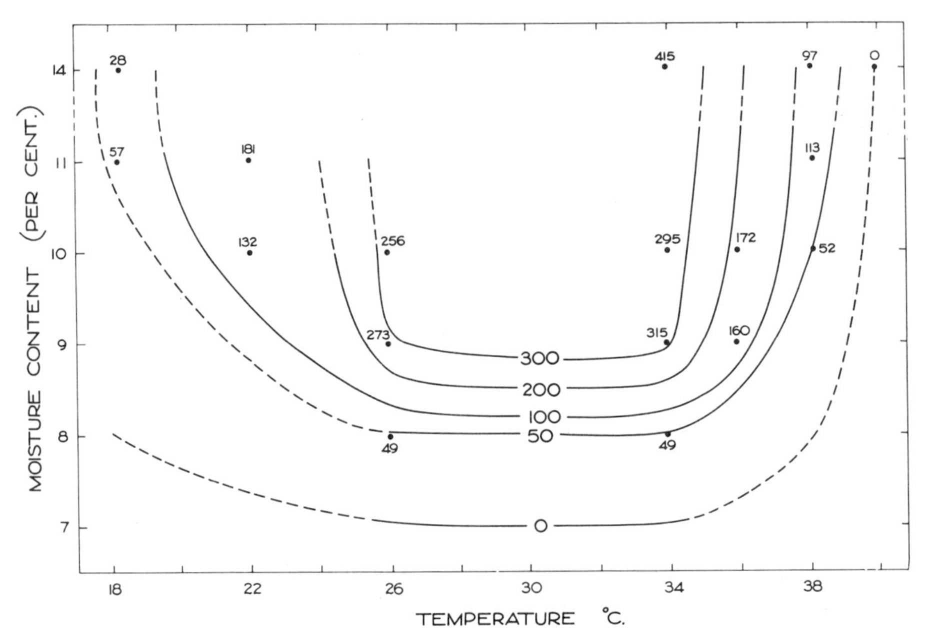

FIG. 7.37.—Lines of equal egg-production for *Rhizopertha dominica* at different temperatures in wheat of different moisture contents. Small figures are experimental values. Note shift of lines to higher temperatures and lower moisture contents than in the corresponding graph for *Calandra oryzae* (Fig. 7.36). (After Birch, 1945*d*.)

this. The graphs also show that the range of relative humidities within which the nymphs developed was reduced as temperature increased. At relative humidities greater and less than those shown in the graphs, all the insects died. This illustrates a point of general importance when dealing with the influence of moisture on development. We are concerned only with that range of humidity within which development can be completed before the animal dies. At humidities beyond this range, moisture is more appropriately considered in relation to longevity (sec. 7.32).

When the speed of development of the nymphs of the tick *Ixodes ricinus* (MacLeod, 1934) was plotted against humidity, the curves at different temperatures showed trends similar to those for *Locusta* (Fig. 7.40). There was an optimal relative humidity at 90 per cent, above and below which the speed of development fell off. The optimum was more definite at 25° C. than at higher temperatures. The optimum for ticks like *Ornithodorus*, which usually live in dry places, was much lower (Cunliffe, 1921, 1922).

With many species of insects this phenomenon of an optimal humidity cannot be observed because development is most rapid in saturated air. The speed of development of the eggs of *Lucilia sericata* (Evans, 1934) and of the eggs of *Musca domestica* Larsen (1943) increased consistently as the

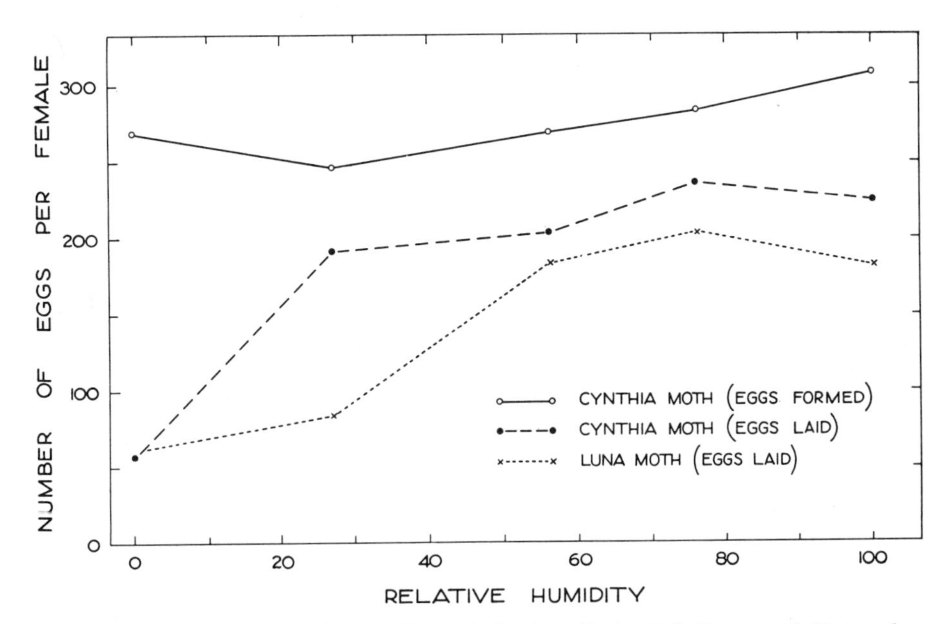

FIG. 7.38.—The fecundity of the Cynthia moth *Samia walkeri* and the Luna moth *Tropaea luna* under favorable conditions, after pupae were kept in their cocoons all their lives at different relative humidities. (After Ludwig, 1942.)

humidity was increased from the lowest humidity at which they will develop to saturated air. The relationship between duration of development and relative humidity at any one temperature was linear for the eggs of these two insects. The relationship between the duration of development and the saturation deficit must also be linear at any one temperature, since at any one temperature the saturation deficit is a linear function of the relative humidity. The linear relationship between the saturation deficit and the duration of development for eggs of *Lucilia* is shown in Figure 7.41. Evans also found that the rate at which eggs lost water at any one temperature was directly proportional to the saturation deficit. The duration of development was therefore linearly related to the rate at which water was lost from the eggs. Larsen (1943) found with *Musca domestica* that the linear relationship between duration of development and saturation deficit held for three temperatures (25°,

30°, and 35° C.) which she studied. The position and slope of the graphs were, of course, different at each temperature.

Nymphs of *Locusta* provide an example of one sort of relationship between moisture and speed of development in which the speed of development is re-

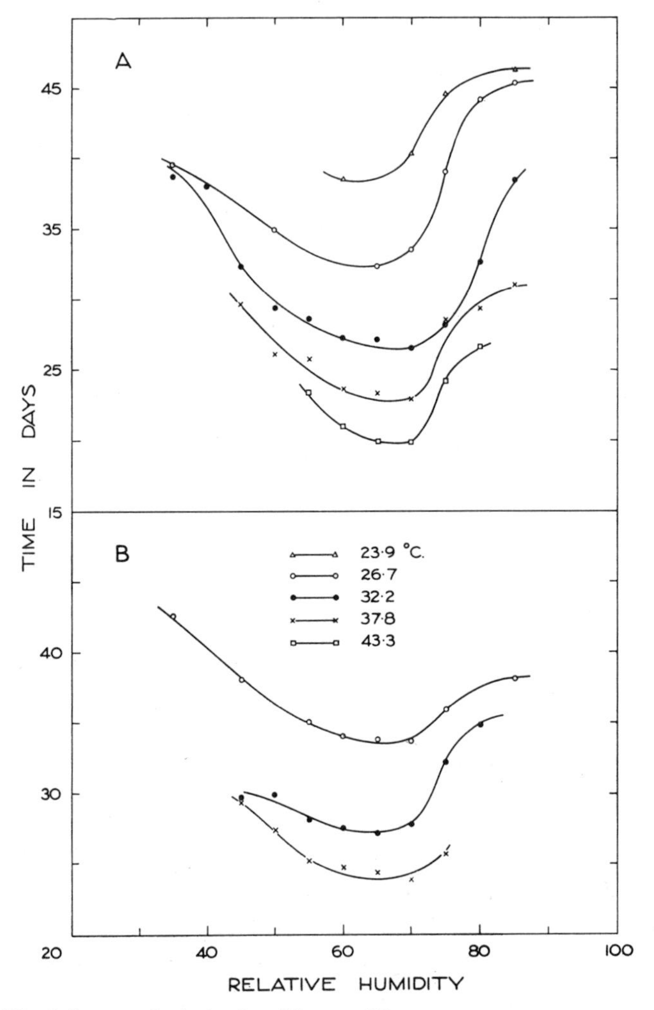

FIG. 7.39.—The influence of relative humidity at different temperatures on the time taken for nymphs of locusts to become adults. *A, Locusta migratoria migratorioides; B, Schistocerca gregaria.* (After Hamilton, 1950.)

tarded at high humidities. The eggs of *Lucilia* are examples of a second type of relationship in which high humidities accelerate the speed of development. There is yet a third category, which includes those insects in which the speed of development is relatively independent of humidity. Ludwig and Anderson (1942) incubated eggs of four saturniid moths at different relative humidities and at different temperatures. They found very little difference in the duration of development at different relative humidities at any one temperature, except for eggs of *Telea polyphemus*, which developed a little more slowly at 0 per cent relative humidity than in saturated air. Other insects which have eggs like this include the bedbugs *Cimex lectularius* and *C. rotunda* (Geisthardt, 1937), the firebrat *Thermobia domestica* (Sweetman, 1938), the bean weevil *Bruchus obtectus* (Menusan, 1934), and the clothes moth *Tineola bisselliella*

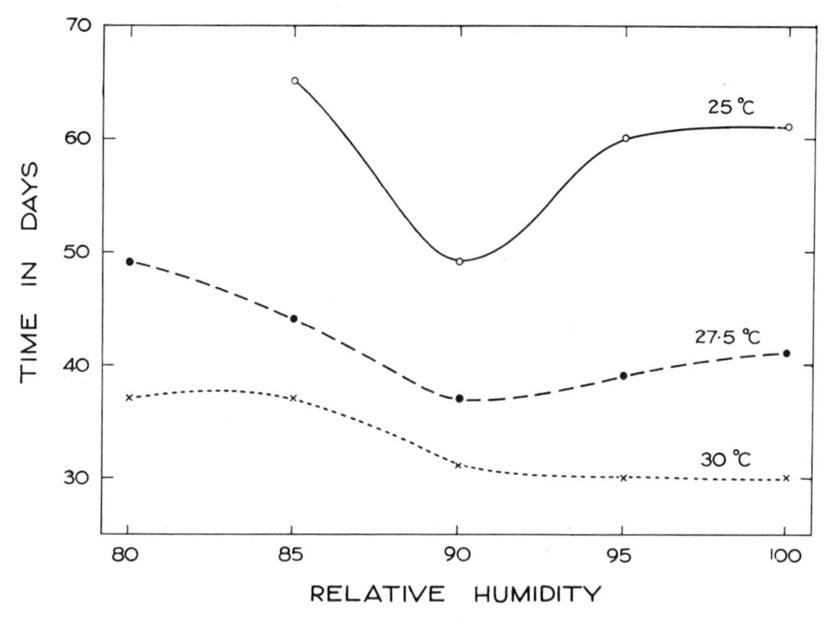

FIG. 7.40.—The influence of relative humidity at different temperatures on the time taken for nymphs of *Ixodes ricinus* to become adults. (After MacLeod, 1934.)

(Griswold, 1944). All these insects occur in dry places, and their eggs have cuticles which are impermeable to water vapor. The same is true of the pupae of the moths *Bombyx mori* (Leclerq, 1946*b*) and *Araschnia levana* (Leclerq, 1946*c*), which develop at the same rate at relative humidities from 20 to 100 per cent. In so far as speed of development depends on humidity, certain difficulties arise in the design of experiments planned to show the influence of temperature on speed of development. The experimenter usually wants to know the influence of temperature when the influence of moisture is invariable. This will clearly not be given by experiments at different temperatures at the one relative humidity, since at different relative humidities the saturation deficit is different, and therefore the rates of evaporation from the insects are likely to be different. The rate of development is more likely to depend on the total water lost during development than on the rate of loss of water. This is because the total water lost takes into account both rate of loss and duration of exposure. In determining the influence of temperature on rate of development, the influence of moisture is therefore most likely to be kept constant throughout the range of different temperatures, when the conditions are such that the animals lose the same amount of water during the experiments at each temperature. Birch (1944*a*) designed an experiment with eggs of *Calandra oryzae* to test this hypothesis. The interaction between temperature and

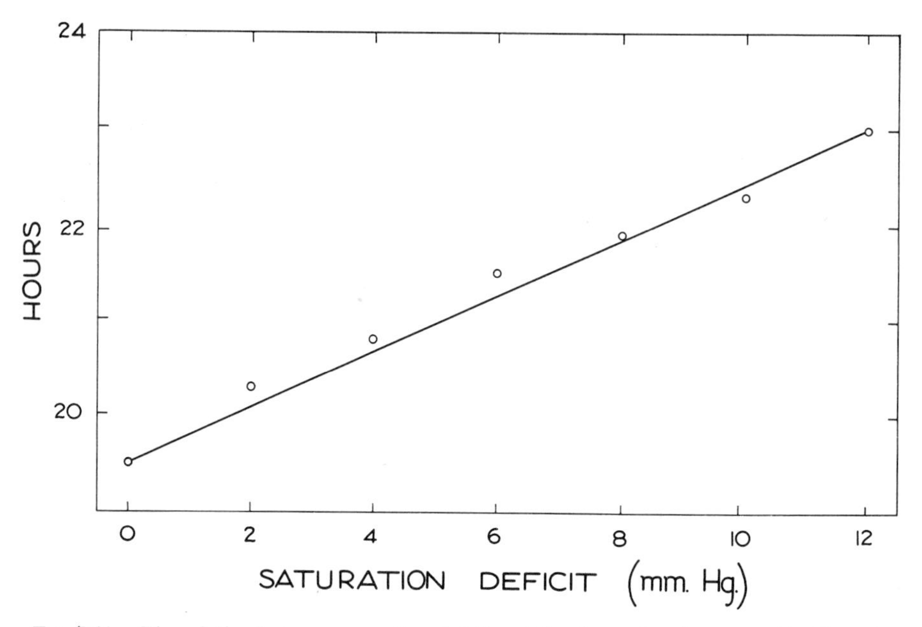

FIG. 7.41.—The relation between saturation deficit and the time taken for eggs of *Lucilia sericata* to hatch at 27° C. (After Evans, 1934.)

"moisture" (defined in this special way) was small, and he concluded that the hypothesis had been confirmed for the eggs of *C. oryzae*.

7.7 SUMMARY

We have seen that animals may respond to moisture in a way which takes them to places where the moisture gives them a good chance to survive and multiply. Some animals move away from moisture to dry places, while others move from dryness to wetter places. But whether they are in dry or wet places, animals are still faced with the necessity for maintaining the water content of their bodies at a relatively constant level. When their water content falls below a certain minimal amount, they die. There is also some evidence, though it is less well substantiated, that animals also die when their water content increases above "normal," as it may in very wet places. There is thus a zone of lethal wetness and a zone of lethal dryness, with a tolerable zone of favorable moistness in between. These are analogous to the zone of lethal high temperatures, the zone of lethal low temperatures, and the medial favorable zone on the temperature scale (Fig. 6.31).

The influence of moisture in the three zones from 0 to 100 per cent relative humidity has been summarized in a hypothetical diagram in Figure 7.42.

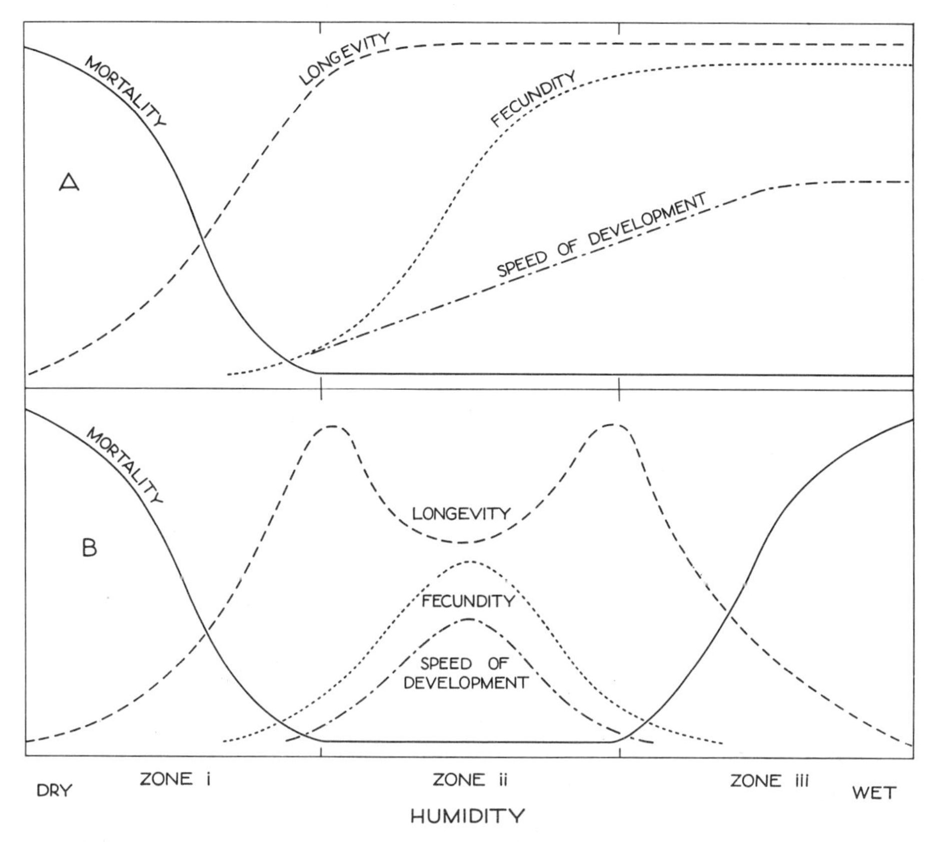

FIG. 7.42.—Hypothetical curves summarizing the influence of moisture on the longevity, fecundity and speed of development of an animal. *A* relates to those kinds of animals which seem not to be harmed by exposure to high humidity. *B* relates to those kinds of animals which are harmed by exposure to high humidity. In zone (*i*) for *A* and *B* and in zone (*iii*) for *B* only, interest centers on the lethal influence of dryness and wetness, respectively. In zone (*ii*) interest centers on speed of development and fecundity. In *B* both these curves have well-defined optima. In *A* the curves approach a maximum at 100 per cent relative humidity. Fecundity in this case is represented by a curve which tends to an asymptote, and speed of development is represented by a straight line.

Figure 7.42, *A*, summarizes the influence of moisture on those animals which appear to be able to tolerate high humidity and Figure 7.42, *B*, summarizes the influence of moisture on those animals which are harmed by high humidity.

In the zones of lethal dryness (zone *i*) and lethal wetness (zone *iii*), interest centers on the influence of moisture on the animal's chance to survive. This has been expressed as a sigmoid curve depicting the death-rate in a population; this is well established for the dry zone, but not for the wet zone. Alternatively, the mean duration of life for the individuals comprising the population is shown as a hyperbola in the dry zone. This curve may be continued as an asymptote (*dotted line*) over the zone of favorable moistness *A*; alternatively, it may be represented, as in *B*, by a U-shaped curve in the zone of favorable moisture, the dip in the curve corresponding to the humidity at which the animals become sexually mature in the shortest time. The curve then falls away again to zero in the wet zone. This is typified by certain locusts.

Within the zone of favorable moisture, interest centers on the influence of moisture on fecundity and speed of development. Here again there are two sorts of relationships, represented by the two sets of curves in *A* and *B*. Fecundity may increase in a sigmoid fashion, reaching a maximum in saturated air, as in *A*. Alternatively, the curve may rise to a peak at a relative humidity less than saturation and fall away to zero before saturation is reached, as in *B*.

CHAPTER 8

Weather: Light

8.0 INTRODUCTION

IN CHAPTER 6, in discussing temperature, we found it convenient to consider the gradient from very low lethal temperatures through the median range of favorable temperatures up to the range within which temperature is lethal because it is too high. Moisture lent itself to similar treatment in chapter 7. But light is different. There are plenty of animals which can develop equally well in total darkness and in continuous light. Nor is the brightest light normally found in nature likely to be lethal.

Light is chiefly important in relation to behavior and as a stimulus for those mechanisms which regulate life-cycles and keep them in step with the seasons. The essential difference between light and temperature or moisture may be illustrated by describing the behavior of caterpillars of the spruce budworm *Choristoneura fumiferana* in respect to light. The caterpillars tend to move toward a diffuse light; therefore, they will move from the center of a tree outward, thereby coming to the tips of the branches, where there may be fresh young leaves to eat (Wellington, 1948). Compare this with the behavior of *Choristoneura* in a gradient of moisture (sec. 7.16). The caterpillars will move along the gradient away from a place where they are likely to be harmed by too little or too much moisture into a zone of favorable moistness. This is not so with light, for the larvae will not be harmed by exposure either to complete darkness or to the brightest light that they are likely to meet in nature. The response to light is nevertheless a valuable adaptation, because it brings the caterpillars to a place where there is likely to be plenty of good food. This is quite characteristic of the adaptations associated with light; and for this reason Fraenkel and Gunn (1940, p. 190) referred to light as a "token" stimulus —a "token" being a sign which represents something else. Light may indicate circumstances which are, for other reasons, favorable or unfavorable; and the animal possessing the appropriate adaptations responds accordingly.

In some cases the stimulus may be provided by a gradient in illuminance.[1]

1. Throughout this chapter "illuminance" is used in preference to the older term "light-intensity." The illuminance of any surface is measured by the luminous flux incident on unit area. The unit of illuminance is the lumen per square foot which is equivalent to the foot-candle of older usage. The unit of luminous intensity of a source is the candle.

The diurnal rhythms in illuminance and in the quality of light may be associated with rhythms in temperature, moisture, food, and so on. Of these, light may be the most consistent and reliable; the animal becomes adapted to respond to a gradient in illuminance or in the quality of the light, and light thus acts as a "token" stimulus, leading the animal to a place where temperature or moisture is favorable or to a place where there is an abundance of food, etc. (sec. 8.3). In other cases the stimulus may be provided by the changing length of day (photoperiod); this serves as a "clock" indicating the seasonal changes in temperature, moisture, food, and so on. The responses to photoperiod may be subtle, but, in general, they are of the sort which serve to "gear" the life-cycle to the seasons (secs. 8.1, 8.2).

8.1 THE INFLUENCE OF LIGHT IN SYNCHRONIZING THE LIFE-CYCLE TO THE SEASONS

8.11 *Breeding Seasons in Vertebrates*

The periodicity of reproduction in vertebrates has been reviewed by Marshall (1942), Bullough (1951), and others. Their general conclusion was that the breeding season is usually determined chiefly, though not solely, by light. With certain birds and mammals and at least with some other vertebrates, the periodicity of breeding seems to be related to an internal physiological rhythm which is under the control of the pituitary gland. But Baker and Baker (1936) pointed out that such a physiological mechanism could never in itself account for periodicity in seasonal breeding without the additional influence of some component of the environment which varies seasonally. A thousand clocks, set off ticking together and showing the same time, sooner or later get out of time. Slight differences between clocks will quite soon result in big differences in the time shown by them. But the striking feature of breeding seasons in animals is that whole populations of a species mature together, frequently with small variability within the population. This could not be achieved by an innate biological timing mechanism alone. Baker and Baker (1936) calculated that if the hypothetical "clock" within an animal varied by only 6 minutes in each year, then a spring-breeding animal of the last northern ice age would reproduce at the opposite time of the year today—that is, in the autumn. The internal "clock" would need periodic adjustment to insure that the members of the population were all "in time" and, second, to insure that they kept the "right time." Any external stimulus which fluctuated with the seasons might serve, but the most reliable one is length of day. It is therefore not surprising to find that in many animals the activity of the pituitary gland is determined by the changing length of the day.

The innate rhythm of the pituitary appears to be more plastic in some animals than in others. When pregnant Southdown ewes were shipped from

England to Australia, they bred for a second time in the same year after their arrival in the Southern Hemisphere. Within two seasons they were completely adjusted to the Southern Hemisphere, and they bred in the southern autumn. In contrast to the sheep, the hooded parakeet and Brown's parakeet from northern Australia bred in the same calendar month (October) when transferred from Australia to the Northern Hemisphere (Marshall, 1942). The sheep is so adaptable in this respect that there is some doubt as to whether it does, in fact, possess any innate rhythm of the pituitary at all. The facts so far known could be equally well interpreted on the hypothesis that the rhythm of the pituitary in sheep is entirely dependent upon light. Confirmation of this hypothesis, however, awaits further investigation (Yeates, 1949). Rowan (1925, 1938) kept the finch *Junco hyemalis* in outdoor aviaries exposed to the weather. In one aviary he had additional light provided by two 50-watt electric bulbs. From October on, the birds in this aviary were exposed to days of increasing length; the lights were switched on at sunset and left on for 5 minutes longer each day. By mid-November the testes of males were increasing in size, and by the end of December they were larger than those of normal birds in early spring, despite the fact that temperatures in the aviary had been below 0° C. The testes of birds in the aviary which was not receiving additional light were quite small. This was the first experimental test of the hypothesis that maturation of the sexual organs in birds depended on the photoperiod. Since then, these results have been confirmed with many other species of birds (Marshall, 1942; Bullough, 1951).

Baker and Ranson (1932) did similar experiments with mammals. They found that the field mouse *Microtus* bred freely when exposed to 15 hours of light each day but virtually ceased breeding when this was reduced to 9 hours. By gradually increasing the length of day experienced by the ferret, Bissonnette (1935) was able to make it mature more rapidly, and he obtained animals which were sexually mature outside the normal breeding season. Similar results have since been obtained by increasing the length of day experienced by the hedgehog, raccoon, a lizard, a turtle, and certain species of fish (Marshall, 1942; Bullough, 1951). On the other hand, Hill and Parkes (1930) found that ferrets still bred in the spring, even when they were kept in darkness except for ½ hour in the light each day.

Animals which become reproductively mature in the autumn through the influence of light would have to respond to a decreasing photoperiod if they were dependent upon the photoperiod to time their cycles. Yeates (1949) showed that this was indeed so for sheep. He exposed sheep to an increasing length of day during autumn; the daily exposure to light was increased from 13 hours in mid-October to 21 hours at the end of January. These animals became nonreproductive 2 months in advance of control animals. The day-length was then reduced so that they experienced a decreasing photoperiod at a time of the year when they would normally be exposed to increasing hours of daylight. By the end of June these experimental animals were receiving only 5½ hours of light per day. The first of them came into breeding condition in May and the rest followed shortly afterward. Decreasing the photoperiod also hastened the arrival of reproductive maturity in the goat (Bissonnette, 1941) and the trout (Hoover and Hubbard, 1937). The trout were made to spawn in August instead of December, which is the usual month for spawning in nature.

Parker *et al.* (1952) pointed out that none of these experiments had been designed to discover the respective influences of light and darkness. In view of the fact that with plants it is the duration of the period of darkness which is most important (Gregory, 1948), it would seem that this would be well worth closer study in animals as well. Hart (1951) experimented with four groups of mature anoestrus ferrets. All four groups were exposed to natural daylight between December 5 and March 30 in the vicinity of Cambridge, England. Group A (the controls) experienced no artificial light. For group B an artificial light was switched on for 1 hour each night near midnight, thus breaking the night into two equal periods of darkness. Group C started off the same way as group B, but the light was left on for a period that was increased by ½ hour every 3 days. Group D experienced periods of light and darkness equivalent to those for group C, except that the light was switched on at sundown instead of in the middle of the night. The treatments and the results are summarized graphically in Figure 8.01. None of the controls came into oestrus before March 30. All the experimental animals came into full oestrus between January 25 and February 9. The important comparisons are between groups A and B and groups B and D. In the case of groups A and B, at the time when oestrus occurred in group B, each lot was experiencing about 13 hours of darkness each day. In the former this occurred as one long period and did not induce oestrus; in the latter it occurred as two short periods, and this brought on oestrus by the eighth week. With groups B and D, the total duration of the daily dark period was different, and oestrus appeared in group D shortly after the dark period had become as short as the dark periods experienced by group B. The general conclusion to be drawn from these experiments is that oestrus was determined either by the absolute shortness of the dark period or by the ratio of durations of light and dark periods. Further experiments would be required to discriminate between these alternatives.

On the other hand, Yeates (1949) concluded, as a result of his experiments with sheep, that the increasing duration of the periods of light was in itself sufficient to bring the animals into a nonreproductive condition. Hart (1951) considered that the increasing duration of daylight which occurs in nature as winter changes into summer is not in itself essential but is merely the means whereby the change-over from a sequence of short light and long dark periods to a sequence of long light and short dark periods takes place.

Well-defined breeding seasons also exist among animals in the tropics, where length of day varies little throughout the year (Baker, 1938). But not much is known about the controlling influences in tropical climates. A number of the

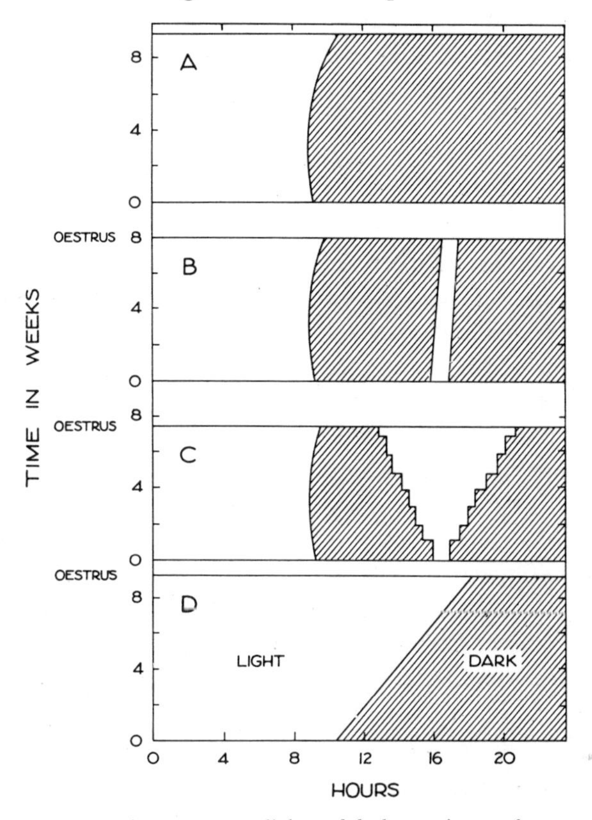

FIG. 8.01.—Charts showing the exposure to light and darkness given to four groups of ferrets. The time of oestrus is shown on the left. *A*, natural daylight only; *B*, same as *A*, but with an additional hour of light at about midnight; *C*, same as *B* at the beginning, but thereafter the artificial light was left switched on for ½ hour more every 3 days; *D*, total light the same as *C* but given at the end of the day instead of in the middle of the night. (After Hart, 1951.)

components of the environment do show quite marked seasonal changes, notably the incidence of tropical fruits, which vary seasonally. Similarly, in dry areas such as in the center of Australia, it would seem unlikely that birds would depend upon length of day for timing their breeding seasons, since favorable seasons are dependent on rain and these seasons do not occur every year. When they do occur, it is not always at the same time of the year. It is generally thought that birds in central Australia breed after rains, but this needs confirmation. The breeding season of some of the frogs of the central desert is determined by rain (Fletcher, 1889; Harrison, 1922), and it is reasonable to suppose that this may apply to other groups as well.

In most experiments in which the photoperiod has been the variable, the response has been independent of illuminance above a threshold of a few lumens per square foot. But Bissonnette (1936) found that starlings were exceptional; the rate of development of gonads was greater when the birds were illuminated by a 40-watt globe as compared with a 10-watt globe; but higher illuminance than that produced by a 40-watt globe did not increase the rate of development.

8.12 *The Influence of Light on the Inception and Completion of Diapause in Insects*

If the chief advantage of diapause is to impose a rhythm on the life-cycle which will match the rhythm of the seasons (sec. 4.6), then the advantage will be greater, the more effective the "clock" which foretells the approach of the unfavorable season. It was mentioned in section 4.4 that some species respond to changes in their food, and others to the seasonal rhythm of temperature. But with many species the seasonal rhythm of the life-cycle seems to be determined by more than merely temperature and food. Food, temperature, moisture, and air pressure are all very much less regular in their seasonal procession and are therefore likely to be less effective as "clocks" than photoperiod. Until recently, remarkably little work had been done on the influence of photoperiod on the inception of diapause in insects; but those who have investigated this subject have been rewarded by some striking discoveries.

Kogure (1933) observed that, in certain bi-voltine races of the silkworm *Bombyx mori*, nearly all the females of the first generation which arose from eggs that were developing during the cool, relatively short days of the spring laid only nondiapause eggs; but females of the second generation, which arose from eggs that developed during the long, warm days of mid-summer, laid eggs all of which entered diapause. The silkworm is a convenient subject for an experimental analysis of this observation because (*a*) there is a close correlation between egg color and the presence or absence of diapause and (*b*) eggs in diapause will resume development if they are suitably treated with acid. Thus Kogure was able to have eggs developing at any time, independently of the season, and he had a ready method of telling whether a particular treatment had produced eggs with or without a diapause. He reared eggs at different photoperiods ranging from 0 to 24 hours of light each day and at different temperatures ranging from 15° to 28° C. The outcome was that eggs which were exposed to less than 14 hours of light each day at 15° C. gave rise to moths all of which laid pale (nondiapause) eggs, while those experiencing 16 or more hours of light each day gave rise to moths of which from 30 to 70 per cent laid dark (diapause) eggs. At 20° C. the results were comparable, but the tendency toward diapause was stronger at every photoperiod; and at higher temperatures (24° and 28° C.) the influence of the photoperiod was entirely masked, and virtually 100 per cent of the eggs gave rise to moths laying dark eggs. The interaction between photoperiod and temperature is illustrated in Figure 8.02, which has been constructed from Kogure's data for the bi-voltine race *shohaku*.

The influence of light was most marked on the developing egg, but the

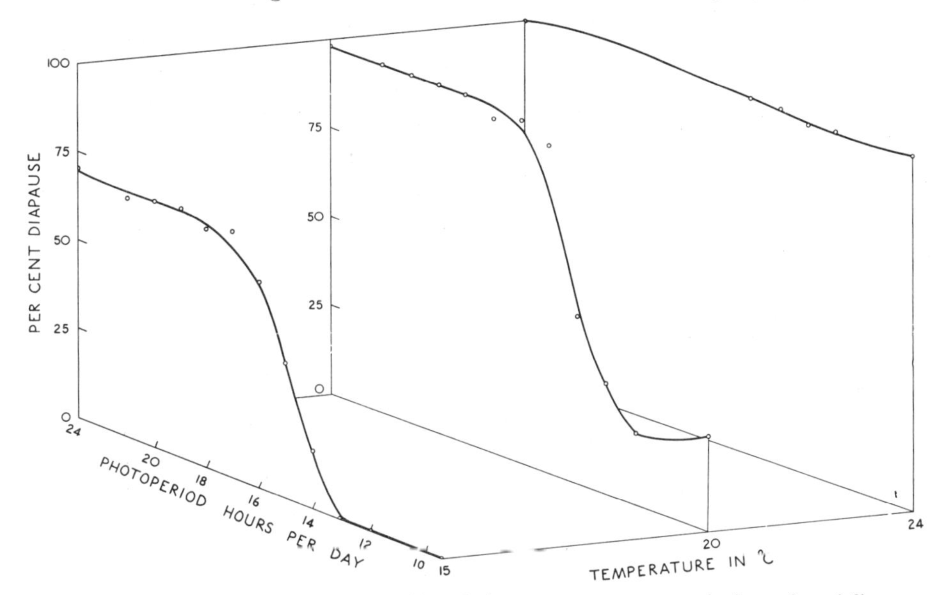

FIG. 8.02.—The influence of photoperiod in relation to temperature, on the inception of diapause in the eggs of *Bombyx mori*. The ordinates indicate the proportion of females laying dark (diapausing) eggs when they had spent their embryonic life exposed to the particular combinations of photoperiod and temperature indicated by the abscissae. (After Kogure, 1933.)

pattern of development (i.e., diapause or nondiapause in the next generation) determined during this stage was also modified by the influence of light and temperature on the larval instars. Eggs of the bi-voltine race *showa* were incubated at 19° C. in darkness. The larvae were reared at 24° C., some in the dark for the first three instars, some in the light for the first three instars, and the others were illuminated for the first, second, or third instar only. The results, summarized in Table 8.01, show that the influence of light on the

TABLE 8.01*

INFLUENCE OF LIGHT ON EARLY LARVAL INSTARS OF *Bombyx*

Treatment of Larvae	Per Cent Moths Laying Dark Eggs
Instar I illuminated	41
Instar II illuminated	38
Instar III illuminated	21
Instars I, II, and III illuminated	45
Instars I, II, and III in the dark	2

* After Kogure (1933).

early-stage larva was similar to that on the egg and that the first and second instars responded more strongly than did the third. The strength of the response of the young larva to light depended on its environment when it was an embryo. For example, with the bi-voltine race *shohaku*, the response was least with those which had been incubated at 15° C. in the dark, at 24° C. in the light, or at 28° C. irrespective of illumination. In the last two groups the "diapause pattern" and in the first group the "nondiapause pattern" had been laid down so firmly during the development of the embryo that it was not modified by any subsequent treatment given to the larva. The response of the young larva to light was greatest when the strength of the "diapause pattern" laid down during incubation had been intermediate between these two extremes, notably with larvae arising from eggs which had been incubated at 15° C. in the light or at 20° C. in the dark. These points are clearly illustrated by the information in Table 8.02, which is a summary of the much more extensive data given by Kogure (1933).

TABLE 8.02*

INTERACTION OF LIGHT DURING FIRST TWO LARVAL INSTARS AND LIGHT AND TEMPERATURE DURING INCUBATION OF EGGS OF *Bombyx* AND THEIR INFLUENCE ON PROPORTION OF DIAPAUSE IN NEXT GENERATION

	Temperature during Incubation of Eggs					
	15° C.		20° C.		24° C.	
Illumination:						
During incubation of eggs	Dark	Light	Dark	Light	Dark	Light
During first 2 larval instars	D L	D L	D L	D L	D L	D L
Per cent moths laying dark (diapause) eggs	0 2	50 86	32 62	93 99	95 99	100 100

* After Kogure (1933).

The responses of the fourth and the fifth larval instars and the pupa to light were relatively weaker than those of the first and second larval instars. It was not evident at all in those cases where the "diapause pattern" had been firmly imprinted during the life of the embryo. But Kogure's experiments seemed to indicate that when the response could be recognized (e.g., in larvae arising from eggs that were incubated at 15° C. in the light or at 20° C. in the darkness), light influenced diapause in the opposite way from eggs or young larvae: a few less of the moths which had spent the fourth and fifth larval instars and the pupal stage in the light laid dark (diapause) eggs than did the ones which had spent these stages in the dark.

The illuminance which was effective varied with the temperature, but at 15° C. the threshold for effectiveness was about 0.014 lumens per square foot, although a significant difference from darkness was also obtained with the lowest intensity tried, namely, 0.004 lumens per square foot. At 20° C. the threshold was higher, lying somewhere between 0.025 and 0.5 lumens per square foot. Red and orange-yellow light with wave lengths exceeding 5,500 A was equivalent to darkness, and violet and blue-green light with wave lengths between 5,100 and 3,500 A was equivalent to white light. These results are confirmed by those given by Dickson (1949) for *Grapholitha* with respect both to the abruptness of the threshold and to the particular part of the spectrum that was effective.

There are several generations of *Grapholitha molesta* during the summer in southern California, but, independent of generation, an increasing proportion

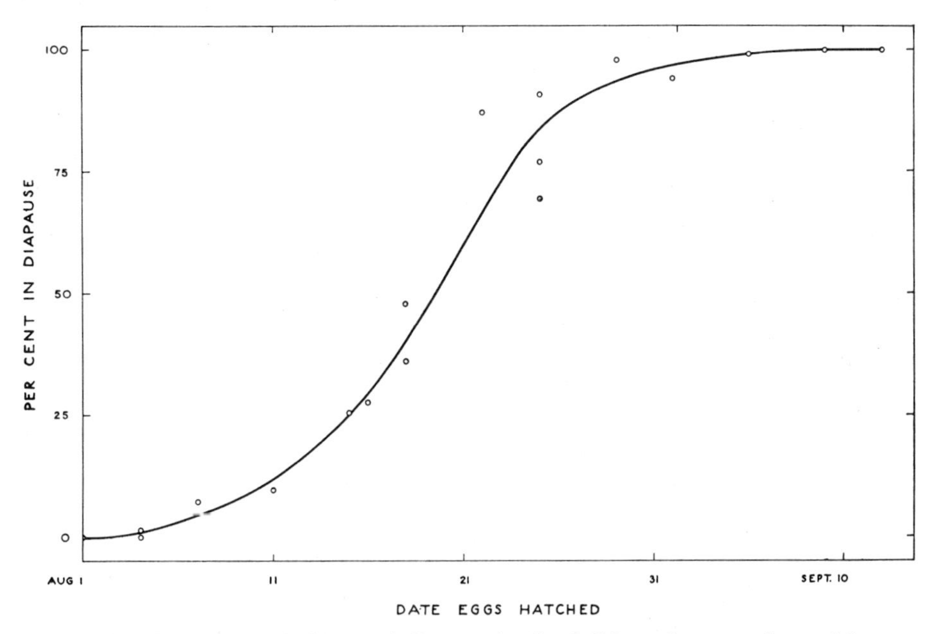

FIG. 8.03.—The seasonal incidence of diapause in *Grapholitha molesta* reared out-of-doors at Riverside, California. (After Dickson, 1949.)

of the larvae that hatch from eggs after the middle of August enter diapause. In this species diapause occurs in the fully grown larva, which spins a dense hibernaculum instead of the flimsy cocoon characteristic of the larva that is going to develop without diapause. Diapause, once induced, may endure for many months. Figure 8.03 shows that a substantial proportion of the larvae enter diapause early in the autumn, long before the temperature is low enough to inhibit active development: in these, diapause has already been determined by the photoperiod experienced during the early larval instars (Dickson, 1949).

Very few larvae entered diapause when they were reared in continuous light or continuous darkness at any temperature between 12° and 30° C.; at intermediate temperatures between 21° and 26° C. the numbers entering diapause remained low, with a short photoperiod of less than 6 hours or a long one exceeding 14 hours, but increased rapidly with photoperiods exceeding 9 hours, and reached a maximum of about 100 per cent for photoperiods between

11 and 13 hours (Fig. 8.04). There was a subtle interaction between light and darkness: the response was not obtained by exposure to alternate equal periods of light and darkness outside the narrow limits of 10–13 hours: a dark period of 11 hours (which, when alternated with light for 11 hours, produced 100 per cent diapause) became ineffective when alternated with light periods outside the limits of 6–15 hours: similarly, a light period of 12 hours became ineffective when alternated with a dark period of less than 9 hours'

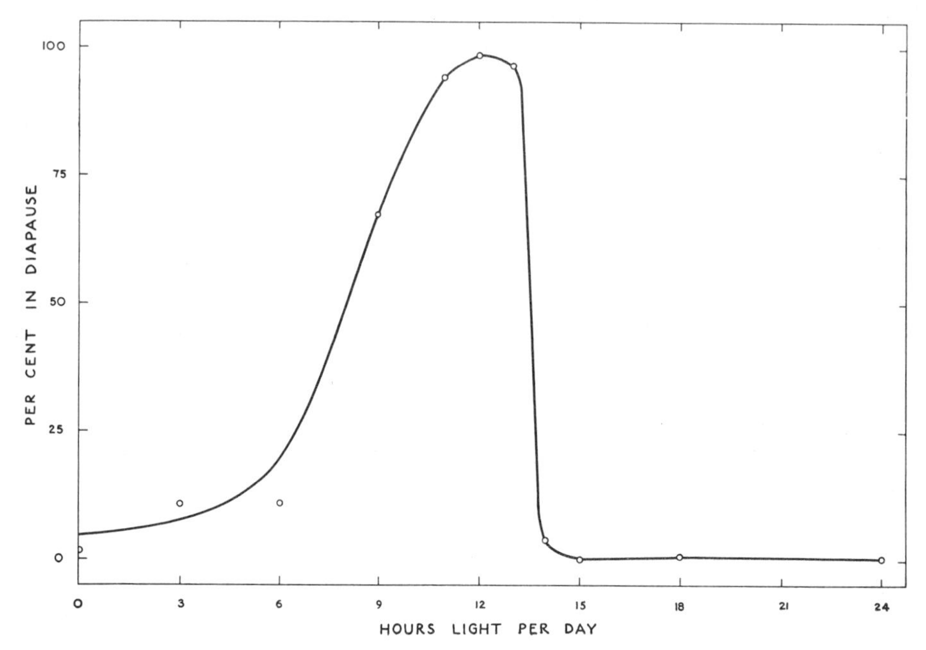

FIG. 8.04.—The influence of photoperiod at 24° C. on the inception of diapause in *Grapholitha molesta*. (After Dickson, 1949.)

duration. The insect did not respond to fluctuations of light and darkness which were too different from the natural diurnal fluctuations; but the full explanation of these results remains baffling and should provide a valuable basis for further work of a more strictly physiological character. The threshold of intensity for *Grapholitha* was about 3 lumens per square foot falling on the outside of the fruit in which the larva was tunneling. As mentioned above, the same part of the spectrum (namely, the band between 4,000 and 6,000 A) was influential with the larvae of *Grapholitha* as with the eggs of *Bombyx:* both the red and the ultraviolet were ineffective.

The influence of the photoperiod was most marked at intermediate temperatures between 21° and 26° C.: at 12° and 30° C. it was almost entirely masked: even with the most influential photoperiod (12 hours' light and 12 hours' darkness), virtually no larvae entered diapause at these temperatures.

The response to the photoperiod varied in a subtle way with the age of the larva. Diapause was irrevocably inhibited by a long photoperiod during the first 4 days of larval life; but the converse did not hold: exposure to a 12-hour photoperiod for the first 4 days of larval life failed completely to induce diapause in larvae which spent the remainder of their larval life under the influence of a longer photoperiod. But diapause was irrevocably determined for 70 per cent of the larvae when the exposure to the 12-hour photoperiod was extended to include the first 8 days of larval life.

The pupal diapause in certain Lepidoptera appears also to respond to light. In nature the noctuid *Diataraxia oleracea* hibernates as a pupa; but when the species was reared throughout the year in a glasshouse, a higher proportion of pupae entered diapause in the spring and autumn than in the summer. When batches of 40–50 larvae were kept at a constant temperature of 24° C. and exposed to various photoperiods, more pupae entered diapause at the short photoperiods than at the long. This is shown in Table 8.03, which is taken from the data of Way *et al.* (1949).

TABLE 8.03*

INFLUENCE OF PHOTOPERIOD DURING LARVAL LIFE ON INCEPTION OF DIAPAUSE IN PUPAE OF *Diataraxia*

Hours of light each day......	24	16	8	4	0
Per cent pupae in diapause....	2	27	100	100	94

* After Way *et al.* (1949).

The tick *Dermacentor* in Massachusetts ceases feeding in the autumn, when the temperature is higher than that prevailing in the spring, when activity is resumed. And since exposure to certain photoperiods may be shown to stimulate activity in hibernating ticks, the inference is that the photoperiod may be influential in the inception of diapause in the autumn (Smith *et al.*, 1941). Similarly, the eggs or larvae of a number of species of mosquitoes are known to enter diapause in the autumn, when the temperature is relatively high; they remain in diapause throughout the winter, even in a heated room, and resume development in the spring, while the temperature is even lower than it was in the autumn, when activity ceased. Such observations have been made for *Aedes triseriatus*, *Anopheles barberi* (Baker, 1935), *Orthopodomyia pulchripalpis* (Tate, 1932), *Anopheles plumbeus* (Roubaud and Colas-Belcour, 1933), and *A. geniculatus* (Legendre, 1934). In *Aedes* diapause occurs in the fully grown embryo. It develops to the point of hatching in the autumn but does not emerge from the egg until the spring. Eggs kept warm during the winter still hatched about the same time in the spring, but they would emerge at any time during the winter if they were exposed to an artificially lengthened photoperiod (Baker, 1935). Similar observations have been made on the larvae of *Anopheles* (Baker, 1935; Kennedy, 1947). A number of other examples relating especially to more recent work were discussed by Lees (1952).

The flowering of certain plants depends on their having been exposed to particular diurnal fluctuations in light and darkness; not the whole spectrum but only a relatively narrow band of it is influential. With most plants this band is situated near the red end of the spectrum and not, as with *Bombyx* and *Grapholitha*, the blue end. With plants this leads to the conclusion that chlorophyll or related green pigments absorb the light which is responsible for the photoperiodic effect. Similar reasoning would suggest that in *Bombyx* a yellow-red pigment is involved; and the close agreement between the color bands that are influential with the egg of *Bombyx* and the larvae of *Grapholitha* indicates that the same or similar pigments are present in both these species. Plants may be classified into three categories according to the influence of light on flowering: short-day, long-day, and independent. Three comparable categories occur in insects, too: in *Grapholitha* diapause is induced by exposure to short photoperiod; in *Bombyx* it arises in response to long photoperiod; and in *Lucilia* and *Listroderes* diapause is quite independent of photoperiod (Dickson, 1949).

8.2 THE INFLUENCE OF LIGHT ON SYNCHRONIZING LIFE-CYCLES WITHIN A BREEDING SEASON

In the examples discussed in section 8.1 the photoperiod was responsible for keeping the "physiological clocks" of different individuals "in time" with one another and also at the "right time." In this sense it synchronizes the life-cycle with the season, so that any individual reaches reproductive maturity at a favorable season of the year. Simultaneously, it causes all individuals in the population to come into reproductive maturity together. Among some marine invertebrates, which have a distinct breeding season, notably certain polychaetes and mollusks, influences other than light seem to determine the time of the year in which the animals breed, but within the breeding season the photoperiod of the moon determines the time of the month when the animals come out to lay their eggs or mate. It is as though the photoperiod, being a very precise timekeeper, takes over from some other stimulus, such as temperature, which must be a rather crude timekeeper, within that period of the animal's life known as the breeding season.

One of the most extraordinary of these periodicities is the swarming of the palolo worm. The Atlantic species *Leodice fucata* swarms once a year in July. The schedule of reproductive maturation of these worms is timed with such precision that millions of them swarm to the surface as sexually mature worms within a couple of days. They release their eggs and sperm practically simultaneously on the surface of the ocean, insuring a very efficient means of shedding and fertilizing of eggs. Many of the details of how this comes about in response to light have been worked out by Clark and Hess (1940*a*, *b*) at the Dry Tortugas Islands off Florida. Figure 8.05 summarizes the time of swarming over many years in relation to the phases of the moon. Most of the swarming occurred at the third quarter of the moon. Major swarms practically always appeared within 5 days of the third quarter of the moon in the month of July, yet some swarms also occurred in the first quarter. The swarms came at these times, even in the absence of bright moonlight; the series of events culminating in swarming must therefore somehow be influenced by the

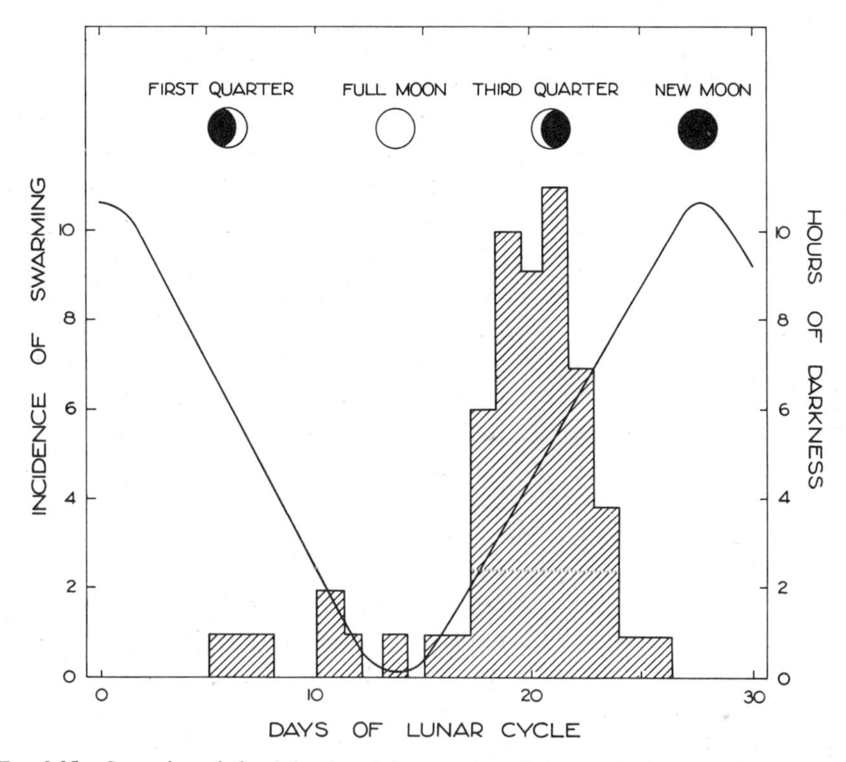

FIG. 8.05.—Swarming of the Atlantic palolo worm in relation to the lunar cycle; summary of records from 1898 to 1939. Hours of darkness (*continuous line*) give the number of hours during which both the sun and the moon were below the horizon. (After Clark and Hess, 1940*a*.)

moon before that time. Indeed, this must be so, in view of the long period of maturation of the worms.

The sequence of events which leads to the massive swarming is somewhat as follows. The worms live in crevices and burrows in rocks and corals to depths of about 150 fathoms. Immature worms are photopositive to illuminance varying from 0.001 to 0.01 lumens per square foot. In higher illuminance they are photonegative. This means that the immature worms are confined to their burrows by their negative response during the day and at night when moonlight exceeds an illuminance of 0.01 lumens per square foot. It also means that the worms emerge from their burrows to feed only during certain parts of the night. The duration of the feeding period will be determined by the illuminance and duration of the moonlight, and both of these are determined by the lunar cycle. Eventually, the worms become sexually mature, the posterior half becoming laden with eggs and sperm. The posterior half of the worm is then referred to as the "epitoke," while the anterior half is called the "atoke." The atokal portion of the mature worm has about the same threshold of response to light as the immature worm, and it, too, shows a periodicity in feeding related to the lunar cycle. This may be an important means of keeping the sexual development in step with the lunar cycle, with the result that they become completely mature at a definite phase of the moon.

At an illuminance of about 0.0005 lumens per square foot, mature worms undergo convulsive movements which result in the severing of the atoke from the epitoke. This will presumably happen in the moonlight, as the worms will be confined in their burrows during the daytime when the light is brighter. Whereas the atoke is photonegative to all but very low illuminance, the epitoke, which is packed with eggs or sperm, is highly photopositive up to an illuminance of at least 50 lumens per square foot. In consequence, they swarm to the surface of the ocean in response to the moonlight. But their responses to light have not yet finished. The rupture of the epitoke, with release of eggs and sperm, can take place at any illuminance or even in the dark, but it is greatly speeded up under high illuminance. In consequence, most of them rupture at dawn with the rising sun. There is an amazing simultaneous release of eggs and sperm on the surface of the ocean, and they sometimes cover acres to a depth of several feet!

The responses of the immature and mature worms to light was worked out in the laboratory, and, as we have seen, they link in well with what is known of the relation of swarming to the phases of the moon. This in itself is circumstantial evidence that moonlight is a cause of periodicity. This hypothesis was also supported by experiments in which Clark (1941) varied the amount of illumination received from the moon by illuminating and shading rocks containing worms. When the duration of light was increased, reproductive maturity occurred earlier than in the controls. When the duration of moonlight was decreased, the time of swarming was later than in the controls. The duration of moonlight is therefore important. If it were the only stimulus, its effectiveness should increase to a maximum about 3 days after full moon, but this does not happen. Clark (1941) suggested that the most important stimulus may be the rate of change of photoperiod from night to night, for this is at a maximum at the first and third quarters of the moon, when swarming occurs.

Data on the reproductive periodicity of a number of other species of polychaetes were summarized by Korringa (1947). Some of them, for example, *Platynereis dumerilii*, swarms twice per lunar cycle on the Atlantic coast of France, once during the first quarter of the moon and again during the third (Fig. 8.06). But the same species near Naples swarms only once, just after full moon. Since the two swarms occur where there are large tides and the one swarm occurs in the Mediterranean, which has negligible tides, Korringa (1947) suggested that the worms may respond to differences in hydrostatic pressure in the Atlantic and to differences in illuminance in the Mediterranean. There is no direct evidence one way or the other, but whereas the hypothesis of hydrostatic pressure might be applicable to the Atlantic, it clearly cannot be invoked to explain swarming in the Mediterranean.

There is more substantial evidence for the hypothesis that the lunar cycle through the tides determines the periodic breeding of the oyster *Ostraea edulis*

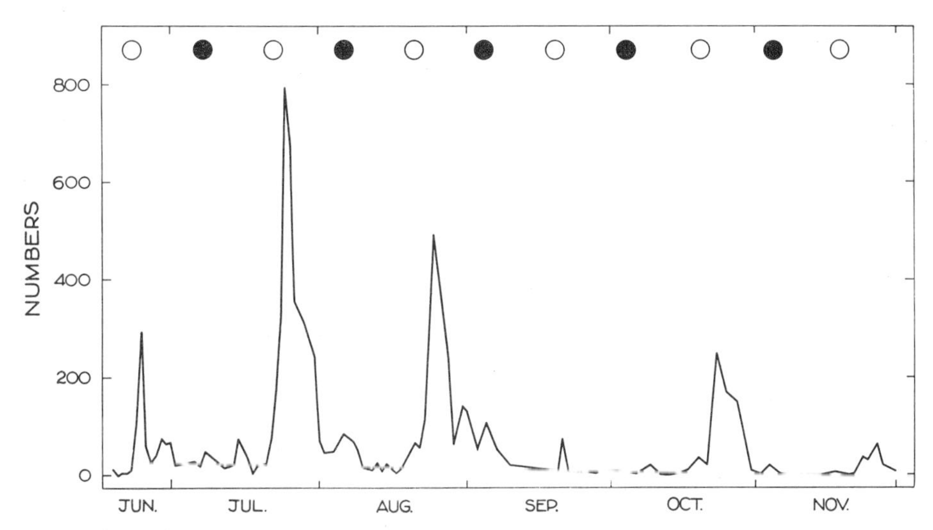

FIG. 8.06.—The relative numbers of swarming *Platynereis dumerilii* caught by Ranzi at Naples, shown in relation to the phases of the moon. (After Korringa, 1947.)

in Holland. Most of the swarming maxima for larvae occur 10 days after full or new moon, that is, about the time of the spring tides (Fig. 8.07). Since swarming occurs at both the full and the new moon, Korringa suggested that moonlight cannot be the regulating influence. Instead, he proposed that the rhythm of the tides influenced the animals through changes in hydrostatic pressure. However, this hypothesis remains untested. Other species of oysters are known to show periodic cycles in their breeding which can be correlated with the cycles of the moon (Orton, 1926; Korringa, 1947).

Periodicities in breeding which appear to be related to the lunar or tidal cycle exist in other groups besides polychaetes and oysters. For example, Fox (1924, 1932) found that the sea urchin *Centrechinus setosus* at Suez had a cycle of development such that the gonads reached their greatest bulk just before full moon and spawning occurred at the full moon. Wheeler (1937) found a rhythm of swarming in three species of prawns at Bermuda. They came to the surface at regular intervals throughout the year, about an hour after sunset. The beginning of swarming coincided with the new moon and reached a maximum on the second night of the lunar month and again on the twenty-sixth night. Marshall and Stephenson (1933) found that the coral *Pocillopora bulbosa* spawned discontinuously on the Great Barrier Reef, with spawning maxima coinciding with the new moon in summer and autumn and the full moon in winter. And Abe (1937) made reference to a coral, *Fungia actiniformis*,

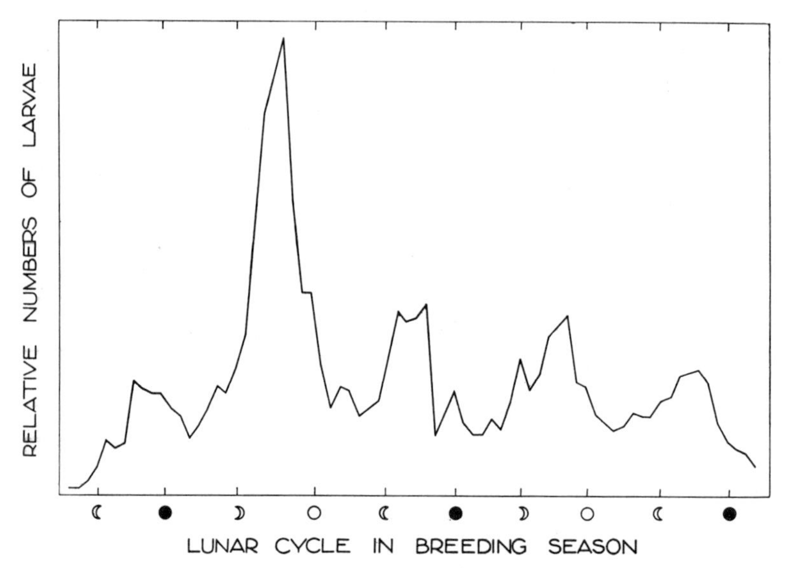

Fig. 8.07.—The relative numbers of larvae of the oyster *Ostrea edulis* in Holland in relation to the phases of the moon from the years 1937 to 1946. The numbers of larvae in any swarm were expressed as a proportion of the total larvae in swarms for the year. (After Korringa, 1947.)

the planulae of which were liberated at the time of the new moon in the Palau Islands of the Pacific Ocean.

In all these examples of periodicity in breeding of marine invertebrates, it is possible to distinguish an annual rhythm in the time from one breeding season to the next. Korringa suggested that this is probably controlled by temperature. Within the breeding season there may be a monthly rhythm corresponding with the tides or the procession of dark and moonlight nights. Then, in addition, there may be a daily rhythm manifested by the simultaneous performance of nuptial behavior, swarming, and the release of reproductive products; this, too, seems to be under the control of the lunar cycle. Korringa pointed out that in some animals these three rhythms are so remarkably adjusted to seasonal and diurnal changes that an entire population may spawn simultaneously during one single hour a year.

Gray (1951) showed that the numbers of the protozoan *Stylonichia* in an English stream reached a maximum 4–7 days after full moon. The chief food of *Stylonichia* is diatoms. Gray pointed out that fluctuations in the abundance of diatoms in phase with the lunar cycle have been recorded in the Illinois River by Kofoid; these reached a maximum 12–14 days after the full moon, depending upon the season. The fluctuations of *Stylonichia* may simply follow the fluctuations of its food in the stream. It is well known that the appearance of winged or wingless aphids at certain seasons of the year is related to the photoperiod, but it is not known whether the photoperiod acts directly on the insect or indirectly through its food. Marcovitch (1924) exposed the parthenogenetic forms of *Aphis forbesi* and the strawberry plants on which they were living to 7½ hours of light each day during February to May; the sexual forms appeared in May instead of November. Davidson (1929) obtained similar results with *Aphis rumicis*. Some of Marcovitch's experiments were done with aphids on the roots of the plants; and, since these gave similar results, it would seem that the sexual forms develop in response to some change in the plant which is induced by the exposure to short days and long nights.

8.3 THE BEHAVIOR OF INVERTEBRATES IN A GRADIENT OF LIGHT

Among the more dramatic mass movements of animals in relation to daily changes in illuminance are the well-known daily vertical migrations of zoöplankton in lakes and seas. Many animals of zoöplankton come to the surface at night and descend to depths of about 5–60 meters in the daytime. Details of the movements of various species are summarized by Allee *et al.* (1949, pp. 554–56), Clarke (1933, 1934), and Russell (1927). The ecological explanation of these migrations has not yet been discovered. The zoöplankton obviously profit from being near the surface, where their food, the phytoplankton, is most concentrated. On the other hand, the surface may be an unsuitable place to linger during the day, and the response to light may be an adaptation to move out of this unfavorable zone. Schallek (1942*a*) observed the movements of the marine copepod *Acartia tonsa* in vertical glass cylinders illuminated solely by diffuse light from a fluorescent tube behind an opal glass screen. The light was held in a position 36° from the horizontal above the top of the cylinders. While the light was on during the day, the copepods stayed at the bottom of the cylinders. When it was off at night, they rose to the surface. They sank again the next day, when the light was turned on in the morning. This is illustrated in Figure 8.08, which shows the proportion of animals in the top third of the cylinders. The first two drops in numbers correspond to the periods of light. When the period of darkness was continued from then onward for a couple of days, the animals began their descent in the cylinders on the first morning of the dark phase, despite the darkness. They rose again at night. They continued this rhythm of migration on a smaller and less perceptible scale the next day. It disappeared after a few days in darkness. By giving more frequent alternations of light and darkness, Schallek was able to cause the animals to move up and down every 4 hours, as though the 4-hour period corresponded to their normal day.

When the diffuse oblique light was replaced by a direct light from an incandescent bulb overhead, their behavior was reversed: they moved up in the light and down in the darkness. When the light was left on, they stayed up. They also moved toward a discrete light source when they were in a horizontal trough with the light at one end. It is difficult to suggest any ecological ex-

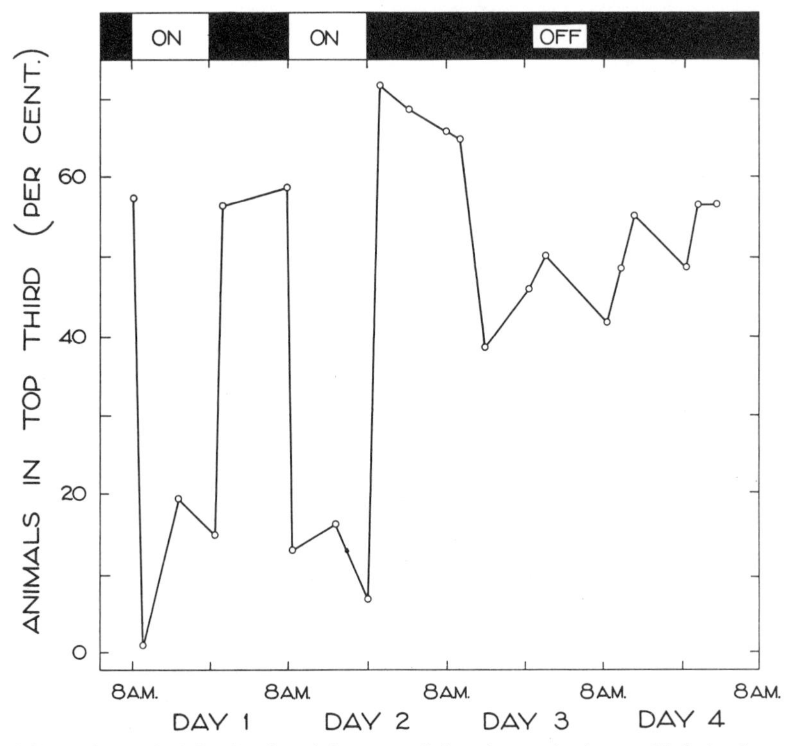

Fig. 8.08.—The vertical distribution of the copepod *Acartia tonsa* in the top third of a glass cylinder of sea water. The cylinder was illuminated obliquely from above by light from fluorescent tubes. The darkened areas indicate the dark periods of the experiment. (After Schallek, 1942*a*.)

planation for this reversal in behavior of the copepod, for the normal lighting experienced by these animals in the sea has components in practically every direction, with, of course, a minimum from below. We shall show later (sec. 8.32) how a similar reversal in the behavior of the caterpillar *Choristoneura* is an adaptation which is useful to the species.

In the mass rearing of certain small insect "parasites" for liberation in the citrus groves of California, it is an essential part of the technique to allow the insects to be attracted to the inner surface of a brightly illuminated window, where they may be easily caught and packaged for distribution. Adult blowflies are usually attracted to light, but when they are hungry, they will enter dark places in search of food. Once they have eaten, they become strongly photopositive again. Certain blowfly traps depend on this behavior. There is no need to add to this list, for the reader will have no difficulty in thinking of almost innumerable examples from nature of how insects and other invertebrates move either toward or away from light. Students of this subject have recognized several different sorts of behavior, to which they have given technical names (Fraenkel and Gunn, 1940). We shall explain these briefly, since they have to be used in the next section.

An undirected movement in response to light is called a "photokinesis"; a directed movement in response to light is called a "phototaxis." The lamprey tends to move more quickly in the light, though its movements remain undirected; it thereby tends to spend most of its time in the darker places. This would be known as "orthokinesis." Similarly, the planarian turns (though not in any directed way) more frequently in light than in darkness, and this would be called "klinokinesis." Apparently, kineses are less common than taxes.

The larvae of the blowflies *Calliphora erythrocephala* and *Lucilia sericata* move away from light toward darkness by alternating lateral deviations of the anterior part of the body. This sort of movement is called "klinotaxis." Animals with a single light receptor generally behave like this, and the usual explanation offered is that the single light receptor makes comparisons of the illuminance at successive moments of time. Imagine a one-eyed man looking through a telescope with ground glass in place of lenses, directing his movements toward a distant light by perpetually scanning the horizon to right and left in order to keep the bright glow in view in the telescope. A slight deviation to right or left would have to be compensated for by a movement in the opposite direction, to bring the light again into line with the direction of travel.

But most animals possess two light receptors and seem able to move to or from a source of light without these "trial-and-error" deviations of the body as they progress. For example, the larvae of *Choristoneura fumiferana* when placed in a gradient of illuminance will move directly toward the light; the larvae of *Ephestia kühniella* in the same circumstances will move directly away from the light. When there are two lights of equal brightness, these species typically move toward or away from a position midway between the two lights. This behavior is called "tropotaxis"; it is usually considered that the animal makes simultaneous comparisons of the brightness of the light on either side of its body (Fraenkel and Gunn, 1940).

The male firefly keeps moving toward the flash of the female, as if it retains an impression of the flash (Mast, 1912). This sort of behavior is called "telotaxis." Animals with this capacity, when exposed to two lights, will move toward one of them, but not along a line between the two, as in the case of "tropotaxis."

The bug *Notonecta*, when moving in the presence of a discrete source of light, will tend to keep the long and transverse axes of its body perpendicular to the path of the light. This is called the "dorsal light-reaction." Other species may so direct their movements, temporarily at least, that their path makes a constant angle with the path of the light. This sort of behavior is called "light-compassing movements." When the light is very distant, the animal traces a straight path; but when light is close, for example, an incandescent globe in a room, the path is curved. Light-compassing movements are especially important in ants and bees in their daily excursions after food (Fraenkel and Gunn, 1940).

In some of these reactions it is probable that the animal responds to illuminance as a stimulus. It may move either up or down a gradient in intensity. A receptor which responds to differences in illuminance can act as a direction-finder by comparing the illuminance in different directions. This is almost certainly the nature of the response in photokinesis. But some animals respond to the direction of the light rays. Telotaxis appears to be of this nature. Similarly, "light-compassing movements" are movements in relation to direction of light rays and not to illuminance. Perception of direction of light rays must involve the formation of an image of the source of light. The separate elements of the compound eye are stimulated from different directions, and this gives rise to a retinal pattern of illuminance which corresponds in some crude sort of way to an image of the light source. For the ant this would be the disk of the sun.

It seems reasonably certain that light-compassing movements involve the formation of an image of the source of light; and photokinesis depends on comparisons of the illuminance coming from different directions with respect to the position of the body. But the explanation for phototaxis has not been so clearly worked out.

8.31 *Experimental Demonstrations of Phototaxes*

Light emanating from a source like an incandescent lamp diminishes in illuminance in inverse proportion to the square of the distance from the source. An animal which moves toward the light source may be moving parallel to the rays of light, that is, in response to the direction of the rays, or it may be moving in response to a gradient in illuminance. It is possible to separate illuminance and direction by using a parallel beam of light in which there is no gradient. There are plenty of animals which move toward a source of light in a parallel beam; these animals are evidently not responding to a gradient in illuminance but to the direction of the light rays (Fraenkel and Gunn, 1940, p. 195). A second type of experiment has been used in which light is directed at right angles to the plane of movement of the animals. The light is then said to be without "effective direction," so far as the animals are concerned, since they cannot move in the direction of the light. Wellington (1948) placed photo-negative larvae of the spruce budworm on a board in a dark room such that movement away from an incandescent bulb on the board (directed light) brought them into a patch of light thrown onto the board from an overhead shielded lamp. They moved directly into the lighted patch thrown onto the board from above, although the intensity was ten times greater facing the patch than it was toward the single lamp. When a second lamp, which was in the center of the patch, was turned on as the larvae moved away from the original light, they then moved away from the second light, following a course which took them away from the two lamps and the patch of light. If the second light was then switched off, the larvae did not move back toward the patch. Wellington's interpretation of this experiment was that the larvae were repelled not by intensity but by the direction (or image) of the source, since there was never any hesitation in crossing the dark-light boundary of the bright patch. This seems the best interpretation to place on the behavior of the caterpillars; but it might have been stated more certainly if the illuminance which reached the eye of the insect from the overhead source had been known. Loeb (1918) did experiments, which were redescribed by Fraenkel and Gunn (1940, pp. 65 and 185), in which he claimed that he was able to make photopositive caterpillars of *Porthesia* move toward weaker light and photonegative blowfly larvae move toward stronger light by rearranging the direction of the weak and strong sources of light. Loeb concluded that the insects moved in response to the direction of the light but were not influenced by the illuminance; but the design of his experiments did not altogether preclude certain other interpretations.

A more critical experiment was described by Spooner (1933), who studied the movement of plankton in a gradient of illuminance from a convergent beam produced by a lens. Photopositive animals moved along the beam from a point of low illuminance through the focus (where illuminance was greatest) on through a gradient from high to low illuminance as they approached the source of light. They evidently responded to the direction of the rays, which they followed, and not to the gradient of illuminance.

These experiments give rise to the hypothesis which was stated by Fraenkel and Gunn (1940, p. 185) as follows: "In taxes photo-positive animals go towards the light source, even if the intensity is lower there, and photonegative ones go away from the source, even if they reach a higher intensity; it looks as if intensity is unimportant." The subject is so complex that a great deal more investigation is still required to verify this hypothesis.

Animals with eyes as complex as those of adult insects, or even as complex as the ocelli of most lepidopterous larvae, may exhibit a variety of reactions to light, even within the species. This is well illustrated by Wellington's (1948) studies on the caterpillars of the spruce budworm *Choristoneura fumiferana*. Two lights of equal brilliance were mounted on a board 1 meter square. The lights were at the end of a line XY, shown diagrammatically in Figure 8.09.

This line was the base of an isosceles right-angled triangle. Individual larvae were released at the apex of the triangle, and the direction of their movements was specified by the angle made to the bisector of the right angle. The bisector itself was called "0° orientation angle." Thus a larva which moved straight ahead to a point equidistant between the two lights had an orientation angle

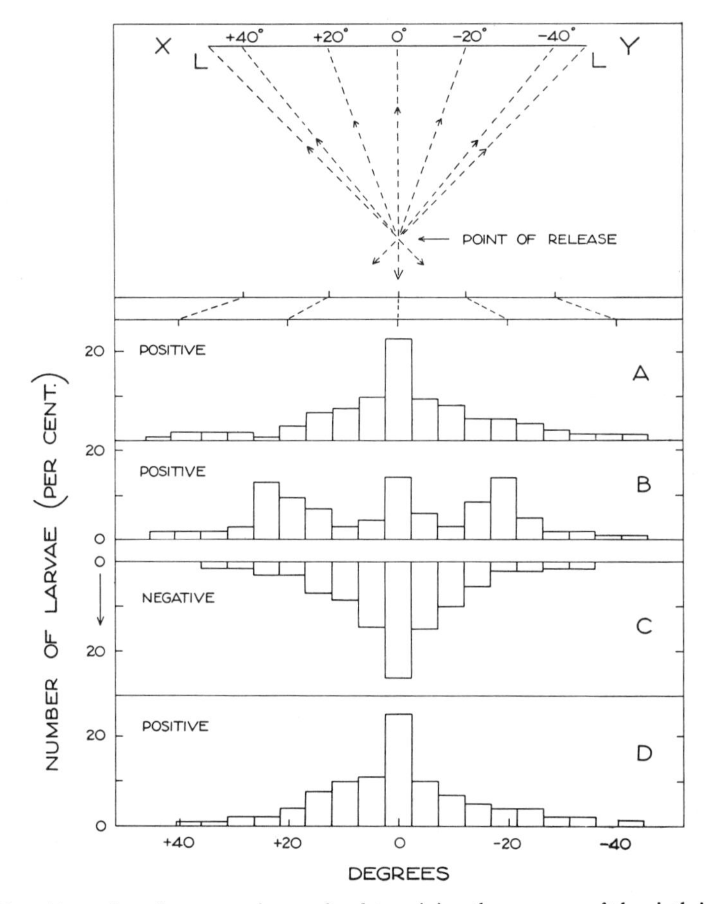

FIG. 8.09.—*Above*, plan of two experiments for determining the response of the sixth-instar larvae of the spruce budworm *Choristoneura fumiferana* to light and showing "orientation-angles." *Below, A*, frequency distribution of larvae between the two lights in the first trial; *B*, fourth trial, showing three peaks; *C*, photonegative larvae which have been starved overnight; *D*, larvae from *C* after having been fed again overnight. (After Wellington, 1948.)

of 0°; but one which moved directly to the left-hand light had an orientation angle of +45°; and one that moved to the right-hand light had an orientation of −45° The distribution of sixth-instar larvae in the different positions (from +45° to −45°) is illustrated by the frequency diagram shown in Figure 8.09, *A*, which shows the distribution taken up by larvae in the first trial. They were all photopositive and tended to distribute themselves with a mode at 0°. By the fourth trial (Fig. 8.09, *B*), secondary peaks arose besides the central one at 0°. These were due to larvae "compassing" about one or the other light; having taken up a position with reference to one of the lights, they moved at a fixed angle to the rays from this light. In the ninth trial all the larvae were photonegative. This is illustrated by the reversed frequency diagram in Figure 8.09, *C*. Again they moved with reference to both lights, so that all larvae moved between the angles of +45° and −45°. The larvae were then put in a "feeding jar" overnight, and when tested the next morning, they were photopositive (Fig. 8.09, *D*), with a distribution similar to that shown in the first test of the previous day. When photopositive larvae were kept without food overnight, they had become photonegative by the next day. When Ringer's solution was injected into these larvae through the anus until the gut became distended, the larvae resumed the photopositive condition. A ligature of silk tied around the fifth abdominal segment produced the same result. Conversely, larvae in which the gut was distended by food and which were therefore photopositive became photonegative when the ganglion of the ventral nervous cord in the fifth abdominal segment was anesthetized with methalyl. These experiments showed that the movements toward or away from light depended upon the ventral ganglion in the fifth abdominal segment and that the natural stimulus for this ganglion came from the pressure exerted by the gut when it was distended with food.

The behavior of the larvae of *Choristoneura* with respect to light depended also on whether the light came from a diffuse or a discrete source. Wellington hung fluorescent tubes masked with white paper above the board, to provide a source of diffuse light. He found that all instars were photopositive to diffuse light, irrespective of the condition of their gut or their behavior when exposed to light from a discrete source. The results of these carefully controlled experiments were confirmed and illustrated by another done more simply out-of-doors. Larvae which were photonegative to discrete light were placed on an inclined rod pointing toward the sky. During clouded periods, they climbed toward the sky; but when the sun was visible, they crawled down the rod and fell off. They were photopositive in the diffuse light from a clouded sky but photonegative to the light coming directly from the sun. An alternative explanation might be that they were negative to the illuminance of the sun shining brightly; but this seems unlikely when one also takes into account the results of the experiments described above. The differences in the behavior of larvae which had been kept without food when they were exposed to diffuse light (to which they were photopositive) and light from a discrete source (to which they were photonegative) might be interpreted as a difference in reaction to low as compared with high illuminance. But the larvae moved away from a discrete source of light into a patch of light which had ten times the illuminance when this light came from overhead (see above), and Wellington concluded that they were responding to the direction of the light rather than its illuminance. To respond to direction in this way, the larva must recognize, however dimly, the source of the light, that is its eye must form, at least, some sort of image. Wellington tested this hypothesis by moving white squares against a black background. The larvae responded in a way which indicated a capacity to perceive the white squares. The perception of a bright disk does not involve any high degree of visual acuity, and it seems likely that the simple eye of the caterpillar might do this. It follows that the larva of *Choristoneura* is unable to perceive a bright disk when the gut is full and pressing on the ganglion of the fifth abdominal segment but becomes able to do so when this pressure is removed, as it is when the larva has been without food for a period. Schallek (1942*a*) listed a number of fresh-water and marine invertebrates which are photopositive in the presence of light from a discrete source but photonegative when they are exposed to diffuse light.

Wellington (1948) described one way in which the behavior of the spruce budworm with respect to light might have adaptive value for caterpillars living in spruce forests. We might imagine a sixth-instar larva in a heavy infestation finding itself on a defoliated twig. The high evaporative rate in the area of the defoliated twig would stimulate the larva to greater activity, so that it would probably drop off on a thread. Being photopositive to diffuse light, it would move upward if it did not meet foliage on the way down. As starvation progressed, it would become capable of compassing by the sun when this was in sight. This could result in a movement around the tree instead of up it, so increasing the chances of finding food. On an overcast day the larva, in the absence of a discrete source of light, would tend to move upward to the top of the tree, where it would stay, provided that the overcast conditions persisted without rain. But if the day was sunny, the larva would become negatively phototactic to the disk of the sun and would move down the tree until the sun was out of sight. This descent would not be followed by a return to any higher level from which the disk of the sun was visible. But since the larva is positively phototactic to diffuse light, it would tend to be held in the outer periphery of the tree, where the chance of encountering fresh food would be better than in the middle. If it were photonegative to illuminance as well as to the disk of the sun, it would be driven in toward the trunk of the tree, where there is little or no food.

Wellington's studies show that the daily changes in the weather, particularly in the incidence of sunshine and cloud and the extent to which larvae happen to be starved or well fed, result in movements of larvae on spruce trees in response to the sort of illumination they are receiving. Light is clearly an important influence in the distribution of the insects on the trees throughout the day. The survival-value of some of these movements is obvious enough. Not many examples have been worked out in such detail as this one, although light is known to influence the movements of a great number of animals.

Clark (1950) counted the numbers of locusts (*Chortoicetes terminifera*) at intervals along transects that started in grassland and continued through belts of woodland. The locusts were most numerous in the grassland and along the margins of the woodland. They became fewer, the further the transects penetrated into the wooded areas. Clark explained these observations in terms of the locusts' responses to light. Being strongly photopositive, the locusts move toward places of high illuminance, and this tends to prevent them from penetrating deeply into a stand of trees. Those which did wander in would soon reach a zone where the illuminance was low and would then return toward the zone of higher illuminance.

The individual animals in a population which have all had much the same history do not necessarily behave in a uniform way in response to light. Figure 8.09, *B*, shows that some of the larvae of the spruce budworm may exhibit light-compassing movements, while others have a direct taxis toward light. Richards (1951) reported a difference in the sign of phototaxis of adult beetles of *Calandra granaria* in a single population. Most of the beetles moved away from a light, but a small number moved toward a light. The difference was shown to be genetic, for Richards made selective matings, for two generations, of beetles which were photopositive and others of beetles which were photonegative. In this way he obtained two populations, one of which was completely photonegative and the other of which was almost indifferent to light.

8.32 *Reversal of Phototaxes*

When the larvae of *Choristoneura* move toward the source of a diffuse light and away from the source of a discrete light, such as an incandescent lamp, or, when the marine copepod *Acartia tonsa* moves away from a diffuse light but toward an incandescent globe, the reversal of the behavior is due to the differences in the stimuli received from the two sources of light. In this section we shall mention a number of examples of similar reversals in phototaxes which may occur while the stimulus received from light remains constant; in these cases the changed behavior may be due to changes in temperature, moisture, food, or even in the age or stage of development of the animal.

Wellington (1948) put larvae of the spruce budworm on the floor of a chamber which had some parts shaded and others illuminated from a diffuse light above. The air in the chamber was kept almost saturated with moisture, so as to eliminate evaporation as a possible influence on the behavior of the larvae in it. Larvae which were all photopositive at the beginning of the experiment were placed on the illuminated part of the floor, the temperature of which was then raised. The larvae remained in the illuminated part of the chamber until a certain critical temperature was reached, when they moved into the dark part of the chamber. The critical temperature was different for different instars (Table 8.04). The striking difference between the reversal temperatures of

TABLE 8.04*

TEMPERATURES AT WHICH 100 LARVAE OF EACH INSTAR OF SPRUCE BUDWORM *Choristoneura fumiferana* CHANGED FROM PHOTOPOSITIVE TO PHOTONEGATIVE WITH RESPECT TO DIFFUSE LIGHT

Instar No.	1	2	3	4	5	6
Temperature (° C.)......	28.6	36.5	36.5	36.8	37.5	37.9

* After Wellington (1948).

the first and second instars was spectacularly shown when larvae were put in a gradient of temperature (Fig. 8.10). The gradient was illuminated in the center

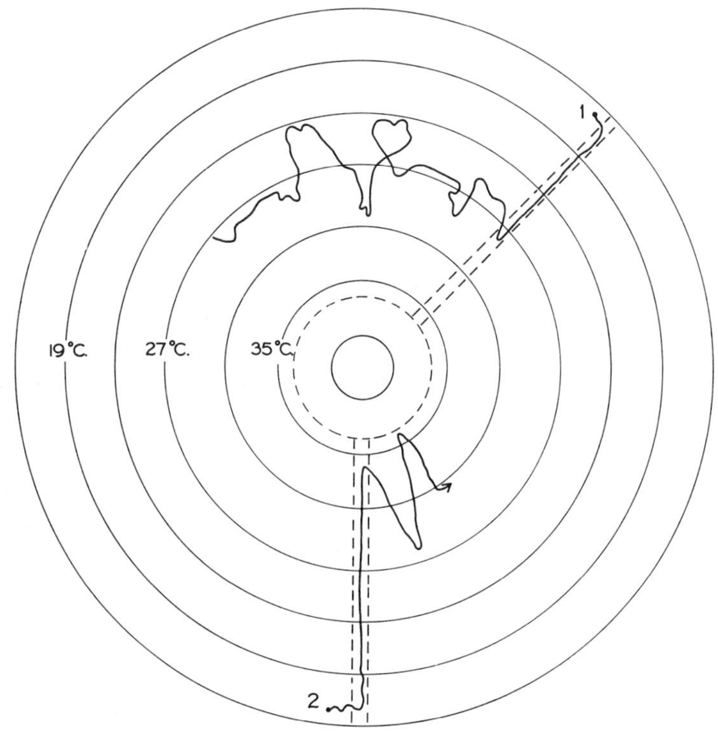

FIG. 8.10.—The reversal of phototaxis of (*1*) a first- and (*2*) a second-instar larva of the spruce budworm at high temperatures. The area within the dotted lines indicates the lighted zone of a temperature gradient. (After Wellington, 1948.)

by a spot of light from an overhead lamp and by two narrow pencils of light from the center to the periphery, provided by two microscope lamps. Larvae were placed in a cool zone near this beam. They immediately traveled to the beam and along it toward the bright spot in the center, until the critical temperature was encountered. The wavy lines in Figure 8.10 show the typical tracks of a first-instar and a second-instar larva. At the critical temperature the larva turned sharply away from the light into the shaded part of the gradient. Temperatures as high as the critical temperature may be experienced by larvae in the field. This was demonstrated by measuring the body temperatures of larvae in the field with thermocouples. Field observations also showed that larvae on the exposed face of a tree either dropped to shaded levels or retreated to the inner parts of a branch during periods of exposure to intense radiant heating from the sun. The reversal of phototaxis at these temperatures caused the animals to move away from the sun to a cooler region, which was more favorable to them.

Reversal of phototaxis at high temperatures is also shown by larvae of *Malacosoma* and *Neodiprion* (Wellington *et al.*, 1951). The tsetse fly *Glossina morsitans* is positively phototactic at normal temperatures, but, at temperatures of 32° C. and above, it becomes negatively phototactic. This, like the similar behavior of the spruce budworm, has survival-value for the animals in nature. They move away from the dangerous heat of the sun to the cooler refuge of shade. In Rhodesia the temperature inside dense patches of evergreen forest along the banks of rivers is generally below that needed to make the flies photonegative. Such places are rarely penetrated by the flies to any considerable depth. The flies play a sort of hide and seek in the tropical forest, moving from light to shade, depending upon the temperature. The complexities of their distribution in the forests can largely be explained in terms of their phototaxes (Jack and Williams, 1937).

Adults of the syrphid *Eristalis tenax* moved toward light at temperatures between 10° and 30° C., but above 30° C. they moved away from the light. The critical temperature at which the reversal in phototaxis occurred was largely independent of illuminance between the limits of 800 and 1,600 lumens per square foot. But with illuminance below 800 lumens per square foot, the temperature at which reversal occurred tended to be higher, the lower the illuminance (Fig. 8.11). The temperature at which reversal of phototaxis occurred was higher for females than for males (Dolley and White, 1951). The reversal of phototaxis at high temperatures has been shown to occur in many more animals than those mentioned here (Dolley and Golden, 1947).

The potato beetle *Leptinotarsa decemlineata* is normally photopositive, but it may become photonegative as it becomes desiccated (sec. 7.34). During dry weather the beetles move down the stems of plants into the soil, where they remain until they have been remoistened by rain. They then become photopositive again and ascend the stems of the plants, where they lay their eggs.

Rudolfs and Lackey (1929) were able to make larvae of the mosquitos *Culex pipiens*, *Aedes sylvestris*, and *A. canadensis* photopositive by feeding them on a mixed diet or on dinoflagellates. But they became photonegative when confined to a diet of ciliates. Larvae of the spruce budworm were photopositive while the gut was distended with food, but they became photonegative in their response to a discrete source of light when they were kept without food (see above). The reverse seems to apply to *Porthesia* larvae, which were photopositive before they had eaten but lost this reaction completely after they had been feeding for a while (Loeb, 1918).

In some animals a reversal of phototaxis occurs as they grow older. This is common among blowflies. Larvae of *Calliphora erythrocephala* are photonega-

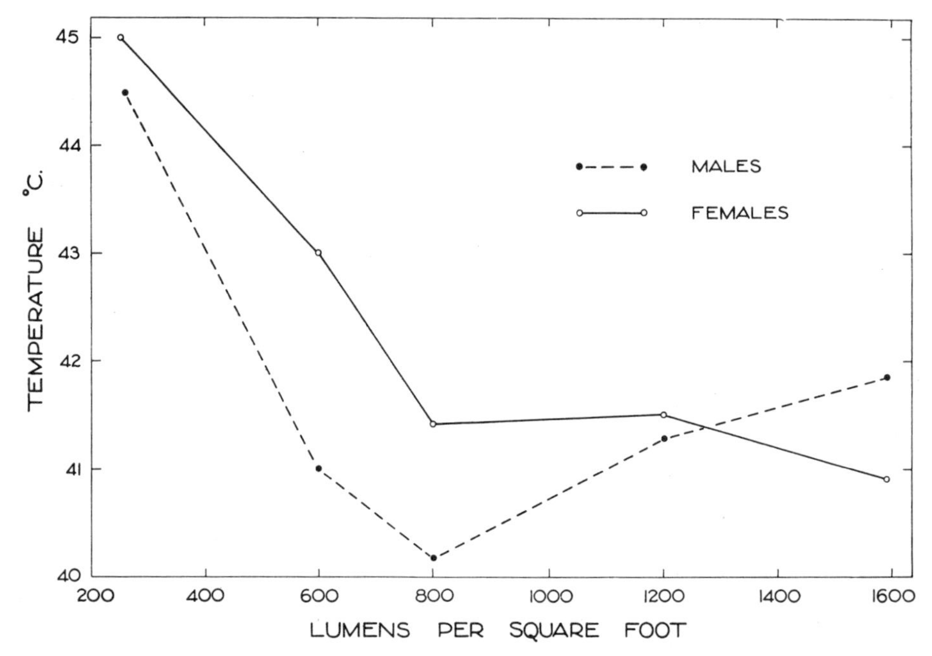

FIG. 8.11.—The influence of illuminance on the temperature at which photopositive adults of the fly *Eristalis tenax* became photonegative. (After Dolley and White, 1951.)

tive all their lives, but the intensity of the response varies with age. This was measured by Patten (1916), who put individual larvae of different ages in the path of a horizontal beam of light from one of two alternative lamps, shown as *X* and *Y* in Figure 8.12. Larvae typically moved along the beam away from the light. Light *X* (or *Y*) was then turned off, and *Z* was turned on to produce a beam at right angles to the original beam. A highly photonegative larva would be deflected from its path at an angle of 90°, moving away from the new source of light (as shown by the arrow in the diagram). A less sensitive larva would trace a path which diverged from the original by a smaller angle. Figure 8.12 shows the average deflection of 50 larvae of different ages (at room temperature). The intensity of the reaction to light varied greatly with age, reaching a maximum about the fourth day of larval life, after which it fell until the seventh day, when it remained constant until pupation. Between the fourth and fifth days, about 15 per cent of the larvae had migrated to pupate; by the sixth day, 60 per cent had migrated; and by the seventh day, practically all had migrated. The drop in the intensity of negative phototaxis thus coincided with migration. Newly emerged adults of the bean weevil were photonegative, but females became photopositive after they had laid eggs. Old adults were also photopositive (Menusan, 1935). Larvae of *Eristalis tenax* were highly photopositive for the first few hours after hatching. They became photonegative

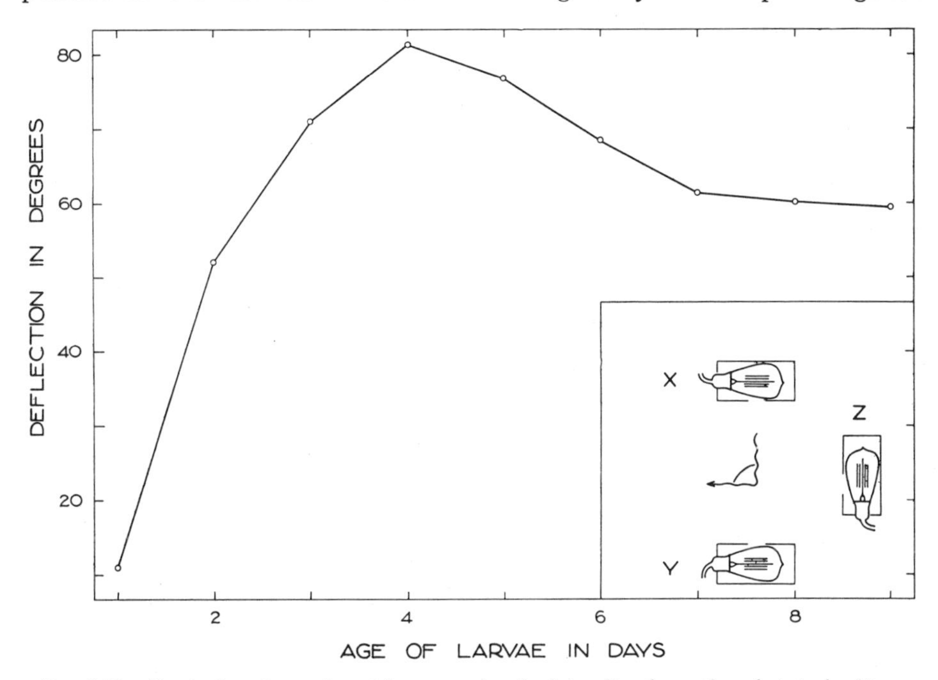

FIG. 8.12.—*Inset:* plan of experiment for measuring the intensity of negative phototaxis of larvae of *Calliphora crythrocephala*. *X* or *Y* is used as a starting light; *Z* is used for deflecting the path of the maggots. Arrow shows path of maggots with maximal sensitivity. The graph shows the angle of deflection of the path of 50 larvae at different ages. (After Patten, 1916.)

very quickly and remained so until they pupated. The newly emerged adults were photonegative and then became photopositive and remained so unless the temperature was unduly high (Dolley and Haines, 1930). Larvae of the lobster *Homarus americanus* were photopositive for a couple of days after hatching; they then became photonegative and remained so until shortly before molting, when they became photopositive. They became photonegative after the first molt, and in the fourth and later stages they remained so (Hadley, 1908). Young *Daphnia magna* were photopositive, but after the release of the first brood they mostly became photonegative (Clarke, 1932).

The reversal of phototaxis has distinct adaptive value, at least in some cases; this is well illustrated in the example of the change of phototaxis of the spruce budworm and the tsetse fly with increase in temperature. The changes associated with age are of adaptive value, at least in some animals in which they are associated with the changed habits of the older animal. This is most clearly seen in the changes of the blowfly from larva to mature adult. In other cases the adaptive advantage of reversal of phototaxis may not be so evident. This may be because they have not yet been sufficiently studied.

8.33 *Illuminance-Preferendum*

When the larva of *Choristoneura* moves away from an incandescent globe, it is probably perceiving a crude image of the globe, and the direction of its movement is probably determined largely by the direction from which the light is coming. At the same time, it is moving down a gradient of illuminance, but this is probably not the chief cause of its progress in that direction (sec. 8.31). When the same larva moves toward a fluorescent tube shielded by white paper, it can hardly be perceiving any image of the source of light, so it must be responding to the gradient in illuminance. This raises the question: Do the larvae of *Choristoneura* prefer a particular level of illuminance? And would they, given the opportunity, move away from places where the illuminance was brighter or duller than this and come to rest in the place where the preferred level of illuminance occurred? In other words, is there, in a gradient of illuminance, a zone which can be recognized as a "preferendum," as with moisture or temperature? Wellington's (1948) experiments with spruce budworm were not designed to answer these questions.

Herrstroem (1949) found that adults of the carabid *Agonum dorsale* preferred a particular level of illuminance. His apparatus consisted of a metal disk in which he had cut a ring-shaped groove. The beetles were allowed to run in the groove, which was kept damp. The groove was closed above with a sheet of glass, on which was placed a "gray key." This was a disk of papers divided into sectors of different thickness. The whole was covered with another sheet of glass. A lamp from above, shining through the paper, illuminated the groove with diffuse light. The thickness of the paper was so arranged that there was a gradient of illuminance around the groove. A similar gradient might have been obtained with rather less trouble by suspending a fluorescent tube under a plate of opal glass, with one end close to and the other end further away from the glass.

Herrstroem recorded the number of beetles in each sector at regular intervals on a series of successive days. A large number of beetles were used. Most of those which had previously been living in weak diffuse light came to rest in those parts of the groove where the illuminance was less than 0.05 lumens per square foot; the mode of the distribution was 0.022 lumens per square foot, and its mean was 0.056 lumens per square foot (Fig. 8.13, *A*). This corresponds to a dull light in which one might just be able to read a typewritten page.

The distributions of beetles which had been exposed during 48 hours before the experiment to an illuminance of 1.27 lumens per square foot are shown in Figure 8.13, *B*. At the close of the first ½ hour, most of the beetles were found where the illuminance was greatest (0.25 lumens per square foot). Later they moved to duller parts of the groove, until at the end of 4 hours there was little difference from those which had been living continuously in a dull light (Fig. 8.13, *A*). The means of the distributions found at the end of 30, 90, and 240 minutes were 0.116, 0.078, and 0.046 lumens, respectively. The "acclimatiza-

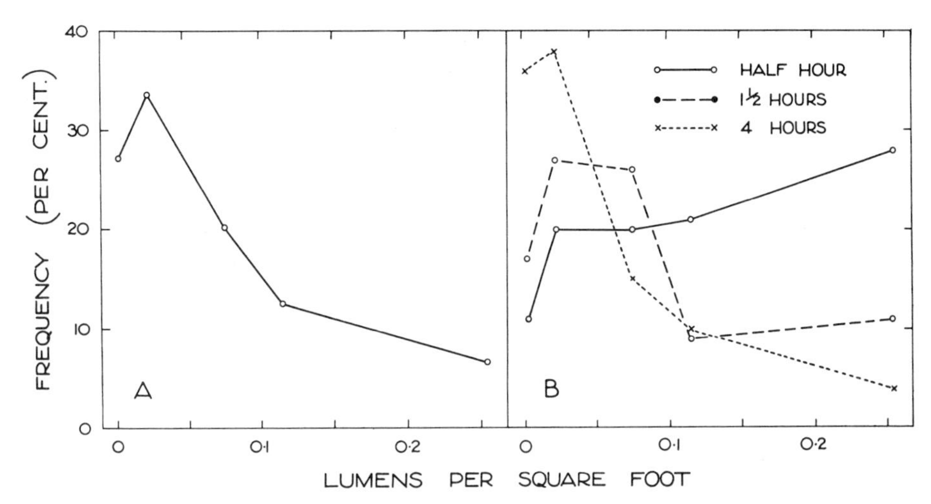

FIG. 8.13. Illuminance preferendum of the beetle *Agonum dorsale*. *A*, after having been in diffuse dull light prior to the experiment; *B*, after having been in bright light for 2 days prior to the experiment. The curves show the lowering of the preferred illuminance after ½, 1½, and 4 hours in the apparatus. (After Herrstroem, 1949.)

tion" which was reflected in the early behavior of the beetles soon wore off, leaving little trace by the end of 4 hours. The beetles used in these experiments were collected in the autumn. In view of the possible influence of acclimatization, it would have been interesting to repeat the experiments with beetles collected during summer. It is not known how widespread the phenomenon of illuminance-preferendum is in nature, because remarkably few species have been tested for it.

8.34 *The Influence of Colored and Polarized Light on Insects*

In experiments in which color is the variable, it is essential that illuminance be kept constant. This introduces technical difficulties; for example, it is impossible to work with high illuminance and cover a broad range of the spectrum. Much of the work which has been done both in the field and in the laboratory on the ability of insects to perceive color suffers from the defect that illuminance was not kept constant. Also many insects can be conditioned to prefer a certain color. This adds to the difficulty of experimenting in this field.

Weiss (1943) exposed some 50 different species of insects to the light from colored lamps. He kept the illuminance constant, while varying the color from ultraviolet to infrared. When the insects were placed about 1 foot away from the colored lamps and allowed to move freely, most of them came to rest where the light was blue-green (4,700–5,280 *A*). Few individuals of any species went to where the wave lengths fell outside these limits. When the insects were put down some 6 feet away from the lights, most of them moved toward the ultraviolet light (3,650–3,663 *A*). In some species a number of individuals also moved toward the blue-green light (4,700–5,280 *A*). The rest of the spectrum from yellow to infrared was not attractive. When only two lamps were used at a time, most of the insects went toward the lamp which was emitting light with a shorter wave length. But this result could be reversed by increasing the illuminance of the longer wave length. For example, the potato beetle *Leptinotarsa decemlineata* and the Japanese beetle *Popillia japonica*, when offered the choice of ultraviolet or red of equal illuminance, chose ultraviolet. But they could be made to go toward the red light in preference to ultraviolet by increasing the illuminance of the red light. Wigglesworth (1950) added a number of species to those discussed by Weiss (1943) and Weiss *et al.* (1941, 1942), in which the shorter wave lengths are the more attractive. But there are exceptions, notably among some butterflies and especially the firefly, which is attracted by flashes of red light with wave length of 6,900 *A*. This is the longest wave length yet recorded as visible to an insect (Wigglesworth, 1950).

Recent experiments have shown that certain insects can detect and respond to polarized light. The proportion of the light from the sky which is polarized depends on the relative position of the sun. Near to and far away from the sun, there is the least polarization; the maximum occurs between these two extremes, where about 70 per cent of the light may be polarized. Besides differences in the proportion of the light which is polarized in the different parts of the sky, there are also differences in the plane of polarization across the sky. The first indication that insects might be able to perceive such differences came from Von Frisch's (1950) remarkable experiments on food-finding by honey bees. Bees which have located a source of food are able to find their way to the food again, and also they are able to communicate the direction and distance of the food from the hive to the other bees in the hive. They appear to find the food again by registering the angle made between the line joining the hive and the food and the line joining the hive and the sun, and then keeping the sun at this angle to their direction of flight. But the ability to do this persisted even when the sun was obscured on cloudy days. This was evidently possible by virtue of a remarkable sensitivity to the differences in the degree of polarization of light in different regions of the sky. Von Frisch demonstrated this by altering the polarization of light visible to the bee in the hive with polaroid sheets and observing the change in the dancing behavior by means of which they communicated direction and distance to other bees. He concluded that the eye of the bee could analyze polarized light. The ability to discern differences in the polarization of light has since been demonstrated for the ant *Myrmica* (Vowles, 1950), in the larvae of, *Neodiprion*, *Malacosoma*, and *Choristoneura* (Wellington *et. al.*, 1951), in *Drosophila melanogaster* (Stephens *et al.*, 1952), and in *Limulus* (Waterman, 1950). Wellington and Vowles suggested that the insects with which they worked orientated themselves by reference to patterns of polarization in the sky.

8.4 THE BEHAVIOR OF INVERTEBRATES IN ALTERNATING LIGHT AND DARKNESS

8.41 *Daily Rhythms of Activity*

It is a common observation that many species of animals are not equally active throughout the 24 hours of the day. Some are primarily nocturnal (a long list of these is given by Park, 1940), others are diurnal, and others, which are referred to as "crepuscular," are active mainly at dawn and dusk. A species may confine its activities of feeding, mating, and moving about from one place to another to these particular periods of the 24 hours. The main environmental components which undergo a daily fluctuation are temperature, moisture, and light. In section 6.22 some examples were given of the influence of the daily rhythm of temperature on the daily activities of insects, such as locusts and grasshoppers. Light may also cause a daily rhythm of activity in some animals.

It is characteristic of animals that they display intermittent bursts of activity, spaced between periods of rest. In the absence of a systematic stimulus, these may be irregular. When the sea anemone *Metridium* was exposed to light constantly, it could be seen sometimes contracted and sometimes expanded; but there was no regularity about its activity. When it was exposed alternately to light and darkness, most, though not all, individuals were expanded while it was dark and contracted while it was light. In other words, their activity assumed a rhythm which had the same period as the stimulus (Batham and Pantin, 1950). In nature the activities of many species are correlated with the diurnal rhythm of night and day, and there are experiments to show that at least sometimes the causal stimulus is light.

Cockroaches move about much more at night than in the daytime, when they tend to remain hidden in crevices and other dark places. Gunn (1940) recorded the daily activity of *Blatta orientalis* by placing single cockroaches in an apparatus which had a floor which tilted when the cockroach walked its length. Smaller excursions were not recorded. Every tilt was registered on a revolving smoked drum. The temperature and humidity in the apparatus were kept constant, and the activity was measured in artificial light and darkness. When the light was on from 7:00 A.M. to 7:00 P.M. and off for the rest of the 24 hours, the greatest activity was recorded in the dark half of the day. In continuous light the rhythm of activity persisted for a couple of days, but then disappeared. Continuous darkness had the same effect. These results indicate that the rhythm of activity is associated with the rhythm of illuminance, though it can persist for a short time in the absence of the external rhythm. By keeping the cockroaches in continuous light for a couple of weeks and then giving them a reversed rhythm of light and darkness, they could be made active in the day hours (in the dark) and inactive in the night hours (in the light). Bentley, Gunn, and Ewer (1941) measured the activity of the beetle *Ptinus tectus* by counting the number of animals moving during 15 minutes. They obtained results which were similar to those obtained by Gunn with the cockroach. The rhythm persisted for a few days in continuous light, and it was reversed by reversing the hours of darkness and light.

The crayfish *Cambarus virilis* is normally active at night and quiet during the daytime. Schallek (1942*b*) measured its activity in a tank of water by means of a celluloid ball which was floating on the surface of the water. This was attached to a writing point on a revolving smoked drum, so that every time the animal moved sufficiently to disturb the surface of the water, a mark was recorded on the drum. Activity was recorded as the number of movements per hour. More complete measurements of activity can be obtained by confining animals in a treadmill and making a continuous record of the revolutions of the mill. However, Schallek's methods were quite adequate for measuring the changes in activity of *Cambarus* during the 24 hours of the day. When the animals were in continuous darkness, the daily rhythm of activity persisted for at least 5 weeks. Figure 8.14 shows the persistent rhythm in darkness. Some of the animals were active only once a day; this was usually, though not always, at noon or midnight; others were active twice a day, at dawn and dusk. When light and darkness were alternated at 4-hourly intervals, the animals were more active in the "dark period" which occurred in the middle of the night, as compared with the "dark period" which came in the middle of the day. The innate rhythm of activity thus appears to be more deeply impressed in *Cambarus* than in the cockroach. Yet it is still a rhythm which is modifiable by different alternations of light and darkness.

Rhythms of activity which are modified by rhythms of light and darkness, yet which persist in constant illuminations, have been shown in the cricket *Gryllus assimilis* (Lutz, 1932), the beetles *Boletotherus cornutus* (Park and Keller, 1932) and *Megalodacne heros* (Park and Sejba, 1935), the millipede *Spirobolus marginalis* (Park, 1935), the laboratory rat (Richter, 1922; Browman, 1936), the deermouse *Peromyscus* (Johnson, 1939), the fieldmouse *Microtus* (Davis, 1933), the Japanese waltzing mouse (Wolf, 1930), and the bat *Myotis* (Griffin and Welsh, 1937). Still further examples are given by Park (1940). The rhythm of *Peromyscus* persisted for 7 days in the dark, but the active period was no longer in the night. Park and Keller (1932) found that the beetle *Boletotherus*, even those which had been reared in the laboratory in complete darkness at constant temperature and humidity from the larval stage, showed a 24-hourly rhythm of locomotory activity. Bats of the genus *Myotis* normally emerge at sunset, and, if food is plentiful, they return to their roost within a couple of hours, sometimes going out again in the early morning. When kept in

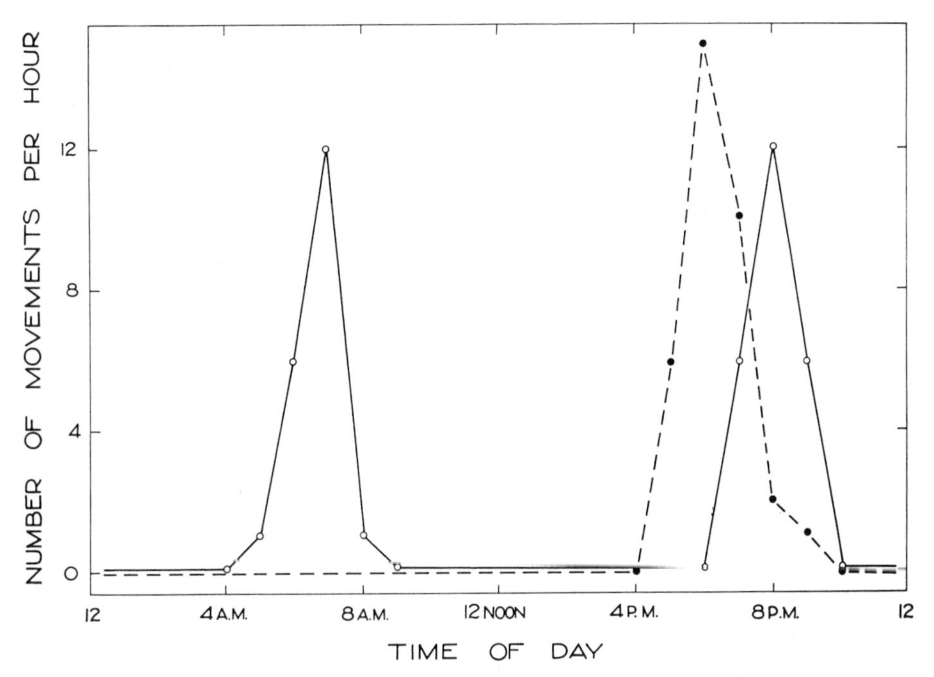

FIG. 8.14.—Activity of the crayfish *Cambarus virilis* in constant darkness. The solid line shows the activity of an animal which was active twice a day; the dotted line shows the activity of an animal which was active once each day. (After Schallek, 1942*b*.)

subdued light in a cage, *Myotis* had the same rhythm of activity, which also persisted for 4 days in darkness (Griffin and Welsh, 1937).

The rhythm of activity of *Cambarus* ceased altogether when the eyestalks were removed; the animals became continuously active. The eyestalk contains both nerve centers and an endocrine organ, the sinus gland. Injections of extract from the sinus gland did not influence the activity of animals from which the eyestalks had been removed. But when the optic nerve of normal animals was severed, they became more active and were comparable to those from which the eyestalks had been removed. Schallek concluded that activity is inhibited in the normal animal in the quiet phase of the rhythm by the action of fibers contained in the optic nerve. It would seem reasonable to suppose that the inhibiting fibers become effective when light falls on the retina. This could doubtless be tested by blackening the eyes, but this experiment was not done.

Although experiments on *Cambarus* indicated a nervous, rather than a hormonal, control of activity, this may not be universal. For example, Kalmus (1938) found that extract of eyestalk caused increased locomotion of *Potamobius*, from which he concluded that *Potamobius* is normally quiet but is activated at night by a substance released from the eyestalk.

These experiments indicate a physiological mechanism through which the stimulus of light may become effective. But the explanation for the rhythms which persist after the stimulus has been removed remains to be discovered. It would seem necessary to seek something which would serve as an "internal clock." An analogous rhythm may be seen in the behavior of the chromatophores of certain crustaceans. In the crab *Uca* the expansion and contraction of the chromatophores depend on a secretion from the sinus gland. The secretory activity of the gland is rhythmical; it can be stimulated by light, but the rhythm persists in constant darkness (Brown and Hines, 1952). In the moth *Plusia gamma* and also in certain other insects and crustaceans a daily migration of the pigments of the compound eyes is correlated with the rhythm of night and day; but the rhythmical migration of pigments persisted even when the animals were kept constantly in the light (Welsh, 1938). The electric potential of the corneal surfaces of the eyes of certain insects showed diurnal fluctuations even when the insects were kept constantly in the dark except for the momentary flash of light which occurred when electrical measurements were being made. The illuminance of this momentary stimulus, which was necessary to cause an electric response in *Dytiscus*, was 1,000 times greater in the day than at night (Jahn and Crescitelli, 1938, 1940; Jahn and Wulff, 1943).

It will be noticed from the foregoing examples that most of the animals for which there is experimental evidence that the diurnal rhythm of activity is determined by light or the absence of light are of the sort which are active at night. One might infer from this that those types which are naturally active in the daytime are more likely to depend on some other stimulus, such as temperature. This may well be so, but there are some experiments with certain species of *Drosophila* which show how light may impose a rhythm on the activity of animals which are active during the daytime.

Dobzhansky and Epling (1944) and Mitchell and Epling (1951) measured the number of *Drosophila pseudoobscura* caught visiting traps during each hour of the day in certain districts in western North America. Timofeeff-Ressovsky (quoted by Pavan *et al.*, 1950) did the same for *D. melanogaster* and *D. funebris* near Berlin. And Pavan *et al.* (1950) made a similar study of certain species of *Drosophila* in Brazil. Figure 8.15 shows the proportions of a day's catch caught between the hours of 5:00 A.M. and 6:00 P.M. in an open wood in Brazil. The flies consisted of about 10 different species, all of which had a similar rhythm of activity. Flies visited the bait chiefly between 6:00 and 8:00 A.M. and 4:30 to 6:00 P.M. During the night they did not visit the baits at all. The trend of activity was correlated with the sequence of changes in temperature and relative humidity during the day. Temperature was low and humidity was high when the flies visited the baits. On cloudy days more flies visited baits during the hours of daylight. Inside a tropical rain forest there was no appreciable drop in activity at the middle of the day, though at the margin of the same forest the activity fell during periods of sunshine. These observations suggest that the differences in activity during a day are adaptations to the changes in tempera-

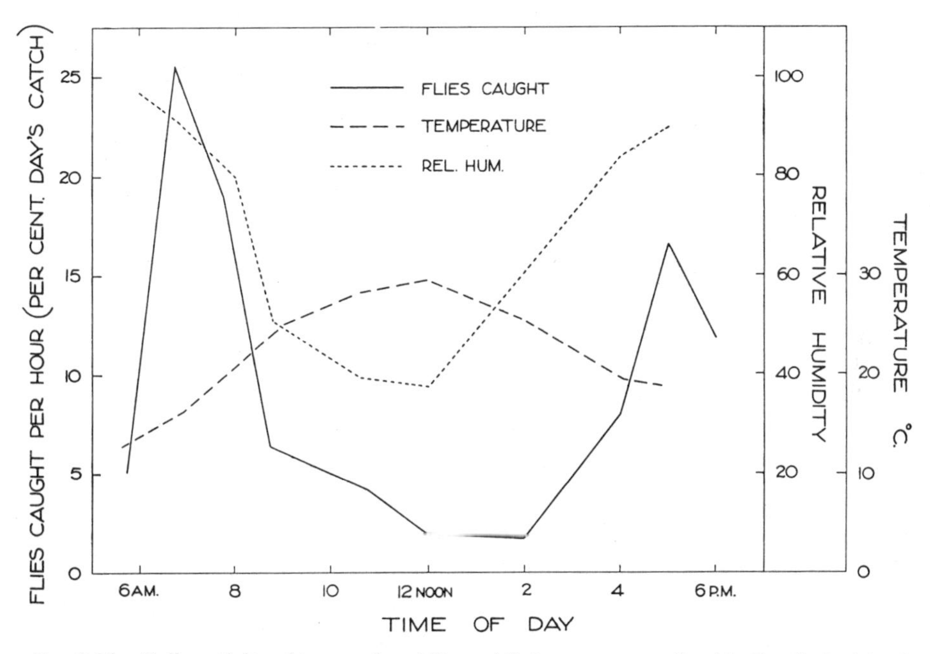

FIG. 8.15.—Daily activity of ten species of *Drosophila* in an open woodland in Brazil. Activity is expressed as the percentage of the total day's catch which were caught at each hour of the day. (After Pavan *et al.*, 1950.)

ture and moisture; flies do not venture out to visit food when they are likely to suffer desiccation. But it does not follow that the changes in activity are initiated by changes in temperature and relative humidity. It is more likely that light is a "token" stimulus which regulates activity. Mitchell and Epling (1951) measured illuminance, as well as temperature and humidity, during the daily cycle of activity of *D. pseudoobscura* in California. They found that the flies visited baits when the temperature was between 50° and 90° F. and illuminance was less than 300–400 lumens per square foot and greater than 1 lumen per square foot. These requirements were met only in the mornings and evenings of clear days and occasionally on cloudy days. Differences in temperature influenced the degree of activity, but not the timing and duration of the active period. This depended on illuminance. Other observations by Dobzhansky and Epling (1944) also indicate that illuminance determines whether the flies are active or not. The periodic activity of *D. pseudoobscura* is quite rigid in comparison with the activity of other species of *Drosophila* in the wet tropics of Brazil. These show a greater plasticity, in that they are active in the middle of the day, when humidity happens to be high.

The dermestid beetle *Attagenus japonica*, which is common on chrysanthemums in Japan, has a well-marked daily rhythm of activity which reaches a peak between 9:00 and 11:00 A.M. Katô (1949) found that *Attagenus* did not fly unless illuminance exceeded 1,000 lumens per square foot and temperature exceeded 15° C. In the direct sunshine the beetles migrated to the undersides of flowers and became motionless, but this was because high temperature made the beetles inactive.

8.42 *Daily Rhythms in Ecdyses in Certain Insects*

Bateman (unpublished thesis) showed that in certain circumstances the pupal molt in the fruit fly *Dacus tryoni* occurred with a diurnal rhythm. When the insects were reared throughout the life-cycle at a constant temperature and illuminated by lights which were switched on at 9:00 A.M. and off at 5:00 P.M. each day, 90 per cent of the pupal molts occurred between 6:00 A.M. and 6:00 P.M. and the mean time for emergence was 1:45 P.M. It was necessary to keep temperature controlled precisely, for even turning on the fluorescent lights in the constant-temperature room, where the experiments were done, caused a rise of a couple of degrees in temperature and when precautions were not taken to prevent this rise, emergence, instead of being spread over 18 hours, became concentrated into the 7-hour period following the turning-on of the lights. Exposure to light during the pupal stage had practically no influence on the rhythm of emergence; it remained the same, no matter whether the pupae were kept in constant light, constant darkness, or alternating light and darkness. The rhythm of emergence was determined by light experienced by the larvae and the adults of the previous generation. This is illustrated quite strikingly in Figures 8.16 and 8.17. Figure 8.16, *A*, shows the distribution of emergence when pupae were kept in the dark and larvae were illuminated from 9:00 A.M. to 5:00 P.M.; and Figure 8.16, *B*, shows emergence when the larvae were illuminated from 9:00 P.M. to 5:00 A.M. In the former experiment the mean time of emergence was 1:00 P.M., but in the latter the mean time of emergence was 8:00 P.M. An even more striking difference is shown in Figure 8.17, *A* and *B;* in these experiments both larvae and pupae were kept in the dark, but in the one case (Fig. 8.17, *A*) the adults of the previous generation were illuminated from 9:00 A.M. to 5:00 P.M., while in the other experiment (Fig. 8.17, *B*) the adults were illuminated from 9:00 P.M. to 5:00 A.M. In the former experiment the mean time of emergence was 12:20 P.M., while in the latter case it was 12:50 A.M. Thus Figure 8.16, *A* and *B*, shows how the light experienced by the larvae may influence the emergence of adults from the pupae and Figure 8.17, *A* and *B*, shows how the light experienced by the adults of the previous generation may influence the emergence of adults from pupae in the next generation.

The adults of *Drosophila melanogaster* emerged between the hours of 6:00 A.M. and 2:00 P.M. when kept at constant temperature and in normal sequence of daylight and night. The rhythm of emergence persisted for some generations when the insects were reared in constant light or constant darkness (Bremer,

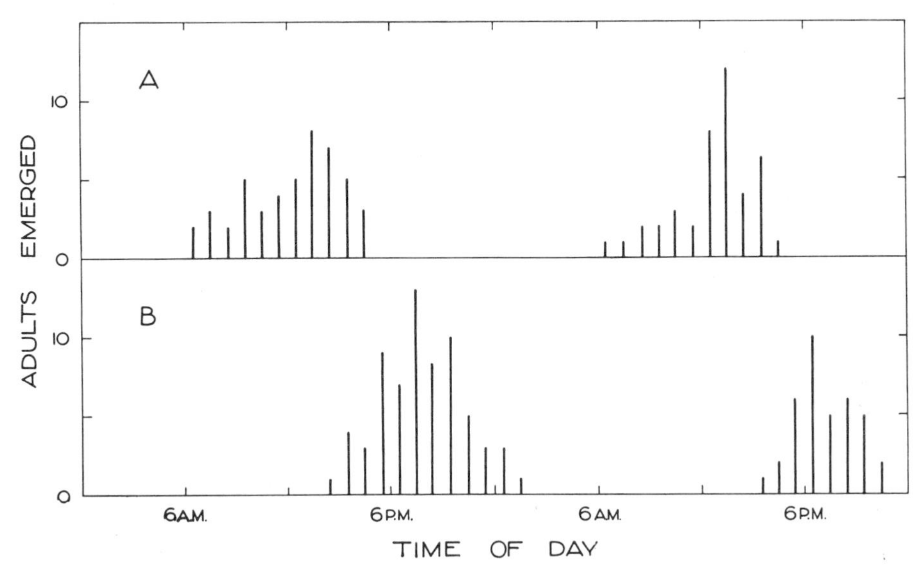

FIG. 8.16.—The influence of exposure to light during the larval stage on the "timing" of the pupal ecdysis in *Dacus tryoni*. *A*, the larvae were exposed to light from 9:00 A.M. to 5:00 P.M. and to darkness at all other times; *B*, the larvae were exposed to light from 9:00 P.M. to 5:00 A.M. and to darkness at all other times. (After Bateman, unpublished thesis.)

1926). Bremer was able to reverse the rhythm, so that the flies emerged during the night, by reversing the alternation of light and darkness. The reversed rhythm also persisted for some generations when the insects were reared in constant light or constant darkness. Although the rhythm of emergence of flies was eventually lost after they had been reared for a number of generations in continuous darkness, Bünning (1935) found that a single stimulus was sufficient to reinstate the rhythm. He kept cultures under constant weak illumination for 15 generations, by which time no rhythm was left. At the end of this time a single brief exposure to darkness resulted in the reappearance of the rhythm. Kalmus (1940) kept a culture of *Drosophila* in darkness until the rhythm disappeared. A single illumination of at least 4 hours then induced a 24-hour rhythm of emergence in the pupae when they were returned to darkness. Lewis and Bletchley (1943) found a peak in emergence of the dung-fly *Scopeuma stercoraria* between 9:00 A.M. and 2:00 P.M. when they were kept exposed to room temperature and natural light and darkness. In constant darkness or in constant light the rhythm persisted but was less pronounced. The moth *Ephestia kühniella* had a peak of emergence between 5:00 and 6:00 P.M. in an insectary. The rhythm was evidently influenced by both temperature and light, for at constant temperature and alternating light it persisted but was much less pronounced, while it was possible to induce a 16-hour rhythm by exposing the pupae to a 16-hour cycle of temperature in place of the normal 24-hour cycle (Scott, 1936). Hueck (1951) found that the emergence of the red spider *Metatetranychus ulmi* from eggs took place only in the light when they were exposed

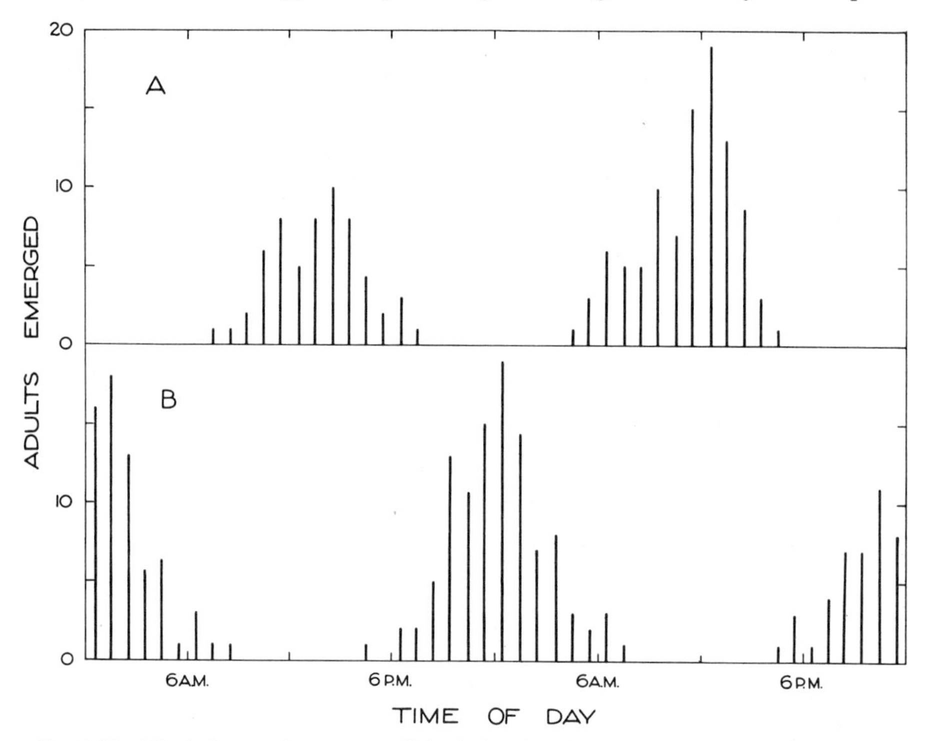

FIG. 8.17.—The influence of exposure to light during the adult stage on the "timing" of the pupal ecdysis in the dark in the next generation, in *Dacus tryoni*. *A*, adults were illuminated from 9:00 A.M. to 5:00 P.M. only; *B*, adults were illuminated from 9:00 P.M. to 5:00 A.M. only. (After Bateman, unpublished thesis.)

to light and darkness. In complete darkness, only 52 per cent hatched, as compared with 75 per cent in alternating darkness and light. In constant blue light, 83 per cent hatched; in continuous red light, only 62 per cent hatched. Since eggs which failed to hatch in the dark had fully formed embryos, Hueck suggested that the light provided the stimulus for the fully developed embryo to break the egg membranes.

8.5 THE INFLUENCE OF LIGHT ON MATING AND OVIPOSITION IN INSECTS

Observations in the laboratory and in the field have shown that the fruit fly *Dacus tryoni* mates only at dusk. Before they mate, the flies behave in a characteristic way; both sexes fly actively, and the males make a high-pitched "call." When the flies were kept in cages at constant temperature but exposed to natural light, the characteristic premating behavior began at dusk and continued for about an hour. Most of the flies took part in this activity, and during this period some of them copulated. But at no other time did any of the flies become sexually active. Both the mating behavior and copulation were confined to the period of dusk. If during the half-hour of dusk, when flies were sexually active, a light was turned on, they rapidly abandoned their activity, only to resume it again when the light was switched off (Myers, 1952; Barton-Browne, unpublished thesis). Barton-Browne kept flies in cages at constant temperature and humidity under an illuminance of 230 lumens per square foot during the daytime. At night the lamps were switched off. At the end of the light period, natural dusk was simulated by screening the lamps, so that the flies were under an illuminance of about 1 lumen per square foot. This induced normal sexual activity. When the hours of light were reversed, so that the flies were illuminated in the night and kept in the dark during the day, sexual activity was successfully induced by screening the lamps at the end of the night period at 5:00 A.M. But the flies did not become sexually active when lights were screened at the end of a period of darkness. The response to low illuminance occurred only when flies had been exposed during the "day" to high illuminance.

Some species of *Drosophila* mate only in the light. When adults of *D. auraria* were kept constantly in the dark, none mated. Two lots of *D. rufa* were kept in cages for 7 days, one in darkness, the other in light. In the first cage only 32 per cent of the females had mated by the seventh day, but in the lighted cage, 72 per cent. Similar results were obtained with *D. simulans* (Spieth and Hsu, 1950). But *D. subobscura* was like *D. auraria*, since none mated when they were kept continuously in the darkness (Rendel, 1945). Wallace and Dobzhansky (1946) counted the number of matings among 22 pairs of *D. subobscura* which were exposed to illuminances of 10, 27, and 85 lumens per square foot. In 2 hours they observed 1, 8, and 2 matings, respectively, and they concluded that for this species there was an optimal illuminance at about 30 lumens per square foot.

The beetle *Aphodius howitti* mates only at dusk. Carne (personal communication) has observed the beetles emerge from dung-pads at sunset, when there is a sudden drop in illuminance to about 2 lumens per square foot. They come together on the surface of the soil and copulate, after which they climb grass stalks, from which they fly off in large numbers within a matter of a few minutes. The hepialid *Oncopera fasciculata* is also crepuscular. Mating follows a characteristic nuptial flight, which takes place for a brief period in the evening. The flights commence abruptly a few minutes after sunset; about 6:00 P.M. in early September and about 6:40 P.M. early in October (southern spring). The illuminance when flight commences usually comes within the limits of 1 to 1½ lumens per square foot. The flights last for about 15 minutes and then come to an end almost as abruptly as they had started (Madge, personal communication). Many species of mosquitos are known to form mating swarms during late afternoon or at dusk. Bates (1949, p. 55) induced swarming in *Anopheles superpictus* by reducing the illuminance to 1 lumen per square foot. Once a swarm had formed, it persisted when the illuminance was increased up to 50

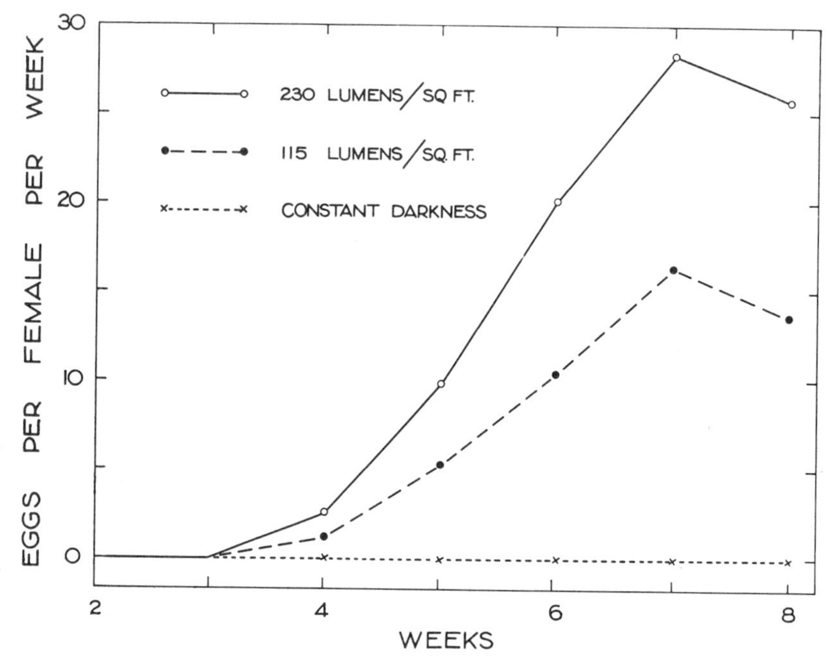

FIG. 8.18.—The fecundity of the fruit fly *Dacus tryoni* at 25.6° C. and 75 per cent relative humidity when fed on a pawpaw mixture and exposed to darkness and two other levels of illuminance. (After Barton-Browne, unpublished thesis.)

lumens per square foot; but when the lamp was turned off, the swarm dispersed immediately. Nielson and Greve (1950) observed that *Aedes cantans* and *Chironomus plumosus* formed swarms at dusk and at dawn, but these swarms consisted entirely of males.

In some insects oviposition may be stimulated by exposure to light of a particular illuminance or to darkness. Barton-Browne (unpublished thesis) kept the fruit fly *Dacus tryoni* in cages under artificial lights, which were switched on for 8 hours to give 230 lumens per square foot, then screened for a brief period to give a dull light simulating dusk, and then switched off, leaving the flies in darkness for 16 hours; temperature and humidity were kept constant throughout. Under these circumstances 97 per cent of all eggs were laid during periods when the lights were on. Isely and Ackerman (1923) found that the codlin moth *Cydia pomonella*, unlike *Dacus*, laid most of its eggs during darkness. Fifty-five moths which were kept in an insectary exposed to natural light laid 93 per cent of their eggs between sunset and dawn. When moths were kept continuously in darkness, eggs were laid more regularly throughout the day and night. When moths were exposed to alternating 2-hourly periods of light and darkness, 72 per cent of the eggs were laid in the dark. Oviposition could be started at any time by putting the moths in the darkness and stopped at any time by putting them in the light. The cotton-boll worms *Earias* spp. and moths of the genus *Amsacta* also lay most of their eggs during darkness or twilight (Pruthi, 1940).

Little is known about the influence of light on the maturation of eggs in insects. Barton-Browne (unpublished thesis) kept the fruit fly *Dacus tryoni* in cages made of "perspex," and exposed them to illuminances of 0, 115, or 230 lumens per square foot; temperature and humidity were kept constant. No eggs were laid in the dark, and considerably more in the brighter light than in the duller one (Fig. 8.18). When the eyes were painted with black paint, no eggs were laid. When the illuminance was increased, the flies copulated more often (though always during the period of dusk), and they ate more food while exposed to the bright light. These two activities together accounted for the increased number of eggs laid under high illuminance compared with the number laid under low illuminance. Unlike the fruit fly, the bean weevil *Bruchus obtectus* laid fewer eggs as the illuminance was increased (Menusan, 1935). This is illustrated in Table 8.05.

TABLE 8.05*

NUMBER OF EGGS LAID BY BEAN WEEVIL *Bruchus obtectus* AT 27° C. AND 90 PER CENT RELATIVE HUMIDITY EXPOSED TO DIFFERENT ILLUMINANCES

Candle-Power of Light 2.5 Feet from Beetles	Total Eggs Laid
120	30
50	35
15	42
Dark	67

* After Menusan (1935).

8.6 THE INFLUENCE OF LIGHT ON GROWTH

We discussed (in sec. 8.1) the influence of the photoperiod on the maturation of the gonads and the incidence of diapause. In this section we mention several other ways in which light may influence growth. The larvae of *Tenebrio molitor*, which usually live in the dark, developed more rapidly when exposed to the light from a 10-watt lamp (Cotton, 1930). Waitzinger (1933) kept some larvae of the silkworm *Bombyx mori* in the dark and others in the light; he found that they developed from the third to the fourth instar in 5 days in the dark, compared to 3.5 days in the light. Kogure (1933) found that when first- and second-instar larvae of *B. mori* were reared in light at temperatures above 30° C., the proportion of larvae which molted three times as compared with four was 7 per cent in the light and 0.5 per cent in the dark. Exposure to light for less than 12 hours at a time was equivalent to darkness. At temperatures below 28° C. very few three-molters were produced in either light or darkness.

The reported necessity of light for the vigorous growth of reef-forming corals may be associated with the requirements for light of the commensal zoöxanthellae. Vaughan (1919) found that, after 43 days in light-tight boxes, 5 out of 18 species of coral were dead and all were colorless from loss of zoöxanthellae. It is thought that the coral benefits from association with zoöxanthellae because they release oxygen (Vaughan, 1919; Verwey, 1930). It is difficult to explain the interesting observation of Nakamura (1941) that cut pieces of the colonial hydroid *Syncoryne nipponica* required light in order to regenerate; not one of 700 pieces regenerated in the dark. When exposed for 70 hours to light from an electric globe, such pieces usually regenerated successfully. Hydroids which had regenerated in the light were reabsorbed if they were subsequently placed in darkness or diffuse light of low illuminance.

8.7 SUMMARY

The influence of light on animals has been discussed in relation to the following qualities of light: illuminance, direction, photoperiod, wave length, and degree of polarization. Illuminance and direction may influence the daily rhythms of feeding, dispersal, and mating. Directed movement in response to light may serve to bring the animal to the right place at the right time. Perception of differences in wave length and degree of polarization may be used by certain invertebrates in finding their way about.

The photoperiod serves in some species to synchronize the life-cycle with the seasons. This has been well shown for birds and mammals, in which the maturation of gonads depends upon photoperiod; and in some insects the inception of diapause depends upon photoperiod. In a number of marine invertebrates, notably polychaetes and mollusks, strikingly well-defined breeding seasons depend upon the periodicity of the moon.

The experiments and observations which have been discussed in this chapter serve chiefly to show how much more there is to be found out about the influence of light on animals. Such studies may turn out to be highly rewarding to the ecologist.

CHAPTER 9

Other Animals of the Same Kind

There is no doubt that a rational study of the struggle for existence among animals can be begun only after the questions of the multiplication of organisms have undergone a thoroughly exact quantitative analysis. . . . We must therefore begin by analysing the laws of growth of homogeneous groups consisting of individuals of one and the same species, and the competition between individuals in such homogeneous groups.

G. F. GAUSE, *The Struggle for Existence* (1934)

9.0 INTRODUCTION

WE SAID in sections 3.0 and 3.132 that species or populations were characterized by a quality called the "innate capacity for increase," or r_m. We defined r_m as the maximal rate of increase attained by a population of stable age-distribution at any particular combination of temperature, moisture, quality of food, and so on, when the quantity of food, space, and other animals of the same kind are kept at an optimum and other organisms of different kinds are excluded from the experiment. In section 3.4 we pointed out that r_m is an abstraction from the actual rate of increase r, which may be observed in real populations either natural or artificial, such as those which are reared in laboratories. The actual rate of increase r may be influenced by variations in a number of components of the environment which are either excluded from or assumed constant in the definition of r_m. In this chapter we consider one special cause of variation in r, namely, the numbers of other animals of the same kind. For this it is necessary to consider not the absolute numbers but rather the numbers per unit area, which is another way of saying the density of the population.

This involves an analysis of the two phenomena "underpopulation" and "crowding." The value of r reaches a maximum when there is neither underpopulation nor crowding and falls away on either side of this point. "Underpopulation" has scarcely been studied seriously by animal ecologists, and in discussing this phenomenon we shall have to be content with just a few examples of the different ways in which the value of r may be reduced because there are too few other animals of the same sort in the environment. On the other hand, "crowding" has received a relatively enormous amount of attention, and this subject we shall be able to discuss quite exhaustively.

We do not believe that this perspective reflects the relative importance of the two phenomena: the reverse may be nearer the truth. General experience and observation of nature have convinced us that many, perhaps most, natural populations suffer either periodically or permanently from the disadvantages of underpopulation. In nature, abundant species are exceptional; most species are rare, and it seems likely that the distribution and abundance of many species are determined partly by the sorts of phenomena which we discuss in the next section under the heading "Underpopulation." On the other hand, "crowding" is relatively rarely observed in nature. When it does occur, it is usually a very different phenomenon from the one which is usually studied in experimental models in the laboratory. For example, it would be difficult to find a natural population which remains crowded in a limited space but with a supply of food maintained consistently at the same level.

One important attribute of a crowd will not be discussed in this chapter, because it is more appropriately considered in relation to food (chap. 11). Large numbers of animals require more food than few do, and therefore crowding may result in an actual shortage of food. In this chapter we are concerned with other aspects of crowding and therefore consider only situations in which food is artificially kept plentiful.

In section 9.1 we show that the value of r may be reduced when the density of the population is low; and we analyze some of the ways in which this result may be brought about.

9.1 UNDERPOPULATION

9.11 *Reduction of r in Sparse Experimental Populations*

Certain laboratory experiments with Protozoa, Insecta, and *Daphnia* have shown that either the rate at which young were produced or the duration of reproductive life was reduced as the density of the experimental populations decreased. For example, Robertson (1921) reported that for certain ciliates the rate of division was greatest at certain optimal densities and that it fell away as the density decreased. This report stimulated much discussion and led to many other experiments with different species of Protozoa. These have been adequately summarized by Allee *et al.* (1949) and Hutchinson and Deevey (1949), who concluded that the reproductive rates of some Protozoa do vary with the size of the inoculum, there being an optimal density at which the rate is fastest. Robertson postulated that a stimulating substance was released into the medium by the reproducing organisms. There is, however, no direct evidence of this. Chapman (1928) and Allee (1931) noticed that the rate of multiplication in colonies of the flour beetle *Tribolium confusum* was slow when there were few beetles in the population but increased to a maximum as the density of the population approached an optimum. Park (1933) showed that this was due to a reduction in the rate at which the beetles produced eggs in the sparse populations, because in these circumstances the sexes met rather infrequently and a female often waited a long time before she was fertilized. Similar results have been reported from laboratory experiments with the Indian meal moth *Plodia interpunctella* (Snyman, 1949) and the flour moth *Ephestia kühniella* (Ullyett, 1945). With *Drosophila*, Pearl *et al.* (1927) found that the adult flies lived longest when there were from 35 to 55 of them living in a 1-ounce vial. Their average life was shorter when fewer flies than this were kept in a vial. Pratt (1943) similarly found an optimal density at which the longevity of *Daphnia magna* was greatest. Neither of these last two observations has been followed up to the point where it might be explained. We do not know of any other experiments which demonstrate a reduction in longevity associated with a sparse population, but this phenomenon can sometimes be observed in natural populations.

9.12 *Reduction of r in Sparse Natural Populations*

In natural populations the reduction in the value of r associated with the increasing sparseness of the population may be carried so far that r becomes negative and the population proceeds to dwindle to extinction. There are two aspects to this phenomenon. One is the final extinction of a population which has been well established in an area but which has been brought down to low numbers by natural vicissitudes or the deliberate destructiveness of man. The other is the failure of a small colony of immigrants to become established in a new area which is favorable in all respects except for the sparseness of the colonizing population. These two phenomena must be going on around us all the time, but they are mostly missed because they are so difficult to see. For obvious reasons, underpopulation is a phenomenon which is not readily measured in nature. It is difficult enough to gather precise information about the numbers of animals that are numerous, and very much more so when they are scarce. Yet the effort should and must be made. This is, at present, one of the most important and most neglected branches of population ecology.

A well-known example of the extinction of an established population relates to the heath hen, which used to live in an area in Massachusetts known as Martha's Vineyard. We take our account from the one given in Allee *et al.* (1949). Observations began in 1880, when the birds were relatively abundant. They were hunted severely, and by 1890 there were only 200 birds in existence. After 1890 the heath hen was protected, and by 1916 the numbers had risen to 2,000—still quite a sparse population for this area. That year a combination of adverse circumstances reduced the population to 50 breeding pairs, and after that the numbers continued to decline until in 1928 only one male could be found. The record of events during this period suggests that the hunting which took place up to 1890 depleted the population to a critical low level from which it did not permanently recover, despite the introduction at this late stage of a conservation program.

We do not know of any other example so well documented as this one. But Allee (1938) estimated that 25 is about the smallest herd of elephants that can persist in South Africa. If the numbers become fewer than this, the animals tend to die out. Similarly, Allee *et al.* (1949) estimated that the minimal number of reindeer that can survive as a single population in the Eurasian tundra is between 300 and 400; when the herd becomes smaller than this, it tends to decrease and eventually dies out. Entomologists, experienced in the eradication of the tsetse fly from selected areas in South Africa, consider that it is not necessary to aim at complete eradication by artificial measures, since a small population will eventually die out if left alone. There is evidently a threshold below which the population cannot survive, but in this case we do not know what this threshold is. Ford (1949, p. 313) was able to measure the change in numbers of three colonies of the butterfly *Maniola jurtina*, two of which were much larger than 1,000 and one was under 500. The survival-rate in the small population was substantially less than that of the larger one. Ford suggested that colonies much smaller than about 1,000 individuals may be liable to extinction.

Another aspect of this same phenomenon is exemplified by the difficulty experienced by immigrants into new territory in establishing themselves as a permanent colony. We know very little about this because it has hardly been studied. Practical entomologists, using the method of biological control against insect pests and weeds, generally accept it as a sound maxim that if there are only limited numbers of a predator, it is best to liberate them all in the same place and at the same time rather than to distribute them more thinly over a wider or more diverse area (Clausen, 1951, p. 8). This is because experience teaches that it is often difficult to build up even in the one place a density of population which will exceed the threshold required to give the colony a reasonable chance of becoming established. There are exceptions to this rule, which only serve to emphasize the complexity of the problems that we are studying. For example, a tachinid fly is reported to have become established in Hawaii after the liberation of only nine individuals (Sweetman, 1936). Taylor (1937) commented as follows on *Pleurotropus parvulus*, which is a successful predator of the coconut leaf miner in Fiji: "It was everywhere evident that if the parasites are already present in every tree (even if there is only one per tree) at the beginning of the generation of the pest they can entirely suppress it before the end of the generation." It is not possible to discover from the literature on biological control just what have been the minimal numbers required to establish a colony in those cases where new species have been successfully introduced into new territories. But it is clear that in some instances small numbers have sufficed, whereas in many others large numbers were required.

The achievements of quarantine services in delaying or preventing the establishment of pests in new territories is further evidence of the truth of the principle which we are discussing. When, as sometimes happens, total prohibition of the import of certain articles would cause intolerable hardship, quarantine authorities often allow their import under conditions designed to reduce to a minimum the numbers of the pest species that come in. This is sound practice, for if the numbers that slip past quarantine are low enough, the odds against a population's becoming established may be high indeed. The rabbit did not slip into Australia past quarantine; it was introduced intentionally! Today it is enormously abundant, ranking perhaps as the major pest of pastures. Yet the first attempt to establish a local population failed, and a second lot had to be brought from England.

Some ornithologists consider that the proportion of nesting success with some birds increases with the size of the colony (Darling, 1938). Conversely, in small colonies this may be so small that r becomes negative and the colony dwindles and dies. For example, it is well known that small colonies of the herring gull *Larus argentatus* are unable to establish themselves in situations where large ones would have no difficulty. Another example is the fulmar, *Fulmaris glacialis*. There is, on the average, a lapse of 4 years (in some few cases 10 years) from the time of arrival of the first birds in a new area to the time that the first egg is laid. During this time further immigrants arrive to swell the numbers of the colony, suggesting that a minimal number has to be present before the colony can begin to breed. There is some evidence that mature birds tend to fly back to colonized areas and that the birds reaching new areas are young and immature. This in itself might account for the slowness of new colonies in commencing breeding (Fisher and Waterson, 1941).

So far in this section we have discussed the failure of sparse populations to maintain their numbers in situations which seemed favorable in all respects except for the low density of the population. And we have argued that this has happened because, somehow or other, in these situations r was reduced to a negative value by the adverse influence of that component of the environment which we have called "other animals of the same sort." The examples discussed so far do not give much indication of how this may have been brought about. If we were to try to write down, drawing first upon experience but enlarging this with imagination, all the sorts of situations in which the value of r might be reduced by the lack of a sufficient stock of other animals of the same sort, we might arrive at something like the headings in the following sections.

9.13 *Reduced Birth-Rate in Sparse Populations Owing to Small Chance of Mating*

When only a few animals are present in a large space, a female, on reaching the reproductive stage, may have a relatively small chance of meeting with a male which has also reached the reproductive stage at that time. In this case many females may remain unfertilized, and the average fecundity may be much lower than it would be in a denser population.

Milne's (1950*b*) careful and thorough study of natural populations of the sheep tick *Ixodes ricinus* provides a good example of this principle. The adult sheep tick, when it is ready to feed and mate, emerges from the matted vegetation close to the soil, where it has been sheltering, and climbs to the tip of a grass stem. If a sheep brushes the grass stem in passing, the tick climbs on the sheep. There it mates if, by good fortune, there happens to be a member of the opposite sex close to it on the same sheep. If no sheep comes by, within 4 or 5 days, the tick is forced, by desiccation, to return to the moist, matted vegetation near the surface of the soil. On the average, it spends about 9 days exposed on the tip of a grass stem. Usually this is made up of several periods of 2 or 3 days each. If it fails to become attached to a host during one of these periods, it dies without leaving any progeny. With sheep grazing at a density of about 1 to the acre, which is usual in some of the areas studied by Milne, many ticks fail to become attached to a host. On one pasture which Milne estimated to contain about 150,000 female ticks, a flock of sheep grazing throughout the season picked up about 30,000 ticks; so that about 80 per cent of the ticks failed to become attached to a host. This proportion will be dependent upon the number of sheep grazing in the area but will be largely independent of the density of the tick population. That is to say, that so long as the number of sheep grazing in the area remains constant, about the same proportion of the total ticks present will be picked up independent of the density of the tick population. But the actual numbers picked up by the sheep and therefore the average number per sheep will depend on the density of the tick population (sec. 13.32).

Not all those that are picked up will be able to find a mate. Some will be on sheep on which there is no member of the opposite sex, or not one that can be found on such a large host as a sheep. Moreover, a male copulates only once, so that each male may fertilize only one female; as the sexes are often picked up in unequal numbers, this leads to considerable wastage, which increases rapidly from this cause as the density of the population decreases. A numerical example will make this clear. First it is necessary to consider the distribution of the sexes on sheep that pick up 2, 3, 4, . . . , n ticks. We know from Milne's observations that the sexes are picked up at random. Therefore, it is appropriate to use the expansion of the binomial $(p + q)^n$ to describe the distribution of the sexes on the sheep. In this expression, p is the probability that any tick which is picked up is a male and q is the probability that it will be a female; p and q are proportions which add up to unity, and n is the number of ticks picked up per sheep. We know also that the sex ratio is unity. Therefore, the binomial becomes $(\frac{1}{2} + \frac{1}{2})^n$. When n is 2, the three terms of the expanded binomial are in the proportion 1:2:1. This means that if we had a sufficiently large sample of sheep, each carrying 2 ticks, we would expect to find 2 males on 25 per cent of the sheep, a male and a female on 50 per cent, and 2 females on the remaining 25 per cent. In this case only half the females would have mates present on the sheep on which they had become attached. The remaining half of the females would have no chance of being fertilized and would be destined to die without leaving any progeny. When there are 4 ticks to each sheep, the probability that a female will have a potential mate present on the same sheep is 62.5 per cent. The theoretical distribution of the sexes and the subsequent steps in the simple calculations necessary to estimate this probability are best followed in Table 9.01. The third column of the table gives the

TABLE 9.01*
STEPS IN ESTIMATION OF PERCENTAGE OF FEMALE TICKS WHICH HAVE POSSIBLE MATES WHEN SHEEP PICK UP 4 TICKS EACH

No. of Ticks per Sheep		No. of Sheep	No. of Possible Pairs	Total No. of Ticks
Females	Males			
0	4	1	0	4
1	3	4	4	16
2	2	6	12	24
3	1	4	4	16
4	0	1	0	4
Total		16	20	64

* The third column of the table gives the expansion of the expression $(\frac{1}{2} + \frac{1}{2})^4$.

expansion of the expression $(\frac{1}{2} + \frac{1}{2})^4$. The bottom row shows that, of the 32 females present on the 16 sheep, only 20 (62.5 per cent) occur on sheep on which there is also a male that may be available to mate with them. Similar calculations may be made for any number of ticks per sheep: Table 9.02 gives them for numbers from 1 to 12.

TABLE 9.02
PROBABILITY THAT FEMALE TICK WILL HAVE POTENTIAL MATE ON SHEEP AS NUMBER OF TICKS ON SHEEP VARIES FROM 1 TO 12

No. of Ticks per Sheep	Probability that Female Will Be Matched by Male on Same Sheep	No. of Ticks per Sheep	Probability that Female Will Be Matched by Male on Same Sheep
1	0.000	7	0.688
2	.500	8	.727
3	.500	9	.727
4	.625	10	.754
5	.625	11	.754
6	0.688	12	0.774

Now we are in a position to consider what happens in a natural population. The sheep will carry different numbers of ticks. The numbers will be distributed about a mean with a particular variance. Suppose we have 100 sheep, and the mean number of ticks per sheep is 10; let us suppose further that this mean has a variance of 0.7. The first and second columns of Table 9.03 give an arbitrary set of figures which fulfil these postulates. We might have chosen any other values for the variance, but this would not materially influence the argument to be developed from this example. The figures in the third column are taken from Table 9.02, and those in the fourth column were found by taking the product of the first, second, and third columns and dividing by 2, since only half the ticks were females. It will be seen from the bottom row of Table 9.03 that, of the

TABLE 9.03
NUMBER OF FEMALE TICKS WITH POSSIBLE MATES WHEN 100 SHEEP HAVE MEAN NUMBER OF 10 TICKS PER SHEEP, WITH VARIANCE OF 0.7

No. of Ticks per Sheep	No. of Sheep	Proportion of Females with Possible Mates	No. of Females with Possible Mates
8	5	0.727	14.54
9	15	.727	49.08
10	60	.754	226.20
11	15	.754	62.20
12	5	0.774	23.22
Total	100	....	375.24

500 females on the 100 sheep, only 375, or 75 per cent, have a chance of being fertilized. Now we may reasonably compare this population with another in which the variance remains constant at 0.7 but the mean varies. When this is done, we get the very interesting results given in Table 9.04.

TABLE 9.04
PROPORTION OF FEMALE TICKS WITH POSSIBLE MATES WHEN MEAN NUMBER OF TICKS PER SHEEP VARIES BUT VARIANCE REMAINS CONSTANT AT 0.7

Mean No. of Ticks per Sheep	Proportion of Females with Possible Mates, i.e., Those "Matched" by Male on Same Sheep	Mean No. of Ticks per Sheep	Proportion of Females with Possible Mates, i.e., Those "Matched" by Male on Same Sheep
1	0.200	5	0.637
2	0.475	10	0.751

It is clear from Table 9.04 that the proportion of females that become attached to sheep on which there is a possible mate for them becomes smaller as the population becomes sparser. Moreover, the decrease is not linear with the density of the population but tends to become relatively greater as the population becomes sparser. Halving the density from 10 per sheep to 5 reduced the proportion of females with possible mates from 0.75 to 0.64; but the actual numbers of females with possible mates on the same sheep would be reduced in the ratio 75:32. On the other hand, halving the density of the population from 2 to 1 tick per sheep reduced the proportion of females with possible mates from 0.48 to 0.20; but the reduction in absolute numbers would be in the ratio 48:10. A decrease to half the density from 10 to 5 per sheep may have reduced the absolute number of mated females to about one-half, whereas reducing the density by one-half from 2 to 1 per sheep reduced the absolute numbers of ticks that may have been mated to about one-fifth. It is not known how low the fecundity must be reduced before the birth-rate becomes inadequate to match the death-rate from the many hazards which beset the tick in the various stages of its life-cycle, but it is clear that there must be a threshold low density below which r will become negative. Populations that are less dense than this will be unable to maintain themselves or to increase; they will dwindle and die out. At slightly greater densities, r, though positive, may be so small that the rate of increase of the population may be extremely slow.

Milne (1950*b*) found plenty of evidence that both these things do, in fact, happen in nature and are, indeed, quite characteristic of natural populations of *Ixodes*. Certain features of the ecology of *Ixodes* would be (and, indeed, have proved to be) most puzzling in the absence of this sort of theoretical background. Milne found instances where apparently suitable pastures had remained uninfested for decades, although separated by no more than a sheep fence from similar ones that had been carrying a dense population of ticks for many years. And he concluded: "In these cases, the few ticks brought on by wild hosts with overlapping territories must fail to provide the critical level of adult population necessary for primary tick establishment. Unusual wild fauna conditions, such as, for instance, a 'plague' of hares might occasionally tip the scale, but the writer finds that spread can invariably be traced to straying or deliberate moving of sheep in sufficient numbers. . . . A good sheep fence can protect a farm from neighbouring ticks, at least for a long period." He was able to confirm this with one striking example. In 1940 a certain field suddenly became infested with ticks, although it had been free from them up to that time, notwithstanding the fact that the adjacent field (belonging to another farm) had been heavily infested for at least 30 years. In 1940 a break occurred in the fence and escaped notice for some time, allowing a number of sheep to stray from the infested to the uninfested field.

On other occasions Milne noticed that "on hills and moorlands there are some pastures with a lower population than they ought to have considering the amount of cover and host-potential available." These are usually populations which are in the early stages of building up their numbers, which is a very slow process with *Ixodes* when it starts with a small colony in new territory. An old shepherd who was also a very good naturalist told Milne of the introduction of *Ixodes* into an area that had long been free of the pest and was able to give an authentic account of the very slow rate of multiplication for a number of years after the first introduction into the new area. Milne's account of this is given below.

Before 1860–70, the tick was unknown in College Valley and for miles around. About that time it was introduced to one pasture of Westnewton Tors, a farm contiguous with Hethpool. It came on a flock comprising sheep bought at May Day from the notorious infested areas of North Tyne in England and of south-west Scotland, "where ticks had always been." Since then, partly by transferring of flocks and partly by sheep-straying due to fence disrepair, the tick had gradually encroached upon pasture after pasture and farm after farm until today about 15 to 20 farms in that area are infested. Their rate of spread was slow. The original focus, Westnewton Tors, has been infested for about 80 years. Farms within a radius of roughly 2 or 3 miles have been infested for 30 to 60 years. Outside that, up to a radius of 5 to 6 miles from the centre, the infestations date from only 5 to 20 years back, and some farms are not infested. When Douglass (the shepherd) first came to work on Hethpool, one of its pastures, the Ewe Hill, was still free of ticks. He unwittingly infested the Ewe Hill by transferring to it a flock from one of his "ticky" pastures during high activity. Over the years he watched the tick population grow on the Ewe Hill until now it is the most heavily infested pasture on the farm.

The above example indicates the broad reality of spread. The mass of this kind of evidence shows that the tick colonizes most surely and its population comes into equilibrium soonest when a whole flock of sheep is transferred quickly from a heavily tick-infested to a suitable tick-free farm during peak activity (as about May Day). . . . Even so, the build-up process is, according to hill-men, comparatively slow. Ticks seemingly do not become very numerous until 15 to 20 years after introduction.

When the population of *Ixodes* is sparse, as when they have been freshly introduced into new territory, relatively few females find mates, and consequently the birth-rate is low. With *Ixodes* the relationship between density of population and the proportion of females likely to find mates could be analyzed with the aid of some very simple mathematics (see above), because of three rather unusual characters of *Ixodes*, namely, (*a*) mating occurs only on the host; (*b*) the ticks are completely passive when seeking a host, and consequently there is a random distribution of ticks and sexes on the hosts; and (*c*) the ticks are largely monogamous. It is unlikely that many other species would fulfil the requirements of this simple mathematical model. With other species which do not require to mate on a host, which are more mobile and less passive in seeking mates, and in which polygamy is possible, the reduction in birth-rate in sparse populations due to a shortage of mates may be less pronounced, except at very low densities; certainly, the relationship will be more complicated to analyze. But it is likely that it occurs frequently in natural populations without having been noticed very often. We give below a few further examples (none so striking as *Ixodes*) which indicate that the phenomenon is at least widespread.

The beetles *Chrysomela gemellata* and *C. hyperici* were introduced from Europe into Australia in the hope that they might control the weed *Hypericum perforatum*. The early attempts to get the beetles established in the field were only partly successful, and there is some evidence that a large colony is more likely to become permanently established than a small one. Some beetles have been observed to fly away from established colonies during the spring. Many of these arrive in other favorable sites, but frequently they fail to become established in these new areas. Although the death-rate among the beetles is high everywhere, the migrants are not all destroyed before their normal reproductive period. A few beetles can often be found in these areas, but it is quite commonplace to find no eggs there. As the eggs, being more prominent than the adults, are the more likely to be found, it is reasonable to conclude that the insects have not laid eggs. This may happen in situations where there are no obvious features of the environment which would lead one to expect that the beetles would fail to establish a colony, except this, that each beetle has in its environment too few others of the same sort, so that its chance of mating may be quite small (Clark, personal communication). Clark has also noticed that in areas which are only sparsely populated the numbers gradually dwindle until the colony eventually dies out. There is circumstantial evidence that this is due to the fact that the beetles may be too widely dispersed to find one another during the mating period: for they are relatively immobile, moving only a few yards at a time when the food supply becomes exhausted or when they are forced into flight by unfavorably high temperatures.

Errington (1940, 1943) has noticed that sparse populations of the muskrat *Ondatra zibethicus* do not breed successfully and multiply. He estimated that the most sparse population that can breed successfully is about one pair to each mile of stream or to each 15 acres of marshland: when they are fewer than this, it seems that the males too often fail to find the females during their short period of oestrus. Ornithologists consider that the proportion of nesting success in some birds increases with the size of the colony (Darling, 1938). For example, social breeders like the herring gull, *Larus argentatus*, are stimulated to successful breeding by sexual displays. In small colonies the stimulus may be so slight as to be ineffective. This explanation is still conjectural, though the inability of small colonies to establish themselves is well known. We have already mentioned in an earlier section (9.11) that in laboratory cultures of three insect species that have been studied, *Tribolium confusum*, *Ephestia kühniella*, and *Plodia interpunctella*, the fecundity becomes less as the population becomes sparser. This is because of infrequency of mating in the sparser populations. An increasing proportion of females either remain unmated or have to wait longer for the arrival of a male as the population becomes sparser. The result is a reduction in average fecundity, which is, of course, reflected in a lower value for r.

9.14 *Increased Death-Rate in Sparse Populations Owing to Predation by Non-specific Predators*

An animal which may be preyed upon by a predator whose numbers are limited otherwise than by the abundance of this particular sort of prey, may have a greater chance of surviving when it is one of many than when there are only a few others with it.

Darwin appreciated this principle when he wrote in chapter 3 of *The Origin of Species:* "We can easily raise plenty of corn and rapeseed, etc., in our fields, because the seeds are in great excess compared with the birds which feed on them; nor can the birds, though having a superabundance of food at this one season, increase in number proportionally to the supply of seed, as their numbers are checked during winter; but anyone who has tried, knows how troublesome it is to get seed from a few wheat or other such plants in a garden: I have in this case lost every single seed. This view of the necessity of a large stock of the same species for its preservation, explains, I believe, some singular facts in nature, such as that of very rare plants being sometimes extremely abundant in the few spots where they do occur; and that of some social plants being social, that is, abounding in individuals even on the extreme confines of their range. For in such cases, we may believe, that a plant could exist only where the conditions of its life were so favourable that many could exist together, and thus save the species from utter destruction." Although this passage from Darwin refers only to plants, the principles seem equally applicable to animal populations.

Birch and Andrewartha (1941) described just such an occurrence in relation to the grasshopper *Austroicetes cruciata*. This species multiplies to plague numbers at irregular intervals; it has one generation a year, and, once a plague has developed, the high numbers may be maintained for several generations. Such a period of dense populations began in 1935 and endured until 1940. There were the usual large number of eggs present at the beginning of this generation, and the nymphs hatched in the spring without unusual mortality. But because of an unusually severe drought during the winter of 1940 which carried forward into the spring, the grass withered and dried everywhere except in the most humid places. Local depressions in the eroded plain or small flats at the base of a rocky slope which collected a little extra water after a light shower of rain carried enough grass to support a few grasshoppers, and those which happened to be in these places survived, while those elsewhere perished from starvation. None of these areas was extensive: some were no more than a square chain or two; and they were not numerous. The birds which prey upon *Austroicetes*, lacking the food to maintain large numbers during the remainder of the year, were not numerous enough to make any appreciable difference to the density of the grasshopper population during the favorable years. But during the drought of 1940 the result was completely different. There were scarcely enough grasshoppers to satisfy the birds, which congregated in the places where the grasshoppers had survived the drought, staying there until they had apparently sought out the very last grasshopper, with the result that a very high proportion of the insects was destroyed in this way. In other words, when the numbers of grasshoppers were high, as during 1935–39, each one had very little chance of being eaten by a bird. But in 1940 when the numbers were low, each grasshopper had an extremely high chance of being eaten. That is, longevity (and therefore r) had decreased as the number of other animals of the same sort in the environment had become fewer (see also secs. 10.33, 13.124).

9.15 *Reduced Death-Rate in Large Organized Herds or Flocks*

In the presence of predators, an animal may have a greater chance of surviving if it is one of a large organized herd or flock than if it is in a small herd or solitary.

Vogt gives evidence of a critical lower limit for successful breeding and multiplication in the Guanay *Phalacrocorax bougainvillei* (Hutchinson and Deevey, 1949). In small colonies the ratio of circumference to population size is larger than in large colonies, and the loss of chicks from the edge of the colony is therefore greater. The example we gave in section 9.1 of reindeer herds in the Eurasian tundra dying out if their numbers decreased below about 300 may be largely due to this cause, since ungulates generally seek to protect themselves against predators by the organized resistance of the herd as a whole. Leopold (1933, p. 86) stated that antelope failed to "come back" even after completely closed seasons for hunting had been attempted, because they had become too few to protect themselves against predators. Herds of less than 12–15 usually do not fight off wolves or coyotes as a herd. When such small herds are attacked, they usually stampede and scatter, so that the loss is great. Herds of more than 15 usually bunch into a defensive group and ward off attackers more effectively.

9.16 *Amelioration of Environment by the Action of Dense Populations*

It may happen that some component of the environment, being unfavorable for survival or multiplication, may be ameliorated by the combined action of many animals where few may fail to achieve this end.

Anyone who has noticed, on the one hand, the relative immortality of a large colony of ants or termites and, on the other hand, has witnessed the great numbers of individuals taking part in the nuptial flights of these insects will appreciate the enormous hazards that attend the establishment of a new colony. We have traveled in southern Australia for a distance of 50 miles while the air remained continuously thick with the nuptial swarms of the small ant *Iridomyrmex refoniger* and have seen comparable numbers of the termite *Coptotermes acinaciformis* in nuptial flights. Every year swarms of this magnitude supply more than enough potential parents to colonize every suitable nook and cranny in the area. Yet at all times in this area one may observe much suitable space and food not being used by these insects. One can only attribute this to a very high death-rate either at the time the single pair of individuals falls to the ground after the mating flight or up to the stage when the colony is still so small as to be invisible to the general observer. But once the colony has become visible in size, it is common experience that it survives for very long periods. It is clear that large colonies have a higher survival-rate than small ones or than single pairs of colonizing individuals.

In the absence of any investigation of this problem, we can only imagine an explanation. For instance, it could be that in this climate the particular component of environment which is unfavorable and which is ameliorated by the activity of a large colony is moisture: for it is well known that during the arid summer the interior of the termitarium is much more humid than either the air or the soil around. This is a result that could not be achieved by a pair or even a small colony. If, in general, this area is unfavorable because of its dryness, then we would expect a newly mated pair to be successful in establishing a permanent colony only in local situations that are much more humid than the general run of places available to them; or, alternatively, they might succeed in a less favorable situation during a spell of unusually humid weather. Once the colony has grown to a certain size, it may be relatively independent of this particular hazard and may live on in the situation which would normally be too dry for a small colony or a pair.

The social insects and other gregarious animals seem to derive security from their numbers because of their organization. But nonsocial animals may also derive similar advantages from the more random activity of large numbers. For example, the grain borer, *Rhizopertha dominica*, cannot be established in laboratory cultures of sound wheat unless a large number of animals are put together in the same jar. On the other hand, quite a small number of insects will suffice to found a colony on cracked or damaged grain. When the insects are placed with the sound grain, they begin to "nibble" it. When many insects are placed in the same jar, this "nibbling" proceeds to the point where at least some grain has been sufficiently "damaged" to become favorable for *Rhizopertha* before all the insects have died. At this stage, r becomes positive, and the survivors begin to multiply. On the other hand, when only a few insects are present with sound grain, the "nibbling" of the few is less effective, and they all die without having made any grain favorable for multiplication (Birch, unpublished data). The flour beetles *Tribolium confusum* and *T. castaneum*, though usually found in flour, can breed in damaged wheat. In the laboratory it is difficult to establish colonies in sound grain when only a few animals are introduced for a start, but it is easy when many animals are used. Here again the larger numbers may make the grain more suitable by damaging it (Birch, 1947).

Allee (1938, 1951) gave a number of examples of animals growing more rapidly in the presence of a number of their own species. The snail *Lymnaea* grew more rapidly in water which had been "conditioned" by other snails. This is possibly due to the dense growth of microflora (on which the snails feed) in the stale, as compared with fresh, water. Allee found that goldfish grew more rapidly when they were kept in water in which other similar goldfish had lived than when they were daily transferred to clean water. The fish are said to "condition" the water by regurgitating food and perhaps also by the secretion of mucus. Increased growth is evidently associated in some way with this "conditioning" of the water. But it is not certain whether it is due to the removal of harmful substances or to the liberation of some "growth substance" into the water.

9.2 THE INFLUENCE OF CROWDING ON THE VALUE OF r

As the density of a population increases from being very sparse, it may be considered to attain an optimal density at which the rate of increase becomes maximal. For a population of stable age-distribution living in certain specified circumstances (sec. 9.0), this maximal rate of increase would be equivalent to the innate capacity of increase (r_m). If the resources of food and space were unlimited, a population of stable age-distribution, having attained this density, would go on increasing indefinitely, with a rate of increase equal to r_m. The curve representing the growth of the population in absolute numbers (not density) would be the familiar exponential curve of the geometric progression. But when space is limited and the numbers in unit space (density) are plotted, instead of absolute numbers, the simple exponential curve would be replaced by a sigmoid one, at least during the early stages, while the colony is spreading into the available space. In nature the condition of the population is often complicated by changing (often diminishing) resources of food; but in this chapter we must exclude food from consideration and concentrate our attention on the situation in which sufficient food is maintained artificially and only space is limited.

This problem was first studied in relation to human populations; later the inquiry was extended to experimental populations of other animals. In 1839 the Belgian mathematician Verhulst suggested the use of a curve he called the "logistic curve" to describe the growth of human populations. The same mathematical expression was independently derived by Pearl and Reed (1920) as an empirical description of the growth of the population of the United States of America. A rational derivation of the curve was given by Lotka (1925), who also extended the theory to include the growth of two species living together in the same space. Volterra (1931), without knowing of Lotka's work, covered the same ground. Pearl (1926) applied the logistic curve to an experimental population; he reared the fly *Drosophila melanogaster* in milk bottles. He also used the logistic curve to describe the growth of populations of yeast cells and human populations. Since then, a number of experimenters have studied the growth of artificial populations of other species reared in a medium where temperature and moisture were kept constant and with a supply of food replenished at regular intervals. It soon became clear that the explanation which had been offered by the earlier authors was too simple. For example, it is now known that the *Drosophila* model is far more complex than Pearl thought it to be. Sang (1950) criticized the application of the logistic theory to this particular model.

The logistic theory is about the simplest that could be derived to describe the growth of a population in a limited space. Its simplicity has many theoretical and practical advantages. This has led to its being widely used as a starting point for general mathematical theories about the "struggle for existence" (Volterra, 1931; Lotka, 1932; Gause, 1934; Kostitzin, 1939; D'Ancona, 1942). Indeed, a just criticism might be that the general application of the logistic theory has gone ahead without a sufficiently critical examination of the theory itself. For these reasons it is worth while to re-examine the logistic theory to see how general it is and what use we may make of it. Fortunately, we have at our disposal the results of a number of recent experiments which have been carefully controlled.

We shall start by reviewing the theoretical derivation of the logistic equations and then consider how well these explain the empirical facts that are available. We shall leave Pearl's original example, *Drosophila*, on one side at first, because we now know that this is an especially complex model. Instead, we shall begin with some simpler examples from populations of unicellular organisms with simple life-histories, returning later to *Drosophila* and others with more complex life-histories.

9.21 *Theoretical Derivation of the Logistic Curve*

When a population of animals is growing in a limited space, the number of animals in unit space (i.e., the density of the population) rises. In due course the presence in the medium of many other animals of the same sort may reduce the fecundity, the longevity, and the speed of development of the individuals comprising the population. The precise way in which this may be brought about will be considered later. For the present it is sufficient to note that this is reflected in a reduced rate of increase of the population as a whole. According to the logistic theory, the reduction in rate of increase which occurs in these circumstances is assumed to be a linear function of density, i.e., the rate of increase r starts off at its maximum (which is r_m) when there is no crowding. Then for each individual added to the population, r_m is reduced by a constant amount; herein lies the basic simplicity of the logistic theory. It should be emphasized, because it is generally not appreciated, that this involves the assumption that every individual makes an equal contribution to the population. But we know that among many animals individuals of different ages have quite a different influence on the growth of the population; the addition of an egg of an insect to a population of insects cannot be considered equivalent to the addition of an adult. In other words, the theory takes no account of the possible influence of differences in ages of individuals in the population. This aspect of the theory will be discussed in detail in section 9.231.

We may express the simple proposition of the logistic theory in symbols. Thus the equation for the geometric progression (which gives the rate of increase, in absolute numbers, of a population growing in unlimited space) is

$$\frac{\delta N}{\delta t} = r_m N. \tag{1}$$

And the rate of increase of the same population growing in a limited space becomes

$$\frac{\delta N}{\delta t} = N(r_m - cN) = rN \tag{2}$$

(of course, to get relative increase, e.g., rate of increase per head, it is necessary to divide by N). In this expression N is the density of the population and c is the constant amount by which, we postulated, the rate of increase was reduced by the addition of one individual to the numbers already in unit space. The rate of increase thus continues to be reduced as a linear function of N as N becomes larger, until, at some maximal value of N, cN approaches r_m and $r_m - cN$ approaches 0.

Equation (2) is the differential form of the logistic curve. It is more instructive to include in the equation the symbol K, which stands for the maximal or asymptotic value which N approaches as t (time) approaches infinity. As N approaches this asymptote, the rate of increase approaches 0. So we can write: when $N = K$,

$$\frac{\delta N}{\delta t} = 0; \qquad \text{and} \qquad r_m - cN = 0;$$

hence

$$r_m - cK = 0$$

and

$$c = \frac{r_m}{K}. \tag{3}$$

Substituting in equation (2), we get

$$\frac{\delta N}{\delta t} = r_m N\left(1 - \frac{N}{K}\right)$$

$$= r_m N\left(\frac{K - N}{K}\right). \tag{4}$$

In fitting empirical data to the curve, it is convenient to use the integral form of the equation:

$$N = \frac{K}{1 + e^{(a - r_m t)}}. \tag{5}$$

In this equation, a is the constant of integration defining the position of the curve relative to the origin (Pearl, 1930, p. 420). The curve has an inflection at the co-ordinates:

$$t = \frac{a}{r_m}, \quad \text{and} \quad N = \frac{K}{2}.$$

Equation (5) may take the form:

$$\log_e\left(\frac{K - N}{N}\right) = a - r_m t. \tag{6}$$

It is derived from equation (5) by the following steps:

$$N = \frac{K}{1 + e^{(a - r_m t)}};$$

$$\therefore \frac{K - N}{N} = e^{(a - r_m t)};$$

$$\therefore \log_e\left(\frac{K - N}{N}\right) = a - r_m t.$$

This is the equation of a straight line with co-ordinates $\log_e (K - N)/N$ and t. The constant a is the value of $\log_e (K - N)/N$ when t is 0; and r_m is the coefficient for linear regression of $\log_e (K - N)/N$ on t. In fitting data to this equation, a value of K is found by trial and error which gives the best fit to the straight line. The procedure is described fully by Pearl (1930).

Two attributes of the logistic theory make it attractive: its simplicity and its "reality." The differential form of the logistic equation (eq. [4]) contains only two constants: r_m has a biological meaning as the maximal value of r for any set of conditions as defined by temperature, moisture, food, and so on, i.e., all the components of environment, excluding other animals (sec. 3.0); it seems reasonable to attribute to K a biological meaning, viz., the density at which the particular space under consideration becomes "saturated" by the animals being studied; and, in equation (2), c also has a biological meaning, in that it is the amount by which r_m is supposed to decrease for each additional animal added to unit space. This and several other assumptions will need to be a sufficiently close description of what the animals actually do if the apparent reality is not to prove fictitious; and simplicity may not be a sufficient recommendation if the theory does not fit the facts.

There are two ways to test this: (*a*) A colony of animals may be reared in a constant space with a constant supply of food. With the data so obtained, the constants of the best-fitting logistic curve may be calculated empirically, and the result examined to see how closely the observed data follow the trend of the calculated curve. We do this in section 9.22. (*b*) The several assumptions on which the logistic theory rests may be examined separately and perhaps compared with the conclusions that may be drawn from the appropriate experiments. We do this in section 9.23.

9.22 *Empirical Tests of the Logistic Theory*

9.221 ANIMALS WITH SIMPLE LIFE-HISTORIES

The growth of populations of *Paramecium aurelia* and *P. caudatum* has been studied under a number of conditions by Gause (1934). We have selected one experiment which was most carefully controlled. It is worth while to describe the method Gause used in this experiment because it illustrates very well what authors mean by a "constant environment" and a "limited space" when they write about experiments like this one. At the beginning of an experiment Gause placed 20 *Paramecium* in a tube containing 5 ml. of Osterhaut's medium, which is a mixture of salts carefully balanced in relation to the physiology of the animal. It was buffered to pH 8 with a phosphate buffer. Each day Gause added a constant quantity of bacteria, which served as food for the *Paramecium.* The bacteria were a pure culture of *Bacillus pyocyaneus* grown on a solid medium. When added to the salt solution, they did not multiply, so that the amount of food available for the *Paramecium* was known and controlled precisely. A standard, uniformly filled platinum loop of fresh bacteria taken from the solid medium was diluted in some Osterhaut's medium, and a fixed volume of this was added each day to the tube containing the *Paramecium* after the old medium had been removed. The quantity of bacteria in one set of experiments was twice that in the others. These are referred to as the "full-loop" and the "half-loop" experiments. The culture was incubated at 26° C. Every second day the culture was washed with bacteria-free Osterhaut's medium, to prevent the accumulation of waste products. The numbers of *Paramecium* in the tube were estimated once each day. The experiments were done in duplicate.

With the temperature and the volume and chemical composition of the medium kept constant, with the waste products removed at frequent intervals, and with food of a uniform quality renewed in equal quantities each day, it would seem that these experiments did indeed provide a "constant environment" in a "limited space." The only obvious criticism of the experiment is the small number of replicates.

The growth of four populations is shown in Figures 9.01 and 9.02. The observed points have been taken from Gause (1934, appendix, Table 4). Because of apparent discrepancies between Gause's diagrams and his tables of data, the curves have been recalculated. These have been drawn as smooth lines in Figures 9.01 and 9.02. There is a certain amount of subjective judgment re-

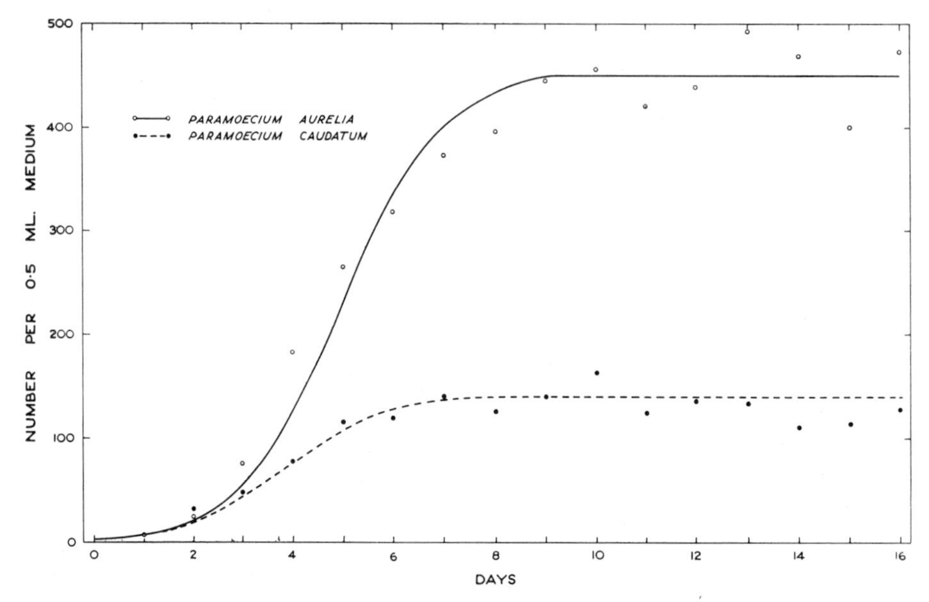

FIG. 9.01.—The trends in densities of populations of *Paramecium aurelia* and *P. caudatum* living in Osterhaut's "full-loop" media. The smooth curves are logistic curves calculated from observed points between 0 and 7 days (*P. aurelia*) and between 0 and 6 days for *P. caudatum* (Table 9.05). (After Gause, 1934.)

quired in the fitting of logistic curves, and for this reason it seems necessary to indicate how these curves were calculated.

The first step in calculating a logistic curve is to choose a value to serve as a first approximation to K. If, as in the present experiments, the data indicate an asymptote, the mean of these observations may be taken as K. The values of the constants a and r_m are then determined by the data on the ascending part of the graph. If the fit is not good enough, it may be improved by arbitrarily excluding one or several points from the calculation of a and r_m. Alternatively, any arbitrary value may be chosen for K which may not necessarily be related to the observed values in this part of the graph. The values of the constants a and r_m are then determined by the data in the ascending part of the graph; but if the fit is not good enough, it may be improved by varying the value of K or by excluding some points from the calculation of a and r_m. The latter is the only method available when the experiments have been stopped short or for some other reason do not enable K to be calculated directly. But when, as in Gause's experiments, this can be done, it would seem reasonable to use this value for K.

In fitting logistic curves to Gause's data, both these methods were tried, but by neither was it possible to get a well-fitting curve that included all the points. The curves in Figures 9.01 and 9.02 were done by the first method, because it is

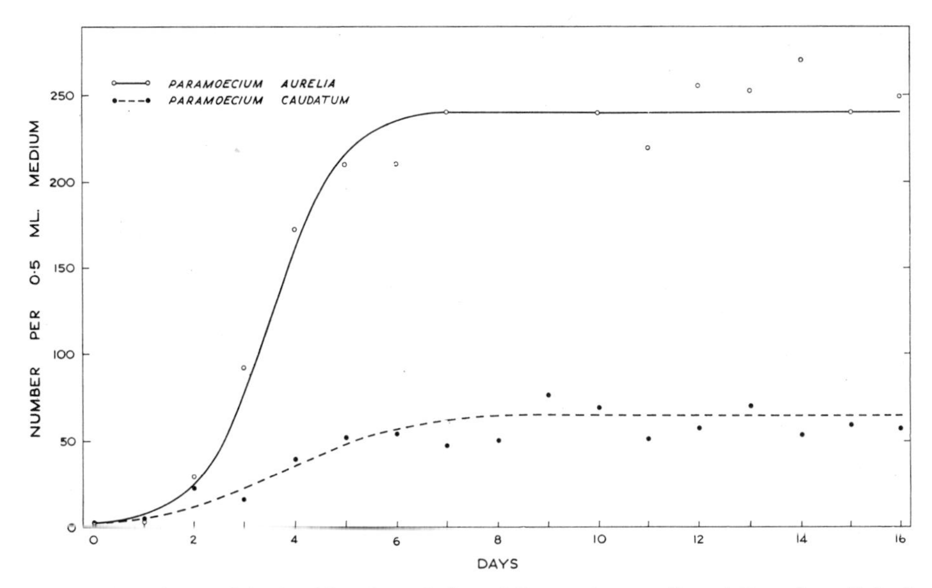

FIG. 9.02.—The trends in densities of populations of *Paramecium aurelia* and *P. caudatum* living in Osterhaut's "half-loop" media. The smooth curves are logistic curves calculated from observed points between 0 and 5 days (*P. aurelia*) and between 0 and 6 days for *P. caudatum* (Table 9.05). (After Gause, 1934.)

fairly clear that the observed points tend to follow an asymptote, once the initial phase of growth has been completed. There is certainly a fluctuation in this part of the graph, and it must be left to subjective judgment to decide where to draw the asymptote through the points; but there was scarcely any difficulty in making this choice, except for *P. caudatum* in the half-loop medium, where the fluctuations were rather greater than in the other experiments.

The growth of *P. aurelia* in the full-loop medium may be given as an example, since the calculated curve was not such a good fit in this experiment as in the others and the calculations involved more subjective judgment. A value of 450 was assigned to K, as this fitted well with the trend of the observed points from 10 to 16 weeks. A logistic curve was then calculated from the data covering the whole period that the population was increasing, i.e., up to 10 weeks ($N = 454$). But the calculated curve was a very poor fit to the observed points. By excluding first the point for $t = 9$ and then those for $t = 9$ and $t = 8$ and so on, it was found that the best fit was obtained when the curve was calculated from the observed points between $t = 0$ and $t = 7$, inclusive. Even so, Figure 9.01 shows that the calculated curve does not conform very closely to the observed points, although it follows the general trend of the observed points perhaps faithfully enough. A better fit, to a very much smaller part of the data, was obtained when K was estimated by the second method. By trial and error a curve that fitted very closely to the first 5 observed points was found when K was taken equal to 525 and a and r_m were calculated from the first 5 observed points only. With this curve, all the observed points from 6 weeks to the end of the experiment at 16 weeks came well below the calculated curve. Such a curve might be justified if we were chiefly interested in the influence of crowding on the value of r and wanted to estimate this precisely for a particular range of densities (sec. 9.241). But at present we are concerned with the general logistic theory; and, in view of the small range covered by this curve and the obvious indication of an asymptote at $N = 450$, the first method was preferred. In each of the four examples shown in Figures 9.01 and 9.02 the numbers of observed points used in calculating the curve are indicated in the legends of the figures. The values of r_m shown in the calculated curves in Table 9.05 are slightly different from those calculated from the same data by Gause (1934, Table 11) and different again as compared with Kostitzin (1939, p. 85). It is unfortunately not possible to tell from their publications how their curves were calculated.

TABLE 9.05

EQUATIONS FOR LOGISTIC CURVES FITTED TO GAUSE'S DATA FOR *Paramecium*

	Full-Loop Medium	Half-Loop Medium
P. caudatum........	$N = \dfrac{140}{1 + e^{3.894-1.015t}}$	$N = \dfrac{65}{1 + e^{3.185-0.855t}}$
P. aurelia..........	$N = \dfrac{450}{1 + e^{5.041-1.022t}}$	$N = \dfrac{240}{1 + e^{5.101-1.460t}}$

The two estimates of r_m for *P. caudatum* are 1.015 and 0.855. If we regard the difference between these two as experimental error, we may combine them for a more reliable estimate of $r_m = 0.935$. Similarly, for *P. aurelia*, $r_m = 1.241$.[1] If the data had been expressed in volumes instead of numbers, the values of the constants a and r_m would have remained the same, but K would have been

1. The higher value for *P. aurelia* is due to the discrepant estimate of r_m for the half-loop medium. This may be an artifact. The equations shown in Table 9.05 were obtained after certain observations had been discarded from Gause's raw data (see above; also legends to Figs. 9.01 and 9.02). The values of a and r_m will vary according to the number of observations retained. This may account for the slight differences in these estimates of r_m and those given by Gause (1934, Table 11); Kostitzin gives still different values. According to his calculations, r_m is very nearly the same for all four curves.

different. Since *P. aurelia* has a volume about 0.39 times that of *P. caudatum*, the transformations of K can be readily made if desired.

The logistic theory implies that the value of $r_m = 0.935$ for *P. caudatum* is the innate capacity for increase of this species when it is living at 26° C. in Osterhaut's medium and feeding on *Bacillus pyocyaneus*. That is to say, it is the infinitesimal rate of increase which could be calculated for a population of stable age-distribution unhampered by limited space, by taking into account the fecundity, longevity, and speed of development of the individuals comprising it. The infinitesimal rate of increase would be converted to a finite rate of increase by the methods described in section 3.11 if this were required for practical reasons. But the infinitesimal rate of increase is more easily manipulated and more useful for theoretical purposes (sec. 9.24). Discounting the discrepant estimate of r_m for *P. aurelia* in the half-loop medium, it would seem that these two species do not differ very much in their innate capacity for increase in the circumstances of this experiment.

9.222 ANIMALS WITH COMPLEX LIFE-HISTORIES

Many workers, having recorded the growth of populations of animals with life-histories more complex than that of *Paramecium*, have tried to fit their results into the logistic theory, often without appreciating the complexity of the particular examples which they have studied. For example, Pearl's (1926) pioneering analysis of the growth of a population of *Drosophila* has been shown by Sang (1950) to include a number of errors because Pearl did not deal with all the complexities of the *Drosophila* culture. In the first place, Pearl did not have a single species but, instead, a predator and prey were confined together in the same limited space, for the yeast in the culture bottles was itself a growing population. It was therefore impossible to control precisely the amount of food available to the flies. As their numbers increased, they ate more, and the yeast grew more slowly. Furthermore, the composition of the yeasts varied as the culture aged. In the second place, the fly, unlike *Paramecium*, has several stages in its life-cycle. These have different requirements for both food and space. The eggs can hardly be considered an "active" part of the population; the larvae and adults certainly are, but each has different requirements: the larvae live and feed in and on the solid medium; the adults live in the space above the medium and feed on its surface; the pupae occupy considerable space but do not require food. It is therefore open to question which stages should be counted in the enumeration of N. There would obviously be an error in lumping all stages together and considering them as contributing equally to the "crowding" of the population. On the other hand, there would obviously be an error in counting only one stage. Pearl counted only the adults, assuming that the immature stages were equivalent to the unborn in a human population.

A third criticism of Pearl's experiments with *Drosophila* relates to the age-distribution of the colony at the beginning of the experiment. The logistic theory requires that the uncrowded population commence its growth with an innate capacity for increase which we have called r_m. But, since the fecundity and expectation of life for the individual inevitably vary with its age, the actual rate of increase, r, will be the same as r_m only if the population has a certain age-distribution known as the "stable" age-distribution (sec. 3.2). Growth curves of experimental populations of insects have usually been obtained from the growth of initially small populations characterized by anything but a stable age-distribution; usually the initial population consists of a single pair or a few pairs of adults. Pearl took the precaution of starting his colonies with a mixture of a few larvae, pupae, and adults; and his experiments were therefore superior to those started from adults only. But the rate of growth at the beginning would probably be quite a lot in excess of r_m, for a population of stable age-distribution would include quite a large proportion of immature stages.

These defects in Pearl's experiments with *Drosophila* may occur in other experiments, and this possibility should be borne in mind when this sort of experiment is being criticized or planned. This is particularly true of the third defect, for we do not know of a single experiment in the literature which was started with a population with a stable age-distribution. The obvious explanation for this is that it presents almost insuperable difficulties. The colony would have to be large in order to include a sufficient number of stages and a reasonable sample of these stages. This is clear from the figures in Table 3.03, which shows the stable age-distribution of the weevil *Calandra oryzae*. It is clear that a close approach to a stable age-distribution would require an initial population of at least several hundreds of weevils; the experiment might then become impracticable because of the large size of the containers that would be needed and the large amount of work involved in attending to the experiment and in sampling. But it is important to remember that the logistic theory applies to populations with the stable age-distribution, even if this condition cannot be fulfilled in practical experiments.

Beetles of the genera *Tribolium* and *Calandra*, which live in flour and wheat, respectively, have often been used for experimental studies of populations. These beetles have a life-cycle which includes egg, larval, pupal, and adult stages; so experiments done with them lack the simplicity of Gause's experiments with *Paramecium*. But they are preferable to *Drosophila*, because all stages of the life-cycle occur side by side and the larvae and adults eat the same sort of food; and since the food is not living, the quantity and quality can be regulated at will. We shall therefore commence our discussion with some experiments that have been done with *Tribolium* and *Calandra*.

Chapman (see Gause, 1931) started colonies of *Tribolium confusum* with a single pair of adults in 16 and 64 gm. of wheat flour at 27° C. The flour was renewed every 17 days. All stages of the life-cycle (eggs, larvae, pupae, and adults) were included in the counts. The experiments were not kept going long enough to find out what happened after the population had completed its first upward surge of growth, but there is just the beginning of a downward trend, which might have been the beginning of an oscillation. Since the results gave

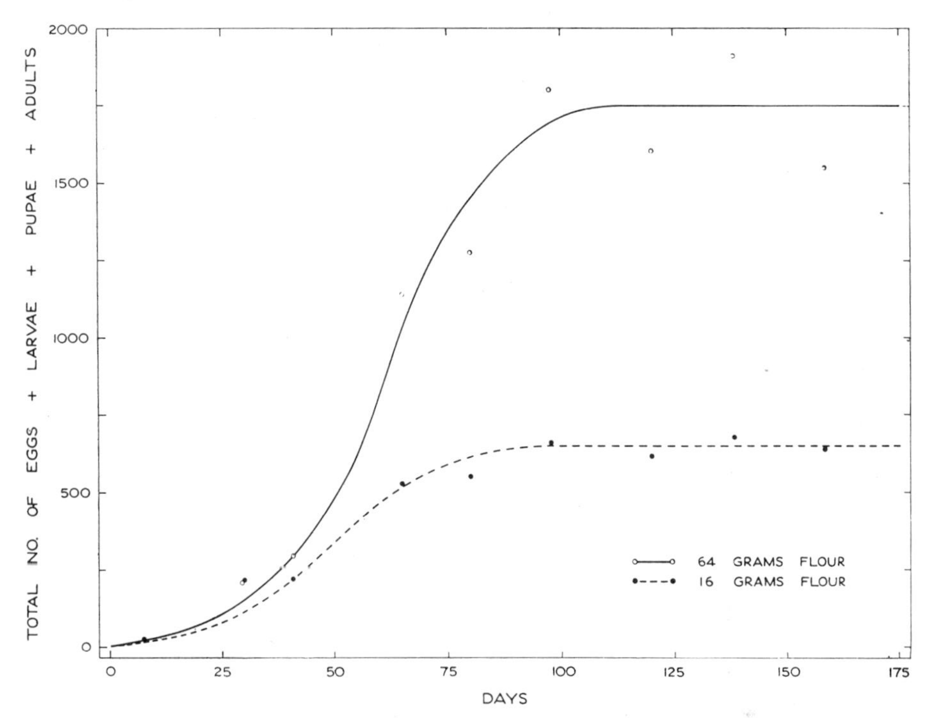

FIG. 9.03.—The trends in the numbers of *Tribolium confusum* in 16 and 64 gm. flour at 27° C. The smooth curves are calculated logistic curves (Table 9.06, p. 366). The observed points represent the combined numbers of eggs, larvae, pupae, and adults. (After Gause, 1931.)

no direct indication of an asymptote, the second method of estimating K was used. This method places full weight on the data in the ascending part of the graph, and the calculated curve followed the trend of the observed points quite faithfully (Fig. 9.03). Of course, we would not expect that the value for K given by this method would necessarily indicate the density of the population which would have been observed, had the experiments been carried on for a longer period. Holdaway (1932) did similar experiments with the same insect and got similar results. Park (1948) reared populations of the related *T. castaneum* in 8 gm. of whole-wheat flour and yeast at 29.5° C. and 60–70 per cent relative humidity. Park's data do not indicate whether or not the early stages of growth conformed to the logistic theory, but they do show quite clearly that in the later stages the density failed to stay steady around an asymptote, as required by the theory. In Figure 9.04, there is no indication of an asymptote. Instead, the density shows a series of long-term fluctuations, with the troughs approaching surprisingly close to zero. Similar oscillations have been found in populations of *T. confusum* (Park, 1948) and *Calandra oryzae* and *Rhizopertha dominica* (Birch, 1953*b*).

In four separate experiments Birch placed a single pair of adults of either the small form or the large form of *Calandra oryzae* in 12 gm. of either wheat or

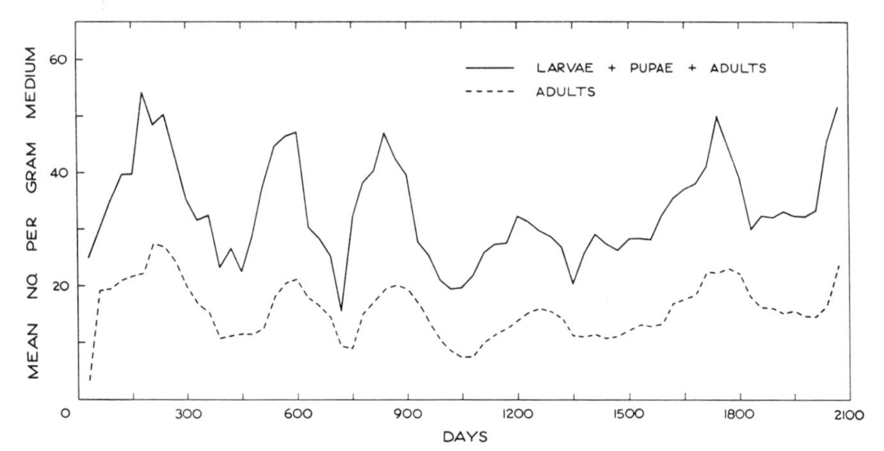

FIG. 9.04.—The trend in the density of a population of *Tribolium castaneum* in a flour-yeast medium at 29.5° C. and 60–70 per cent relative humidity. Details of the initial population-growth phase are not shown. The medium (8 gm.) was replaced every 30 days. Observed points are means of 20 replicates: they represent the combined numbers of larvae, pupae, and adults. (Park, personal communication.)

maize and kept them constantly at 29.1° C. and 70 per cent relative humidity. The grain was renewed to the same quantity every 2 weeks. The growth of the population was measured by counting the number of adults at weekly intervals. Since the immature stages of these insects occur inside the grain, they could not be counted; but it was considered that the density of the adults would reflect the density of the whole population, since Park (1948) had shown that this is so for *Tribolium*. The experiments were continued for 100 weeks, which was equivalent to about 12 generations. With those living in wheat, the first upward surge of growth was completed in about 22 weeks. Thereafter, the density underwent a series of long-term and profound oscillations (Figs. 9.05 and 9.06). In the absence of any tendency for the density to follow an asymptote, a value for K was estimated by the second method. In each case the value for K was chosen such that the ensuing curve would give the best fit for the greatest part of the data relating to the first phase of growth, when the density was increasing continuously; in the case of those living in wheat, it was not possible to find a curve that would follow the data closely enough for the full period of this phase of growth. The curves that were eventually chosen as best follow the observed points for a greater or a lesser part of this phase of the growth of the population (Fig. 9.05). The fit is better in Figure 9.06, which relates to the populations living in maize. The populations in maize were less dense and the oscillations less pronounced.

The results of these experiments with *Tribolium* and *Calandra* may be considered to be highly reliable, for all of them were well replicated: the former 20 times, the latter 15 times. In both cases the replicates were in close agreement,

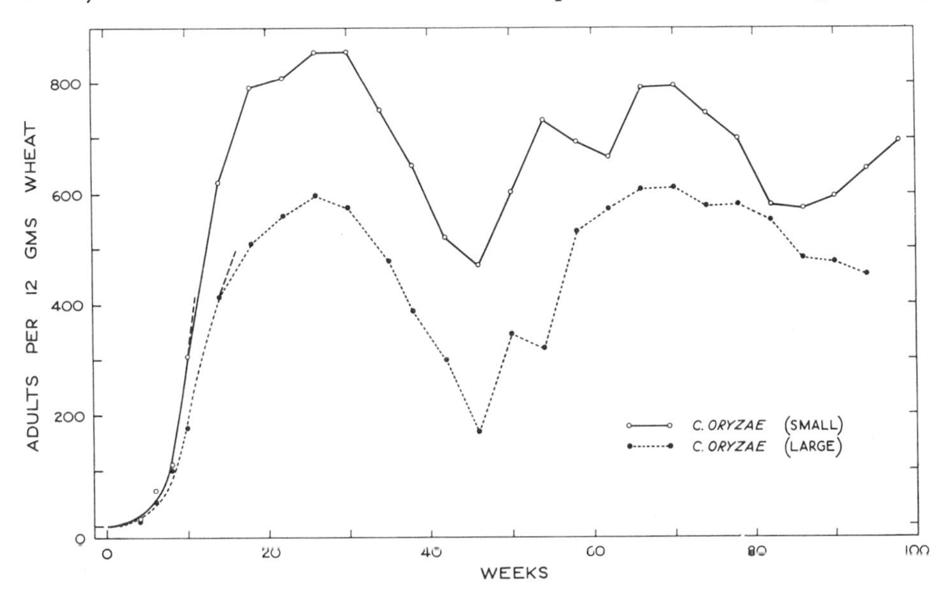

FIG. 9.05.—Trends in the densities of populations of the small and the large "strains" of *Calandra oryzae* at 29.1° C. and 70 per cent relative humidity in wheat. The smooth curves are calculated logistic curves; 12 gm. of wheat were renewed at 2-week intervals. Observed points are means of 15 replicates; only adults were counted. (After Birch, 1953*b*.)

resulting in a small coefficient of variability. In other words, if these experiments were repeated, the probability of again getting similar oscillating curves instead of a steady asymptote is very high indeed. This means that we can confidently rule out accident or artifact in interpreting these results and can be quite sure that these curves faithfully represent characteristic growth trends for populations of these animals grown in the circumstances specified for these experiments. The existence of oscillations like these in a population made up of a single species living at a constant temperature and with a constant supply of food in limited space is of great interest. It shows that fluctuations in numbers can be brought about by the influence of "other animals of the same kind" quite independently of any fluctuations in temperature, food, or the influence of other sorts of animals either predators or non-predators (chap. 10).

We do not know of any other experiments in which oscillations in the final density instead of a steady asymptote have been demonstrated quite so reliably as with *Tribolium* and *Calandra*, but there is a number of other species for which similar oscillations have been indicated. These include *Rhizopertha dominica* (Birch, 1953*b*), *Callosobruchus chinensis* (Utida, 1941*e*), *Trogoderma versicolor* (Park, Gregg, and Lutherman, 1941), *Lucilia cuprina* (Nicholson, 1950), *Daphnia magna* (Pratt, 1943), *Daphnia obtusa* (Slobodkin, 1954), and *Simocephalus vetulus* (Frank, 1952). The results of one of Pratt's experiments with *Daphnia* are shown in Figure 9.07. This particular population was grown

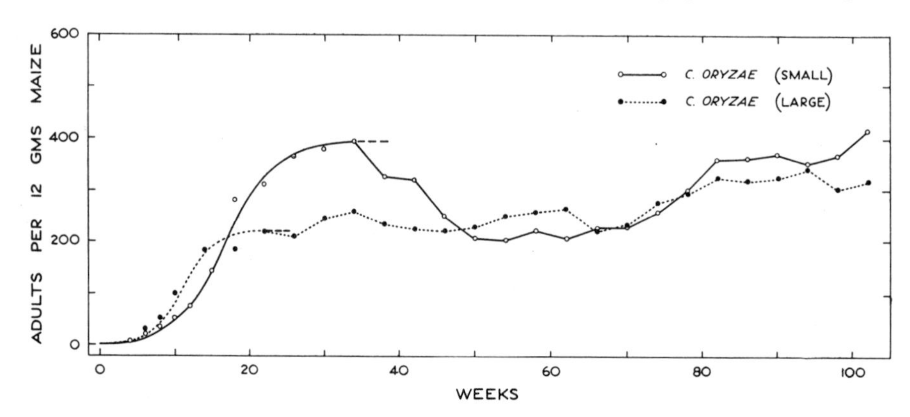

FIG. 9.06.—Trends in the densities of populations of the small and the large "strains" of *Calandra oryzae* at 29.1° C. and 70 per cent relative humidity in maize. The smooth curves are calculated logistic curves; 12 gm. of maize were renewed at 2-week intervals. Observed points are means of 15 replicates; only adults were counted. (After Birch, 1953*b*.)

at 25° C., and its food was the alga *Chlorella pyrenoidosa*. Only part of the cycle of oscillations has been shown in Figure 9.07; equally profound oscillations continued until the end of the experiment at 220 days. On the other hand, we do not know of any experiments with animals with complex life-histories that have unequivocably demonstrated the presence of an upper asymptote during the later phases of growth of a population living in a "constant environment" in "limited space."

9.223 CONCLUSIONS INDICATED BY EMPIRICAL STUDIES

It may justly be concluded, from the empirical studies that we have examined, that the logistic theory fails to account for some of the information provided by the experiments. It conforms rather more closely to the data for animals with simple life-histories, like *Paramecium*, and less closely for animals with complex life-cycles, like insects. With the former, the empirical curve representing the growth of the population follows a sigmoid course during the first phase, when the density is increasing continuously, and it approaches an asymptote during the second phase, while the density remains fairly steady. With such animals it is possible to find a logistic curve which follows the general trend of the growth of the population through both the first and second phases of its growth, provided that some of the intermediate observations relating to the later part of the first phase are discarded. In other words, for animals with simple life-cycles, the logistic curve provides an imperfect, yet sufficiently close and useful, description of the experimental data.

With animals with complex life-cycles, the logistic theory is seen to be still more imperfect. Yet, by discarding all the information provided by the experiment about the second phase of growth, it is still possible to fit a logistic

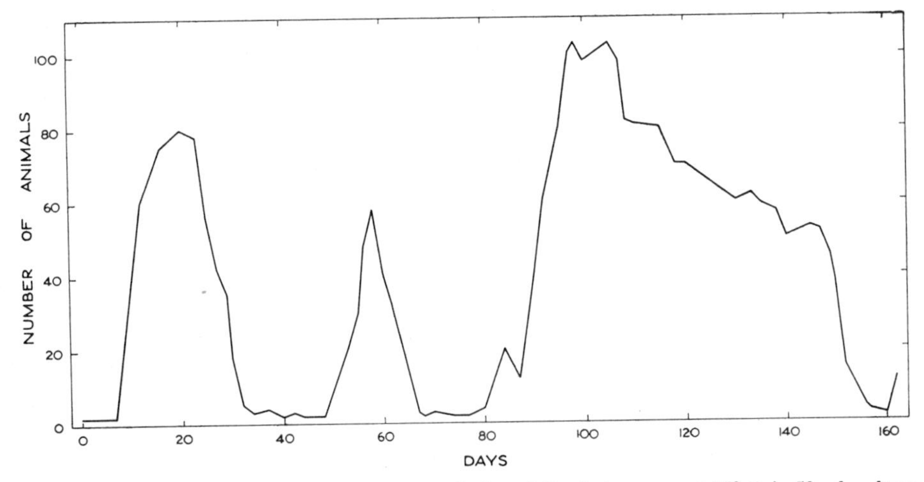

FIG. 9.07.—The trend in the density of a population of *Daphnia magna* at 25° C. in 50-ml. culture of *Chlorella pyrenoidosa*. The medium was replaced every 2 days. The graph shows the trend in numbers in one of 21 replicates. (After Pratt, 1943.)

curve to at least part of the first phase, during which the population is undergoing its initial increase in density. This, as we shall see in section 9.24, means that the logistic theory, severely limited though it may be, remains a useful "tool" in population ecology.

The experiments we discussed in section 9.222 were exceptional, in that they were kept going for a relatively long period after the initial phase of sigmoid growth had been completed. Most other experiments which we know about were discontinued either before or just after this phase had been completed, and we cannot tell whether these populations would have maintained themselves steadily about an asymptote like *Paramecium* or would have undergone extreme oscillations like *Tribolium*, *Calandra*, and *Daphnia*. More experiments are still needed, but we may be sure that the growth of populations of at least some animals with complex life-cycles fails to agree with the logistic theory in the latter phases of their growth: for they undergo profound oscillations where the theory requires that they remain steady about an asymptote. This makes it desirable to look more closely into the assumptions on which the logistic theory is built, to see how well each one agrees with evidence that can be got from experiments.

9.23 *Criticism of the Assumptions Implicit in the Logistic Theory*

When we say that the growth of a population may be represented by a logistic curve, we imply the following provisions:

a) That the population has a stable age-distribution.

b) That its density has been measured in appropriate units.

c) That there is a real attribute of animals corresponding to our concept of r_m, the innate capacity for increase.

d) That the relationship between density and the rate of increase is inversely linear, i.e., that the rate of increase is decreased by a constant amount for each individual added to the numbers already present in unit space.

e) That the depressive influence of density on rate of increase operates immediately, i.e., that the rate of increase is slowed down to the full extent demanded by the theory as soon as each additional animal is added to the population and not only at some later date; or, in still other words, that there is no "time lag" associated with the influence of density on rate of increase.

9.231 THE AGE-DISTRIBUTION OF EXPERIMENTAL POPULATIONS

We have already pointed out in section 9.222 that the theory of the logistic curve involves the assumption that the population commences its growth with an innate capacity for increase r_m and that this cannot happen unless the initial population has a stable age-distribution. A second feature also implicit in the theory is that the age-distribution remains unaltered during the growth of the population. This must be so, as far as the theory is concerned, since it will be remembered that we supposed that every individual added to the population decreased the rate of increase of the population by a certain constant amount, c. But this can hold in practice only if groups of individuals are added in proportion to their representation in a population of stable age-distribution. If the proportion of the different age-groups which were added varied from time to time, then the "average" contribution of each individual would clearly no longer be constant. If, however, the proportions of the different age-groups added represented a sample from a stable age-distribution and this never varied, then the population would, of course, retain its initial stable age-distribution.

The first point we have already commented on (sec. 9.222); none of the experiments that have been done to test the logistic theory has made use of populations of stable age-distribution. This may be due partly to the technical difficulties of working with the large numbers that would be required, but it also seems that the need to fulfil this condition had been overlooked until Leslie (1948) pointed out that the logistic theory required that the population should have a stable age-distribution. There is no way of telling precisely how greatly this error may contribute to discrepancies in the final results, but we would expect it to be least in species like *Paramecium*, in which the nonreproductive stages occupy relatively little of the life-cycle, and most in species like insects, in which the immature stages occupy a relatively large proportion of the total life-cycle. For example, a population of insects made up entirely of adults in the reproductive stage, being quite untrammeled by limited space, would at first increase much more rapidly than a population with a stable age-distribution which would include a high proportion of immature stages. At a slightly later stage in its growth, this population might come to have a much lower rate of increase. Certainly, its rate of increase would not be equivalent to r_m, as required by the logistic theory; it would probably fluctuate quite widely on either side of r_m.

Theoretical considerations by Leslie (1948) show that if there is no time lag in the influence of density, a population of stable age-distribution will retain its stable age-distribution when living in limited space only under one set of conditions, namely, when the birth-rate is independent of density and the changes in rate of increase are due entirely to increase in the death-rate and this increase is independent of age. If, on the other hand, the birth-rate is dependent on density but the survival-rate is not, then the population will assume the stationary or "life-table" form of age-distribution (sec. 3.2). In some species of insects the density is known to influence both birth-rate and survival-rate. Presumably in these species a growing population would assume an age-distribution intermediate between these two.

Very little is known about the age-distribution of experimental populations and how they may change with the growth of the population. This may be partly due to the difficulty of "aging" most animals. With insects an approximation may be made by classifying them into stages of the life-cycle, e.g., egg, larva, pupa, and adult. This is, of course, not fine enough, but it is better than nothing. This is the method Bodenheimer (1938) used to give an approximate description of changes in the age-distribution of a colony of *Drosophila* (Fig. 9.08). This colony began as a single pair of adults. The diagrams in Figure 9.08 show by the width of the bands the proportion of eggs, larvae, pupae, and adults from the fifth to the thirty-fifth day. On the fifth day, 52 per cent of the population were in the egg stage, 30 per cent were young larvae, 17 per cent older larvae, and only 2 per cent adults. This "youthful" population is represented by a pyramid-shaped figure. By the tenth day the average age of the individuals in the population was greater, as shown by the urn-shaped figure. The return of the pyramidal figure for the fifteenth day is associated with the development of a second generation. Of course, this population does not represent the changes in the age-distribution of a population growing in conformity with the logistic theory, but only the changes associated with the early stages of growth of a family in which the procession of generations is still prominent. But these effects must have been prominent in many experiments that have purported to test the logistic theory. We do not know of any case of an experimental population tending to assume a stable age-distribution.

We have seen that the theory requires the existence of a stable age-distribution throughout the growth of the population. Likewise, the existence of an upper asymptote is dependent upon the continuance of the stable age-distribution during this phase of the population's history. A population of the same species, having the same density K and living in the same circumstances, but

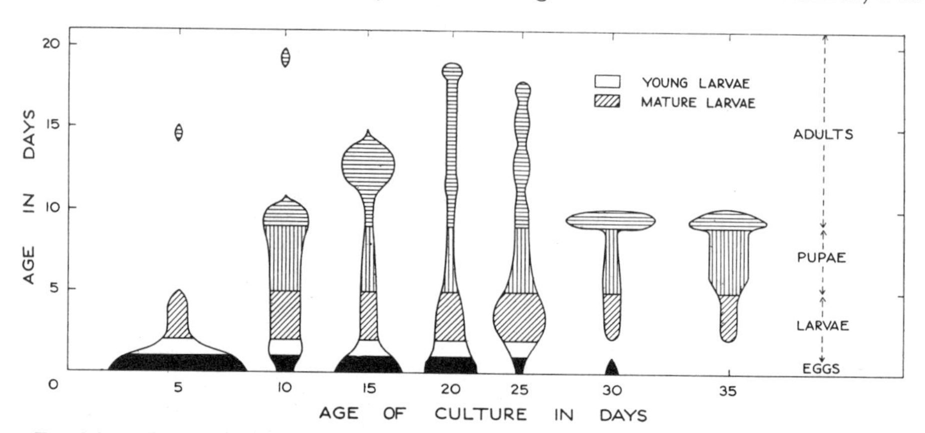

FIG. 9.08.—The age-distribution of *Drosophila melanogaster* cultured on banana-agar at 27° C. The population was initiated with a single pair of adult flies. (After Bodenheimer, 1938.)

having a different age-distribution (either with an excess of old or an excess of young relative to the stable age-distribution), would not in any circumstances remain steadily at this density. Instead, it would at first fluctuate on either side of K. If, as time passed, the age-distribution approached the stable age-distribution, then the density would approach K by a series of damped oscillations and eventually become steady at this level (Leslie, 1948).

It would be of interest to know the way the age-distribution changes both during the growth phases of an experimental population and during the oscillations which follow. This might provide a clue to the explanation of oscillations themselves: the crests may be associated with a predominantly young population, and the troughs with a predominantly old one. Errington (1946) considered that fluctuations in the natural populations of the muskrat may be caused in this way.

When it is considered how limited are the circumstances in which Leslie's theoretical populations will maintain a steady density equivalent to the asymptote of a logistic curve, it is not surprising to find that oscillations are the rule and steady densities the exception in experimental populations. For example, in the populations of *Calandra* described in section 9.222, in order that the oscillations should damp out, the population, having started from a single pair of adults, would need to assume the stable age-distribution; this would not happen unless the birth-rate remained independent of density, which seems unlikely.

9.232 THE CHOICE OF UNITS FOR REPRESENTING THE POPULATION

The unit of the population is the individual. Provided that the size of the individual is independent of the density of the population, the constants a and r_m in the logistic equation would have the same value whether the population had been measured by numbers, weight, or volume; but K would depend on the unit of measurement. Certain species of insects become smaller when they are reared in a crowd (sec. 9.235). The addition of a large animal in the early stages might be expected to influence the rate of increase rather differently from the addition of a small animal in the later stages. Equivalent results might come from the addition of the same weight of animals rather than from the same number of animals. Under these circumstances it might be more logical to consider mass or volume rather than numbers. This is a precaution that is rarely observed, but the error that is introduced by counting instead of weighing is usually quite small.

9.233 THE "REALITY" OF r_m

There are two senses in which we need to discuss the "reality" of r_m. In the first place, it presupposes that species do have definite innate capacities for increase which, in certain circumstances, can be regarded as constant. We consider this capacity to be characteristic of the species in much the same way as we might consider a morphological or taxonomic feature. Changes in the innate capacity of species are said to occur in some animals from season to season (Clements and Shelford, 1939, p. 187), but these examples have scarcely been investigated sufficiently to rule out the possibility of disease or some other influence causing the change. In practice, differences in birth-rates or death-rates between two or more populations living in similar circumstances are more usually regarded as evidence of genetic difference. It is then reasonable to regard the innate capacity for increase of a species as something which can be defined and regarded as characteristic of each species. We have called it r_m.

In the second place, we have to consider the objection that r_m may be a rate of increase which, though incorporated in the logistic equation, is never realized in experimental populations. The theory of the logistic curve involves the assumption that the population commences growth with a rate of increase r_m, but the duration of this rate must be infinitesimally small. The theory does, however, imply that if at any stage in the growth of the population the density were reduced to the optimal level, the population would indeed grow at the rate r_m until crowding increased again. This is what is implied by making r_m a constant in the logistic equation. Strictly speaking, this can be true only if the age-distribution of the population remains stable during its history, and we have shown in section 9.231 that this is unlikely. We are therefore left with the necessity of finding out empirically just how much error is involved in regarding this particular assumption of the logistic theory as approximately true.

By definition, r_m is the rate of increase of a population of stable age-distribution when the depressive influence of crowding is negligible. This suggests an empirical test of the "reality" (or reliability) of the concept r_m as a constant in the logistic equation. The value of r_m may be calculated directly by the methods of chapter 3, using data from life-tables and age-specific fecundity tables; and it may be calculated quite independently of this by fitting a logistic curve to the records of the growth of a population, as in section 9.22. The second method allows several independent estimates of r_m to be made, by varying either the space in which the population grows or the quantity (but not the quality) of the food. Of course, it is important to see that all other components of the environment, such as temperature, moisture, and so on, which may influence fecundity, longevity, or speed of development, are carefully controlled and reproduced as closely as possible in all the experiments, because, as was seen in chapter 3, there is a whole array of values of r_m corresponding to variations in these components of the environment. It is unfortunate that so few tests of this sort have been made. We know only those which we quote below, namely, Gause's experiments with *Paramecium* and his analysis of Chapman's data for *Tribolium* and Birch's experiments with *Calandra* and *Rhizopertha*.

Table 9.05 (sec. 9.221) shows the values of r_m in logistic curves fitted to data for populations of *Paramecium caudatum* and *P. aurelia* growing in identical circumstances except for the amount of food available to the colony. In one set of experiments this was exactly double the amount used in the other set. For *P. aurelia* the two estimates r_m are within about 15 per cent of each other; for *P. caudatum* the difference is nearer 40 per cent (but see the footnote at the end of sec. 9.221). Chapman reared populations of *Tribolium confusum* in identical circumstances except for the amount of flour available to the colonies: two adults were placed in 16, 32, 64, and 128 gm. of flour, which was renewed at regular intervals. The results are summarized in Table 9.06. Gause (1931), in

TABLE 9.06*

EQUATIONS FOR LOGISTIC CURVES FITTED TO DATA FOR FOUR POPULATIONS OF *Tribolium* GROWING IN DIFFERENT QUANTITIES OF FLOUR BUT AT SAME TEMPERATURE AND RELATIVE HUMIDITY

Quantity of Flour (Gm.)	Equation for Logistic Curve	Quantity of Flour (Gm.)	Equation for Logistic Curve
16...........	$N = \frac{650}{1 + e^{4.269-0.091t}}$	64..........	$N = \frac{1,750}{1 + e^{4.940-0.082t}}$
32...........	$N = \frac{1,025}{1 + e^{4.320-0.083t}}$	128.........	$N = \frac{5,000}{1 + e^{5.190-0.050t}}$

* After Gause (1931).

presenting these data, was chiefly interested in the value for K, and he did not discuss r_m. Our chief interest is in the value of r_m. The table shows that the values of r_m for the first three experiments are within 10 per cent of one another, which must be accepted as quite close agreement. The fourth one is discrepant. It is rather likely that this is due to underpopulation. In the early stages, while there were only 2 beetles in 128 gm. of flour, the density was below the optimum for copulation and egg-laying. Such an optimum has been shown to exist for *Tribolium* (sec. 9.11). The data for *P. aurelia* and *T. confusum* show that a number of independent estimates of r_m, made by varying the quantity of food available, are reasonably similar. Unless the same error is being repeated in each case, these results would tend to confirm the "reality" of r_m.

Birch (1953*a*) estimated r_m for *Rhizopertha dominica* and for the large and small "strains" of *Calandra oryzae* by the methods of chapter 3. He reared the insects uncrowded in grain at carefully controlled temperature and moisture, and recorded longevity, fecundity, and speed of development. From these data, life-tables and age-specific fecundity tables were constructed, enabling r_m to be calculated directly. Then he reared populations of the same insects in 12 gm. of the same sort of grain at the same temperature and moisture. The populations were started with single pairs of adults, and only adults were counted; it was explained in section 9.222 why this was considered sufficient. Logistic curves were found which adequately fitted at least part of the data in the ascending part of the curve (Figs. 9.05 and 9.06). Table 9.07 allows a comparison

TABLE 9.07*

TWO INDEPENDENT ESTIMATES OF INNATE CAPACITY FOR INCREASE (r_m) FOR EXPERIMENTAL POPULATIONS OF GRAIN BEETLES

Species	Grain	Temp. °C.	Moisture Content of Grain (Per Cent)	Innate Capacity for Increase (r_m)	
				Derived from Life-Table and Age-Fecundity Table	Derived from Logistic Curve
C. oryzae (small "strain")	Wheat	29.1	14	0.772	0.55
R. dominica	Wheat	32.3	14	.686	.58
R. dominica	Wheat	29.1	14	.578	.50
C. oryzae (large "strain")	Wheat	29.1	14	.564	.52
C. oryzae (small "strain")	Wheat	32.3	14	.501	.48
C. oryzae (large "strain")	Maize	29.1	13	.436	.46
C. oryzae (small "strain")	Maize	29.1	13	0.417	0.30

* After Birch (1953*b*).

of the values of r_m calculated by these two entirely independent methods. Since the populations were not started with a stable age-distribution, the constants calculated from the data did not truly represent r_m but some other value of r; but the error in this case may not have been large. For each example shown in Table 9.07, the estimate of r_m from the logistic curve is somewhat smaller than that from the life-table and age-fecundity table. Even so, the agreement between the two estimates is remarkably close, especially when it is remembered that the value of r_m of the logistic equation is derived from the curve of growth of adults only in the population. The agreement is closest in those populations in which the growth trends fitted the logistic equation for a large part of the initial sigmoid trend. These results tend to support the concept that there is an attribute of animals corresponding to the innate capacity for increase, r_m.

9.234 THE LINEAR RELATIONSHIP BETWEEN DENSITY AND RATE OF INCREASE

It would seem to be impossible to devise satisfactory experiments to test thoroughly the hypothesis that the rate of increase is a linear function of density, because the problem is so complex. But the rate of increase may be broken down into simpler components. A number of experiments have been done with the purpose of measuring the influence of density on longevity, fecundity, and speed of development. These are attributes of individuals: in their collective aspect they represent the death-rate and the birth-rate of the population. We shall review some of these experiments, taking them in this order.

The influence of crowding on longevity has been studied for a number of insects. The outcome of crowding may depend on the age of the insect and particularly on the stage in the life-cycle of those species which have a metamorphosis in the life-history. If all stages occur together, as, for example, with the flour beetle, *Tribolium*, living in flour, it may be necessary to consider how the crowding by one stage, say the larvae, may influence the longevity of another stage, say the eggs or the adults, which may be present at the same time. This aspect of the ecology of *Tribolium confusum* has been thoroughly investigated by Park and his school. The results of some of their experiments may be summarized in the following diagram. The source of the information is indicated by the reference on the arrow. The dotted lines from adults to adults and from larvae to eggs indicate that these experiments have not been

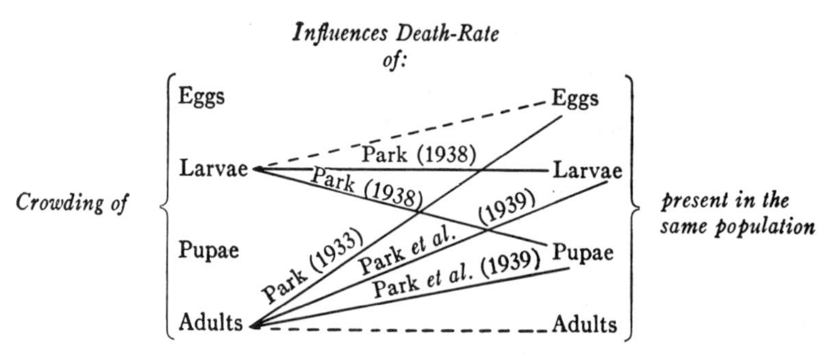

done, but we expect that such results could be demonstrated. Park (1938) and Park, Miller, and Lutherman (1939) demonstrated that fewer larvae and pupae survived in flour that had been "conditioned" by having many *Tribolium* live in it than in fresh flour. On the other hand, when larvae alone were kept in fresh flour, which was renewed frequently to prevent conditioning, the survival-rate among the larvae was almost independent of the density. When the survival-rate of larvae in "conditioned" flour was related to the density of the population of larvae living in it (i.e., to the density of the larval population that had "conditioned" the flour), this relationship was found to be curvilinear. In cultures of *Tribolium* the adults eat the eggs that they can find. The greater the density of adults in the culture, the less is the chance that an egg will survive. Crowding by adults thus increases the death-rate in the egg stage, but in this case also the relationship is not linear; with increasing densities of adults but constant densities of eggs, the proportion of eggs eaten each day increased up to a maximum and then remained steady at about that level. On the other hand, when the density of the adults was kept constant but the density of eggs varied, the proportion of eggs eaten varied directly with the density of the eggs. The results are shown in detail in Table 9.08. This is the

TABLE 9.08*
MEAN NUMBERS OF EGGS CONSUMED BY ONE ADULT MALE *Tribolium confusum* IN CULTURES CONTAINING VARIABLE NUMBERS OF EGGS AND ADULT BEETLES

No. of Beetles per Culture	No. of Eggs per Culture		
	50	150	350
1	0.02	0.95	....
2	.06	....	....
4	.28	.70	1.80
8	.13	.88	2.51
16	.11	.48	1.50
32	.09	.41	0.99
64	0.05	0.20	1.08

* After Boyce (1946).

result that would have been expected, always provided that the greatest density of eggs was still insufficient to satiate the adults (Boyce, 1946). With *Tribolium*, all stages in the life-cycle occur together in the flour. This, as the diagram above illustrates, makes for great complexity in the interactions between stages in the life-cycle.

With *Drosophila*, the adults live above the culture medium, the eggs are on the surface of the medium, the larvae burrow into it, and the pupae are either in the medium or on the walls of the bottle; neither the larvae nor the adults are cannibalistic, so that neither the eggs nor any other stage is likely to be eaten, no matter how dense the population may become. There would seem to be no effect comparable to the conditioning of the medium by *Tribolium*. As would be expected, the relationships between the stages is much simpler for *Drosophila* than for *Tribolium*. This is shown in the following diagram:

Influences the Death-Rate of:

Crowding of { Larvae ——→ Larvae } *present in the same population*
{ Adults ——→ Adults }

With *Drosophila melanogaster*, the combined death-rates of larvae and pupae is nearly proportional to the density of the larvae between the levels of 2 to 120 larvae per 4 ml. of medium (Fig. 9.09). This results from the reduction in the nutritive value of the medium in crowded cultures (Sang, 1949). Adult *Drosophila* lived longest when there were about 35–55 flies in a 1-ounce vial.

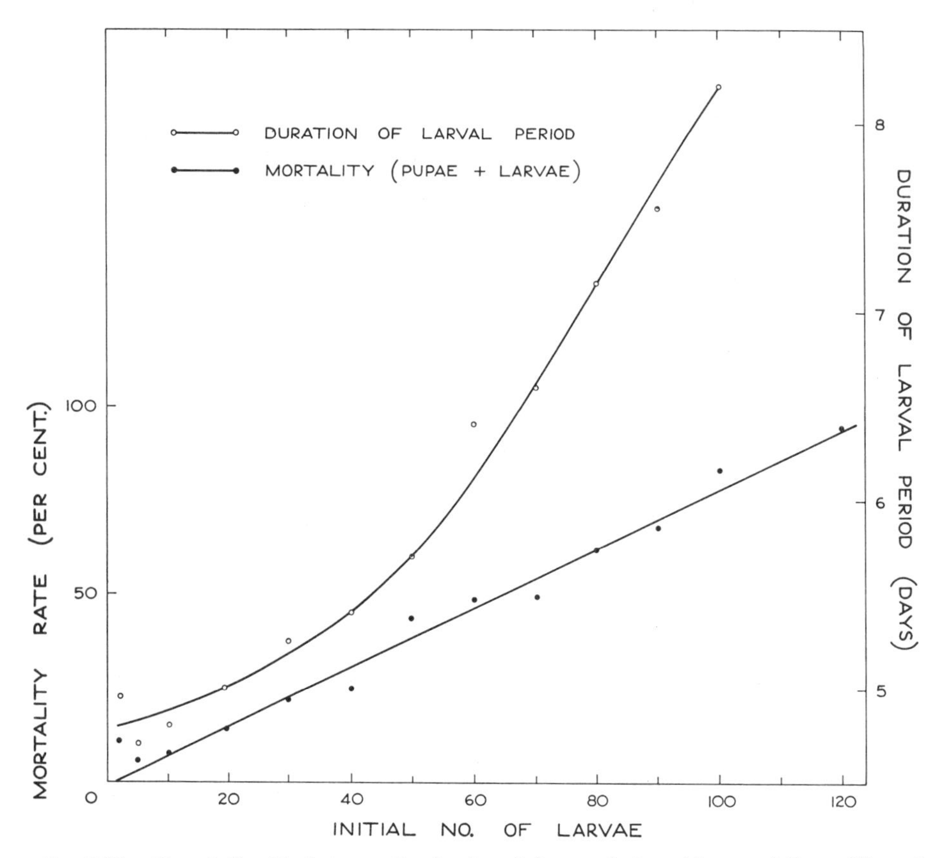

FIG. 9.09.—The relationship between the density of the population of larvae of *Drosophila* and (*a*) the death-rate among larvae and pupae taken together and (*b*) the duration of the larval stage. In these experiments 4 ml. of culture medium were "seeded" with different numbers of larvae and incubated at 25° C. (After Sang, 1949, Table 1.)

Above this density, longevity decreased with increasing density; Figure 9.10 shows that the relationship was curvilinear (Pearl, Miner, and Parker, 1927; Bodenheimer, 1938). But Sang (1950) considered that these results may not indicate what happens in an ordinary culture, because the circumstances are so different. This merely serves to emphasize the extreme difficulty in testing the hypothesis of linearity implicit in the logistic equation. The interactions in experimental populations are usually more complex than those tested in analytical experiments of the sort being described.

Adults of the grain moth *Sitotroga cerealella* and the grain borer *Rhizopertha dominica* live among the grains of stored wheat, but their larvae live inside the grains. The different stages thus have entirely different requirements for space and relatively different requirements for food. Thus they are more like *Drosophila* than like *Tribolium* in the way that the different stages influence the survival-rate of one another. Crombie (1944) measured the survival-rate of larvae at different densities of larvae for these two species (Fig. 9.11). The

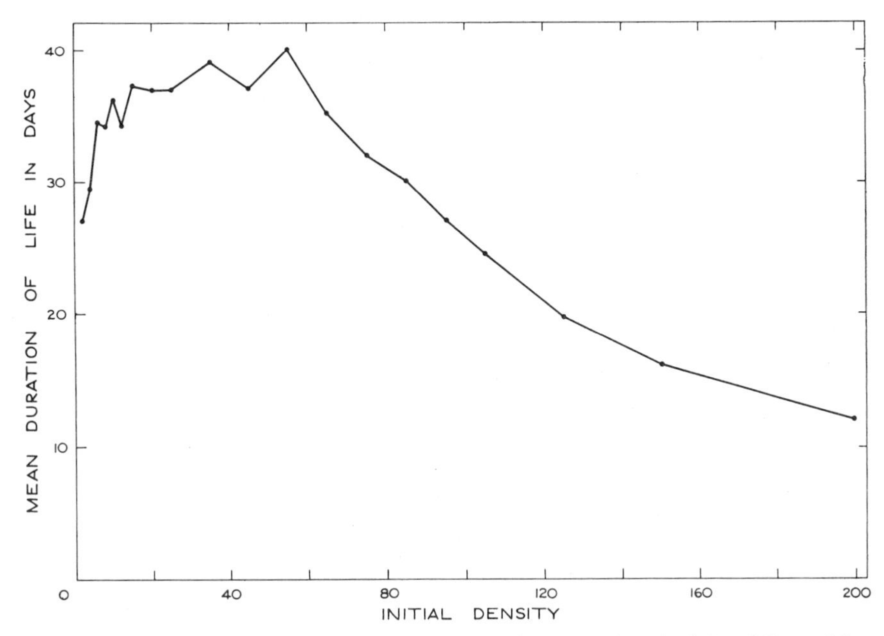

FIG. 9.10.—The relationship between the density of the population of adults of *Drosophila* and the mean longevity of adults. They were provided with fresh food daily. (After Pearl, Miner, and Parker, 1927.)

relationship was curvilinear at low densities but became nearly linear at medium and high densities, of the order usually found in experimental populations.

Utida (1941*c*), in his experiments with the beetle *Callosobruchus chinensis*, found that the relationships between density of larvae and death-rates of larvae and pupae were linear. Snyman (1949), working with the moth *Plodia interpunctella*, found a curvilinear relationship between the density of larvae and the death-rate of larvae. Blowflies, living on carrion, resemble *Drosophila*, in that the larval and adult stages have different habits. Ullyett (1950) worked with several different species of blowflies. The details differed between species, but in each case he found a curvilinear relationship between the density of larvae and the death-rates for larvae and pupae (Fig. 9.12). These results are like those Snyman found for *Plodia*.

Little is known of the influence of density on the longevity of animals other than insects. Pratt (1943) studied *Daphnia magna* and found that the longevity was increased with increasing density up to an optimum and then it decreased (Fig. 9.13). This is like the data for *Drosophila* (Fig. 9.10); in both, there is maximal longevity at an optimal density; in neither case is there any part of the graph which could be considered to approximate a straight line.

As with the above examples relating to mortality, so with the following ex-

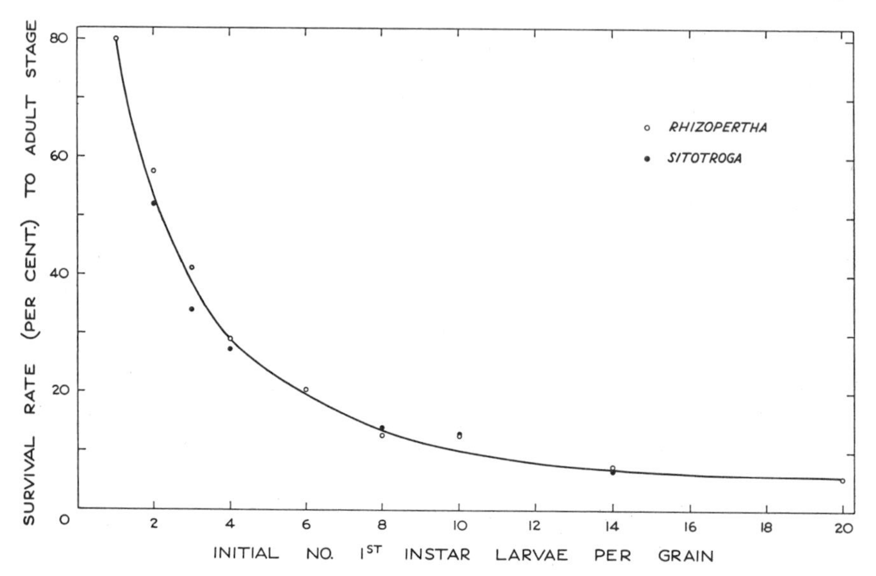

FIG. 9.11.—The relationship between the density of the population of larvae of *Sitotroga* and *Rhizopertha* and their survival-rates measured as the proportion of first-instar larvae which survived to become adults. They were reared in grains of wheat. (After Crombie, 1944.)

amples relating to the influence of density on fecundity, most of the experiments have been done with insects. The relationship between the net fecundity of *Tribolium confusum* and the density of adults is shown in Figure 9.14. As the density increases, the fecundity reaches a maximum (as the adverse influence of underpopulation disappears) at an optimal density and then decreases (MacLagan, 1932*b*). Above the optimal density the graph is distinctly curvilinear. Park (1933) showed that this was probably the resultant of two opposing influences. On the one hand, the frequency of copulation increases with increasing density, and this increases the fecundity. On the other hand, with increasing density, relatively more eggs are eaten; and at relatively high densities, fewer eggs are laid. Birch, Park, and Frank (1951) compared the two species *T. confusum* and *T. castaneum* and found that, for any particular density above the optimum, the proportional reduction in the net fecundity was about the same for both species. They increased the density of males and females separately and found with both species that when most of the population were males, more were required to bring about the same proportional reduction in net fecundity than when most of the population were females. Both sexes eat eggs to about the same extent, but the presence of many females

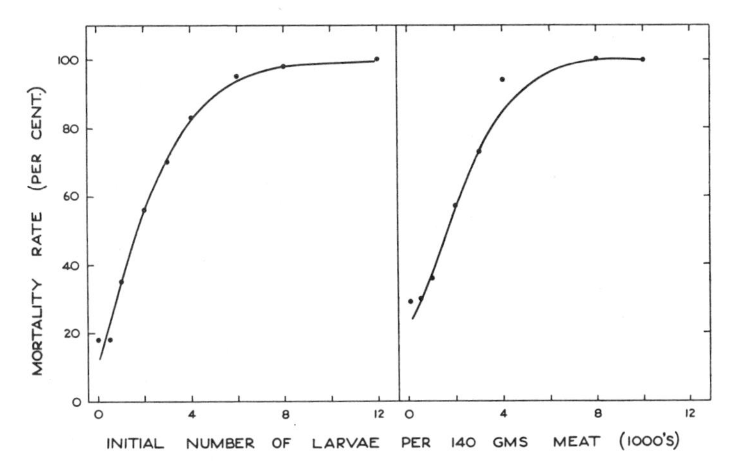

FIG. 9.12.—The relationship between the density of the population of larvae of two species of blowfly (*Chrysomyia*) and the mortality-rate among larvae and pupae taken together. They were reared on meat at 27° C. (After Ullyett, 1950.)

causes a greater reduction in the number of eggs laid than does the presence of the same number of males.

In a number of other species the relationship between density and fecundity has been variously found to be either curvilinear or linear. It was found to be curvilinear for *Rhizopertha* (Crombie, 1942; Fig. 9.15), but linear for

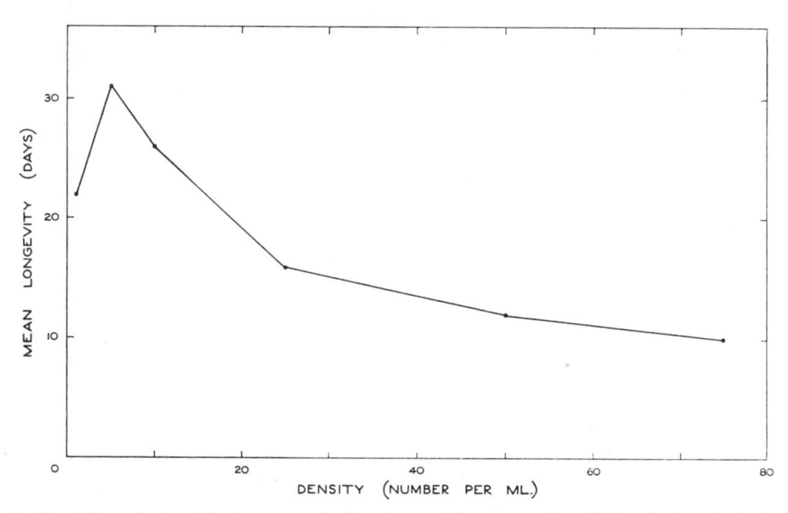

FIG. 9.13.—The relationship between the density of the population of *Daphnia* and the longevity of the adults. They were reared in 50 ml. of pond water at 25° C. with *Chlorella* for food. (After Pratt, 1943.)

Oryzaephilus surinamensis (Crombie, 1943), *Callosobruchus chinensis* (Utida, (1941*d*), and *Calandra oryzae* (Richards, 1947). Much work has been done with *Drosophila* also, but, as the food in these experiments consisted of a living culture of yeasts, they did not fulfil the requirements of a constant supply of food; these experiments are more relevant to chapter 10, where we discuss the growth of populations of predator and prey living together in a limited space. In growing cultures when the supply of food is limited, the relationship be-

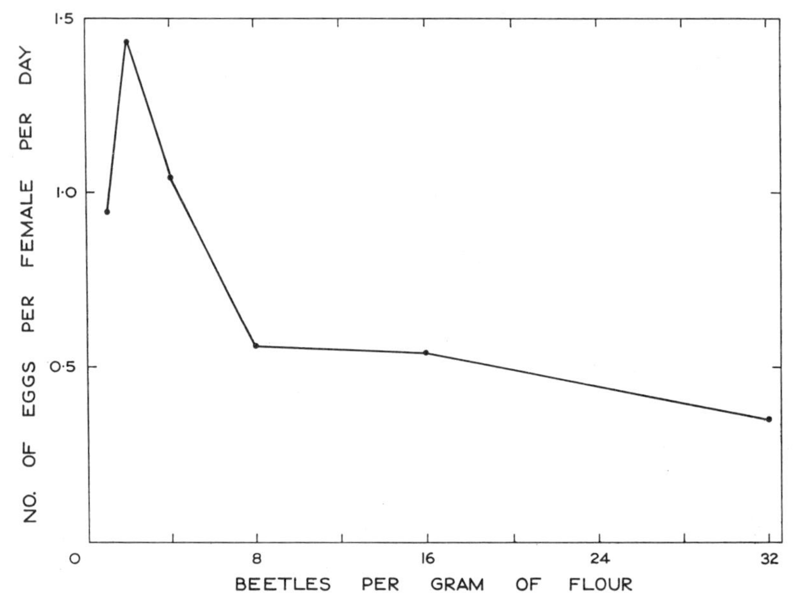

FIG. 9.14.—The relationship between the density of the population of adults of *Tribolium confusum* in flour and their fecundity. (After MacLagan, 1932*b*.)

tween density and fecundity is curvilinear (Fig. 9.16); but when food is adequate, fecundity is largely independent of density (Robertson and Sang, 1944). This shows that in *Drosophila* crowding by adults per se has little influence on fecundity. In this respect *Drosophila* resembles *Tribolium* when crowded with males, but not with females.

The third and final component of r_m which can be investigated in this way is the speed of development. Density has been found to have no influence on the speed of development of the larval stages of several species of blowflies (Ullyett, 1950), of *Calandra oryzae* (Birch, 1945*c*), *Callosobruchus chinensis* (Utida, 1941*a*), and *Tribolium confusum* in "fresh" flour, i.e., flour that has been renewed frequently so that it did not become "conditioned." The only known exceptions to this rule are the larvae of *Tribolium* in "conditioned" flour (Park, 1938; Fig. 9.17) and the larvae of *Drosophila* in the usual culture of living yeast (Fig. 9.09). The former represents only an indirect influence of density on speed of development; and the latter is not due to the influence of crowding as such but to a shortage of food, the growing yeast population having failed to replenish itself after being "eaten out" by the dense population of *Drosophila* (Sang, 1949). Both with *Tribolium* in "conditioned" flour and with *Drosophila*, the relationship between density and speed of development is approximately linear.

With sufficiently detailed information about the longevity, fecundity, and

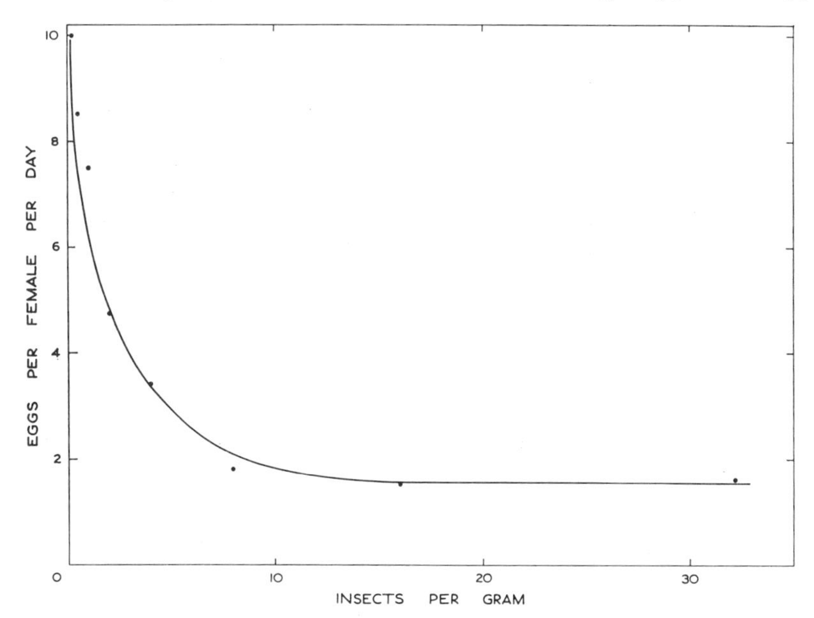

FIG. 9.15.—The relationship between the density of the population of adults of *Rhizopertha* and their fecundity. They were living in wheat at 30° C. and 70 per cent relative humidity. (After Crombie, 1942.)

speed of development of a species, its rate of increase may be calculated for the particular environment for which these quantities have been measured (chap. 3). It follows that, with sufficiently detailed information about the influence of density on these three components of *r*, it would be possible to evaluate the influence of density on the rate of increase. No one has ever attempted this enormous task: none of the experiments we have just been discussing represents more than a very beginning of the work that would be required. But several experiments have been done which aimed to measure directly the influence of density on the rate of increase within a rather narrow part of the whole field.

For example, Pearl (1926) started colonies of *Drosophila* with varying numbers of flies and counted the average number of surviving progeny per female at the end of 16 days. He had to keep the period fairly short, or else the density would have changed too much between the beginning and the end of the experiment. This and other obvious defects of this sort of experiment make it advisable to interpret the results as no more than an approximate guide to the influence of density on rate of increase. The results of Pearl's experiments indicate that the relationship between density and number of progeny per female was hyperbolic (Fig. 9.18). Utida (1941*a*, *b*, 1942) did a similar experiment with the beetle *Callosobruchus chinensis*. With this species the adults live for about 10 days, while the immature stages occupy about 4

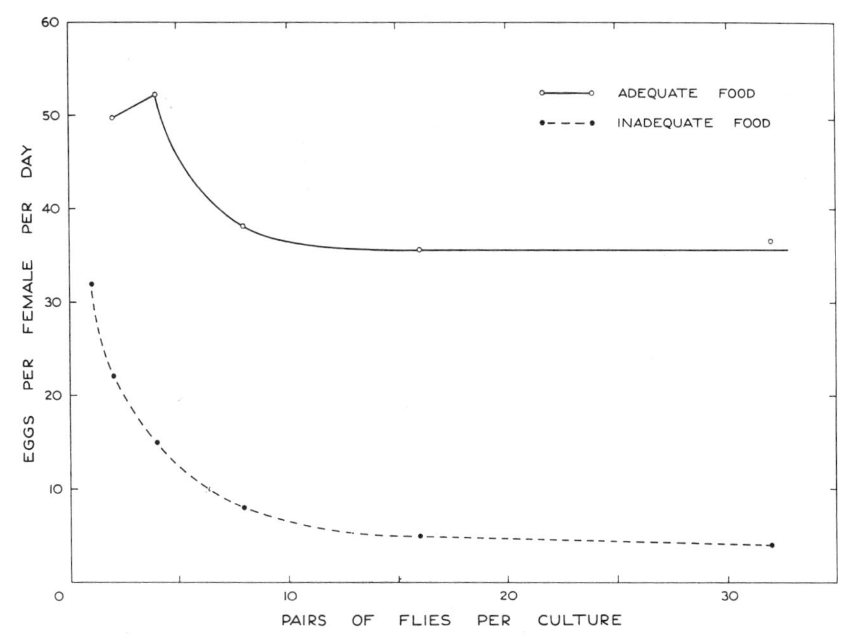

FIG. 9.16.—The relationship between the density of the population of adults of *Drosophila* and their fecundity, both when food was maintained in adequate supply, irrespective of the density of the population, and when the supply of food was limited. (After Robertson and Sang, 1944.)

weeks. It is thus possible to record the progeny of a complete generation by counting the adults that emerge. The curve which relates density at the beginning of the generation to the number of progeny per female per generation approximates to a hyperbola (Fig. 9.19). Pratt (1943) did some similar experiments with *Daphnia magna* and found that after a certain period the relationship between density and live progeny per female became hyperbolic.

9.235 THE INSTANTANEOUS RESPONSE IN THE RATE OF INCREASE TO CHANGES IN DENSITY

Equation (2) (sec. 9.21), was written: rate of increase in absolute numbers is equal to

$$rN = \frac{\delta N}{\delta t} = (r_m - cN)N.$$

It is obvious, since this equation does not have any modifying factor associated with cN, that the equation implies that the quantity in parentheses is always less than r_m by a quantity exactly equal to cN, i.e., that the quantity cN changes as soon as N changes. If this were not to be implied, the equation would contain a modifying factor for time associated with cN. But then the equation would no longer describe a simple sigmoid curve like the logistic. In this section we have to consider whether it is likely that rate of increase in

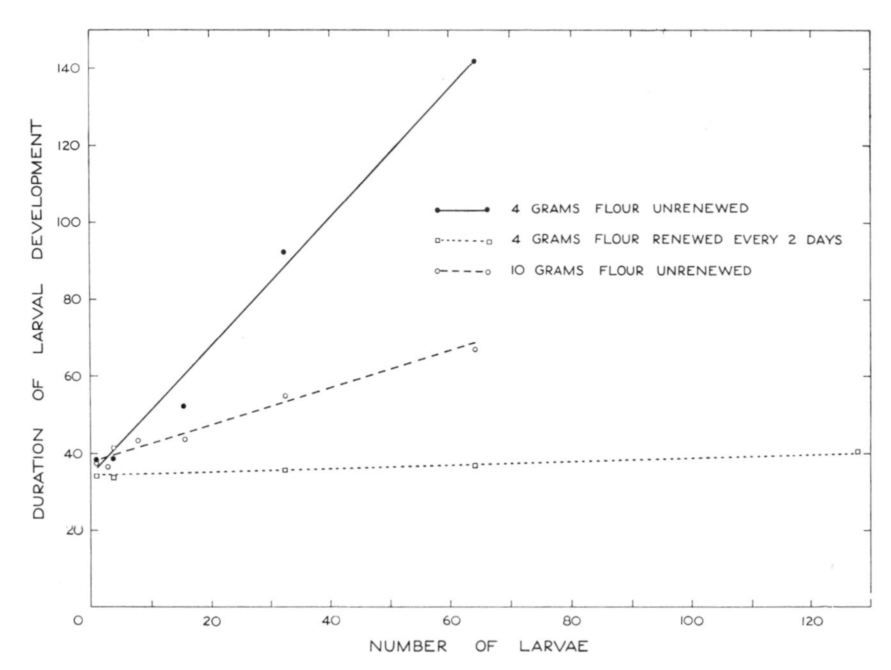

FIG. 9.17.—The relationship between the density of the population of larvae of *Tribolium confusum* and the duration of the larval stage. They were reared in flour at 28° C. and 48 per cent relative humidity. (After Park, 1938.)

experimental populations growing in limited space does, in fact, react immediately to changes in density, as implied by the logistic equation. There are three aspects to this problem:

a) A number of experimenters have measured the interval which elapses before longevity, fecundity, or speed of development begins to react to changes in density. As in the preceding section, this sort of experiment with the individual components of r is practicable, whereas the more comprehensive one, which attempts to measure the same thing in relation to r as a whole, is technically too difficult.

b) Certain attributes of animals with complex life-cycles make it unlikely (on a priori grounds) that rate of increase would react immediately to changes in density.

c) In the case of animals with simple life-cycles (like *Paramecium*) there is internal evidence, provided by the growth of the experimental population itself, that rate of increase may react immediately, or nearly so, to changes of density.

We shall discuss these three aspects of this problem in this order.

With those species in which crowding reduces the average length of life of

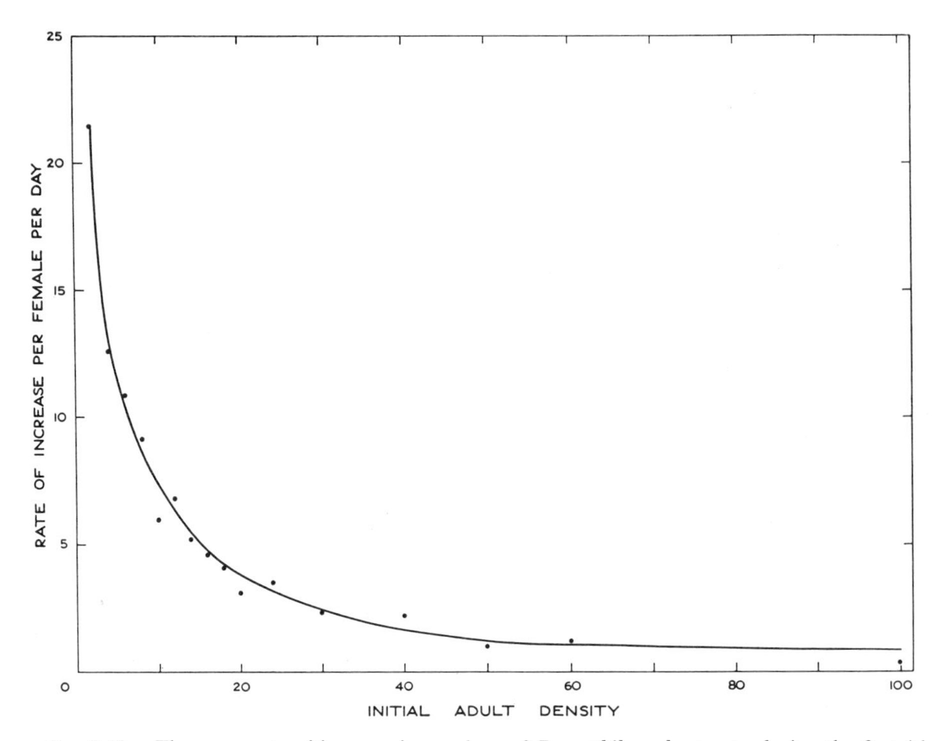

FIG. 9.18.—The mean rate of increase in numbers of *Drosophila melanogaster* during the first 16 days after the colony was founded. They were all reared in similar cultures at 25° C., but each colony was started with a different number of flies. (After Pearl, 1926.)

the individual and therefore the survival-rate in the population, the immediate cause of death may be different with each species. With *Tribolium confusum* the insects die because the flour has been "conditioned" (sec. 9.234). In this case it is inconceivable that the addition of another individual to the culture should immediately decrease the survival-rate in the population; for both the "conditioning" of the flour and the killing by "conditioned" flour take time. The individuals that are destined to die because another has been added to the population must first sicken; and, indeed, their death is likely to be quite a protracted affair. Empirical evidence for this is supplied by Park's (1938) experiments. He found that when larvae of *T. confusum* were crowded in flour that was not changed frequently, relatively more of them died without reaching the pupal stage than from another lot which had been crowded to the same extent in flour that had been kept "fresh" by frequent changes. But he also observed another effect of crowding: of those that pupated, relatively fewer survived to become adult among the former than among the latter; in this case the outcome of crowding during the larval stage was not fully evident until much later in the pupal stage.

With the adult stage also, the death-rate may not rise as soon as crowding

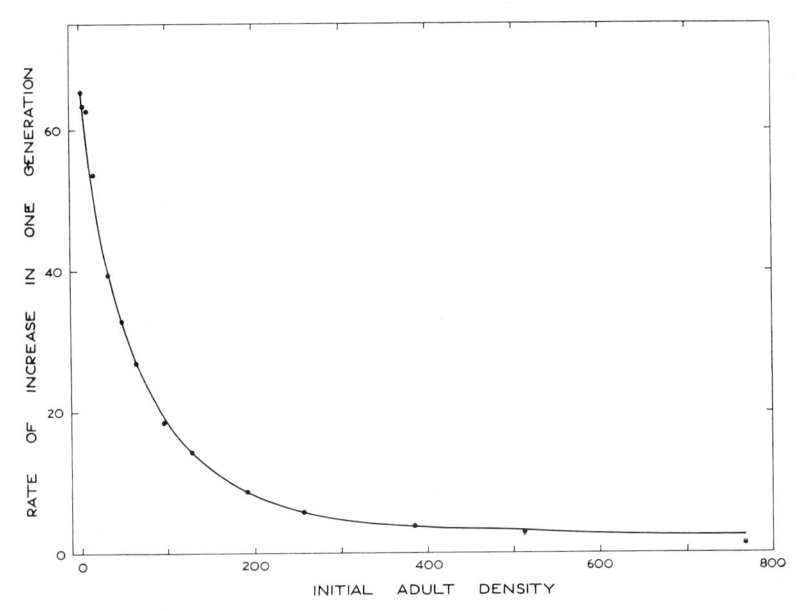

FIG. 9.19.—The number of living offspring per generation of the Azuki bean weevil *Callosobruchus chinensis* from colonies which had been started with different numbers of adults. They were all reared at 24.8° C. and 74 per cent relative humidity. (After Utida, 1941*a*.)

begins, but only after an interval. Thus Pearl, Miner, and Parker (1927) found that the life-expectancy of adult *Drosophila melanogaster* when 16 days old, was influenced by the density experienced during an earlier period of their lives. Flies which started life at a density of 35 per 1-ounce vial were transferred to a density of 200 on the sixteenth day of their lives. Their length of life was greater than that of another group which spent the first 16 days at a density of 200.

Ullyett (1950) crowded larvae of blowflies in circumstances rather different from those in the sort of experiments we have been discussing in this chapter. A fixed quantity of food was supplied at the beginning of the experiment and was not renewed. In these circumstances crowding resulted in a shortage of food, and the larvae suffered from starvation and malnutrition. More of the pupae from crowded cultures failed to become adults than from the less crowded cultures in which the larvae had plenty of good food (Fig.

9.20). This showed that the influence of starvation on the survival-rate was not immediate; but it is doubtful whether starvation is ever important in well-planned experiments designed to test the logistic theory; in such experiments food should be kept in adequate supply and only space should be limited.

In Ullyett's experiments the larvae that had been crowded, and therefore starved, gave rise to small flies, which laid fewer eggs than normal flies which had not been crowded as larvae (Fig. 9.21). This shows that the influence of

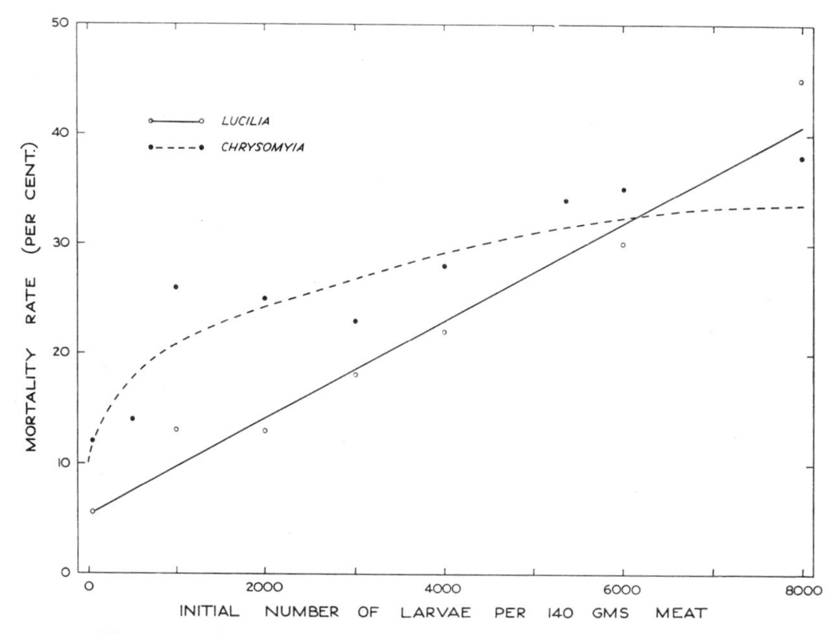

FIG. 9.20.—The relationship between pupal mortality and initial larval density of the blowflies *Lucilia sericata* and *Chrysomyia albiceps*. (After Ullyett, 1950.)

starvation on fecundity need not be immediate; but again it is doubtful whether starvation is relevant to the present discussion. However, it is known that adults from crowded cultures of *Calandrd oryzae* (Birch, 1953*b*), *C. granaria* (Richards, 1948), and *Tribolium confusum* are smaller than normal adults. It is almost certain that this is not due to shortage of food. These small adults, like the blowflies of Ullyett's experiments, probably lay fewer eggs. A similar effect has been shown for the moth *Plodia interpunctella* (Snyman, 1949). When the larvae were crowded, the adults laid fewer eggs. Similarly with *Drosophila*, the adults that emerged when the larvae were reared in a crowd were smaller and laid fewer eggs than those which had not been crowded as larvae. Like Ullyett's blowflies, this was due to shortage of food and malnutrition, for in the ordinary *Drosophila* culture the flies feed on living yeast and the food is not therefore supplied at a constant level, as it should be (Robertson and Sang, 1944; Chiang and Hodson, 1950).

There is, however, at least one circumstance in which fecundity responds immediately to changes in density. When adults of a number of grain beetles were crowded together, there was an immediate reduction in the rate at which eggs were laid (Crombie, 1943). It has been suggested by MacLagan (1932*b*) in respect to *Calandra oryzae* that this may be due to the increased "jostling" by the insects, which disturb one another at oviposition. Whenever crowding influences the behavior of the animal in some such way as this, one would

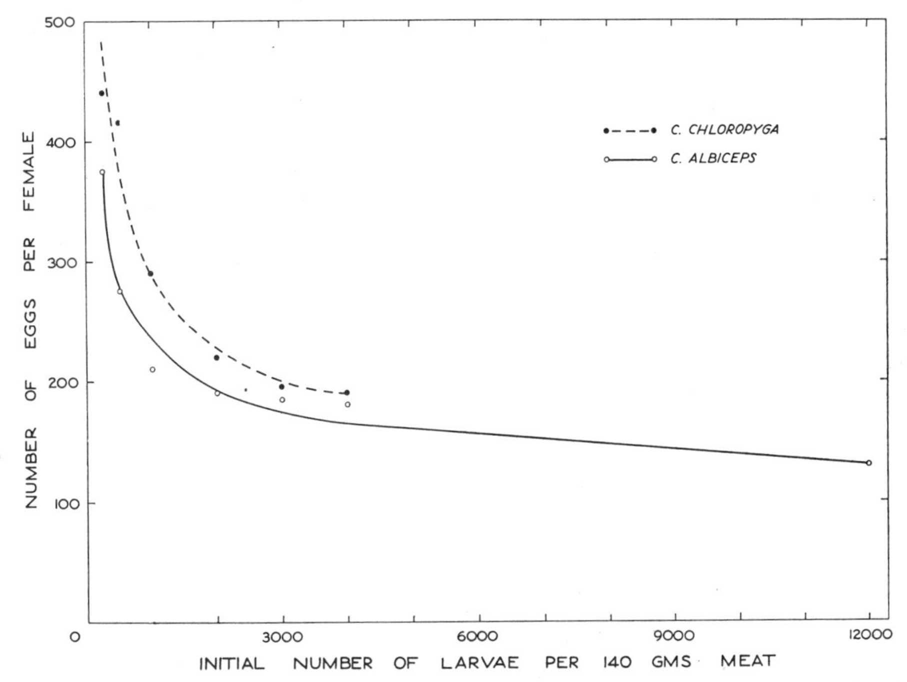

FIG. 9.21.—The relationship between initial larval density and the number of eggs laid per female by the blowflies *Chrysomyia chloropyga* and *C. albiceps*. (After Ullyett, 1950.)

expect the fecundity to react immediately to changes in density. But it is difficult to imagine any other circumstance in which the reaction would be immediate. It has already been pointed out (sec. 9.234) that density is not known to influence the speed of development of insects to any important degree. And this may be neglected in considering "time lag."

We have seen in the preceding paragraphs that changes in the birth-rate and death-rate, which are a consequence of changes in the density of the population, are often not evident at the time when crowding occurs but become manifest only at a later stage in the development of the animal or the population. These are qualities which require to be demonstrated for each species in particular, because there is no other way by which we can infer that they are likely to hold for most species with complex life-cycles. There is however, a quality possessed in general by all animals with complex life-cycles, which makes it seem likely that for such animals there will usually be a delay between the time that density changes and the time when the rate of increase begins to react to the changed density. This can be readily appreciated by comparing the simple life-history of *Paramecium* with the complex one of *Drosophila*.

In the logistic equation, N stands for the number of individuals, and it is implicit in the equation that each individual is equivalent to any other in its requirements for food and space. That is, the variability in the rate of increase brought about by the addition of any individual chosen at random should fall within the limits proper to a homogeneous population. There should not be, in the population, groups which are significantly different from one another. Now a population of *Paramecium* may be homogeneous in this sense, or nearly so; but a population of *Drosophila* certainly is not. It includes eggs, larvae, pupae, and adults. Each stage makes a different contribution to density, and, at least with larvae and adults, this varies with their age also. The addition of n eggs to the population is not the same as the addition of the same number of larvae or adults. At a particular stage of its growth the population may be able to accommodate, without any immediate reduction in the rate of increase, the addition of n eggs, whereas the addition of the same number of adults might result in a large reduction in the rate of increase. Let us suppose these n eggs were added to the population. In due course they became adults, and the rate of increase was depressed accordingly. The interval that elapsed between the laying of the eggs and the emergence of the adults is a measure of the "time lag" which occurred in the influence of density on the rate of increase. This is a particular example of a more general quality possessed by animals with complex life-histories. Whenever one stage in the life-cycle differs significantly from another in its contribution to density, then a "time lag" will be generated in the way illustrated for *Drosophila*.

On the other hand, when the population is homogeneous, or nearly so, as may be the case with *Paramecium*, there is no reason to expect a "time lag" from this cause. Referring back to section 9.221 and Figures 9.01 and 9.02, it will be seen that the theoretical logistic curve follows the observed data fairly closely for most of the time. Unless there have been compensating errors (which seems most unlikely), this must be taken to indicate that the *Paramecium* population behaves in each detail in the way postulated by the logistic curve. This may be taken as internal evidence that, with animals with simple life-histories, the response of rate of increase to changing density is immediate or nearly so.

9.236 CONCLUSIONS RELATING TO THE "REALITY" OF THE LOGISTIC THEORY

The growth of an experimental population of animals with complex life-cycles does not, as a rule, conform closely to the theoretical curve except for a small part of the data: in some instances most of the first phase of steadily increasing density may be represented by part of a logistic curve, and in others no more than a portion even of this phase can be so fitted. In order to find a curve that will fit even this limited part of the data, it is usually necessary to assume quite an unrealistic value for K which may not be related to the observed data in that part of the curve. Therefore, these populations, unlike those of *Paramecium*, do not provide any internal evidence for the reality of the assumptions on which the logistic theory is based. Consequently, conclusions must be based on the "external" evidence, which was discussed in sections 9.231–9.235.

It was shown in section 9.233 that the innate capacity for increase (r_m) may be accepted as a real and consistent attribute of a species. If an experimental population started with a stable age-distribution and maintained this as it grew and, in fact, its growth conformed to the logistic theory, then the value of the constant r_m in the logistic equation which could be fitted to such data would truly represent the innate capacity for increase. But experimental populations have always been started with populations with other age-distributions. Therefore, the constant given by such data cannot theoretically be the true innate capacity for increase but a quantity appropriate to some other age-distribution. This, of course, will not be constant but will vary with time as the age-distribution of the population changes. The discrepancy between theory and fact may not in this case be large enough to be important, since Birch (1953*b*) found that the values of r_m for *Calandra* and *Rhizopertha* determined by the methods of this chapter and of chapter 3 agreed fairly closely (Table 9.07).

Section 9.234 resulted in no clear-cut conclusion regarding the relationship between density and the rate of increase. A number of experiments indicated that one or another of the components of r might be a linear function of density for certain species, but rather more indicated a curvilinear (perhaps hyperbolic) relationship. On the whole, it would seem likely that the graph relating density to rate of increase would not very often be a straight line.

In section 9.231 it was pointed out that we did not know of any experiments that had been started with a population of stable age-distribution. This defect is likely to result in a curve which is too steep at first, which climbs to an excessive density, and then enters upon a series of fluctuations. In special circumstances the curve will approach an asymptote by a series of damped oscillations; but we cannot assess the chance that these circumstances will occur (sec. 9.231). These conclusions are based on theoretical reasoning (Leslie, 1948); but, at the same time, this is a just description of experimental populations of *Calandra* and other species which had been started with a single pair of adults instead of a population of stable age-distribution.

Nevertheless, this may not be the full explanation of the discrepancy between the observed growth curves for these populations and the theoretical logistic curves. In section 9.235 it was shown that occasions on which the rate of increase responded immediately to changes in density must be exceptional, if they occur at all, in experimental populations of animals with complex life-cycles. The presence of a "time lag" in this relationship might be expected to produce deviations from the theoretical curve rather like those which occur when the colony is founded with a population which contains a greater proportion of individuals of reproductive age than would be found in the stable age-distribution. The rate of increase, at first, will be much higher than would be expected by the theory: later it may be lower. This will result in a curve which will at first be too steep; it will tend to rise quickly to an unduly high density; this may be followed by a decline, thus generating a series of oscillations.

A striking, and perhaps unexpected, aspect of the oscillations in experimental populations of *Calandra*, *Tribolium*, and other species is their long period. For example, *Calandra oryzae* (small form) completed a cycle from peak to peak in about 45 weeks when the population was living in wheat at 29° C. (Fig. 9.05). Since a generation occupies, on the average, about 8 weeks, the cycle has a period of 5.4 generations. With *Tribolium castaneum* in flour at 29.5° C. the period between peaks is 44 weeks. At 28.5° C. a generation takes 8 weeks (Fig. 9.04; Leslie and Park, 1949). The cycle thus has a period of about 5.5 generations. Utida (1941*e*) found similar oscillations in the density of an experimental population of the beetle *Callosobruchus chinensis*. The cycle occupied a period of about 10 weeks; a generation occupied about 4 weeks. The cycle thus occupied about 2.5 generations. The long periods of these oscillations suggest that they may not have any simple explanation; the presence of a "time lag" in the reaction to density and the changing age-distribution are obvious causes, but there may be others.

We started section 9.23 with the statement that the chief attractions of the logistic theory were its simplicity and its "reality." By "reality" we meant that a realistic biological meaning could be attributed to each of its constants. Our analysis has shown that this may be true enough for animals with simple life-cycles like *Paramecium;* but for animals with complex life-cycles, like most insects, it would appear that the seeming "reality" of the logistic constants is largely fictitious, for they do not correspond at all closely to biological realities. Even the constant r_m of the logistic equation has been shown to represent the true innate capacity for increase only when the experimental population has a stable age-distribution and grows in accordance with the other postulates of the logistic theory: we have seen that this is not so for species with complex life-cycles.

Because of the theoretical shortcomings of the logistic equation, several workers have suggested alternative expressions. All these have had more constants and have been more complicated; and some of them have had constants to which no biological meaning could be attributed. So that these expressions, in addition to being more difficult to manipulate, are "meaningless" as well. Other types of sigmoid curve have been fitted to some of Gause's data for *Paramecium* (sec. 9.221) by Feller (1940), who found one expression that gave a fit at least as good as the logistic. The "Gompertz" curve is another sigmoid function that might be used to describe the growth of an experimental population (Winsor, 1932). Still others were suggested by Baas-Becking (1946). An alternative approach to the construction of a simple equation with a few constants is the complex equation with many constants. For example, the two constants r_m and c of the logistic curve (eq. [2]) might be broken down into components for longevity, fecundity, and speed of development; factors might be added for "time lag" and so on. An approach along these lines has been made by Stanley (1932). The resulting equation is exceedingly complex. For the experimenter it possesses the merit that he can alter one term at a time, such as the rate of egg-laying, and he can then follow the consequences of this in terms of the growth of the population. The theoretical model can be used as a plan for experimental populations and the results compared with the theory.

Mendes (1949) has also taken the basic equation of the logistic curve and elaborated it so that, instead of three constants, it includes about a dozen. In this case the constants apply particularly to a population of coffee-berry borers, and each constant has a biological meaning applicable to this species of insect. The disadvantages of these elaborate equations is that they are unwieldy. None is superior to the logistic in this regard. There is every advantage in seeking simplicity in an equation, provided that it can be shown to be a reasonably good description of the observed data. The importance of this is well put by Pearl (1930, p. 408): "He [the mathematician] furthermore knows that by putting as many constants into his equation as there are observations in the data he can make his curve hit all the observed points exactly, but in so doing will have defeated the very purpose with which he started, which was to emphasize the law (if any) and minimize the fluctuations, because actually if he does what has been described he emphasizes the fluctuations and probably loses completely any chance of discovering a law."

In section 9.22 we fitted logistic curves to data got by allowing colonies of animals to multiply in "constant environments" and "limited space" and then examined them to see how well the theoretical curve fitted the empirical data. In doing this, we judged the logistic equation by whether the resulting curve gave an adequate description of the data, or part of it. We were not seeking the expression which gave the best possible fit, irrespective of simplicity. Surprisingly enough, in the light of the discussion in section 9.23, we found that, at least in some cases, the logistic curve could be made to fit at least part of the data quite closely. The general conclusion to this section, therefore, is that, despite its theoretical limitations, the logistic curve remains a useful tool for the ecologist. Because of its limitations, too much reliance should not be placed on it in particular cases until it has been verified empirically for each case.

9.24 *Applications of the Logistic Theory*

When, for a particular set of data it may be shown empirically, or may be reasonably inferred, that the data, or part of it, follow sufficiently closely along a logistic curve, or portion of it, then the logistic equation provides the basis for several interesting and useful deductions about the population during that period of its growth which is faithfully described by the logistic curve. These relate to the influence of crowding on the rate of increase, r (sec. 9.241), and the relationship between innate capacity for increase, r_m, and the upper limit to density (sec. 9.242).

9.241 THE INFLUENCE OF CROWDING ON THE RATE OF INCREASE, r

Any component of environment which influences longevity, fecundity, or speed of development must thereby influence the rate of increase in the population. These components of the rate of increase reach a maximum when each animal in the population has neither too many nor too few other animals of its own kind around it; but the rate of increase may become zero or negative (in which case it becomes a rate of decrease) in very sparse or very dense populations. Although this has been repeatedly demonstrated qualitatively, it seems that the experiment that would measure quantitively the influence of density on rate of increase has yet to be devised. It is not necessary to elaborate all the reasons for this; chief among them may be the need to keep the experimental animal as one of a population of constant density, when the population, by its very nature, is constantly changing its density (sec. 9.234).

When a particular set of data may be considered to fit the logistic theory, then the equation provides an indirect method of measuring quantitatively the influence of density on rate of increase. From equation (2) we have:

$$\frac{\delta N}{\delta t} = N(r_m - cN).$$

At the upper asymptote $\delta N/\delta t = 0$; $N = K$. By substitution, this gives

$$r_m - cK = 0;$$

$$\therefore c = \frac{r_m}{K};$$

$$\therefore cN = \frac{r_m N}{K}.$$

But cN is the quantity by which r_m is reduced at any density N. Therefore, if we write r as the rate of increase at any density, N, we have

$$
\begin{aligned}
r &= r_m - cN \\
&= r_m - \frac{r_m N}{K} \\
&= \frac{r_m(K - N)}{K}. \qquad (7)
\end{aligned}
$$

This method of using the logistic equation to calculate the actual value of the rate of increase r_N for any particular density N was applied to data for *Paramecium caudatum* in half-loop medium (sec. 9.221, Fig. 9.02, and Table 9.05). The particular equation that was fitted to these data was

$$N = \frac{65}{1 + e^{3.185 - 0.855t}}.$$

Therefore, from equation (7), we obtain

$$r = \frac{0.855(65 - N)}{65}.$$

By substituting successively the values for N in the second column of Table 9.09, the quantities in the last column were obtained. This method gives values for r which approach r_m as N becomes very small and which approach

TABLE 9.09*
WAY IN WHICH LOGISTIC EQUATION MAY BE USED TO MEASURE INFLUENCE OF DENSITY (CROWDING) ON RATE OF INCREASE, r

Time (Days)	Density, N, = Numbers per Unit Volume as Given by Logistic Equation	Rate of Increase, r, as Given by: $r = \frac{r_m(K - N)}{K}$
0	2.0	0.83
1	5.8	.78
2	11.9	.70
3	22.7	.56
4	36.3	.38
5	48.7	.21
6	57.0	.10
10	64.7	.00
20	65.0	0.00

* Data for *Paramecium caudatum* from Gause (1934).

0 as N approaches K. Note that equation (7) gives $r = r_m$ when $N = 0$, whereas we know that r approaches r_m as a maximum at some optimal density greater than 0 (sec. 9.1). There is also a practical weakness in equation (7), because it does not provide for negative values of r. This is because the logistic

theory does not allow for oscillations in the density but only a steady increase, which gradually approaches an asymptote, K.

In deriving equation (7), we have taken Gause's concept of "environmental resistance" and twisted it to suit our own purpose. This phrase was coined by Chapman (1931), who undoubtedly meant it to refer to all components of the environment which might influence longevity, fecundity, or speed of development, i.e., he meant it to refer to weather, food, predators, etc., as well as that component of environment which we have called "other animals of the same kind." Gause took the phrase and used it with a much narrower meaning, within the limits of the logistic theory. He abstracted all these other components of the environment which might influence the rate of increase and assumed that the animal possessed a "potential" rate of increase as indicated by the constant r_m in the logistic equation. It realized this "potential" rate of increase only in a "perfect environment" in which there was no crowding. The addition of more animals to the "environment" conferred on it a "resistance" which prevented the full "potential" rate of increase from being realized. Gause measured "environmental resistance" as N/K. It follows that the "degree of realization of the potential rate of increase" is given by $(K - N)/K$. We think that this phrase and the concepts which it describes are chiefly of historical interest now. The profound separation between the animal and its environment which Chapman had in mind is a mistake because this does not happen in nature. Although the narrower concept of Gause has a more precise meaning, it seems not to be very instructive.

9.242 THE RELATIONSHIP BETWEEN r_m AND THE UPPER LIMIT TO DENSITY

The numbers attained by the experimental populations described in section 9.22 either remained constant (Fig. 9.02) or fluctuated profoundly (Fig. 9.05). The former is characteristic of simple animals like *Paramecium*, and in this case a logistic curve may be fitted which approaches an asymptote, K, which may be quite close to the observed numbers. The latter is characteristic of more complex animals like insects. In these cases a logistic curve which fits part of the observed data during the early phases of the growth of the population may approach an asymptote which may bear little relation to the observed data for the later stages of the growth of the population. We shall consider the former sort of population first. From equation (7) we may write

$$K = \frac{r_m}{c}. \tag{8}$$

For animals like *Paramecium*, with simple life-cycles, the asymptote K of the theoretical curve is close to the numbers that were observed in the population during the later stages of its growth. Thus we may deduce from equation (8) that the upper asymptote, K, which represents the limiting density for the population, is determined solely by the values of r_m and c, which are, respectively, the animal's innate capacity for increase and the density component of the logistic equation. This means, for example, that the upper limit to density is quite independent of the absolute size of the colony: this was verified empirically for *Tribolium* by Chapman (1928).

The theoretical dependence of the size of the upper asymptote upon the value of the innate capacity for increase of the species and the density coefficient, c, is worth stressing, since ecologists have sometimes implied that there is no possible connection between the upper limit to numbers and the capacity of a population to increase in numbers. Darwin rightly pointed out in *The Origin of Species* that there was no correlation between the reproductive rate (birth-rate) of a species and its commonness and rareness. But it has sometimes been erroneously implied that there is therefore no reason to expect a relationship between the capacity to increase and its numbers in nature. This is certainly not a corollary of Darwin's statement. The innate capacity of the species to increase, r_m, is a statistic which takes into account the very point which Darwin wished to stress: that death-rate as well as birth-rate must be considered when numbers in the population are under consideration. Some species, for example, marine animals with planktonic larvae, which have tremendous birth-rates have also tremendous death-rates, and so their capacity to increase bears little relation to the birth-rate.

The expression $K = r_m/c$ may be put into words thus: the upper limit to numbers in a population growing in conformity with the logistic theory is proportional to the innate capacity for increase, when the limiting density coefficient, c, is constant. It is indirectly proportional to the limiting density coefficient, c, when the innate capacity for increase is constant. Thus, in a comparison between two or more populations to which the logistic theory was applicable, a direct correlation between r_m and K would be expected only if c was the same for both populations. This argument will be clearer from a consideration of some hypothetical situations.

Situation A.—Let us imagine a population X which gives rise in evolution to an isolated population Y. Let us also suppose that population Y carries a gene or genes which confer upon it an innate capacity for increase which is greater than that of the parent-population X. We shall suppose that there are no other differences. The two species have the same rate of reduction in growth of population as density increases. In other words, the value of c in the simpler form of the logistic equation has not changed in the evolution of Y from X. Now the upper limit to numbers of the population of Y (the asymptote K) must be higher than that of X by an amount proportional to the difference between their innate capacities to increase.

The relationship between c, r_m, and K in this example can be shown graphically by dividing r_m into its two components—a birth-rate and a death-rate. The outcome of increasing density is to reduce the actual rate of increase by decreasing the birth-rate or increasing the death-rate or both. Assuming linear relationships between density and birth-rate and death-rate, the situation can be represented by the graph in Figure 9.22, *A*. The birth-rate and death-rate of population *X* at different densities is represented by the lines b_x and d_x. As the density of the population increases, the birth-rate exceeds the death-

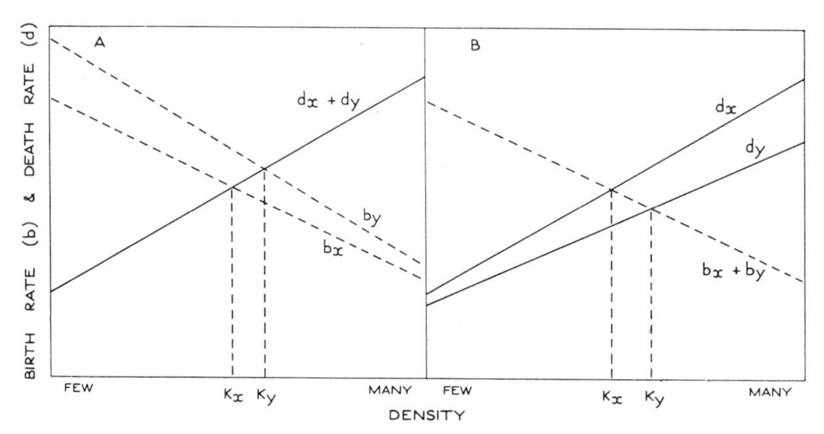

FIG. 9.22.—Showing the effect of increased birth-rate (*A*) and decreased death-rate (*B*) on the asymptotic density, when the proportional effect of density on these components remains unaltered.

rate until a density is reached at which b_x equals d_x. This is the density K_x, where the two lines intersect.

Consider now the second population *Y*, which has evolved a higher innate rate of increase. We shall examine the outcome of this on its upper asymptote K_y. Figure 9.22, *A* and *B*, represents the various ways in which the higher value of r_m may have been brought about. It could be due to a higher birth-rate and nothing else (Fig. 9.22, *A*). This is represented by the line b_y. It runs above b_x but is not parallel to it, as we have assumed that the *proportional* change with density is the same in both populations. The lines are separated by a constant proportion and not by a constant amount. The curve for the death-rate of population *Y* is the same as that of population *X*. The significant feature of this graph is that b_y intersects the line $d_x + d_y$ at a higher density, K_y, than compared with the corresponding intersection for population *X*. Furthermore, the death-rate in this population as it approaches its asymptote is higher. The corollary to be drawn from this example is that a population with a higher birth-rate has both a higher upper limit to its numbers and a higher death-rate as it approaches this upper limit.

A second way in which the innate capacity for increase of population *Y* might have become greater than that of population *X* is shown in Figure 9.22, *B*. The birth-rates of the two populations are identical, but the death-rate of population Y is less than that of population X. As before, K_y exceeds K_x; but in this case the death-rate of population Y at the density K_y is lower than that of population X at its asymptotic density, K_x. This is the opposite of the position in Figure 9.22, *A*.

Situation B.—In the previous examples we have assumed that the two populations varied only in their innate capacity for increase while the density factor, c, remained the same. It is, of course, conceivable that one population

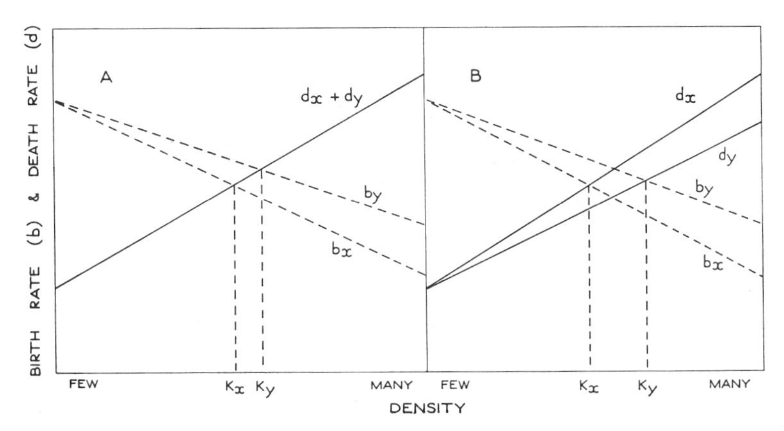

FIG. 9.23.—Showing variations in the asymptotic density which result from *A*, a reduction in the relative influence of density on birth-rate; and *B*, a reduction in the relative influence of density on both birth-rate and death-rate.

may differ from the second in its density factor, c, while the two populations may have identical values for the innate capacity for increase, r_m. Figure 9.23, *A*, represents a situation in which the density factor c is smaller in population Y than in X. In this example this happens to be due to the smaller influence which density has in reducing the birth-rate. It is assumed that the death-rate remains the same in the two populations. The outcome of this is to raise the value of K. This would be increased still higher if the influence of density on death-rate were also relatively less (Fig. 9.23, *B*).

Situation A is summarized in Figure 9.22, *A* and *B*, which indicates a correlation between r_m and K. Situation B is summarized in Figure 9.23, *A* and *B*, which indicates the absence of such a correlation. In nature either might happen, but it may be that in natural populations the possible combinations are limited so that not all conceivable theoretical possibilities occur. Much more experimental data are necessary than are available to establish these points. It is, however, of interest to compare what data are available.

Table 9.10 shows the values of r_m and K for *Paramecium aurelia* and *P. caudatum* reared in the way described in section 9.221, in which the only difference between the two experiments was that one population was limited to half the amount of food maintained for the other. The values shown in Table

TABLE 9.10*

VALUES OF r_m AND K IN CALCULATED CURVES FOR GROWTH OF *P. aurelia* AND *P. caudatum* IN "FULL-LOOP" AND "HALF-LOOP" MEDIA

	FULL-LOOP			HALF-LOOP		
	r_m	K		r_m	K	
		No.	Volume		No.	Volume
P. aurelia........	1.021	450	176	1.460	240	94
P. caudatum......	1.015	140	140	0.855	65	65

* Data from Gause (1934).

9.10 are abstracted from Table 9.05. The upper asymptotes are expressed as volumes as well as numbers. Since the two species vary in size, it is more reasonable to make comparisons between total volumes. We may compare the values of r_m and K obtained in either the "full-loop" or the "half-loop" medium, but we are not justified in making comparisons between experiments in which the quantity of food was different. The reason for this is simply that the double quantity of food in one series compared with the other series necessarily permits a higher number of organisms to exist in it, and therefore a higher value for K would be expected, and, of course, the table shows this. However, within the limitations of the two valid comparisons, there is clearly a correlation between r_m and K in these experiments. The data are too limited for any further analysis than this. Unfortunately, we know of few other experiments which might serve to test the hypotheses discussed in this section. Information which can be drawn upon from animals with complex life-histories will be discussed below.

The growth phase of the populations discussed in section 9.222 is not followed by constant numbers but by oscillations of quite large amplitude. We now consider whether the densities represented by the peaks of these oscillations bear a relation to the innate capacity for increase, r_m.

a) Calandra oryzae and Rhizopertha dominica.—Populations of the grain beetles *Calandra oryzae* and *Rhizopertha dominica* have been studied at different temperatures in different foods and at different relative humidities. In this way the innate capacity for increase of the populations has been varied at will. The values of r_m in these different circumstances have been calculated from the life-tables and age-specific fecundity-rates by the procedure outlined in chapter 3. The growth of populations from the time that they began as a single pair of newly emerged beetles has been followed by counting the number of adults present. The insects were provided with 12 gm. of grain as food, which was replenished at 2-week intervals. The experiments were continued for periods up to 3 years. Growth curves of some of the populations are shown in Figures 9.05 and 9.06, which illustrate how the densities about which numbers oscillated varied for each experiment. In Figure 9.05, for example, the density of the large "strain" of *C. oryzae* was never as high as the maximal density of the small "strain," nor was the density of the small "strain" as low as that of the large "strain" in the troughs of the oscillations. Rather than compare the mean densities about which numbers appear to oscillate, it seems more realistic to compare the maximal numbers reached in the population. An estimate of maximal numbers is based on the mean of 15 replicates at two stages in their histories when the growth curves reached peaks. These values for seven series of experiments have been plotted against the values of the

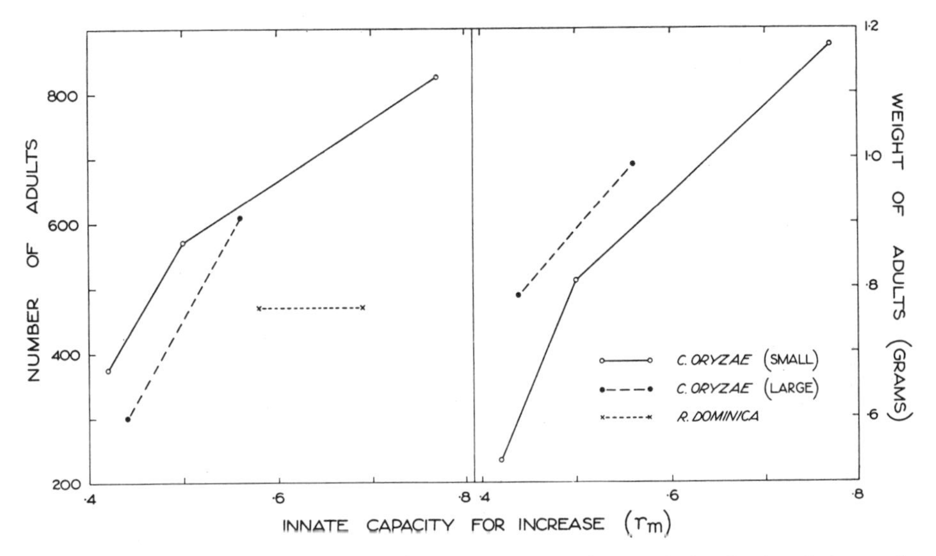

FIG. 9.24.—The upper limit to density of populations of three grain beetles in relation to the innate capacity for increase, r_m. *Left:* the densities of the populations are expressed as numbers of adults living in 12 gm. of grain; *right:* the densities of the populations are expressed as the weight of the adults in 12 gm. of grain. (After Birch, 1953*b*.)

innate capacity for increase in Figure 9.24, *left*. For the small and the large "strain" of *C. oryzae*, the greater the innate capacity for increase, the greater the maximal numbers of animals in the population. More data are available for the small strain, and this shows that the relationship is not linear but that the curve flattens out as r_m increases. Clearly, there is a limit to the numbers of animals which can occupy a limited space, and this is evidently approached as r_m reaches its maximal possible value. Data for the large "strain" are not sufficient to indicate how the curve may flatten as r_m increases. In contrast to these two insects, the two values of maximal numbers for *R. dominica* show no trend with increase in the value of r_m; perhaps this is because these values may lie on the flattened part of what is a curvilinear relationship. The results as a whole suggest a relationship between r_m and the maximal numbers in the population, but the relationship is different for the different species. Figure 9.24, *right*, shows that when the densities of the populations at their maxima are expressed in units of weight instead of numbers, the general slope and shape of the curves are the same, but the relative positions of the curves for the two "strains" are reversed. At their maxima the small "strain" had greater numbers than the large "strain," but, since the individuals were relatively much lighter, the populations of the small "strain" weighed less than the populations of the large "strain" at their maxima.

b) *Tribolium confusum and T. castaneum.*—A relationship between r_m and the upper density is clearly evident in Holdaway's (1932) data for *Tribolium*

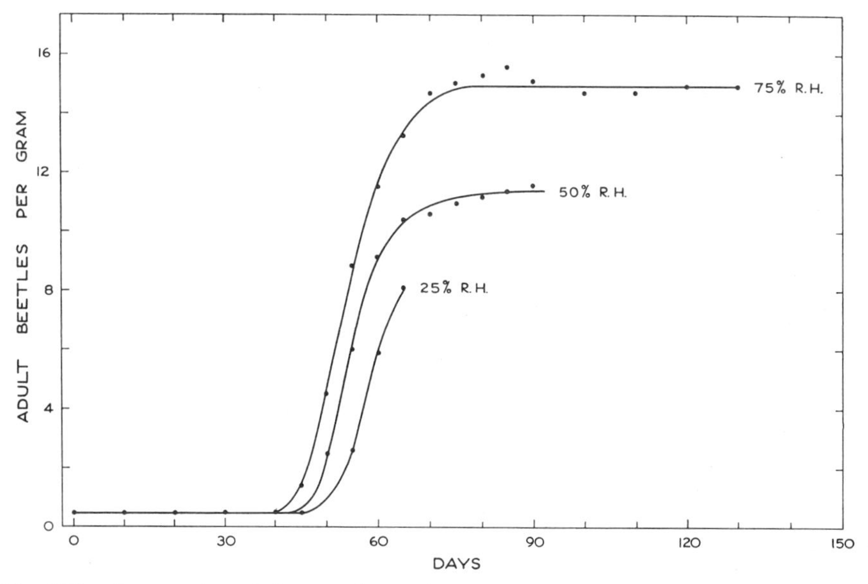

FIG. 9.25.—A comparison of the growth-rates in populations of adults of *Tribolium confusum* at relative humidities of 25, 50, and 75 per cent at 27° C. (After Holdaway, 1932.)

confusum at different relative humidities. Figure 9.25 shows the curves for the growth of populations based on the counting of adults at three relative humidities—75, 50, and 25 per cent. Although the values of the innate capacity for increase have not been determined precisely for these three humidities, the value of r_m must be lower at 25 per cent relative humidity than at 50 per cent and lower at 50 per cent than at 75 per cent. Figure 9.25 shows that the maximal densities vary in this order. Holdaway considered that this was due to lower net fecundity and slower speed of development at the lower humidities. The lower net fecundity is due to more eggs being eaten at lower humidities. This therefore provides a nice example of a correlation between moisture in the environment, the innate capacity for increase, and the upper limit to numbers. Park (1954) has summarized his extensive data on populations of *T. confusum* and *T. castaneum* in a way which confirms and extends the conclusions we have drawn from Holdaway's data. Table 9.11 shows the mean total population of flour beetles for monthly censuses covering a period of over 2 years of oscillations in numbers. Both temperature and moisture had a marked influence on the numbers in the populations; the mean density was greater at 70 per cent relative humidity than at 30 per cent for each temperature, except at 24° C. for *T. confusum.* The influence of moisture was very much greater for *T. castaneum* than for *T. confusum.* The numbers of *T. confusum* tended to be greatest at 34° C. and those of *T. castaneum* were greatest at 29° C.:

TABLE 9.11*
MEAN TOTAL POPULATION OF ADULTS, LARVAE, AND PUPAE OF *T. confusum* AND *T. castaneum* EXPRESSED AS NUMBERS PER GRAM FOR MONTHLY CENSUSES OF POPULATIONS FOR PERIOD OF OVER 2 YEARS

Temp. (° C.)	Relative Humidity (Per Cent)	*T. confusum*	*T. castaneum*
34	70	41.2	38.3
34	30	23.7	9.5
29	70	32.9	50.1
29	30	29.7	18.8
24	70	28.2	45.2
24	30	30.7	2.6

* After Park (1954).

One feature of particular interest in considering the upper density is the relation of the upper density to the amount of food used. It is a striking feature of experimental populations of *Tribolium, Calandra,* and *Rhizopertha* that the food which is replenished at regular intervals is never completely used by the populations. Even at the highest densities, there is always plenty of medium left over in the *Tribolium* cultures when it is replenished. It is not known, however, what the food value of this is. But with the other two beetles, the maximal amount of whole grain used during the 2 weeks was always a fairly small proportion of the total grain available. In cultures of the small "strain" of *Calandra oryzae,* for example, 3 gm. on the average, were utilized out of 12 gm. available during the 2-week period when the numbers were at the high level of 800 adults, with the associated immature stages, which were not counted. This shows that the attempt to keep food constant in these experiments, as well as temperature and moisture, has been reasonably successful. It seems unlikely that the limiting factor is the value of the food, such as that reported by Richards (1928) for yeast cultures, since it is a common observation that if a number of beetles are left in a tube with a certain amount of grain which is not replenished, they will eventually reduce the wheat to a dust. In the population experiments in which the food is replenished, there are usually numerous whole grains as well as husks left in the culture after 2 weeks. It is difficult to understand why the population does not increase to the point where all the food will be consumed during the interval between changes. Evidently, density imposes limitations directly on population increase in the confined space which have little or nothing to do with the amount of food that is available; there may be some cause, such as cannibalism of larvae by larvae, which is of much importance, or perhaps there is a number of causes. This problem will hardly be solved until more is known about the qualitative food requirements of the insects and the way in which the food value of the remaining food drops when insects are feeding on it. Even so, it is not difficult to see in principle how the upper limit to numbers may be determined by the influence of density on birth-rate and death-rate rather than by the amount of food present. This has already been illustrated in another context in Figure 9.23, *B* which illustrates a simple hypothetical model in which the upper density is dependent upon the birth-rate and death-rate. We do not suppose that the hypothetical model corresponds with the experimental models we have been discussing; it merely indicates one possible way of interpreting these results.

9.243 CONCLUSION

The experimental study of the growth of a population in a limited space where temperature and food are kept constant reveals a complex set of interactions which influence the density and form of growth of the population. Some aspects of this problem have been investigated more thoroughly than others. It is clear that there are big differences in detail between species, and therefore generalizations, even within the limited fields of experimental populations, are difficult to make.

The approach of this chapter has been to consider relatively simple situations and to stress the need for an appreciation of additional complexities in other populations. The logistic theory may be applied strictly to only one set of limited circumstances; this has often been overlooked. The circumstances in which the growth of a population would be expected, on theoretical grounds, to conform to the logistic theory are these: when the population at the beginning has a stable age-distribution; when the effect of crowding is to reduce the birth-rate and survival-rate in such a way that the influence of density on the rate of increase is linear; and when the influence of density is instantaneous, i.e., without lag. Accordingly, as the circumstances of the experiment and the responses of the animals depart from these requirements of the theory, so might the growth form of the population be expected to depart from the form of the logistic curve.

On the whole, the growth of populations does follow an S-shaped curve, but the postulated asymptote of the logistic curve may be replaced by oscillations, especially with animals which have complex life-histories such as holometabolic insects. However, the particular limitations of the logistic theory should not blind us to its possible usefulness and the possibility of further developments in this theory as more knowledge accumulates from experiments.

The experimental populations discussed in this chapter are abstractions from nature. Those in which populations were reared for long periods of time differed from natural populations in one important respect: the food was replenished at regular intervals by the experimenter. In nature the food of animals is derived from either plants or other animals which themselves form populations in which the numbers fluctuate. The experimental population of the sort described in this chapter is thus a simplification, because food is kept constant by the experimenter. There are, of course, advantages in analyzing the situation in which food does not vary. The study of the growth of an insect population on a growing tree, for example, presents all sorts of technical difficulties. We shall, of course, wish to use the information obtained from experimental populations in the interpretation of populations in nature, but in so doing, it is best to be cautious, for it would be naïve not to recognize that the experimental population is an abstraction from nature.

9.25 *The Reduction of r in Crowded Natural Populations*

In sections 10.322 and 10.323 we mention how predators and diseases may be more likely to cause a high death-rate in a dense population than in a sparse one. In section 11.22 we show how an absolute shortage of food may be accentuated by the presence of too many animals having to share the same limited stock of food. And elsewhere we discuss other indirect aspects of crowding. In this section we have to consider only the direct outcomes of crowding: how the presence of too many animals of the same kind in an area may reduce the chances of each one of them to survive and multiply by virtue of the direct influence which they exert on one another.

Certain experiments dealing with this problem were discussed in sections 9.22 and 9.23. These were all done with insects. We do not know of any parallel examples from natural populations of insects. Certain other sorts of animals commonly suffer from an absolute shortage of space. This may be observed among rodents, birds, and fishes and other vertebrates which recognize "territory" and often seem to need, or demand, room out of all proportion to their numbers (secs. 10.322 and 12.31). It may also be observed among certain marine invertebrates, such as intertidal barnacles.

In the district around Lake Vyrnwy, Wales, during 1937 and 1938 the voles *Microtus agrestis* were numerous. During 1938 they declined to very low numbers, and during 1939 they became still fewer. It is characteristic of this species that the numbers decline abruptly after they have been high for a year or two. Chitty (1952) made a thorough study of the ecology of the vole in this area during this period and concluded that the abrupt change from large positive to large negative values for *r* could hardly have been caused by weather, shortage of food, migration, predators, or disease. In fact, there were no obvious differences between the "good" vole years and the "bad" vole years. Chitty considered that there were probably five related causes for the decline in numbers which was observed during 1938 and 1939. We give these in his own words: "(1) Strife during the breeding season resulted in (2) the early death of the young and physiological derangement among the adults. (3) The later progeny of these adults survived, but (4) were abnormal from birth and thus more susceptible to various mortality factors. (5) These constitutional defects, in more severe form, were transmitted to the next generation."

Chitty's hypothesis is interesting because it suggests that *r* may be reduced not only by an increased death-rate from fighting, especially among the young, but also by a reduced birth-rate due to "physiological derangement" which persists for several generations, even though the population may, in the meantime, have become quite sparse. The last point has not been confirmed by experiment. But Clarke (1953) showed that when strange voles were placed in a cage where a colony was already established, the residents attacked the strangers viciously. Those which experienced the disadvantage of fighting on strange territory (and it is a very grave one) suffered, in addition to their wounds, marked increase in the weights of adrenals and spleen and involution of the thymus. These changes in the adrenals and thymus may have been caused by a disturbance in the normal functioning of the adrenopituitary system. Chitty (1952) remarked that "voles seem to be much less tolerant of crowding at the height of the breeding season than they are in autumn or winter."

Chitty considered that his hypothesis was supported by published observations on the ecology of the snowshoe hare, *Lepus americanus.* We give his argument in his own words:

> The best evidence about the course of a decline is given by Green and Evans (1940). Their study began at the end of a period of increase, maximum numbers being found in the 1933 estimates of spring abundance. On one square mile of the area, however, the peak year was 1935, and it is possible that spring densities of population during 1933–5 were, in fact, rather similar over the rest of the 5 square miles. Whatever the truth of this it is certain that at least between 1935 and 1938 there was a rapid and continuous decline; also that this was largely because of an annually increasing death-rate among the young.
>
> We must therefore account for the fact that although abundance (and presumably competition) was much reduced, yet the population continued to decline for at least three and possibly five consecutive years. Green and Evans were unable to account for this decrease in terms of the external environment, but found the population was suffering from severe physiological derangements typified by low levels of blood-sugar and liver glycogen and known as "shock disease" (Green and Larson, 1938). Christian (1950) gives reasons for believing that shock disease may be a direct result of stress among crowded populations. However, the decrease from 3 to 5 years after peak numbers had been passed cannot be explained by the amount of overcrowding during the period of maximum juvenile mortality. It seems more likely that, as in the vole, the descendants of crowded animals were suffering from physiological derangement derived *in utero* from their parents. A progressive reduction in viability in each of the generations born during and after the period of peak numbers fits the observed facts well. On this assumption an acceleration in the total juvenile death-rate is an inevitable result of the dying out of the older, least affected breeders.

Working with natural populations of great tits, Kluijver (1951) found an inverse relationship between density and fecundity. He suggested that this was due to fighting, which resulted in a "disturbed" physiology. An inverse relationship between density and rate of increase has been shown for the muskrat and the bobwhite quail in natural populations (sec. 10.322).

CHAPTER 10

Other Organisms of Different Kinds

Battle within battle must be continually recurring with varying success. . . . It is good to try in imagination to give any one species an advantage over another. Probably in no single instance should we know what to do. This ought to convince us of our ignorance of the mutual relations of all organic beings; a conviction as necessary as it is difficult to acquire.

DARWIN, *The Origin of Species*

10.0 INTRODUCTION

IN THIS chapter we shall be making quite a lot of use of information accumulated by biologists who have set out to study, and who have interpreted their results in terms of, "interspecies competition." This approach inevitably involves the conception of the "environment of a population." We have shown in chapter 2 that this way of analyzing environment leads to difficulties which can be avoided by the simple device of concentrating attention on the environment of the individual. For example, we tend not to think of *multispecific populations* in which the individuals and species are competing for, say, food which is in short supply, but rather to think of an *animal* in whose environment are many other animals which consume the food that it requires and thus, by reducing the amount of food available to it, reduce its chance to survive and multiply. In other words, we concentrate our attention on the amount of a necessary resource (in this case food) available to the animal that we are studying, and we take an interest in all those components of the environment (in this case other animals) which tend to modify the amount of food (etc.) available. We believe that this approach can lead to a clear insight into the principles of population ecology. In chapter 9 we discussed the abstract and relatively simple situation in which all the other animals were of the same kind. In this chapter we consider certain more complex situations in which there are also present, in the environment, animals of different kinds, as well as other sorts of organisms.

10.01 *The Categories of Other Organisms*

Not all the other animals (and other sorts of organisms) which are within, or may come within, touch, sight, sound, or smell of the particular animal being studied are necessarily to be considered as within its environment; conversely, some which never come within touch, sight, sound, or smell may be important components of the environment. It is therefore important to start this chapter by describing the ways in which other organisms (excluding those of the same kind) may come into the environment of the animal being studied. Later in this chapter these different categories will be illustrated by examples drawn from nature and from laboratory experiments. Some of these are rather complex, and for the present the exposition will be clearer if we use a hypothetical example which includes the essential features of nature but is rather simplified with respect to detail.

Suppose we have to study the ecology of a particular species of beetle which can usually be found in meadows of a particular sort. In its larval stage the beetle is a grub, which lives in a burrow in the soil in situations where the vegetation is rather open; and it feeds upon the leaves of a particular species of thistle. The other plants in the meadow are grasses and clovers. We have to find out by observation and, where necessary, experiment the kinds of other organisms that may be in or near the meadow, whether they influence this grub's chance to survive and leave progeny behind it, and, if so, in what way they do this.

a) We may find an animal, perhaps a caterpillar, which lives on the surface of the ground and eats the same thistle. If the caterpillar is present only during spring and the grub only during midsummer, so that the two may never meet, the result may still be the same as if they were there together, jostling for an opportunity to eat the same leaf: the presence of the caterpillar results in less food for the grub. Or we may find that there is another grub (perhaps a closely related species) which eats not thistle but grass and makes similar burrows in the soil in similar situations. If there are plenty of thistles and grass in the meadow but relatively few suitable places for burrows, then the presence of the second grub may reduce the amount of space available for burrows for the first one. These two animals (the caterpillar and the second grub) and others like them come into the category of *nonpredators which require to share the same resources*.

b) There may be in the meadow another animal, perhaps a sheep, which does not reduce the amount of food available to the grub (since it does not eat thistles), does not alter the amount of space available for burrows, does not seek out the grub to eat it, yet may still kill the grub by trampling on it as it seeks its own food. Or there may be another kind of animal, perhaps a rabbit, which does not eat thistles but grazes the clovers and grasses so closely that it causes an increase in the area suitable for burrows for the grub. These two animals, the sheep and the rabbit, come into another category of nonpredators. They are *nonpredators which do not require to share any of the resources that are used by the grub*, yet influence its chance to survive and multiply, the sheep by directly interfering with the grub itself, the rabbit by modifying some other component of the grub's environment. Scavengers like the dung beetles that live in the meadow may ameliorate the environment, and in this they are like the rabbit. Some examples of mutualism (see Allee *et al.*, 1949, pp. 245 ff.) would come into this category, too. But others might not.

c) There may be in the meadow a parasitic fly which seeks out the grub and lays an egg on it so that its larva will grow up inside the grub and devour it. The life-cycle of this fly may be so nicely adapted to that of the beetle that it does not need, nor can it use, any other kind of prey. Or there may be in this meadow a bird which seeks out the grub and devours it. The bird must find food throughout the year and therefore needs to have an alternative sort of prey for that period of the year when the grub is not present. These two animals, the fly and the bird, are *predators* in the environment of the grub. The fly is *an obligate*, the bird *a facultative*, *predator* of the grub. Or there may be pathogenic organisms (bacteria, fungi, etc.) which may kill the animal.

d) There may be in the meadow another kind of animal, perhaps a cricket, which does not prey upon the grub or require to share any of its resources or in any other way come into any of the categories *a*, *b*, or *c* above. It is present at a different time of the year from the grub and serves as food for the bird at that time of the year when the grub is absent. The cricket is *an alternative prey for a facultative predator* of the grub and, as such, may be an important component in its environment.

e) The groups of organisms included in categories *a*, *b*, *c*, and *d* may be considered to enter *directly* into the environment of the animal whose ecology is being studied. There is another group of organisms which, for the sake of making a distinction, may be considered to enter *indirectly* into the environment. This is best explained in relation to Figure 10.01. In this diagram we have placed the grub *A*, which is the subject of our study, in the center, and on the same level we have drawn animals which represent the four "direct" categories *a*, *b*, *c*, and *d*. Associated with each one of these is a group of organisms which, in their turn, in relation to the animals which they influence directly, may be classified into categories *a*, *b*, *c*, and *d*. These have been placed above each of the animals in categories *a*, *b*, *c*, and *d*, in order to indicate that they enter only indirectly into the environment of the central animal, the grub *A*. These vertical chains could theoretically be extended indefinitely. The discussion of such complex relationships is the delight of the biocoenologist. But their practical importance may not be very great. We include all such organisms in the one broad category *e*.

We have left out of this classification other organisms which are chiefly important because they serve as food for the animal that is being studied, for example, the thistle on which the grub *A* feeds. Similarly, when we study the ecology of a carnivorous animal, we regard the animals which it eats as "food" (chap. 11) rather than as a category of other organisms in its environment.

Other more subtle examples may suggest themselves to the reader, for example, the flagellate living in the gut of the termite which seems to serve partly as an aid to digestion and partly as food. This and other examples of mutualism will be discussed in section 11.23.

There is no difficulty in imagining how an animal may live quite close to another without in any way influencing its chance to survive and multiply. In

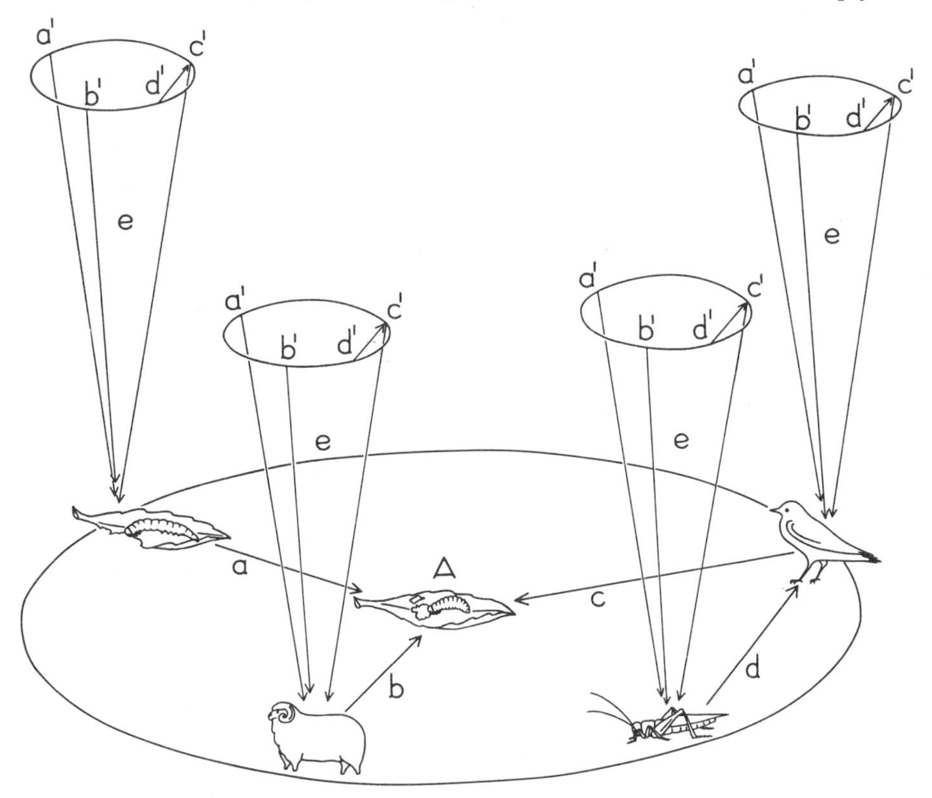

FIG. 10.01.—A diagrammatic representation of the categories of other animals. Those on the upper planes are considered to come into the environment less directly than those on the lower plane. For further explanation see sec. 10.01.

nature it is easy enough to find examples in which this seems to be the case—at least, examples in which, if the relationship exists at all, it must be indirect and quite unimportant. In the laboratory it is possible to demonstrate the same phenomenon in an artificial medium. For example, Gause (1935*a*) made up a liquid medium in a glass vessel into which he put a yeast and a bacterium and then added colonies of *Paramecium caudatum* and *P. bursaria*. The latter species fed entirely on the yeast near the floor of the vessel, and the former almost entirely on the bacteria suspended in the fluid. The two species seemed not to interfere with each other in any important way, and Gause concluded that the two species were living together in the same medium quite independently of each other.

When we come, later in this chapter, to discuss the environments of real animals, we shall find them often much more complex than is indicated by this classification, which is based on a simplified hypothetical example. For example, the relationship between the two flour beetles, *Tribolium castaneum* and *T. confusum* (sec. 10.212) is like that between the sheep and the grub in our example but is also more complex, because *T. castaneum*, unlike the sheep, requires the same sort of food as does *T. confusum*. Some of its activities belong to category *a* and some to category *b*. This means that a particular animal cannot always be placed unequivocably in one category or another, because it may influence the animal that we are studying in more than one way. But its activities, be they ever so diverse, can be so classified, and it is helpful to do so.

In the case of the two species of flour beetles we can be reasonably sure that they require the same, or very nearly the same, food. Much more often in nature it is difficult to decide whether two species do, in fact, require the same food (or the same space, etc.) or only seem to. For example, the larvae of two closely allied species, *Drosophila mulleri* and *D. aldrichi*, both live in the ripe fruit of the cactus *Opuntia lindheimeri* in Texas. In this sense they occupy the "same space." Also they both eat ripe fruit and to this extent require "the same sort of food." But Wagner's (1944) experiments showed that there were at least 9 different yeasts which were sufficient for *D. mulleri*, whereas *D. aldrichi* could complete its development on only 6 of these. When maggots of the two species were reared separately but on the same sort of yeast, their survival-rates were different. We would like to know whether, in nature, the maggots feed indiscriminately on all the yeasts which happen to be present. If the larvae discriminate between the different sorts of yeast, then we must conclude that there is a difference in their requirements for food, despite the seeming similarity.

Whenever two species do, in fact, require to share the same resources, to whatever degree, it is clear that each forms part of the environment of the other. In our example the caterpillar was treated as part of the environment of the grub because we had set out to study the ecology of the grub, though, of course, the grub is equally part of the environment of the caterpillar. This is why most ecologists have preferred to analyze this sort of relationship in terms of "interspecific competition." But it is the essence of our method that we look at the situation from the point of view of one species at a time, and we describe our attitude by saying that we are studying the ecology of a particular species. This is the essential difference between our method and that of those other ecologists who have made "competition" the keynote of their studies. Our fields are equally broad, but we look at them with a different perspective.

10.02 *Some Criticisms of the Competition Concept*

Although many, perhaps most, ecologists have preferred to interpret the relationships between nonpredatory animals in terms of competition, and a few have even included predators, there have also been some who have disagreed from this point of view. It will be instructive to refer briefly to some of the criticisms which they have raised.

Many of the ideas about competition stem, in the first instance, from Darwin's *The Origin of Species*. This was a book chiefly about evolution, but it inevitably overlapped into the field of ecology. Competition has a very different perspective in these two allied disciplines. Darwin was well aware of this and was careful to avoid using the word loosely. Instead, he used Malthus' phrase "struggle for existence" and defined it very carefully: "I use the term struggle for existence," wrote Darwin, "in a large and metaphorical sense, including dependence of one being on another, and including (which is more important) not only the life of the individual, but success in leaving progeny. Two canine animals in a time of dearth, may be truly said to struggle with each other which shall get food and live. But a plant on the edge of a desert is said to struggle for life against the drought, though more properly it should be said to be dependent on the moisture. A plant which annually produces a thousand seeds, of which on an average only one comes to maturity, may be more truly said to struggle with the plants of the same and other kinds which already clothe the ground. . . . In these several senses, which pass into each other, I use for convenience' sake the general term of struggle for existence."

This quotation makes it quite clear that Darwin knew that the advantages possessed by some species have no connection with "competition" (which literally means "together seek"). An antarctic species has an advantage over a tropical species which might happen to be transported to antarctic regions, by virtue of its capacity to survive low temperatures. The antarctic species will survive in preference to the other, without competition entering into the picture at all. Many major changes in evolution are of this nature. It is not necessary to invoke competition to explain the replacement of reptiles in the early Tertiary by birds and mammals. As Simpson (1949) showed, the facts are better fitted by the reverse proposition that the birds and mammals replaced the reptiles because the reptiles had dwindled and become extinct as a result of other causes, possibly climate. The relationship between species which utilize the same food and space, which is called "competition," exists only when food or space is limited in relation to the numbers of organisms utilizing it or when one species harms the other by fighting or in some other way (sec. 2.121). This emphasis is necessary, since "competition" is sometimes loosely used to cover all "struggles" which the organism may have with its environment. In addition to this, the word has also an unfortunate implication of emotion or purpose, and for this reason Dobzhansky (1950*a*) suggested that it is best avoided in discussions of the causes of evolution.

This is equally sound advice in relation to ecology. Thompson (1939) drew attention to some of the undesirable implications in the way in which the word "competition" has been used by ecologists: "No doubt there are cases in which organisms do struggle among themselves for food or territory. But if a caterpillar on an isolated plant completely devours its food supply and then dies of starvation, it is difficult to see how this can legitimately be called competition. If a colony of twenty caterpillars comes in the same way to the same end, the case is not notably altered; and even if nineteen die and one more lucky one survives, 'competition' still does not seem the right word to use." It is better to seek an explanation in terms of the limiting resource, which, in this case, is food (sec. 11.22). The failure of either the one or the twenty caterpillars to come through was not due to "competition" but simply to the lack of enough food for even one caterpillar. In the case where one "lucky" one survived, there was at least enough food for one, and we may suppose that there was enough food for more than one at the beginning. But not all the food was "effective," since some of it was eaten by caterpillars which did not get enough to become mature. Doubtless the one which came through did so because of some advantage it possessed, such as a faster rate of development, which enabled it to get more than its "share" of food. The race goes to the swift. This may be called "competition" by analogy with a sporting event, but we consider it more helpful to interpret this in terms of our concept of "effective food" rather than in terms of "competition."

Ullyett (1950) also questioned the appropriateness of the word "competition" as applied to the relationship between two or more species of blowflies which feed on the same food without apparently interfering with each other in any other way. If the maggots are numerous, there may not be enough food to go around, and only a few of them get enough to survive. Ullyett compared this situation to the slums of a large city where families are usually large, even though food is scarce. He wrote: "It can hardly be said that children in these families 'compete' for the available food. Yet they suffer from the shortage very much in the same way as do the overcrowded progeny of the blowflies. Mortality is higher and the general size smaller than in better-fed sections of humanity. The problem, therefore, must be regarded as one of nutritional requirements and their satisfaction or otherwise by the provisions of nature rather than in the light of a specific 'struggle.' The phenomenon which really acts as the controlling agent is, basically, sheer starvation. If, on the other hand, we use the word 'competition' in a metaphorical sense or in that which is intended when referring to a sporting event, such as a race which goes to the swift, then it is true that *Lucilia* and *C. chloropyga* do compete for the available food."

We mentioned in chapter 2 that some authors (e.g., Nicholson, 1933) have used "competition" to describe the relationship between the eater and the eaten. One needs only to replace "compete" with "together seek" (its literal meaning) in such a sentence as "The lion competes with the zebra," to see how misleading this usage is. The lion may seek the zebra and eat it, but the lion and the zebra do not have anything in common that they can seek together, any more than do the rabbit and the grass which it eats. From the point of view of the zebra, the lion is a predator; from the point of view of the lion, the zebra is food. And it is in this way that we would treat them in this book. Once again it is interesting to note that the distinction between predation and competition was made quite clearly by Darwin in the third chapter of *The Origin of Species.*

10.1 MATHEMATICAL MODELS

The mathematical model is a theoretical device which has been used to build up hypotheses about what happens when two species live together and either require to share the same food or occupy the same space, or else one preys upon the other. The best-known of these are called the "Lotka-Volterra equations," because they were derived independently and at about the same time by Lotka (1925, 1932) in the United States and Volterra (1926, 1931) in Italy. Volterra's interest in the matter arose from a practical problem in Italian fisheries brought to his notice by D'Ancona; Lotka's interest was more academic. The Lotka-Volterra equations relating to nonpredators which require to share the same resources will be examined in section 10.11; and in section 10.12 those which apply to the relationship between predator and prey will be discussed.

The former are derived directly from the equation for the logistic curve; the latter, though not derived directly from this equation, are based upon the same or very similar premises. The limitations of the logistic equation were discussed in section 9.23. It was pointed out that there are sound theoretical grounds for expecting that experimental data would not fit this curve very closely; and this expectation has been borne out by all the experiments that have been done. On the other hand, by certain devices, such as choosing an arbitrary value for K (which may be quite unrealistic) or by arbitrarily discarding some of the data, it is possible to fit the data for at least that part which relates to the early stages of the growth of the population to a curve which forms part of a logistic curve (Figs. 9.05 and 9.06). These limitations to the logistic theory mean that the Lotka-Volterra equations which are based on it must suffer at least the same limitations. We shall see below that they also have other shortcomings and in particular, that the theory relating to nonpredators contains a fundamental contradiction which invalidates its principal conclusions. Unfortunately, it is especially this branch of the theory that has had a great influence on ecological thought. Hutchinson and Deevey (1949), in a general review of population ecology, expressed the following view: "The generalization implicit in cases (2) and (3), that two species with the same niche requirements cannot form mixed steady-state populations in the same region has become one of the chief foundations of modern ecology." And Mayr (1948, p. 212) wrote: "Individuals of two species with identical ecological requirements would be subject to the same competition for space and food as if they were members of a single species. However, since the two species are genetically different, one of them will undoubtedly be slightly superior to the other in a given habitat. Natural selection will discriminate against the less efficient individuals and thus eventually eliminate the less efficient species. This was deduced mathematically by Volterra and proven experimentally by Gause (1934) and others." (See also quotation from Crombie, given in sec. 10.311 and from Lack in sec. 10.213.) For this reason it is necessary to criticize this model rather more fully than it would otherwise merit (see also Andrewartha and Birch, 1953).

10.11 *Nonpredators*

In section 9.21 the logistic equation was expressed in the form:

$$\frac{\delta N'}{\delta t} = r'_m N'\left(1 - \frac{N'}{K'}\right). \qquad (1)$$

In this equation N' represents the number of animals in unit space at time t, K' represents the maximal number of animals that can exist in the particular circumstances of food, temperature, etc. (which require to be defined for each case), and r'_m is the innate capacity for increase in these same circumstances.

The extension of this theory to the circumstances in which two species live together was made by Volterra (1926, 1931) and Lotka (1925, 1932). It was expounded and popularized by Gause (1934 and other papers) and Crombie (1945 and other papers). The expositions by Gause and Crombie have been accepted and quoted widely by ecologists, and we shall base our description and criticism of the model on them. For the present the discussion is to be restricted to species which are supposed to require the same sort of food or space, etc., when living together, but neither requires to eat the other. According to Gause, when two such species A and B live together in a restricted space or with only a limited amount of food between them, the rate of increase in the density of the population (numbers per unit space) of species A may be expressed by the equation:

$$\frac{\delta N'}{\delta t} = r'_m N'\left(1 - \frac{N' + M'}{K'}\right). \qquad (2)$$

In this equation, N' is the number of species A at time t; K' is the maximal number per unit space achieved by this species when it is living alone (i.e., K' is the value taken by K' in eq. [1] when this equation refers to species A);

M' is a number such that $M' = \alpha N''$, when N'' is the number of species B at time t, and α is a coefficient which represents the average influence of species B on the multiplication of species A, relative to the influence of species A on itself. In other words, M' represents the number of species B expressed as the number of species A which would have the equivalent influence. The following long quotation from Gause (1934, p. 45) leaves no doubt on this point:

The unutilized opportunity for growth of the first species in the mixed population can be better understood if we compare it with the value of the unutilized opportunity for the separate growth of the same species. In the latter case the unutilized opportunity for growth is expressed by the difference (expressed in a relative form) between the maximal number of places and the number of places already occupied by the given species. Instead of this for the mixed population we write the difference between the maximal number of places and that of the number of places already taken up by *our species together with the second species* growing simultaneously.

An attempt may be made to express the value M' directly by the number of individuals of the second species at a given moment, which can be measured in the experiment. But it is of course unlikely that in nature two species would utilize their environment in an absolutely identical way, or in other words that equal numbers of individuals would consume (on the average) equal quantities of food and excrete equal quantities of metabolic products of the same chemical composition. Even if such cases do exist, as a rule different species do not utilize the environment in the same way. Therefore the number of individuals of the second species accumulated at a given moment of time in a mixed population in respect to the place it occupies, which might be suitable for the first species, is by no means equivalent to the same number of individuals of the first species. The individuals of the second species have taken up a certain larger or smaller space. If N'' expresses the number of individuals of the second species in a mixed population at a given moment, then the places of the first species which they occupy *in terms of the number of individuals of the first species*, will be $M' = \alpha N''$. Thus, the coefficient α is the coefficient reducing the number of individuals of the second species to the number of places of the first species which they occupy.

Substituting $\alpha N''$ for M' and completing the second equation for species B, we have

$$\frac{\delta N'}{\delta t} = r'_m N' \left(1 - \frac{N' + \alpha N''}{K'}\right), \tag{3}$$

$$\frac{\delta N''}{\delta t} = r''_m N'' \left(1 - \frac{N'' + \beta N'}{K''}\right). \tag{4}$$

These equations were originally given in a different form by Volterra (1931, pp. 9 and 79), from which the present forms may easily be derived. It is quite clear not only from the original presentations of Lotka and Volterra but also from the expositions by Gause and Crombie that α and β are defined as constants which are independent of N' and N''.

When the density of neither population is changing, that is, when

$$\frac{\delta N'}{\delta t} = 0 = \frac{\delta N''}{\delta t},$$

it is said that "the system has reached equilibrium." If there were no restrictions on the numerical values that might be taken by the quantities $(N' + \alpha N'')$ and $(N'' + \beta N')$, then there would be three different ways for the system to approach equilibrium (Fig. 10.02). These are: (*a*) N' and N'' may both remain above zero and approach asymptotes which will be less than K' and K'', respectively. (*b*) N' may approach zero as N'' approaches the asymptote K''. (*c*) N' may approach the asymptote K' as N'' approaches zero. If all three of these end-points were possible, the following four inequalities would define the circumstances determining each end-point.

(i) When

$$\frac{\alpha}{K'} < \frac{1}{K''} \quad \text{and} \quad \frac{\beta}{K''} < \frac{1}{K'},$$

then

$$\lim_{t \to \infty} N' = \frac{K' - \alpha K''}{1 - \alpha\beta} \quad \text{and} \quad \lim_{t \to \infty} N'' = \frac{K'' - \beta K'}{1 - \alpha\beta},$$

both these quantities are greater than zero; the former is less than K' and the latter less than K''.

(ii) When

$$\frac{\alpha}{K'} < \frac{1}{K''} \quad \text{and} \quad \frac{\beta}{K''} > \frac{1}{K'},$$

then limit $N' = K'$ and limit $N'' = 0$ as $t \to \infty$.

(iii) When

$$\frac{\alpha}{K'} > \frac{1}{K''} \quad \text{and} \quad \frac{\beta}{K''} < \frac{1}{K'},$$

then limit $N' = 0$ and limit $N'' = K''$ as $t \to \infty$.

(iv) When

$$\frac{\alpha}{K'} > \frac{1}{K''} \quad \text{and} \quad \frac{\beta}{K''} > \frac{1}{K'},$$

then limit $N' = 0$ or K' and limit $N'' = 0$ or K'' as $t \to \infty$, depending on the relative values of N' and N'' at the beginning.

This, then, is the theory as it relates to nonpredators. We consider the model to be unrealistic for the following reasons:

1. The assumptions of the logistic theory which we have already criticized as unrealistic (sec. 9.23) are repeated twice over in this theory—in equations (3) and (4).

2. There is an additional assumption that α and β are constants which are independent of the numbers of each species. Even assuming that α and β can be assigned any strict biological meaning, it is unrealistic to suppose that they will be independent of the density of either species.

3. The solution of equations (3) and (4) poses a very arbitrary biological situation which amounts to a contradiction. It makes evident the restricted nature of the logistic theory and the virtual impossibility of attributing any real biological meaning to the constants α and β. In order to appreciate this

criticism, it is necessary to grasp the following two points: (i) Equation (3) is derived from equation (1), and the symbols N' and K' are common to both. They must therefore retain precisely the same meaning in both equations. Thus if N' and K' enumerate animals of a certain kind in equation (1), they must

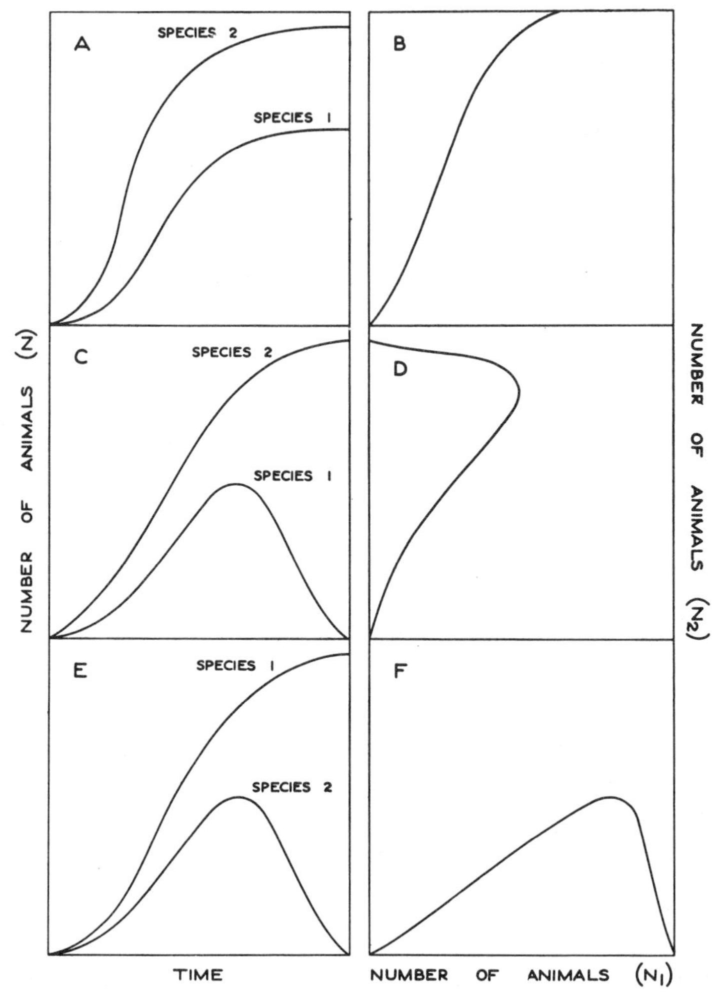

FIG. 10.02.—Curves representing the three possible ways in which the populations may grow when two species are confined together in a restricted space. In *A*, *C*, and *E*, numbers are plotted against time; in *B*, *D*, and *F*, numbers of the first species are plotted against numbers of the second species.

also enumerate animals of precisely the same kind in equation (3). (ii) In order that equation (3) may be manipulated, N' and $\alpha N''$ must be additive. That is to say, $\alpha N''$ must enumerate precisely the same entities, "qualities," or "effects" as N'.

Now if one follows Gause's definition of α (see above), then point (ii) raises no initial difficulty, because, by definition, $\alpha N''$ enumerates precisely the same sort of *animals* as N' and the two can be added together. But a contradiction becomes apparent when one compares the solutions of equations (1) and (3). The solution of equation (1) shows that N' approaches the limit K' as t approaches infinity and $\delta N'/\delta t$ approaches zero as a limit but never becomes negative. Translating this to biological language, we may say that K' represents the greatest number of animals of this particular sort which the unit volume of this particular place can accommodate when the initial number was less than K'. In the general solution of equation (3), $N' + \alpha N''$ may exceed K'. This means that K' is no longer regarded as a true maximal density; instead, when the two species are living together, the same space may now accommodate (at least temporarily) a number of animals which is greater than K', even when these animals have been transformed to a common sort of animal by multiplying N'' by α.

This conclusion contradicts the conclusion implied in equation (1). If more than K' entities can be accommodated when two species are living together, then there is no logical reason why more than K' entities should not be accommodated (at least temporarily) when only the one species is present. But equation (1) is so limited that this cannot happen in the situation it attempts to represent. Provided that we accept equation (1) as the basis of the model and Gause's definition of α and β, we must discard as unreal any solution of equation (3) which required $N' + \alpha N''$ to exceed K'. Equation (3) may be written in the form

$$\frac{\delta N'}{\delta t} = r'_m N' - r'_m N' \left(\frac{N' + \alpha N''}{K'}\right),$$

from which it is immediately clear, since r'_m and α must, by definition, be positive and $N' + \alpha N''$ must not exceed K', that $\delta N'/\delta t$ has a limit at zero; it may never become negative. This means that there is no provision for the density of the population of species A (enumerated by N') to become less; it may increase or it may remain stationary, but it may not decline. Exactly the same reasoning may be applied to equation (4). Therefore, these equations cannot be used to predict any state of equilibrium which requires either N' or N'' to approach zero. The last three inequalities must be discarded on these grounds, leaving only the first, from which it must be concluded that, once the two species have come together, they must continue to live together forever, because the model does not permit either to die out. Reduced to these terms, the model is so unreal as to hold practically no interest for a biologist.

When Gause's definition of α is followed, the ratio $(N' + \alpha N'')/K'$ in equation (3) may be regarded as measuring the influence of the populations of both kinds of animals on the rate of increase of N', with N' representing here (as in eq. [1]) the density of the population of species A. It has been suggested to us that there may be an alternative way of defining α as a coefficient, such that when the number of animals of species B is multiplied by α, the product $\alpha N''$ comes to be expressed in units of the influence of species A on the rate of increase in N'. But since (from paragraph ii above) N' must be expressed in the same units as $\alpha N''$, it follows that $\delta N'/\delta t$ must be expressed in units of the same "influence." It is not rational to suppose that N' may enumerate an "*influence*" or a "*deleterious effect*" on one side of the equation and *animals* on the other side. It is only necessary to translate this statement to biological language to see that this definition of α has no strict biological meaning. The translation would have to read something like this: $\alpha N''$ refers to the influence of animals of species A on the rate of increase of the influence of animals of species A. This is, of course, quite meaningless, and so we discard this definition of α.

It is difficult to conceive of any biological meaning to the constants α and β which is sound, and this is why it is so difficult to translate these equations into biological language. Gause's definition is so simple as to be restrictive, and the alternative definition is meaningless. But these conclusions should not surprise us. It is a false abstraction to suppose that a number of one sort of animal may be multiplied by a constant to convert that number into a number of another sort of animal. The biological situation is too complex to be represented in such simple mathematical terms.

In many of the experiments done by Gause (1934), Crombie (1945), and others, one species, after multiplying for a time, decreased and then eventually died out. By ignoring the limitations to the theory which we have discussed above, it was claimed that these experiments confirmed the hypothesis, which thus came to be widely accepted and to have a great deal of influence on ecological thought. This is unfortunate, for these mathematical models are more likely to mislead than to help in the interpretation of observations and measurements made on natural populations (see also sec. 10.311).

10.12 *Predators*

What was probably the first mathematical model to represent the relationship between the densities of the populations of host and parasite living together was constructed by Sir Ronald Ross (see Lotka, 1925). He established a system of equations to represent the spread of malaria in human populations. Martini developed a similar set of equations in 1921 in relation to measles and scarlet fever (Lotka, 1925, p. 79). For more recent developments in this field the student may consult Jordan and Burrows (1945). The study of the way that disease may spread in human populations is essentially ecology, but it is a rather specialized study, and the methods used are not directly applicable to the study of hosts and parasites, prey and predator, in the senses in which ecologists usually understand these words. The best-known mathematical models dealing in a general way with populations of predators and prey are those of Lotka (1925) and Volterra (1926, 1931) and of Nicholson and Bailey (1935).

In defining the category "predators" (sec. 10.01), we made little mention of the different ways in which carnivorous animals may capture and eat their prey. We shall discuss this in section 10.32. For the present it is more important to be quite clear about the qualities of the hypothetical animals that enter into the equations of the mathematical models which we are considering. In the Lotka-Volterra models the "animals" have the following qualities:

a) They move about at random to one another in the same sense that it may be said that the molecules of nitrogen move at random with respect to one another and to the molecules of oxygen in a mixture of these two gases.

b) The predators are infallible and their appetites insatiable. Every "encounter" with a prey results in its capture, and every prey that is captured is eaten. This goes on indefinitely, no matter how frequently the encounters may occur, i.e., irrespective of the densities of the populations of predators and prey.

c) The populations of both predator and prey have all the qualities necessary for them to conform to the logistic theory (sec. 9.21).

The Lotka-Volterra models are based on arguments with which we are familiar from our consideration of the logistic theory in chapter 9. The rate of increase of a single species of prey living by itself in unlimited space may be written

$$\frac{\delta N'}{\delta t} - r'_m N'.$$

In the absence of prey, the predators would die from starvation. The rate at which the population is decreased by deaths may be regarded as a negative rate of increase and written thus:

$$\frac{\delta N''}{\delta t} = d'' N'',$$

where d'' is a negative quantity and is a death-rate (a relative quantity related to the number of animals in the population).

If both prey and predator are living together in a limited space, r'_m will be reduced by an amount which depends on the density of the population of the predators. Similarly, the coefficient d'' will be increased by an amount depending on the density of the population of the prey species. By virtue of the qualities attributed to these animals (see above), the changes in r'_m will be proportional to N'', and those in d'' will be proportional to N'. We may therefore write the two equations thus:

$$\frac{\delta N'}{\delta t} = (r'_m - C'N'')N' \tag{6}$$

and

$$\frac{\delta N''}{\delta t} = (d'' + C''N')N'', \tag{7}$$

where C' and C'' are constants. Volterra regarded C' as a measure of the "aptitude of the prey to defend itself" and C'' as a measure of "the effectiveness of the means of offense of the parasite." Equations (6) and (7) are the fundamental equations of Lotka and Volterra. There has been a number of modifications and extensions made to them which we need not go into here. They are

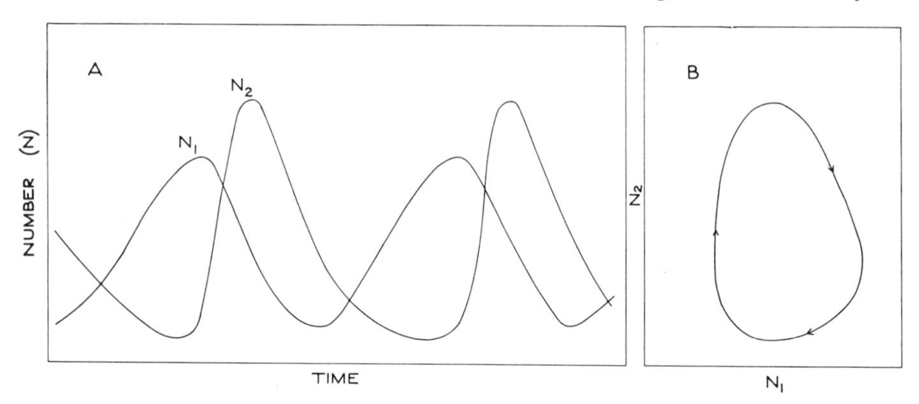

FIG. 10.03.—Hypothetical curves showing the oscillations in the numbers of prey (N_1) and predator (N_2) according to the Lotka-Volterra equations. *A*, numbers plotted against time; *B*, prey plotted against predators.

characterized by the so-called "periodic solution" (Lotka, 1925; Volterra, 1926). This means that the trends in the densities of the populations of prey and predator change direction in a systematic way. The predators increase while the prey is decreasing to a point where the trends become reversed. Then, for a period, the prey increases while the predator is decreasing until the trends are again reversed. Volterra called this the "law of the periodic cycle," because the fluctuations in the numbers of the two species are periodic. This is illustrated in Figure 10.03, *A*, where the fluctuations in the densities of prey and predator are plotted against time in the usual way. In Figure 10.03, *B*, we have abstracted from time and plotted the same data with N' and N'' as the co-ordinates. The closed loop which is shown in this diagram follows from the nature (and extreme simplicity) of equations (6) and (7), which, in turn, follows from the unnatural simplicity of the hypothetical animals in this model. Lotka and Volterra have shown that quite small increases in the complexity of the qualities of the hypothetical animals involving quite minor changes in equations (6) and (7) result in the curve cutting either the abscissae or the ordinate, which means that one species becomes exterminated, leaving the other in complete possession of the field. Gause (1934) called this a "relaxation oscillation," to distinguish it from the "classical oscillation," in which the closed loop indicates that both predator and prey live together indefinitely.

The height and breadth of the loop measure the magnitude of the fluctuations in the populations of prey and predator. The point within the loop about which the oscillations occur is called the "singular point." Mathematically, it is the point which satisfies the condition $\delta N'/\delta t = 0$ and $\delta N''/\delta t = 0$. Theoretically, its co-ordinates are the values of N' and N'' such that all oscillations disappear and the prey and predator live together indefinitely, with the densities of the two populations invariable. The "singular point" in the Lotka-Volterra models is thus analogous to the "steady density" which is a fundamental concept in the Nicholson-Bailey models.

In order to describe hypothetical animals in their models, Nicholson and Bailey (1935) coined a special terminology in which certain ordinary words have been used with quite special meanings (just as throughout their paper "competition" has a most unusual meaning—see secs. 2.121 and 10.02). We shall try to explain these as we meet them below. The hypothetical animals of the Nicholson-Bailey models have the following qualities:

(i) The predator, when it encounters the prey, does not devour it but instead lays an egg (or eggs) in it; the prey is thus eaten by the next generation of the predator. (In the introduction to their paper, Nicholson and Bailey speak of "animals" in a completely general sense, but in the special part where the models are being developed they seem to refer only to a very special sort of predator, namely, the sort of holometabolous insect which the student of "biological control" knows as an "internal parasite"—see sec. 10.321.)

(ii) The prey is distributed uniformly over an area which is itself uniform and constant with respect to weather, food, etc. The ease with which the prey can be found does not vary with the density of the population.

(iii) Although the individual predator may be endowed with highly specialized instincts for finding the prey, still the population of predators searches "at random." This assumption has been the subject of controversy, perhaps because it is difficult to grasp the abstract idea of a population searching. But its meaning can be expressed rather more concretely. Since "random," like "big" or "fast," has no meaning except when it relates one thing to another, the first step is to see what the searching is random to. It is clearly stated not to be random to the prey, so it must be random to the other members of the predator's population. This means, for example, that a particular predator will choose a course in searching quite independently of whether the same ground is being, or has been, covered before. And when, in the course of its search, it finds a prey that is "pre-occupied," because it has already been found and oviposited in, its behavior is such that the outcome (for populations of both the prey and the predator) is the same as if this "pre-occupied" prey had not been there to find. That is, the predator must either be possessed of an infallible instinct which restrains it from laying an egg in a prey that already has been oviposited in; or else, if it lacks this restraint, then (*a*) the predator must be provided with an unlimited stock of eggs and unlimited time in which to lay them, and (*b*) the presence inside one prey of an excess of eggs of the predator must not prejudice the survival and healthy development of the usual number of the predators in this prey.

All these conditions follow quite clearly from Nicholson and Bailey's definition of "random searching by the population" and from their use of this concept in developing the "competition curve." It is an unusual usage of "random," but it is in this sense that it must be understood and criticized in the Nicholson-Bailey models (sec. 10.221).

(iv) The appetites (i.e., the capacity for oviposition) of the predators are insatiable, irrespective of the density of the population of prey.

(v) The predator has an "areal range" which is independent of the density of the population of the prey. The "areal range" is simply the product of the distance traveled by the predator during its lifetime, while searching, multiplied by twice the distance from which it can perceive a prey. A constant value for "areal range" implies that the predator travels just as far in search of prey when these are abundant as when they are scarce!

(vi) The predator has an "area of discovery." This is not an *area* at all, but a *proportion*. It is that fraction which has for its numerator the number of prey found by a predator during its lifetime and for denominator the number of prey in the "areal range" of the predator. (Presumably these must remain constant during the lifetime of the predator.) A constant value for "area of discovery," together with a constant value for "areal range," implies that a predator finds the same proportion of the total population of prey, irrespective of whether these are very scarce or very abundant!

Nicholson and Bailey (1935) criticized the Lotka-Volterra models on three major grounds:

a) In most of their equations they assume that the "reaction" when predator encounters prey is instantaneous. With real animals this is not so, and the time which elapses before the result of the "encounter" is evident may in some cases be as long as a generation of the predator. It is true that Volterra (1931) did allow for a delay in his equation for the interaction of one species of predator with one species of prey, but not in the more complex situations where several species were supposed to be living together.

b) It is implicit in the Lotka-Volterra models that each individual in the populations of prey and predator is exactly equivalent to every other individual of the same species. No allowance is made for the different age-groups or, in the case of insects, the different stages in the life-cycle, e.g., egg, larva, pupa, and adult of the holometabolous insect.

c) Lotka and Volterra used the methods of the calculus.

Nicholson and Bailey claim to have produced a more realistic model by using the discontinuous method of variation and by allowing not only for the age-distribution in the populations but also for the delay between the time of encounter and the time when the result becomes evident. This is incontrovertible: their "animals" seem less unreal than the "gas-molecule" animals of Lotka-Volterra. But after considering the qualities set out in paragraphs i to vi above, may we reasonably conclude that the hypothetical animals of the Nicholson-Bailey models seem very like real animals? Moreover, in both the Lotka-Volterra and the Nicholson-Bailey models it is implicit that the populations of prey and predator are living in a limited space where all the other components of the environment, such as food for the prey, temperature, places to live, etc., are not only uniform throughout the space but also invariable with time. So, no matter how closely we may judge the hypothetical animals of the models to resemble real animals, the models themselves would still not tell us about nature, because their animals live in quite unnatural environments. So long as this is borne clearly in mind (which, alas! it so frequently is not), the model may continue to serve the useful purpose of inspiring work on laboratory experiments with populations and furnishing valuable basic training for ecologists.

The Nicholson-Bailey models are built around two basic concepts: the "competition curve" and the "steady state." These can be derived by simple reasoning and simple algebra from the qualities attributed to the hypothetical animals in the models. The competition curve relates the density of the population of predators to the proportion of previously undiscovered prey found by each predator. It follows particularly from quality iii, namely, that the "population searches at random," and may be derived as follows (see Nicholson and Bailey, 1935, p. 555):

"Let u_1 be the number of objects originally present in a unit of area and let u be the number undiscovered after an area s has been traversed.

"The number of previously undiscovered objects found in traversing ds is uds.

"This must equal the decrease $-du$ of the number of undiscovered objects.

$$\therefore\ -du = uds;$$

and since

$$u = u_1 \text{ when } s = 0$$

$$\therefore\ u = u_1 e^{-s}$$

that is

$$s = \log\left(\frac{u_1}{u}\right)$$

"The fraction of the original objects which is discovered is $\frac{u_1 - u}{u_1}$ i.e., $1 - e^{-s}$."

This is represented by the "competition curve" which was published by Nicholson (1933).

The "steady state" was defined as follows (see Nicholson and Bailey, 1935, p. 557):

"For a parasite-species with a given area of discovery there is clearly a particular density at which it effectively searches a fraction of the environment equal to the fraction of hosts that is surplus, and at this density, exactly the surplus of hosts is destroyed. Also there is a particular density of a host-species which is just sufficient to maintain the above density of parasites from generation to generation. When the densities of the interacting animals have these particular values it is clear that they must remain constant indefinitely in a constant environment. This situation will be referred to as the *steady state*, and the densities of the animals when in this state as their *steady densities*."

If the "areal range" and the "area of discovery" of the predator varied with the density of the population of prey (as they almost certainly do in nature), this conclusion would be true for only one particular density of prey; but, by *assuming* these qualities constant, this conclusion becomes, *by definition*, a general conclusion and can become the foundation for all the subsequent equations which enter into the models.

The "steady state," like the "singular point" of the Lotka-Volterra equations, is a mathematical concept. It cannot be imagined except as a "point" with no dimensions. The slightest departure on either side of the point induces oscillations. In certain circumstances, when age-distribution is ignored, the oscillations continue indefinitely with a constant amplitude, as in the "classical oscillations" of Volterra. But in other circumstances and when age-distribution is taken into account, the oscillations increase in amplitude with time, as in the "relaxation oscillations."

In the mathematical model there is no reason why this should not continue indefinitely, with the densities of the two populations eventually alternately approaching infinity and zero. But in nature this does not happen. Instead of deducing from this that the mathematical model has failed to conform to what is known of nature, Nicholson and Bailey (1935, p. 589) deduced the opposite:

> Each female is the centre of diffusion of her offspring, so when the density of a species becomes very low as a result of violent oscillation the zones of diffusion from the various females can seldom overlap and may frequently be separated from one another by great intervals. Thus the interacting animals exist in numerous disconnected small groups, within each of which interspecific oscillation follows its course independently of that in the other groups. Since the numbers of animals in these groups are very small compared with species-populations, the great reduction in numbers soon produced by increasing oscillation frequently exterminates the groups, but meanwhile some hosts will have migrated into the surrounding previously unoccupied country and have established new groups there. It follows that:
>
> Conclusion 51. A probable ultimate effect of increasing oscillation is the breaking up of the

species population into numerous small widely separated groups which wax and wane and then disappear, to be replaced by new groups in previously unoccupied situations.

Animals in nature are distributed unevenly like this: it is a fact that can readily be verified by observation (sec. 13.02). But it is not a conclusion that may be deduced from, nor an observation to be fitted into, a mathematical model which assumes at the beginning that its animals are distributed evenly over a uniform area.

Nicholson and Bailey varied the circumstances in their models through a wide range of details, chiefly with respect to variations in the numbers of species and prey living together and in the respective life-cycles and "habits" of predator and prey. They expressed their results in the form of 59 specific conclusions. We are not concerned with the details here. In sections 10.221 and 10.223 we shall discuss certain experiments and observations which are relevant to the assumptions on which these models are built and to the conclusions reached in them.

10.2 EXPERIMENTAL POPULATIONS

In this section, after describing briefly some laboratory experiments which demonstrate very neatly the two ways in which nonpredators may enter into the environment, we shall proceed to describe certain experiments and observations which relate to the assumptions and the conclusions of the mathematical models which have been discussed in the preceding section. Most of these come from experiments that have been designed in relation to the mathematical models; but others, notably those of Park and his school, have had a wider purpose. We also include some measurements and observations of natural populations which have a direct bearing on this subject, while leaving for section 10.3 the more general aspects of "other animals" as a component of the environment of an animal living in nature.

10.21 *Nonpredators*

10.211 EXPERIMENTAL DEMONSTRATION OF THE TWO MAJOR CATEGORIES OF NONPREDATORS

Ullyett (1950) reared the larvae of the two species of blowflies, *Lucilia sericata* and *Chrysomyia chloropyga*, on a limited quantity of meat. The number of newly hatched larvae placed on 140 gm. of meat varied from 100 to 10,000. The death-rate was measured for each species at each density, both when the species were reared separately and when they were reared together. When the two species were reared together, the experiments were started with equal numbers of each. So the comparison to be made is the death-rate of, say, *Lucilia* when all the other animals in the meat are of the same kind and when every second one belongs to a different species, in this case *Chrysomyia chloropyga*. Ullyett could find no evidence that either species interfered with the other in any way except that they both required exactly the same sort of food. The experimental results confirmed this in a very interesting way. The larvae of *Chrysomyia* are heavier than those of *Lucilia*. When the densities of the two populations were expressed in terms of weight of larvae per unit quantity of meat, it was found that the death-rate among *Lucilia* was the same for any particular degree of crowding, independent of whether the other animals were all of its own kind or half of them were *C. chloropyga*. This was not so when the degree of crowding was expressed in terms of numbers rather than weight. But this was only to be expected, since weight is obviously the correct unit for comparison in this case. The larvae of *C. chloropyga* in these experiments are nonpredators that require to share the same food as *L. sericata* without interfering with it in any other way (category *a*, sec. 10.01).

In a parallel series of experiments Ullyett compared *Lucilia sericata* with a different species, *Chrysomyia albiceps*, with quite different habits. In these experiments the death-rate of *Lucilia* was much greater when half the other animals around it were *C. albiceps* than when all were of its own kind. As the densities of the populations increased, the death-rate among *Lucilia* increased until it reached 100 per cent, i.e., none survived to become adult. Although *C. albiceps* undoubtedly needs to share the same food as *L. sericata*, this is a relatively unimportant part of its relationship to *Lucilia*. Ullyett observed that the larvae of *C. albiceps* attack and eat those of *Lucilia*. This happens to some extent when both populations are quite sparse but to an increasing degree as the densities increase. Despite these "predatory" habits, the larvae of *C. albiceps* are not to be considered strictly as predators because, in the absence of *Lucilia*, they can do quite well on the meat alone and, in the presence of *Lucilia*, meat is still their chief source of food. They come into the environment of *Lucilia* chiefly as nonpredators which interfere directly with *Lucilia*, i.e., they are most important with respect to category *b* of section 10.01. It is characteristic of real animals, as distinct from the hypothetical ones, with which we have so far been chiefly concerned, that they manifest complexities of this sort. This will be demonstrated several times in the course of the examples which we shall be considering in the remainder of section 10.2 and in section 10.3.

10.212 EMPIRICAL "TESTS" OF MATHEMATICAL MODELS FOR NONPREDATORS

We showed in section 10.11, with respect to the Lotka-Volterra models for nonpredators, that there was a fundamental contradiction between the basic equations and three of the four conclusions derived from them. Consequently, no empirical test of the conclusions can reasonably be attempted except perhaps for the first inequality, which predicts that both species shall continue to live together indefinitely. But, ironically, in all the experiments giving this result, the animals have been found to have differences in habits which place them outside the scope of the theory, which, by definition, is restricted to animals having "similar ecological requirements" (sec. 10.213).

In most of the experiments with animals which seem to conform to this definition, one species or the other usually dies out, often after having first increased in numbers for a period; but sometimes the decline sets in from the beginning. The only conclusion with respect to the Lotka-Volterra models which can be drawn from these experiments is that the fundamental equations (3) and (4) fail to provide even an approximate description of the empirical results, since they do not provide for negative values for $\delta N/\delta t$, although it has been supposed that they do cover this eventuality. One or more of the assumptions on which they were based must therefore be wrong. This is not surprising when it is considered that these include all the assumptions of the logistic theory (sec. 9.23) made twice over with respect to species A and B.

Many of the experiments which we shall discuss in the next section, notably those by Gause and his colleagues and Crombie and others, were designed expressly to test the conclusions of the mathematical models. Since we do not regard the models as valid, we need not be concerned with the conclusions that these authors reached. But the experimental results are interesting in relation to what happens when two species whose requirements for food or some other resource overlap to some degree or who interfere with each other in some other way are forced to live crowded together in a glass jar or some other artificial cage in the laboratory.

10.213 EXPERIMENTAL DEMONSTRATIONS OF THE RELATIONS BETWEEN NONPREDATORS WHEN THEY ARE ARTIFICIALLY CROWDED

Birch (1953*a*, *b*) reared *Calandra oryzae* small "strain," *C. oryzae* large "strain" (sibling species, since they do not interbreed), and *Rhizopertha dominica* in wheat and in maize at a number of different temperatures and moisture contents and used the results to calculate the innate capacity for increase for each species at each combination of temperature and moisture (sec. 3.4). He then reared populations of each species in small glass vessels containing a constant amount of wheat or maize at the same temperatures and moisture contents, observing all the usual precautions for experiments designed to show logistic growth (sec. 9.22). He found a close association between the numerical value of r_m and the maximal density achieved by the population (Fig. 9.24). The next step was to discover what happened when populations of the two species were allowed to grow side by side in the same small glass vessel. In these experiments one pair of adults of each species was introduced into each vessel. The grain was renewed at regular (2-week) intervals, as is usual in this sort of experiment. Each treatment was replicated 15 times; a striking part of the results was the close similarity between the replicates. So we can say with assurance that if the experiment were repeated, similar results would be obtained (Birch, 1953*c*).

TABLE 10.01*

RELATIONSHIP BETWEEN INNATE CAPACITY FOR INCREASE, r_m, AND CAPACITY TO SURVIVE IN A CROWD AS ILLUSTRATED BY PAIRS OF SPECIES OF GRAIN BEETLES

SPECIES: SUCCESSFUL ONE GIVEN FIRST IN EACH PAIR	r_m	GRAIN	TEMP. (° C.)	MOISTURE (PER CENT)	NO. WEEKS FOR UNSUCCESSFUL SPECIES TO DIE OUT		FIGURE
					Mean	Range	
C. oryzae (large "strain").. / *C. oryzae* (small "strain")..	0.436 / 0.417	Maize	29.1	13	100	74–150	10.04
C. oryzae (small "strain").. / *C. oryzae* (large "strain")..	0.772 / 0.564	Wheat	29.1	14	69	46–95	10.05
C. oryzae (small "strain").. / *R. dominica*..............	0.772 / 0.578	Wheat	29.1	14	....	38–>190	10.06
R. dominica.............. / *C. oryzae* (small "strain")..	0.686 / 0.501	Wheat	32.3	14	37	26–47	10.07

* In the first column the successful species is given first in each pair, followed by the one that died out. Note (from second column) that in every instance r_m was greater for the successful species. The sixth column is based on 15 replicates. The last column refers to the diagrams in which the growth and decline of the populations are traced in detail. Data from Birch (1953*c*).

The detailed results are given in Table 10.01 and Figures 10.04–10.07, from which it is clear that the species which was favored by the particular combination of temperature, moisture, and food, to the degree that it had the higher innate capacity for increase, survived to the end, while the one with the lower

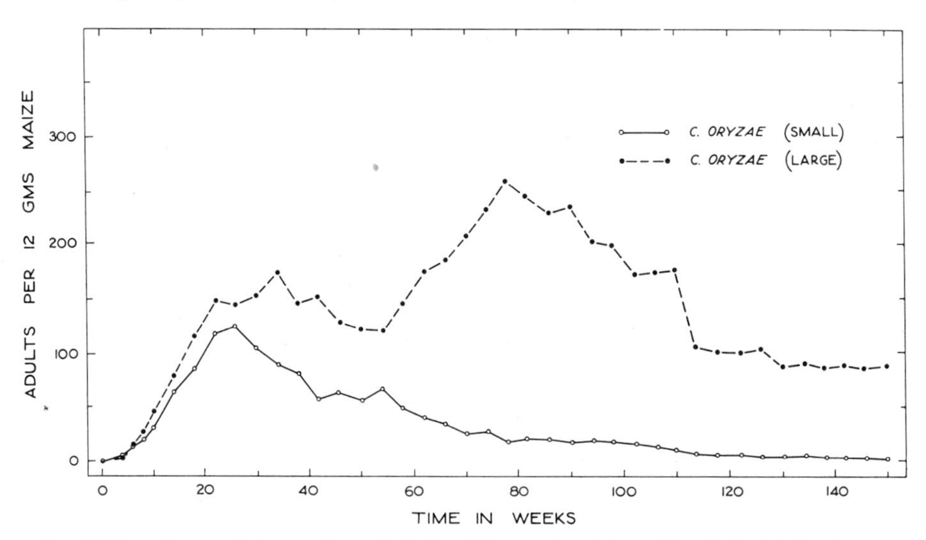

FIG. 10.04.—The trends in numbers of adults of the small and large "strains" of *Calandra oryzae* living together in maize of 13 per cent moisture content at 29.1° C., when the innate capacity for increase, r_m, of the large "strain" was greater than that of the small "strain." (After Birch, 1953*c*.)

value for r_m eventually died out. The species with the lower rate of increase always multiplied for a while before beginning to decline. Sometimes the decline was irregular, and often it was prolonged over many generations (300–400 days) before the species eventually died out. It is noteworthy that success or failure (persistence or extinction) could in every case have been foretold from a knowledge of the qualities of the species when they were living in the same circumstances but by themselves. This may indicate that the relationship between the species when living together is relatively simple and that each influences the other's chance to survive and multiply chiefly by eating its food. A change in temperature of 3° C. was sufficient to reverse the relative values of the innate capacity for increase of the small "strain" of *C. oryzae* and *R.*

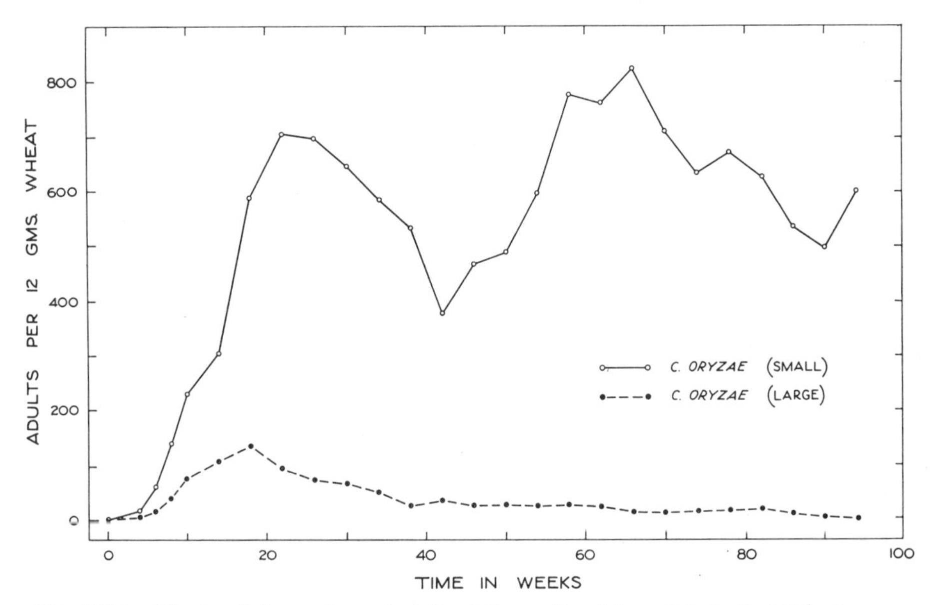

FIG. 10.05.—The trends in numbers of adults of the small and large "strains" of *Calandra oryzae* living together in wheat of 14 per cent moisture content at 29.1° C., when the innate capacity for increase, r_m, of the small "strain" was greater than that of the large "strain." (After Birch, 1953*c*.)

dominica and to reverse the outcome of the experiments when these two species were crowded together. A change from wheat to maize caused the same sort of result with the two strains of *C. oryzae.* Moore's (1952*a*) experiments with two species of *Drosophila* gave similar results. He found that at 25° C. *D. melanogaster* was the successful species when associated with *D. simulans*, but at 15° C. *D. simulans* became the successful species. Park's (1948, 1954) experiments with two species of *Tribolium* provided another example of this phenomenon (see below, later on in this section).

Birch did not vary the relative numbers of the two species at the beginning, but Park, Gregg, and Lutherman (1941) did this in a similar series of experiments with the beetles *Gnathocerus*, *Tribolium*, and *Trogoderma*. Populations of the beetles (usually two species at a time) were reared in small vessels containing a mixture of flour, rolled oats, and yeast; temperature, moisture, and food were not varied. In many of the experiments the final outcome depended on the relative densities of the two populations at the beginning. For example, *Tribolium* failed to become established when a few were introduced into a vessel that already contained a fairly dense population of *Gnathocerus;* and, vice versa, *Gnathocerus* failed to establish itself in the presence of an already dense population of *Tribolium*. Parallel results were obtained when similar experiments were done with *Gnathocerus* and *Trogoderma*.

Gause (1934, 1935*b*) cultured *Paramecium aurelia* and *P. caudatum* as single

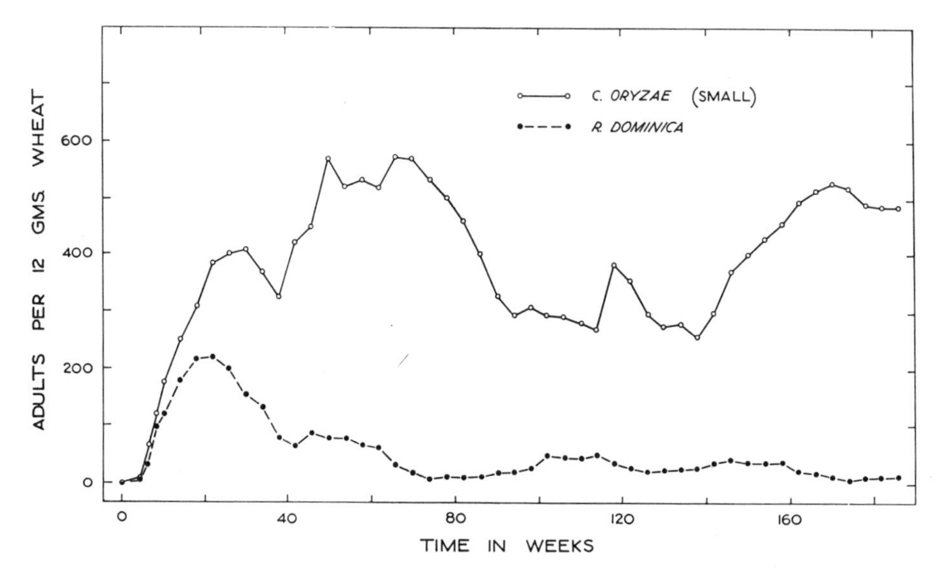

FIG. 10.06.—The trends in numbers of adults of the small "strain" of *Calandra oryzae* and *Rhizopertha dominica* living together in wheat of 14 per cent moisture content at 29.1° C., when the innate capacity for increase, r_m, of *C. oryzae* was greater than that of *R. dominica*. (After Birch, 1953*c*.)

species and as mixed species in Osterhaut's medium containing a regulated supply of *Bacillus pyocyaneus* as food. The growth of a single species is shown in Figure 9.02. When the two species were started off together, *P. caudatum* eventually died out, leaving a dense population of *P. aurelia* (Fig. 10.08). The success of *P. aurelia* seemed to be due to its higher innate capacity for increase in the particular circumstances of the experiment. This was partly because it is relatively insensitive to the excretions from the bacillus which was supplied as food. When a less virulent strain of bacillus was used, *P. caudatum* showed the higher innate capacity for increase and survived, while *P. aurelia* died out. There was no evidence of any direct interference of one species by the other. In all these experiments the relationships of the pairs of species are relatively simple and seem to consist chiefly of the need to share the same food. So they would come into category *a*, section 10.01.

The results of these experiments may be generalized with the help of a *hypothetical* model (Fig. 10.09). This diagram is not intended to represent nature but merely to generalize the results obtained in the artificial circumstances of these laboratory experiments. Consider two species such that the ranges of, say, temperature at which they can develop overlap but do not coincide. There will be a certain range of low temperature at which only one species (say species A) can survive; at the other end of the range the reverse will hold. Another way of saying this is that r_m for species B is zero at certain low temperatures at which r_m for species A is positive, and, conversely, r_m for

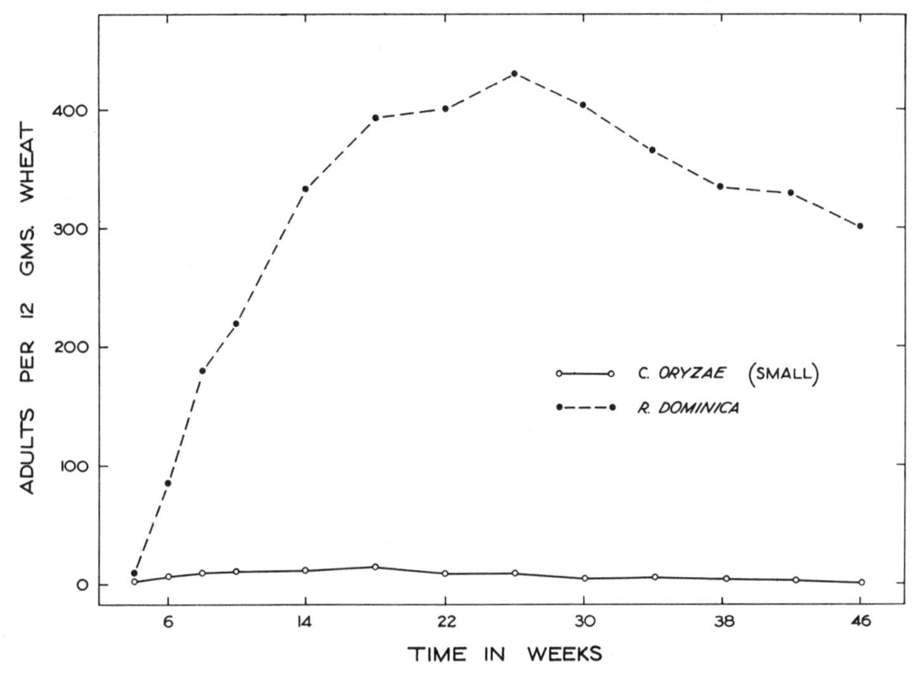

FIG. 10.07.—The trends in numbers of adults of the small "strain" of *Calandra oryzae* and *Rhizopertha dominica* living together in wheat of 14 per cent moisture content at 32.3° C., when the innate capacity for increase of *R. dominica* was greater than that of *C. oryzae*. (After Birch, 1953*c*.)

species A is zero at certain high temperatures when r_m for species B is positive. In the zone of overlap, r_m will be positive for both species but is likely to be the same for both species only over a very narrow range of temperature. Now suppose these two species come together (where the opportunity for multiplication is continuous but food and space are limited) in a series of situations which constitute a temperature gradient. In all those situations where the magnitudes of r_m for the two species are sufficiently different, one species only will survive, and the other will be exterminated. Only species A will occupy the situations where temperature is low; only species B will occupy those where temperature is high. There will be an intermediate zone, where r_m is about the same for both species, in which the result may depend on the relative densities of the two populations when they happen to come together. We may suppose that in this zone about half the situations will contain species A and half species B. The distribution of each species will be smaller than it would have been, had only one of them been there; and in the intermediate zone where both occur, each will be only half as abundant as it would have been, had the other not been there. Since it is based upon laboratory experiments, an essential feature of this hypothetical model is that opportunity for multiplication continues until space

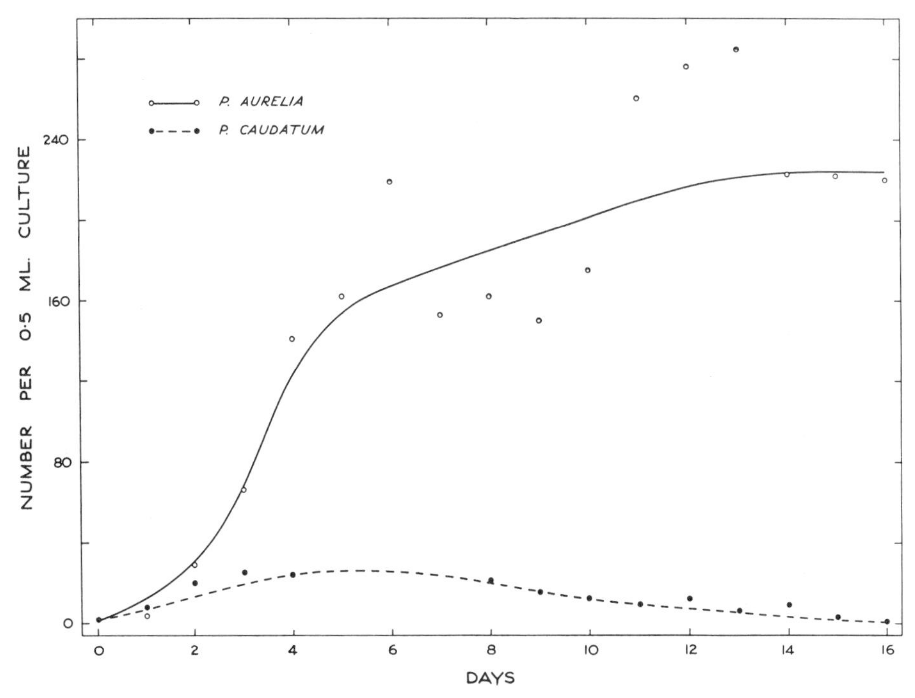

FIG. 10.08.—The trends in numbers of *Paramecium caudatum* and *P. aurelia* growing together in buffered Osterhaut's medium with "half-loop" supply of bacteria as food (see Fig. 9.02 for growth of populations when these species were reared separately). (After Gause, 1934.)

or some other resource becomes limiting. We think this rarely happens in nature. But, within these limitations, the model provides an example of how the component "other animals" may modify the influence of weather on the distribution and abundance of a species.

The relations between the two species of flour beetle, *Tribolium castaneum* and *T. confusum*, discovered by the very thorough experiments of Park (1948) seem to be more complex than those between any of the species that we have been discussing so far. In these experiments small numbers of the two species, though not always the same number of each species, were placed in small tubes containing a limited amount of flour. The flour was renewed at regular intervals, as usual. The changing densities of the two populations were followed until, invariably, one or the other species had died out. Each treatment was replicated a number of times. Usually there was no disagreement between replicates as to which species eventually died out, but occasionally the "wrong" species would survive in some replicates. This in itself indicates a greater complexity than in Birch's experiments with *Calandra* and *Rhizopertha* (see above), where invariably there was complete agreement between replicates.

Carrying the investigation further, Park found that his cultures were naturally infested with a sporozoan parasite, *Adelina*. Then, having found a

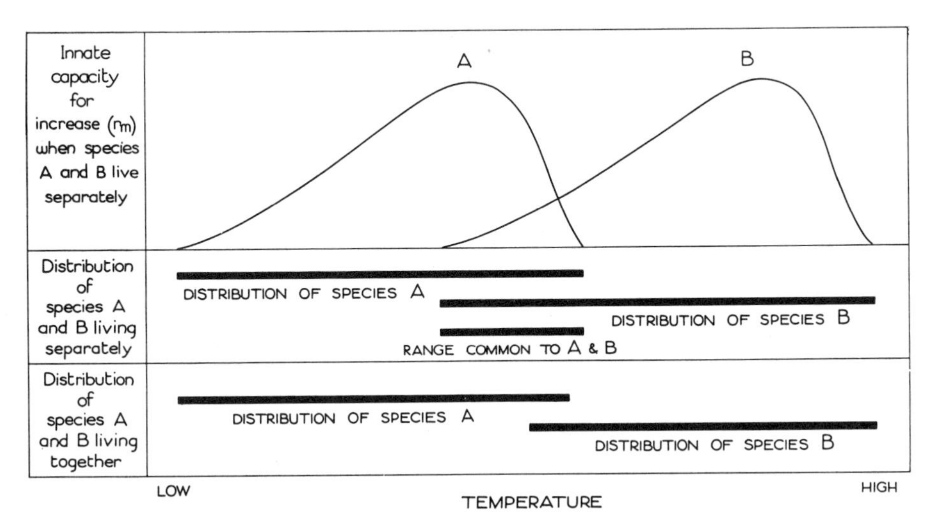

FIG. 10.09.—Hypothetical diagram illustrating one way in which two nonpredators which require to utilize the same resources might reduce the area of each other's distribution with respect to some other component of environment—in this case temperature. In the diagram, temperature increases from left to right.

method for obtaining populations that were free from this parasite, he was able to show that, irrespective of the relative numbers of the two species at the beginning, *Tribolium confusum* was usually the one to survive when *Adelina* was present (Fig. 10.10, *B*); but when *Adelina* was absent, it was usually *T. castaneum* that survived.

These experiments with *Tribolium* were all done at 29.5° C., and at a relative humidity of 70 per cent. In later experiments Park (1954) used three temperatures, 24°, 29°, and 34° C., and two humidities, 30 and 70 per cent. This time *T. confusum* survived in (and *T. castaneum* disappeared from) 71 per cent of the replicates at 24° C. and 70 per cent relative humidity. At 30 per cent relative humidity and 29° C., *T. confusum* survived in 87 per cent of the replicates. At 30 per cent relative humidity and 34° C., *T. confusum* survived in 90 per cent of the replicates. But when the humidity was raised to 70 per cent at these temperatures, *T. castaneum* survived in 86 per cent of the replicates at 29° C. and in all the replicates at 34° C. Thus the final outcome (in terms of the species which survived in the majority of the replicates) was reversed either by altering the temperature from 29° to 24° C. while keeping the humidity constant at 70 per cent, or by altering the humidity from 70 to 30 per cent while keeping the temperature constant at either 29° or 34° C. These results are like those described above for *Calandra* and *Rhizopertha*, illustrating how the success of one species relative to another can be reversed by altering

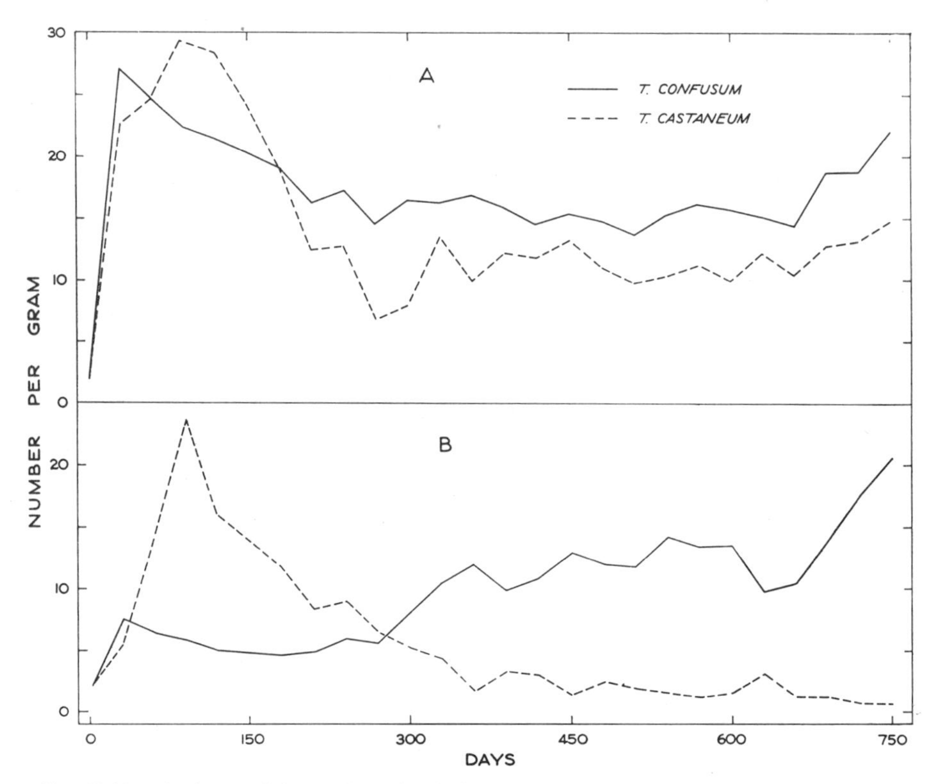

FIG. 10.10.—*A:* the trends in numbers of *Tribolium confusum* and *T. castaneum* when each species was reared by itself in 8 gm. of medium (yeast to which flour had been added) which was renewed regularly. The temperature was kept constant at 29.5° C. and the relative humidity at 60–70 per cent. Each population began as two pairs of newly emerged adults. The curves represent larvae, pupae, and adults combined. *B:* the only difference from *A* was that the two species were reared together in the same container. (After Park, 1948.)

the temperature and the humidity. But they differ in another way, since the same species did not survive in all replicates in any one experiment.

Park had previously shown that these species of *Tribolium* produce an excretion which "conditions" the flour, making it unfavorable for the development of the immature stages (sec. 9.234). And Birch, Park, and Frank (1951) studied the fecundity of the two species living alone and together, with *Adelina* present in each case. The net fecundity of *T. confusum* was less than that of *T. castaneum*, and when crowded with adults of their own species, the extent of the reduction in fecundity was the same for both species. When, however, the imagines of *T. castaneum* were crowded with adult *T. confusum*, the fecundity of *T. castaneum* was greatly reduced. Interspecies crowding did not have any effect on *T. confusum* different from that produced by crowding it with its own species. There was thus evidence that the presence of many *T. confusum* could be the direct cause of a pronounced reduction in the fecundity of *T. castaneum*. In terms of our categories of other animals in section 10.01, *T. confusum* has some qualities which place it in category *a*, because it undoubtedly requires to share the same food as *T. castaneum*, but it also interferes both directly and indirectly with *T. castaneum*, so that it also has some qualities which place it in category *b*. The latter seem to be more important with respect to its influence on *T. castaneum*'s chance to survive and multiply.

Crombie's (1945) experiments indicated a similar, though less complex, relationship between *Rhizopertha dominica* and *Sitotroga cerealella*. The two species were reared in small vessels containing a limited amount of wheat, which was renewed at regular intervals. Many different experiments were done in which an arbitrary, though not always the same, number of each species was placed in the wheat and the changes in the densities of the two populations were followed until one species (in this case always *Sitotroga*) had died out. In one experiment the population of *Rhizopertha* was started with one pair of adults and that of *Sitotroga* with 100 eggs, which were taken as the equivalent of one pair of adults. Figure 10.11, *A*, shows the growth of the two populations when they were living alone and Figure 10.11, *B*, shows what happened when the two species were living together: both populations increased in density during the first 75 days but thereafter, while *Rhizopertha* continued to increase, *Sitotroga* decreased until, at the end of 300 days, none was left. Another experiment was started with 400 *Rhizopertha* and 100 eggs of *Sitotroga;* another was started with 2 *Rhizopertha* and 10,000 eggs of *Sitotroga;* in a number of others the initial densities were intermediate between these two extremes. The final result was the same: *Rhizopertha* survived, and *Sitotroga* died out, irrespective of initial numbers.

Some earlier experiments (Crombie, 1944) had shown that the larvae of each species decreased the probability of the survival of the other in direct proportion to its own numbers. Crombie attributed all the deaths which were associated with crowding to the chance that a larva feeding in the grain might be eaten along with the grain. Apparently, the larvae of *Sitotroga* suffered more severely in this way than did those of *Rhizopertha*. The analysis was not carried further; but with respect to this quality at least, we can say that *Rhizopertha*'s chief importance in the environment of *Sitotroga* is as an animal which interferes with and destroys it (category *b* of sec. 10.01), although, since it shares the same food, it also has some (less important) qualities which belong to category *a*.

In all the experiments that have been discussed so far, one species has come eventually to dominate the available space, and the other has died out completely, usually after having first increased for a period before beginning to decline. But even with the severe crowding which is usual in laboratory experiments, it is possible for two species to live together indefinitely, provided that one does not actively harm the other (i.e., does not belong to category *b* of sec. 10.01) and that the two differ a little in their requirements for food. For ex-

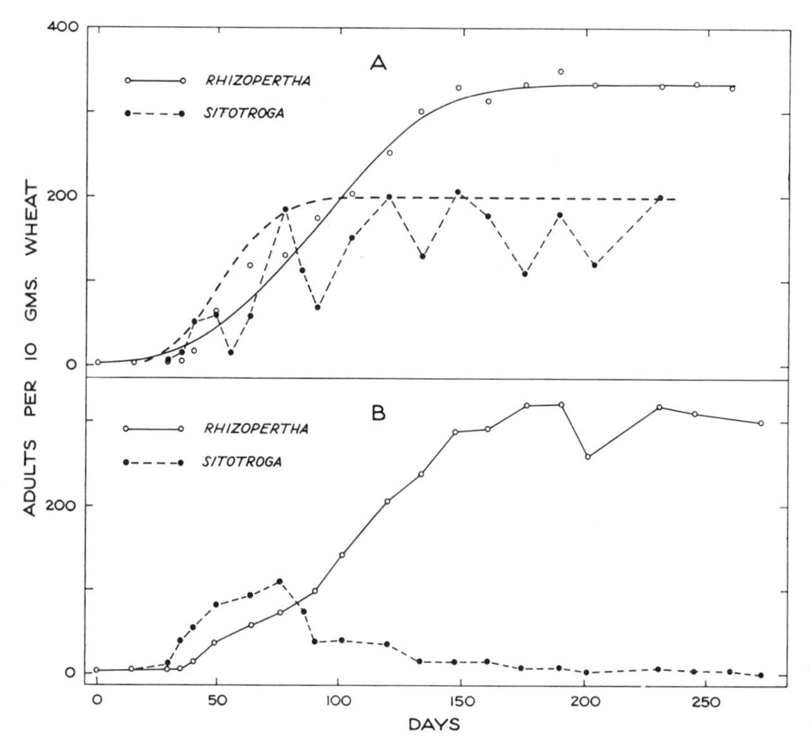

FIG. 10.11.—*A:* the trends in numbers of *Rhizopertha dominica* and *Sitotroga cerealella* when each species was reared by itself in 10 gm. of cracked wheat which was renewed at regular intervals. Temperature was kept constant at 30° C. and relative humidity at 70 per cent. Each population began as a single pair of adults. The curves represent adults only. The theoretical curves were calculated from the logistic equation. *B:* the only difference from *A* was that the two species were reared together in the same container. (After Crombie, 1945.)

ample, Crombie (1945) reared populations of the two beetles *Rhizopertha dominica* and *Oryzaephilus surinamensis* in limited and constant quantities of broken wheat, which was, as usual, renewed at regular intervals. The two species have slightly different habits. The larvae of *Rhizopertha* live and feed inside the grain; the larvae of *Oryzaephilus* live and feed from outside the grain. The adults of both species mostly live and feed from outside the grain. Apparently, these differences were sufficient to allow them to live together indefinitely in the same jar of wheat (Fig. 10.12).

Alternatively, two species which cannot live together in a truly homogeneous medium may be able to do so after a very slight measure of heterogeneity has been introduced into the medium. For example, *Paramecium aurelia*, living in a tube of Osterhaut's medium containing yeast, will feed on the suspension in the upper layers of the fluid, whereas *P. bursaria* in a similar medium will browse on the bottom. Similarly, *P. caudatum* will feed mostly in the upper layers. Gause (1935*b*) showed that when either *P. caudatum* or *P. aurelia* were living in a tube with *P. bursaria*, both populations would survive indefinitely.

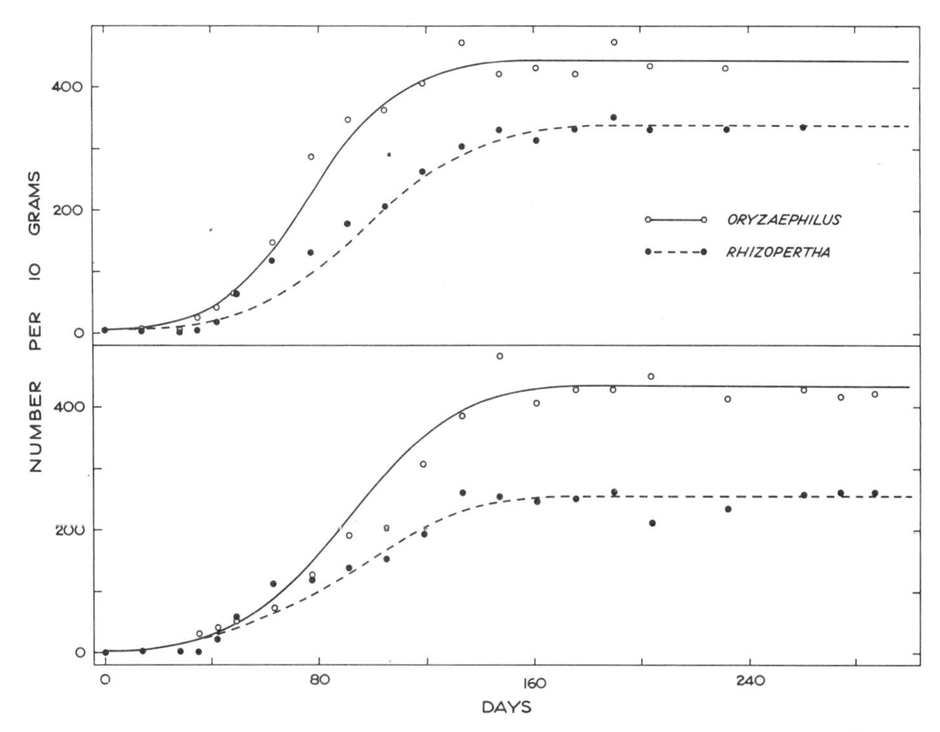

FIG. 10.12.—*Above:* the trends in numbers of *Oryzaephilus surinamensis* and *Rhizopertha dominica* when each species was reared by itself in 10 gm. of cracked wheat which was renewed at regular intervals. The temperature was kept constant at 30° C. and relative humidity at 70 per cent. Each population began as a single pair of adults. The curves represent adults only. *Below:* the same as above except that the two species were reared together in the same container. (After Crombie, 1945.)

But when *P. aurelia* and *P. caudatum* were placed together in the one tube, one would eventually die out, leaving only the other surviving (see above). Crombie's (1946) experiments with *Tribolium confusum* and *Oryzaephilus surinamensis* provide an analogous example. Both species may be reared separately in flour, though, in nature, *Oryzaephilus* breeds in the dust that accumulates from stored wheat. When populations of the two species were confined together in the same vessel of flour, invariably *Oryzaephilus* died out and *Tribolium* survived. The explanation for this is that each species eats the eggs of the other but that *Tribolium* is more active in this regard than *Oryzaephilus;*

also the larvae and adults of *Tribolium* eat the pupae of *Oryzaephilus*, whereas the reverse does not happen. Crombie found a simple way to change the relationship between these two species. He placed small glass tubes in the flour. Provided that these were large enough for the larvae of *Oryzaephilus* to crawl into and pupate, but small enough to keep *Tribolium* out, the two species were able to live together indefinitely in the same jar of flour. This result was brought about by changing what had been a uniform mass of flour into one that was nonuniform, at least to the extent of containing a sufficient number of "refuges" for the "weaker" species.

The results of these experiments, and others like them, have often been generalized as if they could be related directly to nature (sec. 10.311). For example, Lack (1946 and earlier papers), starting from Gause's hypothesis (as stated by Lack), that "two species of the same ecology cannot persist together in the same region," made a special study of "ecological isolation" among three groups of birds, the passerines of Britain, the finches of the Galapagos Islands, and the birds of prey of Europe. He did not find one pair of species for which he could be sure that they had similar requirements for any resource except, occasionally, food that was superabundant. Although Lack demonstrated that between all these species of birds there is virtually no competition for resources in short supply, he still sought an explanation for his results in terms of the mathematical models of Lotka and Volterra and the laboratory experiments of Gause. So he postulated that the present distributions and behavior of these species of birds is the outcome of competition between them in the past. The correct inference to be drawn from these observations is that the mathematical and experimental models are quite unlike nature.

The animals in these laboratory experiments live in circumstances in which the environmental variations of nature are excluded. A test tube of Osterhaut's medium, a jar of wheat, or a tube of flour is just about as homogeneous a distribution as might ever be found. Yet we have seen that quite small differences in the behavior of species that do not actively harm each other and, even when one of them does actively harm the other, quite minor irregularities in the medium allow similar species to live together. In the light of this experience, one wonders where, in nature, to look for a distribution so uniform as to allow the results of these experiments to be repeated among natural populations.

Most of these experiments have been done with species especially selected for the similarity in their requirements, and yet one is impressed by the long time required for the complete annihilation of the "weaker" species. In some of Park's experiments with *Tribolium* and Birch's experiments with *Calandra* and *Rhizopertha*, more than 4 years were required. These are short-lived species, with a generation completed in a couple of months. A similar number of generations with a longer-lived animal, a bird, for example, might occupy many years. In these experiments food, temperature, moisture, and all other important components of the environment are kept as constant as possible. With *Calandra* a shift of a few degrees in temperature was sufficient to transfer the advantage from one species to the other. The same thing happened with *Tribolium* when the parasite *Adelina* was eliminated. A slight change in pH was sufficient to transfer the advantage from one species of *Paramecium* to another (Gause, 1936). Merrell (1951) showed that a periodic fluctuation in the quality of the food was all that was required to allow *Drosophila melanogaster* and *D. funebris* to live together indefinitely in a limited space where all other components of environment were kept as constant as possible. The food was changed at regular intervals as usual; while it remained fresh, one species multiplied and the other decreased; when it became stale, the advantage was reversed. The numbers of the two species fluctuated rhythmically between wide limits, but neither died out. In nature, erratic, as well as rhythmical, fluctuations in all the components of the environment are the rule, and there is not, so far as we can see, any reason to expect, when the distributions of two similar species overlap or coincide, that one would be likely to be favored at the expense of the other continuously for long enough for either to be exterminated. Indeed, so far as the results of these laboratory experiments go, the reverse would seem to be indicated (sec. 10.311).

The laboratory experiments which we have been discussing in this section are indeed interesting and instructive in their proper perspective; there may be special situations in nature to which they can be related directly; but, in general, any extrapolation to nature should be done with the utmost caution and with full awareness that the experimental models apply to homogeneous media in constant conditions. It is a safer procedure to examine natural populations without preconceived ideas based on laboratory models. This is not to deny them a valuable role in population ecology but to put them in proper perspective.

10.22 *Predators*

Volterra worked out his mathematical models after D'Ancona had called his attention to an increase in the numbers of *Selachians* in the Adriatic Sea after the war of 1914–18. These are the main predators of the bottom fauna in this sea. When fishing was resumed after the war, the numbers of *Selachians* decreased. There is nothing unusual in this, for it may truly be said that the most characteristic feature of natural populations is their tendency to fluctuate. Sometimes, when a predator has been recognized as an important component of the environment, it has been claimed that the fluctuations have been caused by the interactions of predators and prey. But fluctuations occur just as characteristically in the absence of predators as when they are present. For example, parallel fluctuations in the numbers of bacteria and protozoans in soil at Rothamsted (Cutler, 1923) have been quoted as an example of fluctuations due to the interaction of predator (Protozoa) and prey (bacteria). But more recent experiments (Taylor, 1936) demonstrated similar fluctuations in the numbers of bacteria in the absence of Protozoa. These fluctuations were all the more striking because they occurred even when the temperature and moisture content of the soil were held constant. Similarly, Bodenheimer (1938) has observed fluctuations in the numbers of fish in the Adriatic Sea which he has explained in terms of hydrological changes. A natural population of *Thrips imaginis*, lacking any association with a predator, fluctuated continuously and more or less rhythmically during the 14 years that records were kept (Davidson and Andrewartha, 1948*b*; see also sec. 13.114). It is clear that the mere observation of oscillations in natural populations, whether of prey or predator or both, provides no evidence of a causal relationship between predators and oscillations nor yet confirms either the premises or the conclusions of the mathematical models. To some, this statement may seem so obvious that it is a truism. But a study of the literature should convince them that the conclusions of the mathematical models have often been accepted and believed without a proper appreciation of the need for empirical confirmation.

Experimental "tests" of the mathematical models have been of two sorts. In the first place, numerous workers, notably Salt, Flanders, and Ullyett and others, have studied the behavior of predators, and we may compare their accounts of the behavior of real animals with the qualities assumed for the hypothetical "animals" of the models. In the second place, several workers have reared populations of prey and predator together and compared the results with those predicted by the models. And Varley (1947), having measured the changing densities in populations of species comprising a more complex association of prey and predators in nature, tested the observed results against those predicted by the appropriate model of the Nicholson-Bailey series. In the next three sections we shall discuss these experiments in this order.

10.221 EXPERIMENTAL DEMONSTRATIONS OF THE QUALITIES OF PREDATORS IN SEEKING THEIR PREY

There is a great wealth of data from observation and experiment relating to the behavior of predaceous insects when searching for their prey. Useful summaries may be found in Flanders (1947) and Clausen (1940). For the present, we are concerned only to discover what support these experiments offer for assumptions made in the mathematical models; in particular, we would test the assumptions that (*a*) the predators search at random and (*b*) the area searched by a predator and the proportion of prey found are independent of the densities of the populations of prey and predator (cf. "areal range" and "area of discovery" of the Nicholson-Bailey models).

We need not spend time considering whether the experiments verify the assumption of the Lotka-Volterra models that predators in seeking their prey move at random to the prey, for it is common knowledge—and a glance at any one of the long list of papers reviewed by Flanders (1947) would show it—that animals have instincts which guide them in search of their food. But the conditions of the Nicholson-Bailey models with respect to random search require only that the searching by each predator be random to that of all the other predators in the population. Nicholson and Bailey specifically state that it is not necessary for the searching to be random to the prey; indeed, if the "competition curve" is to hold good, it is necessary that the predator have an infallible instinct which restrains it from laying an egg in a prey that has already been oviposited in. The only alternative to this, if the competition curve is to be true, is for the predator to have an unlimited supply of eggs and time to lay them; and the number of healthy predators that can develop from the body of one prey must be independent of the number of eggs laid in one prey, no matter how great the excess.

We know of at least one species in which such an infallible instinct occurs: Jones (1937) stated that superparasitism[1] never occurs in *Telenomus ullyetti* because the female of this species always refrains from ovipositing in a host that has already been parasitized. In many other species the instinct is present but not so strongly developed. Ullyett (1945, 1949*a*, *b*, 1950) made a thorough experimental study of this instinct in *Microbracon hebetor*, *Cryptus inornatus*, and *Chelonus texanus*. In all three species he found a strong tendency to avoid oviposition in a host that had already been parasitized. This instinct prevailed when the ratio of prey to predator was broad, and very little superparasitism occurred; but when the ratio was narrow, the urge to lay eggs conflicted with the instinct to avoid prey already containing eggs, with the result that the restraint broke down and excessive numbers of eggs were laid in the prey. Similar results were found for *Trichogramma evanescens*, *Ibalia leucospoides*, and *Collyria calcitrator* (Salt, 1934) and *Diadromus collaris*, *Angitia cerophaga*, and *Apanteles plutellae* (Lloyd, 1940). It would seem that most species which place their eggs inside the body of the prey have this instinct developed to a greater or lesser degree. On the other hand, Varley (1941) could not detect this instinct in *Eurytoma robusta*, *Habrocytus trypetae*, *Torymus cyanimus*, and *Eupelmella vesicularis*. These are all species which lay their eggs on the outside of the body of the prey. It would seem that species which do this do not as a rule recognize the presence of eggs that have already been laid on the prey, and they will continue to lay eggs on prey as they find them, irrespective of how many eggs may already be there. Flanders (1947) showed that the number of eggs produced per day and during the lifetime of many species of the parasitic Hymenoptera is quite characteristic of the species. Moreover, oögenesis depends upon a stimulus from the presence of prey in the vicinity, and in the absence of the appropriate stimulus fully developed eggs may be resorbed (Flanders,

1. This term is defined in sec. 10.321.

1942). This and other evidence lead to the conclusion that the number of eggs and the time avaliable for laying them is quite strictly limited. We may conclude, therefore, with respect to this particular aspect of "random searching by populations," that some species may come fairly close to it but many do not.

The other aspect of this question—namely, whether the movements of a predator while searching are random to the movements of the other predators in the same population—has received relatively little attention. Ullyett's ex-

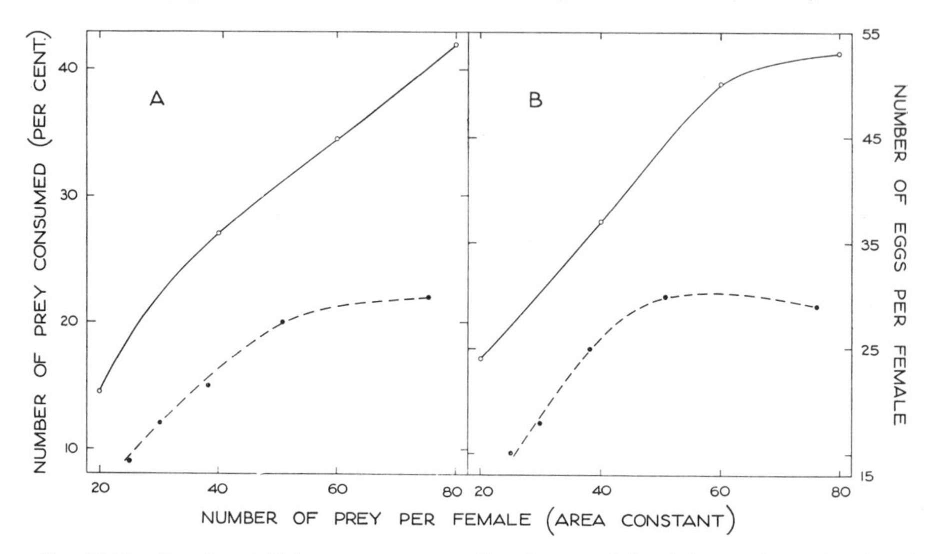

FIG. 10.13.—Females of *Chelonus texanus* were allowed to search for their prey (eggs of *Ephestia*) which were distributed over the same area in every experiment. In the experiments represented by the continuous lines a single female searched each time, and the number of prey in the area was varied. In the experiments represented by the broken lines, the number of prey in the area remained the same, but the number of predators searching for them was varied. *A* shows the proportion of the prey found, and *B* the number of eggs laid by one predator. (After Ullyett, 1949*a*.)

periments, which we quoted above, throw some light on this matter also. In one series of experiments he placed one female of *Chelonus texanus* in a Petri dish with 20, 40, 60, . . . , 200 prey (eggs of *Ephestia*). In another series of experiments a number of female *Chelonus*, varying from 2 to 6, were placed in a Petri dish of the same area with 150 prey in it. In Figure 10.13, *A*, the number of prey attacked per female is plotted against the concentration of hosts per female. It is clear that at every concentration of prey the predator found a consistently larger proportion of the prey when only one female was searching the area than when two or more were in the Petri dish together. Not only were more prey found, but more eggs were laid also (Fig. 10.13, *B*). Similar experiments with *Cryptus inornatus* gave similar results. Ullyett (1949*b*, p. 296), commenting on these results, says: "It is evident that any sudden increase in the population of the parasite within a given area and with a given host population does not give a corresponding increase in parasitism of the host (and therefore in its control) which is commensurate with the sum of the *potential* capacities of the individual females. Some factor or factors step in to modify this capacity when more than one female is present in the area. One of these factors is undoubtedly the partly mechanical one produced by interference between the individual females during the search for hosts and oviposition. . . . Observation of the parasite behaviour during this phase showed that it can be a very real factor in the modification of results." These experiments indicate that the

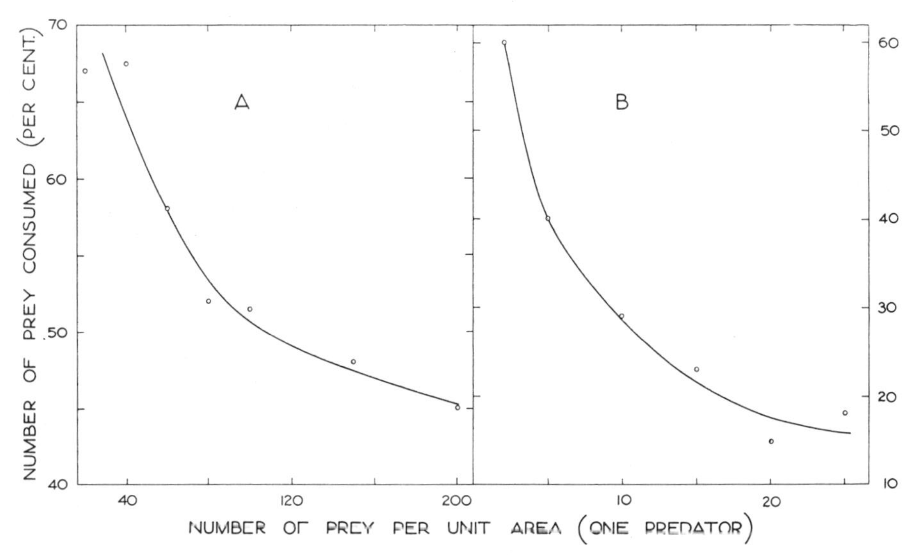

Fig. 10.14.—*A:* one predator (female of *Chelonus texanus*) was allowed to search for prey (eggs of *Ephestia*). The area was the same in every experiment, but the number of prey was varied to give differing *densities* of prey. *B:* the same experiment repeated, using *Cryptus inornatus* as predator and prepupae (in cocoons) of *Loxostege sticticalis* as prey. (After Ullyett, 1949*b*.)

"searching by the population" was not "random" in the way postulated for the Nicholson-Bailey models, in these two species at least.

Ullyett's experiments are also instructive in relation to the constancy of the "area of discovery" postulated by the Nicholson-Bailey models. The models postulate that a predator with a constant "areal range" will find a constant proportion of the prey, independent of the density of the population of the prey. The data for Figure 10.14, *A*, were got by varying the number of prey in a constant area, i.e., by varying the density of the population of prey while keeping artificially constant the area to be searched. A solitary female of *Chelonus* was used each time, thereby eliminating the influence of "nonrandom searching by a population" discussed in the preceding paragraph. The figures relate to the numbers of prey found by one predator in 24 hours. The proportion of prey found (i.e., the "area of discovery") was not constant but varied with the density of the prey; the relationship is curvilinear. Figure 10.14, *B*, shows the same phenomenon for *Cryptus inornatus*. In this the area was again constant, a cage $3\frac{1}{4} \times 3\frac{1}{4} \times 2\frac{1}{2}$ inches. Solitary females were used as before, and the prey (cocoons of *Loxostege*) were scattered on the floor of the cage. Both the proportion of prey found and the proportion used varied with the density of the prey. That is, the "area of discovery" for this species is not constant either.

We do not know of any critical experiments dealing with the "areal range" of a predator, so it must suffice to quote several relevant observations from nature. Rosenberg (1934) measured the dispersal of *Aphelinus mali* in an apple orchard. Early in the summer, about 100 *Aphelinus* were liberated on each of about a dozen apple trees widely spaced throughout the orchard. He found that "the parasites tend to remain within a foot or two of the point of emergence from their host, if plenty of Aphids are present." Nevertheless, later in the summer when the aphids had become scarcer, the *Aphelinus* were found in numbers at least four trees away from the nearest place where they had been liberated. This is a clear indication that the distance traveled by an *Aphelinus* in search of its prey depends upon the numbers of aphids in its vicinity. These examples could be multiplied, for there is plenty of evidence that in nature the area searched over by a predator depends very much on the density of the population of prey. In other words, predators are not characterized by a constant "areal range."

10.222 EMPIRICAL "TESTS" OF LOTKA-VOLTERRA MODELS

Gause (1934, 1935*a*, 1936) was probably the first to make an empirical test of a Lotka-Volterra model of prey and predator. Keeping temperature and certain other components of environment constant, he reared prey (*Paramecium caudatum*) and predator (*Didinium nasutum*) together; an oat medium, frequently replenished, formed an adequate food supply for the prey. The predator always exterminated the prey, irrespective of the relative densities of the two populations at the beginning. After all the prey had been destroyed, the predator died out from starvation (Fig. 10.15, *A*). The periodic oscillations predicted by the Lotka-Volterra models were not realized. Doubtless, the real animals of these experiments differed from the stereotyped hypothetical animals of the models in not moving at random to their prey and in not having insatiable appetites; but, in addition, Gause noticed a special difference. The predator (*Didinium*) was able to multiply intensively even when the prey were scarce. It did this at the expense of size. This meant that the multiplication of the predators was not a linear function of the density of the population of prey and that the departure from linearity was relatively great when the density of the population of prey was low. A similar explanation has been offered for the extermination of the fly *Phormia groenlandica* by the wasp *Mormoniella vitripennis* (Wladimirow; quoted in Gause, 1934).

Whatever the explanation, this has invariably been the outcome of all experiments in which the fundamental conditions of the Lotka-Volterra models have been fulfilled, namely, that the two populations should occur together in a space that is uniform and constant. In other words, the mathematical models have not been confirmed by the empirical tests. We know of no exception to this rule, although some experiments by Gause (1935*b*) and Gause, Smaragdova,

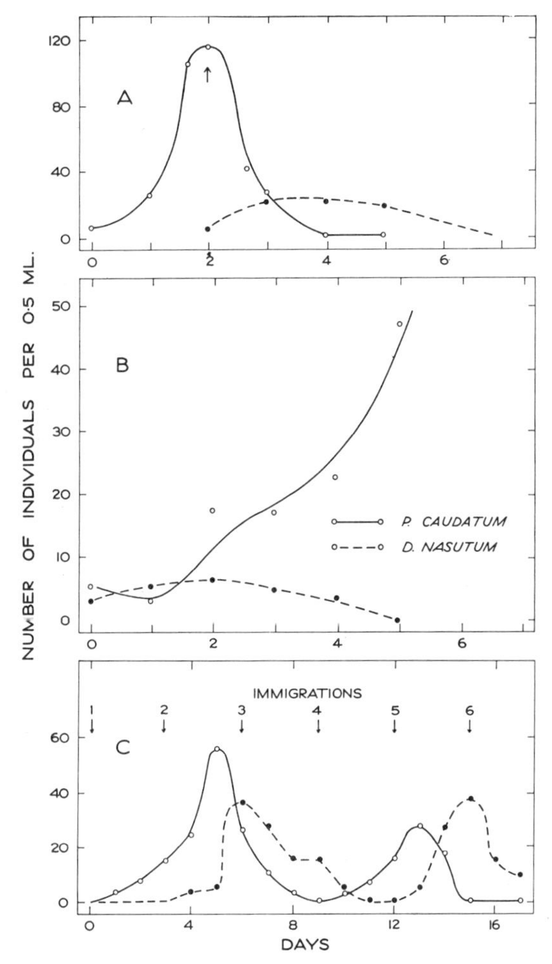

FIG. 10.15.—The trends in the numbers of *Paramecium caudatum* and *Didinium nasutum* when the two species were reared together in the same container on a medium made from an infusion of oats. *A*, no sediment in the medium; *B*, with sediment to serve as a refuge for the prey; *C*, with periodic immigrations of predator and prey. The three experiments were not done in identical circumstances. The scales for the different diagrams are not the same. (After Gause, 1934.)

and Witt (1936) provided what may seem at first sight to be an exception. The prey were yeasts, either *Saccharomyces exiguus* or *Schizosaccharomyces pombe*, and the predators were *Paramecium bursaria*. In experiments which ran up to 25 days, both species survived together, and the densities of the populations oscillated. A certain amount of the yeast sedimented on the bottom of the culture, and these cells escaped attack by the *Paramecium.* This enabled the yeast to survive. The predators, finding less and less to feed on in the liquid medium, decreased in size, in this way retaining their numbers to some extent, even though food was scarce. While the predators were relatively scarce, the yeast cells began to multiply rapidly and so provided food in the liquid medium for the *Paramecium.* It is possible, of course, that a continuation of the experiments might have resulted in the predator's becoming extinct at one of the low points in its fluctuations. This is not an example of an oscillation due simply to the interaction between prey and predator of the type postulated by Lotka and Volterra. The oscillations are due to special qualities possessed by prey and predator, making them quite unlike the hypothetical animals of the models: the sedimentation of the yeast and the ability of *Paramecium* to retain numbers at the expense of size when food is scarce.

Gause (1934) repeated his experiment with *Paramecium* and *Didinium*, but this time he provided a sediment, on the floor of the tube, into which the prey could retreat and into which the predator would not follow them. He found that the prey were exterminated from the clear medium above the sediment, and then the predator died out for lack of food, exactly as in the earlier experiment. But in this experiment the *Paramecium* which had taken refuge in the sediment continued to multiply and reinvaded the clear medium above. The final outcome was a dense population of prey and no predators (Fig. 10.15, *B*).

The experiments of Gause, Smaragdova, and Witt (1936) with two species of mites illustrated the same phenomenon, with a slight variation in detail. The prey was *Aleuroglyphus agilis*, the predator *Cheyletus eruditus.* The two populations were brought together in small vessels containing either millet or flour or a mixture of the two. In no case did both populations persist together, nor were there any indications of periodic oscillations in the densities of the populations. Either the predator or the prey died out, depending on the relative densities of the two populations at the beginning and the nature of the medium. The rate of consumption of prey by predator was slowest in the flour, most rapid in the millet, and intermediate in the mixture of the two, presumably because the predator was not able to search so well in flour, which thus is analogous to the sediment in the experiment with *Paramecium* and *Didinium.* The flour did not provide such an effective refuge for *Aleuroglyphus* as the sediment did for *Paramecium.* Eventually the prey were exterminated, although this took longer in flour than in millet.

Flanders (1948) devised an experimental model to demonstrate the phenomenon of protective refuges. The prey were larvae of *Ephestia kühniella*, the predators were braconid wasps. The two populations lived together in a cage, on the floor of which was an open Petri dish containing grain to a depth of 12 or 24 mm. The ovipositor of the braconid could probe to a depth of 6 mm. Those larvae which happened to be living between the surface and 6 mm. were exposed to the predator; those situated more deeply than this were secure. In every generation sufficient prey escaped to carry the population on. The adults were, of course, "immune." Flanders said that the populations of prey and predator persisted in the cage together indefinitely.

In still other experiments Gause periodically added prey to the experimental population so that it was never allowed to die out. On these occasions, as was to be expected, he was able to maintain populations of the predator and prey living together indefinitely and to produce oscillations at will (Fig. 10.15, *C*). Gause's own comment on these experiments was: "The periodic oscillations in the numbers of predators and prey are not a property of the predator-prey interaction itself, as the mathematicians suspected, but apparently occur as a result of constant interferences from without in the development of this interaction" (Gause, 1934). He would surely be the first to deny the misleading comment made by Volterra and D'Ancona (1935) on these experiments that "they are a very important experimental verification of the mathematical calculations." The chief importance of these experiments is that they are nice laboratory models of phenomena which are commonplace and important in nature.

Gause himself pointed out that the *Paramecium* which sheltered in the sediment from *Didinium* were analogous to the immune hosts that develop in a population being ravaged by disease. Certain stages in a complex life-cycle often carry an innate "immunity." For example, we have observed laboratory colonies of *Calandra oryzae* which were accidentally parasitized with the wasp *Aplastomorpha calandrae*. This wasp lays eggs in the larvae of *Calandra* inside grains of wheat, but it does not attack the adult weevil. Cultures became so heavily parasitized that no weevils were able to mature to the adult stage. Nevertheless, the original adults persisted in the culture unmolested by the parasites.

But the principle is much wider than this. In most natural populations there are individuals that are better protected than others either by virtue of the place where they happen to be living or because they have some protective quality or instinct better developed than their fellows (secs. 12.0 and 12.2). The scientific study of diversities of this sort is likely to be far more fruitful than the construction of mathematical models in which it is postulated that animals and their environments have an unnatural and stereotyped uniformity.

10.223 EMPIRICAL "TESTS" OF THE NICHOLSON-BAILEY MODELS

De Bach and Smith (1941) used *Musca domestica* as prey and *Mormoniella vitripennis* as predator in an experiment which was designed to be a test of the Nicholson-Bailey model, in which a single specific predator interacts with a single host. The particular conclusions which can be justifiably drawn from this ingenious experiment are so dependent upon the way in which the experiment was done that it is necessary to indicate precisely the experimental procedure. Pupae of the fly were distributed in a quantity of barley through which the parasite had to search. Further restrictions were added to these populations. The length of the host's generation and its power of increase were "artifically" limited, and the setup was such that all deaths of hosts were "eliminated" except those that were due to parasites. Without these restrictions, the experimenters considered that the experiment would have been unwieldy. The procedure was as follows: 36 pupae of the fly were distributed in 2 quarts of barley; 18 parasites were added. After the parasites had searched for 24 hours, the number of parasitized pupae was counted. This gave the number of hosts parasitized within an artificial "generation." A "generation" is thus defined as 24 hours of pupal life. In the first 24 hours, on an average, 20.6 hosts were parasitized. This left 15.4 undiscovered hosts. The host was then given a "power of increase" of 2 in a "generation." Hence the next "generation" started with 31 hosts. The original "generation" of parasites was presumed to have died. Only their "offspring" (21 parasites) were available for the next "generation." The procedure was followed through for seven successive "generations." In each "generation" the host was given a "power of increase" of 2, i.e., the unparasitized hosts were doubled for the beginning of the next "generation." The results of these manipulations are shown in Table 10.02. A fall in number of hosts was followed by a fall in number of parasites. The hosts then increased, to be followed by the parasites. The trend suggests the beginning of an oscillation, though only about two-thirds of one complete oscillation is represented.

TABLE 10.02*
NUMBERS OF PARASITES, *Mormoniella vitripennis*, AND HOSTS *Musca domestica*, IN SEVEN "GENERATIONS"

"Generations"		1	2	3	4	5	6	7
Initial no. of parasites	18	21	18	15	11	9	11	14
Initial no. of hosts	36	31	26	22	23	29	37	47

* After De Bach and Smith (1941).

De Bach and Smith calculated the curves for the particular Nicholson-Bailey model in which the prey has a "power of increase" of 2 and completes one generation every 24 hours and in which the populations of prey and predator start with 36 and 18 individuals, respectively. The experimental points fitted quite well to the first section of the calculated oscillation. They concluded that, "so far as the data of this experiment go, they follow with remarkable fidelity the theoretical conclusions of Nicholson and Bailey (1935)."

The somewhat considerable difficulty of regarding this particular experiment as a test of Nicholson and Bailey's theory is that most of the characteristics of the host and the parasite were arbitrarily attributed to the animals by the experimenters. The host's rate of increase is arbitrarily fixed and does not vary with the density of the population of hosts. Populations neither of hosts nor of parasites have any natural age-distribution. The length of life of both host and parasite are limited to 24 hours, and there is no overlapping of generations at all. The only aspect of the experiment which is not fixed by the experimenters is the particular capacity of the parasite to find hosts of various densities. It does not serve as an empirical test of the Nicholson-Bailey model, since all the premises of the model were supplied arbitrarily by the design of the experiment. It is unlikely that populations of *Musca* and *Mormoniella* living together in a limited space would display trends similar to those found in this very abstract experiment. It is more likely that they would conform to the results of Gause's experiments, when either predator or prey died out promptly without showing any tendency to oscillate.

Varley (1947) counted the numbers of the gallfly, *Urophora jaceana*, in a natural stand of knapweed for 2 successive years. Although he studied a natural population, he interpreted his results in terms of the hypothesis of Nicholson and Bailey. So this work is better discussed in this section than in the sections on natural populations. The samples were taken weekly for part of the summer and autumn and less frequently for the rest of the year. Records were kept of the numbers of *Urophora* that died throughout the year, and a careful analysis was made of all the causes of death. The investigation revealed a most interesting complex of other animals associated with *Urophora*, but only one important specific predator. Incidental predation by mice eating the galled flower heads of knapweed and "winter disappearance" accounted for a large proportion of the total deaths. These formed a substantial component of "nonspecific mortality" (i.e., deaths from causes which did not discriminate between prey and predator), which accounted for over 90 per cent of all the deaths of *Urophora*. Nevertheless, having accepted axiomatically that the numbers of *Urophora* could be "regulated" only by a "delayed density-dependent factor"[2] and having decided that *Eurytoma* was the only important "delayed density-dependent factor" present, Varley set out to measure the influence of *Eurytoma* on *Urophora* by substituting his observed values in the appropriate model of the Nicholson-Bailey series.

Varley referred to his results in the following words: "In this contribution to insect ecology the theory of balance of animal populations, formulated by Nicholson (1933) and Nicholson and Bailey (1935), is used for the first time in the interpretation of the results of a field survey. The conclusions are sufficiently striking to claim the attention both of ecologists and economic zoologists, and their importance goes beyond that of the insect material on which they are based. . . . The clarification of a complex situation achieved in this way may provide the economic entomologist with a new and powerful tech-

2. The essentials of Varley's definition of this new term are contained in the following quotation from his paper (p. 140): "A parasite acts as a delayed density-dependent factor if its fecundity or its effective rate of increase is strongly correlated with host density."

nique." This work has also been accepted by several critics as a successful demonstration of a Nicholson-Bailey model (e.g., Elton, 1949, p. 17; MacFadyen, 1949, p. 539).

In view of the widespread acceptance of the claims made for this paper and because of the methodological importance of the issues which it raises, we must explain in some detail why we cannot accept these claims.

a) It is difficult to see how the data from a natural population can be used in this way directly to test the predictions of equations which are built up from the premise that the animals are living in a "universe" that is uniform and invariable with respect to the places where the animals may live, temperature, food, and everything else except predators. If the observed data failed to agree with the predictions of the models, we would say that this might be because some of the premises, namely, with respect to constancy of "environment," had not been fulfilled. If the observed data did happen to agree with the predictions of the equations, it might be correctly inferred that, since some of the premises had not been fulfilled, this agreement was specious because it was due to some fortuitous compensating error in the remaining premises.

b) There are three errors of logic in the interpretation of the empirical results. These are:

(i) The data that were collected provide empirical estimates for all but one of the quantities needed to solve the equations, namely, nonspecific mortality; a value was arbitrarily assumed for this unknown. This was done by substituting in the equations different values for nonspecific mortality until one was found which gave a calculated value for the "steady state" which agreed closely with the census data for the two years of study. This procedure is described in the first paragraph of page 180 of Varley's (1947) paper and in his Table 14. Varley concluded: "This application of Nicholson and Bailey's theory to the study of the steady state has produced a series of calculated values for the population densities of the gall fly and of *Eurytoma curta* which agree closely with the values found in the census." But this agreement is artificial, because the calculated values for the "steady density" were made to agree with the census data for the 2 years by arbitrarily choosing the appropriate value (92 per cent) for nonspecific mortality.

(ii) In a Nicholson-Bailey model the chance that the densities in any two consecutive generations, chosen at random, would fall one on either side of the "steady density" depends on the period of the oscillations. Nicholson (1933, p. 161) stated that four generations is a minimum for the period of the oscillations, and he showed that this occurred only when the "power of increase" exceeded 50. The "power of increase" of *Urophora* was taken as 18, and the oscillations must therefore have a period exceeding four generations. If the period exceeds four, the chance that any two consecutive generations will fall one on either side of the "steady density" is less than the chance that they will both fall on one side. Varley's calculated value for the "steady density" of *Urophora* falls between the two observed densities and therefore provides no evidence that the fluctuations in the numbers of *Urophora* and its predator conform to a Nicholson-Bailey model.

(iii) Throughout this paper the premises and conclusions of the Nicholson-Bailey models are accepted on faith, and most of Varley's conclusions depend on this. For example, the two "major conclusions," which are set out on pages 180–81 of Varley's paper, follow implicitly if one grants the truth of the mathematical models. But in this case they could have been stated without knowing any of the estimates of death-rates, etc., with which the empirical part of this paper is chiefly concerned. To quote these conclusions to verify these mathematical models would be to argue in a circle.

c) The variability in the original samples collected in the field was great, and the estimates of the statistics required to construct a Nicholson-Bailey model consequently lacked the necessary precision. We are indebted to Dr. A. Milne for the opportunity to read, in advance of its publication, a criticism of the statistics used by Varley. We do not have the space to discuss the matter thoroughly, but we have taken the following points from Milne's criticism:

(i) The statistics which were built into the models were derived (by the process of taking ratios, sums, or products) from simpler quantities which were got from counts of the numbers of insects in samples without recourse to "normalizing" transformations for the very skewed distributions. The methods used in calculating these "derived" statistics underestimated their variances in many cases. For example, Varley, in the computation set out under his Table 2, used what appears to be an incorrect method for estimating the variance of the ratio of the number of flies which emerge from the flower heads to the total number of flies developing in the flower heads. He gave the standard error of this ratio as 0.049. But Milne recalculated the same statistic, using the more accurate method:

$$V\left(\frac{a}{b}\right) = \frac{1}{b^2}\left[V(a) + \frac{a^2V(b)}{b^2}\right],$$

where a = mean number of gallflies that emerged, b = the mean number of larvae and (or) pupae per flower head, and V = variance. He found the standard error to be 0.163, which was three times that given by Varley. Several steps were involved in the calculation of certain other statistics, such as the fecundity of the predator, the "area of search" of the predator, and the "rate of increase" of the prey. It may be expected that errors introduced by inaccurate methods of calculating their variances would be greater with these more complex statistics than with the simple ratio which we illustrated above.

(ii) Although Varley estimated the variances of most of the statistics he used in his calculations, he ignored these variances when it came to "testing" the model. For example, the "proof" that *Eurytoma curta* was a "density-dependent factor" depended largely on a calculated value of 7 for the ratio of the fecundity of the species in 1935 to its fecundity in 1936. That rounded figure of 7 was obtained merely by taking the ratio of the appropriate sample means, viz., $63.0 \div 8.4 = 7.5$. But variance must not be ignored in drawing conclusions from data which depend on samples. Varley's estimates of the mean fecundities of the gallflies present in 1935 and 1936 were 63.0 ± 23.0 and 8.4 ± 2.6, respectively. The standard errors are probably underestimated for the reasons given above; but if his statistics are accepted, the fiducial limits of the ratio may be calculated from the following expression (Finney, 1950):

$$\text{Fiducial limits} = \left[\frac{a}{b} \pm \frac{d}{b}\sqrt{\frac{(1-g)s_1^2}{n_1} + \frac{a^2 s_2^2}{b^2 n_2}}\right] \div (1-g).$$

The variances of a and b are given by s_1^2/n_1 and s_2^2/n_2, respectively; d is the statistic given in Table V_1 of Fisher and Yates; and g is calculated as follows:

$$g = \frac{ds_2^2}{b^2 n_2}.$$

If we substitute $n_1 = n_2 \geqslant 120$ in the above expression (we are not told the actual values of n_1 and n_2, and this is the most favorable allowance that can be made), the fiducial limits of the ratio $(63 \pm 23) \div (8.4 \pm 2.6)$ would be 2 and 20 at the 5 per cent level of probability. Clearly, Varley's hard figure of 7 for the "fecundity-ratio" of *Eurytoma* is not to be trusted. Nor is his similarly calculated ratio of 3 for *Urophora*. According to his method, this ratio should have been 5, since here it is the larva and not the adult that is relevant. The fiducial limits, at the 5 per cent level of probability, for this ratio are 3 and 9. Yet the difference between the hard figures 7 and 5 (or 3) is the evidence upon which the interaction between *Eurytoma* and *Urophora* is claimed to be density-dependent!

Varley's estimate of 18 as the "power of increase" of the prey is the mean of two ratios, 21 and 14, which were estimated in 1935 and 1936. By the methods used above, it may be shown that the fiducial limits, at the 5 per cent level of probability, for the estimated rate of increase in 1935 were 13 and 43 and for 1936 they were 8 and 25. The other statistic which entered into the final "test" of the Nicholson-Bailey model was the "area of search" of the predator. This was estimated as 0.25. Varley did not give the variances from which the fiducial limits of this statistic might be estimated; by its nature it was not likely to have been estimated more precisely than the others.

Without taking up more space in detailed criticism we can say in general terms that Varley did not manage to estimate the necessary statistics precisely enough to make us reasonably sure that *Eurytoma curta* was really operating as a "density-dependent factor" or to convince us that the data which he had collected conformed to a Nicholson-Bailey model.

We conclude this section by repeating that neither of the experiments which have, from time to time, been said to provide an empirical verification for the Nicholson-Bailey models do, in fact, verify them.

10.3 NATURAL POPULATIONS

In this section we are chiefly concerned to amplify the categories of other animals, outlined in section 10.01, with examples drawn from natural populations. The discussion, particularly of predators, could go on almost indefinitely, but it has been limited to a few examples which have been chosen to illustrate the major principles. It should be clear from what was said in section 10.01 and elsewhere that the categories describe "types of interaction" rather than types of animals, and one animal may have several sorts of activities which belong in different categories. Usually in such cases one activity outweighs the others in importance, and not much confusion is created by thinking of an animal as belonging to a particular category. This is perhaps desirable, because it makes for facility in writing, but it should be remembered that strictly it is the activities and not the animals that are being classified.

10.31 *Nonpredators*

The rabbit, in Australia, eats much the same food as the sheep. The preferences of the two species may be quite different, but there are no foods eaten by the sheep which the rabbit will not eat, though there may be a few that serve the rabbit but not the sheep. Their requirements in other matters may be quite different, and it is well known in the famous semiarid pastures of the interior that during the hot, dry summer the sheep may wander as far as 5 miles from water in search of food, whereas the rabbit is confined to an area within a mile or two of water. Notwithstanding these differences in the requirements of the two species, the presence of many rabbits in an area means fewer sheep, as the farmers know to their cost. In ecological terms we would say that, despite many differences, the requirements of these two species overlap sufficiently to make the rabbit an important component of the environment of the sheep: it is a nonpredator of category *a* (sec. 10.01).

In South Australia the scelionid wasp *Scelio chortoicetes*, during its larval stage, feeds upon the eggs of the grasshopper *Austroicetes cruciata*. The nymphs and adults of the grasshopper are eaten by a number of different sorts of birds (Andrewartha and Birch, 1948). The eggs are present from midsummer to early spring, the active stages during spring and early summer, so there is no common requirement which the birds and *Scelio* "seek together" in any literal sense. Yet in studying the ecology of *Scelio,* we would recognize the birds as nonpredators of category *a;* for the eating of the adult grasshopper, which might produce the eggs that *Scelio* needs for food, is equivalent to eating the eggs themselves. Usually, the grasshoppers are widespread and very numerous relative to the numbers both of birds and of *Scelio.* There is plenty of food for all; and if this were the only circumstance we knew of, we would rate the birds as quite an unimportant component in the environment of *Scelio.* But at infrequent intervals extraordinary circumstances occur, when drought kills all the grasshoppers in the area except for a few which happen to be living in specially favorable local situations; these are not at all plentiful, and they are usually quite small areas which, because of their topography, are rather less arid than the surrounding country. The birds, because of the shortage of food elsewhere, congregate in these places and may feed on the grasshoppers until there is scarcely one left (Birch and Andrewartha, 1941). This results, indirectly, in a scarcity of food for *Scelio* and a consequent great, though temporary, reduction in the density of its population in the area. How important these events, and especially the part played by the birds, are in determining the status of *Scelio* as a rare or common species in the area will depend on the sorts of considerations discussed in sections 14.1 and 14.2. It happens, in this case, that the animal which is eaten has a complex life-cycle and different stages of the life-cycle are eaten by different predators. Consequently, this example may be more subtle than some, but otherwise it is typical of the relationships which exist when carnivores enter into one another's environments as nonpredators requiring to share the same food.

A good example among herbivores is provided by four species of caterpillars which feed on the leaves of *Pinus sylvestris* in central Europe; this has been well documented by an impressively long list of records going back to 1880, which were published by Schwerdtfeger (1941). These records were discussed by Varley (1949), and he replotted them on a logarithmic scale (Fig. 10.16), which is more instructive because it shows the proportional changes in numbers especially during the long periods when the insects were relatively scarce. Schwerdtfeger recorded that during a mass outbreak, especially of *Bupalis,* the pines might be defoliated, resulting in a shortage of food for this and the other species. But Figure 10.16 shows that, taking all four species together, outbreaks of this magnitude may have occurred less often than 10 times in 60 years. There are in each of the curves many small maxima, indicating that with all the species there were occasions when a small but increasing population ceased increasing and began to decrease; and most of these occurred when none of the species was numerous enough to indicate a shortage of food for any of them.

These four species, though not closely related taxonomically, live on the same tree and, so far as can be determined, have identical requirements for food.

Each one enters into the environment of every other simply as a nonpredator requiring to share the same resource. This resource is hardly ever in short supply, and Varley (1949), in reviewing these data, rightly concluded that "the population densities of each species are separately controlled by different factors at such low population densities that direct competition for food is very uncommon." He then went on to attempt an explanation in terms of a Nichol-

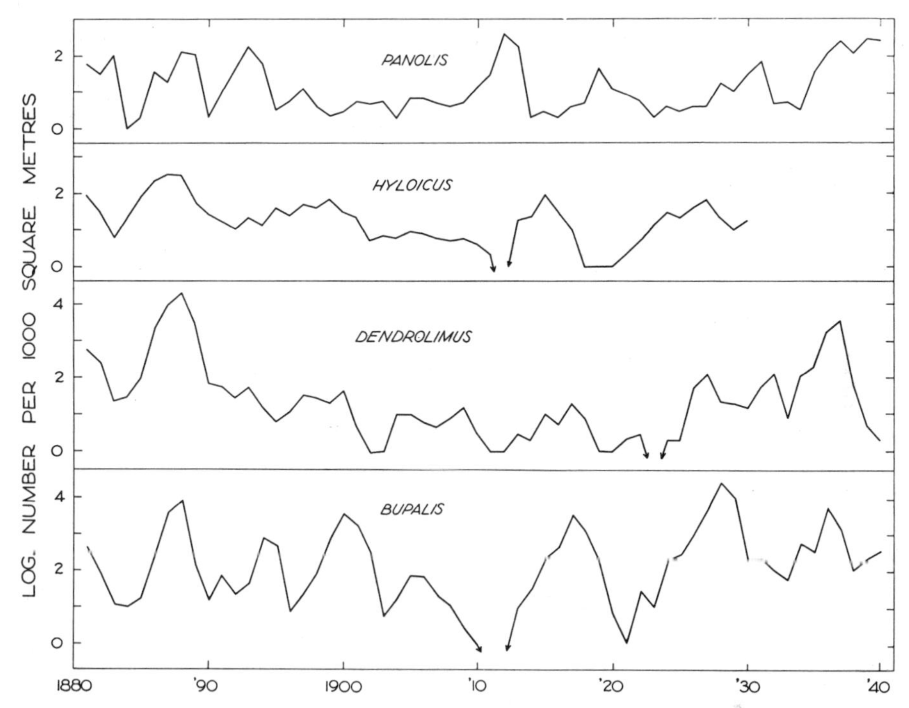

FIG. 10.16.—Annual fluctuations in the numbers of four species of moths in a pine forest at Letzlingen, Germany. The caterpillars feed on the leaves of the pine trees. The densities of the populations were estimated by counting the overwintering stages (larvae or pupae) in the leaf litter on the ground below the trees. A logarithmic scale was used for the ordinates. (After Schwerdtfeger, 1941.)

son-Bailey model. That he failed in this is not surprising, for the more one examines these well-documented examples from nature, the more it becomes clear that the postulates of the mathematical models are inadequate and their conclusions misleading when applied as yardsticks to natural populations. It is true that in this case food is hardly ever in short supply, and, on the face of it, we would be inclined to say that these four caterpillars are not important components of one another's environments.

As a source of food, from the ecological aspect, a piece of carrion or a fallen log is relatively simple compared to the living animals and plants that serve as food for most creatures. The relationships of the animals living in a carcass may be nicely analyzed, as, indeed, they have been in relation to the Australian sheep blowfly *Lucilia cuprina* by Fuller (1934), Waterhouse (1947), Nicholson (1950), and their colleagues. This species is abundant in Australia wherever sheep occur, being responsible for over 90 per cent of the myiasis observed in sheep; it is rare or absent from places where there are no sheep (Waterhouse, 1947). In the laboratory, *L. cuprina* may be reared quite readily on meat (Nicholson, 1950), but it would seem that in nature the only important breeding place is on the living sheep; only a negligible proportion come from carrion and other breeding places. There is a clear explanation for this in the experiments of Fuller (1934) and Waterhouse (1947). Waterhouse exposed the carcasses of 27 freshly killed sheep at intervals throughout the year and recorded the species and numbers of flies that completed development in them. The carcasses were exposed on deep trays in the open and were freely available to the natural populations of blowflies in the field. A carcass, on the average, produced about 10,000 adult flies of all kinds, but there was great variability, from 109 to 67,000. These numbers seem small in relation to a carcass weighing about 100 pounds when it is recalled that Nicholson (1950), in favorable circumstances in the laboratory, was able to rear about 800 *L. cuprina* from 50 gm. of meat (about the equivalent of 7,000 from 1 lb.) and Ullyett (1950) got comparable results with *L. sericata* and several other species. In Waterhouse's experiments there was always a superabundance of eggs laid, and the low numbers emerging were due to the high death-rates among the larvae.

The death-rate among *Lucilia cuprina* was relatively higher than with most other species. Waterhouse observed that the females of this species were strongly attracted to the carcasses while they remained fresh; the flies laid many eggs, but hardly any survived to become adults. None emerged from six carcasses exposed during winter (June to August); eight adult flies were reared from 12 carcasses exposed during summer (November to February); four other carcasses which were exposed during spring (September to October) and autumn (March to May) gave rise to numbers between 11 and 45, with a total of 100: but five others, exposed also during March to May, failed to produce any adults of *L. cuprina*. In another series of experiments special flat trays were used which were especially designed to favor *L. cuprina*. Slightly higher numbers survived, but still the greatest number from one carcass was 645. In the Canberra district this species spends the winter as a larva or pupa; consequently, no eggs would have been laid on carcasses exposed during this season. But the scarcity of *L. cuprina* from carcasses during spring and autumn and its virtual inability to survive in them during summer, when it was breeding freely in living sheep near by, was almost entirely due to the presence in the carcasses of other species which are better adapted for life in a crowd.

During the winter almost the only species of blowfly breeding in carcasses around Canberra is *Calliphora stygia*. Their numbers are usually few relative to the size of the carcass, so that there is no shortage of food. With the approach of spring, when the temperature of the soil reaches about 15° C., *Lucilia cuprina* and *L. sericata* emerge from hibernation; shortly afterward *C. augur* appears in numbers. From this time on, so many eggs are laid on each carcass that there is not enough food to go round. Experiments show that when *L. cuprina* is alone, the first result of crowding is to reduce the size of the individuals without greatly reducing the proportion of survivors, but, with further increases in the density of the population, the number of survivors is reduced until eventually none survives (secs. 9.235 and 11.22). When *L. cuprina* shares the carcass with *C. stygia* or with *L. sericata*, the result is much the same, and up to quite an intense degree of crowding the survivors include some of each species. Apparently, the most important way in which *C. stygia* or *L. sericata* enters into the environment of *L. cuprina* is as a nonpredator which requires to share the same food (category *a* of sec. 10.01).

When *Lucilia cuprina* shares the carcass with *Calliphora augur*, the result is similar, except that, as the intensity of crowding is increased, there comes relatively soon a density at which all the *L. cuprina* die and all the survivors are *C. augur*. So far as can be observed directly, *C. augur* does no more than eat the food which *L. cuprina* requires. In nature it usually arrives at a freshly killed carcass as soon as, or perhaps before, *L. cuprina*. Its eggs hatch immediately they are laid, whereas those of *L. cuprina* usually require about 12 hours. This gives the larvae of *C. augur* so much start, and it may be that this is adequate explanation for the superiority of one species over the other. If this were the only relationship between the two species, *C. augur* would, like *C. stygia*, be classified simply as a nonpredator requiring to share some of the same resources. But certain of Fuller's experiments suggest that there may be more to it than this, and it is possible that *C. augur* modifies the environment of *L. cuprina* in some other way, in addition to merely eating its food (category *b* of sec. 10.01). One way in which it might do this is suggested by Waterhouse's observation that the temperature of a heavily infested carcass may be as much as 20° C. higher than the surrounding air, and it may be that the two species differ in their capacities to thrive at high temperature.

As the summer advances and the temperature of the soil reaches 21° C., *Chrysomyia rufifacies* emerges from hibernation and soon becomes the most abundant species in the area. This species is attracted to the carcass after the others have become established and decomposition is well-advanced. Its larvae are tough-skinned, robust, and vigorous relative to the others. They are well-suited to the high temperatures of summer and develop rapidly. It is their presence in large numbers which is chiefly responsible for the virtual nonappearance of *Lucilia cuprina* among the survivors from carcasses exposed during summer; *Calliphora stygia* and *C. augur* also suffer in the same way, but the latter not quite so severely. There is no doubt that *Chrysomyia rufifacies* eats the food that *L. cuprina* needs, and this may be an important part of its influence on *L. cuprina*. But in a carcass crowded with both species, *L. cuprina* may be observed to be irritated and disturbed: many leave the carcass before they have had enough food to complete their development, and they perish. Also *Chrysomyia rufifacies* attacks and eats the other species. Its influence on *L. cuprina* is thus threefold: it eats the food that the other needs, it interferes with *L. cuprina* to such an extent that many are driven forth to die of starvation, and it also attacks and eats *L. cuprina*. Since it can be reared on a diet of carrion alone, it is not a typical predator, and perhaps it is best regarded as essentially a nonpredator with qualities appropriate to both categories *a* and *b* of section 10.01 (see also experiments described in sec. 10.211). Blowflies on carrion thus provide an example of the relationships between nonpredators which feed on a common food; but they differ from the caterpillars of the preceding example in two important ways: (*a*) their food (carrion) is more strictly limited because, unlike that of the caterpillars' (living pine tree), it does not grow; (*b*) each species of caterpillar merely ate the food which the others required, whereas the blowflies not only did this but interfered actively with one another as well. The interference of one species by another is characteristic of the behavior of some birds when they arrive at food together. Jackdaws and crows fight for voles; and eagles and vultures fight over carcasses. Other birds even fight for the possession of tree tops from which to sing. In seeking a common requirement, one or another may succumb in the struggle, even though the resource sought is not necessarily in short supply (Udvardy, 1951); this, however, is not at all common among birds (sec. 10.311).

The relationships between the animals living in the sewage-filter beds studied by Lloyd, Graham, and Reynoldson (1940) and Lloyd (1943) were rather like those between the animals living in carrion. The communities differed in different beds, depending on management (e.g., one set of beds was "rested" more often than another) and the nature of the material flowing through them; for example, in one set of beds this contained a substance that was toxic to some species. In one particular set of beds which we have chosen as an example, the group of animals which was interesting because of their interrelationships comprised:

An enchytraeid worm, *Lumbricillus lineatus*
Four midges of the family Chironomidae, *Metriocnemus longitarsus*, *M. hirticollis*, *Spaniotoma minima*, and *S. perrenis*
And two moth midges of the family Psychodidae, *Psychoda severina*, and *P. alternata*

The food for these animals consisted of masses of algae, which formed an encrusted layer on the surface of the beds, and a slimy zoöglaea of bacteria, fungi, and Protozoa in the depths. From time to time the encrusted algae would slough away from the stones near the surface and sink. This augmented the stocks of food for the animals living in the depths, while making food rather scarce near the surface. Two of the above species, the worm *Lumbricillus lineata* and the larva of the midge *Metriocnemus longitarsus*, lived chiefly in the topmost foot of the beds and fed on the encrusted mass of algae. Once a year during spring, their numbers, especially those of the worm, increased so greatly relative to the growth of the algae that they caused it to disintegrate and settle into the depths of the beds, where it was added to the amount of food already available to the animals living in the depths. In the absence of these "surface-scourers," the beds would become choked and probably unsuitable as places for the other species to live permanently. After this surface layer had been sloughed off in this manner, food temporarily became very scarce for both *L. lineatus* and *M. longitarsus*. Since the sloughing was chiefly due to the worm, which increased in numbers during winter and especially during warm winters, *L. lineatus* was obviously important in the environment of *M. longitarsus*, partly as a nonpredator requiring to share the same resource, but chiefly as a nonpredator which reduced, without itself consuming them, the stocks of food available to *M. longitarsus* (category *b*).

The larvae of the fly *Metriocnemus longitarsus* feed chiefly on the encrusted algae, but in the course of their feeding they destroy (largely by eating) any eggs or small larvae of the other species that they happen to meet. This species bred throughout the year, with continuously overlapping generations. So larvae were always present in the surface layers of the beds, but they were usually most abundant from December to March and again in June. Since the other three species of Chironomidae laid their eggs near the surface, their chance of survival during this stage depended quite largely on the density of the population of *M. longitarsus*. Accordingly, the timing of the life-cycle of the other species relative to the density of the population of *M. longitarsus* was important. Thus *M. hirticollis* had its major flight during June, when the population of *M. longitarsus* was usually dense, so that the death-rate among eggs of this species was usually high unless some adverse circumstance (usually weather) had reduced the numbers of *M. longitarsus*. Similarly for *Spaniotoma minima*, except that this species, having a shorter life-cycle during summer, had a better chance of recovering from an initial setback. These two species, and especially *M. hirticollis*, were relatively rare, but occasionally (during the course of these investigations it happened twice in eight years) an unusually hot, dry period during June so reduced the numbers of *M. longitarsus* that the survival-rate among the eggs of *M. hirticollis* and *S. minima* was high, and these species became temporarily quite abundant. Since *M. longitarsus* did not usually live or feed in the depths of the beds where *M. hirticollis* and *S. minima* lived (once the eggs had hatched), it did not require to share, to any important extent, any of the resources needed by the other two species, although the stocks of food for the latter were sometimes augmented by algae which sank into the depths from the surface layers where *M. longitarsus* was living. Nor is it strictly a predator, though it may eat some of the eggs that it destroys. It is best regarded, like the hypothetical sheep of section 10.01 category *b*, as a nonpredator which interferes with or destroys the animal of whose environment it forms a part, without necessarily needing to share any of the same resources.

The larvae of the two moth midges, *Psychoda severina* and *P. alternata*, lived in the depths of the beds, and the adults probably always laid at least some of their eggs in the depths, thus avoiding the hazards associated with the presence of *M. longitarsus* in the surface layers. They had to share their food with the larvae of *M. hirticollis* and *S. minima*. Moreover, they suffered the same sort of interference and destruction from these species that these species suffered from *M. longitarsus*. So we say that the psychodids had in their environments the two species of other animals, *M. hirticollis* and *S. minima*, which are nonpredators requiring to share the same food but which, in addition, are also directly destructive.

The abundance of the psychodids depends partly on the abundance of *M. hirticollis* and *S. minima*, which, in turn, is related to the abundance of *M. longitarsus*, whose numbers are related to the abundance of *Lumbricillus lineata*. So both *L. lineata* and *M. longitarsus* are important components in the environments of *P. severina* and *P. alternata*. They come in indirectly and thus belong to category *e* of section 10.01.

Nonpredatory other animals are important components in the environments of these species, influencing them in the ways which we have mentioned. But other components of the environment, especially fluctuations in the weather, are important for these species also. We shall return to this example in sections 13.21 and 13.22 and show how the interactions of the other components may also be taken into account.

We shall close this section with an example which has been selected because it shows nicely still another sort of interaction between the components of environment: how the relationships between a group of species may be changed dramatically by the place in which they happen to be living. It was mentioned above that *Lucilia cuprina* breeds less in carrion than in the living sheep. Waterhouse (1947) collected at random 26 "flyblown" sheep in the vicinity of Canberra, brought them into an insectary, and recorded the species and numbers of flies reared from them. From 16 sheep *L. cuprina* was the only species reared. In each of the other 10, *L. cuprina* was easily the most abundant, but there were also present varying numbers of the other species which are found in carrion, including *L. sericata*, *Calliphora stygia*, *C. augur*, and *Chrysomyia rufifacies*. Ninety per cent of all the flies bred from these 26 sheep were *L. cuprina*. This is a very different picture from what prevails in carrion (see above). In the living sheep *L. cuprina* seems to have the advantage; this may be partly because it gets in ahead of the others (see below), but there may be other reasons which have not yet been discovered.

The two species *Lucilia cuprina* and *L. sericata* are almost sibling species, because they are almost indistinguishable morphologically. For many years the distinction was not recognized; but with experience and with long series of specimens, they may be told apart. Both are attracted to fresh carrion and may be reared in it. Both may also be reared from the living sheep. But, as was shown by Mackerras and Mackerras (1944), it is doubtful whether *L. sericata* would very often in nature become established in a living sheep except when the way had been prepared for it by *L. cuprina*. This is best illustrated by describing a typical experiment. We give it in the authors' own words (p. 15):

> Eight ewes which had soiled or moist crutches were exposed to a pure culture of mature *L. sericata* for a week. Oviposition occurred on two sheep; on one a strike did not follow, on the other on which oviposition had occurred on the 6th day small maggots were present on the 7th day. Several hundred *L. cuprina* were then added to the insectary and within four hours every sheep had egg-masses on the breech. A definite strike occurred on each animal. Some eggs were collected from each breech and were allowed to develop into flies and the curious fact emerged that in seven out of eight both *L. cuprina* and *L. sericata* were present while the eighth contained *L. cuprina* alone. Apparently the act of oviposition by *L. cuprina* had in some way stimulated *L. sericata* to lay eggs on areas which they had previously ignored.

When carrion is being considered, *L. cuprina* is no more than a rather ineffectual and much suppressed sharer of other species' food. In the living sheep, however, it has an entirely different status: in its own right it is the dominant, i.e., the most abundant, species; considered as a component in the environments of the other species, it is very important as a nonpredator which ameliorates the environment for them, as in category *b* (sec. 10.01).

A similar principle is illustrated by the beetles *Calandra oryzae* and *Rhizopertha dominica*, which may infest wheat in storage. In Australia the former is ubiquitous and abundant at all times. The latter is usually quite rare and is seldom noticed as a pest; but in exceptional circumstances it may increase greatly in numbers and become a serious pest. During the war of 1914–18 and again during 1939–45 wheat was stored in unusually large quantities for unusually protracted periods, and on each occasion *R. dominica* reappeared as a pest. Usually the wheat became infested by *C. oryzae* first, and *R. dominica* followed. Figure 3.03 and the experiments discussed in section 10.213 show that *R. dominica* requires a higher temperature than *C. oryzae*. It is likely that the wheat was at first too cold for *R. dominica* to thrive in but that, after it had been infested for some time by *C. oryzae*, its temperature rose to a level that favored *R. dominica*, which increased rapidly, tending in some cases to displace *C. oryzae*. As the infestation developed, the wheat in the depths of the bulk became too hot for either species, and the infestation became restricted to the wheat near the surface (Birch, 1946*a;* Wilson, 1946). Both these insects are to be classified as nonpredators of category *a* (sec. 10.01), since each eats the food that the other requires. But *C. oryzae* ameliorates the environment for *R.*

dominica and thus possesses, in addition, qualities which belong to category *b*.

These examples could be added to almost indefinitely, because there are so many animals that share a common food or some other resource to a greater or a lesser extent and the variation in detail in their relationships is great. If, however, we were asked to nominate just one pair of species whose requirements were identical, then it is doubtful whether we could name one example that had been rigorously determined. This, as we have seen, is of little moment to us in our efforts to understand the distribution and abundance of animals in nature. We are more concerned to know, not whether the overlap is complete, but whether there is any area of overlap and, if so, how this influences the animal's chance to survive and multiply. But the other question has received a lot of attention. So far as evolution and speciation are concerned, this interest may be legitimate; in ecology it stems often from an uncritical extension to nature of the conclusions of the mathematical models, which, as we have seen elsewhere, should certainly not be used in this way. The conclusion having been accepted as an axiom, the evidence is then sought, often along the following lines. Where two species have recently come together across an ecological barrier and one of them has died out or seems likely to, it is said that this must be because the two have identical or very similar requirements. Or when two species which seem alike in their requirements are apparently living together, a search is made for some difference in their requirements or habits. Naturally, such a search is almost always successful.

10.311 THE SPECIAL CASE OF SPECIES SUPPOSED TO HAVE VERY SIMILAR REQUIREMENTS

Gause (1934), having done several experiments in which two nonpredatory species with similar requirements were kept together in a restricted space where temperature and the supply of food were kept as constant as possible and having found in these experiments that one species always died out, leaving the other to thrive alone (sec. 10.21), looked around to see if he could find a similar phenomenon in nature. He called attention to the gradual replacement in Russian fishing waters of the crayfish *Potamobius astacus* by the related species *P. leptodactylis;* but he did not investigate the ecology of these two species to see what was, in fact, the explanation. Crombie (1945) confirmed Gause's experimental results, using different species, and (p. 393) extrapolated the results to nature with the following general statement: "It is easy to demonstrate theoretically that two species with identical ecological niches cannot survive together in the same environment unless density-independent factors keep the population low enough to eliminate interspecific competition." The ideas implicit in this sentence depend on the mathematical model which we criticize in section 10.11 and on the results of certain laboratory experiments in which all the components of environment which could be controlled were kept as constant as possible. Neither is very relevant to what may happen in nature. In this section we examine the evidence provided by the study of natural populations to see whether it supports the view that "two species with identical ecological niches cannot survive together in the same environment."

The chance that the establishment in an area of a newcomer, from across some geographical barrier, will result in the extermination of any of the old-established species must be quite small. One has only to consider the wholesale dispersal of insects and other animals across the world by means of commerce during the last hundred or so years in order to appreciate this point. "Introduced" species make up a substantial part of the insects in almost every part of the world. If it were at all usual for the invading species to cause the extinction of the old-established ones, then one might have expected to hear more about it. On the contrary, the examples given below stand out as being rather unusual. In none of them has it been demonstrated that food or any other essential resource was in short supply, and for several of them the reverse is strongly indicated.

The Mediterranean fruit fly *Ceratitis capitata* was introduced and became established in the fruit-growing districts around Sydney in eastern Australia sometime during the last century. For many years it remained abundant and a serious pest. At the present time *C. capitata* would seem to have become extinct around Sydney, for it has not been found, despite diligent searching. Its place has been taken by another species of the same family, *Dacus tryoni*, which is indigenous to Queensland, farther north. This is now abundant in the same area and living in almost the identical circumstances which previously were characteristic of *C. capitata*. Throughout this whole period a diversity of fruit has continued to be grown in suburban back yards (a favorite breeding place) and fruit has been grown and marketed commercially to supply the needs of the million or so people who live in the city of Sydney. From this it is clear that there was not at any stage any shortage of food for the fruit flies. Knowing their biology well, we feel confident that nothing else that they require has been in short supply either. There must be some other explanation which has not yet been discovered.

Similar phenomena, some of which have been "explained" in terms of competition, include the replacement in Britain of the indigenous and erstwhile abundant red squirrel, *Sciurus vulgaris*, by the introduced gray squirrel, *S. carolinensis*. Similarly, the black rat, *Rattus rattus*, is disappearing in temperate Europe and is being replaced by the brown rat, *R. norvegicus* (Crombie, 1945). When the two species *Plasmodium malariae* and *P. vivax* are injected into man, only one of them is said to persist (Mayne and Young, 1938). Lack (1944) provided a number of examples from birds. The curlew *Numerius arquata* had a more southerly range in Europe than the whimbrel *N. phaeolus;* but the curlew is extending northward and replacing the whimbrel in these regions. Similarly, the whimbrel is extending southward and replacing the curlew in the south. Other examples are more conjectural. The European chaffinch, for example, is alleged to have made two separate invasions of the Canary Islands, which resulted in the different species found on the island today. In Europe the chaffinch breeds freely in both coniferous and broad-leaved woodlands. But on the island of Gran Canaria, *Fringilla teydea*, which is presumed to be the earlier arrival, breeds only in coniferous woods. On the same island *F. coelebs canariensis*, which is presumed to be the later arrival, breeds in the chestnut and laurel forests below and the tree-heath forests above the pine belt. It does not breed in the pine belt in between. On the island of Palma, *F. teydea* is not found, but the Palman form of *F. coelebs* breeds not only in chestnut and laurel but also in pine forests. Lack (1944) inferred from these observations that *F. coelebs* can breed in both pine and broad-leaved forests when it is the only occupant, but in Gran Canaria it is unable to breed in pine forests because they are occupied by another species. So far as the case is stated, there is no direct evidence that the two species could not live together if they were put together. It is assumed that they must have come together at some time in the past and become segregated as a result. In high mountain lakes of Sweden and Norway, trout which have been introduced have thrived and provided good sport for fishermen. In those lakes which have also been stocked with char the trout population sometimes has decreased to a small number (Svärdson, 1949). Svärdson also stated that when trout and char were reared together in the laboratory, the char maintained relatively greater numbers. Pheasants and partridges increased in numbers in parts of Hungary in recent years both in regions occupied by both birds and in those where they occurred separately. But after a time, while the pheasants continued to increase, partridges grew scarce. This may have been due to the pheasant's preference for laying her eggs in foreign nests when she can find any. When birds are so numerous that the nests are close together, pheasants have no difficulty in finding partridge nests. When this happens, partridges suffer great losses at hatching, since pheasant eggs are bigger and this hinders the partridge from hatching all her own clutch properly (Peterfal, 1950, quoted by Udvardy, 1951).

Whether in any of these cases the result may be properly attributed to a shortage of some essential resource or to direct interference of one species by another still remains to be found out. It may perhaps be taken as a reasonable hypothesis that the decline of one species was causally related to the presence of the other. But even this should not be taken as very likely unless it is supported by empirical evidence. Consider, for example, the case of the butterfly *Pararge aegeria*. During the nineteenth century this species was widely distributed throughout southern England and also in the north of England and in Scotland. More recently, especially during the early part of the twentieth century, its numbers became generally very much fewer, and at the same time its distribution contracted. It disappeared entirely from large areas in south-eastern England, where it had previously been abundant, and in Scotland its distribution became restricted to two small areas in western Argyllshire and Inverness-shire. There is evidence, though not yet very definite, that now, in the middle of the twentieth century, it is beginning once again to increase both its abundance and the area of its distribution. The larvae of *P. aegeria* feed on grass, and the adults are weak fliers. Sufficient is known of the ecology of this species to make it reasonably certain that these fluctuations have happened quite independently of any other animal (including man) in the area where it occurs (Downes, 1948).

Dobzhansky and Pavan (1950) restated the general conclusions of the Lotka-Volterra mathematical models and Gause's experimental models in terms that are relevant to natural populations in the following generalization: "Sympatric species or strains either occupy different habitats within the same territory, or exploit the same habitat in different ways, which in the last analysis, amounts to the same thing." This is an alternative way of stating Crombie's generalization, except that this statement makes no allowance for rare species: it implies that all species and strains normally live crowded up to the limits of the resources available to them. We have seen that the reverse is true of the four species of caterpillar studied by Schwerdtfeger (1941); and, since rareness is a more usual attribute than commonness (chap. 14), we would say that Dobzhansky and Pavan's statement may be true for special circumstances but not as a generalization.

The same idea was pursued by Lack (1944, 1946, 1947). Darwin wrote: "As the species of the same genus usually have, though by no means invariably, some similarity in habits and constitution, and always in structure, the struggle will generally be more severe between them, if they come into competition with each other, than between the species of distinct genera." On the basis of the first part of this statement, Lack argued that one would expect species that are close to one another phylogenetically to have similar ecological requirements. So he studied the ecology of the passerine birds of Britain, the finches of the Galapagos Islands, and the European birds of prey, with the specific objective of finding out whether related species within these groups were separated from one another ecologically, in other words, whether they were prevented from competing with one another by virtue of not needing or preferring the same sorts of food, nesting sites, and so on.

Lack concluded from these studies that the distributions of closely related species sometimes did not overlap, being separated, perhaps, by geographic barriers of one sort or another. If their distributions did overlap, then it was usually possible to recognize some consistent differences either in the sorts of food that they ate or in the sorts of places they chose for nests, or so on. When this failed, Lack was sometimes able to point to differences in size, which he accepted as evidence in itself of differences in their requirements. Even so, there were some species for which none of these criteria held, and Lack expressed the faith that further study would reveal differences between these also—as most assuredly it will.

Some of the differences observed by Lack may be mentioned: for example, the genus *Corvus* has four species in Great Britain. The carrion crow, *C. corvea*, and the hooded crow, *C. cornix*, are found in open woodland, parks, heath, moorland, and on sea cliffs. The former is in the southeast of England, and the latter in the northwest, with a small zone of overlap. The rook, *C. frugilegus*, lives among farms, and the raven, *C. corax*, overlaps with it to some extent, but it is much larger in size; the size difference suggests to Lack that it might have differences in habits which have not yet been observed. Among the British

TABLE 10.03*

NUMBER OF CASES OF ECOLOGICAL OR GEOGRAPHIC DIFFERENCES IN DISTRIBUTION OF PAIRS OF RELATED SPECIES OF BRITISH PASSERINE BIRDS

	Cases
Geographic separation	3
Separation by habitat	18 or more
Separated by feeding habits	1
Separated by having different feeding zones in the same general habitat	3
Size differences which are assumed to have ecological significance	5
Separated by different winter ranges	2
Apparent ecological overlap	5–7

* After Lack (1944).

TABLE 10.04*

NUMBER OF CASES OF ECOLOGICAL OR GEOGRAPHIC DIFFERENCES AMONG CLOSELY RELATED FINCHES ON GALAPAGOS ISLANDS

Geographic separation . . . . . 2 cases
- (i) *Geospiza scandens* and *G. conirostris*
- (ii) *G. fuliginosa* and *G. difficilis*

Separation by habitat . . . . . 2 cases
- (i) *G. difficilis* (humid forests) and *G. fuliginosa* (arid zones)
- (ii) *Camarhynchus heliobates* (mangroves) and *C. pallidus* (inland)

Separation by feeding habits . . . . . 2 cases
- (i) *G. scandens*, *G. magnirostris*, *G. fortis*, and *G. fuliginosa* are all ground finches which feed in the same habitat
- (ii) *C. crassirostris* is a tree finch, which feeds on leaves, fruit, and buds, whereas the other species of *Camarhynchus* are mainly insectivorous

Separation by size of beak and presumably by feeding habits . . . . . 2 cases
- (i) The tree finches *G. magnirostris*, *G. fortis*, and *G. fuliginosa*
- (ii) The ground finches *C. psittacula*, *C. parvulus*, and *C. pauper*

* After Lack (1947).

passerine birds, Lack found several cases of ecological overlap; but he states that this is extremely rare in land birds of remote islands. His findings for British passerine birds and the finches on the Galapagos Islands are summarized in Tables 10.03 and 10.04, which we quote directly from his papers. In these papers "habitat" seems to refer simply to the sorts of surroundings in which the birds are usually found, viz., woodland, meadow, etc. Among the nonpasserine birds in Britain, there appear to be many examples of closely related species living together in the same sort of place; but Lack (1945) suggested that closer analysis of these cases may reveal differences. For example, the shag *Phalacrocorax aristotelis* and the cormorant *P. carbo* are closely related species which appear to overlap widely in their ranges. Both nest on cliffs and both feed on fish. But Lack stated that the cormorant nests chiefly on flat, broad cliff-ledges and feeds chiefly in shallow estuaries and harbors, while the shag nests on narrow cliff-ledges and feeds mainly out at sea. Amadon (1947) studied the honey creepers (Drepaniidae) in the Hawaiian Islands and showed that these birds vary according to food and "habitat" in much the same way as do the Galapagos finches. He also explained many of the differences between species in terms of "competition."

There are plenty of examples of closely related mammals which appear to live in the same sorts of places but differ in their feeding habits. Lack (1947) quoted the four species of seals in the Ross Sea in the Antarctic. The crabeater, *Lobodon carcinophagus*, feeds exclusively on euphausid crustacea; the Weddell seal, *Leptonychotes weddelli*, feeds mainly on fish; the leopard seal, *Stenorhinchus leptonyx*, feeds mainly on penguins, seals, and also fish and cephalopods; the Ross seal, *Ommatophoca rossi*, feeds mainly on cephalopods.

Ecological characteristics of closely related species of insects have not been examined as closely as those of birds. Shelford (1907) found five species of tiger beetles on the shores of Lake Michigan; each species lived at a different distance from the shore. Dobzhansky and Pavan (1950) found many species of *Drosophila* in an area of jungle in Brazil; but their collections along a 200-meter transect showed local concentrations of the different species at different points along the transect. Samples taken at distances of only 10 meters often showed different relative frequencies of different species. They were also able to show differences in the food-preferences of some of the species.

Da Cunha, Dobzhansky, and Sokoloff (1951) studied food-preferences in species of *Drosophila* occurring in the transition zone of the Sierra Nevada of California. They measured the attractiveness of baits, consisting of suspensions of different yeasts. The various yeasts used were differentially attractive to the species. For example, the highest relative frequencies of adults of *D. azteca* were observed on the three yeasts isolated from the gut of that species. The lowest frequency of this species was obtained on a yeast isolated from *D. persimilis*. This sort of difference was obtained with a number of the species studied. However, the difference is not at all rigid. The common species of *Drosophila* were attracted to some extent to all the yeasts used in these experiments. Thus, although these species probably have a good deal of their food in common, they also utilize different foods, depending upon the species. On the other hand, an analysis of the habits of the adults of closely related hover flies by Diver (1940) did not reveal any obvious differences between the species *Syrphus torvus*, *S. ribesii*, and *S. vitripennis*. These three occur widely in woods, gardens, marshes, fields, and other places in Europe. Apart from the greater abundance of *S. vitripennis* in gardens as compared with *S. ribesii*, there appears to be no gross difference in the habits of the adults of these three species. Unfortunately, the investigation did not include the habits of the larvae.

What conclusions may be drawn from these studies? It would seem to be generally true that closely related species of birds either live in different sorts of places or else use different sorts of food. This is true, only more so, of distantly related species; but no one seriously suggests "competition" as a cause for this. Why, then, should it be necessary to invoke competition to explain the same phenomenon among closely related species, especially when there is virtually no empirical evidence for it? Lack's (1944, 1947) interpretation of these data may be summarized as follows: The physical differences between the various "habitats" cannot account for the different distributions of birds, since they are relatively independent of their "physical environment." By this he evidently means that the birth-rates and death-rates are not likely to depend on the sort of place in which the species lives. Food-preferences are not sufficient to account for the differences, since in most cases the majority of closely related species living in different sorts of places eat much the same sort of food. He concludes that "the only reasonable hypothesis is that these habitat differences are brought about by competition between species." But there is no direct evidence that one species ever makes a serious or effective attempt to drive another from its "habitat" or even to take up residence in a "habitat" unfamiliar to the species. In other words, there is no evidence that "competition" occurs between different species today. He therefore supposes that it must have occurred in the past. Related species are assumed to have differentiated in geographic isolation from one another and then to have come together. Where their distributions overlapped, "competition" ensued, with the result that one or the other, but not both, survived in the zone of overlap. One species would become the successful occupant of one part of the territory, and the other might be successful in another part. Lack then visualizes the members of each species gradually evolving differences in behavior with respect to that part of the territory in which they are successful, so that, after a time, they rarely attempt to breed outside the place which has become their specific "habitat." It is well known that birds rarely make mistakes in choosing the sort of place that is characteristic of their species.

The difficulty in this hypothesis is that, by the very nature of the case, it can hardly be proved or disproved, because we have no evidence that "competition" took place in some past epoch after the species had become distinct. An alternative hypothesis would seem to warrant more consideration than has been given it. If we assume with Lack that related species have differentiated in geographic isolation, the chances are that they would also have developed different habits and preferences, so that when they were later brought into the same territory, they would select different sorts of places in which to live. It is not obligatory to suppose that these preferences were developed as a result of "competition."

In the Galapagos Islands, *Geospiza fuliginosa* may be found on one island on which there are none of the closely related *G. difficilis;* on another island *G. difficilis* is present without *G. fuliginosa;* on a third island both species may be found. On the first two islands each species may be found living in a wide range of different sorts of places, and there is no considerable difference between them. But on the third island each species is more narrowly restricted; there is approximately the same range of places to choose from, but these are divided between the two species, and neither is found in the sort of place that is characteristic of the other. This would seem to be a case where the species are not kept apart by inherited preferences; for, if the birds living on the third island are, in fact, the same species as those living on the other two, then they would seem not to differ in this respect. There is no need to invoke "competition in the past" to explain these observations. If "competition" is indeed the cause, it must be operating in the present, which would seem to afford an opportunity to study this elusive phenomenon. It would first be necessary to show by the appropriate experiments that the species on the third island are, in fact, the same, not only in appearance but also in behavior, as those on the first two islands. Then the way would be open for a study of how "competition" keeps the two species apart on the third island. The chances are that the explanation, when discovered, may be something quite different.

A similar phenomenon was described by Beauchamp and Ullyott (1932) with respect to the two planarians, *Planaria montenegrina* and *P. gonocephala.* According to these authors, when *P. montenegrina* is the only species in the stream, it occurs in water ranging in temperature from 6.6° to 17° C.; when *P. gonocephala* is the only species in the stream, its range is from 8.5° to 20° C. or more. But when they occur in the same stream, *P. montenegrina* occurs in that part of the stream in which the water is below about 13° C., whereas *P. gonocephala* occurs in that part of the stream which is warmer than 13° C. Beauchamp and Ullyott (1932) concluded that "this limitation of the ranges of the two species by the presence of each other proves that interspecific competition is occurring; *P. montenegrina* is the more successful between 13–14° C. and *P. gonocephala* above this temperature." There is, of course, no "proof" at all in these observations that "competition" is the cause of the different distributions of the two animals. This would seem to be another example which would be worthy of closer examination. The hypothesis, which tends to be stated as fact, could possibly be tested quite simply by experiments in the laboratory with the two species at different temperatures, combined with analysis of actual numbers of planarians in the streams and the amount of food available for them. Neither experiment has been done. Until they have been, Beauchamp and Ullyott's observations have little weight as evidence for "competition" in nature.

One of the few good examples recorded of the interference of one species of bird by another is provided by Pitelka's (1951) observations on the humming birds *Calypte anna* and *Selasphorus sasin* in Woolsey Canyon, California. The territories of the two species overlap to some extent, and where this occurs, neighboring males of the two species show aggressive displays and may chase each other. In this way a *Calypte* male is usually successful in preventing a *Selasphorus* male from occupying its territory. From these and other observations Pitelka concluded that the presence of *Calypte* males in Woolsey Canyon resulted in there being fewer *Selasphorus* males than there would have been if they alone occupied the canyon.

Lack (1946) recorded some examples among birds, and there must be many more among insects, which appear to be exceptions to the general rule which he was trying to establish, that species with the same requirements do not exist together. Five species of predatory birds in Europe live in the same sorts of places and feed principally on the same food, the vole *Microtus arvulus*. Lack concluded that these species evidently do not "compete" for food. Most of the time *Microtus* is superabundant, and there is little reason to suppose that the birds then go short of food. But *Microtus* is sometimes scarce, and then there may not be enough for all the individuals of all five species. In these circumstances Lack would expect only one species to survive; so he postulated that when the vole is present in low numbers, each predator turns to a different prey; in this way there would be no "competition," even when the main food supply was limited. Lack suggested that whenever a number of different species are found eating the same food, this denotes a "superabundance" of that particular food, for example, euphausiid Crustacea appear to be "superabundant" in the summer, when they form the staple food of several species of marine mammals and birds in the Antarctic. Many English woodland birds feed on geometrid caterpillars, and many arctic predators feed on lemmings when they are abundant. We suggest that natural populations rarely consume a large proportion of the food available to them (chap. 13).

We have discussed Lack's studies of birds in some detail because this work is so well documented. But we are forced to conclude that his interesting results do not in any way demonstrate that "competition" between birds in nature is at all commonplace or usual. On the contrary, his results seem to show that it hardly ever occurs. Where he finds species together, there is evidence that their food is "superabundant," or else they live on different foods. When they are separated, there is no evidence that they do invade one another's territories.

Another way of testing whether the mathematical or experimental models are like nature would be to compare the number of closely related species found in a restricted uniform area with those occurring in a wider, more diverse one. Elton (1946) approached this problem by recording the number of species in each genus occurring in 55 areas that were specially selected because they were "uniform" and their boundaries could be recognized. His areas included, for example, 50 miles of the River Wharfe, the estuary of the River Thames at Southend, the Reindeer Peninsula, West Spitzbergen (invertebrates and birds), and so on. Some of the data were collected over a number of years, during which changes doubtless occurred. Elton compared these data with the number of species per genus given in extensive lists, such as, for example, the check list of the insects of the British Isles. Unfortunately, he did not make allowance for the random distribution of species in samples of different sizes, and his conclusions were not relevant to the point that we are discussing. But Williams (1947) re-analyzed Elton's data and showed that Elton's restricted situations contained rather more species per genus than would have been expected in samples of those sizes drawn at random from the faunal lists used by Elton. This is what we would have expected, because closely related species are likely to have requirements which resemble each other's more closely than they resemble those of unlike species, and they are more likely to be found sharing the same bioclimatic zone rather than scattered through widely different zones.

At first sight, this would suggest that in nature exactly the opposite is happening from what was found in Gause's highly artificial experiments, and this would not be surprising either. But in looking closely at the origin of these data, we see that, however interesting they may be in other respects, they tell us nothing about the hypothesis that we set out to examine. In order to get reasonably large numbers in his samples, Elton had to choose areas that were far too broad and not sufficiently uniform: there was room for too much subdivision and specialization within each area (Bagenal, 1951). Elton's data apply to many different groups of animals and plants, and they generally confirm Lack's results with birds; but neither provide any confirmation, in particular, of the relevance of the Lotka-Volterra models to nature or, in general, of the importance of competition. On this whole topic we agree with the statement of Mayr (1948, p. 212): "So far there is only scanty direct proof for the assumption that populations are kept in check by competition for space and food."

10.32 *Predators*

The practical study of predation in nature has been carried forward actively by two diverse groups. On the one hand, there are the students of wildlife, who work chiefly with game and fur-bearing animals, and, on the other hand, there are entomologists, working actively on the "biological control" of insect pests. Conservation is the keynote of most wildlife studies, though exceptionally, as with the rabbit in Australia, the practical objective may be to control a pest. Biological control, as the name implies, relates to the use of predators to reduce the numbers of a pest. Because the circumstances of life for predators of vertebrates and invertebrates are so different and because they have been studied from such different perspectives, the two branches of the subject have developed along rather divergent lines. Certain common principles may be discerned deep below the surface, and we shall examine these in due course; meanwhile, the two aspects of this subject are best discussed separately.

10.321 PREDATORS OF INVERTEBRATES

The students of "biological control," like most other ecologists, have not been backward in coining a special set of technical terms for their own use. However helpful these may be to the specialist, it will be best not to introduce them into our discussions of general principles. But we mention them and briefly indicate their meanings, because this may be the best way briefly to indicate the complexity of the phenomena exhibited by predators of invertebrates. There is a major distinction made between the "predator" which wanders actively among its prey, devouring those which it captures, and the "parasite" which lays an egg (or eggs) in or on the prey (now called the "host"): the searching is done by the adult and the feeding is restricted to the larval stage. "Parasites" are divided into "endoparasites" and "ectoparasites," depending on whether the larva feeds from inside or from outside the host. Then a parasite may be "primary," "secondary," "tertiary," or so on, depending on whether its host is a herbivore, a carnivore which feeds on a herbivore or a carnivore which feeds on a carnivore or so on; these are also called "hyperparasites" and the phenomenon is called "hyperparasitism." When the secondary parasite is the same species as the primary (that is, when the adult parasite lays more eggs in one host than it can support to maturity), then these are "superparasites" or, more properly, the phenomenon is called "superparasitism." Parasitism may be "singular" or "multiple," depending on whether one or several species "parasitize" the same host, and parasites may be "specific" or "general," depending on whether they can use only one or several species of hosts. In this section we shall continue, as elsewhere in this book, to include all these categories in the single term "predator."

The first ten volumes of Thompson's (1942——) parasite catalogue is restricted to insect predators (of insects and spiders) which are of the sort that lay their eggs in or on the body of the host; the records have been collected almost entirely from the *Review of Applied Entomology*, so that the prey includes only species considered to be of economic importance. The catalogue is thus doubly restricted to predators of a particular sort preying on the relatively small proportion of species that are of economic importance. At an approximate estimate the catalogue has about 10,000 species of predators listed in it. It would be difficult to guess how many names might be in a comparable list covering all the known species of insects. We made another simple test with the aid of this catalogue. Having chosen at random and from several different countries the names of some 20 insect pests which we know are kept in check only by the repeated application of insecticides, we looked them up in the catalogue and found, without exception, that each had several predators associated with it. From these simple researches into Thompson's parasite catalogue we can deduce two well-known facts:

a) There are an enormous number and variety of species of insects that prey on other insects.

b) There are many species of herbivorous insects which remain numerous enough to be serious pests, despite the presence in their environments of a number of predators. On the other hand, the mere existence of Thompson's catalogue is tangible evidence of the success of "biological control" in a sufficient proportion of cases to justify the expenditure of substantial sums of money on it.

Sweetman (1935) listed 10 examples of pests which had been brought under "highly beneficial control" and 13 others brought under "adequate control" by the introduction of a new predator or predators into the area where they lived. Each example was subjected to a critical scrutiny before it was included in the list, and as a result of this extensive study, Sweetman reached certain conclusions which we quote: "Undoubtedly, the decline in the ravages of a number of insect pests has been attributed wrongly to the artificial manipulation of parasites and predators, when this decline really was due to other factors. Frequently high percentages of parasitism and numerous specimens of beneficial insects have been observed and from such observations and data the conclusion drawn that the pest would have destroyed the crop from the commercial viewpoint if a particular enemy of the pest had not been introduced or liberated. Such evidence will not withstand critical analysis but frequently no better data are available. . . . Apparently many workers have overlooked the fact that a number of destructive agents may be present in an environment, any one of which in the absence of others, would have reduced the ravages of a particular pest." In the final analysis, the influence of the predator on the abundance of the prey must be expressed in terms of the relative abundance of the two species. The abundance of a predator, as with any other sort of animal, depends upon the way the four major components of environment influence its chance to survive and multiply.

Weather often influences the prey and the predator differently: any component of the innate capacity for increase (r_m; see sec. 3.1) may be influenced in this way by any component of the weather. For example, Burnett (1949) studied the influence of temperature on the greenhouse white fly *Trialeurodes vaporariorum* and its predator *Encarsia formosa*. The relative influence of temperature on the birth-rate and speed of development for the two species is shown in Figure 10.17. The third component of r_m was also studied, and the trend with temperature was similar to that shown in Figure 10.17 for the other two components: between 27° and 30° C. there was little difference in the life-

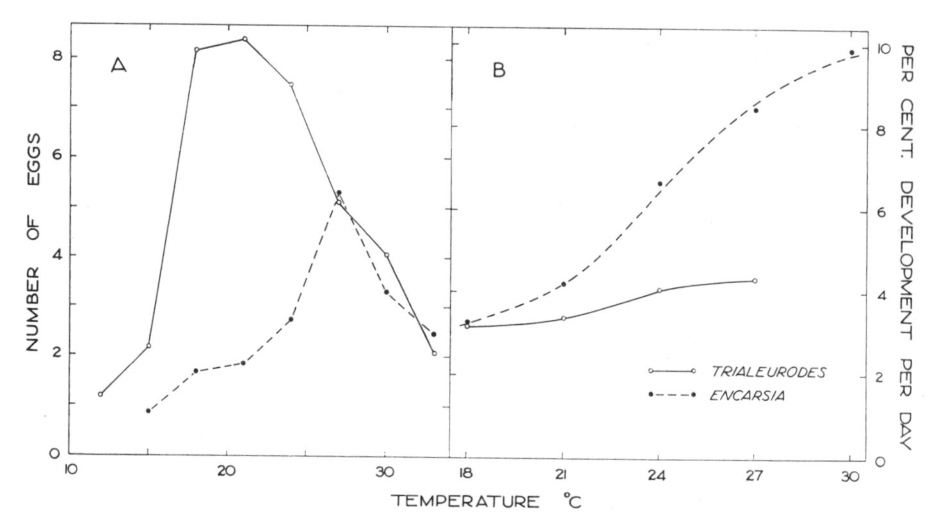

FIG. 10.17.—The influence of temperature on *A*, the daily rate of egg-production, and *B*, the speed of development of a predator *Encarsia formosa* and its prey *Trialeurodes vaporariorum*. Note that the relative advantage possessed by the prey at low temperature disappears at high temperature. (After Burnett, 1949.)

span of adult females of *Trialeurodes* and *Encarsia*, but at temperatures near 20° C. the whitefly lived longer and consequently laid many more eggs than its predator. Burnett did not present these data in the form of age-specific tables, so it is not possible to calculate r_m precisely, but it is clear from inspection of the data that the relative changes in the value of r_m with temperature are quite different for the two species: with *Trialeurodes*, r_m would be at a maximum in the range 18°–21° C.; with *Encarsia*, in the range 25°–28° C.

When the two species were placed together in a greenhouse at 18° C., *Encarsia* increased so slowly relative to the rate of increase of its prey, despite the superabundance of food (larval *Trialeurodes*), that it offered no check to the increase of its prey. When the temperature in the greenhouse was 24° or 27° C., the rate of increase of *Encarsia* relative to that of *Trialeurodes* was so great that food soon became scarce (Fig. 10.18). Local colonies of the prey were exterminated, but the population in the greenhouse as a whole persisted at a low density, because there were always some prey that the predators failed to find. This was partly because *Trialeurodes* was constantly colonizing new situations and partly because the behavior of the two species was slightly different: they tended to prefer different parts of the tomato plants and to respond differently to gradients of humidity, light, and so on.

Hefley (1928) found that atmospheric humidity influenced the survival-rate of the hawk moth *Protoparce quinquemaculata* and its predator *Winthemia 4-pustulata*, differentially. Relatively more of the prey survived at low humidities and relatively more of the predators at high humidities. Ahmad (1936)

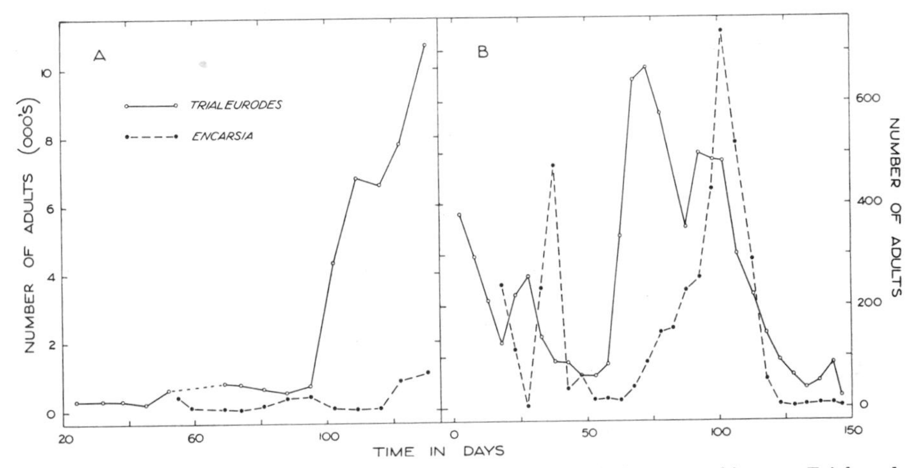

FIG. 10.18.—The trends in the numbers of the predator *Encarsia formosa* and its prey *Trialeurodes vaporariorum* when they were reared together in a greenhouse in which there were four tomato plants spaced several feet apart. In *A* the temperature was maintained at 18° C.; in *B*, at 24° C. Note that the predator never multiplied appreciably at 18° C., despite the presence of plenty of food. (After Burnett, 1949.)

measured the net increase per generation in *Ephestia kühniella* and its predator *Nemeritus canescens* and found a differential response to both temperature and humidity. For example, with *Ephestia* the net increase was greatest (78.3) at 23° C.; it fell to 42.1 at 15° and to 25.4 at 30°. With *Nemeritus* the net increase was greatest (38.7) also at 23° C. but fell relatively more steeply at lower temperatures, being only 3.3 at 15°. The relative change in speed of development also favored the prey as the temperature fell from 23° to 15° C.

Sometimes quite subtle differences in climate are sufficient to tip the balance in favor of one species or the other. Clausen (1951) referred to the well-known case of *Centeter cinerea*, which is a predator of the Japanese beetle *Popillia japonica*. In Japan this predator maintains high numbers relative to its prey, and every year, on the average, destroys about 80 per cent of the beetles before they can lay eggs. In 1922 it was introduced with high hopes into North America, where *P. japonica* is a serious pest; but the result was disappointing. In this new situation the great majority of *Centeter* emerge so far in advance of *Popillia* that they die without leaving any progeny behind them; only a few stragglers, the last to emerge, are still present when the first of the *Popillia* appear. This has been sufficient to carry the population forward to the next generation, but every year for more than 25 years since *Centeter* was first established in North America, the death-rate has been so high that the numbers have never increased to the point where they might make any considerable difference to the numbers of *Popillia* in the area.

With *Metaphycus helvolus*, predator of the black scale *Saissetia oleae*, the influence of weather is more direct. Since its introduction into the coastal districts of southern California in 1937, this species has been responsible for a great and consistent reduction in the numbers of its prey, and Clausen (1951) included it in his list of "fully effective parasites." But in the areas more remote from the coast, frosts periodically kill a high proportion of *M. helvolus* without causing a corresponding death-rate among *Saissetia*. In between such catastrophes, *Metaphycus* reduces the numbers of *Saissetia* to a low level; but periodically, after a severe frost, the numbers of *Saissetia* increase for a while without serious check, and the population may temporarily become quite dense. Many other examples might be quoted in which weather is, in one way or another, influential, to a greater or lesser extent, in determining the abundance of a predator relative to its prey, for there are many aspects to this complicated phenomenon. We shall illustrate only one other aspect, leaving the reader to fill in further details for himself.

In the apple orchards near Perth in Western Australia the aphid *Eriosoma lanigerum* has become quite rare since the introduction into the area of the encyrtid predator *Aphelinus mali*. Previously, *Eriosoma* had been abundant and a major pest. Usually there is a resurgence of the aphid population each spring, but later in the summer they become so few that there is difficulty in finding any; they are exterminated completely from most trees, but there are always some trees where a few aphids happen to survive. Both *Eriosoma* and *Aphelinus* become quiescent during winter, but *Eriosoma* is active at a slightly lower temperature than *Aphelinus*, so it resumes breeding somewhat earlier in the spring than the predator, with the result that each year at this season its numbers increase temporarily. This increase is never carried very far, because the predator, having a relatively short life-cycle, completes a number of generations during the favorable season and always overtakes the prey before the summer is far advanced.

Picard (1923) described a similar phenomenon in France in relation to the cabbage butterfly *Pieris rapae* and its predator *Apanteles glomeratus*. In the more northern regions the larvae of *Pieris* are not present after October; neither is *Apanteles* active after this date, so the life-cycles of the two are nicely synchronized. Farther south the larvae of *Pieris* may be found feeding throughout the winter, but *Apanteles* ceases activity about November. Picard considered that this is an adequate explanation of the relatively dense population of *Pieris* found in the Midi, compared with the sparser populations characteristic of the area farther north.

Many carnivorous species themselves are preyed upon by predators, which may effectively reduce the densities of their populations below the level at which they can exert much influence on the densities of the populations of the herbivores on which they prey. In the early days of biological control this point was often overlooked, and the secondary predators were sometimes introduced along with the primary ones. For example, in Australia the scale insect *Saissetia oleae* is preyed upon by the encyrtid *Metaphycus lounsburyi*, which was introduced into this country about 1925. It seems to have had little influence on the abundance of *Saissetia*, and this has been attributed to the presence of *Quaylea whittieri*, which had been introduced some time before; *Quaylea* is a predator of *Metaphycus;* it has also to be counted as one of the other animals in the environment of *Saissetia*, belonging to category *e* of section 10.01.

The braconid *Angitia cerophaga* was introduced into New Zealand about 1936 because it was known to be an important predator of the cabbage-riddler, *Plutella maculipennis* in Europe. The introduction nearly failed in the early stages because the pteromalid *Eupteromalus* sp. found its way into the field cages in which *Angitia* was being reared and preyed upon it to such an extent that very few survived. When *Angitia* became established naturally in the field, *Eupteromalus* continued to prey upon it, and one series of samples taken over a number of months indicated that as many as 60–70 per cent of the *Angitia* were being destroyed from dense populations, rather fewer from sparser ones. Now that *Angitia* has become firmly established in New Zealand, it is rated as a "valuable" predator of *Plutella* (Robertson, 1948); nevertheless, *Plutella* remains a pest of some economic importance, and it is not possible to say whether it would be very much less abundant in the absence of *Eupteromalus*.

De Bach (1949) found a complex of predators that preyed on the mealybug *Pseudococcus longispinus* in southern California. He followed the trends in their populations throughout one season by counting, at monthly intervals, the numbers of each species trapped in corrugated cardboard bands around the trunks of citrus trees. Among the 13 species recorded, only 3 or 4 were very influential, the most important being the two lacewings, *Sympherobius californicus* and *Chrysopa californica*, and the ladybird, *Cryptolaemus montrouzieri*, which, during the period of this study at least, seemed to be relatively free from predators of their own. There was also present, but of rather less importance, the encyrtid *Anarhopus sydneyensis*, which was preyed upon by the calliceratid *Lygocerus* sp. The predator *Lygocerus* was handicapped by cold relatively more than its prey, with the result that *Anarhopus* was most numerous during spring and early summer (Fig. 10.19). On the other hand, Figure

10.19 also shows that *Lygocerus* was favored when a high proportion of the mealybugs contained larvae of *Anarhopus*, even though the absolute abundance of *Anarhopus* may be less. Thus *Lygocerus* presses most heavily on *Anarhopus* at the time when *Anarhopus* is pressing most heavily on *Pseudococcus;* this makes *Lygocerus* relatively a more important component in the environment of the mealybug than might appear at first sight.

It is partly because of complexities of this sort that authorities on biological control continue to disagree about the importance of these predators of predators ("hyperparasites"). For example, Clausen (1933), speaking to the Fifth Pacific Science Congress, said: "It is self-evident that whenever possible the

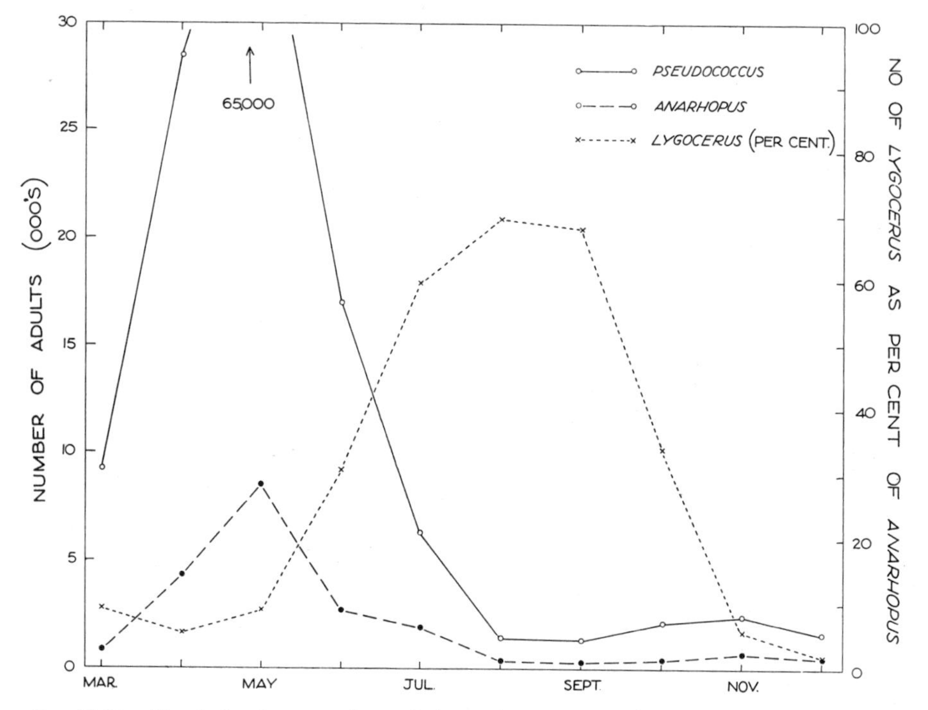

FIG. 10.19.—Trends in the natural populations of *Lygocerus* sp., *Anarhopus sydneyensis*, and *Pseudococcus longispinus* in a citrus grove in California. The first preys on the second, and the second on the third. (After De Bach, 1949.)

importations should comprise only pure colonies of adults or of dormant stages, the latter insectary-reared, if necessary, to insure freedom from hyperparasites. The bulk shipment of field-collected host material should be resorted to only as a last resort and with adequate safeguards. This procedure very evidently involves increased work abroad and greater expenditures, but the greater security afforded warrants its adoption." And Smith and Armitage (1931) wrote: "This insect [the lacewing *Chrysopa californica*] would be one of the most important enemies of mealybugs, particularly the citrophilous species, if it were not destroyed in large numbers by parasites." On the other hand, Flanders (1943) has re-examined the evidence relating to *Quaylia whittieri* in California and *Saronotum americanum* in Hawaii (two cases in which "hyperparasitism" has been held to be very influential) and has come to the conclusion that in neither instance does the presence of the "hyperparasite" have much to do with the failure of the predator to "control" its prey. With these species, as with *Lygocerus* (see above), the prey is living inside the body of another insect, and this makes the search for the prey doubly difficult. This has been advanced, reasonably enough, as a theoretical ground for expecting the predator of a predator to have less chance of being an important component of environment than a predator of a herbivore.

When all the other components of a predator's environment are relatively favorable, that is, when there is no other important check to increase, food may become the limiting resource which largely determines its abundance. Even so, it does not always happen that the predator presses heavily on the prey. The reader should turn to section 11.12 for a discussion on the accessibility of food. The food of an insect predator may be inaccessible in three ways.

First, the prey may be widely scattered relative to the powers of dispersal of the predator. With insects, because of their complex instincts and adaptations, it happens, even more than with other sorts of animals, that the population of any substantial area is likely to be broken up into discrete groups or colonies, following in part the distribution of food plants or suitable places in which to live. Also it often happens with insects that all members of a population live exposed to the attacks of any predator which happens to find them. This is especially true, for example, of scale insects, aphids, or leaf-eating caterpillars living on shrubs or trees either in an artificial place like an orchard or more naturally in a wood. It is a matter of common observation and experience that a predator, provided that there is no other check to its increase, will go on multiplying until the population (or colony) of prey in a local restricted situation has been exterminated. In the meantime, some individuals of prey will have escaped and may have founded another colony, for the instinct for dispersal is strong and almost universal among animals (chap. 5). The predator may also have the instinct for dispersal strongly developed, and usually, sooner or later, some of the predators find the new colony. Whether this happens, on the average, soon or late may be very important in determining the status of the prey as a rare or common species. Though, of course, the respective innate capacities for increase of the two species must also be taken into account. To quote from Smith (1939): "In measuring the capabilities of an entomophagous insect as a biological control factor one cannot then ignore the nature of the host's dispersion. A parasite's efficiency depends upon a combination of the two qualities, host dispersion and its own power of discovery."

Clausen (1951) listed a number of examples in which the introduction of a predator had resulted in the speedy reduction in the numbers of the prey to the point where they were no longer a serious pest. In each case, both the powers of dispersal and the innate capacity for increase of the predator relative to the prey were shown to be great. In certain cases where the predator had relatively poor powers of dispersal and consequently did not bring about a great or permanent reduction in the numbers of the prey, it has been possible to achieve this by breeding the predator artificially in insectaries and dispersing it artificially in the orchards. The case of great historical interest—and it still remains one of the most striking—concerns the introduction into California in 1888–89 from Australia of the ladybird *Rodolia cardinalis*, which preys upon the cottony-cushion scale *Icerya purchasi*, which had at that time become so numerous that it threatened to destroy the citrus industry in southern California. A total of 129 adult *Rodolia* were placed on a heavily infested tree over which a cage had been built, in December, 1888. By April, 1889, they had increased greatly and had destroyed all the scale insects on the tree. The cage was removed, and within 3 months the beetles had dispersed through the whole orchard and had virtually exterminated all the *Icerya* in it. This success was repeated throughout the whole of southern California, and, ever since, the numbers of *Icerya* in this area have remained low. Smith (1939) emphasized that this result could not be explained only in terms of the high powers of dispersal of *Rodolia*, because *Icerya* happens to have rather poor powers of dispersal and tends to assume an extremely colonial type of distribution, which favors the predator. The power of dispersal of the predator relative to that of the prey, together with a high rate of increase in the predators, provides the full explanation for the extraordinary influence that this predator exercises over the abundance of its prey.

As an example of a closely related predator which was much less influential, Clausen quoted *Cryptolaemus montrouzieri*. This ladybird was also introduced from Australia to California in 1891–92 with the purpose of controlling certain species of mealybugs. At first, it seemed successful, particularly in the more humid coastal areas, in reducing the numbers in dense populations of prey; but once these had become rather less plentiful and the numbers of the predator had decreased correspondingly, they seemed to be unable to reduce the density of the population any further. This species does not disperse readily (Smith and Armitage, 1931), whereas the prey disperses more readily than *Icerya*. Nevertheless, *Cryptolaemus* was used for many years very successfully in California for the control of mealybugs: when reared in insectaries and dispersed artificially through the orchards early in April at the rate of about 10 to 20 beetles per tree, it multiplies and has usually destroyed most of its prey by about midsummer. Another ladybird, *Rhizobius ventralis*, which was introduced into California about the same time, has become permanently established there but has never had much influence on the density of the population of its prey (the scale insect *Saissetia oleae*) because of the poor powers of dispersal of the larval *Rhizobius* relative to the dispersal of the young scale insects shortly after hatching from the egg (Smith, 1939). In none of these examples has the prey, or any portion of it, been concealed from the predator in any material way. Those individuals among the prey which survive being eaten long enough to contribute progeny to the next generation do so not because they are less vulnerable than their fellows but purely because they are more fortunate: no predator happened to come in time to the place where they were living. In considering such examples, the emphasis is properly placed on the relative powers of dispersal of the predator and prey.

The second way in which the food of an insect predator may be inaccessible differs from the first not fundamentally but only in emphasizing a different aspect of the same phenomenon. The senses of insects limit their perception of objects to those that are quite close to them; so the predator which, as a result of dispersal, has come into the vicinity of its prey still has to search for it. The female *Trichogramma* may recognize its prey (the egg of the moth *Sitotroga*) from about ¼ inch away (Laing, 1938), but *Brachymeria euphloeae* may perceive its prey (the pupa of *Oreta carnea*) from 4–12 inches away (Schneider, 1939). An unidentified species of *Apanteles* is described by Flanders (1947) as being able, once it has perceived the track of the tortricid caterpillar on which it preys, to "follow it tenaciously until contact is made and an egg laid."

When the prey is concealed in the way that larvae boring in a tree or living in the soil or inside the body of another insect may be concealed, then its chance of being eaten by a particular predator may depend, as before, on the relative powers of dispersal of the prey and predator but, in addition and perhaps more importantly, on the power of concealment of the prey relative to the searching capacity of the predator. We have already seen (see above; also Fig. 10.19) how the chance that *Anarhopus* will be found by its predator *Lygocerus* depends largely on the abundance of *Anarhopus* relative to its prey, *Pseudococcus*. This is because the larval *Anarhopus* lives inside the mealybug; *Lygocerus* has to search for it here and can do so only by probing in the bodies of the mealybugs with its ovipositor. When a high proportion of the mealybugs do not contain larval *Anarhopus*, this results in a lot of wasted time and effort for *Lygocerus* and relative immunity for *Anarhopus*. Similar situations have been described by Flanders (1943) for *Quaylea whittieri* and by Compere (1925) for *Eusemion californicum*. Both these species are predators of the larval stage of *Metaphycus lounsburyi*, which lives inside the body of the scale insect *Saissetia oleae*. Compere's account of *Eusemion* may be quoted: "With the individuals under observation, in some instances five minutes or more was spent in drilling before the ovipositor penetrated for an exploratory examination. After the ovipositor is inserted the parasite works the organ in all directions until satisfied of the presence of a primary larva. When a satisfactory primary host is located, an egg is deposited within its body. If the secondary fails to locate a victim, the ovipositor is withdrawn and the female moves to another scale."

The third way in which the food of an insect predator may be inaccessible may be regarded as a special extension of the second. In certain circumstances or always with certain species, a proportion of a population may be afforded absolute protection from a particular sort of predator by the places in which they are living. We described in section 10.222 a neat laboratory model which was designed by Flanders to illustrate this point.

The same phenomenon, with many variations in detail, may often be observed in nature. For example, the moth *Laspeyresia molesta* has several generations during the summer. Early in the season the larvae may be found mining in the young twigs of peach; later most of them are in the fruit. Once a larva has penetrated deeply into the fruit, it is safe from the predator *Macrocentrus ancylivorus*, whose ovipositor will not penetrate so far. But the larvae mining in the twigs and the very young larvae living just under the skin of the fruit are not so protected, and the predator may find many of these and lay eggs in them (Steenburgh and Boyce, 1937).

Often the protection conferred upon the prey by the place where it happens to be living depends not on the physical limitations of the predator but on its behavior. Flanders (1947) quoted a number of examples of this. The gypsy moth occurs equally in dense woodland and the more open edges of forests or clearings, but the predator *Hyposeter disparis* shows a strong preference for dense woodland and seldom finds those individuals of *Lymantria* which happen to be living near the edges of clearings (Muesebeck and Parker, 1933). In greenhouses the aphid *Myzus persicae* will live on most aspects of its host plant, but the predator *Ephedrus nitidus* attacks only those aphids on the more exposed parts of the plant. In this instance the immunity gained by the *Myzus* living in the more shady situations avails them little because there is another predator, *Aphidius phorodontis*, which prefers these aspects (McLeod, 1937). The sawfly *Cephus pygmaeus* lives in the stems of either wheat or barley, but the predator *Collyria calcitrator* goes much more readily into fields of wheat than of barley (Walker, 1940).

The Japanese beetle *Popillia japonica* is widely distributed in eastern North America, but only the ones—and they are relatively very few—which happen to live close to certain flowering plants which provide food for the adults of *Typhia popilliavora* are vulnerable to the attacks of this predator (Gardner, 1938). There is a number of predators which prey upon *Lucilia cuprina* and related species of blowflies when they are living in carrion (Fuller, 1934). But virtually none of these seeks out the maggots when they occur on the living sheep (Waterhouse, 1947). This is one of the reasons why *L. cuprina*, which breeds freely on the living sheep, is a common species in Australia wherever sheep occur and is rare or absent from places where carrion is the only source of food.

The failures in biological control have probably outnumbered the successes, but the successes have sometimes been spectacular, leaving no room for doubt that the newly introduced predator exerts a dominant influence on the abundance of the prey. In more natural communities, that is, in those composed of indigenous species or at least where men have not deliberately introduced foreign predators, there has, until recently, been no way of demonstrating the influence of the predator. With the invention and widespread use of chlorinated hydrocarbons as insecticides, it has become possible to gather experimental evidence on this point. Two interesting papers by Pickett *et al.* (1946) and Pickett (1949) drew attention to the selectiveness of certain insecticides and how their continued use resulted in the increased abundance of certain pests. De Bach (1946), taking advantage of this property of DDT, devised what he called "An insecticidal check method of measuring the efficacy of entomophagous insects" and subsequently used this method successfully with the mealybug *Pseudococcus longispinus* and the scale insect *Aonidiella aurantii* and their respective predators. But even before this method had become available, it had been generally accepted that in nature there are many species which are dominated by predators just as thoroughly as any that have been brought under biological control artificially. In Australia the indigenous cottony-cushion scale is a rare species, and no one would seriously doubt that this is chiefly due to the presence in Australia of the indigenous predator *Rodolia*. And, by inference, it is accepted as likely that many others among the rare species which, being of no economic importance, have not been studied, remain rare largely because of the influence of predators in their environments. There would seem to be less grounds for agreement among students of vertebrate predators.

10.322 PREDATORS OF VERTEBRATES

With vertebrates, especially the higher ones, learning and intelligence are relatively more important than with the invertebrates, and it becomes necessary to consider the special skills of the hunter (e.g., fox or wolf), the means of defense developed by the hunted (large ungulates), social behavior especially among prey (e.g., muskrat or "territorial" bird), and the readiness to feed on a wide variety of prey which is characteristic of most vertebrate predators. For this reason it is often difficult to separate cause and effect in predator-prey relationships among vertebrates.

From records kept by the fur trade and notes kept by trappers and naturalists in the area, Elton (1942) was able to trace the fluctuations in the numbers of the Labrador vole (*Microtus enixus*) and its chief predator, the arctic fox (*Alopex lagopus*) back for more than 100 years. Similar data, but based on an annual questionnaire sent out by the Northwest Territories Administration, Ottawa, have been summarized in a series of papers beginning with Chitty and Elton (1937) and culminating with a paper by H. Chitty (1950). These records are unique for the length of the period that they cover and the thoroughness with which they have been gathered. They show large fluctuations in the numbers of voles and lemmings and strikingly similar fluctuations in the numbers of the chief predators, the arctic fox and the snowy owl. It used to be thought that the predators caused the fluctuations in the numbers of the prey, and this interpretation may still be accepted by some students (Dymond, 1947). But Elton (1942, p. 205) doubted whether this is an adequate explanation: "Without at present putting the matter any higher than this, we can see that there are inherent properties of the population dynamics of voles that may eventually explain their cycle as a self-contained system that is not so much dependent on other animals like predators and parasites as we at first supposed." And Chitty's (1950) explanation reversed the original causal relationship completely: "The fur-trade of the Canadian Arctic is chiefly dependent upon the white or arctic fox (*Alopex lagopus*) which has a short-term fluctuation in numbers generally of three or four years. The main reason for the fox cycle is the cycle in its principal prey, the lemmings (*Lemmus* and *Dicrostonyx* spp.). The periodic disappearance of lemmings affects another of their predators, the snowy owl (*Nyctea nyctea*) which is often forced by starvation to migrate hundreds of miles to the south."

For the same reason it is difficult to make generalizations about predator-prey relationships which will cover vertebrates as a whole; and Errington (1946) in his comprehensive review of this subject chose to discuss a number of broad groups separately. The following three conclusions may be abstracted from the general substance of Errington's review:

a) Predators are most likely to be influential in determining the abundance of the prey when the prey belong to one of the lower groups, e.g., reptiles or fishes, or among higher groups when the predator is one of the highly skilful canids (dog, fox).

b) Predators are least likely to be influential when prey and predator belong to intermediate groups (rodents, birds) and especially when the prey has an awareness of "territory" leading to internecine strife which may outweigh the influence of predators. In such cases, what may seem to be a relatively enormous destruction of prey by predators may be quite misleading, because it does, in fact, have only a negligible influence on the average abundance of the prey in the area.

c) With vertebrates, when predators are important, there is usually also a strong interaction between the two components of environment, "predators" and "place to live."

Errington did not mention specific studies of predators of reptiles, and we do not know of any. He referred to a number of studies with fish, in which it was shown that the influence of the predator was largely determined by the proportion of the prey that happened to be living in places where the predator could not or would not penetrate. (For analogous situations with invertebrates see sec. 10.321, the third way in which food may be inaccessible to a predator.) Virtually no predators penetrate the thick, submerged plant growths where the pygmy sunfish (*Elassoma zonatum*) lives, but those sunfish which venture into the clearer water are vulnerable to a number of predaceous fish (Barney and Anson, 1920). White (1937, 1939) observed that kingfishers (*Megacerle alcyon*) and mergansers (*Mergus* spp.) preyed very successfully on young salmon (*Salmo salar*) in streams that were shallow and clear; when most of the streams were in this condition, the salmon's chance of survival was relatively small: in areas where the birds were undisturbed, very few fish survived except in deeper pools and other places offering greater protection; in areas where the birds were killed or kept away by men, more fish survived in the less sheltered places, and the numbers in the whole area were correspondingly greater. Thompson's (1941) studies of the fish populations in several lakes in the Illinois River Valley indicated that, in the absence or scarcity of certain predaceous species, the prey "tend to multiply beyond their food resources; they become emaciated and fail to attain sizes that are attractive to anglers." In addition to these cases in which predators seem to be quite influential in determining the abundance of the prey, Errington also mentioned several examples from among reptiles and fishes in which predators seem not to be important. Instead, "self-regulating mechanisms," analogous to those found in many rodents and territorial birds, seem to keep the population from exceeding an upper limit well below that which the resources of food and space in the area might support.

Among rodents, the muskrat (*Ondatra zibethicus*) and its chief predator the mink (*Mustela vison*) have been intensively studied (Errington, 1943). The muskrat leads a semiaquatic life. Wherever there is shallow water, along the banks of drains, around the shores of swamps, along the edges of ponds, and so on, it will construct its burrows and galleries, emerging from these to feed on grain and grasses; corn (*Zea mays*) is a favorite food. At all times but especially during the breeding season, which corresponds approximately to summer, the muskrat shows a keen awareness of territory and will not tolerate crowding beyond a certain limit. So, once the best and most secure places have been occupied to capacity by the more fortunate, the fiercer, or the more determined individuals, the others are either killed (especially the young) or driven forth to seek homes in less secure places or left with no homes at all. Some of these the predators take, but all or most of them are doomed to die, predators or no.

The muskrat forms a staple food of the mink (*Mustela vison*), but it may be eaten by a number of other predators, and it may also be killed by men who seek its fur. The mink and these others will feed quite readily on a number of other sorts of prey if these are more readily available than muskrats, so they will not, except in circumstances of special scarcity, need to seek out the muskrats that are living in the more secure places. Even so, most of these are virtually invulnerable to the mink; it is those that have been driven out and failed to find adequate shelter that are eaten not only by the mink but also by many other lesser predators.

Errington also showed, during many years' study of the same populations living naturally in the same area in Iowa, that there was a close inverse relationship between the density of the population at the end of winter and the rate at which it increased during summer. That is, the muskrats tended to increase more rapidly when they were few, so that the carrying capacity of the area was usually reached by the end of the season, whether at the beginning there had been few or many. Since the mink determines neither the amount of cover nor the rate at which it becomes fully occupied, Errington concluded that predation had very little influence on the average abundance of muskrats in any particular area.

Similarly with the bobwhite quail *Colinus virginianus*, Errington (1945) concluded that predators were not important. This species is relatively tolerant of its own kind during winter, when coveys of 15 or 30 birds, often comprising several family groups, come together and live in the shelter of low brush adjacent to fields and meadows which are the source of their food—insects and seeds. They are adept at seeking shelter at the slightest warning, so that, if sufficient shelter is available, their chance of being taken by predators during the winter is quite small. But when there are more quail than the area provides adequate shelter for, then they are eaten by predators. The horned owl (*Bubo virginianus*) is the chief one, but many lesser ones join in the hunt, depending on how accessible the food is. All the predators, including the owl, normally have many other sorts of food, so that they do not press the hunt to the point where a quail is likely to be destroyed if it has adequate shelter.

During summer the quail breed, building nests in the shelter of grassy tussocks in the meadows. But by this time they have become highly intolerant of one another, and each male defends his chosen territory with the utmost vigor. This means that the rate of increase in a particular area during summer bears an inverse relationship to the density of the population at the end of winter. These two phenomena—the relative invulnerability of well-sheltered quail during winter and the influence of the territorial instinct in determining fecundity during summer—make predators relatively unimportant. The analogy with the muskrat is close, and Errington used these species as the basis for a more general theory.

According to this theory, predators are regarded as merely one of several "intercompensatory mechanisms" which, together, prevent the prey, in any area, from exceeding, except very temporarily, a characteristic level of abundance which is determined primarily by the amount of "cover" in the area and which Errington calls the "carrying capacity" of the area. The population may fall much below this level of abundance under the influence of other components of the environment; with muskrats, drought seems to be one of the most influential. But predators have little to do with this phase of the cycle. Using our terminology, we would say that this theory places little emphasis on the predators in the environment and attributes most importance to the quality and abundance of suitable places in which the prey may live. The theory also stresses the influence of other animals of the same kind, in limiting fecundity (through the territorial instinct) and thus the rate of increase. Anticipating what has still to be said in chapter 12, we interpolate here that the place to live often requires to be evaluated in terms of the other components of environment, in this case predators especially. This means that this theory depends upon the implicit proviso that there are no changes in the sorts of predators in the environment. For example, Errington (1943) himself showed that muskrats which were secure from mink might be vulnerable to red fox. This proviso does not seriously limit Errington's theory.

Throughout this review, Errington (1946) emphasized the difficulty in trying to generalize too widely with respect to the predators of vertebrates:

> In stating a broad rule that the lower vertebrates are greatly more subject to limitation through food or predation than are the strongly territorial (or, at any rate, intraspecifically intolerant) mammals and birds, we must not lay undue emphasis upon phyletic classification. Much variation in behaviour is to be witnessed within related groups. Moreau and Moreau (1938) wrote "that in one species (of east African bishop bird, *Euplectes hordeacea*), the specific territory size can, on present evidence, be regarded as a limiting factor in population density, while in another closely allied and with generally similar habits (*E. nigroventris*) this possibility is excluded," the territories of the latter being "almost indefinitely compressible."

Errington recognized that at least some of the higher ungulates which are preyed upon by canids were exceptional also. In this connection he wrote (Errington, 1946, p. 158):

> Intercompensations in rates of gain and loss are evidently less complete in the life equations of the ungulates, however, than in the muskrats. There is vastly more reason that I can see for believing that predation can have a truly significant influence on population levels of at least some wild ungulates. A rather famous case of antelopes (*Antilocapra americana*) increasing in Oregon after the reduction of the abundant coyotes, which preyed upon the young, looks circumstantially convincing (Riter, 1941). . . . Most examples of predation upon wild ungulates showing a reasonably clear evidence of population effect have one thing in common: the predators involved had special abilities as killers—indeed were usually *Canis* spp. members of a subhuman group inferior as mammals only to man in adaptiveness and potential destructiveness to conspicuous, relatively slow-breeding forms. . . . Like man, they may to some extent be diverted from a particular prey species by the availability of another more preferred (Murie, 1940, 1944)—a matter of special population significance in that predation losses inflicted by canids seem less marked by intercompensations than losses from general predation. Against canids as against man, ungulate prey may often find a good measure of security in first-class habitats even when poorer grades are untenable. Clarke

(1940, p. 84) wrote of a species, much reduced through human exploitation ". . . the places where the musk-oxen have survived down to the present on the mainlands are those where they were most numerous 100 years ago."

One would expect sheep to be among the most vulnerable of ungulates to predators, and we do, indeed, find that in nature they tend to occur in places that afford special protection from predators. Thus the native North American mountain sheep *Ovis dalli* and *O. canadensis* live in rough hills and mountains, where they can find special protection against eagles, wolves, and other flesh-eaters (Murie 1944, pp. 95–143). Some of the larger sorts of ungulates, which, by virtue of their formidable size and herd instincts, are better able to protect themselves against the most sagacious predators, may live in more open places. And these, over which, it would seem, predators are rather less influential, have, according to Errington (1946, p. 157), "a propensity to increase up to the limit of the food supply and to the extent of actually starving, thus conforming to the Malthusian thesis more literally than do the general run of higher vertebrates."

These examples illustrate the principle which seems to apply to vertebrates that whenever predation is important, there is usually also a strong interaction between the two components of environment—"predators" and "place in which to live."

10.323 PATHOGENS

The ecologist who has to measure the survival-rates and understand the causes of deaths in natural populations may sometimes find that disease is the most important of all, overshadowing, at least temporarily, all other components of the environment. The viruses and organisms responsible for disease —the bacteria, fungi, Protozoa, nematodes, insects, mites, and so on—must all be considered as components in the environments of their hosts. In our classification of the components of environment we group these pathogens and parasites with predators, but we discuss them separately, because the methods used for studying them are so different.

The practical aspect is exemplified by the artificial use of "milky disease" (*Bacillus popillae* and *B. lentimorbus*) of the Japanese beetle (Steinhaus, 1949, p. 675). In certain parts of the world, notably North America, quite extensive laboratories have been set aside for the study of the diseases of insect pests. Some early successes, notably with insect pests of forests in Canada, have given rise to the hope that this method may prove a valuable complement to the older methods of biological control (Cameron, 1950). And, with vertebrates, the recent large-scale attempt in Australia to reduce the number of rabbits by introducing the South American virus which causes the disease myxomatosis represents still another development. This is a complex problem: the virus is spread chiefly by insects; already a large number of species of mosquitos, sand flies, and other biting insects have been established or reasonably suspected as vectors. Solution of the problem requires knowledge not only of the virus but also of the rabbit and a diverse number of vectors. The ecology of each requires to be studied and the interactions of the disease with them all (Ratcliffe *et al.*, 1952). In the more academic aspects of the investigation of disease, a great deal of thought and time has been put into the construction of mathematical and empirical models, but we shall not go deeply into these.

The construction of theoretical models to represent the way disease may influence the numbers of animals in a natural population was initiated by Ross in his study of malaria and by Martini in the study of the spread of measles and scarlet fever in man (sec. 10.2). Since then, there has been extensive development of both theoretical and empirical models in the study of disease, but a review of this very large field is quite beyond the scope of this book (See Abbey, 1952; Costa Maia, 1952; Serfling, 1952).

Empirical models by Topley (1926), Greenwood *et al.* (1936), and others on mice infected with bacteria and viruses have shown that, in experimental populations, disease typically spreads in waves. The number of animals which become infected, following the introduction of the disease into the population, rises to a maximum and then falls, perhaps to quite a low level. The epizoötic (a term corresponding to "epidemic" in human populations) may be preceded by an increase in the virulence of the organisms in their repeated passage through susceptible hosts, and the susceptibility of hosts themselves may undergo changes. As a result, the disease may spread rapidly (i.e., an epizoötic develops), particularly if there are many susceptible individuals living close together. Once an epizoötic has begun, it can be maintained in experimental populations by adding susceptible uninfected animals at regular intervals. The severity of the epizoötic and the phase and amplitude of its waxing and waning could be modified by varying the rate at which susceptible uninfected animals were added to the population. The natural history of disease in human populations has parallels with these models. Examples were given by Burnet (1940), who compared the spread of an epidemic to the spread of a forest fire, which flares up in different places but does not recur until fresh undergrowth has grown up. There is some evidence, too, that the empirical models may be representative of what happens in populations of insects infected with disease. However, this speculation remains to be confirmed (Steinhaus, 1949, p. 184).

Both in the laboratory and in nature the disease spreads most rapidly and the epizoötic develops most vigorously when the density of the population of hosts is high. Darwin noticed that "when a species, owing to highly favourable circumstances, increases inordinately in numbers on a small tract, epidemics—at least this seems generally to occur with our game animals—often ensue." Naturalists before and since Darwin have associated disease with the fall in numbers of various species of animals after they have been very abundant. But it does not follow, because the animals have been very numerous and dense and disease is prevalent, that the disease is the cause of the fall in numbers. Elton (1942) drew attention to this in the case of the vole in England. He was unable to find any direct causal relationship between disease and the fall in numbers which ended each cycle. His analysis of work on the Continent shows this to be inconclusive also, despite the age-old tradition in some places that disease ended the outbreak of voles. In Canada a variety of diseases are common during the rapid decline in numbers of the hare *Lepus americanus* after years of abundance (MacLulich, 1937); similar observations have been recorded for the ruffed grouse *Bonasa umbellus* in Canada (Clarke, 1936); and in Great Britain when the red grouse *Lagopus scoticus* reaches very high numbers, the "crash" which follows is frequently (though not always) associated with high incidence of strongylosis (*Trichostrongylus perigracilis*). Dane, Miles, and Stoker (1953) found that the incidence of the virus disease "puffinosis" among juvenile shearwaters (*Puffinus puffinus*) during epizoötics of this disease on Skomer Island was greatest in those parts of the island where a high concentration of birds was associated with extremely dense cover. In some crowded areas the incidence of disease and the death-rate were low, but in all of these the cover was not dense. Dense cover has the effect of crowding the birds, since they are all forced to use the same few tracks and they tend to crowd together to use the few favorable sites, such as large boulders, for wing-flapping exercises.

The most spectacular outbreaks of disease in insects occur when the population of hosts is dense. For example, the virus wilt disease of the alfalfa caterpillar *Colias philodice eurytheme* in California reaches the proportions of an epizoötic when the caterpillars in fields of alfalfa are especially numerous and if the humidity is high at the same time. In spring when few larvae are about, the disease is rarely in evidence; but in late summer and early autumn the proportion of infected caterpillars increases, with resulting widespread epizoötics when the numbers of caterpillars are at a maximum (Steinhaus, 1949).

The incidence of disease in insects is greatest when temperature and humidity are high, and this may be the way that disease prevents a species from becoming established in wet regions or limits its abundance in wet years (sec. 7.4). For this reason, California is less favorable than Florida for the spread of fungal diseases among insects. In Japan the coffee scale *Coccus viridis* is attacked more by fungi in the lower slopes of the island than at high altitudes and on the lower branches of trees, where humidity is higher, than on the higher branches. These and other examples are quoted by Steinhaus (1949).

Although the immediate effect of disease is to reduce the number of hosts, the ultimate effect may be quite different. Ullyett and Schonken (see Steinhaus, 1949, p. 669) studied the influence of the fungus *Entomophthora* sp. on the numbers of caterpillars of the moth *Plutella maculipennis* in the field in South

Africa. The sudden appearance of the disease was followed by a rapid fall in the number of larvae, and these low numbers were maintained for about 10 weeks. During this time the weather was favorable for the disease. The disease then disappeared, and the number of larvae increased, but this time they became much more numerous than before. Whereas the original population was not very destructive, the new population was sufficiently dense to cause much damage. Ullyett and Schonken considered that this new high density was due to a large-scale destruction of predaceous insects during the epizoötic. The reverse effect has also been recorded; the increase of disease in the cutworm *Euxoa ochrogaster* was associated with an increase in the proportion of the parasite *Meteorus vulgaris* in relation to the number of cutworms. Thus the predators were unusually abundant compared to the number of prey in the season following the epizoötic, and they exercised "very effective control" (Steinhaus, 1949, p. 701).

10.33 *Other Categories of Other Organisms*

At the beginning of section 10.321 we said that the most fundamental way to think about predation is to consider the predator as an animal, like any other sort, whose numbers depend on all the components of its environment, though not equally on all. It is clear that a predator is not likely to influence the abundance of its prey to any considerable extent unless all the other components of its environment are favorable enough to make food the chief limiting resource. The other categories of other animals which have to be mentioned in this section are therefore those which enter into the predator's environment, making it more or less abundant than it would otherwise be and thereby indirectly influencing the prey's chance of survival. The two sorts which we shall mention briefly are the predators of predators (category *e* of sec. 10.01) and the alternative foods of predators (category *d* of sec. 10.01). These are obvious; but, the complexity of nature being what it is, the reader will doubtless think of many others.

Predators of predators were mentioned a number of times in section 10.321, and no more need be said about invertebrates except that obviously there is need for critical investigations to find out how generally animals in this category may be important in nature. With vertebrates, the occurrence of this sort of animal in many environments is obvious. This explains the attention given to food chains, for example, Darwin's cats, mice, bumblebees, and clover seed and Elton's well-known pyramid of numbers. But with the demonstration by Errington and others that the importance of predators in relation to vertebrates is far less general than was previously thought to be the case, there is a need to re-examine much that has been written about these relationships. Doubtless, critical investigations would show that there are particular cases among vertebrates, as with invertebrates, in which a predator largely influences the abundance of a predator and thereby becomes an important animal in the environment of the second predator's prey.

An example of the important influence that may be exerted by an alternative food for a predator, which is striking because it is so well documented, concerned the zygaenid moth *Levuana iridescens* and its predator, the tachinid fly *Ptychomyia remota* in Fiji (Tothill *et al.*, 1930). It is considered that *Levuana* is not indigenous to Fiji, but it has been there a long time. Prior to 1925 it was an extremely serious pest of coconuts. "There must be very few Lepidopterous pests that are able to defoliate trees over an extensive area as completely and as rapidly as did *Levuana* before it was controlled by introduced parasites; and yet there is not the slightest doubt that *Levuana* outbreaks were in the majority of cases terminated primarily by the complete exhaustion of the food supply over a circumscribed area of infestation" (Tothill *et al.*, 1930, p. 144). This quotation not only graphically summarizes the destructiveness of *Levuana* but also indicates how poorly it disperses. A moth rarely attempts sustained flight and usually lays all her eggs on palms adjacent to the one from which she originated. Breeding goes on without interruption throughout the year, a generation occupying about 45 days; but, owing to overlapping of generations, a natural population characteristically contains all stages of the life-cycle together as contemporaries. Before 1925 the population in any typical large area would be made up of a number of more or less discrete colonies, each containing all stages of the life-cycle at one time and gradually increasing in density, without greatly expanding in area, until eventually it had eaten all the available food; then the colony would die out almost completely for lack of food.

In 1925 the predator *Ptychomyia remota* was introduced from Malaya. This species has all the qualities of an effective predator: it has an innate capacity for increase greater than that of its prey, and its powers of dispersal are incomparably greater than those of *Levuana*. Its success was immediate and thorough: within a year *Levuana* had become, and has remained ever since, a rare insect. Not even the famous example of the introduction of *Rodolia* into California was more spectacular than this. Yet *Ptychomyia* owed its effectiveness to one additional circumstance which was not required for *Rodolia*.

The adult fly *Ptychomyia* will not oviposit on any stage in the life-cycle of *Levuana* except a caterpillar which has reached or exceeded a late stage in the third instar. The life of the adult fly is relatively short, about 10 days. Now one outcome of the heavy pressure of the predator on the prey was to eliminate or reduce the overlapping of generations which had been so characteristic of the original populations of *Levuana*. Sooner or later, particularly as the density of the population became less, a time came when none of the prey was present in a stage suitable for the predator. In one island where no alternative food existed for *Ptychomyia*, the death-rate among the predators was so high that *Levuana* experienced a respite and was able to increase to large numbers again. Elsewhere in Fiji two introduced species, *Plutella maculipennis* and *Eublemma* sp. and probably one or more native ones living naturally in the "bush," provide alternative food for *Ptychomyia*. It maintains its numbers on these during emergencies, returning to its favored prey *Levuana* as soon as this becomes available again. These species which serve as alternative foods for *Ptychomyia* are very important animals in the environment of *Levuana:* in their absence, but with *Ptychomyia* present, *Levuana* might not be so abundant in Fiji as it was before 1925, but certainly it would be much more abundant than it is now.

The leaf-mining beetle *Promecotheca reichei* in Fiji has in its environment the two predators *Elasmus hispidarum* (Hymenoptera) and the mite *Pediculoides ventricosus* (Taylor, 1937). Before *P. ventricosus* was introduced into Fiji, a natural population of *P. reichei* usually contained all stages of the life-cycle living together, and in these circumstances *Elasmus* usually kept the numbers of *Promecotheca* quite low. The presence of *Pediculoides* changes the situation dramatically. The mites are able to multiply rapidly while the leaf miners are abundant, but only very slowly when the prey is scarce. In a dense population of leaf miners the mites become very abundant, killing out all stages of the leaf miner except adults. Then the mites, too, die out for want of food. The adults of *Promecotheca* lay eggs, and the population begins to increase. But it is a relatively long time before there are present any larvae in a suitable stage for *Elasmus* to lay eggs in them. In the meantime, most of the *Elasmus* also perish without leaving progeny behind them. Thus *Promecotheca*, freed from both its predators, continues to multiply and soon becomes temporarily very abundant —until the mites again appear and the cycle is repeated.

In this example, special interest is centered on the mite. In relation to *Promecotheca*, it may be considered simply as a predator (category *c*); or it may be considered as entering the environment indirectly (category *e*) as a nonpredator of category *a'* in the environment of *Elasmus* (Fig. 10.01). In this case there is no doubt which is the more important aspect of its relations to *Promecotheca*. Its presence results in the increased abundance of the leaf miner, and this is readily explained in terms of the relationship which we call category *e* (sec. 10.01).

In temperate climates the life-cycles of most animals display a seasonal rhythm. Many species of insects in which a vulnerable stage in the life-cycle occurs only at one season of the year may be preyed on by birds without this influencing their ultimate abundance very much, for the simple reason that the birds are too few. Lack (1946, p. 125) wrote: "In the Galapagos at the beginning of the rainy season, nearly all the species of Darwin's finches eat young leaves and buds, also nectar in tree-flowers. Some weeks later, when the young leaves, buds and flowers have passed, nearly all the finches take sphingid caterpillars, which may be very abundant. A little later the fruits ripen, and most species turn to them. Then the grasses produce seed, and several species find food there." Lack was not specifically interested in the ecology of the sphingid caterpillars, and he did not tell us how the feeding of the birds influenced the abundance of the insects; but, for the sake of our example, we may be sure that, had there been fewer birds, there would have been fewer caterpillars eaten, and there would have been fewer birds if there had been an absence or greater scarcity of suitable food at some other season of the year. We have often observed this with respect to the grasshopper *Austroicetes cruciata* in South Australia. The nymphs and adults of the grasshopper are present during September to December, and during this time a number of species of birds use them as staple food. Usually the birds are so few and the grasshoppers so numerous that the result can scarcely be noticed except in exceptional circumstances when the grasshoppers happen to be present in very low numbers as a result of drought or some other cause not related to the birds (Andrewartha and Birch, 1948).

Finally, to add one more point to the differences in predator-prey relationships between invertebrates and the higher vertebrates, we would recall that it is just the existence of a diversity of alternative foods and the readiness of the predator to avail itself of the most accessible sorts which forms one of the major planks in the platform which Errington erects to prove that predation has little influence on the average abundance of (for example) the bobwhite quail. In the absence of alternative food, the horned owl might have to hunt more assiduously for the quail, and this might mean a relative reduction in their numbers. If this were so, the species which provide alternative food for the owl are important components in the environment of the quail: their presence increases the quail's chance to survive and hence the abundance of its population.

In the present chapter we have tried to illustrate the categories into which we put the other organisms of different kinds which may occur in the environment of the animal whose ecology we wish to study. We have tried to make some helpful generalizations, while recognizing the limitations imposed on generalizing by the differences in behavior between the lowly and the higher sorts of animals; the evolution of instinct and intelligence is at the root of this.

CHAPTER 11

Food

. . . animal numbers seldom grow to the ultimate limit set by food-supply, and not often (except in some parts of the sea) to the limits of available space. This conclusion is also supported by the general experience of naturalists, that mass starvation of herbivorous animals is a comparatively rare event in nature, although it does occasionally happen, as with certain moth caterpillars that abound on oak trees in some years and may cause complete defoliation. With predatory animals it probably happens more often.

ELTON (1938, p. 130)

11.0 INTRODUCTION

FOOD, as a component of environment, may influence an animal's chance to survive and multiply by modifying its fecundity, longevity, or speed of development. Both the quality and the quantity of food are important. The study of the qualitative aspects of food belongs properly to physiology and has been widely pursued in relation to the dietetics and nutrition of domesticated animals and man. The arts of animal husbandry and medicine we must not pursue here; but we do need to discuss how the quality of food may influence the rate of increase, r, in natural populations (sec. 11.3).

The study of the amount of food available for a natural population, or, to state the problem more precisely, the study of an animal's chance of finding an adequate supply of food throughout its lifetime, plainly belongs to ecology. It is unusual for an extensive population to consume all or most of the stock of food in its area; those species that do so regularly or often are few compared to the many that rarely, if ever, consume more than a small proportion of their stock of food (secs. 2.121, 9.1, 10.31, 11.12). It does not follow that the amount of food is rarely of major importance in determining the numbers of natural populations; but usually the relationship is more subtle than the simple one represented by an absolute shortage of food over an extensive area (secs. 11.11 and 11.12). Nevertheless, this does occur with certain species, some of which are of great practical importance, for example, in relation to the "biological control" of insect pests and weeds and the management of "wildlife" (secs. 10.321 and 10.322). When absolute shortage of food is limiting in this way, the relationships which are especially worthy of study are the ways in which the various components in the environment of an animal may influence the abundance of its food; we discuss "weather" in this connection in section 11.21 and "other animals of the same kind" in section 11.22.

11.1 THE DISTRIBUTION AND ABUNDANCE OF FOOD IN RELATION TO THE BEHAVIOR OF THE ANIMAL

Under this major heading we discuss first those situations in which the behavior of the animal, especially in relation to the distribution of its food, is of prime importance (secs. 11.11 and 11.12). Then we consider some rather more complex interactions between food and the other components of environment (sec. 11.2).

11.11 *Territorial Behavior in Certain Vertebrates*

The breeding population of the great tit *Parus major* in the area studied by Kluijver (1951) fluctuated in numbers from year to year but did not exceed a maximum, which was determined largely by the behavior of the birds themselves in relation to the resources of the area. At the end of each summer, owing to the natural increase during the breeding season, the area was usually quite densely populated. During autumn, pairs were formed, and each pair selected and defended a territory, not by overt violence, but by characteristic "threat displays" and by song. This was sufficient to drive out all intruders, and those birds which failed to establish themselves in a territory at this time left the area altogether. With the beginning of winter, territorial boundaries were no longer recognized; the pairs separated, and the birds sought food individually or in flocks over a relatively wide area. Territories were resumed again briefly in the spring during the mating season but were ignored again as soon as there were offspring to be fed. The maintenance of territorial behavior through autumn, until the excess of locally bred birds had been driven away and the chance of more immigrants arriving had become quite small, resulted in the density of the breeding population for the next year being virtually determined in the preceding autumn. This emphasizes the importance of territorial behavior as an adaptation which prevents overcrowding. The special advantage of being freed from territorial behavior during winter and when there are young ones to be fed is that the birds can search as widely as possible when food is least plentiful or when it is needed in greatest quantities.

In many completely natural situations the number of breeding pairs of great tits may be limited by the number of suitable places for nests; but in the area where Kluijver worked he had installed more nesting boxes than were ever occupied, so it was clear that the size and distribution of the territories were determined largely by other stimuli. Whatever these may have been, the behavior of the birds was nicely attuned to the resources of the area where they were living, for Kluijver showed that food was usually in ample supply; only during catastrophes, such as unusually severe frost, did the birds suffer a shortage of food.

The bobwhite quail *Colinus virginianus*, unlike *Parus*, does not take up territories during the autumn. At the end of the breeding season the entire population of the area comes together in gregarious groups or coveys, which persist throughout the winter. With the approach of spring, these disband, and the males take up territories which they defend, often by real fights. Those that fail to establish themselves in territories wander away and suffer the usual fate of such wanderers. The successful ones remain and raise their young. The family stays together throughout the breeding season, and with the approach of autumn several families may join together in one covey. A quail's chance of survival during the winter depends largely on the presence of protective cover associated with a source of food. During the summer, food becomes relatively more important, and the territorial behavior of the birds during the breeding season is a mechanism which tends to prevent overcrowding. Territorial behavior has been observed in a large variety of birds. The Australian magpie *Gymnorhina dorsalis* is unusual, in that a relatively large territory, perhaps as much as 100 acres, belongs to and is defended by a group which may contain between 6 and 20 birds (Serventy and Whittell, 1951).

Rodents do not take up individual territories, like *Parus* or *Colinus*, or communal territories, like *Gymnorhina*. Nevertheless, many species show behavior which is analogous. The muskrat *Ondatra zibethicus* has been thoroughly studied by Errington (1943, 1944, and other papers; see also sec. 10.322). Communities of muskrats show a keen awareness of crowding and, especially as the breeding season approaches, show an increasing intolerance of one another's company. Unlike *Parus*, which relies on seemingly harmless threats to drive away its rivals, the muskrat is a fierce cannibal, so the sorting and weeding-out of the population is accompanied by much bloodshed. But the final outcome is much the same: there is finally left behind in the area a population the density of which has been largely determined by the particular degree of crowding which the animals will tolerate. This may be related not only to the amount of protective shelter in the area but also to food. In this way the stock of food in the area, along with the amounts of other necessities, may determine the upper limit to the density of the population. But the relationship is not direct; it depends on an intricate pattern of behavior which is an innate quality of the species. This sort of behavior is known only for vertebrates and is especially well developed in birds and rodents (sec. 12.31).

11.12 *The Paradox of Scarcity amid Plenty*

Most species are rare, relative not only to the places which they may occupy but also to the stocks of food in the areas where they live. In other words, most natural populations (considering not small localized situations but the whole distribution or a substantial part of it) regularly consume only a small proportion of the food in their area. Territorial behavior may provide a partial explanation for this in those species which possess it; but it is lacking from many vertebrates, and it is absent or rare in the invertebrates. An obvious explanation, and one which may often fit the facts, is that some other component of the environment—possibly, though not necessarily, predators—may hold the population in check. But this is not a universal explanation, for we may recognize certain sorts of situations in which a shortage of food may be chiefly responsible for the failure of the population to increase, even though, at first sight, food would seem to be plentiful, because only a small proportion of the total is consumed by the animals. We give several examples. The reader, from his experience of nature, will be able to add many variations on the same theme.

The small hymenopteron *Lygocerus* sp. requires as food the larva of another hymenopteron, *Anarhopus sydneyensis*, which lives inside the body of the mealybug *Pseudococcus longispinus*. The adult *Lygocerus* seeks the larva of *Anarhopus*, and whenever she finds one she lays an egg in it. But her only method of searching is to probe each mealybug with her ovipositor; this takes time and energy (sec. 10.321); and the number of mealybugs which she may search in this way during her lifetime is quite limited. When mealybugs are abundant but relatively few of them contain a larva of *Anarhopus*, then *Lygocerus* may spend much time on fruitless search, and its rate of increase may be limited by a shortage of food. Only a small proportion of the food in the area is consumed, so the shortage of food is not absolute; it occurs because the food is concealed inside the bodies of the mealybugs and most of it remains undiscovered. Paradoxically, the amount of food found by a *Lygocerus* may be greater when the absolute number of *Anarhopus* in the area is fewer, provided that there are also many fewer mealybugs and a higher proportion of them contain *Anarhopus* (De Bach, 1949, p. 21).

We have noticed an analogous situation in our studies of *Thrips imaginis* (Davidson and Andrewartha, 1948*a*, *b*). This small insect cannot develop healthily and lay eggs unless its diet includes pollen. In the vicinity of Adelaide, *T. imaginis* is found breeding throughout the year in the flowers of many different sorts of plants. During the spring the thrips increase greatly in numbers; our daily samples showed that it was usual to find about 190 thrips in one small rose, as well as comparable numbers in all the diverse flowers of weeds and other plants which, in this climate, abound in the springtime. During the summer, which is hot and arid, and during the winter, flowers of any sort are relatively few and widely scattered; the few that are found contain very few thrips; our records showed about 5 thrips per rose. With such small numbers it is clear that there is no absolute shortage of food, for only a small proportion of what is present is eaten. The flowers are so sparsely distributed relative to the powers of dispersal of the thrips that the chance that a thrips moving away from one flower will find another must be quite small; with low numbers of thrips, the chance that a flower will be found is also small.

The flowers in which the thrips live are ephemeral, and as they die, the thrips need to seek new food. Also the innate urge to disperse is strongly developed in them, and they often fly away from a fresh uncrowded flower with apparently no other stimulus than this (sec. 5.1). During the spring, when flowers are abundant and widespread, a thrips may find fresh food close to where it was bred; even those which disperse more widely have a good chance of coming to rest in the vicinity of food; and relatively few must fail to find it. But during the summer and winter, when flowers are few and widely scattered, many thrips must die of starvation, not because there is no food for them but because they fail to find it. The high death-rate which we observe at these seasons of the year must be attributed, in part at least, to shortage of food, even though relatively little of what is present is consumed. In these examples the inaccessibility of a proportion of the food depends upon the statistical improbability of its being found, and it would occur whether the food were distributed uniformly or not. But there is another sort of situation in which lack of uniformity in the distribution of the food becomes important.

The beetle *Chrysomela gemellata* feeds, both as an adult and in the larval stage, on the leaves of *Hypericum perforatum.* In the vicinity of Bright, Victoria, this plant grows in open grassland and under the trees in forests of *Eucalyptus* and *Pinus*. The beetles may fly into the wooded areas in search of food, but they do this only when there is a shortage of food in the adjacent open grassland. Later, when they have finished feeding for the time being but before they are ready to lay eggs, the beetles, in response to some stimulus which is not fully understood, leave the wooded areas. They rarely lay eggs on the *Hypericum* growing in the forests, which in this way become inaccessible or unavailable as food for the larvae of *C. gemellata* (Clark, 1953). Several analogous examples are described for predaceous insects in section 10.321. In each case a proportion of the food which is distributed differently from the remainder is missed, not because it is physically inaccessible but because the animal, by virtue of its special behavior, is inhibited from going to the place where the food occurs.

In a favorable situation, that is, where *Hypericum* is growing fairly densely in the open, *Chrysomela* may go on increasing in numbers until the plants have been completely defoliated. Most of the insects then die of starvation; but in every case observed by Clark (1953) a few pupated and survived to emerge as adults. These ate what little regrowth had occurred and then also died of starvation, unless, in the course of their wanderings, they came across a fresh source of food. Since their powers of dispersal by crawling and flying are slight, few of the insects succeeded in moving more than 3 chains from where they had originated. So when the *Hypericum* was defoliated over an extensive area, this usually resulted in the virtual extermination of the colony of *Chrysomela* associated with it, no matter how much food there might be at distances greater than 3 chains. Not many animals, even the very small ones, would exhibit such poor powers of dispersal as this; but the principle holds that whenever the food in an area gives out, for whatever cause, the chance that fresh sources of food will be found depends not only on the distribution and abundance of the food but also on the powers of dispersal of the animals seeking it. Dispersal is discussed in chapter 5; in section 11.2 we discuss certain principles which underlie the distribution and abundance of food in nature.

11.2 THE DISTRIBUTION AND ABUNDANCE OF FOOD IN RELATION TO OTHER COMPONENTS OF ENVIRONMENT

11.21 *The Interaction between Weather and Food*

Whether an animal be herbivorous or carnivorous, the ultimate limit to its distribution must be related to the distribution of certain plants: the butterfly *Pieris rapae* occurs only in places where there are cruciferous plants, which are the only sorts on which it will breed. The distribution of *Apanteles*, which preys on *Pieris* and no other sort of animal, is similarly restricted by the distribution of these plants. It is so obvious as to be hardly worth mentioning that the distribution of plants is determined largely by climate. The zonation of the vegetation, due to the temperature gradient, which may be observed as one proceeds from low to high latitudes or altitudes, is familiar to all ecologists.

Australia may be divided into a series of bioclimatic zones which are determined largely by moisture (Davidson, 1936*c;* see also Fig. 11.01). With a great expanse of almost flat land surrounded by water, the rainfall is greatest near the coast and falls away to less than 5 inches in the central deserts. The bioclimatic zones are arranged in a series of concentric strips, with the aridity of each one increasing as its distance from the coast increases. This long and consistent gradient is strikingly reflected in the vegetation types; this is especially true if one makes a northerly transect from the south coast to the interior (Wood, 1937; Andrewartha *et al.*, 1938; Andrewartha, 1940). Three phenomena stand out most convincingly: (*a*) Differences in vegetation types, associated with differences in climate, can be recognized and their boundaries demarcated, often with surprising exactness. (*b*) As one approaches a boundary, the plants which are characteristic of this zone begin to thin out and be intermingled with plants which belong to the next zone. (*c*) After the boundary has been crossed, the plants of the last zone will dwindle and disappear except for isolated communities which may persist in local situations. This is due to the heterogeneity of the terrain: a flat at the base of a stony hill in an arid zone, receiving not only the water which falls on it as rain but also that which runs off

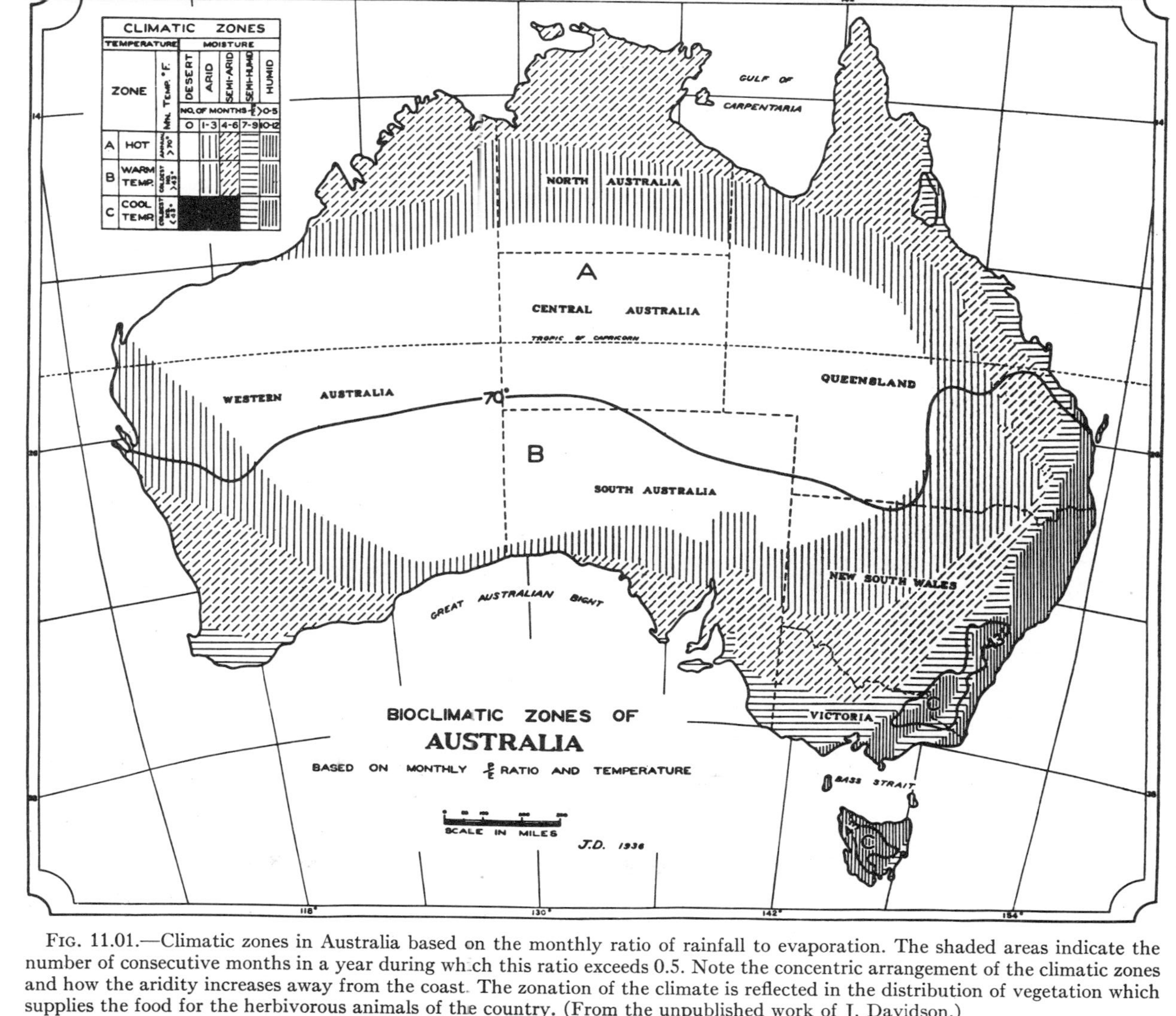

FIG. 11.01.—Climatic zones in Australia based on the monthly ratio of rainfall to evaporation. The shaded areas indicate the number of consecutive months in a year during which this ratio exceeds 0.5. Note the concentric arrangement of the climatic zones and how the aridity increases away from the coast. The zonation of the climate is reflected in the distribution of vegetation which supplies the food for the herbivorous animals of the country. (From the unpublished work of J. Davidson.)

from the stony hill, is moister than the surrounding countryside and may support the plants which are appropriate to a more humid zone.

In other words, there is, in this part of Australia, spread out for him who runs to read, a demonstration of how moisture (rainfall and evaporation) may influence not only the distribution but also the abundance of the different plant species, every one of which is, of course, food for some species of animal. For example, it is well known, and the explanation for it is well understood, that in most parts of Australia and western Canada more rain means more wheat, whereas in most parts of England more rain means less wheat. The animal ecologist, pursuing the interactions between weather and the food of the particular animal which he is studying, will find many leads in the literature not only of plant ecology of the more academic sort but also in the more applied disciplines of agronomy and economic geography. When these interactions are uncovered, they may turn out to be quite subtle.

We followed the fluctuations in a natural population of *Thrips imaginis* near Adelaide for 14 years and showed that the numbers present during the spring (when the insect is most numerous) were largely determined by the amount of food present at that season, which, in turn, was largely determined not only by the amount of rain which had fallen during the preceding autumn but also the dates on which it fell (Davidson and Andrewartha, 1948*b;* see also sec. 13.116). This relationship depends on the climate of the area, which, in turn, has largely determined the sorts of plants that grow there. The climate of Adelaide is characterized by a hot and arid summer lasting about 6 months. During this season all vegetation (except that artificially irrigated in gardens) becomes dormant; annuals die, leaving dormant seeds behind, and the perennials in this area are mostly xerophytes in which dormancy at this time of the year is natural. There is a growing period during winter and a massive flowering of virtually all the vegetation in the spring. This superabundance of food for *T. imaginis* occurring at a time when the weather is also favorable makes for rapid multiplication; and mass outbreaks involving quite stupendous numbers of thrips occur from time to time.

The distribution of *Thrips imaginis* includes all the southern coastal region and extends along the eastern coast beyond Sydney. The species is quite numerous around Sydney during the spring, but it never becomes so abundant as it does farther south. The best explanation for this lies in the way that climate influences the supply of food in the two places. The climate of Sydney is more equable, with rainfall distributed more evenly throughout the year and the temperatures remaining higher during the winter (Fig. 11.02). In these circumstances, there is not the same tremendous concentration of flowering in the spring, and *T. imaginis* is never relieved of the relative shortage of food so thoroughly as occurs in Adelaide for a brief period each year.

11.22 *The Interaction between "Other Animals of the Same Kind" and Food*

In section 11.12 we discuss situations in which only a small proportion of the total stock of food was eaten, yet nevertheless shortage of food set a limit to the rate of increase in the population. In such circumstances an animal's chance of getting enough food to survive and multiply may depend on the distribution of the food and on its own behavior, but scarcely at all on the number of other animals seeking the same food. On the other hand, when there is an absolute shortage of food, i.e., when all the food in the area has been consumed and there is not enough to go around, the number of other animals may be of major

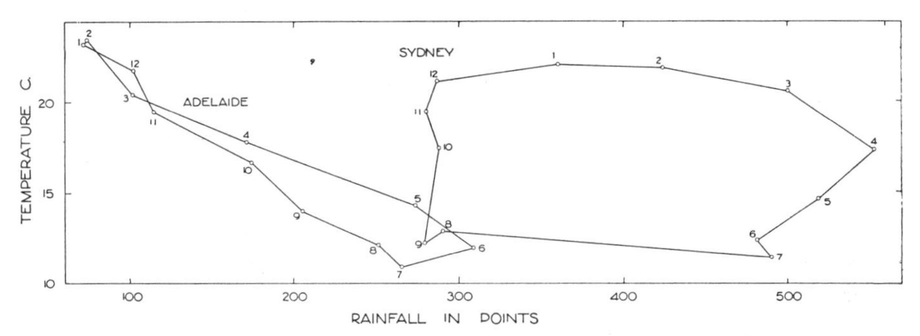

FIG. 11.02.—Climographs based on monthly rainfall and monthly temperature at Sydney and Adelaide. Note the aridity (low rainfall, high temperature) during summer (November to March) at Adelaide and the more even distribution of rainfall at Sydney. At Adelaide the plants grow during the winter, and the growing season culminates in a great burst of flowering during the spring. This is less pronounced at Sydney.

importance. This sort of situation is rather unusual in nature over broad areas, though it may happen sometimes as a temporary phenomenon in the smaller local units of a larger population. Elton (1949, p. 16) wrote of this: "Indeed, so apparent is this that it has been for many years a convention among botanists to treat dynamic vegetation systems as though the animals were not having any influence upon the energetics of the plants at all; or only to bring this idea in where the inroads are of a very conspicuous kind, as with rabbit or stock or deer grazing." However, it is easy to create a shortage of food artificially in the laboratory, and a number of nice experiments have been done along these lines (Nicholson, 1950; Ullyett, 1950).

The blowfly *Lucilia* lays its eggs on carrion at a specific (early) stage in its decomposition. With plenty of food, the maggots grow quickly, reaching a weight of about 60 mg. before pupating. It is not essential for maggots to grow so big, and when food is scarce, they may still pupate, provided that they can attain a weight of about 25 mg. (Ullyett, 1950). When few eggs are laid on a piece of meat of limited size, few large flies are produced; with a moderate number of eggs, a larger number of smaller flies emerge. With increasing numbers of eggs laid on a fixed weight of meat, the number of flies emerging increases up to a maximum and then begins to decrease, eventually becoming zero. In Nicholson's (1950) experiments this maximum was reached when 25 females were allowed to lay their eggs on 50 gm. of meat. With increasing numbers, fewer and fewer maggots were able to get enough food to grow to the weight at which pupation became possible, until, with 150 females laying eggs on 50 gm. of meat, there were so many maggots that none got enough food and all died as larvae. All the food was eaten, but not one individual was contributed to the next generation.

These results have been interpreted in terms of "intraspecific competition," and there is no doubt that the "competition" between the maggots may be intense and most striking to witness. But the numbers of animals which survive in these experiments may not be closely related to the intensity of the competition (secs. 2.121 and 10.02). We think it is better to seek an explanation in terms of limiting resources and to start by asking the question: How is the amount of food influenced by the number of maggots present? The superficial answer to this question for the artificially simple case presented by these experiments with blowflies is: Not at all, since each experiment was started with the same amount of meat, 50 gm. But let us pursue the matter a little further and consider not the total amount of meat placed in the cage but rather that part of it which was consumed by maggots which survived to contribute progeny to the next generation. This depended on the number of maggots in the cage.

When few maggots were present, there was a large amount of food per maggot, and some food remained uneaten. The food which was not eaten was wasted because it was surplus. As the number of maggots feeding on the 50 gm. of meat was increased, a certain number was reached such that there was just enough food for each maggot to complete its development. With the population at this density, one might say that the whole 50 gm. of meat was "effective"; there was no uneaten surplus and none "wasted" on maggots which failed to get enough food to reach maturity. As numbers increased beyond this point, the amount of food per maggot was not sufficient for all the maggots to reach maturity. Some of those which developed faster got more than their share and so were able to reach maturity. But the others failed to get enough food to complete their development. The food eaten by these may be said to have been "wasted," while the remainder (i.e., the part of the 50 gm. eaten by those maggots which completed development) may be regarded as "effective" food. As numbers increased still further, a smaller and smaller part of the stock of food was effective, until, with 150 females laying eggs on 50 gm. of meat, there were so many maggots that none of the food was effective—although all of it was eaten, yet no maggot got enough food to complete its development. We have illustrated the relationship between the number of other animals of the same sort and the amount of effective food in Figure 11.03. The hypothetical curve labeled "dead food" refers to the relatively simple situation which we have been discussing up to the present. The curve is low when there are few animals, because there is a large uneaten surplus, which is wasted; it is low when there are many animals, because much is eaten by animals which fail, through shortage of food, to reach maturity; this portion also must be regarded as wasted, in the sense that it is not used to contribute progeny to the next generation.

When the food, instead of being something dead like carrion, is a living plant, the relationship becomes more complex, but the principle is the same. The chief difference lies in this: with carrion it is possible to do an experiment (like those discussed above) in which the total stock of food can be made independent of the number of animals feeding on it; with the living plant this is not possible, because the total stock of food varies with the number of feeding animals. This could be readily demonstrated by the following simple experiment. A number

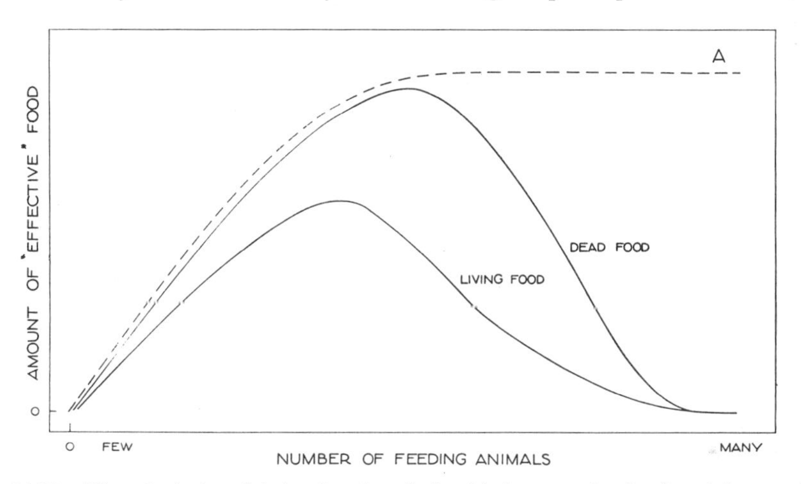

FIG. 11.03.—Hypothetical model showing the relationship between the density of the population and the proportion of their total stock of food which is likely to be "effective," in the sense that it may be used in the production of progeny for the next generation. For further explanation see text.

of growing plants may be divided at random into two equal groups. Those in group A are left to grow unmolested for (say) 50 days and weighed at the end of this period. From those in group B each day leaves are removed which weigh one-fiftieth of the weight of the leaves which were on the plant at the beginning. These are weighed and their sum added to the weight of the plant at the end of the experiment, on the fiftieth day. The plants in group B, together with the leaves taken from them, will weigh less than the plants in group A, for, as the well-known horticultural aphorism says, "pruning stunts." Indeed, every horticulturist knows that the quickest way to produce a big tree is to leave it unpruned, and the surest way to strengthen a weak branch is to leave it unpruned while cutting back the others. Likewise, any plant that is grazed, whether by the farmer's cattle or by insect pests, will make less growth than one which is not so grazed. The general ecological principle which these examples indicate may be stated as follows: when the food is a living plant, the size of the *total stock of food* depends upon the number of animals feeding on it; and the relationship would be something like that indicated by the hypothetical curve, labeled "living food" in Figure 11.04. By contrast, the curve for total stock of dead food is a horizontal straight line, because the total stock of food is independent of the number of animals that may come to feed upon it. The principle of effective food may be applied in exactly the same way as when the food was

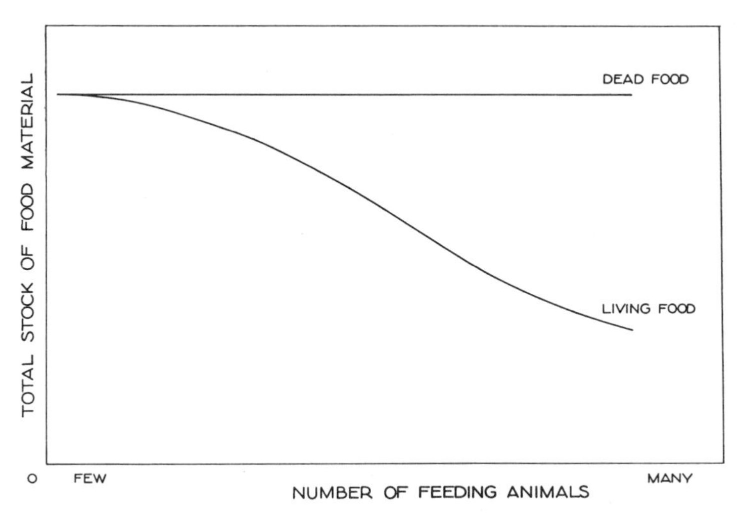

FIG. 11.04.—Hypothetical model showing the relationship between the density of the population and the total stock of food likely to be available to the population. The quantity of dead food does not change with the density of the population feeding upon it. But the quantity of living food is likely to be less, the more heavily it is grazed or preyed upon.

carrion; but in Figure 11.03 we have made this curve descend more abruptly, to indicate that its slope reflects the dual influence of numbers of animals not only on "effective food" but also on "total stock of food," whereas the curve for carrion reflects only the former.

When the food is a population of living animals, there are still other complexities to be considered, for example, the destruction and replacement of the stock of food must be considered in terms of discontinuous variation, the unit usually being one whole animal. At first sight, this may seem to be a fundamental difference, but a moment's reflection will show that it is not. When the food is a growing plant and *n* leaves are eaten, then the remainder take over and make their appropriate contribution to the increase in weight of the plant. When the food is a growing population of animals and *n* individuals are eaten, then the remainder take over and make their appropriate contribution to the increase in numbers of the population. The situations are essentially the same, and there is nothing to be gained by making a fundamental ecological distinction between living plants and living animals as food. So in Figure 11.04 a single line represents living food, whether this be plant or animal.

The concept of effective food may be usefully applied when the food is a population of living animals—with one exception. Many species of insect predators of the sort known to entomologists as "internal parasites"—that is, those in which the female seeks out the prey and lays an egg *inside* it—can recognize those individuals which have already been attacked and they avoid them (sec. 10.221). Other species which lack this faculty may possess an adaptation which insures that one and only one predator shall survive in one prey, no matter how many eggs may have been laid in it. In these cases the whole stock of food (all that is found) is used without waste, and there is no point in invoking the concept of effective food. In Figure 11.03 the broken line, which represents this special case, occupies the same position on the graph as might be occupied by a curve for "total stock of food" on the same scale.

Whenever the numbers in a population become so great that there is an absolute shortage of food, it is likely that the food, whether plant or animal, will be destroyed: if plant, it may be killed; if animal, the population may be wiped out, though not, as a rule, before some individuals have moved off, with the chance of founding new colonies elsewhere. In nature the complete destruction of a source of food rarely happens, except locally. When an animal species is found to be responsible for the destruction of a great proportion of its food over broad areas, investigation usually reveals that the eating species has high powers of dispersal and rate of increase relative to those of the eaten species; and the usual picture is of a high rate of destruction of food going on concurrently in most of the local situations where the food is distributed (see quotation from Nicholson, sec. 5.0). Although unusual among animals, such species are often of great economic importance, being harmful when their food is valuable to us and beneficial when their food is a weed or pest of some kind.

Among the insects, these are the species which have been successfully used for the biological control of weeds and insect pests. The moth *Cactoblastis cactorum*, which occurs naturally feeding on *Opuntia* in America, was introduced into Australia in 1925 at a time when the prickly pear was at its peak in that country. Some 30 million acres of potentially good agricultural land was so densely covered by *Opuntia* spp. as to be valueless; it was estimated that the weight of prickly pear per acre exceeded 500 tons. There was a further 30 million acres where the prickly pears were growing not quite so densely. At this time the area occupied by *Opuntia* spp. was increasing by about 1,000,000 acres each year. During the next 5 years (i.e., from 1926 to 1930) some 3,000,000,000 *Cactoblastis* were liberated in this area (Dodd, 1936). They multiplied and spread: at first, they consumed and destroyed all the food in local situations and then spread farther afield. "During the years 1931 to 1934 the collapse of the main body of prickly-pear continued in very much the same manner as that described along the Moonie River. The last big tract of dense pear in Queensland . . . succumbed to *Cactoblastis* in 1934. At this time, it was estimated that 90 per cent. of the primary pear of the State had disappeared; . . ." (Dodd, 1940, p. 140). This quotation refers to dense stands of *Opuntia;* of course, local colonies persisted, and new ones are constantly starting from seeds. But *Cactoblastis*, because of its high powers of dispersal and rapid rate of increase, soon finds them and destroys them. The result is that *Opuntia* is now relatively scarce over vast areas where it was previously inordinately abundant. Several examples of the successful introduction of predaceous insects to reduce the numbers of insect pests were mentioned in section 10.321, and it will be recalled that all the more "successful" ones had high powers of dispersal and high rates of increase relative to those of the prey. The beetles *Chrysomela gemellata* and *C. hyperici*, which were introduced into Australia to control St.-John's-wort were less successful than *C. cactorum*, partly because of their low powers of dispersal and rates of increase (Clark, 1953; sec. 11.12).

The complicated behavior of vertebrates makes it more difficult to generalize about them. Leopold (1943, in Allee *et al.*, 1949, p. 706) considered that the Kaibab plateau in Arizona in its natural condition supported about 4,000 deer. These would have made little impression on their supply of food, and he estimated that in 1918 the area might have supported 30,000 deer without detriment to their resources of food. From 1907 onward, large numbers of coyotes, pumas, and wolves were destroyed by shooting; the wolves were exterminated by 1926. The deer multiplied rapidly, reaching a maximum of 100,000 by 1924–25. A drastic shortage of food during the next two winters resulted in the death by starvation of some 60,000 deer. At the same time, the feeding of so many animals permanently reduced the total stock of food in the area (in accordance with the general law illustrated in Fig. 11.04), and the numbers continued to decline to 10,000 by 1940. Darwin (1839, p. 101, in the 1952 reprint) commented on the apparent shortage of food for the large herds of ungulates he observed in Africa: "I confess it is truly surprising how such a number of animals can find support in a country producing so little food. The larger quadrupeds no doubt roam over wide extents in search of it; and their food chiefly consists of underwood, which probably contains much nutriment in small bulk. Dr. Smith also informs me that the vegetation has a rapid growth; no sooner is a part consumed than its place is supplied by a fresh stock." A more recent comment on animals in the same region is given by Stevenson-Hamilton (1937, p. 259): "The deliberate elimination of practically all the larger carnivorous animals from the Zululand game reserves resulted, within a few years, in the numerical increase of the 'big game' to such a degree that the reserve could no longer support its needs, and the animals spread out far and wide over the surrounding settled country. . . . In a certain district of the Sabi Game Reserve, a season's intensive trapping of the smaller carnivora—genets, mongooses, civets, wild cats and jackals—resulted in such a plague of bush rodents that the local natives lost most of their grain crop." Errington (1946, p. 157) considered that ungulates are more likely to eat out their stocks of food than are most other sorts of vertebrates. The quotation from Stevenson-Hamilton shows that this may happen also with small rodents if their predators are artificially destroyed.

The food supply of predaceous vertebrates depends to some extent upon the behavior of the animals on which they prey. Populations of species like the muskrat which regulate their density relative to the supply of food and other resources by the complex adaptation known as "territorial behavior" (sec. 11.11) regularly produce each breeding season a "surplus of foot-loose wanderers" which are especially vulnerable to predators (Errington, 1943). The size of the surplus varies greatly from year to year, and this might be expected to influence the rate of increase in the populations of predators. Chitty (1950) considered that the well-known fluctuations in the numbers of the white fox *Alopex lagopus* and the snowy owl *Nyctea nyctea* in the Canadian Arctic may be due to fluctuations in the supply of their chief food, the lemmings, *Lemmus* spp. and *Dicrostonyx* spp. Little is known about the fundamental causes in the fluctuations in numbers of these rodents.

11.23 *The Interaction between "Other Organisms of Different Kinds" and Food*

An animal's chance of getting enough food may be reduced by the presence of the other sorts of organisms in its environment. If these belong to category *a* of section 10.01, they may simply consume the same sort of food and thereby reduce the total stock. Or they may belong to category *b*, in which case they may spoil or destroy some or all of the stock of food incidentally, without themselves consuming it. Still more complex relationships are possible. For example, Ullyett (1947) found that an epizoötic of the fungus *Entomophthora* so reduced the numbers of the caterpillar *Plutella* that the predators (*Angitia* spp. and others) suffered severely from a shortage of food where it had been abundant before the outbreak of disease. Phytopathological microörganisms may similarly reduce the stock of food available for a herbivorous animal. The activity of these pathogens may be largely determined by weather, adding still further to the complexity of this particular interaction. Most of the more important relationships with other organisms were discussed generally, with reference to food as well as other resources in the environment, in chapter 10, and there is no need to say more about them.

On the other hand, an animal's chance of getting enough food may be enhanced by the presence in its environment of other organisms of certain kinds—not including those that are themselves the food. Some of these may be regarded as coming into category *b* of section 10.01, but there are certain others which may not have been covered by the discussion in chapter 10. Consider, for example, the dung beetle living in a field of clover. The ultimate source of the beetle's food is the clover, but it is of no use in this form. The cow, which also lives in the field, eats the clover and the beetle breeds in the dung dropped by the cow; the presence of many cows in the field means much food for the beetles and therefore, in the absence of other checks to increase, more beetles. Similarly, there are many species of small insects, mostly Diptera and Hymenoptera, which live on the "honeydew" secreted by aphids, coccids, and other Homoptera. And many other similar cases might be mentioned. Certain Protozoa live in the gut of the termite and play an essential part in the digestion of cellulose, which forms a large part of the termite's food (Cleveland, 1928). Similarly, microörganisms in the rumen of the ox and the caecum of the horse aid in the digestion of cellulose and are said to synthesize vitamins. The pathogens which deplete the stock of food for the animal which eats living plant or animal (see above) may add to the stock of food for the scavenger.

Rawlings (1948) described the association between the wood wasp *Sirex noctilio* and a fungus, *Stereum* sp. The larva of *S. noctilio* lives as a borer in the main trunks or large branches of trees, usually of the genus *Pinus*. The adult female has a stout ovipositor with which she drills a hole through the bark and into the cambium, where she deposits her eggs. She always carries in two glands associated with the ovipositor a stock of the fruiting bodies (oidia) of the fungus, and she injects an inoculum of this into the cambium each time she drills a hole for oviposition. If the tree is in a suitable condition, the fungus will spread (killing the cambium) in a strip several inches wide and several feet long, up and down, from the original site of the inoculation. The young larva always feeds for a period in this region. If the fungus fails to become established, the larva also fails to survive. How the fungus is essential for the survival and growth of the larva is not known; it may add something to the quality of the food which would otherwise be lacking.

11.3 THE QUALITY OF FOOD

Since so many books have been written about the composition of the food of man and the domesticated animals and the influence of the diverse constituents which are now known to be essential for healthy growth and reproduction, it would be trite to belabor the point here that the quality of the food available to animals in nature may influence any one of the three components of r and thus the animal's chance to survive and multiply. We shall therefore take for granted most of what might have been said in this section and shall mention only a few examples (mostly from among the insects, which is the group that we know best) of how the quality of food may influence the duration of life, the speed of development, or the fecundity of particular species. Different races within a species may also differ in these respects (secs. 15.111, 15.12; Table 15.03).

11.31 *Longevity*

The spruce budworm *Choristoneura fumiferana* in Canada occasionally multiplies to vast numbers, constituting plagues which may do severe damage to wide areas of forest. This is more likely to happen in situations where there are numerous balsam firs in flower. The chief reason for this is the higher survival-rate among the young larvae in these circumstances. The spruce budworm does not feed during the first instar; they are in the second instar when they emerge from their hibernacula in the spring and seek food for the first time. They strongly prefer the staminate flowers of the balsam fir to foliage of either fir or spruce, and they will eat the young foliage more readily than the old. The survival-rates among young larvae feeding on pollen and young foliage were equally high, but a tree in full bloom provides an abundance of good food which may otherwise be lacking. Relatively few of the young larvae which were constrained to feed on old foliage survived. Those that did grew more slowly, and the adults laid fewer eggs. So the advantage conferred by the presence of flowers may not be due entirely to the higher survival-rate (Blais, 1952). In nature *Thrips imaginis* breeds only in flowers. In the laboratory they could be reared with little difficulty on the entire stamens of *Antirrhinum;* but when the anther with the pollen was removed, so that the food consisted only of the filament, none survived to become adult (Andrewartha, 1935).

The larvae of the gipsy moth *Lymantria dispar* in New England are usually found on the foliage of oak, but they also occur on a number of other shrubs and trees, especially if oaks are scarce. A few may even be found on pine, although in experiments it was found that all the young stages and most of the older larvae died when they were segregated in a cage with no food other than pine (Brues, 1946). The gipsy moth is naturally polyphagous; most monophagous species cannot live on any sort of food other than that which is natural to them. Often this may be due to their behavior in refusing to eat the material offered to them rather than to its inadequacy as food.

The species *Ephestia kühniella*, *Stegobium paniceum*, and *Oryzaephilus surinamensis* are commonly found in barns and storehouses, feeding on such foods as flour, whole meal and bran, and, in addition, *Oryzaephilus* may be found in dried fruit. They are restricted to this sort of food because they cannot survive except on a diet that is rich in carbohydrate. On the other hand, the beetles *Ptinus* spp., *Tribolium* spp., and *Lasioderma serricorne* also occur in storehouses, feeding not only on flour and other materials rich in carbohydrates but also in such materials as fish and meat-meals, dried yeast (*Ptinus* and *Tribolium*), and tobacco (*Lasioderma*). This is because these beetles require little, if any, carbohydrate in their diet (Fraenkel and Blewett, 1943*b*). The survival-rates of *Ephestia elutella* when reared on a number of "natural" foods were measured by Waloff (1948). The results are set out in Table 11.01.

TABLE 11.01*
DURATION OF LARVAL DEVELOPMENT AND SURVIVAL-RATES AMONG LARVAE OF *Ephestia elutella* WHEN REARED ON "NATURAL" FOODS AT 25° C. AND 70 PER CENT RELATIVE HUMIDITY

Food	No. of Survivors (Per Cent)	Duration of Larval and Pupal Stages Combined (Excluding Period Spent in Diapause)
Wheat	87	50
Wheat embryo	100	42
Tobacco	10	120
Figs	10	134
Cacao beans	9	87
Beans	6	...†
White flour	25	137

* After Waloff (1948).
† None completed diapause.

All the species of insects which have been tested require vitamins of the B group, but their specific requirements for the different ones vary (Fraenkel and Blewett, 1943*a*). For example, *Ephestia elutella* made no growth in the absence of riboflavin, whereas *E. kühniella* survived on a diet which lacked riboflavin, but it grew more slowly than normally (Fraenkel and Blewett, 1946). Some insects, for example, *Lasioderma*, *Stegobium*, and *Oryzaephilus*, do not require the B vitamins in their food because they get them from symbiotic microörganisms which live in special cells in their bodies. Fraenkel and Blewett (1944) deprived these beetles of their symbionts by washing the eggs in a bactericidal solution. The larvae normally get their supply of symbionts from the outside of the egg from which they hatch, but washed eggs gave rise to symbiont-free larvae. All those which were offered a diet lacking in the B vitamins died; but those which were fed on a diet containing vitamins of the B group survived normally. The larvae of *Drosophila* normally get their vitamin B in yeasts in the medium on which they are feeding. Unless vitamin B was provided in this or in some other way, they died (Bacot and Harden, 1922). There is some evidence that different species of yeasts have a different food value for *Drosophila;* experiments described in section 15.111, in which different yeasts were fed to *D. pseudoobscura*, indicate that the rate of increase of the different chromosomal types varied with the yeast in their diet. All insects require a sterol in their diet. For example, *Dermestes* cannot utilize plant and yeast sterols, and, as a result, it can develop only on food of animal origin, such as furs and hides, on which it often is a serious pest (Fraenkel and Blewett, 1943*c*).

11.32 *Fecundity*

Many species of insects with a complete metamorphosis in the life-cycle store up enough food during the larval stage that the adult may produce its full quota of eggs without any food; moths of the family Hepialidae are like this. Others store up enough protein, but the adults require to take in water and carbohydrate; most species of the family Agrotidae are like this. Still others require a full diet as adults in order to produce eggs. For example, *Thrips imaginis* require pollen. Andrewartha (1935) confined a number of newly transformed adults in small glass vials with a variety of foods and found that only those which received pollen laid many eggs (Table 11.02). Blowflies of the genera

TABLE 11.02*
INFLUENCE OF FOOD (WITH AND WITHOUT POLLEN) ON FECUNDITY AND LONGEVITY OF ADULT *Thrips imaginis* AT 23° C.

FOOD	NO. OF INDIVIDUALS		MEAN NO. OF EGGS LAID DURING 20 DAYS
	Beginning	Alive after 20 Days	
Leaf of *Trifolium repens*	22	2	0.8
Leaf of *T. repens* plus pollen	19	15	158
Leaf of *Plantago lanceolatus*	19	3	0.2
Leaf of *P. lanceolatus* plus pollen	16	9	156
Filament of *Antirrhinum*	9	9	0.3
Filament of *Antirrhinum* plus pollen	7	7	120

* After Andrewartha (1935).

Lucilia and *Calliphora* produced eggs normally when the diet included extracts of liver or muscle or some other animal protein. They produced no eggs when fed only on cane sugar and water; but the males produced normal sperm when fed on this diet (Mackerras, 1933). The housefly *Musca domestica* also requires protein in its diet if it is to produce any eggs (Glaser, 1923). The same is true of many mosquitos, and they get their protein from blood. The adult beetles of the family Meloidae and Cerambycidae eat a lot of pollen, and this no doubt provides them with necessary protein (Brues, 1946, p. 105). The fruit fly *Dacus tryoni* laid few eggs on a diet of pawpaw and honey compared to the many laid when they fed on a protein hydrolyzate of yeast which was rich in proteins and amino acids.

Evans (1938) found that aphids *Brevicoryne brassicae* produced more young when their diet was rich in protein. He grew six cabbages under a bright light. Their leaves were rich in protein, and the aphids feeding on them produced an average of 445 young during 24 days. He grew six other cabbages under a dull light. Their leaves were poor in protein, and the aphids feeding on them produced an average of 160 young during 24 days (Table 11.03). The correlation coefficient between protein-nitrogen and number of young was 0.85 ($P = 0.01$) and that between nonprotein-nitrogen and number of young was 0.67 ($P = 0.05$). In another experiment aphids were confined in a dull light on the petiole of leaves which were exposed to bright light. They produced the same number of young as did the others which were feeding in the bright light. These experiments indicated that the important difference in the treatment of the two lots of aphids was the difference in the proportion of protein in their diets.

TABLE 11.03*

NUMBER OF YOUNG PER FEMALE WHEN *Brevicoryne brassicae* WERE ALLOWED TO FEED ON CABBAGES GROWN IN BRIGHT AND DULL LIGHT, AND NITROGEN CONTENT OF LEAVES

DULL LIGHT			BRIGHT LIGHT		
No. of Young	Nitrogen (Per Cent)		No. of Young	Nitrogen (Per Cent)	
	Protein	Nonprotein		Protein	Nonprotein
80......	0.17	0.06	252...........	0.20	0.05
97......	.15	.05	304...........	.22	.04
113......	.18	.05	337...........	.29	.08
114......	.16	.05	478...........	.21	.07
182......	.21	.11	669...........	.21	.04
243......	0.25	0.06	760...........	0.34	0.09
Mean 160......	0.19	0.06	445...........	0.25	0.06

* After Evans (1938).

11.33 *Speed of Development*

The larval stage of the buprestid *Melanophila californica* is a grub which lives in the sapwood of *Pinus ponderosa* in northwestern North America. The eggs are laid just under the bark scales. The first-instar larva hatches after about 15 days and tunnels into the region of the cambium. If the tree happens to be in a certain condition, the larva will feed and grow actively, completing its larval development during one summer. This condition cannot be specified precisely, but it occurs in trees which are unhealthy and about to die. If the tree should not be in this condition, the young larva may continue to live in its tunnel near the cambial layer for as long as 4 years, feeding only a little and scarcely making any growth. If the tree continues in an unfavorable (i.e., healthy) condition, the larva eventually dies without developing beyond the first instar. But should the tree (or part of it) enter this favorable condition, then all the small larvae present, irrespective of their age from a few days up to 4 years, will start feeding actively and grow rapidly, usually completing the larval stage in one summer. It is not known precisely what changes in the quality of the food are responsible for this dramatic change in the development of the larvae (West, 1947).

A method which is sometimes used in the laboratory to test the suitability of different diets for insects is to compare the speed of development of the immature stages. On unsuitable diets they require longer to complete their development. For example, the larvae of *Tribolium confusum* required about 28 days to complete the larval stage at 25° C. and 95 per cent relative humidity when the food was "high-extraction" flour and from 32 to 50 days with yeast or meat meal as food; the difference was chiefly due to the lack of carbohydrates in the last two materials. On the other hand, the larvae of *Ptinus tectus* developed rather slowly on a diet in which carbohydrates predominated; when yeast was added to whole-meal flour, the speed of development was increased by about 20 per cent.

11.34 *Selection of Suitable Food*

This section is concerned with the adaptations which help certain animals, especially insects, to find suitable food and to select the food which is appropriate to themselves from among all the other sorts which may be near. The most striking adaptations may be found among certain species of insects which are strictly monophagous—that is, they are able to use for food one sort, and one sort only, of plant or animal, as the case may be. A disadvantage of monophagy is that it restricts the animals to one species of food; the chief advantage is that it permits them to become closely adapted for using that sort of food (secs. 15.12, 15.2). There is no close correlation between the nutritional requirements of leaf-feeding insects and the plants they choose; indeed, there seem not to be any great differences in the nutrients in the leaves of different species. Preferred plants may be suitable, not primarily because of any particular nutrients they provide, but because they are present at the right time of the year and in other ways they provide a suitable place in which the insect can carry out its life-history (Dethier, 1954).

The two sibling species of the weevil *Calandra oryzae*, which are known in literature merely as large and small "strains," occur quite commonly in storehouses around Sydney, Adelaide, and Perth. The two strains are virtually monophagous, because the small strain is found only in wheat and the large strain only in maize. Both strains may be reared artificially on either grain. But with the large strain the innate capacity for increase was greater in maize, and with the small strain r_m was greater in wheat (Birch, 1953*a;* see also sec. 10.213). When they were confined in vials with both sorts of grain, the large strain laid more eggs on maize than on wheat, and the small strain laid more eggs on wheat than on maize. But their preferences were not strong enough to account for the clear-cut difference in distributions in nature. It is not known what senses may account for the way that the small strain unerringly selects wheat, and the large strain, maize. Crombie (1941) showed that *Rhizopertha dominica* could smell grain from a little distance, but, once it had alighted on the grain, it was stimulated to lay eggs by its sense of touch.

The larvae of the butterfly *Danaus plexippus* feed almost exclusively on milkweeds of the genus *Asclepias*. The butterfly instinctively lays eggs on milkweed and virtually on no other sort of plant. Since the eggs are placed so unerringly on the appropriate plant, one might not expect to find in the larvae any special adaptation for recognizing its food. But they possess olfactory receptors on the antennae and maxillae. Unless these are stimulated by the smell of *Asclepias*, the caterpillars make no attempt to eat but will keep crawling until such time as they come upon a leaf of *Asclepias* or die from starvation. When a caterpillar does come upon a leaf, it waves its head from side to side very close to the surface. If it receives the appropriate stimulus, it "tastes" the leaf. The taste may then stimulate the caterpillar to continue feeding.

Dethier (1937) "paved" the floor of a large cage with square pieces of leaf cut from *Asclepias* and a number of other plants. Caterpillars of *Danaus* were placed in the cage; 85 per cent of them came to rest on the *Asclepias*. The milkweed was then painted with a number of strong-smelling substances; and the caterpillars came to rest indiscriminately anywhere on the floor of the cage. The first experiment was repeated, but with a wire screen covering the leaves so that the caterpillars could smell the leaves but could not touch them; as a caterpillar came over a leaf of *Asclepias*, it usually took a zigzag course, and half of the caterpillars came to rest over these leaves. When the other leaves were coated with latex from *Asclepias*, the caterpillars behaved as if these leaves were *Asclepias*. Leaves of *Asclepias* were then coated with substances like sodium chloride or hydrochloric acid which were odorless but distasteful to the caterpillars; the caterpillars recognized the leaves and started to chew them, but did not continue to eat. The smell was "right" but the taste was "wrong."

Wireworms (larvae of beetles of the genus *Agriotes* and other related genera) feed on the roots of grasses and crop plants. A wireworm moves at random through the soil until it comes near a root and recognizes the presence in the soil of substances like asparagine and aspartic acid which are excreted by the root. The wireworm continues to move at random, except that it undertakes turning movements which keep it in the soil where these substances can be recognized. In due course it comes within touch of the root. It is stimulated to eat the root by the presence of certain sugars, fats, polypeptides, and other substances (Crombie and Darrah, 1947; Thorpe *et al.*, 1947).

Dethier (1947) mentioned a number of caterpillars which were known to select their food by recognizing a specific chemical compound. For example, the larvae of the butterfly *Papilio anax* readily "sampled" pieces of filter paper soaked in rue (methylonyl ketone). This substance is present in a number of aromatic species of the families Umbelliferae and Rutaceae, on some of which the larvae are known to feed in nature. Dethier tried several others from which the larvae had not been recorded; in every case they "sampled" the leaves but did not eat them, probably because they were too tough. Dethier also offered the larvae the leaves of several species of Rutaceae which lack rue; the larvae were not stimulated to "sample" them.

Most larval insects need to be very close to their food before they can smell it. But in the adults of certain species the olfactory organs are highly sensitive. The dung beetle *Geotrupes sylvaticus* responded to the odor of skatol gas when the concentration was as low as 2.3×10^{-8} M; and they usually moved straight toward dung from a distance of 50 cm. The corresponding distance for *Scarabaeus* was 10 meters; for *Drosophila melanogaster*, 23 cm.; and for *Calliphora erythrocephala*, 6 cm. (Fraenkel and Gunn, 1940).

Some adult Hymenoptera display remarkable ability in finding their prey. The ichneumonid *Rhyssa persuasoria* recognizes larvae of *Sirex* through several centimeters of wood, and *Megarhyssa* finds the larvae of *Tremex* that are similarly situated. When the adults of the ichneumonid *Pimpla ruficollis* first emerge, their ovaries are small, and they need to feed on the nectar from flowers of certain Umbelliferae before their eggs will develop to maturity; this takes about 3 weeks. The prey of *Pimpla* are the pine-feeding caterpillars of *Rhyacionia buoliana*. While its ovaries were immature, *Pimpla* was repelled by pine oil; when the ovaries were mature, *Pimpla* was attracted by pine oil (Thorpe and Candle, 1938). This beautiful adaptation led *Pimpla* away from the pine tree when it was seeking food for itself and back to the pine when it was seeking food for its larvae and was ready to lay its eggs.

Certain insects select specific tissues in the specific plant or animal that they have chosen for food. The leafhopper *Eutettix tenellus* sucks the fluids from the phloem of certain species of Polygonaceae. The fluids are least acid in the phloem and most acid toward the periphery of the parenchyma. It seems that the mouth parts of the leafhopper may be orientated with respect to this gradient. Leafhoppers were placed on a membrane which they could pierce, to feed on droplets of water containing sugar in solution. Some drops were at pH 8.5 and others at pH 5.0. A leafhopper which happened to push its mouth parts into an acid drop would explore it for a while and then move elsewhere. One that found an alkaline drop would stay there to feed (Fife and Frampton, 1936).

It seems that the "search" may start with movements that are random with respect to the food but not necessarily random with respect to other stimuli. Insects tend to move in gradients of temperature, moisture, and light (secs. 6.21, 7.16, and 8.3). This may bring the insect to a place where food is likely to be found. When the insect comes close enough to see or smell its food, it moves directly toward it.

Gast (1937) found that the earthworm *Lumbricus terrestris* preferred the following leaves in the following order: large-toothed aspen, white ash, basswood, sugar maple, and red maple; red oak was not eaten at all. Van der Drift (1951) found that the millipede *Julus scandinavicus* preferred red oak to any other, whereas *Cylindrojulus silvarum* preferred pine. Lindquist (1942) tested 15 species of snails to discover what sorts of food they preferred; most of them rejected leaves of beech and oak for leaves of hazel, elm, and ash.

Räber (1949) experimented with models of birds and mice, to discover how the tawny owl recognized its prey. He found that models of birds were "captured" when they were quite immobile but models of mice were "captured" only if they had moving legs. From this he concluded that birds are recognized by shape but mice by shape and movement. At dusk when the tawny owl is hunting, most birds are likely to be still, whereas mice are likely to be moving.

Besides taxes and instinct, learning may be important with birds. Even closely related species living in the same area may use quite different foods (sec. 10.311).

The examples of this section illustrate how some animals seek out suitable food by means of instincts, taxes, and so on. This is a field of study on the borderland of ecology and behavior which warrants more study than it has been given in the past.

CHAPTER 12

A Place in Which To Live

We have three butterflies which are limited by geological considerations, being inhabitants only of chalk downs or limestone hills in south and central England, and they may reach the shore where such formations break in cliffs to the sea. These are the Silver-spotted Skipper, Hesperia comma, *the Chalk-hill Blue,* Lysandra coridon, *and the Adonis Blue,* L. bellargus. . . . *The two latter insects are further restricted by the distribution of their food plant, the Horse-shoe Vetch,* Hippocrepis comosa, *and possibly by the occurrence of a sufficiency of ants to guard them. Yet any of the three species may be absent from a hillside which seems to possess all the qualification which they need, even though they may occur elsewhere in the immediate neighbourhood. This more subtle type of preference is one which entomologists constantly encounter, and a detailed analysis of it is much needed. A collector who is a careful observer is often able to examine a terrain and to decide, intuitively as it were, whether a given butterfly will be found there, and that rare being, the really accomplished naturalist will nearly always be right. Of course he reaches his conclusions by a synthesis, subconscious as well as conscious, of the varied characteristics of the spot weighed up with great experience; but this is a work of art rather than of science, and we would gladly know the components which make such predictions possible.*

FORD, *Butterflies* (1945*a*, p. 122)

When two sticklebacks meet in battle, it is possible to predict with a high degree of certainty how the fight will end: the one which is farther from his nest will lose the match. In the immediate neighbourhood of his nest, even the smallest male will defeat the largest one. . . . The vanquished fish invariably flees homeward, and the victor, carried away by his successes, chases the other furiously, far into its domain. The further the victor goes from home, the more his courage ebbs, while that of the vanquished rises in proportion. Arrived in the precincts of his nest, the fugitive gains new strength, turns right about and dashes with gathering fury at his pursuer. A new battle begins which ends, with absolute certainty, in the defeat of the former victor, and off goes the chase again in the opposite direction. The pursuit is repeated a few times in alternating directions, swinging to and fro like a pendulum which at last reaches a state of equilibrium at a certain point. The line at which the fighting potentials of the individuals are thus equally balanced marks the border of their territories. The same principle is of great importance in the biology of many animals, particularly that of birds.

LORENZ, *King Solomon's Ring* (1952, pp. 26–27)

12.0 INTRODUCTION

A GOOD case might be made out for the proposition that among vertebrates the behavior patterns associated with territory are more fundamental even than those associated with sex. With invertebrates, "territoriality" as such hardly occurs, but the behavior associated with the animal's choice of a place to live seems to be deeply rooted and complex. So it is probably true of animals in general that the seeking of a special place to live in is one of their most fundamental characteristics. Since this character, like all others, is subject to natural variation within a population and since any area which may support a natural population is inevitably variable with respect to the sorts of places where the animals may live, it follows that some individuals are always better placed than others not only with respect to the usual hazards associated with weather, other animals, food, and so on but also with respect to many less tangible requirements which the naturalist can only guess at.

The quotation from Ford which appears at the head of this chapter was written from the point of view of the collector; but the ecologist who wishes to have a quantitative appreciation of the distribution and abundance of the species that he is studying must also strive to emulate "that rare being, the really accomplished naturalist"; he must learn the habits and requirements of the species that he studies, so that he can say for each one what are the sorts of places it can live in and why. In this chapter we shall discuss this problem from three aspects: for want of a better way to discriminate between them, we call these "relative," "absolute," and "quantitative" aspects. By the "relative aspect" we mean that certain sorts of places may be valuable chiefly because they provide shelter against the extremes of weather, or protection against a predator, or so on; in other words, the way they influence the animal's chance to survive and multiply depends on the influence of other components in the environment. This is discussed in section 12.2. On the other hand, the suitability of a certain sort of place may depend less on its interaction with other components of environment than on its own innate qualities and the behavior of the animal. This is discussed in section 12.1. But irrespective of whether the places where animals may live are chiefly important for their relative or their absolute qualities, it is always important to know how numerous such places may be and how they may be distributed in a particular area. This is what we mean by the "quantitative aspect," and we discuss it in section 12.3.

12.1 SOME EXAMPLES OF THE PLACES WHERE ANIMALS MAY LIVE AND THE SUBTLE BEHAVIOR OF ANIMALS IN CHOOSING A PLACE TO LIVE

Elton (1927, p. 39) wrote: "Most animals have some more or less efficient means of finding and remaining in the habitat which is most favourable to them . . . most animals are, in practice, limited in their distribution by their habits and reactions, the latter being so adjusted that they choose places to live in, which are suitable to their particular physiological requirements or to their breeding habits." As an example he mentioned how the African lion chooses its lair with great attention to a number of rather subtle requirements.

At the other extreme there are numerous insects and other small animals which merely produce a great superabundance of individuals with a powerful urge for dispersal. They launch themselves into the air and are wafted wherever the wind may blow (sec. 5.13). The extent of their freedom to choose a place in which to live would seem to lie in their capacity to recognize a good place when they happen to have alighted on it and to stay there if the place satisfies their requirements. The aphids, scale insects, and mites do this. The reputations which some of these species have as pests of horticultural crops bear witness to the efficacy of this means of finding a place to live. But most animals display more complex behavior than this in seeking a place in which to live.

The requirements of the small case-bearing moth *Luffia ferchaultella* (Psychidae) have been worked out fully by McDonogh (1939). It may be found in southern England, living on the trunks of trees and on stones where lichen is growing. The limits of its distribution are near to the 62° F. isotherm for July. It is found in some places where the mean temperature for July is below 62° F., but these are places with an unusually large amount of sunshine. It is absent from some places where the mean temperature for July is 62° F. or higher, but these are mostly places where the temperature during winter is unusually low. Within the area bounded by this isotherm, *Luffia* occurs from sea-level up to an altitude of about 400 feet, but not in higher places. Outside these limits there are trees and stones which seem in every other way suitable for *Luffia*, but none occurs there, presumably for some reason associated with temperature. McDonogh's explanation for the northern limit was that places where the mean temperature for July is below 62° F. are likely, during winter, to experience temperature low enough to kill *Luffia*, even those individuals which had found the best shelter.

The caterpillars thrive when they have as food the lichen *Lecanora*, which is composed of a green alga, *Pleurococcus*, and the hyphae of a fungus which was not identified. Any small area of bark chosen at random might have growing on it (*a*) neither *Pleurococcus* nor *Lecanora*, (*b*) either one or the other predominantly, or (*c*) both patchily. McDonogh counted the numbers of larvae in November and again in April on patches of bark supporting weak and strong growths of the alga and the lichen. In Table 12.01 the results are given as the average number of larvae on 6 square inches of bark. Although the caterpillars were present in low numbers in places where there was only *Pleurococcus*, they were numerous only in places where *Lecanora* predominated and was growing densely.

There were, however, certain sorts of places where *Lecanora* was growing densely which were characteristically devoid of *Luffia*. The height above ground-level and the position of the tree with respect to the edge of the wood are important. On the latter point McDonogh wrote: "The larvae are not found on trees in a wood if the tree is more than 20 yards from the edge, unless the

TABLE 12.01*

AVERAGE NUMBER OF LARVAE OF *Luffia ferchaultella* FOUND ON SAMPLE AREAS OF 6 SQUARE INCHES OF BARK ON HORSE CHESTNUT WHEN THESE SUPPORTED FOOD IN VARYING AMOUNTS AND OF VARYING QUALITY

NATURE AND QUALITY OF FOOD	AVERAGE NO. OF LARVAE IN 6 SQUARE INCHES		
	November	April	Mean
60% *Lecanora*, thick growth	2.2	2.8	2.5
80% *Lecanora*, patchy growth	2.0	2.4	2.2
90% *Lecanora*, thick growth	5.6	5.0	5.3
No *Lecanora*	0.3	0.7	0.5
80% *Pleurococcus*, thick growth	0.1	0.0	0.1
Both types present but scattered	0.3	0.6	0.5
Both types present and thick	0.5	0.6	0.6

* Data from McDonogh (1939).

undergrowth is very thin. Trees surrounded by thick bushes do not have the moth on them. . . . Besides this screening effect due to trees and undergrowth there is another caused by sudden rises in the ground-level. A group of trees situated on the top of a hill will more often than not be uninfested by the moth, though the trees are apparently suitable. . . . Gentle undulations do not affect the distribution. The optimum type of country for the moth is open park land such as at Richmond Park or Windsor Great Park, where there are plenty of trees situated in unscreened positions."

McDonogh also counted the number of larvae on the trunks of horse chestnut trees at various heights above the ground. The figures in Table 12.02 refer

TABLE 12.02*

MEAN NUMBER OF LARVAE OF *Luffia ferchaultella* ON TRUNK OF HORSE CHESTNUT TREES AT VARIOUS HEIGHTS ABOVE GROUND

HEIGHT ABOVE GROUND	MEAN NO. OF LARVAE PER 6-INCH SQUARE				
	North	East	South	West	Mean
6 inches	3.75	3.75	19.75	3.00	7.56
2 feet	0.75	2.50	7.50	0.75	3.00
4 feet	1.50	2.00	5.00	0.50	2.25
5 feet	0.50	1.75	3.00	0.75	1.50
Mean	1.63	2.50	8.81	1.30	3.58

* Data from McDonogh (1939).

to the mean number of larvae on 36 square inches of bark that was well covered by *Lecanora*. The larvae were most abundant near the ground and on the south side, and scarcely any were found more than 8 feet above the ground.

The absence of *Luffia* from the upper reaches of the trunk and from trees which are deeply shaded may be explained by their reactions to light. McDonogh's experiment with light is best described in his own words: "A set of green screens was made to cover part of a tree trunk, so that the light intensity over that part of the surface was reduced. They were in three degrees of density and are referred to here as light, mid- and dark green screens. At the beginning of each experiment a known number of active larvae was placed under each of the screens. After 24 hours the position of the larvae under the screens was noted. Two areas 4 and 6 inches square were marked under the centre of each screen. The number of larvae found in the squares was taken as a measure of the effect of the screens on the movement of the larvae compared with the movement of a similar set of larvae in a control area without the screens." The results of this experiment are given in Table 12.03.

TABLE 12.03*

NUMBER OF LARVAE OF *Luffia ferchaultella* (AS PER CENT OF TOTAL) WHICH REMAINED IN EACH LEVEL OF ILLUMINANCE 24 HOURS AFTER BEGINNING OF EXPERIMENT

ILLUMINANCE (AS RATIO OF FULL LIGHT IN OPEN)	TYPE OF SCREEN	NO. OF LARVAE (AS PER CENT OF TOTAL)	
		4-In. Square	6-In. Square
0.023	Dark green	7.4	37.2
.047	Mid-green	31.8	56.2
.070	Light green	37.8	60.4
0.117	None (control)	26.8	56.2

* Data from McDonogh (1939).

The elm hardly ever provides a favorable place for *Luffia* to live. The willow and the oak nearly always harbor at least a few, and often quite dense populations live on them. Sometimes the beech and the horse chestnut also provide good places for *Luffia* to live. Young trees of any species which have not attained a girth of about 8 inches rarely support many *Luffia*. The differences between species and the inadequacy of young trees may be associated with the requirements outlined above, but there is also the matter of crevices which are required by the pupae. "While the maximum number of larvae appears to be usually very near to the ground, the pupae tend to be higher up the tree. Irregular distribution is commoner on smooth bark than on well-creviced trees. The amount of alga and lichen controls the distribution of the larvae, but the pupae tend to be more affected by the distribution of cracks in the bark which act as places for pupation."

This unusually thorough documentation of what a particular species requires from the places where it can live shows how subtle these requirements may be. Recapitulating, we note that *Luffia* can live in places where (*a*) the mean temperature during July is not below 62° F. and the temperature during the winter is not low enough to kill the larvae; (*b*) there is an abundance of a particular lichen, but not if the algal component is present alone; (*c*) there is the right amount of light; (*d*) there are suitable crevices for pupation. And, finally, even places having all these attributes have little chance of being occupied unless they happen to be within a mile or two of a population of *Luffia*, for the adult of this species is wingless. The larvae disperse by floating in the air with the aid of silken threads, but McDonogh considered that about 1½ miles was about the greatest distance that they were likely to travel.

Many holometabolous insects spend most of their lives in or on the plant or animal (or debris such as log, carcass, or dung) which serves both as food and as a place in which to live; and it is scarcely practicable with them, during this stage of their life-cycle, to analyze these two components of environment independently. Some species may leave the place where they have been feeding and seek another sort of place when the time comes to pupate, but other species may not. It is interesting that, in species with a facultative diapause intercalated into a multi-voltine life-cycle, the diapausing and nondiapausing individuals usually seek different sorts of places. The former often go farther or search more persistently, with the consequence that they end up in more sheltered places in which they spend the winter, or the summer, as the case may be. This can be readily observed in the codlin moth *Cydia pomonella;* in this species one or two generations in which few individuals enter diapause may be succeeded by a generation in which all enter diapause. Diapause, when present, occurs at the close of the final larval instar. The inception of diapause is largely determined by photoperiod. Those larvae which have experienced long days as they were developing are satisfied to spin a flimsy cocoon under almost the first piece of bark they find, no matter how inadequate the shelter it provides; they pupate at once and emerge as adults without delay. Those larvae which have experienced short days as they were growing show great persistence in pushing or chewing their way into tight cracks or crevices in the bark, where they spin a heavy, dense cocoon in which they spend the winter in diapause.

The beet webworm *Loxostege sticticalis* in Montana has one or two generations during summer, and the winter is spent as a diapausing prepupa. Exceptionally, a small proportion of the second generation may go on to produce a third generation in the one summer (Pepper and Hastings, 1941). The proportion of the first generation which enters diapause may vary from less than 1 per cent to more than 60 per cent. The larvae, on reaching maturity, wander away from the plant and burrow several inches into the soil, constructing a strong silk-lined cell. Those which are going to pupate without delay usually choose loose soil for this purpose, but those which have been determined for diapause tend to wander farther and usually seek out firm soil or sod in which to construct their cells; this is irrespective of what generation they may belong to.

The mite *Bryobia praetiosa*, living on apple trees in South Australia, passes through several generations during summer and then spends the winter as a diapausing egg. All nondiapausing eggs are laid on the backs of the leaves, but all diapausing eggs are laid on the main branches of the tree. It is likely (by analogy with other species) that the mite responds to photoperiod. The striking feature is that exposure to a certain length of day during the early stages of the mite's development determines not only the sort of egg which is to be laid but also the behavior of the mite in seeking a place to lay it. The survival-value of the adaptation is clear, but this makes it no less remarkable.

Fisher and Ford (1947) studied a colony of the moth *Panaxia dominula* (Arctiidae) which occupied about 20 acres of fenlike marsh at Dry Sandford near Oxford. Part of the marsh was wooded, but most of it was covered by reeds and herbaceous plants, including comphrey, *Symphytum officinale*, which was the chief food of the larvae, and several sorts of nettles, which were also suitable for food. The marsh was bounded by woodland and agricultural land, into which the moths seemed never to penetrate, notwithstanding that they flew powerfully, were often observed circling around and above trees in their chosen area, and certainly were quite active in dispersing throughout the 20 acres in which they lived. There was another small area at Tubney, about 1½ miles away, which was suitable for *Panaxia*, but Fisher and Ford found clear evidence that very few, if any, moths found their way from Dry Sandford to Tubney (sec. 5.01). Moreover, entomologists have collected in this vicinity for many years, and their testimony (reliable because *Panaxia* is a large brightly colored day-flying species) confirms that *Panaxia* is not to be found straying beyond the confines of the specialized territory where the colony lives.

This unmitigated adherence to such a small area (in the case of Dry Sandford a mere 20 acres) on the part of a large moth capable of powerful flight indicates a keen response to the boundaries of the area where there are suitable places for it and its larvae to live and an altogether remarkable inhibition of the usual tendency for at least some individuals to disperse widely from the place where they originated (sec. 5.6).

McCabe and Blanchard (1950) made a thorough study of the ecology of three species of deer mice, *Peromyscus maniculatus*, *P. californicus*, and *P. truei*, in an area of about 25 square miles near Berkeley, California; their report is especially interesting for the information they give about the sorts of places where these three species may live. The area studied was an outlying ridge of the Contra Costa hills, with a general elevation of about 1,500 feet and with several peaks approaching 2,000 feet above sea-level. The ridge was dissected by many valleys (canyons), the floors of which might be as much as 1,000 feet below the crests of the adjacent hills. Several different types of vegetation were recognized, and their distributions seemed to indicate that they required different amounts of moisture. The transition from one type to another was often abrupt, and the boundaries were thus often clearly defined. The crests of the moister hills carried woods of *Pinus* or *Eucalyptus*. Grass and *Artemisia* occupied the drier hilltops. The stream beds were lined by narrow bands of moisture-loving trees, *Umbellularia californica*, and oaks also occurred at the lower levels. The hillsides, especially the moister aspects, were characteristically covered by a dense tangled scrub known locally as "chaparral." In different parts of the area the chaparral abutted on all the other vegetation types, and often the transition from chaparral to forest, to grassland, or to woodland was abrupt, giving a well-defined edge to the chaparral. Artificial margins to the chaparral also occurred, perhaps along an old road or a track or some other sort of clearing.

Throughout the whole area the mice were restricted to the margins of the chaparral, whether natural or artificial. Assiduous trapping in the body of the chaparral failed to discover a single mouse; no residents were found in the wooded areas or in the grassland or, indeed, anywhere else except along the very margins of the chaparral. In some places, especially on the more arid slopes, the margins were poorly defined, as the chaparral gradually thinned out and became invaded by grass or *Artemisia;* at the same time, some plants of the chaparral would penetrate the grassland or the area occupied by *Artemisia*. From these and other places where the margins to the chaparral were poorly defined, *Peromyscus* was absent also.

The places most favored by *Peromyscus maniculatus* occurred on the moister slopes, where a luxurious growth of chaparral, composed chiefly of *Bacharis*, gave way abruptly to a narrow band of giant herbs, chiefly Umbelliferae, which, in turn, gave way to grassland. The ground under the *Bacharis* was bare except for a thin covering of dead leaves from *Bacharis*. This obviously suited *P. maniculatus*, which has the habit of seeking shelter in holes and small crevices in the ground and is not able to make very good progress running over a rough surface; but there were doubtless other more subtle qualities which made these the most suitable places for *P. maniculatus* to live. Elsewhere the chaparral manifested clear-cut margins where it met forest or woodland or even a glade with a few trees in it. These margins also provided suitable places for *P. maniculatus;* but they were less likely to be occupied than the other sort of margin, and when they were occupied, the populations were less dense. Especially in the wetter spots, the ground under the chaparral where it joined forest or woodland was covered with a "duff" of dead leaves, accumulations of fallen twigs and branches, and a tangle of low herbage. The presence of the duff on the ground was disadvantageous to *P. maniculatus*, but doubtless there were other ways in which this sort of margin failed to meet the requirements of this species.

The other two species seemed never to seek shelter underground but always in the debris of vegetation on the surface. Consequently, places where the ground was covered with duff suited them quite well. They were most likely to be found and their populations more likely to be dense along the edges of the chaparral, where it met woodland of *Pinus*, *Eucalyptus*, or *Quercus*, or around the edges of a glade. It was characteristic of the places favored by *Peromyscus truei* and *P. californicus* that the chaparral would be composed of a greater variety of species, its margins would be more sinuous and less linear than those which were favored by *P. maniculatus*, and the ground would be covered by duff. Their requirements were more stringent than those of *P. maniculatus*, for they were always very few except in the places that suited them precisely. On the other hand, *P. maniculatus* was almost ubiquitous, being present, if not abundant, wherever the chaparral manifested a well-defined edge.

In some of the areas best suited to *Peromyscus maniculatus* this species contributed more than 95 per cent of all the mice captured. In the places which were best suited for *P. truei* and *P. californicus*, these two might contribute as much as 75 per cent of the total catch, with *P. maniculatus* making up the rest. The ratio of *P. truei* to *P. californicus* varied from 3:1 to 9:1. No place was found where the population consisted exclusively of one species. The places where the relative numbers of the three species were most nearly equal were also the places where the absolute numbers were lowest.

In a number of experiments the trapped mice were destroyed instead of being released in the same place. This never resulted in any change in the relative proportions of the different species in that area. From these and other observations McCabe and Blanchard concluded that neither competition nor any other sort of interaction between the species was important in determining either the absolute numbers of any species or the relative numbers of the three species in an area. On the other hand, McCabe and Blanchard observed what was at least the rudiments of territorial behavior in these mice, and this, in relation to the number of good places to live, seemed to be what chiefly determined the density of the population of each species. Since any particular area seemed to carry a characteristic number of each species, it may be safely inferred that each species had its own special requirements with respect to the place where it would live.

The spotted skunk *Spilogile interrupta*, like other small vertebrates, requires in its territory one or a number of "dens." Crabb (1948) described a den as "any location or cover which the animal uses of its own free will for rest or seclusion." He studied the sorts of places which were used as dens and concluded: "The first consideration seems to be the exclusion of light. Without exception every den or semblance of a den met this requirement. Sometimes the den was only a place to curl up or stretch out in, such as under the corner of a well platform or a shock of small grain, but wherever darkness prevailed there the spotted skunk seemed to feel most at home." The den also provides shelter from the weather and protection from predators, including men and domestic animals. Crabb considered that the number of places (hollow logs, burrows, spaces under woodpiles, and so on) which would meet these requirements around the farms and fields of southern Iowa were almost unlimited. But, of course, some would be very much better than others.

In certain months of the year the number of rats in residential blocks in Baltimore was found to be proportional to the amount of broken paving, which is a measure of shelter available. At other times shelter was not limiting, but food or some other component of environment was (Davis, 1953).

Leopold (1933) described the qualities required in a "deer range" in the Lake states. The deer require a cedar swamp for "yarding" during deep snow; for hiding and sleeping, they need evergreen thickets, preferably on the point of a saddle on a hill; for play, open places are needed; and for fawning, the doe must be near water where she can satisfy her thirst during nursing without having to travel too far. Leopold made the generalization that "a range is habitable for a given species when it furnishes places suitable for it to feed, hide, rest, sleep, play, and breed all within reach of its cruising radius." Birds often include in their requirements not only a suitable site for a nest but also places for singing, for roosting, and, in resident species, a secure place for sleeping during winter. The crossbill and the tree pipit require tall trees for singing posts; the nightingale and the wren sing from vantage points much nearer to the ground (Lack and Venables, 1939). The special requirements of the bobwhite quail and the muskrat have been studied quite thoroughly, but we shall leave the discussion of these two species until section 12.3.

12.11 *Specialized Behavior Associated with the Choice of a Place To Lay Eggs*

With insects it is usually, though not always, the adult which is specialized for dispersal, and it is in this stage that are found the most remarkable adaptations for seeking out the right sorts of places to deposit eggs. In striking contrast to *Panaxia*, which remains so firmly attached to its home territory (sec. 12.1), the two butterflies *Danaida archippus* and *Pieris rapae* are remarkable for their widely ranging flights in search of plants on which to oviposit. The larvae of *Danaida* live and feed on the cotton bushes (*Asclepias* spp.). In southern Australia these shrubs grow on roadsides, in parks, and on wastelands; they are distributed widely but often quite sparsely. Yet it is unusual to find one during summer that does not carry at least a few larvae of *Danaida*. Similarly, the larvae of *P. rapae* are found only on plants of the family Cruciferae; and one is impressed by the high probability that even remotely situated plants of this family will be found and oviposited on by this widely ranging butterfly (see also sec. 11.34).

The mosquito *Anopheles culifacies* lays its eggs in rice fields when the rice is not very high; when the rice plants reach a height of about 1 foot, the mosquitos disappear. The female while laying eggs flies a tortuous course just above the surface of the water. If it is prevented by the presence of tall rice plants from doing this, then it will not lay eggs in that place. An otherwise suitable place can be made quite unsuitable by simply inserting all over it glass rods which project about a foot above the surface (Macan and Worthington, 1951). There is doubtless survival-value in this odd behavior; it would be interesting to know whether this habit prevents the mosquito from laying eggs in places where the larvae would have little chance to survive. In North America there are 8 species of mosquitos whose larvae are found only in rain-filled rot-holes in trees (Jenkins and Carpenter, 1946). There are 3 species in Britain similarly restricted to tree-holes (Elton, 1949). It has been suggested that the gravid females of these species are attracted to the holes and stimulated to lay their eggs in them by the presence of some organic substance which is absent from water accumulated in other sorts of places. But this is not known; nor is it known what other qualities make the tree-hole a suitable place for these species to breed in.

The larva of the beetle *Lyctus brunneus* lives in freshly seasoned timber, provided that it contains a sufficiency of starch and not less than 8 per cent of moisture. Starch is absent from the true wood of all sorts of trees, but it is present, often in adequate concentrations, in the sapwood of certain broad-leaved species. The beetles seek a place where the vessels of the xylem (which run in the sapwood) have been cut across or otherwise exposed, and place their eggs inside the cavities of the vessels. Apart from rare accidents, the eggs are never laid anywhere else. The eggs of *Lyctus brunneus* are about 180 μ wide. The beetles usually choose vessels about this size; they will not put eggs in wide vessels where they would be a loose fit, and, of course, they cannot put them in vessels that are narrower than the egg. Consequently, the timber of coniferous trees and the true wood of broad-leaved species do not have any eggs laid in them, nor does the sapwood from broad-leaved trees, in which the vessels of the xylem are too narrow. The first two would not contain enough starch to support the larvae, but the last may; in that case its unsuitability as a place for *Lyctus* to live depends entirely on the behavior of the beetles in refraining from ovipositing anywhere except in vessels of such a size that the ovipositor can just be neatly pushed into them (Parkin, 1934; Gay, 1953).

Frogs of the species *Rana temporaria* in England emerge from hibernation some weeks before spawning begins. They frequently leave the pond where they hibernated and travel, often quite a considerable distance, to another pond to spawn. They usually place their eggs in a few restricted areas within the pond. The whole phenomenon is rather puzzling because often the abandoned pond looks quite suitable, and the new pond often appears far more uniform than the patchy distribution of eggs would indicate: just what it is that makes a place suitable for *Rana* to lay its eggs remains a mystery (Savage, 1934, 1935).

Birds usually have specific requirements for nesting sites and nesting materials. In Britain certain species of the Corvidae and Falconidae nest only in the canopies of tall trees and may be found only in woods which contain at least one tall tree; each species is characterized by a minimal height below which it will not build. But the nightingale and the wren tend to keep in the secondary growth, well below the canopy, for all their activities during the breeding season. Many species of tits, woodpeckers, and other birds of the woodland require holes in trees for their nests. That is why they are often scarce or absent from young woods and more common where there are old trees. The scarcity of jackdaws and starlings in woods composed entirely of coniferous trees is probably due to the absence of good holes for nesting in these woods. The absence of the nightingale from coniferous forests has a different explanation: it requires the dead leaves from deciduous trees for its nest (Lack and Venables, 1939). In California and Arizona there are two species of woodpecker which make nests by carving a hole in the large upstanding stem of the giant cactus *Cereus giganteus*. In the absence of the cactus, they will also make nests in trees. The elf owl *Microphallus whitneyi* nests exclusively in the abandoned nests of the woodpeckers in the cactus, never in any other tree. This is often quoted as an example of the dependence of one species on another (Elton, 1927, p. 48; Allee *et al.*, 1949, p. 362), but it would seem that none has described the subtle difference that the owl discerns between an abandoned woodpecker's nest in *Cereus* and one in some other kind of tree.

12.12 *Some Clues from Studies of "Succession"*

Studies of the "ecological succession" which may go on in such habitats as a fallen log, a carcass, or a heap of dung belong properly to community ecology. Unfortunately for our purpose, they rarely indicate, except in very broad terms, what any one species requires. We mention several such studies which do indicate how the different species in the succession may have quite different requirements of the places in which they live.

Savely (1939) studied the animals that lived in rotting logs of pine and oak in a forest in North Carolina during successive stages in the logs' decomposition. The first species to enter the fallen pine logs were 22 species of beetles, chiefly Cerambycidae and Scolytidae; these lived and fed almost exclusively in the phloem. The bark soon became loosened from the wood, and the space between wood and bark became packed with chewed-up wood. This favored the growth of fungi. Once the fungi had become established, the log became habitable for a further 37 species of insects and mites which fed on fungi and decayed wood. The greater number of species (29 out of 37) were Coleoptera, but the species having the greatest numbers of individuals were Collembola, Acarina, and Diptera. By the second year the phloem-feeding species had disappeared, the wood was becoming soft and was invaded by 11 new species, among which the termite *Reticulitermes flavipes* was prominent. After this, no marked changes occurred in the occupants of the log as it gradually became incorporated into the soil.

Mohr (1943) made a study of "succession" among animals living in cattle droppings at Urbana, Illinois. Altogether, 150 species were found in the dung at one stage or another from when it was fresh to when it finally became incorporated into the soil. The particular insects breeding in the dung at any stage were largely determined by the microörganisms which decomposed it; but the animals also played a part in determining the "succession." The maggots of *Cryptolucilia* and *Sarcophaga* produced galleries and openings in the dung which were later used by certain staphylinid beetles. It seemed that the beetles depended upon the galleries made by the previous inhabitants, for they used them continuously without making any new ones for themselves.

Woodroffe and Southgate (1950) described the animals which lived in the nests of sparrows in England. At first, while the nest is being more or less permanently occupied by the birds, the other inhabitants are nearly all ectoparasitic arthropods, insects, ticks, and mites. When the nest has been abandoned by the birds, it is invaded by scavengers; these are largely moths of the families Tineidae and Oecophoridae, beetles of the families Ptinidae and Dermestidae, certain silverfish, and mites. Many of the species are the same as those found in warehouses and stores, feeding on grain and other sorts of human foodstuffs. After a period the nest decomposes into a humus-like mass, and the foregoing species are supplanted by another group, largely composed of the sorts of animals that are usually found in soil. A similar sequence may be observed in a carcass (Fuller, 1934; Waterhouse, 1947; see also sec. 10.31). Some examples of succession of animals associated with plant succession are summarized by Odum (1953, pp. 190–94).

12.2 SOME EXAMPLES OF THE INTERDEPENDENCE OF "PLACE TO LIVE" WITH OTHER COMPONENTS OF ENVIRONMENT

12.21 *Temperature*

For many species of terrestrial invertebrates which live in the cool, temperate regions of the Northern Hemisphere the size of the population which resumes activity in the spring may depend largely on the number which has been able to survive exposure to cold during the winter. Special adaptations, such as diapause and the cold-hardiness associated with it and the tendency to seek well-protected situations in which to spend the winter, are important. But some individuals find better winter quarters than others, and the survival-rate may depend partly on the severity of the winter and partly on the proportion of the population which happens to find adequate quarters. Salt (1950, p. 285) expressed this principle well: "Low winter temperatures are a limiting factor in the survival of many species of insects and often restrict their geographical range. Although the few species which hibernate in very exposed situations can survive the formation of ice in their tissues, the remainder are killed if they become frozen. The latter depend for their survival on the insulating protection of their hibernacula (soil, plant debris, snow, ice, etc.) and on their ability to undercool. The combination of these two protective factors may be sufficient that mortality as a result of freezing is a rarity in some species. More often the protection is incomplete, with the result that those individuals in the more exposed hibernacula and with the least ability to undercool perish. In severe winters this fraction may be temporarily large and may seriously deplete the population. If a species has been expanding its range into colder regions, one severe cold period can eradicate it in such areas, restricting it once again to its normal range."

Mail (1930, 1932) and Mail and Salt (1933) measured the temperature of the soil at several depths in Minnesota and Montana and related this to the probable survival-rate of certain insects which spend the winter in the soil.

From experiments in the laboratory with cold-hardy adults of the Colorado potato beetle *Leptinotarsa decemlineata*, Mail and Salt (1933) concluded that even a brief exposure to −12° C. would kill virtually all the beetles; that more than half of them would die if they were exposed for some hours to −7° C.; and that even prolonged exposure of days or weeks to −4° C. would kill very few beetles. From their own experiments and the observations of others, they concluded that the depth to which the beetles burrowed into the soil at the onset of winter depended on the soil; in loose, sandy soil the majority would be between 14 and 24 inches, but in harder soils they might all remain in the top 8 inches, concentrated just above the "plow line."

The temperature experienced by the beetles during winter depends not only on their depth below the surface of the soil and the prevailing temperature of the air but also on the amount and permanence of snow lying on the ground during the cold weather. Snow is a good insulator, and the presence of 6 inches of snow on the ground results in quite moderate temperatures in the soil below it (Figs. 12.01 and 12.02). The data for Figure 12.01 (*lower curve, with snow*) were collected at Bozeman, Montana, on February 9, 1932. Although it was an unusually cold day, beetles that were hibernating more than 12 inches below the surface would have been quite safe. Most of those which had remained in the top 12 inches, either fortuitously or because they had been prevented by hard soil from burrowing more deeply, would have been in danger during this cold spell. The data for Figure 12.01 (*upper curve, no snow*) were collected at St. Paul, Minnesota, on January 1, 1928, in soil from which the snow had been scraped away; the weather was not unusually cold. In the absence of snow, the soil was much colder, and only those few individuals which happened to have gone deeper than 2 feet would have had much chance to live through the winter.

In the regions where Mail and Salt worked, the occurrence of warm "chinook" winds may cause the snow to disappear quite suddenly even in midwinter. Then the temperature of the soil down to a depth of several feet may fall quite dramatically. Figure 12.02 illustrates one such occurrence; beetles hibernating in the top 2 feet of soil would have had little chance of surviving this cold spell. On the other hand, Mail (1932) observed that during the winter of 1931–32 snow remained almost continuously from December 1 to March 31 on the area where he had his instruments, and the temperature of the soil even near the surface fell no more than a few degrees below zero. At no depth was it cold enough to kill many of the overwintering adults of *Leptinotarsa*.

This species is permanently established in Montana, and it must be concluded that during the more severe winters the population is carried forward because at least some individuals happen to find winter quarters in specially

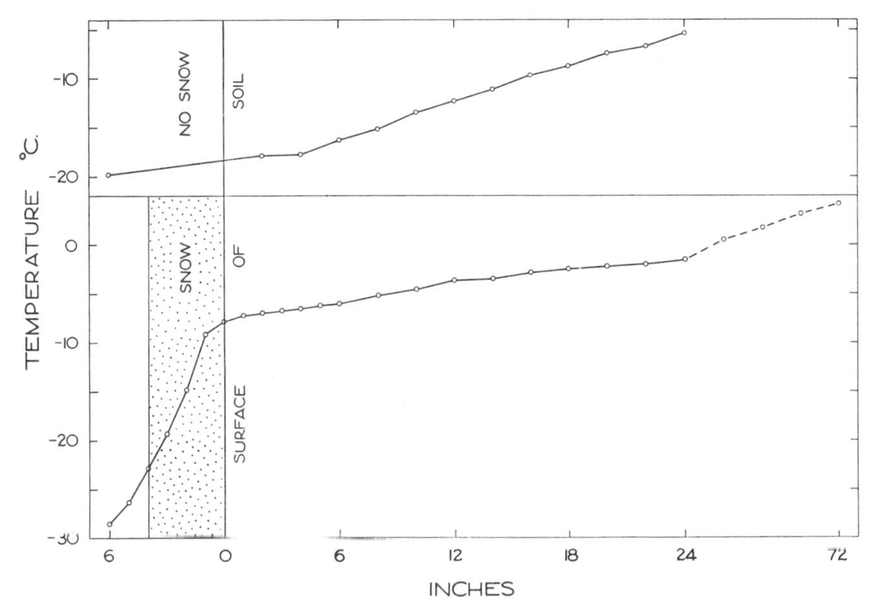

FIG. 12.01.—Temperature of the soil at various depths below the surface: *upper curve*, in the absence of snow; *lower curve*, when snow was present. Note the steep gradient through the 4 inches of snow and the consequent less severe temperature in the soil below the snow. (After Mail, 1930.)

favored situations. The beetles may survive near the surface in places where the snow is more permanent; in more exposed places they have little chance of surviving unless they are buried deeply in the soil. The beetle has not extended its distribution northward into Alberta beyond about the 54th parallel, despite the presence of plenty of food there. This is probably because in this region the temperature of the top 2 feet of soil is likely to fall below −7° C. at least once during most winters (Mail and Salt, 1933).

The interaction between temperature and place to live is nicely illustrated by these studies. The important qualities of the place where *Leptinotarsa* may spend the winter are the nature of the soil, which may determine how deeply the beetles will burrow, and the amount and permanence of the covering of snow. At Bozeman (latitude 45°) *Leptinotarsa* may have a good chance of surviving the winter in situations where the snow is permanent during the winter; in other places, only those individuals which happen to be buried more than 2 feet below the surface have much chance of surviving the winter. At Beaverlodge (latitude 55°) the prevailing atmospheric temperature is lower, and even the most sheltered place may be too cold for *Leptinotarsa*.

During 1924 the noctuid *Euxoa segetum* was abundant in the vicinity of the Lower Volga, doing much damage to crops. It is unusual for this species to be so abundant in this area, though outbreaks occur rather frequently in adjoining districts. Sacharov (1930), in seeking an explanation, explored several hy-

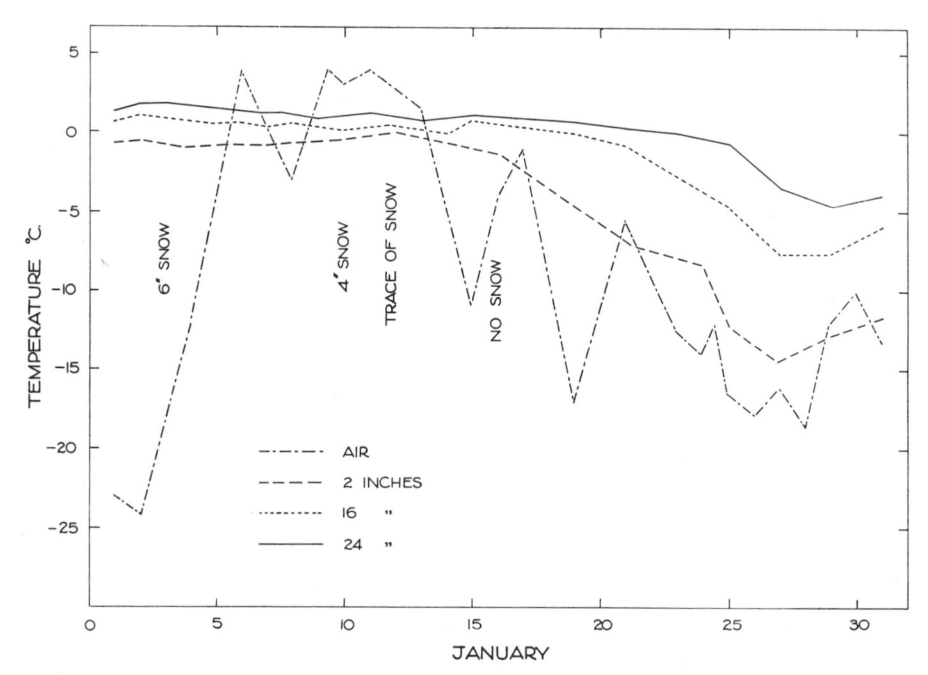

FIG. 12.02.—Temperature of the air and of the soil at several depths below the surface of the soil. Note how, at first, the presence of snow kept the soil warm during a very cold spell of weather and how, after the snow had disappeared, the temperature of the soil fell. (After Mail, 1930.)

potheses relating to food, predators, and the weather during summer but found that none was consistent with the facts. He then examined the survival-rate of overwintering larvae in relation to records of atmospheric temperature, the amount and permanence of snow, and the depth to which the larvae burrowed into the soil; and he found a nice explanation in terms of the interaction between temperature and the places where the larvae spent the winter. There are two generations of *Euxoa* during summer. The larvae of the second generation, maturing during autumn, accumulate large reserves of fat, and their tissues contain relatively little water. Such larvae burrow into the soil to a depth of 6 or 8 inches, often well before winter. But other more tardy individuals of the second generation require to continue feeding later, and these may be trapped near the surface and killed by exposure to cold during winter. Even the others which are relatively cold-hardy because they are in diapause (sec. 6.314) and relatively well protected because they are deeper in the soil may be killed by exposure to cold if the soil does not remain covered by an adequate depth of snow. In one experiment Sacharov scraped the snow away, so that the surface of the soil remained bare, and found that 77 per cent of the diapausing larvae which were overwintering about 7 inches below the surface were killed by the cold.

In the laboratory Sacharov found that diapausing larvae were not harmed by 18 hours' exposure to −6° C. but were killed by exposure to temperatures below −8° C. He related these findings about the behavior and physiology of *Euxoa* to his observations of their distribution and abundance. His conclusions are best given in his own words:

> If we study, further, the January isotherms of the region, we will see that the regions most affected by the insect are the coldest, but at the same time they have the deepest snow cover, which serves to protect the hibernating caterpillars from frosts. In the south-eastern steppe and semi-desert districts winter temperatures are higher than in the forest-steppe zone; but the very thin snow-cover and continual dry east winds cause freezing of the soil to a greater depth, resulting in the killing of caterpillars by autumn and winter frosts.
>
> An example of the fluctuations of soil temperature in connexion with the snow-cover is presented by the observations made in Saratov district for the last three years. During the winter of 1923–24 the snow cover was 12.6 cms. in December, 19 cms. in January, 29.6 cms. in February, and the soil temperature did not drop below −5.5° C. During this winter the deep snow cover and the moderate freezing of the soil were favorable to the hibernation of the caterpillars, and, as a result, very heavy infestation and damage were recorded in the summer of 1924. In the next winter, 1924–25, the snow cover during December, January and February did not exceed 4.5 cms. for each of these months, and the soil temperature at a depth of 25 cms. dropped in December to −12° C., that is, to a level at which the majority of the caterpillars, according to our laboratory experiments, should die. Indeed the damage recorded in 1925 was very slight. Since no parasites or diseases of caterpillars have been observed, this reduction in numbers can be ascribed only to the deep freezing of the soil in December. Further, the winter of 1925–26 was similar to that of 1923–24, as far as the soil temperature was concerned, but the snow cover was as in the winter of 1924–25. One might have expected an outbreak of *Euxoa* next summer, but this did not happen, simply because more than one year is necessary for the insect to increase in numbers to the extent of becoming a pest.

The temperature inside plants or on the surface of leaves, bark, etc., where small animals may live may be quite different from that of the surrounding air. The temperature of one leaf may vary quite markedly from that of another, depending on their exposure to radiation and wind (Wellington, 1950). All stages of the beetle *Dendroctonus brevicornis* live inside "galleries" which they cut into the bark of pine trees. Miller (1931) inserted thermometers into the bark of a tree on its southern and northern aspects. The bark on the south side was warmer and that on the north side cooler than the air in the shade, and the difference between the two was about 8° C. The insects which happened to be living on the south side might be expected to develop more rapidly than those on the north side. Henson and Shepherd (1952) inserted thermocouple junctions into the mines made by the caterpillars of *Recurvaria milleri* in the needles of lodge pine. During the day when the sun was shining, the temperature inside the leaf was as much as 7° C. higher than that of the air in the shade. There was a close linear relationship between the amount of radiation falling on the leaf and the difference between the temperature of the leaf and the air. This depended not only on the extent to which the leaf was shaded but also on the angle made by the leaf with the sun's rays. The temperature of leaves in the most shady situations differed little from that of the air in the shade, so there was a difference of as much as 7° C. in the temperature experienced by different individuals of *Recurvaria*, depending on whether they happened to be living in leaves that were well shaded or exposed to much radiation.

Uvarov (1931) suggested that bubonic plague survived in the steppes around the Caspian Sea because the fleas on the ground squirrel *Spermophilus* were protected against the cold of winter and the dry heat of summer in the burrows of the squirrel. Similarly, Buxton (1932*b*) found that rat holes in Palestine provided a suitably cool and moist place in which *Xenopsylla cheopis* could live, although the temperature and humidity recorded at midday in meteorological screens near by would have been fatal, at least to the larvae.

12.22 *Moisture*

In warm, temperate zones the hot, dry summer may be the most hazardous season for small terrestrial animals; and we can recognize, in the species from these regions, adaptations which lessen the dangers from desiccation. Diapause is important (sec. 4.6), and so is behavior in seeking the most favorable places to aestivate. Still, chance plays a large part, and an unusually severe summer may kill all but the few individuals which happen to be in the best places. This happened to the grasshopper *Austroicetes cruciata* in parts of South Australia during the summer of 1938–39.

This grasshopper aestivates as a diapausing egg. The females can, but in nature they rarely do, lay their eggs in soil that is loose or soft. Instead, they invariably choose soil that is hard and compact; bare, windswept patches are favored, and they have even been observed boring into the edge of a macadam road. The female seems to experience difficulty in starting the hole unless she can keep the tip of her abdomen pressed firmly against the surface of the soil. A female may often be seen to back up to a suitable obstacle, such as the side of a stone or a small bush, flex her posterior legs upward, grip the stone about half an inch from the surface of the ground, and use this leverage to press the tip of the abdomen firmly against the surface of the soil while starting to bore. The result is that many eggs are laid around the edges of such obstacles, though many are also laid out in the open; minute cracks and other minor irregularities in the soil may be exploited in the absence of stones or small bushes.

During the summer of 1938–39 some districts where *Austroicetes* occurs went for 89 days with virtually no rain; this was an extraordinary drought. In one district where it was estimated that at least 90 per cent of the eggs had died, none was found alive except among those that had been laid around the edges of large, flat stones. Many had died even in these situations, and it was specially noticeable that, with very few exceptions, the eggs in any one pod (usually about 20) were either all dead or nearly all alive (Andrewartha, 1939; and unpublished notes). The inference is quite plain. The only eggs which had any chance of surviving this drought were those which had been laid around the edges of stones large enough to provide "runoff" from the few scant showers of rain that occurred; even so, only those few survived which, by virtue of local vagaries in the contours of the stone or the soil, received rather more than their share of the small amount of water thus provided. Farther south, where the drought was less severe, the survival-rate varied from 36 to 78 per cent, but it was still noticeable that small local differences in topography largely determined which eggs should survive and how many.

The distribution of the cricket *Gryllulus commodus* in South Australia is strikingly limited by the distribution of the type of soil known as "rendzina." In this region these soils mostly occur in long narrow strips, because they form the floors of narrow valleys separated by low, sandy hillocks which were once coastal sand dunes. These soils support permanent populations of *Gryllulus*, because, with the weather experienced in this region, the eggs have little chance of surviving unless they are laid in a soil with a very high water-holding capacity. The full explanation for this was worked out by Browning (1952, unpublished thesis).

With *Gryllulus* the life-cycle occupies one year. The eggs are laid during April or May (southern autumn); diapause-development is completed early in winter, but the eggs do not hatch until early summer, usually during November. In some years a high proportion of the eggs give rise to nymphs, but more often only a few survive in the most favorable situations. Most of the deaths occur during spring, when the eggs are nearly ready to hatch. The chief cause of death is desiccation due to the drying-out of the soil around the eggs.

Compared to hardy eggs such as those of *Austroicetes cruciata* (sec. 7.233), the eggs of *Gryllulus* are poorly waterproofed. Browning found that none hatched unless they were kept in contact with free water or in an atmosphere saturated with water vapor; in one experiment 1 day's exposure to a relative humidity of 90 per cent at 29° C. killed 75 per cent of the eggs. In nature the eggs are laid about $\frac{1}{4}$ inch below the surface of the soil, and Browning's experiments show that an egg is unlikely to survive if the soil around it dries out to the "wilting point" even for 1 day.

The climate in this region is of the Mediterranean type; that is, the winter is mild and humid and the summer is warm and dry. The soil remains continuously wet during winter but dries out, at least on the surface, for several months during summer. The chief hazard for the eggs is that the surface layer of soil in which they occur may become dry before they hatch. The risk is less in soils of high water-holding capacity. The rendzinas are heavy black soils, very sticky when wet, with a high proportion of clay and organic matter near the surface; they have a high water-holding capacity.

Occasionally the crickets become very numerous and do a lot of damage to pastures; this is likely to happen after 1 or 2 years in which the rainfall during winter was high, especially if it lasted well into the spring. During the drier, more normal years, when most of the eggs die without hatching, the survival-rate varies widely from place to place. This is associated with variations in the quality of the soil. The rendzinas in this region are highly variable, especially with respect to the important quality of water-holding capacity, which is associated with heaviness and blackness. In the more severe years the higher survival-rates are associated with the heavier and blacker patches of rendzina. In the kinder years the survival-rate may also be high on the lighter variations of this soil type, though scarcely ever on any of the other soil types which occur in this region.

Although the association of *Gryllulus* with the rendzinas is close and almost general, the few exceptions are important because they confirm the theory. Browning found one area of several acres in which the crickets were thriving on a light soil (meadow podsol); the local topography was such that a high water table kept the surface soil in this area moist even in the absence of rain. Elsewhere *Gryllulus* occurs in numbers on almost any sort of soil in gardens around dwellings where the soil may be watered during summer.

In much the same region where *Gryllulus* is found, the distribution of the moth *Oncopera fasciculata* is virtually restricted to the lighter soil types, scarcely ever being found on the rendzinas. The soils on which *Oncopera* is found may be broadly grouped into volcanic soils and meadow podsols. The former are mostly deep, friable, and well drained; they may dry out rather severely during summer but never become excessively wet, even during the wettest of winters. The latter are mostly found in lower situations; the subsoil is heavy and the surface soil may become water-logged during winter.

The life-cycle of *Oncopera* occupies 1 year. The moths are present for a few weeks early in spring; the eggs hatch during October or November, and the larvae are present from then until late in the winter (July–August). The larvae establish themselves in vertical burrows in the soil, from which they emerge during the night to feed. They remain relatively dormant during the summer, which is dry, and feed and grow most actively during the winter, which is mild. A large proportion, amounting at times to virtually the whole population of an extensive area, may die during the summer from excessive dryness or from drowning during winter. The risk of desiccation is considerable for the larvae in both the volcanic soils and the meadow podsols but is greater for the former. The risk of drowning is negligible for larvae on volcanic soils but is quite great for those on meadow podsols. Notwithstanding these twofold risks, exceptional weather may allow a high survival-rate on either sort of soil, and, if this recurs for several consecutive years, dense populations may be built up over quite extensive areas. The multiplication and decline of this species are determined largely by the survival-rate during the larval stages, and this depends mainly on the interaction of weather (chiefly moisture) and the places where the larvae may live (Madge, 1953, unpublished thesis).

In their pristine condition the plains in the region where *Oncopera* is found carried certain perennial grasses with a tussocky habit of growth. The larvae of *Oncopera* tend to extend their burrows upward into the crowns of these tussocks, a habit which enables them to escape drowning in the wetter situations. With the development of the land for agriculture, these grasses have disappeared, and the *Oncopera* living in the modern pastures have no opportunity to practice this habit. But Madge came across one farm on which, because of certain accidental circumstances in the establishment and subsequent management of the pastures, there was a meadow in which a dense low growth of clover was studded, at intervals of several feet, with tussocks of cocksfoot, with crowns rising some 6 or 9 inches above the level of the ground. It also happened that this meadow carried a dense population of *Oncopera*. During the autumn the larvae were present at the rate of four or five per square foot in the flat parts, and each tussock also contained four or five larvae; their burrows extended down among the roots and up into the crowns (Fig. 12.03). Heavy rain fell during May and June, and the soil became water-logged (Fig. 12.04). So far as could be ascertained, all the larvae whose tunnels were in the flat ground between the tussocks were drowned, but hardly any of those which happened to be living in the tussocks were killed.

In northern England the life-cycle of the sheep tick *Ixodes ricinus* occupies 3 years. About 6 months after emerging from the egg, the larva goes in search of its first meal in the spring; it climbs to the tip of a grass stem and awaits the passing of a sheep or some other small animal. If it has the good fortune to be picked up by a host, it engorges itself with blood and after a few days drops to the ground. Being incapable of walking more than a few inches, it is unable to seek for a good place to live; but if it is fortunate enough to have dropped into one where the shelter is adequate, it may survive the 12 months which must elapse before it is ready to look for its next meal. During this time it molts and grows into a nymph. At the appropriate season (April–May) the nymph again climbs a grass stem and awaits the passing of a sheep. The nymph, after engorging with blood, drops to the ground. If it survives during the next 12 months, it becomes adult. In the next spring the adult climbs to the tip of a grass stem, seeking not only a meal of blood but also a mate on the sheep (sec. 9.13). Being fully engorged and having mated, the female drops from the host and burrows, like the larva and nymph, into the vegetation. If she survives, she lays her eggs about 2 months later (MacLeod, 1934; Milne, 1952).

During the three years of its life the tick spends, altogether, about 3 weeks feeding and about an equivalent period on the tips of grass stems waiting to be

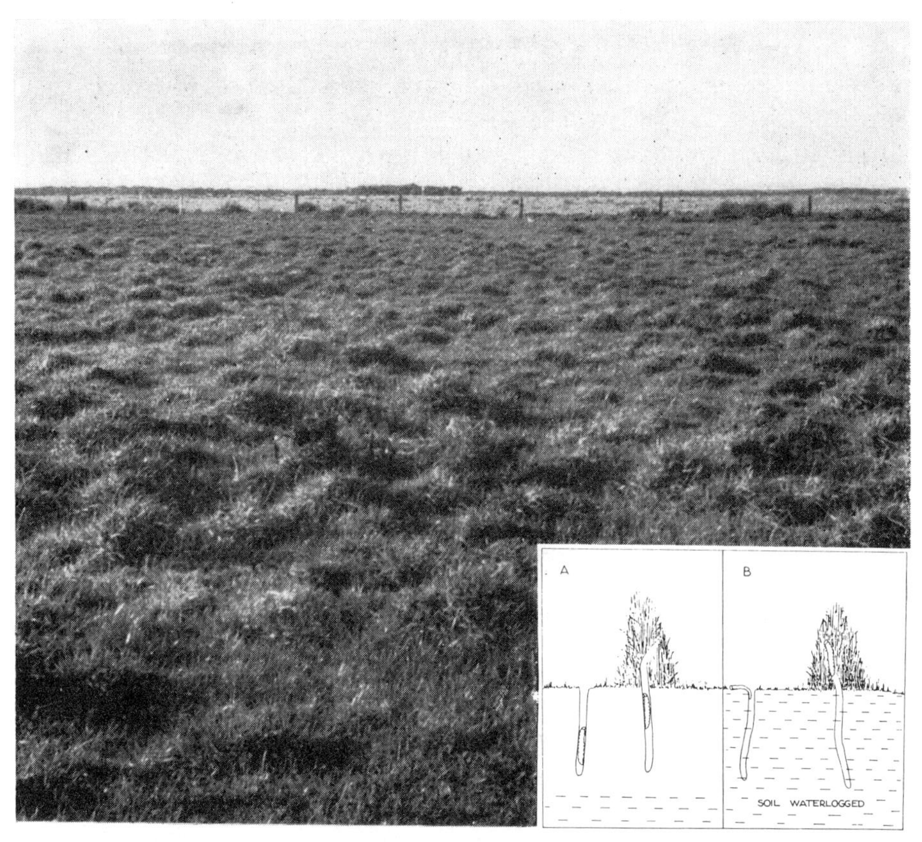

FIG. 12.03.—Two sorts of places which may be occupied by larvae of *Oncopera fasciculata* in a meadow containing tussocks of cocksfoot. *Inset:* diagrammatic section to show the larvae in their burrows. *A*, in the absence of flooding, larvae may survive in both sorts of places, *B*, when the soil becomes water-logged, the larvae whose burrows are in the open may drown, whereas those whose burrows ascend into the tussocks may escape drowning. (After Madge, unpublished thesis.)

picked up; the rest of the time is spent sheltering amid the vegetation close to the ground. Many ticks must die in each generation from starvation; predators, chiefly shrews and birds, take a heavy toll; but more fundamental than either of these hazards is the need to conserve the supply of water in the body during the long intervals between meals (Milne, 1950*a*, *b*). In its younger stages *Ixodes*, in a starved condition, may absorb water from the surrounding air if the relative humidity exceeds 88 per cent; in its later stages a tick requires more humid surroundings, and an unfed adult cannot absorb water from air that is drier than 92 per cent relative humidity (sec. 7.213). Consequently, the ticks after dropping from the host have little chance of surviving until the next meal is due (or to lay eggs, as the case may be) unless they happen to find a place in which the air is likely to remain saturated with water vapor, or nearly so, for most of the time.

According to Milne (1944, 1946), the distribution of *Ixodes* in northern

FIG. 12.04.—Part of the meadow illustrated in Fig. 12.03 during a spell of wet weather. The soil was water-logged, and all the larvae whose burrows were in the open places between the tussocks were drowned. (After Madge, unpublished thesis.)

England is strictly limited to rough pastures on hills and moors. The vegetation in these pastures is coarse, and there is a dense, almost permanently moist, mat of semidecayed vegetable matter, moss, and grass, covering the surface of the soil to a depth of several inches. In this material the relative humidity is close to 100 per cent for most of the year, and the ticks find adequate shelter from desiccation. Milne (1950*a*) did a number of experiments to find out the precise location of the ticks in this mat (Table 12.04). On a number of different occasions he placed about 40 newly molted adults on the surface of three different sorts of turf and allowed them to crawl down into it. Once they had settled down, they did not change position throughout the season, and the data in Table 12.04 summarized their distribution in the mat during the rest of the summer and the winter. Most of them penetrated well into the mat, where they were likely to be surrounded perpetually by saturated air. Not all parts of the pasture were equally favorable in this way, and Milne (1950*b*)

TABLE 12.04*

VERTICAL DISTRIBUTION OF *Ixodes* IN VEGETATION ON ROUGH UPLAND PASTURES AS INDICATED BY RELATIVE NUMBERS FOUND AT VARIOUS DEPTHS IN VEGETATION

TYPE OF PASTURE	LAYER	NO. OF TICKS AS PER CENT OF TOTAL	
		Males	Females
Rough grass (mat 2½ inches)	Grass	3.9	4.8
	Upper mat (top 1½ inches)	84.4	81.2
	Lower mat	10.7	14.0
	Soil	1.0	0.0
Heather (mat 2½ inches)	Heather	0.0	0.0
	Upper mat (top 1½ inches)	96.4	82.3
	Lower mat	3.6	15.2
	Soil	0.0	2.5
Bracken (mat 1 inch)	Bracken	0.0	0.0
	Mat	100.0	98.7
	Soil	0.0	1.3

* Data from Milne (1950*a*).

showed that the abundance of ticks was correlated with the amount of cover. In the short swards, which are more characteristic of the lowland pastures, the vegetation offers no place where the moisture is adequate to support the tick throughout the year. Consequently, *Ixodes* never becomes established on the lowlands, despite the repeated introduction of flocks of infested sheep into these areas.

In Scotland the biting midge *Culicoides impunctatus* is known to have a "patchy" distribution. Kettle (1951) not only demonstrated the patchiness

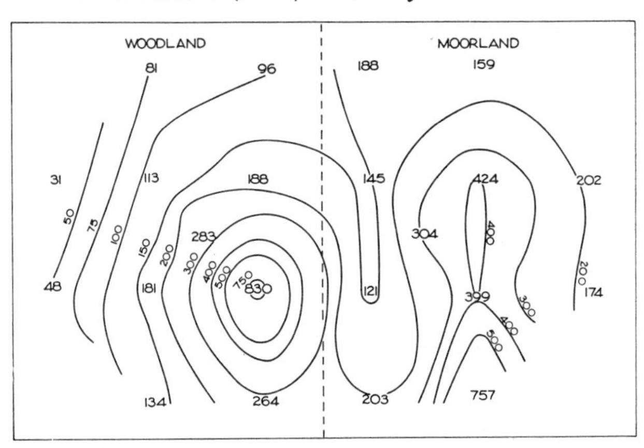

FIG. 12.05.—The distribution of adult females of *Culicoides impunctatus* in a 5-acre field of woodland and moorland at Bannachra. For further explanation see text. (After Kettle, 1951.)

quantitatively but also provided an explanation for it in terms of the sorts of places in which the larvae can live. He placed 20 traps uniformly over an area of about 5 acres and recorded the number of *Culicoides* caught each week for 18 weeks. Half the area was moorland, and half was woodland. The numbers caught in each trap were expressed as a percentage of the total for the area, and results were plotted on a plan of the area (Fig. 12.05). Lines were drawn joining places where the midges were trapped in equal numbers, and these indicated two distinct centers of concentration, from which it was postulated that all the midges in the area were coming from two distinct, probably quite small, breeding grounds. This was proved for the woodland area by taking samples of soil and counting the larvae in them. The breeding ground turned out to be a small area of rather boggy soil carrying *Sphagnum* and *Juncus*. It was flanked by higher ground, supporting bracken or woodland with undergrowth of various shrubs. No larvae were found in the higher ground; apparently the breeding of *Culicoides* is restricted to places where the soil remains moist. The distribution of the adults is limited by the distribution of suitable breeding grounds and the distance they may fly away from the place where they were bred. Kettle estimated this to be about 80 yards.

12.23 *Other Animals*

In many of the examples used in chapters 9 and 10 to illustrate the influence of other organisms on an animal's chance to survive and multiply, it was necessary, in order to keep the discussion realistic, to take into account the sort of place where the particular animal was living. (Indeed, one of the chief criticisms to be leveled against the mathematical models of predator and prey is that they imply that all individuals of the prey may be found equally readily by the predator. This assumption makes these models quite unrealistic.) This emphasizes the close interdependence between "place to live" and "other organisms." It also makes it unnecessary to say very much in the present section (since the matter has already been largely covered in these earlier chapters), and we shall be content merely to describe briefly several examples, selected because they are especially apt. Most of them have been chosen to illustrate how the suitability of a place in which to live may depend on the protection which the place provides from predators.

Udvardy (1951) quoted a description from Turcěk (1949) of how, in the forests of the North Carpathians, the birds nesting in the canopies of trees were relatively safe from predators except when the trees were defoliated by the feeding of numerous caterpillars of the moth *Lipara dispar*. In every case which he observed, predators killed the birds which were nesting in the trees that were defoliated.

In central North America the chief predator of the muskrat *Ondatra zibethicus* is the mink *Mustela vison*. A muskrat which is well established (that is, it is living peacefully, free from serious internecine strife) in a burrow well situated with respect to surrounding water may be unlikely to fall a victim to a mink. But one which is homeless or poorly situated with respect to the place where it is living has small chance of surviving. Water, preferably sufficient to cover the entrance to the burrow, is necessary for security, since drought is one of the most serious hazards experienced by muskrats in Iowa. A quotation from

Errington (1939) illustrates the way in which moisture may influence the muskrat's chance to survive in the presence of predators and how, when droughts occur, some individuals have a better chance of surviving than others, because they happen to be living in places which offer greater security than those occupied by less fortunate individuals:

> The effect of drought on vulnerability of muskrats to mink is so pronounced nevertheless that a few examples should be appropriate in this paper. For both summers of 1936 and 1937 there were on Round Lake close to one adult muskrat per 2.7 acres and in both years the spring mortality from mink was conspicuous. In 1936, the mink pressure slackened in May but was resumed in July as the water level of the marsh went down. Exposure of the bank burrows along about 300 yards of the southeast shore was followed by the killing by mink, largely between July 22 and August 1, of apparently all but one of eight muskrats believed to be resident there. The one individual known to have escaped was living in a newly built lodge about 60 yards from shore. Similar mortality in other exposed shore habitats was also detected but nearly the entire population of Round Lake was already living in much greater security in lodges deeper in the marsh. In contrast, the summer of 1937 was a season of high water in northwest Iowa, and examination of mink prey items and 168 faecal passages gathered from Round Lake did not disclose any evidence of mink pressure upon muskrats from May to early October. The vulnerability of the muskrats of the Dewey's Pasture potholes, which went dry from June to August, 1936, seemed to become critical just after the disappearance of the surface water. Mink faeces deposited before this time rarely contained muskrat remains; then, with the exposure of the muskrats, mink diet ran strongly to this item for a week or two.

The red fox *Vulpes regalis* rarely preys upon muskrats that are protected by the presence of water around their burrows; but when drought lowers the level of water and exposes the muskrats, the fox may, by virtue of its superior skill, exceed even the mink in the severity of its preying. A quotation from Scott (1947) illustrates these points: "The muskrats lived in a marsh habitat[1] at Wall Lake and in a river habitat[1] at Moingona. Field observations indicated that the foxes tended to avoid wading through water, and muskrats appeared to be secure while in water. There was very little evidence of muskrats feeding on land within the area at Moingona, possibly because there was very little cultivated land adjacent to the river. The remains of muskrat were detected but twice in the fecal material (from foxes) from the Moingona area. At Wall Lake the muskrats were left exposed on land in the 34-acre occupied area when the water levels were lowered by drought during June, July and August of 1940. The unusually severe and apparently somewhat uncompensated predation upon these muskrats has been described by Errington and Scott (1945)." Errington (1943) considered that the foxes killed more muskrats than would have been destroyed by mink in the same circumstances. He estimated that the presence of this family of foxes in the area (which was somewhat fortuitous) had resulted in the deaths of some 75 muskrats which might have survived in the absence of the foxes.

1. In this sentence "habitat" seems to refer not to actual space occupied by the animals but to the sort of country in which they were living, e.g., marshes or rivers in general.

The next example, of the scale insect *Saissetia oleae*, is of special interest because the provision in a district of suitable places protecting it from its predator may result not in an increase in the numbers of the scale insect but in a decrease. When the scale insect *S. oleae* is living in a place where there is a mild and humid climate, it will breed continuously throughout the year; generations overlap, and the population consists of individuals in all stages of development. In places where the summer is hot and arid, breeding may be restricted to the winter. For example, in groves of citrus in the Piru district of Ventura County in southern California, *Saissetia* spends the summer in the adult stage; eggs are laid and hatch in the autumn, and the immature stages are present during winter (September–April); there is only one generation each year (Flanders, 1949). But in the same district, on an oleander which is kept well watered during summer so that it grows vigorously and produces a dense growth near the ground, there may be places among the lower branches where it is moist and cool enough during summer to enable *Saissetia* to continue breeding throughout the hot weather.

The small encyrtid *Metaphycus helvolus* preys on *Saissetia* in its immature stages but is not able to pierce the armor of the mature scale. Now Flanders has observed that the predators, by virtue of a certain pattern of behavior, tend not to penetrate into the dark recesses among the lower branches of these oleanders, so that all stages of *Saissetia* living there are secure from *Metaphycus*. (Compare this natural situation with the model that Flanders made in the laboratory with *Ephestia* and *Habrobracon*, sec. 10.222.) Most of the young *Saissetia* hatching among the lower branches climb upward and settle on the higher branches; they provide a continuous supply of food for *Metaphycus* throughout the year. Very few survive, but, no matter how thoroughly *Metaphycus* seeks them out and eats them, Flanders' observations suggest that the predators may never go seriously short of food because a fresh supply of prey is constantly moving up from below.

In the groves of citrus quite a different sequence of events takes place. No *Saissetia* is hidden away in a place where *Metaphycus* will not go. During winter (September–April) the entire population is in a stage suitable for *Metaphycus*. In the mild climate of the Piru district the predator may have several generations during this period, and it has been known virtually to exterminate *Saissetia* from local areas which may be quite extensive. In these circumstances *Metaphycus* dies out also. In due course the area may be recolonized by *Saissetia*, and in the interval before it is found again by *Metaphycus* the scale insects may become very numerous indeed.

Flanders has suggested that if there are planted adjacent to a grove of citrus, preferably upwind with respect to the prevailing winds, a number of oleanders and if these are kept well watered, then the last event in this sequence may not occur. The populations in the grove of citrus both of *Saissetia* and of *Meta-*

phycus will constantly be replenished from the populations on the oleanders. They will not die out altogether, but neither will they become abundant. Paradoxically, the occurrence in the district of places which are especially suitable for *Saissetia*, both because the humidity permits it to breed continuously and because it is secure from its predator, may result, not as might have been expected, in an increase in the numbers of *Saissetia* but, instead, in a decrease.

12.3 THE DISTRIBUTION AND ABUNDANCE OF PLACES WHERE AN ANIMAL MAY LIVE

12.31 *The Concept of "Carrying Capacity" in the Ecology of Animals Which Recognize Territories*

According to Leopold (1933, p. 136), the requirements of the bobwhite quail *Colinus virginianus* may be summarized under four headings: (*a*) For nesting, it requires well-drained ground covered with moderately thin grass or brush, but with bare ground near by so that the young may dry out after rain. This probably accounts for the frequency with which nests are found near paths and roads. (*b*) The food of the young consists mostly of insects, but as they grow older they also eat seeds and berries; insects make up about 20 per cent of the diet of the adults during summer; seeds are the most important constituent during winter. (*c*) For protection from bad weather and as a place in which to hide from predators, thickets of scrub or tangles of vines are required, especially during winter. (*d*) For sleeping, an elevated place is best, because it enables the birds to take wing readily if they are attacked by a predator. Where meadow, field, scrub, and woods meet, the quail is likely to find an ideal place to live.

Imagine an area of 1 square mile on which these four sorts of country were equally represented but each consisted of solid blocks, occupying one-quarter of the area (see Fig. 12.06, *A*). It is probable that the central point, where all four areas meet, would provide a good place for one covey of quail, but there would be none elsewhere on the square mile. Now imagine the same area of each sort of country arranged patchily as in Figure 12.06, *B*. On this square mile there would be five places where a covey of quail would find all its requirements. Errington would say that the "carrying capacity" of the second arrangement was higher than that of the first. The concept of carrying capacity was developed by Errington (1934) and Errington and Hamerstrom (1936), in the first place, with special reference to bobwhite quail and was later used by Errington (1946) in a more general way to apply to populations of other vertebrates, especially birds and rodents, in which "territorial behavior" was manifest in at least some degree.

Errington and Hamerstrom (1936) found that for the bobwhite quail the carrying capacity of an area was largely determined by the amount and suitability of places for hiding and sheltering during winter, because in most of the places that they studied the presence of farmland insured plenty of food during both winter and summer. During summer there was also plenty of "cover" almost everywhere, with the result that there was room for many more quail during summer than during winter. With the approach of autumn, much of this cover disappeared, and the quails could find adequate shelter only among the

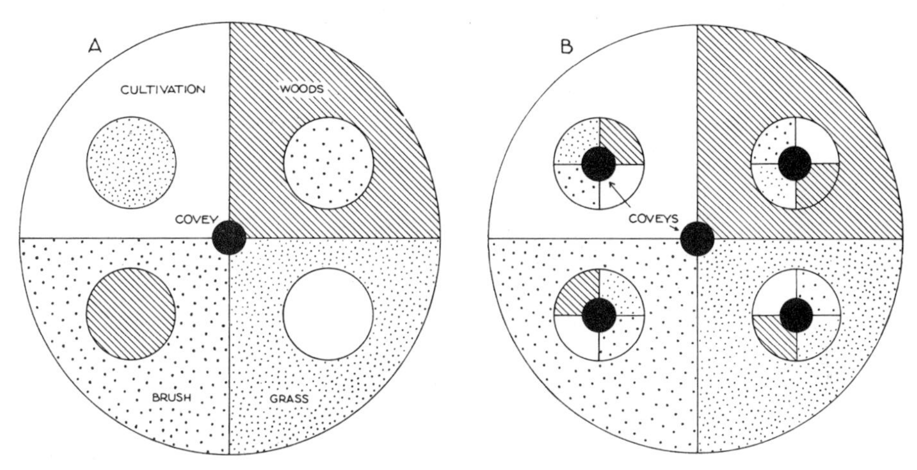

FIG. 12.06.—Hypothetical distributions of woodland, brushland, meadow, and field in two areas, to illustrate the concept of "carrying capacity" for a species like the bobwhite quail, which requires all four sorts of vegetation in the place where it lives. There is the same area of woodland, brushland, meadow, and field in *A* as in *B;* but in *A* their distribution provides a place for only one covey, whereas in *B* there are places for five covies. (Adapted from Leopold, 1933.)

vegetation growing in gullies, woodlots, or along watercourses. If the natural increase during summer had resulted in numbers in excess of what this amount of vegetation might shelter during the winter, then the surplus was driven out to be destroyed, largely by predators. For 6 years Errington and Hamerstrom counted or estimated the numbers of quails overwintering in 20 different areas in Iowa and Wisconsin. They found that the maximal numbers surviving in a particular area were remarkably constant, and they considered this to indicate the "carrying capacity" of the area. In 4 areas which were studied more intensively, this varied from one quail to 5 acres to one quail to about 65 acres; but the number, whatever it might be, was characteristic of each area (Errington, 1934).

Errington and Hamerstrom (1936) were not able to recognize with precision what were the qualities in the vegetation that made for high or low carrying capacity. There was no doubt that drastic changes in the vegetation providing shelter during the winter changed the carrying capacity substantially; sometimes even minor changes caused marked alterations in the carrying capacity. They wrote (p. 396): "We do know, on the other hand, that carrying capacity has been both raised and lowered by changes in cover conditions. Environmental manipulation, for one purpose or another, may often be of profound consequence to bob-white populations, but we see no way of predicting its effect in advance. To be sure, we may predict with reasonable certainty that a strong spread of fencerow and roadside brush in central Iowa may be followed by an increase of quail and any wholesale reduction of existing cover would probably mean decidedly fewer quail, but there is a great deal about this that we do not know, especially as to details. Leads as good as any perhaps may be given by the data from territories number 11, number 17 and others where relatively slight environmental modifications have made territories unattractive, if not lethal. The cleaning out of one small but strategically located patch of roadside growth apparently changed number 13 from a habitable to a lethal territory. The burning of a single brush-pile, likewise strategically located, patently left territory number 11 with an altered status, so far as wintering quail were concerned." The development of the country for agriculture has, of course, profoundly altered its carrying capacity for quail. Errington and Hamerstrom, taking into account the probable relative distribution of areas providing food and shelter, expressed the opinion that the carrying capacity of this particular region was probably at a maximum about 1880.

In some of the areas studied during some years, the numbers remained below the estimated carrying capacity. Sometimes the number would be below the carrying capacity at the beginning of the winter, despite the great opportunities for multiplication during the summer; sometimes catastrophic shortage of food, severe weather, the presence of other nonpredatory animals, or harrying by men during winter reduced the numbers below the carrying capacity. Also, when the observations were carried on for a longer period (one area was studied continuously for 15 years), the carrying capacity seemed to fluctuate in a way that was not obviously related to changes in the vegetation. These matters will be mentioned again in section 13.34.

The principle of the carrying capacity is also well illustrated in Errington's (1944, 1948) accounts of the ecology of the muskrat *Ondatra zibethicus*. During the winter the muskrat is relatively tolerant of its fellows, and larger numbers can overwinter together than would be tolerated in the same space, once the breeding season has begun. The number left behind in the place where the population spent the winter, after the surplus have been driven forth in the spring, depends largely on how many the muskrats will themselves tolerate in those particular quarters. What proportion of those which are driven out at this time survives to breed depends on their chances of finding adequate quarters before they succumb to predators or one of the numerous other hazards that beset a wandering muskrat. Thus "carrying capacity" has much the same meaning as the number of places in the area where a muskrat may live securely.

The muskrat is semiaquatic; it builds burrows or "lodges" which open below the surface of the water. They are best suited by an expanse of shallow water which does not fluctuate in depth. Marshes are good; so are drains, if they are fed by tiles, especially if they run near cultivated fields which provide good food. Streams also provide places for muskrats to live; but, according to Errington, a stream becomes increasingly suitable, the more it resembles a marsh. Fluctuations in the level of the water may be harmful: floods may drive the muskrats out of their burrows; droughts may uncover the burrows and expose the muskrats to their predators (sec. 12.23). In some situations the level of the water, being dependent on the weather, is likely to fluctuate rather abruptly. This makes it more difficult with the muskrat than it was with the bobwhite quail to think of the "carrying capacity" as a relatively stable quality of a particular area. Nevertheless, Errington (1946, p. 147) has observed that: "The numbers of breeding pairs tolerated in a specific habitat[2] tended to be similar from year to year, irrespective of the less drastic environmental changes, such as moderate fluctuations in water and food."

Temporary changes in carrying capacity may be brought about by the weather, especially fluctuations in rainfall. In Iowa the high level of water sustained in certain marshes as a result of an unusual amount of rain reduced the carrying capacities of these marshes. On the other hand, a series of wet seasons resulted in a big increase in the numbers of muskrats breeding in the state (Iowa) as a whole because the heavy rain filled up ponds, ditches, and marshland which had previously been dry (Errington, 1944).

Changes of a more permanent nature have been brought about by artificially draining, ditching, or damming in areas of marshland and other suitable country. Dymond (1947) referred to the construction of dams, dikes, and canals in the delta of the Saskatchewan River, Manitoba, which diverted water into dried-up marshes and thereby greatly increased the number of suitable places for muskrats to live: the number of muskrats in the area increased from 1,000 to 200,000 in 4 years. Errington (1948) referred to a number of other projects which had permanently increased the carrying capacities of particular areas: "But manipulation of water surely always will be one of the principal techniques in management of muskrat environment, whether the project involves the damming of a ravine for ponds (Allan, 1939), the establishment of a private muskrat 'ranch' (Grange, 1947), or something as ambitious as the restoration of the Lower Souris marshes of North Dakota (Henry, 1939)."

According to Errington (1946, p. 147), food is often less important than "cover" in determining the number of muskrats that an area may support. Nevertheless, when the amount of cover is artificially increased, the point may be reached where food becomes limiting. Errington (1948, p. 598) wrote in connection with this:

2. In this sentence "habitat" seems to mean the actual space occupied by the particular group of muskrats, e.g., a particular marsh in Iowa.

. . . mention should be made that manipulation of water levels on marshlands needs to be directed towards insuring against too much water as well as against too little. The broad-leaved cattail (*Typha latifolia*), a most outstanding native food for marsh-dwelling muskrats of northern United States, seems unable for long to tolerate a depth of more than about 4 feet of water, and it has happened that, when the deep-water stands started dying, those of the shallows likewise died, thus sharply lowering the attractiveness and habitability of given marshes for muskrats. . . . I would estimate that the best cattail marshes allow ascendancies of muskrats to reach nearly twice the populations as do even superior marshes grown to other vegetation. Specifically, the highest wintering density recorded during the Iowa investigations was about 35 per acre for a sizeable area of a cattail marsh, compared with maxima of between 15 and 20 per acre for bulrush (*Scirpus* spp.) marshes. In less extreme cases, one seemingly may expect about as many 20-per-acre populations in cattails as 10-per-acre populations in bulrushes.

The mourning dove *Zenaidura macroura* in North America used to live along the edges of forests, flying to the prairie to feed and to the forest to nest. As the country was developed for agriculture, there came a great increase in the area where forest and farmland met and a corresponding increase in the numbers of mourning doves. But at the present time, in some areas at least, the more intensive use of land for farming and the failure to replace shelter belts and windbreaks is reducing the number of places where *Zenaidura* may live. McClure (1943) recorded the number of *Zenaidura* on three small farms in Iowa during the years 1938, 1939, and 1940. On all the farms and especially on two of them, shade trees were dying and not being replaced during this period. There was a marked reduction in the number of doves from 1938 to 1940, especially on the two farms where more trees died (Table 12.05). McClure

TABLE 12.05*

NUMBER OF NESTINGS AND NUMBER OF YOUNG RAISED BY *Zenaidura macroura* DURING THREE SUCCESSIVE YEARS IN AREAS WHERE NUMBERS OF TREES WERE DIMINISHED

	FARM A			FARM B			FARM C		
	1938	1939	1940	1938	1939	1940	1938	1939	1940
Nestings	102	51	37	78	42	31	51	44	46
Successful nestings	47	20	15	36	15	10	34	22	21
Young raised	80	34	26	62	25	17	61	38	38

* After McClure (1943).

attributed the decrease in the numbers of doves to the reduction in the number of places where nests might be built. The English sparrow, *Passer domesticus*, increased in numbers at all the farms during this period, but the increase was most striking at farm B, where the decrease in doves was also greatest. The sparrow and the dove require much the same sort of food, and a possible (though by no means certain) explanation of this observation would be that the departure of the doves, due to scarcity of places for nesting, left more food available for the sparrows. Considerably more data would be required to verify this hypothesis.

In section 11.11 we mentioned Kluijver's (1951) observations of the great tit *Parus major* in an area of woodland where the places for nesting had been artificially increased so that there were always some boxes left unoccupied; also there was usually plenty of food. Since neither food nor places for nesting were limiting, it may be inferred that the carrying capacity of this area (that is the number of territories which the birds maintained there) depended upon some other more subtle requirement which the birds needed from the place where they chose to live. Although this could not be demonstrated quite so objectively with respect to the bobwhite quail and the muskrat, it is clear from Errington's writings that he had recognized the same phenomenon with these species, too. On the other hand, Kluijver (1951) mentioned certain other areas where the introduction of nesting boxes had been followed by an increase in the numbers of great tits, from which it may be inferred that the carrying capacity of these areas had been chiefly limited by the number of places where a nest might be built. This is quite like the very simple situation we described for the imaginary bees in our hypothetical example in section 2.121.

12.32 *The Concept of "Outbreak Centers" in the Ecology of Locusts and Other Species Which Do Not Recognize Territory*

A few individuals of the locust *Chortoicetes terminifera* may be found during summer (and perhaps at other seasons) if one searches diligently enough, in the right sort of country, scattered throughout most of the continent of Australia (Fig. 12.07). Over most of this enormous region the death-rate is so high that one doubts whether an isolated population would survive there for long. This point cannot be resolved because *Chortoicetes*, like other well-known locusts, migrates freely over great distances in its solitary, as well as its gregarious, phases (sec. 5.4) and areas which may be unsuited for permanent habitation are continuously receiving immigrants from distant places, where the locusts breed more successfully. Occasionally the immigrants come in immense gregarious swarms, and their progeny, if they encounter favorable weather in the area which they have invaded, may breed there for a few generations and constitute a major "locust plague" until such time as catastrophe, in the form of drought and starvation or disease, overtakes them and the plague comes to an abrupt end (Davidson, 1936*a;* Andrewartha, 1937, 1940; Key, 1942, 1943, 1945).

The locusts experience similar hazards in the "outbreak areas," whence these plagues originate, and, during periods of unfavorable weather, r may remain negative, leading to a consequent decrease in the numbers of the locusts in these areas, too. But the outbreak areas are situated in a zone where the climate is kinder to the locusts and catastrophes may be both less severe and less frequent; also opportunities for increase may be greater and more frequent in this climate (see theoretical discussion in chap. 14). Nevertheless, the locusts do not survive equally well or multiply equally readily throughout the whole of the zone experiencing this more favorable climate; they survive and multiply best in certain areas where the particular distribution of certain kinds of soil and vegetation fulfils their rather specialized requirements for egg-laying, feeding, and sheltering. The special qualities of an outbreak area were investigated by Clark (1947*a*), the distribution of outbreak areas in eastern Australia by Key (1945, and earlier papers), and in South Australia by Andrewartha (1940).

In the outbreak area studied by Clark (1947*a*), *Chortoicetes* developed

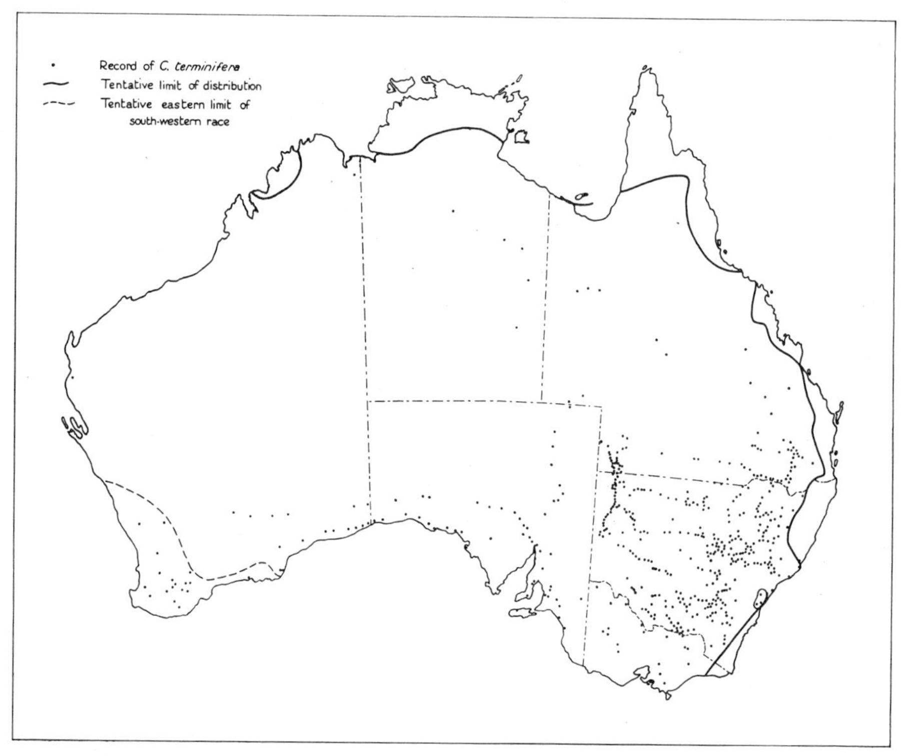

FIG. 12.07.—Localities from which individuals of *Chortoicetes terminifera* have been collected. (After Key, 1954.)

actively only during the warmer parts of the year, from September to May. A few of the older nymphs and adults hibernated successfully, but usually the winter was spent in the egg stage. Between September and May, one, two, or three generations might develop, depending on the amount and distribution of rainfall. Rain was needed to maintain moisture in the soil for the eggs to develop and hatch and to promote the growth of grasses for the nymphs and adults to eat. At least some eggs might remain alive for 3 months during summer in soil that was too dry to promote development, but exposure to drought more prolonged than this would kill most of the eggs. No ordinary drought would kill the hardy, tussocky perennial grasses which dominated local situations in the area, but they would not produce any succulent growth during a severe dry spell. The annuals would die off, leaving only seeds, and the areas which had been dominated by them would remain largely bare while the drought lasted. Much rain would produce a wealth of succulent grass for the locusts to eat, but even after light rain the hardy perennial species would put forth a little succulent growth. Excessive rain (probably not a very frequent hazard in this climate) might engender outbreaks of fungal or bacterial disease among the locusts.

In addition to food, the nymphs and adults required shelter. They would crawl into a tussock of grass to escape from a cold wind at night; during the hottest part of the day they would rest in the shade of the tussock or climb it on the shady side to a position where the breeze might cool them. After unusually good and soaking rains at the right time of the year, there was plenty of shelter provided by a luxuriant blanket of grasses and other herbage not only in the situations normally supporting perennials but also in other areas, where, at other times, the vegetation would at best be low and sparse or even absent. Except after such unusual rains, the only places where adequate shelter obtained were areas which supported perennial grasses with a tussocky habit of growth.

The adults also required, for oviposition, soils of a particular degree of compactness; and it was necessary that these soils should carry, at the time when the eggs hatched, at least sufficient vegetation to provide food and shelter for the early nymphal instars. Soils that were compact enough for oviposition did not support the sort of vegetation which would insure food and shelter in dry weather; and vegetation which adequately provided these requirements was never found on soil which was suitable for oviposition. So it turned out, on investigation, that the distribution and abundance of certain types of soil constituted the most essential qualities of an outbreak area.

The distribution of soils in the area was summarized by Clark (1947*a*, p. 10): "The Bogan-Macquarie outbreak area has a flat topography. Practically the whole of the present surface is alluvial in origin. The slight local differences in level occurring throughout the area rarely exceed a few feet. However, to them are related major changes in soil type and vegetation. The soils of the higher ground are compact. In general those at the lower level are self-mulching. The latter are regarded as a more recent alluvium than the former. The outbreak area as a whole consists of a mosaic of compact and self-mulching soils. Rarely can more than a few miles be travelled in a straight line in any direction without a major change in soil type being encountered." The heavy self-mulching soils vary in texture from clay loams to clay, as well as in other features; they are all chernozems or rendzinas. The compact soils of the higher levels vary in texture from sandy loam to loam.

All but the most compact of these were favored by *Chortoicetes* for oviposition. The nymphs, hatching from the eggs, might in dry weather find themselves short of food and shelter; but at other times, hatching after rain, they usually found in the low vegetation on these areas sufficient food and shelter to carry them through the first two or three instars. But this vegetation was likely to dry off quickly if no more rain came; moreover, the older nymphs and adults required more adequate shelter than the low sparse vegetation which grew on the compact soils (Fig. 12.08). Their chance of survival in most generations,

FIG. 12.08.—The sort of place favored by *Chortoicetes terminifera* for oviposition. The eggs are mostly laid in the bare ground near the margins of the vegetation. The low sparse vegetation is largely *Chloris truncata* and *Medicago* spp. (After Clark, 1947*a*.)

and especially during dry weather, depended on their being able to find more reliable sources of food and better places to shelter. The nymphs cannot fly, so they had to find their requirements within a few hundred yards of where they were hatched from eggs.

The characteristic vegetation growing on the heavier soils at the lower levels included perennial grasses, such as *Stipa*, *Eragrostis*, *Chloris*, and *Danthonia*, the tussocks of which provided good shelter for the locusts. These species and others associated with them on these soils of high water-holding capacity remained green and succulent for a relatively long time after rain and thus provided a relatively secure source of food for the locusts (Fig. 12.09).

FIG. 12.09.—The sort of place where the older nymphs and adults of *Chortoicetes terminifera* are likely to find adequate supplies of food and plenty of shelter even during dry times. (After Clark, 1947*a*.)

Wherever these two soils with their concomitant vegetations occur close together, especially if the boundary between them is abrupt, the locust is likely to find its best chance to survive during dry weather and multiply when the weather becomes more clement (Fig. 12.10). This is especially true if the two soils which come into juxtaposition represent the extremes of their classes, because then the vegetation is likely to remain true to type even through extremes of drought. In the area studied by Clark these two soils were distributed as in a mosaic; consequently, there were in this area a great many places where a locust would have a good chance to survive and multiply; the area was extensive—Key (1945) estimated that it contained 10,000 square miles.

These highly suitable places were called "outbreak centers" by Key (1945).

Just as outbreak centers may have a characteristic distribution and abundance in a particular outbreak area, so may outbreak areas be distributed in a particular way in a broader zone whose boundaries are determined by climate. Key plotted the limits of the zone in eastern Australia where the weather would be about as kind to *Chortoicetes* as the weather in the Bogan-Macquarie out-

FIG. 12.10.—A place that is especially suitable for *Chortoicetes terminifera* because of the abrupt transition from the area of compact soil, with its low sparse vegetation which is suitable for oviposition, to the area of heavy self-mulching soil, carrying vegetation which provides a reliable source of food and shelter. (After Key, 1945.)

break area. Then, by studying the distribution of soil and vegetation, he mapped the boundaries of a number of other outbreak areas. Figure 12.11 shows that these are irregularly distributed and that the whole of the area within the uniform climatic zone is far from equally likely to promote the multiplication of *Chortoicetes terminifera*.

Figure 12.11 also includes two outbreak areas in South Australia, which were mapped by Andrewartha (1940). These occur in a broad inland area of semi-desert, where the mean annual rainfall is about 8 inches, compared to 18 inches in the area studied by Clark. The country in the outbreak areas in the desert is characterized by a special topography; the lack of a well-developed drainage system and the presence of local features, such as flat watercourses, flats, and depressions, result in a series of local situations moister than the surrounding countryside, where perennial tussocky grasses persist and where the locust may

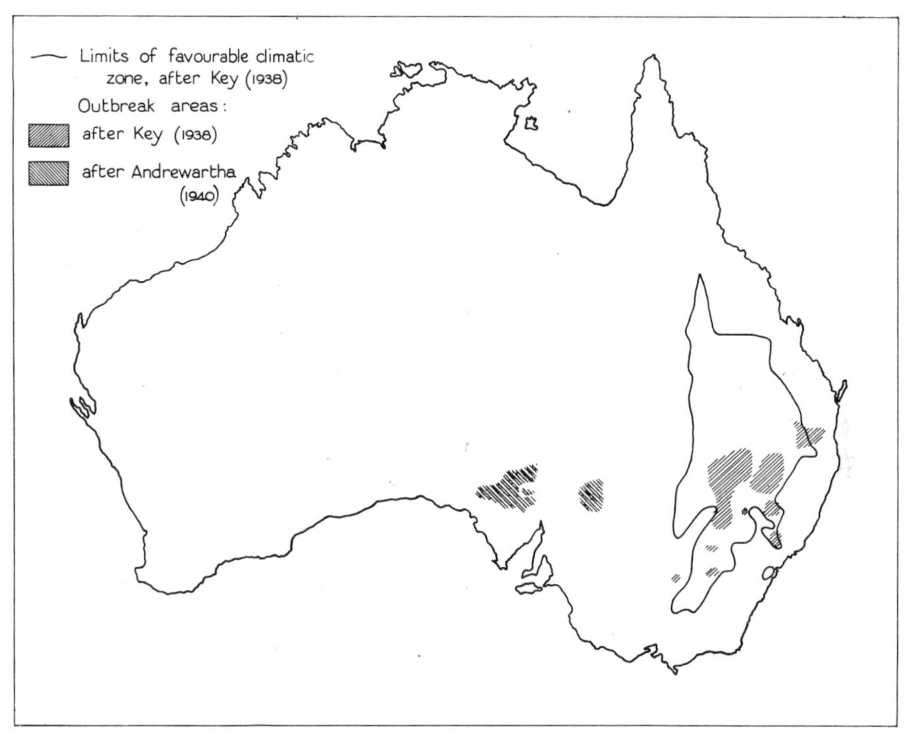

Fig. 12.11.—The distribution of outbreak areas for *Chortoicetes terminifera* in eastern and South Australia. (Modified from Key, 1945; and Andrewartha, 1940.)

find food and shelter to enable it to survive during bad times (Fig. 12.12). Elsewhere in the desert, although the climate is little different, the country lacks the essential physiographic features and suitable places, for *Chortoicetes* may be rare or absent. Although the average annual rainfall does not exceed 8 inches, it is highly variable, and the desert occasionally experiences a run of wet years which enables the locusts to multiply. This happens only rarely. The first plague of locusts recorded for South Australia is known to have reached the coast in 1845, less than 10 years after the colony was founded. Since then, plagues have recurred at intervals of about 30 or 40 years. In eastern Australia, plagues have been at least three or four times more frequent than this (Key, 1938). This reflects the difference in climate in the areas where the plagues originate.

Also it is reasonably certain that in eastern Australia the development of the country for sheep-grazing (the chief industry in the area that we have been discussing) has resulted in an enormous increase in the number of places where *Chortoicetes* has a good chance to survive and multiply. It follows that, to a large extent, the locust plagues of this region are man-made; if they have not become more frequent, they must, at least, have become greater since the

FIG. 12.12.—The sort of place where *Chortoicetes terminifera* breed during good years and survive during bad in the outbreak areas situated in the semidesert parts of South Australia. (After Andrewartha, 1940.)

country was settled by Europeans. It will be best to quote the evidence for this directly from Clark (1947*a*, p. 69):

Several years of intensive field observation and discussions on the development of this part of New South Wales, led the author to conclude that most of the outbreak centres have been produced as a result of human activities, including wholesale ring-barking and clearing of timber, grazing, and the introduction of the rabbit to Australia. In combination with severe droughts, and damage due to trampling by the passage of stock to and from watering places and on stock routes, rabbits have undoubtedly played an important part in facilitating soil erosion, thus rendering considerable sections of the country favourable as oviposition habitats[3] for very many years to come.

Some evidence was gained to suggest that the introduction of sheep to the area led to the establishment of pasture species, some of which are favourable to *Chortoicetes*, e.g., *Hordeum leporinum*. The area was first used for cattle grazing, and several of the older property holders, whose statements may be considered reliable, informed the author that when the country was first settled the pasture was much sparser on both the light and heavy soil types than it is today. It was not until after the introduction of sheep that relatively dense pastures

3. "Habitat" in this passage refers to the special sort of place which serves not for the whole life of the animal but for one of its particular functions, e.g., oviposition, feeding, sheltering.

developed, mainly as a result of extensive tree destruction. Thus not only have oviposition nuclei been created, but the food-shelter habitat has been improved and extended as a result of human occupation of the country.

In western Australia and South Australia the development of certain tracts of country for agriculture has similarly resulted in a great increase in the area suitable for the grasshopper *Austroicetes cruciata*. The ecology of this species is discussed in section 13.12.

PART IV

The Numbers of Animals in Natural Populations

CHAPTER 13

Empirical Examples of the Numbers of Animals in Natural Populations

Fluctuations occur in every group of animals and in every habitat that has been investigated: an impression not derived solely from deliberate research upon fluctuating populations, since it has also been widely supported by ecologists studying populations for other purposes and from other points of view. These fluctuations are rather forcibly expressed in human affairs through the irregularity from year to year of pests, epidemics, and of various animal resources, notably in the sea. . . . Although the amplitude of fluctuation is often very great, scarcity alternating with high abundance every so many years, two things that we might expect do not often happen. The first, complete destruction of vegetation by herbivorous animals, has already been mentioned. The second is complete destruction over any wide area of either predators or prey. The factors controlling the limitation of these fluctuations are therefore of great interest, since they are the factors that critically affect the survival or extinction of species, ...

ELTON (1938, pp. 130–31)

13.0 INTRODUCTION

ALL the examples which are discussed in this chapter have several qualities in common: (*a*) each one represents an attempt, by one means or another, to measure the number of animals constituting a natural population; (*b*) most of them have been continued through several generations, some up to 15 years or more; (*c*) in each case the investigator has had a more or less thorough knowledge of the behavior and physiology of the species to guide him in his selection of facts about the particular natural population which he has chosen to study; (*d*) in each case the investigator has also studied the environment and related the observed numbers (or more often changes in the numbers) to measurements of the environment. In other words, these are all particular examples of the general method which we extolled in section 1.2—that is, all except the examples given in section 13.4. There is an extensive literature about the so-called "regular cycles" in the numbers of fishes, game birds, rodents, and other fur-bearing animals; and nearly all of it suffers from the grave disadvantage that information about numbers has been gathered without much reference to the biology of the animals themselves; this was mostly because the knowledge did not exist. Most of the examples in this chapter are species which, for some reason or another, are of importance to man. They have been selected not for this reason but because they have been studied intensively. In some examples man has radically altered their ecology; but whether this has happened or not, the principles which emerge are the same.

Whereas it must surely be generally true that an animal's chance to survive and multiply depends on all four of the major components of its environment (sec. 2.2), yet it is also true that there are plenty of cases, perhaps the majority, in which one or several of the components are of chief importance; if their influence is understood, the others may, in practice, be safely disregarded.

13.01 *Methods of Sampling*

If the chief purpose of an investigation is to explain the influence of environment on the animal's chance to survive and multiply, it may be sufficient to measure relative changes in the density of the population without troubling to ascertain the absolute numbers in the population that is being studied. Simple trapping may suffice for this purpose (sec. 13.113). In the absence of a method of trapping which can be relied upon to catch a sufficiently constant proportion of the population at all times, it may be necessary to make an estimate of the absolute numbers. Errington (1945), in his study of the bobwhite quail (sec. 13.34), found it practicable to enumerate every bird in the area. But with most species it is not practicable to count the individuals on any area large enough to be of use in a practical study, and some method of sampling becomes necessary. With relatively sedentary species, like the arthropods in soil, which were studied by Salt and Hollick (1946; see sec. 13.02), the numbers may be counted precisely in an adequate number of small samples chosen from a larger area. With more mobile species and those which are not so easily seen or counted even in a small area, the special method known as capture, marking, release, and recapture is most useful if it is practicable (sec. 13.33). On the other hand, even mobile species, if they are easily seen, may be counted on a number of small areas or "quadrats"; this method was used successfully by Clark (1947*a*) to estimate the relative numbers of locusts on different quadrats.

In the case of the wireworms mentioned in section 13.02, it was sufficient, having chosen a suitably large and uniform area of grassland, to take a number of samples at random over the whole meadow. But suppose one wanted to estimate the numbers of a certain caterpillar in a certain area of mixed woodland. If this caterpillar were known to live only on oak trees, then it would be a rather futile and wasteful procedure to sample all the trees in the wood at random. It would be better to count the number of oaks in the area and then take a random sample of oaks on which to count the caterpillars. This means, in more general terms, that it is usually necessary to place a restriction on the randomness of the sampling within the whole area chosen for study, by taking into account the patchy distribution of suitable places for the animals to live in.

13.02 *Tests for Randomness in the Distributions of Animals in Natural Populations*

We mentioned in section 2.12 and elsewhere that the places where each sort of animal may live are usually distributed unevenly and that this imparts a fundamental patchiness to the distributions of animals. Furthermore, the animals are usually distributed quite unevenly through the places that are suitable; some places will be crowded, while similar ones contain unexpectedly few individuals. The prevailing scarcity of most species results in many suitable places being quite empty, purely by chance. Other causes of uneven distributions have also been mentioned. For example, with living plants the ordinary processes of growth, senescence, and decay and of succession may result in constantly changing distributions of food and suitable places to live (sec. 5.0). Also the presence of an active predator tends to accentuate the patchiness of the distribution of its prey (secs. 5.3 and 10.321). We give later several examples in which the degree of unevenness in the distribution of the animals in a natural population has been estimated quantitatively.

The best way to find out whether the individuals of a particular species living in an area are distributed over it more or less evenly than they would be if their distribution were random is to partition the area into a number of stations or "quadrats" and count or estimate the numbers occurring in each quadrat. It is necessary to choose a large number of small quadrats, so that the chance that a particular individual will occur in a particular quadrat is small. Then if the animals are distributed at random over the whole area, the frequencies of the quadrats containing 0, 1, 2, 3, . . . , x individuals should tend toward a Poisson series. The frequencies in a Poisson series are given by the expression

$$Ne^{-m}\left(1,\ m,\ \frac{m^2}{1\times 2},\ \frac{m^3}{1\times 2\times 3},\ \ldots,\ \frac{m^x}{1\times 2\times 3\times \ldots x}\right),$$

where N is the total number of animals counted, m is the mean number per quadrat, and the series 0, 1, 2, 3, . . . , x, represents the number of animals that may be found in any one quadrat. It follows that the frequency of quadrats containing x individuals is given by

$$N\,e^{-m}\left(\frac{m^x}{x(x-1)(x-2)\ldots 1}\right).$$

In a Poisson series the mean equals the variance, so that

$$\frac{\Sigma(x-\bar{x})^2}{\bar{x}(n-1)} = 1.$$

If, in an observed series, $\Sigma(x - \bar{x})^2/\bar{x}(n - 1)$ is significantly less than unity, this indicates that the variance of the distribution must be less than that appropriate to a Poisson series, which means that the individuals in the population are distributed over the area more evenly than would have been expected from a mere random scattering. Conversely, if this quantity is found to be significantly greater than unity, then it may be concluded that the variance in the population that was sampled was greater than that which is appropriate to a Poisson series; in other words, the animals were distributed less evenly (that is, with greater patchiness) than would have been expected from a mere random scattering. If the number of samples is large (several hundred or more), the theoretical Poisson series may be calculated, and the observed distribution compared directly with the theoretical one. If, as more often happens, the number of samples is smaller, it is appropriate to make use of the fact that, for a Poisson series, $\Sigma(x - \bar{x})^2/\bar{x}$ is distributed as χ^2, with the number of degrees of freedom one less than the number of samples. Alternatively, for larger samples it is convenient to make use of the fact that the variance of the expression $\Sigma(x - \bar{x})^2/\bar{x}(n - 1)$ (for a Poisson series) is $2n/(n - 1)^2$ and therefore observed values of $\Sigma(x - \bar{x})^2/\bar{x}(n - 1)$ may be considered to differ significantly from unity, at the 5 per cent level of probability, if the difference exceeds $2\sqrt{2n/(n - 1)^2}$.

Salt and Hollick (1946) divided a seemingly uniform square yard of grassland into 81 "quadrats," each 4 inches square, and counted the wireworms (larvae of *Agriotes* sp.) in each quadrat. The soil was dug up, taken away, and washed through a series of sieves by a special method which separated nearly all the larvae, even the smallest, from the soil. The results are shown in Tables 13.01 and 13.02. The value of $\Sigma(x - \bar{x})^2/\bar{x}(n - 1)$ calculated from

TABLE 13.01*

NUMBERS OF WIREWORMS IN 81 QUADRATS IN SQUARE YARD OF SOIL UNDER GRASS

0	3	8	9	4	2	1	2	3
7	3	7	6	3	1	3	1	3
3	1	8	1	4	3	4	3	6
0	1	4	2	11	5	8	1	3
5	1	6	11	9	7	3	9	6
3	2	7	13	2	7	8	9	14
2	0	7	8	15	8	12	7	5
8	3	8	5	6	12	5	3	2
7	8	10	10	11	15	8	10	10

* Each square represents a quadrat in the relative position in which it occurred in nature. After Salt and Hollick (1946).

these results was 6.24. With 81 samples, a value of this magnitude indicates a highly significant departure from the theoretical Poisson series: the wireworms were definitely not distributed at random in the square yard; there were many quadrats with many or few wireworms and not enough with moderate numbers. In other words, the wireworms showed a strong tendency to occur in patches.

Salt and Hollick also studied the distribution of wireworms in one area of ¼ acre of grassland and another of 8 acres. From the former, 21 samples were taken from each of 16 "stations" over a period of 3 years. The samples consisted of cylinders of soil, 4 inches in diameter and 12 inches deep. The "stations" were separated from one another by 5 yards in one direction and 15 in the other. In the 8-acre area, 20 stations were established, and these were sampled at monthly intervals for 28 months. In both areas the wireworms were distributed less evenly than would be expected on the hypothesis of random distribution, and the discrepancies were highly significant.

TABLE 13.02

DATA FROM TABLE 13.01 REARRANGED TO SHOW FREQUENCIES WITH WHICH DIFFERENT NUMBERS OCCURRED

No. of Wireworms	No. of Quadrats	No. of Wireworms	No. of Quadrats
0	3	9	4
1	8	10	4
2	7	11	3
3	14	12	2
4	4	13	1
5	5	14	1
6	5	15	2
7	8	16+	0
8	10		
Total	....	461	81
Mean	....	5.69	....
$\frac{\Sigma(x - \bar{x})^2}{\bar{x}(n - 1)}$	....	2.48	....
$2\sqrt{\frac{2n}{(n - 1)^2}}$	....	0.318	....

In the ¼-acre area there were relatively fewer wireworms where the loam was deeper and fewer in the wetter situations. In the 8-acre meadow there were fewer where the soil contained a lot of organic matter, and these were usually also places where *Lolium* was growing thickly. This indicates that at least some of the patchiness of the numbers of wireworms in these larger areas was associated with variability in the soil with respect to its suitability as a place for wireworms to live. The small area of 1 square yard might be expected to be less variable than the larger areas measured in acres. So one might expect that most of the patchiness in the numbers of wireworms in an area as small as this might be attributed to the behavior of the insects.

Cole (1946*a*,*b*) placed a large number of boards of the same size on the ground in several sorts of woodland. He examined the boards at intervals and recorded the numbers of the different species of animals that were living under each one. In one experiment he collected 127 centipedes (*Lithobius forficatus*) from 1,152 boards; they were distributed among the boards as in Table 13.03. For these data $\chi^2 = 1{,}258.6$; $n = 1{,}151$. With such large values of

TABLE 13.03*
DISTRIBUTION OF 127 CENTIPEDES (*Lithobius forficatus*) UNDER 1,152 BOARDS PLACED ON GROUND IN WOOD

No. of Centipedes per Board	No. of Boards	
	Observed	Poisson Series
0	1,039	1,031.8
1	101	113.7
2	10	6.3
3	2	0.2

* After Cole (1946*a*).

n, one makes use of the fact that $(\sqrt{2\chi^2} - \sqrt{2n - 1})$ is distributed normally with unit standard deviation about a mean of zero. In this case $(\sqrt{2\chi^2} - \sqrt{2n - 1}) = 2.09$. So the deviations from the theoretical distribution, though small, must be judged significant at the 5 per cent level of probability. There were too many boards with 0, 2, or 3 centipedes and too few with 1; this indicates a patchiness in excess of what would be expected on the hypothesis of random distribution of the centipedes throughout the wood. Similarly, with the isopod *Trachelipus rathkei* and the beetles of the family Carabidae, Cole found that their distributions were significantly nonrandom and patchy. In fact, the only animals in these studies of which the distribution was not significantly nonrandom were certain spiders. Cole also analyzed data from Beall (1940) for the caterpillar *Loxostege sticticalis* and for fleas on rats and found in each case that the distributions differed significantly from the Poisson series and that the difference was due to a patchiness in excess of what would be expected to occur by chance.

Gilmour *et al.* (1946) measured the distribution of the blowfly *Lucilia cuprina* in an area of grassland near Canberra. It was a circle 8 miles in diameter. The country was undulating; most of the original vegetation of savannah woodland had been destroyed, and the area had been developed for grazing sheep; a plantation of pines encroached into the area in one sector. One hundred and two traps were spaced evenly, ¾ mile apart, over this area, as indicated in Figure 13.01. The traps were examined daily, early in the morning before the flies became active, and the numbers of flies in the traps were recorded. The same experiment was repeated three times during the summer of 1941–42, on

December 7, January 18, and March 13. The traps were in the same situations on each occasion.

The authors of this paper very kindly lent us their original records, and we have analyzed them, first, to see whether the individuals of *L. cuprina* occurring naturally on this area were distributed at random over it and, second, to see whether the concentrations of flies recurred in the same local situations in each experiment.

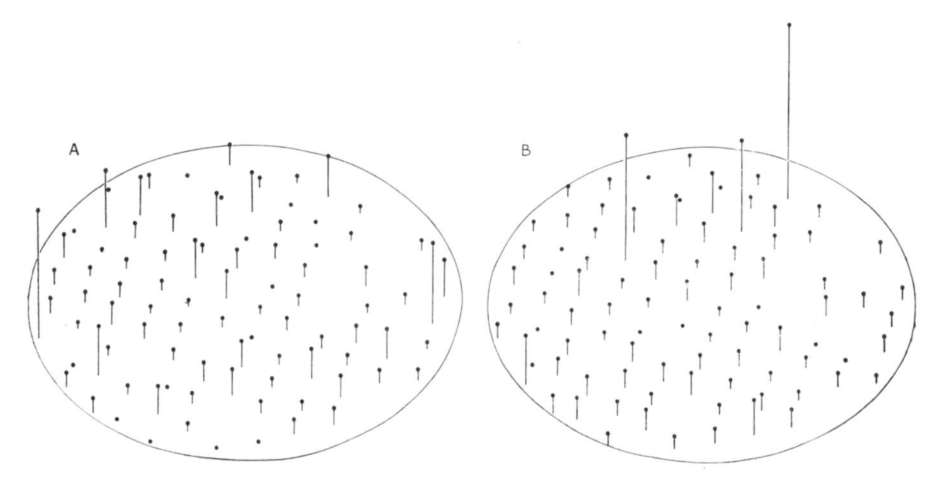

FIG. 13.01.—The distribution of the blowfly *Lucilia cuprina* in an area of grassland near Canberra. The diameter of the circle was 8 miles. Traps were distributed evenly over the area, as indicated by the columns. The height of each column is proportional to the number of flies caught in the trap. Two experiments are represented: *A* was started on December 7; *B*, on March 13. (Data from Gilmour *et al.*, 1946.)

Gilmour *et al.* (1946) had already shown that the number of flies caught in the traps represented between 0.001 and 0.0001 of the total population on the area, from which we may conclude that the chance that any one fly would be caught in a trap was quite small. Therefore, if the flies were distributed at random over the area, the numbers caught in the traps might be expected to conform to the Poisson series. We calculated the theoretical Poisson series for each experiment and compared the theoretical frequencies with the observed ones. The results are set out in Table 13.04. If the flies had been distributed over the area truly at random, values of χ^2 as large as those that may be calculated from the bottom row of the table would be expected less often than once in 1,000 trials. We can therefore say with a high degree of assurance that *Lucilia cuprina* were not distributed at random over this area of approximately 50 square miles. The values of $\Sigma(x + \bar{x})^2/\bar{x}(n + 1)$ show that the departures from randomness were due to excessive patchiness. The same may be inferred by inspection of the frequencies set out in the body of the table; for example, in the second experiment, too many traps caught from 0 to 12 flies or from 23 to 113 flies, and too few traps caught the intermediate numbers of flies. Similar discrepancies between the observed and theoretical frequencies occurred in experiments 3 and 4.

The statistics shown in Table 13.04 establish beyond reasonable doubt that the flies were nonrandomly distributed over the experimental area at the time

TABLE 13.04

OBSERVED FREQUENCIES (TRAPS) COMPARED WITH APPROPRIATE POISSON SERIES FOR EACH OF THREE EXPERIMENTS WITH *Lucilia cuprina* BY GILMOUR *et al.* (1946)

Experiment No. 2			Experiment No. 3			Experiment No. 4		
No. of Flies per Trap	No. of Traps Obs.	No. of Traps Expected	No. of Flies per Trap	No. of Traps Obs.	No. of Traps Expected	No. of Flies per Trap	No. of Traps Obs.	No. of Traps Expected
0– 12	52	11.64	0– 3	55	18.26	0– 4	54	19.80
13– 14	7	13.48	4	10	15.12	5	9	13.44
15– 16	6	18.13	5	8	17.17	6	6	15.13
17– 18	1	19.04	6	3	16.25	7	8	14.59
19– 20	3	16.06	7	3	13.17	8– 9	5	21.55
21– 22	4	11.11	8–64	23	21.66	10–88	16	14.29
23–113	28	11.55	....	....	....	....	....	....
Total	207	207	....	99	99	....	81	81
P.........	0.001			0.001			0.001	
$\frac{\Sigma(x-\bar{x})^2}{\bar{x}(n-1)}$..	22			13			22	

of each experiment. An inspection of Figure 13.01 suggests that the distribution of flies had changed markedly from one experiment to the next. To test this hypothesis, we correlated the numbers of flies caught in each trap during the different experiments. The values obtained were $r_{23} = 0.18$, $r_{24} = 0.08$, $r_{34} = 0.11$. None of these values was significant; hence it may be concluded that there was no consistency in the way that the flies were distributed over the experimental area at the times of the different experiments. In other words, the concentrations of flies occurred in different places at different times.

An important feature of these experiments was that 40,000 flies which had been reared in an insectary and marked with a dye were liberated near the center of the circle one day before the trapping began for each experiment. Some of the marked flies were subsequently recaptured in the traps. In each experiment there was a tendency for the traps that caught most wild flies to catch most marked ones also, and vice versa. From this it may be inferred that the local situations where the wild flies were most numerous were also the places that were most attractive to the marked ones. From the correlations given in the preceding paragraph, we know that these centers of concentration were differently situated in December, January, and March. They were therefore unlikely to be associated with permanent physical features like streams, dams, sheds, shade belts, and so on. Nor was there any evidence that they were associated with flocks of sheep.

Dobzhansky and Pavan (1950) measured the relative density of various species of *Drosophila* in a forest in Brazil. Baits were set at 10-meter intervals along two transects, each 200 meters long, which intersected in the middle at right angles. The baits attracted flies in the immediate vicinity, and, at regular intervals during the day, flies that came to the baits were collected by sweeping a net over the baits. The numbers collected reflected the density of flies in the vicinity of the baits. The results for two groups of species are illustrated in Figure 13.02. The lines at right angles represent the transects, and the height

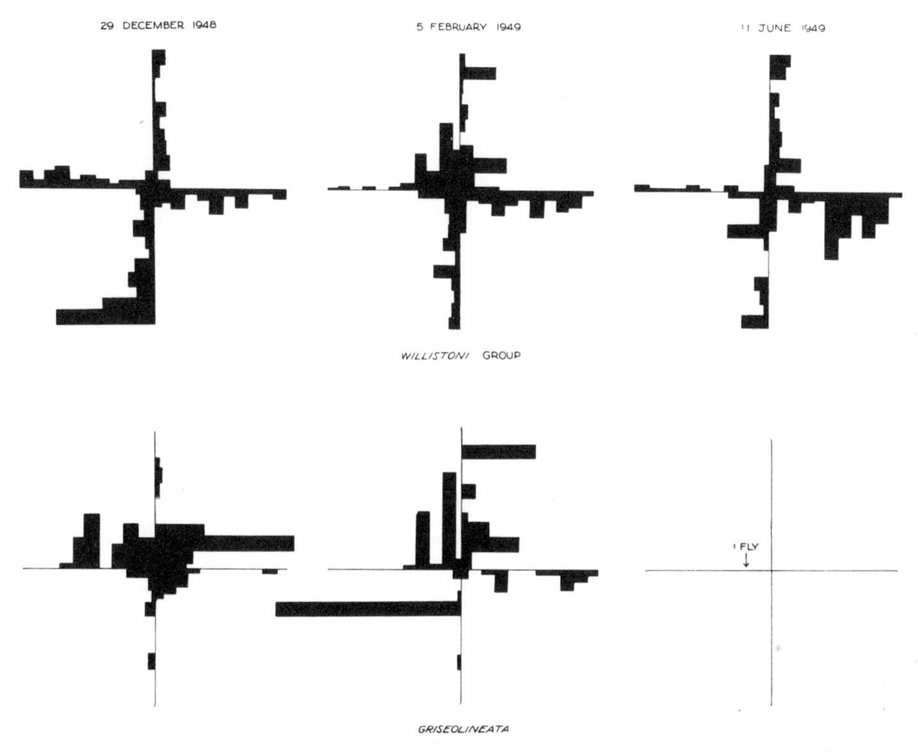

FIG. 13.02.—The distributions of two groups of sibling species of *Drosophila* in a forest in Brazil. Two rows of traps were arranged in the form of a cross, and the experiment was repeated on dates indicated. For further explanation see text. (After Dobzhansky and Pavan, 1950.)

of the black columns along these lines are proportional to the percentage of flies collected at each point. Tall columns indicate that a large proportion of the total flies along the transect occurred at this point; low columns signify that the species were rare. These and other results showed that the distributions of the flies in the forest were extremely patchy. The pattern of patchiness varied from month to month; in places where one species was numerous, the others were often scarce; in other words, the pattern varied for each species. These points are illustrated in Figure 13.02. The causes of patchiness among these populations may have been quite complex. Dobzhansky and Pavan

(1950) suggested that it was associated with irregular distributions of fruits of different sorts which attracted the different species of flies. The flies are also sensitive to moisture; this varied as the transect crossed a damp depression or a dry slope. The flies are also sensitive to light; this varied, depending largely on the amount of vegetation. But whatever may be the chief causes of patchiness, this is a good example of patchiness within a relatively small area. We have no doubt that the phenomenon would have been more marked if the transects had extended for miles instead of meters.

Many sorts of insects and other animals concentrate in small areas for courting or mating. Barton-Browne (personal communication) has observed adults of the fruit fly *Dacus tryoni* congregating at dusk toward the end of an orchard nearest the setting sun. During the brief period of dusk (about 30 minutes) the flies court and mate. They do this at no other time of the day, when they are more widely distributed throughout the orchard.

So far the examples all refer to relatively small areas chosen because they possessed a certain uniformity; the floor of a forest, the soil in a meadow, even the 60 square miles of grassland studied by Gilmour *et al.*, may all be regarded as small uniform areas, arbitrarily chosen from the much larger area which is the distribution of the whole species. Let us now consider a much broader sort of area. Take, for example, the distribution of certain butterflies in Great Britain. Ford (1945*a*, p. 268), drawing from his wide experience of many species, wrote: "Colonies of a species may be localised by a variety of conditions. They may be cut off on islands, or they may be restricted to a particular type of habitat, such as mountains, marshes, moors or woods, and any of these may be completely isolated from places of a similar character. It is probable that in Great Britain the course of civilisation is increasing the isolation of colonies, . . ."

The numbers in each local colony fluctuate, perhaps between wide extremes, and not always synchronously. This adds another component to the variability in the relative abundance of a species from place to place when broad areas are concerned. A quotation from Ford (1945*a*, p. 268) illustrates this point:

> ". . . fluctuations in numbers and in areas of distribution take place throughout the whole population of certain butterflies, but isolated colonies of a species are especially susceptible to such fluctuations, whether as a part of the more general trend or in response to local conditions. In an isolated habitat the food supply may partly fail, circumstances may for a season or more favour the parasites or the diseases of an insect, or bad weather may occur at some critical stage in the life-cycle. Any of these events may greatly reduce an isolated population and cause it to fall far below the numbers which the area could support in favourable conditions. When these return, a recrudescence of the colony may be expected, so that the colony will fluctuate over a long or short period according to circumstances.

Records of the relative abundance of the butterfly *Euphydryas aurinia* in one isolated colony in Cumberland are available for 55 years. These were summarized by Ford (1945*a*, p. 268) as follows: "The species was quite common in 1881, and gradually increased until by 1894 it had become exceedingly abundant. After 1897, the numbers began to decline, and from 1906 to 1912 it was quite scarce. From 1913 to 1919 it was very rare, so that a few specimens only could be caught each year as a result of long-continued search, where once they were to be seen in thousands. From 1920 to 1926, a very rapid increase took place, so that by 1925 the butterfly had become excessively common, and so it remained until we ceased our observations in 1935."

The colony in this place, though greatly depleted at times, survived. Elsewhere colonies have been known to become extinct. And this is a risk that is quite important for small colonies that fluctuate widely in numbers. Another quotation from Ford (1945*a*, p. 135) illustrates how this may add to the patchiness of the distribution of the species at any one time and adds to the variability from time to time: "This butterfly [*Euphydryas aurinia*] occurs widely in suitable places in England, though it largely avoids the eastern counties. However, it used to occur near Deal in Kent, and it has recently reappeared in Hertfordshire after a long absence. Formerly it existed in a few places in East Anglia, where it seems to be extinct, and there are one or two localities for it in Yorkshire."

The best example that we can give of a population distributed with excessive uniformity comes from the work of Holme (1950) with the small lamellibranch *Tellina tenuis*, which lives in vertical burrows in the sand on beaches in Britain. Holme studied the distribution of *T. tenuis* in certain selected areas of beach in the estuary of the River Exe. He chose a number of "stations" situated between high- and low-water marks and counted the numbers of *T. tenuis* in from 2 to 12 quadrats of 0.01 square meter from each station. Altogether, there were 33 samples, which is scarcely enough for an adequate test. The values of $\Sigma(x - \bar{x})^2/\bar{x}(n - 1)$ at three stations, where 4, 12, and 7 samples were taken, were, respectively, 0.13, 0.31, and 0.09. With so few samples, none of these figures is independently significant. But all the 33 samples may be pooled to give an estimate of χ^2. This was found to be 6.437, which, with 22 degrees of freedom, corresponds to a probability of less than 0.01. The observed distribution is significantly different from the Poisson series, but in this case the variance is too small; such a degree of uniformity in spacing would have happened by chance less often than once in 100 trials.

It is not known whether *Tellina tenuis* has some means of "defending a territory" or whether this uniformity in its distribution comes about in some other way. In an area as uniform as a few square yards of beach, "territorial behavior" on the part of the animals living there might be expected to lead to a high degree of uniformity in their distribution. With birds living in an area where resources of food, nesting sites, shelter, and so on may be unevenly distributed, territorial behavior may not lead to any marked uniformity in the distribution of the birds.

Allee *et al.* (1949, p. 366) and Hutchinson (1953) mentioned several other examples in which the pattern of distribution of certain natural populations had been studied. It is most unusual to find one in which the distribution is random. It is almost equally unusual to find one in which the departure from randomness is in the direction of excessive uniformity. It is generally true, with very few exceptions, that natural populations are distributed nonrandomly and that the departure from randomness is in the direction of excessive patchiness. This has to be taken into account when devising methods for sampling natural populations and for testing the significance of differences between means (see Bliss and Fisher, 1953). Also, of course, it emphasizes the unreality of those mathematical models which assume that populations of animals are distributed either at random or uniformly through any considerable area of country.

13.1 OF FOUR NATURAL POPULATIONS IN WHICH THE NUMBERS ARE DETERMINED LARGELY BY WEATHER

The numbers of many animals are largely determined by weather. In chapters 6, 7, and 8 we have shown how the main components of weather, which are temperature, moisture, and light, may influence the animal's chance to survive and multiply. In this section we describe how weather determines the numbers in certain natural populations. Studies of this sort must be based on a sound knowledge of the physiology and behavior of the animal, but it is not necessary to add much of these details to this chapter, since this has already been done in earlier chapters.

In the first example in this section, weather is shown to influence the animal's chance to survive and multiply directly and also indirectly through its influence on the amount and distribution of food. In the second example, weather is important, in addition to its direct influence, indirectly through its association with food and diseases. In the last two examples the association between weather and the numbers in the populations is clear enough, but the relationships have not been explained so fully as in the first two examples. Weather may also come into the discussions of the examples in sections 13.2 and 13.3, but it does not have such a dominant place as in those selected for discussion in this section.

13.11 *Fluctuations in the Numbers of* Thrips imaginis *in a Garden in South Australia*

A natural population of *Thrips imaginis* in the grounds of the Waite Institute near Adelaide, South Australia, was studied by Davidson and Andrewartha (1948*a*, *b*). This insect belongs to the order Thysanoptera, family Thripidae. It is indigenous to southern Australia, and, like many other species in this region, its opportunities for multiplication have increased greatly since the country was developed for agriculture. But this era of change is largely over now.

13.111 BIOLOGY OF *Thrips imaginis*

The life-cycle of *Thrips imaginis* is not complicated by diapause, nor is there any season of the year when it becomes quiescent. Breeding and development go on continuously, but it will be shown later that both the birth-rate and the survival-rate fall to low levels during the height of the summer and during the winter, so that r remains negative during these seasons of the year.

The fully grown insect is about 1 mm. long, and many hundreds of them can find room in one rose. In nature they frequent the flowers of roses, fruit trees, a wide variety of garden plants, and weeds. In the laboratory, adults lived for 77 days at 24° C. on a diet that lacked pollen; but they laid scarcely any eggs; 9 thrips averaged 20 eggs each. When pollen was added to the diet, the average length of life was decreased to 55 days, but the average number of eggs increased to 209. On a diet which included pollen the mean length of life varied from 250 days at 8° to 46 days at 23° C. The mean number of eggs per thrips varied from 192 at 12.5° to 252 at 23° C. Eggs were produced fairly evenly throughout adult life at the rate of 1.4 per day at 12.5° and 5.6 per day at 23° C. (Andrewartha, 1935).

In nature the eggs are usually imbedded in the soft tissues of the flower, and the two nymphal stages are commonly found in the flowers alongside the adults. In the laboratory it proved impossible to rear the nymphs on a diet that lacked pollen; but when pollen was added to the diet, they could be reared quite readily. The duration of the life-cycle from egg to adult was 44 days at 11° and 9 days at 25° C. (Andrewartha, 1936). A generation may be completed in a few weeks during the warm period of the year but may require many months during the winter (Davidson, 1936*b*).

Not only is the speed of development greatly retarded during the winter, but the death-rate, especially among the nymphs, may be high because few flowers would last long enough for the nymphs to complete their development at this season of the year. The stages best adapted to survive the winter are the adults and the pupae. The latter are sometimes found in the flowers alongside the active stages, but it is more usual for the fully grown nymph to leave the flower and pupate among the litter at the base of the plant or in the soil just below the surface. The egg, imbedded in the tissue of the plant, is surrounded by moisture; the nymphs and adults may replenish their supplies of moisture by sucking sap; but the pupae, which do not feed, are likely to be fatally desiccated during hot, dry weather. This may be one of the major causes of death during the summer.

The behavior of the adult with respect to dispersal was not observed directly. Nevertheless, there was clear evidence that the species has a strong innate tendency toward dispersal; this evidence was summarized in section 5.1. This behavior is advantageous during the spring, when there are suitable flowers everywhere. But a strong tendency to disperse when flowers are very sparsely distributed may result in a high death-rate among the adults during the height of the summer and during the winter (sec. 11.12).

During this investigation samples of flowers containing *Thrips imaginis* were examined almost daily for 14 years. We occasionally found a few other species of animals in the flowers. These included several other species of thrips, an occasional beetle, a few stray aphids or mites. All told, they would add up to a fraction of 1 per cent of the numbers of *T. imaginis* in the flowers. A few adult *T. imaginis* might occasionally be caught in spiders' webs, though certainly this was an infrequent occurrence. A few pupae might have been eaten by foraging beetles, spiders, and mites, etc.; but we gained no evidence of this, and, if it occurred at all, it must have been on quite a small scale. Indeed, *T. imaginis* is remarkable for the virtual absence of other sorts of animals from its environment.

13.112 CLIMATE AND VEGETATION IN THE AREA WHERE THE POPULATION WAS LIVING

The climate of Adelaide is broadly like that of the region near the Mediterranean Sea. The winter is mild; the mean minimal and maximal temperatures for July (coldest month) are 7.5° and 15° C., respectively. The summer is hot and rather severely dry; the mean minimal and maximal temperatures for January (hottest month) are 16.3° and 30° C., respectively. The little rain that falls during summer is quite inadequate to moisten the parched earth except very temporarily (Fig. 6.04). The more permanent vegetation consists mainly of xerophytic shrubs and trees, which remain dormant during summer; the annuals, which constitute nearly all the herbaceous vegetation of the area, spend the summer as seeds. Most of the rain falls during the winter; usually there is sufficient to maintain the soil permanently moist for about 7 months of the year (Davidson, 1936*c;* Fig. 11.01). This is the growing season for plants, and it culminates in a great outburst of blossoming in the spring.

The ending of the summer and the beginning of the growing season is quite abrupt, and one can usually recognize the "break" of the season quite unequivocably. During the 15 years between 1932 and 1946 the earliest "break" came on February 18 (1946) and the latest on June 3 (1934). When the break came early, the seeds germinated quickly, and the plants completed relatively more of their vegetative development during the autumn before growth was slowed down by the lower temperatures of winter. In these circumstances the plants flowered earlier in the spring and more abundantly. For example, one annual *Echium plantagineum*, which is commonplace and of which the flowers are highly favored by *Thrips imaginis*, was observed in full bloom on August 3 in 1946 but not until October 3 in 1936; other species responded similarly. It will be shown later (sec. 13.115) that this has a big influence on the numbers of *T. imaginis* present during the spring.

13.113 METHOD OF SAMPLING

Simple trapping was practiced, and only adults were recorded. Simple trapping may indicate relative changes in abundance but provides no way of estimating the absolute numbers in the area drawn upon by the traps. In this case the traps consisted of 20 roses picked at random from a long hedge in the garden; the roses were always chosen at the same stage of development. The particular variety, "Cecil Brunner," lacks anthers, so they are useless as breeding places for *Thrips imaginis*. But the thrips are strongly attracted to them, and they are excellent traps for the adults. From April, 1932, to December, 1946, the roses were collected at 9:00 A.M. every day except Sundays and certain holidays; from 1939 to 1946 daily samples were collected each year during the period September to December, except that no records are available for 1944.

Simple trapping gives an adequate estimate of relative abundance only if the traps catch a sufficiently constant proportion of the population independent of variations in numbers and other circumstances. In the present experiments there was clear evidence of rather pronounced short-term fluctuations superimposed on a steady trend either up or down, depending on the season (Fig. 13.03). This suggested that the numbers which we were catching on any one day depended partly on the numbers in the area and partly on the activity of the thrips in seeking out the flowers. We have no elaborate information about the behavior of the *Thrips imaginis* to guide us in seeking an explanation of this varying activity. But experience suggested that the thrips were likely to be more active on a fine warm day. Such weather often occurs in Adelaide when the barometer is falling.

The hypothesis was tested by fitting the data to a partial regression of the general form

$$\log \frac{Y}{Y_t} = b_1(x_1 - X_1) + b_2(x_2 - X_2) + b_3(x_3 - X_3) \ . \ . \ . \ ,$$

in which Y and Y_t refer to the numbers of thrips; $x_1, X_1, x_2, X_2, \ldots$, refer to certain components of the weather; and $b_1, b_2, b_3, \ldots$, are coefficients which measure the strength of the association of the dependent variate $\log Y/Y_t$ with the independent variate $(x_1 - X_1), (x_2 - X_2), \ldots$, independent of their association with each other. The numbers of thrips were converted to logarithms because it was more instructive and realistic to consider relative changes in their numbers than absolute changes. The dependent variate was expressed as the logarithm of a ratio, and the independent variates as differences, because it was necessary with all of them to eliminate trends with time before testing the hypothesis about activity. In the case of the thrips, the trend was due chiefly to the growth of the population; with the independent

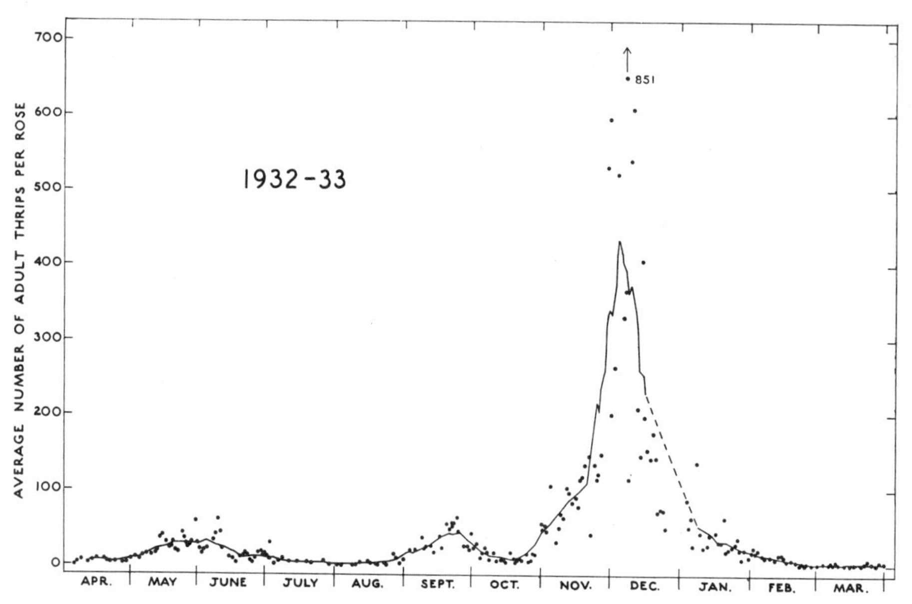

FIG. 13.03.—The individual daily records and the trend throughout 1 year in the number of *Thrips imaginis* per rose. The points represent daily records. The curve is a 15-point moving average. (After Davidson and Andrewartha, 1948*a*.)

variates, the trends were due to the progression of the seasons. The trends were eliminated by fitting the data to equations of the general form

$$Y = a + bx + cx^2 + dx^3.$$

Hence in the equation for partial regression given above, Y_t, X_1, X_2, X_3, . . . , refer to values taken from the calculated smooth trend lines and Y, x_1, x_2, x_3, . . . , refer to the actual daily records of thrips, temperature, rainfall, and so on (Fig. 13.04).

In the equation for partial regression $\log Y/Y_t$ may also be written $(\log Y - \log Y_t)$, which serves to emphasize the fact that in this method we are comparing departures from the usual, or expected, number of thrips with departures from the usual, or expected, weather. Thus if there are unexpectedly many thrips in the roses on days that are unusually warm, this will be indicated by the value of the appropriate coefficient. Similarly, if there are unusually few thrips found on unusually wet days, this will be measured too.

Moreover, if wet days are also sometimes cold, this method will tell to what extent the low numbers recorded on wet days are associated with rain independently of temperature and to what extent with low temperature independently of rain.

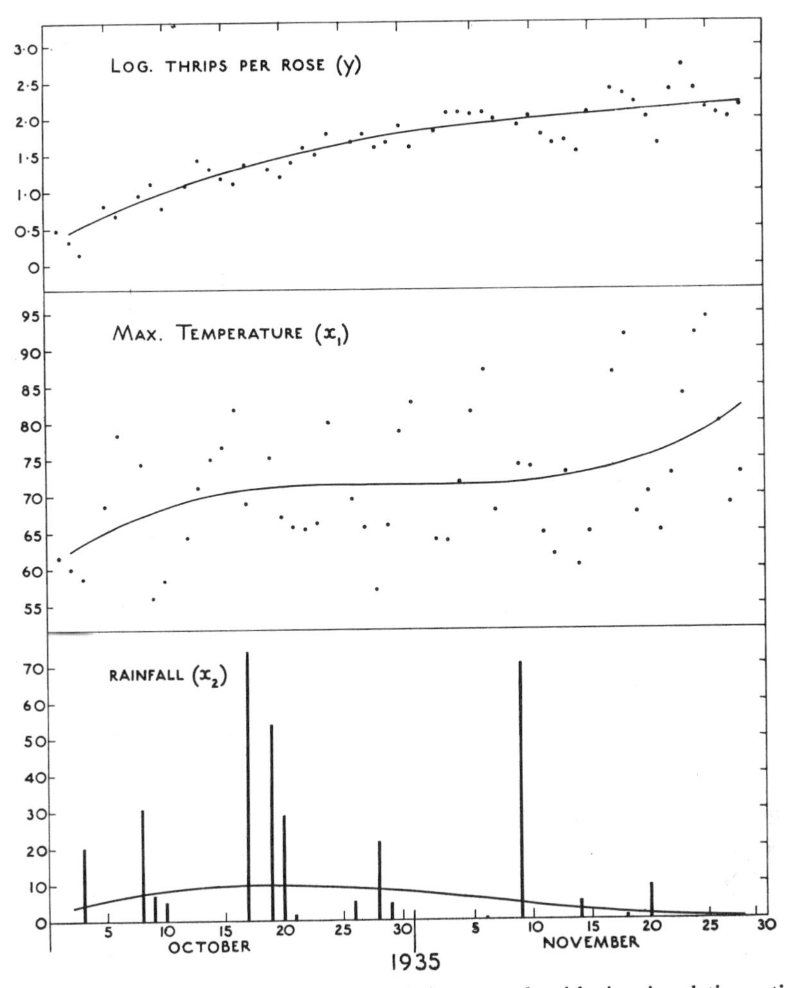

FIG. 13.04.—The curves which were used to eliminate trends with time in relating activity of thrips, log y, to maximum temperature, x_1, and rainfall, x_2. For further explanation see text. The curves were calculated from the expressions:

$$Y = 1.6195 + 0.04288\,\xi_1 - 0.005235\,\xi_2 + 0.00006393\,\xi_3;$$
$$X_1 = 71.805 + 0.3359\,\xi_1 + 0.008789\,\xi_2 + 0.003627\,\xi_3;$$
$$X_2 = 5.715 - 0.2056\,\xi_1 + 0.06132\,\xi_2 + 0.002255\,\xi_3;$$

in which X_1 is in degrees Fahrenheit; X_2 in units of 0.01 inches of rain; and Y in units of log thrips per rose. (After Davidson and Andrewartha, 1948*b*.)

The first partial regression which was formulated included three independent variates—for rainfall, temperature, and barometric pressure. The last was shown to have no influence independently of temperature and rainfall. So it was excluded from the regression. Since the roses may be attractive to *Thrips imaginis* for several days before they are picked, the next regressions included six independent variates, namely, temperature and rainfall for each of the three days preceding the collection of the roses. The regression as a whole was significant, but only those terms relating to the day immediately before sampling were significant in their own right. So a third regression was calculated with only two terms. It was

$$\log Y - \log Y_t = 0.007925(x_1 - X_1) - 0.001969(x_2 - X_2),$$

where x_1 was the maximal temperature in degrees Fahrenheit and x_2 was rainfall in units of 0.01 inch for the day immediately before that on which the sample was taken.

From this equation it may be shown that, on the average, the samples overestimated the numbers of thrips by 2.5 per cent for every degree the temperature exceeded the "usual" as indicated by the trend line. Similarly, the samples underestimated the numbers by 6.6 per cent for every 0.1 inch of rain. The regression accounted for only 10 per cent of the total variance, so there was probably much "activity" of the thrips which we had failed to explain. Nevertheless, this equation may be used to correct the crude results given by the samples. All the figures used in the calculations discussed in section 13.115 have been so corrected. Details of the method may be seen in Davidson and Andrewartha (1948*b*).

13.114 THE DATA

Samples were taken on 2,291 days throughout the 14 years, and, of these, 1,252 samples were taken during the period from September 1 to December 30, which is the season when the thrips multiply to a maximum each year. Altogether, during the whole investigation some 6,000,000 thrips were recorded from the samples.

These data were analyzed to discover how closely certain components in the environment of a *Thrips imaginis* were associated with the maximal numbers attained each year. The method of partial regression was used. The dependent variate was a quantity which represented the number of thrips; the independent variates were quantities which measured certain components of the weather. The first step was to calculate from the data a quantity which would represent the abundance of the thrips during the period of each year when they reached their greatest numbers. Figure 13.05 shows that the numbers of thrips in the roses were always low at the beginning of the spring; they increased during the next few months and then declined more or less abruptly. Although the daily fluctuations were pronounced, the trend, as indicated by a 15-day running average, was clear, and the maximum of the curve could be clearly recognized. The date on which the maximum occurred was usually close to November 30, but it varied from November 15 in 1939 to December 13 in 1945. The dates on which the maxima occurred each year are indicated by arrows in Figure 13.05. The maximum of the smoothed curve might have served as the criterion we were seeking; but, for reasons which were explained in Davidson and Andrewartha (1948*b*), we chose instead the average for the 30 days preceding the maximum. The mean logarithm of the numbers of thrips in one rose for these 30 days for each of the 14 years are shown in Table 13.05.

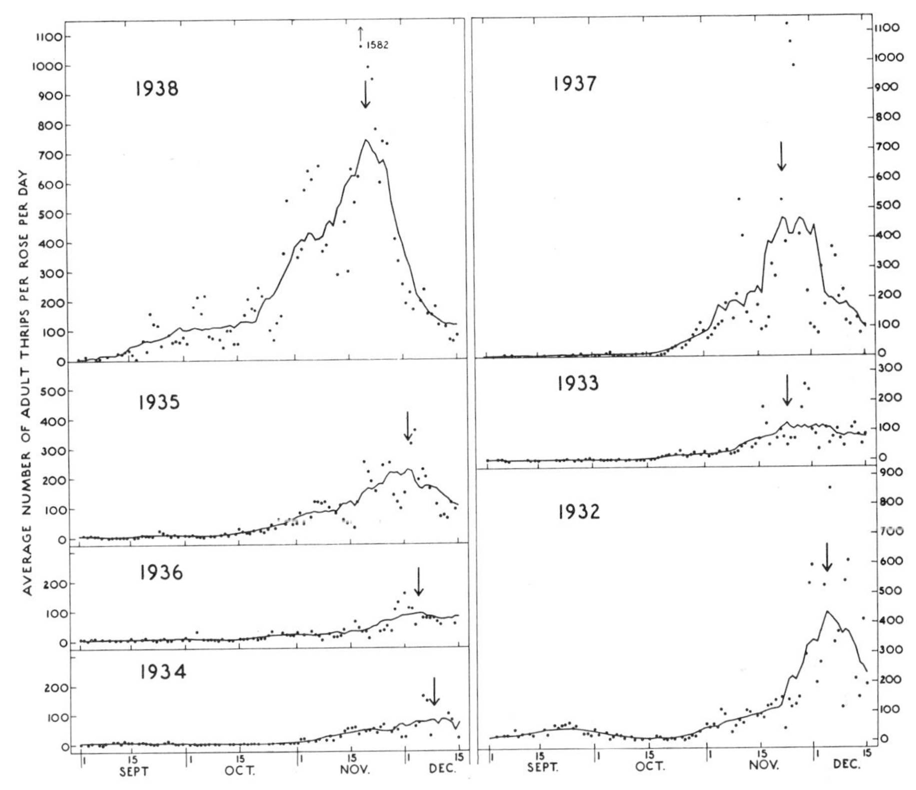

FIG. 13.05.—The numbers of *Thrips imaginis* per rose during the spring each year for 7 consecutive years. The points represent daily records; the curve is a 15-point moving average. The arrow indicates the date on which the curve reaches the maximum for each year. Note that the same general trend was repeated during the spring each year, with variations in the height attained by the curve and date on which the maximum was attained. (After Davidson and Andrewartha, 1948*a*.)

These are the values which constituted the dependent variate (log y) in the regression, which is discussed in section 13.115.

In our choice of independent variates for the regression we were guided by our knowledge of the biology of *Thrips imaginis* and the climate of the area where the population was living. In this climate, r for *T. imaginis* is likely to be negative at all seasons of the year except during spring and perhaps for a brief period during autumn in some years (Figs. 13.03, 13.05). The explanation for this may be given briefly as follows: During the summer, except for the early part, the places where pupae occur are likely to be so arid that few may be fortunate enough to survive; also, suitable flowers for breeding become scarce and sparsely scattered, and this adds to the hazards of life for adults, reducing their chances both of surviving and of leaving offspring behind them. During the winter, food continues to be sparsely distributed; also with the prevailing low temperature the thrips develop slowly, and many nymphs die simply because they fail to complete their development before the flowers in which they are living wither. With rising temperatures in the spring, the thrips develop more quickly; the soil, wet from winter and remoistened by spring rains, permits a high rate of survival among pupae. In some years flowers continue to be scarce at first, but it is a striking feature of this region that the blossoming of a high proportion of the natural vegetation is condensed into a brief period in the spring; the annuals, which preponderate, must set seeds before the summer drought sets in; and the perennials also tend to produce most of their flowers at this season. Consequently, the thrips flourish and multiply at this time until, as summer develops, the soil dries out and the flowers wither and disappear. The thrips which had been so abundant become few again.

Even when the thrips are most numerous, the flowers in which they are breeding do not appear to be overcrowded except perhaps locally or temporarily: while the thrips are multiplying, the flowers increase even more rapidly. Then, when the population begins to decline, the flowers become less crowded still. This may be partly because the survival-rate among pupae depends upon the moisture in the top half-inch of soil, whereas the plants draw their water from a greater depth. Also, as the flowers begin to thin out and the distances between suitable breeding places become greater, r may be reduced by this cause long before the flowers become absolutely scarce (sec. 11.12).

Considerations of this sort led to the hypothesis that the numbers achieved by the thrips during each year were determined largely by the duration of the period that was favorable for their multiplication. When this period was prolonged, the thrips would ultimately reach higher numbers; when it was briefer, the decline would set in while the numbers were still relatively low.

It was explained in section 13.112 that the flowers of a wide variety of annual plants are of chief importance for *Thrips imaginis*. The date on which these plants burst into flower depends largely on the stage in vegetative growth reached by the end of the winter. The date on which blossoming finishes depends largely on the amount of rain falling during the spring. The beginning of blossoming is more variable than its ending, but both may influence the duration of the favorable season for *T. imaginis*.

Therefore, in seeking quantities which might be associated with the numbers of thrips in the spring, we looked first for one which would represent the opportunity for growth during autumn and winter afforded to the annual plants which were chiefly important in the ecology of *Thrips imaginis*. This was done by the familiar method of summing "effective temperature" in units of day-degrees. Assuming arbitrarily a threshold of 48° F., the number of day-degrees of effective temperature appropriate to each day were calculated from the expression

$$T = \frac{\text{Maximal daily temperature} - 48}{2}.$$

By reference to certain climatological studies of this region (Davidson, 1936*c;* Trumble, 1937) we devised an arbitrary method of determining the "break" in the season; this was taken as the date on which the seeds began to germinate. Starting from this date and proceeding to August 31, daily effective temperatures were summed. The 14 yearly totals are given under the heading x_1 in Table 13.05.

TABLE 13.05*
INDIVIDUAL VALUES OF QUANTITIES POSTULATED AS LIKELY TO INFLUENCE NUMBERS OF THRIPS PRESENT DURING SPRING

Year	Date of "Break" of Season	Independent Variates†				Dependent Variate‡ log y
		x_1	x_2	x_3	x_4	
1932	Mar. 27	12.09	4.37	6.09	13.77	2.08
1933	Apr. 11	10.42	4.39	7.04	12.09	1.57
1934	June 3	6.33	6.55	7.07	10.42	1.65
1935	Mar. 14	13.92	5.48	7.15	6.33	2.03
1936	Apr. 7	11.54	3.94	6.48	13.92	1.54
1937	Mar. 28	12.37	4.37	7.16	11.54	2.04
1938	Feb. 19	18.74	2.03	7.29	12.37	2.51
1939	Feb. 24	18.34	2.71	6.82	18.74	2.72
1940	Apr. 7	10.89	2.38	7.85	18.34	2.10
1941	Apr. 3	12.96	6.48	6.81	10.89	2.40
1942	Mar. 30	12.62	6.05	7.28	12.96	2.57
1943	Apr. 7	9.63	4.36	6.96	12.62	1.79
1944	Apr. 2	....	....	....	....	....
1945	May 4	8.33	6.76	6.42	9.65	1.93
1945	Feb. 18	17.69	3.28	6.43	8.33	2.28

* The numbers of thrips are given as log y in this table. After Davidson and Andrewartha (1948*b*).
† x_1 = Total effective day-degrees (in units of 100 day-degrees) from the "break" of the season to August 31. x_2 = Total rainfall (in inches) for September and October. x_3 = Total effective day-degrees (in units of 100 day-degrees) for September and October. x_4 = As for x_1 but for the year preceding the one when the samples of thrips were taken.
‡ y = The geometric mean number of thrips in one rose for the 30 days preceding the maximum of the smoothed curve (see explanation in text).

Rain falling during the spring would tend to sustain the plants and thus prolong the period favorable for the multiplication of thrips. Also rain falling at this time might be expected to increase the rate of survival of pupae. So the total rainfall for September and October was included in the hypothesis; it appears as x_2 in Table 13.05.

Although, as a rule, the spring is warm enough to promote the breeding of *Thrips imaginis*, the temperature may be lower than optimal. Therefore, we included the temperatures during the spring in the hypothesis and calculated as x_3 the sum of the daily effective temperatures for September and October.

Finally, there was the possibility that the influence of a "good" season might be carried forward to the next by some means or another, perhaps through a greater "carry-over" of thrips or of seeds. To test this possibility, x_4 was added to the quantities in Table 13.05. It is the same quantity as x_1, but the series has been displaced by 1 year.

13.115 ANALYSIS OF THE DATA

The method of partial regression described by Fisher (1948, p. 147) is well suited to the analysis of data of this sort, because each coefficient of partial regression expresses the degree of association between its independent variate and the dependent variate when all other terms *included in the regression* are held at their means. In other words, it measures the degree of association which each one has with the dependent variate independently of all the others. Regression measures degree of association; causal relationships need to be inferred on biological grounds. This is generally true of all correlations, but the point may be driven home in relation to the present study by supposing that some other quantity *not included in the regression* were correlated closely with, say, x_2. Then if this hypothetical quantity were substituted for x_2 in the calculations, it would be found to have the same degree of association with y as x_2 had. But we would still be left to decide which one (if either) was causally related to y; and this would have to be done from our knowledge of the biology of the animal and its environment. The biology of *Thrips imaginis* had been investigated thoroughly in preparation for this ecological study; and the climate of Adelaide was well known from the work in climatology that had been done at the Waite Institute.

Before giving the results of the analyses, it is necessary to explain why the numbers of thrips were transformed to logarithms. We would expect that for a certain increment in, say, temperature the numbers of thrips would be multiplied by a constant factor; we would not expect their numbers to be increased by the addition of a constant number. This may be expressed in symbols: the expected relationship would not be of the form $Y = k + ax$ but would more likely be of the form $Y = ka^x$; or, if several "causes" were being investigated, it would be

$$Y = ka_1^{x_1} \cdot a_2^{x_2}. \tag{i}$$

Taking logarithms of this expression gives

$$\log Y = \log k + x_1 \log a_1 + x_2 \log a_2. \tag{ii}$$

This equation is, indeed, the general equation for partial regression, which is usually written in statistical textbooks in the form

$$Y = a + b_1x_1 + b_2x_2. \tag{iii}$$

Therefore, it was appropriate merely to transform the numbers of thrips to logarithms and proceed with the analysis by the standard methods of partial regression. This was done, giving the equation

$$\log Y = -2.390 + 0.1254x_1 + 0.2019x_2 + 0.1866x_3 + 0.0850x_4. \qquad \text{(iv)}$$

From equation (iv) the expected (or theoretical) values for y may be calculated. These are shown in Figure 13.06. The agreement with the observed values is

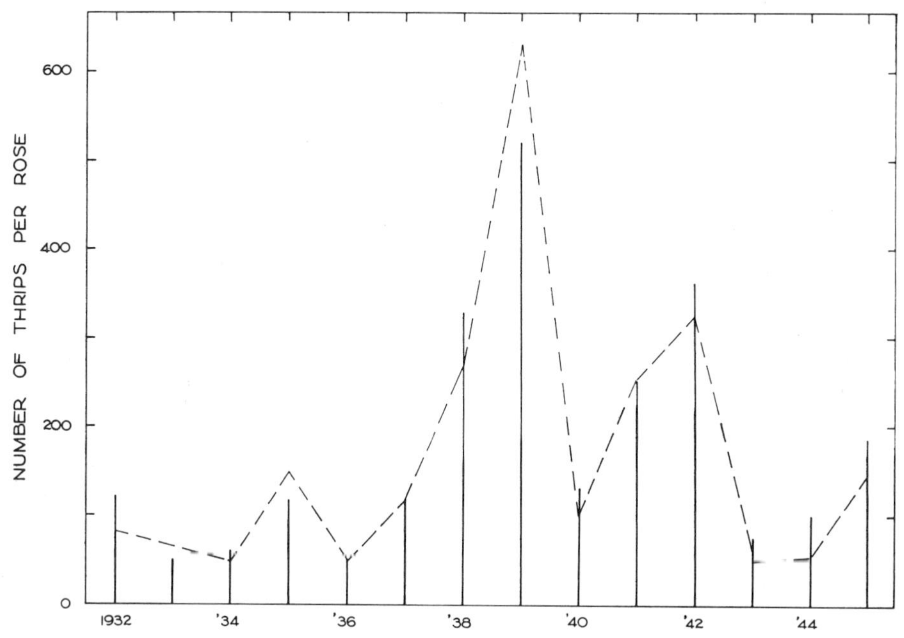

FIG. 13.06.—The columns represent the geometric means of the daily counts of thrips per rose. The curve represents the theoretical values for the same quantities calculated from the expression:

$$\log Y = -2.390 + 0.125x_1 + 0.202x_2 + 0.187x_3 + 0.058x_4.$$

The independent variates x_1, x_2, x_3, and x_4 are the quantities given in Table 13.05. Note that they are altogether different quantities from those designated x_1 and x_2 in Fig. 13.04. (After Davidson and Andrewartha, 1948*b*.)

remarkable. But it is more instructive for our present purposes to present the results in tables setting out the analysis of variance (Table 13.06*b*) and the significance which can be attributed to the regression and the individual terms in it (Table 13.06*a*).

The relative magnitudes of the coefficients $\log a_1$, $\log a_2$, and likewise the percentage increases associated with unit increases in $x_1, x_2, \ldots$, depend upon the arbitrary choice of units for $x_1, x_2, \ldots$ Therefore, the magnitudes of these quantities provide no guide to the relative importance of $x_1, x_2, \ldots$, in determining the magnitude of $\log y$. This difficulty may be overcome by expressing the coefficients in "standard measure" thus:

$$\beta_1 = b_1 \sqrt{\frac{S(x_1 - \bar{x}_1)^2}{S(y' - \bar{y}')^2}},$$

TABLE 13.06*a*

REGRESSION OF LOG y (NUMBER OF THRIPS PER ROSE) ON x_1, x_2, x_3, AND x_4 (SEE TABLE 13.05)

VARIATE	REGRESSION COEFFICIENTS		S.E.	t	P
	β	b			
x_1	1.224	0.1254	0.0185	6.79	< 0.001
x_2	0.848	.2019	.0514	3.93	$< .01$
x_3	0.226	.1866	.1122	1.66	$< .2 > 0.1$
x_4	0.511	0.0580	0.0185	3.08	< 0.02

Increase per unit increase in x_1 = 41.7 per cent,
Increase per unit increase in x_2 = 67.1 per cent,
Increase per unit increase in x_3 = 62.0 per cent,
Increase per unit increase in x_4 = 18.9 per cent.

where b_1 stands for log a_1 and y' stands for log y. The values of β_1, β_2, . . . , are given in the first column of Table 13.06*a*. They show that x_1 was easily the most important of the independent variates and that x_2 was the next most important. The untransformed regression coefficients, b_1, b_2, . . . , are also shown in Table 13.06*a*, because they are used in the tests of significance. The analysis of variance is set out in Table 13.06*b*.

TABLE 13.06*b*

ANALYSIS OF VARIANCE

Source of Variance	Degrees of Freedom	Sum of Squares	Mean Square	Variance (Per Cent)
Regression	4	1.5537	0.3884	78.4
Residual	9	0.2736	0.0304	
Total	13	1.8273		

$z = 1.2739$; $P < 0.001$.

It was not practicable to make a similar precise analysis of the numbers of thrips which were recorded during the periods when they were at a minimum. At times the thrips averaged less than one per flower; and parallel samples taken from other sorts of flowers in which the thrips were breeding indicated correspondingly low numbers. The sampling variance of such low numbers is large, making quantitative analysis difficult.

But there are, nevertheless, several general conclusions which may be stated with reasonable certainty. Since, during times of low numbers, the thrips were always very few in every flower which was examined, we can say categorically that they were not short of food in the sense that all the available food was consumed; on the contrary, only a small fraction of the food was used. On the other hand, food certainly was scarce (or perhaps better to say "sparse") in the sense we discussed in section 11.12. It is obvious that shortage of food (in this sense) cannot be interpreted as a "density-dependent factor."[1] We can

1. The phrase "density-dependent factors" recurs in this chapter. We use it with the precise meaning in which it is defined in sec. 2.12 in a quotation from Elton (1949).

also be reasonably certain, as we stated in section 13.111, that predators and diseases are virtually absent from the environment of *Thrips imaginis*. Nor was there any other observation made which could be taken to indicate the operation of a "density-dependent factor."

On the other hand, by analogy with what is known to happen during the season of maximal abundance, we can hazard a well-informed guess as to what might be determining the numbers during the season of minimal abundance. During the summer the rate of increase becomes and remains negative, largely because of the lack of moisture. Aridity causes a high rate of deaths among pupae and a scarcity of places for breeding. There is a brief respite in the autumn, after the rains come, and r sometimes becomes positive again for a brief period. During the winter, r becomes negative once more, and the chief cause of this is low temperature. The numbers might therefore be expected to decline further, the longer and more arid the summer and the longer and colder the winter.

13.116 CONCLUSIONS TO BE DRAWN FROM THE ANALYSIS OF THE DATA

During 14 years the numbers of this natural population of *Thrips imaginis* fluctuated with a consistent seasonal rhythm (Figs. 13.03 and 13.05). The maximal numbers were usually reached about the end of November; there was a minimum at the end of the summer about March–April, followed by a small increase, which reached its peak about May or June; then, finally, another decrease, during which the lowest numbers of the year would be registered, usually toward the end of July or during August.

The maximal numbers varied from year to year, as illustrated in Figure 13.06, but at no time were the thrips numerous enough to consume more than a small fraction of the food or occupy more than a small fraction of the total places available to them. So their numbers were never limited by shortage of food in the absolute sense. Nevertheless, the degree of concentration of breeding places was important because the thrips do not multiply rapidly except when they are closely surrounded by suitable flowers for breeding. Indeed, it was shown that a substantial part of the variation in the maximal numbers from year to year could be explained by reference to the duration of the period when breeding places were distributed densely over the area. This was largely determined by weather. Other causes for the fluctuations in the maximal numbers of thrips from year to year were also related to the weather; and, altogether, 78 per cent of the variance was explained by four quantities which were calculated entirely from meteorological records.

This left virtually no chance of finding any other systematic cause for variation, because 22 per cent is a rather small residuum to be left as due to random sampling errors. All the variation in maximal numbers from year to year may therefore be attributed to causes that are not related to density:

not only did we fail to find a "density-dependent factor," but we also showed that there was no room for one.

The evidence with respect to what determined the numbers during the season of minimal abundance is not quite so definite. It was shown beyond doubt that food never acted as a "density-dependent factor" and with reasonable certainty that no other "density-dependent factor" was operating. It was suggested, by analogy with what is known of the influence of weather on maximal numbers, that weather also determined minimal numbers in a way that was essentially independent of density. The maximal and minimal numbers for the seasonal cycles were studied in these ways, but the average for the whole year was not calculated, because this would have little meaning and would not be relevant to the discussion of "density-dependent factors."

Those who believe that the numbers in natural populations can be "regulated" only by "density-dependent factors" argue that, in the absence of "density-dependent factors," populations must either go on increasing without limit or else become extinct. And we have, indeed, been asked by our colleagues, in discussing these data, why *Thrips imaginis* does not become extinct and why *T. imaginis* does not go on increasing without limit. The answer to the first question is that in many local situations the thrips do die out every year. But in the broader area which we were studying, there were always some less severe places where there was never time enough for the population to become extinct. In them, r was negative also, and the population was declining there, as elsewhere, but not rapidly enough to die out before the weather changed and permitted the population to start increasing again.

To make this point doubly clear, consider what happens near the margins of the distribution of *Thrips imaginis*. As one travels north from Adelaide, one eventually approaches the central desert. But long before the desert is reached, there comes a region where *T. imaginis* may not be found, although there is little doubt that if a population of *T. imaginis* were to be set down in a favorable place in this region during spring, it would survive for a while. But it would not survive for long, probably not even through its first year, for in this region the summer is more arid and more prolonged than in Adelaide; less relief is provided by the autumn; and droughts may also occur during the winter. In this area, therefore, a population of *T. imaginis,* after the spring had passed, would be likely to decrease so rapidly, and the decline would be likely to continue for so long, that the population would have little chance of surviving long enough to take advantage of the arrival of spring. Consequently, *T. imaginis* is not found there. But in the area that we studied there were always many local situations which were sufficiently favorable even in summer for *T. imaginis* to survive there.

The answer to the second question as to why *Thrips imaginis* does not go on multiplying without limit can be given with equal simplicity. It just does not have time to do this, for the favorable season of the year, which is the spring, is invariably followed by an unfavorable period in the summer, when hot, dry weather (but not "density-dependent factors") knocks the numbers back.

The dogma of "density-dependent factors" is unrealistic on at least two major counts: it ignores the fluctuations of r with time, which may be induced by seasonal and other fluctuations in the components of environment; it also ignores the heterogeneity of the places where animals may live. This empirical study of a natural population of *Thrips imaginis* has shown that if these two facts are recognized, it is not necessary to invoke "density-dependent factors" to explain either the maximal or the minimal numbers occurring in a natural population.

13.12 *The Distribution and Abundance of* Austroicetes cruciata *in South Australia*

The ecology of the grasshopper *Austroicetes cruciata* was studied by a team of workers at the Waite Institute during the period 1935–42. The work was summarized by Andrewartha (1944*b*). Particular aspects of the work were reported by Andrewartha (1939, 1943*a*, *b*, 1944*a*, *c*), Andrewartha, Davidson, and Swan (1938), Birch (1942), Birch and Andrewartha (1941, 1942, 1944), and Andrewartha and Birch (1948).

13.121 THE LIFE-CYCLE OF *Austroicetes cruciata*

The life-cycle of *Austroicetes cruciata* is characterized by an intense obligate diapause in the egg stage (secs. 4.3 and 4.6). Consequently, there is one and only one generation each year; also the same stage in the life-cycle recurs at the same season each year, with remarkably little variation in the dates on which they begin and end. Diapause is usually completed by about midwinter (June). The speed of development of the eggs thereafter depends largely on the temperature experienced during the latter part of the winter and early spring (sec. 6.234). Variability in the temperature at this season induces variability in the date on which the eggs hatch; but, for one district, it was estimated, from the records of temperatures for the years 1891–1940, that the peak of the hatching varied by no more than ± 10 days about the average date for 80 per cent of the years. The other 20 per cent of occasions included departures of up to 20 days on either side of the average date. During the few years for which direct observations are available, the date varied from August 25 to September 16.

This is the season of the year when the herbage attains its greatest development (sec. 4.6), and, except on the occasion of an unusually severe drought, the nymphs usually find plenty of food. The speed with which they develop is therefore determined almost entirely by temperatures. The nymphal development is condensed into a rather brief period, and the adults lay their first batch of eggs within a few days of the final molt. According to our observations, the nymphal stage occupied from 41 to 54 days, and the date on which the first eggs were laid varied from October 30 to November 3.

By November, summer is already advancing, and the grass is beginning to wither. As their food disappears, the adults of *Austroicetes cruciata* begin to die from starvation long before most of them would have died from old age. The date by which they had virtually disappeared varied from November 20 to December 3. From then until the following spring (September) the whole population remained in the egg stage. During the whole of the summer the eggs remained firmly in diapause, which confers upon the eggs a remarkable capacity to withstand the severe drought characteristic of the summer in this area (sec. 7.233).

The regularity of the life-cycle with respect to the seasons and the fact that virtually only one stage was present at a time made it easier to ascertain the probable influence of weather on the grasshopper's chance to survive and multiply (sec. 13.124). The behavior of the females in seeking places for oviposition was described in section 12.22; the distribution of soil that is suitable for oviposition has an important influence on the distribution and abundance of *Austroicetes cruciata* (Andrewartha, 1944*b*).

13.122 CLIMATE AND VEGETATION IN THE AREA WHERE *Austroicetes cruciata* LIVES

The grasshopper *Austroicetes cruciata* lives in an area where the climate is more arid than that in the region where *Thrips imaginis* is found in abundance (sec. 13.112). For example, the mean annual rainfall in the district of Hammond, which is in the heart of the distribution of *A. cruciata*, is 11.7 inches; about 62 per cent of this falls during the cooler half of the year (May to October). The growing season in this district lasts, on the average, for about 5 months. During the warmer half of the year (November to April) 4.6 inches of rain falls, on the average. But during the same period the evaporation from a free-water surface amounts to about 31 inches. The soil remains parched and the plants dormant (Figs. 13.07 and 11.01).

The rainfall in this region is variable with respect to both the amount that falls each year and its distribution throughout the year. So the average values do not indicate the full severity of the hazards experienced by *Austroicetes cruciata*. From the records for 33 stations in the area where *A. cruciata* lives, it was calculated by Cornish (in Andrewartha, 1943*a*) that, on the average, once in 20 years the total annual rainfall would be nearly double the mean and once, on the average, during the same 20 years it would be little greater than half the mean. In other words, the rainfall for the wettest year in any group of 20 years is likely to be 3.4 times as great as the rainfall for the driest year of the same 20-year period; the extremes in any 10-year period are likely to differ from each other in the ratio 2.5:1, and for any 5-year period in the ratio 1.8:1. These figures give only an indirect measure of the variability of the rainfall; but since both excessive dryness and excessive wetness may, in certain circumstances, result in the deaths of many *A. cruciata*, it is clear that variability of this magnitude must add to the hazards of life for the grasshopper.

The temperature during the winter in the area where *Austroicetes cruciata* lives is low enough, especially at night, to promote diapause-development but not to harm the eggs. But the summer may be extremely hot. The mean maximal temperature in the shade for January (the hottest month) at Hammond is 32° C. The mean maximal temperature of the soil at about ¾ inch below the surface, where the eggs of *A. cruciata* spend the summer, is likely to be at least 10° C. higher than this (sec. 6.234); and, on unusually hot days, the maximal temperature of the soil at this depth must exceed 50° C. But there is evidence, both from experiments in the laboratory and from observation in nature, that very few eggs ever die from the direct influence of high temperatures, provided that they do not at the same time lose a fatal amount of water.

The pristine vegetation of this country was woodland and scrub, with a sparse understory of xerophytic shrubs and grasses. There must have been relatively little area suitable for the multiplication of *Austroicetes cruciata*. Most of the original vegetation has now been destroyed, and its place has been taken by pastures composed of a variety of hardy grasses and other low-growing herbs which serve as food for grasshoppers. Most of them are annuals, which are adapted to take advantage of the brief growing season characterizing this region: they become established from seed during winter while the soil is moist; they put forth a great burst of growth during spring, and complete the production of seeds in the brief period that elapses before the summer drought sets in. Barley grass, *Hordeum murinum*, is typical and is one of the most important foods of the nymphs. The adults depend more on the grasses which stay green a little longer. The growth of several perennial species of *Stipa*, *Danthonia*, and others follows the same seasonal pattern as the annuals, but they shoot afresh each year from a dormant crown; also they will put forward a small showing of green leaves at any time after a few points of rain. These and some of the longer-lived annuals are important as food for the adults. Also in isolated local situations small areas sown to wheat or lucerne may continue to support a local population after grasshoppers have mostly disappeared elsewhere.

13.123 THE NUMBERS OF *Austroicetes cruciata*

The numbers of *Austroicetes cruciata* fluctuated between wide extremes. From 1935 to 1939 the swarms were too obvious to need counting; the individual swarms did not seem to increase in density, nor did the boundaries of the area affected by swarms expand, but there was a perceptible increase from 1935 to 1939 in the proportion of this area occupied by the swarms. By the end of October, 1940, the grasshoppers had become so scarce that hardly any could be found, although we searched most thoroughly. Virtually this entire generation died without laying an egg, and in 1941 the numbers were still extremely low. A few sparse populations were found, but the great majority of the places which were examined seemed to be quite devoid of *A. cruciata.*

The distribution of *Austroicetes cruciata* was determined precisely, but the numbers only qualitatively. We traveled along roads and tracks running north and south and east and west, stopping every 5 or 10 miles in places which seemed suitable for the grasshoppers and recorded their numbers in five categories which we called "absent," "solitary," "dense solitary," "loose swarm," and "swarm." Apart from local fluctuations such as the disappearance of swarms from certain districts after the hot, dry summer of 1938–39, there was no appreciable change in the boundaries of the area where the grasshoppers were numerous during the period 1935–39. We also went back as far as 1891, searching the technical publications and newspapers for references to plagues of *A. cruciata.* There was no reference to swarms outside the boundaries of the area indicated by our surveys, although there was a number of references to swarms inside these boundaries. The results are shown in Figure 13.07.

We also studied the causes of deaths in the different stages of the life-cycle. We were successful in this, partly because the life-cycle of *Austroicetes cruciata* is so regular with respect to the seasons, but also because we fortunately encountered at one time or another during the course of the investigations extremes of weather during summer, winter, and spring.

13.124 THE CAUSES OF DEATHS AMONG NATURAL POPULATIONS OF *Austroicetes cruciata*

The egg, being present during summer, autumn, and winter, is exposed to a number of different sorts of hazards from weather at different seasons; but the only one that matters is the risk of losing a fatal amount of water during summer. The fully turgid diapausing egg at the beginning of summer contains about 2.4 mg. of water. It can lose more than half this and still remain alive; but if its water content falls below about 0.75 mg., the egg is likely to die. The membranes which inclose the tissues of the egg are remarkably impermeable to water passing outward, and eggs that were stored for seven months at 20° C. and 55 per cent R.H. lost water to the air at the rate of about 0.01 mg. per day. In another experiment, four batches of eggs were stored for 210 days at 30° or 35° C. at 55 per cent and 75 per cent. R.H. The death-rate varied from 32 to 59 per cent. Although the passage of water outward through the egg membranes to dry air is so slow, it can nevertheless be taken in rapidly from damp soil. Eggs which had been desiccated until they were flaccid were placed on moist paper at 25° C. During the first 20 hours they absorbed water at the rate of 0.21 mg. per hour, for the second 20 hours at the rate of 0.10 mg. per hour, and after 60 hours they had become fully turgid again.

In nature during the summer the diapausing egg is losing water most of the time but regains some after each substantial shower of rain. If the rain is sufficient to keep the top inch of soil moist for several days, the egg will have time to replace all the water that it has lost. During the summer of 1938–39 the drought was unusually severe, and in certain districts a high proportion of

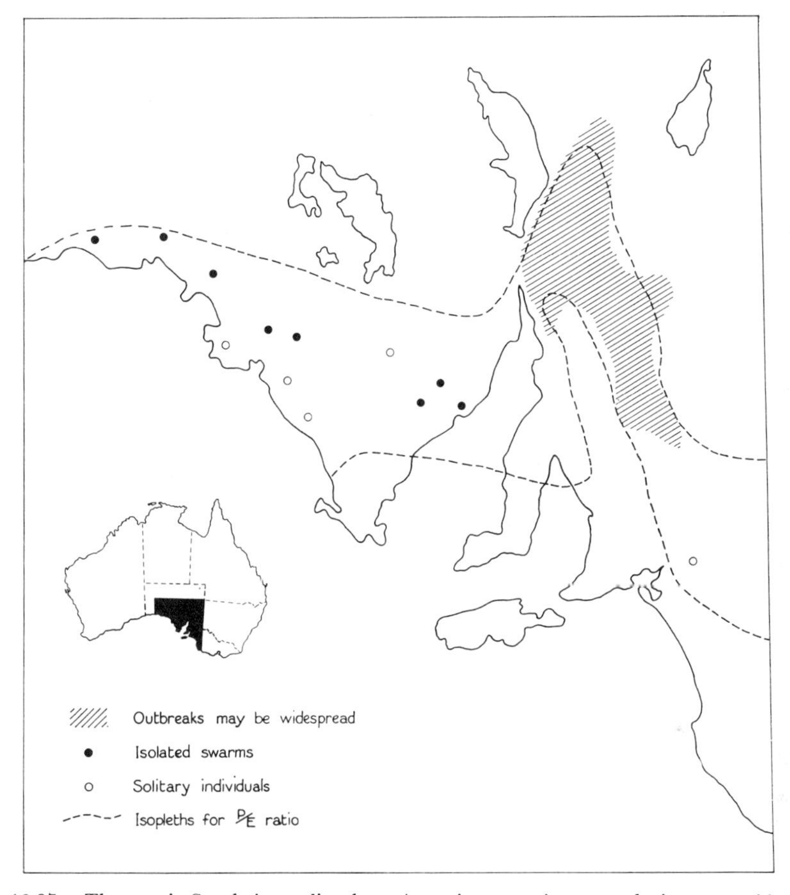

FIG. 13.07.—The area in South Australia where *Austroicetes cruciata* may, during a run of favorable years, maintain a dense population. For further explanation see text. (After Andrewartha, 1944*b*.)

eggs lost a fatal amount of water. We measured the proportion of eggs that died in the two districts of Hawker and Hammond, which are separated by about 40 miles. At Hawker the proportion of dead eggs varied from 94 to 88 per cent depending on the local situation, and at Hammond from 75 to 54 per cent (sec. 12.22). We measured the intensity of drought associated with these death-rates by estimating from records of rainfall, temperature, and atmospheric humidity the amount of evaporation that would have occurred from a free surface of water during the longest spell without rain. At Hammond there were 80 consecutive days without rain, at Hawker 89; the estimated evaporation during these periods was 26.2 inches at Hammond and 40.6 inches at Hawker. Assuming that the severity of drought at Hammond during the summer of 1938–39 was just about critical, we estimated the number of occasions that this had happened during the 50 years between 1891 and 1940. We found that at Hammond there had been 11 such occasions, giving a probability of 22 per cent for any one generation of a high death-rate from drought in the summer. If a more severe criterion were chosen, say, an estimated evaporation of 40 inches, then this probability would be about 10 per cent (Table 13.07). The figures in Table 13.07 show that the risk from drought during summer is

TABLE 13.07*
PARTICULAR YEARS BETWEEN 1891 AND 1940 FOR WHICH DROUGHTS AS SEVERE AS THOSE AT HAMMOND AND HAWKER DURING 1938–39 WERE RECORDED FOR THREE DISTRICTS IN AREA WHERE *Austroicetes cruciata* LIVES

Year	Hawker		Hammond		Orroroo	
	Duration of Longest Dry Spell (Days)	Total Evaporation (Inches)	Duration of Longest Dry Spell (Days)	Total Evaporation (Inches)	Duration of Longest Dry Spell (Days)	Total Evaporation (Inches)
1892/93	107	37.2	...	...	109	34.9
1897/98	133	45.0	...	...	113	44.1
1898/99	...	...	91	39.1	...	...
1900/01	...	...	81	32.7	119	37.6
1905/06	110	39.6	117	40.1	105	36.0
1914/15	...	...	73	30.7	...	...
1915/16	116	39.6	141	51.4	96	28.4
1918/19	102	42.9	...	...	106	41.9
1921/22	114	37.9	99	41.3	76	26.5
1922/23	...	...	88	40.6	69	29.9
1924/25	...	...	78	33.3	...	...
1925/26	96	40.7	...	...	...	...
1928/29	154	71.1	154	50.5	85	32.3
1932/33	74	28.8	124	49.4	...	...
1938/39	89	40.6	80	26.2	...	...
1939/40	66	34.1	...	...	...	...
No. years evaporation > 26.2 in.	11		11		9	
No. years evaporation > 40.6 in.	5		5		2	

* After Birch and Andrewartha (1944).

greater in some parts of the area where *Austroicetes cruciata* lives than in others; also the difference in severity between Hawker and Hammond in 1938–39 was fortuitous, because, on the average, the risk is equally great in both these districts. This is characteristic of the sort of variability which is nearly always found when natural populations are studied thoroughly.

Most eggs which survive the summer develop normally and survive the winter. But another hazard may occur when newly hatched nymphs are ready to emerge from the egg pods. The egg pod of *Austroicetes cruciata* is closed by a "cap" of the same material that forms its walls. The female also rakes loose soil over the top of the pod, covering it to a depth of about 1 mm. The soil becomes compact during winter. If, at the time when the nymphs are ready to emerge, the surface of the soil is dry, the cap may be cemented on so firmly that the nymphs cannot push it off. We sometimes found nymphs which had been trapped in this way and had died from starvation, but we never found any extensive situation where the death-rate from this cause exceeded a few per cent. Nevertheless, it is conceivable that a drought coinciding with the main hatching of the nymphs might cause many of them to be trapped in the egg pods. So we estimated the probability of such an event. Experiments in the laboratory showed that newly emerged nymphs could live for about 5 days without food. Experiments in nature showed that 0.02 inch of rain would soften the soil sufficiently to allow the nymphs to break out from the pod. Using methods which were described by Birch and Andrewartha (1944) and Andrewartha (1944*b*) and taking the 50 years between 1891 and 1940, we estimated the frequency with which the main hatching of *Austroicetes cruciata* would have coincided with a 6-day drought. This was found to be about 0.012, indicating that a high death-rate from this cause might happen about 12 times in a thousand years. The nymphs encounter much greater risks than this from other sources as they become older.

When the nymphs first hatch, they are not likely to be short of food because this is the season of the year when the grass is growing most strongly. During the period 1935 to 1939 when swarms of *Austroicetes cruciata* were present, the later stages sometimes experienced a shortage of food, not because there was insufficient grass in the area but because it dried up before the grasshoppers could complete their development. The nymphs did not usually suffer much from shortage of food, but the duration of life and the fecundity of the adults were largely determined by the date when the grasses dried up. As the grasses withered, the grasshoppers died from starvation. But the sheep which shared the pastures with the grasshoppers could thrive on dry grass; and it was the aim of the farmer to have standing in his paddocks enough "dry feed" to carry his flocks through the summer. Perhaps the surest evidence that the grasshoppers on no occasion between 1935 and 1939 ate more than a small proportion of their potential stock of food lies in the records of the numbers of sheep living on the same pastures during this period; the number of sheep increased from 469 thousand in 1935 to 712 thousand in 1939.

It was fortunate for our study that 1940 was a year of severe drought, with a heavy death-rate among nymphs during the spring. By measuring the severity of this drought and consulting past meteorological records, we were able to estimate the frequency with which nymphs might suffer a heavy death-rate from drought.

Many eggs were laid in the spring of 1939, and everything pointed to a generation in 1940 with probably even greater numbers. But little rain fell during the growing season of 1940, and when the nymphs hatched during the first week in September, there were already signs that food would be scarce later on. The dramatic ending to the five years of plagues was described by Birch and Andrewartha (1941, p. 95):

> In 1940 the nymphs emerged mostly during the first week of September. There was very little food available, due to the lack of rain during winter. For example, at Orroroo the total rain from April to August was 2.44 inches (mean for this period 6.09 inches). Annuals were scarce or absent from most situations; speargrass had only a few short green shoots near the base. During September and October rain occurred as light falls widely spaced. Temperature and atmospheric saturation deficit were above average. An examination of records for 15 representative towns showed that no rain fell between 21st August and 12th September; on that date between 11 and 68 points were recorded; light falls varying from 1 to 30 points were registered on 20th September; on 28th September falls between 3 and 40 points were recorded; falls ranging between 4 and 34 points were recorded on 15th October, and further falls varying from 1 point to 17 points on 29th October. No further rain was recorded until 9th November. These falls of rain were insufficient to maintain growth of speargrass and the grasshopper numbers were reduced as a result of starvation.
>
> Survey trips were made on 3rd September, 16th September, 3rd October, 29th October and 18th November. On the 3rd September hoppers were present in plague numbers in most situations examined; in one or two local situations there was evidence that death from starvation had already occurred on a large scale. On the 3rd October the hoppers were found in plague numbers in only a few favoured local situations; it was noticed that their development had been unusually slow. By 29th October the grasshoppers had practically disappeared; only two small situations were found where they were numerous. By 18th November they had disappeared almost completely. During an extensive survey which covered many districts no situation was found where there was more than an occasional grasshopper. In some places the grasshoppers had disappeared from areas close to wheat crops which were still green. Observations by many local residents provided the explanation for this. As the grasshoppers had died out over the countryside at large, the birds, which normally prey upon them, had been forced to seek out the grasshoppers in the local situations where the latter still survived. The congregation of starlings, crows and other birds had been so great that in most places the grasshoppers had been almost completely exterminated.

From this account it is clear that the primary cause of the calamity was lack of rain during winter and spring. We analyzed the records for rainfall, temperature, and atmospheric humidity for the 50 years 1891–1940 and found that droughts as severe as this one had occurred 7 times at Hammond, 5 times at Hawker, and 3 times at Orroroo. The probability of a high death-rate among grasshoppers from the three causes we have so far analyzed—a hot, dry summer, drought at the time of hatching, and drought during the nymphal stages—has been summarized in Table 1.01.

We next consider a fourth possible cause of death. The known variability of the rainfall in this area makes it likely that excessive wetness may sometimes cause the deaths of many *Austroicetes cruciata* by permitting the spread of disease among them, but we did not observe this in South Australia. Andrewartha (1944*b*) described indirect evidence that this had happened in Western Australia during July, 1938, but it was not possible to estimate the probability of such an event in the area that we studied. There is no way, other than those which we have discussed, in which weather is likely to cause a high death-rate among a natural population of *A. cruciata*.

The sheep is the only other animal of any importance that requires to eat the same food as *Austroicetes cruciata*. The farmer tries to keep a reserve of "dry feed" to maintain his flock of sheep during summer, so it is unlikely that he would ever willingly allow so many sheep on the pasture that they would appreciably reduce the stock of food available to *A. cruciata*. In special cases, as in 1940, when the usual "spring renewal" of food did not take place and the total stock of food in the area was very much less than usual, the presence of the sheep may have accentuated the shortage of the food for the grasshopper. But even in these circumstances the primary cause of the shortage was the weather.

The predators of *Austroicetes cruciata* include a scelionid, a bombyliid, and a dermestid which eat the eggs and a sphecid which preys on the nymphs and adults. None was abundant, and none made any appreciable difference to the numbers of the grasshoppers. Birds of various sorts prey on the nymphs and adults. The presence of *A. cruciata* during spring may assist the birds to provide for their young, but the number of birds in the area depends less on the number of grasshoppers than on the amount of food they can find during the remaining 9 months of the year when the grasshoppers are absent. The relative influence of the birds on the number of grasshoppers increases as the grasshoppers become fewer. This is the reverse of what is expected from a "density-dependent factor."

13.125 GENERAL CONCLUSIONS

From 1935 to 1939 the grasshoppers were present in large and increasing numbers. It is likely that they would have increased more rapidly if the females in each generation had lived long enough to lay all the eggs which they are potentially able to produce. Dissections indicated that this was about 240 eggs, the equivalent of about 12 pods. In nature the food dried up, and the grasshoppers died of starvation before they could lay all their eggs. We made no precise estimate, but we think that the average would have been about 2–4 pods per female. The precise number does not matter. The important point is that the limit to fecundity was determined by the duration of the period when succulent food was present. Thus there is a sense in which it is true that fecundity was determined by the supply of food. But the stocks of food ran out, not because they were eaten by the grasshoppers but because, in the absence of rain to keep the soil moist, the grass changed its condition, becoming useless as food for the grasshoppers. In the closing stages the little green food that remained was sparsely scattered over many plants. So, in the end, the scarcity of food which overtook the grasshoppers was more of a relative scarcity (in the sense of sec. 11.12) than an absolute one. At a slightly later stage there would, of course, be virtually no green food left anywhere, but the last grasshopper had usually died of starvation before this stage had been reached. The supply of food was therefore quite unrelated to the numbers of the grasshoppers, and it cannot in any sense be construed as a "density-dependent factor."

It was shown in section 13.124 that the records of the numbers of sheep living on the same pastures provided independent evidence that, during the period when the grasshoppers were abundant, they did not eat more than a small proportion of the total herbage growing in the pastures. It was also shown in section 13.124 that none of the causes of deaths of *Austroicetes cruciata* in natural populations could be recognized as a "density-dependent factor." Weather was most important and was clearly not "density-dependent"; the influence of predatory birds, which become important in special circumstances, is in many ways the opposite to that of a "density-dependent factor."

The rate of increase *r* was not determined by "density-dependent factors"; yet the grasshoppers did not go on increasing in numbers to the limit of their resources of food. (Places in which to lay eggs and all the other spatial requirements of the grasshoppers were in great excess of the demands made on them.) The explanation in this case is obvious and simple. Calamity in the form of the drought of 1940 overtook the population long before this stage was reached. Virtually all the grasshoppers died without leaving progeny behind them; and nearly all the deaths were due to starvation. But, once again, the shortage of food did not depend on the numbers of grasshoppers. The food was in short supply because, from lack of moisture, the grass withered before the nymphs could complete their growth. The result would have been the same, had there been few or many grasshoppers present at the beginning.

The grasshoppers died out completely from most local situations during the calamity of 1940, but they did not become extinct from the area as a whole; a few survived in isolated places here and there. It was shown in section 13.124 that similar hazards recur with certain frequencies which can be estimated from the meteorological records. This has been going on for a very long time, and the continued persistence of *Austroicetes cruciata* in this area shows that the probability of complete extinction during any one vicissitude must be exceedingly small. It has been argued that such small probabilities cannot be explained except in terms of "density-dependent factors." We think that this view fails to allow for the great heterogeneity of places where *A. cruciata* may live in this area and for the chance that a few individuals here and there will have the good luck to find themselves in places where life can be sustained. This may be a little more likely to happen if the population is large at the time that catastrophe strikes, but this does not make the weather a "density-dependent factor" in the usually accepted sense of this phrase. Indeed, the number of survivors is likely to be determined largely by the interaction of the two components of environment which we have called "weather" and "place to live"; the number of predatory birds may also be important; this does not depend on the number of grasshoppers but on the amount of food which the birds can find in the area during the 9 months when the grasshoppers are absent. We showed in section 1.1 how, as the climate becomes increasingly severe, the probability of complete extinction from a substantial area increases. Indeed, this is what happens in the areas which we regard as being outside the distribution of *A. cruciata*.

The area which we had mapped (secs. 13.123; Fig. 13.07) might be taken to represent the distribution of *Austroicetes cruciata*, or at least that part of it where they have a good chance of becoming numerous from time to time. If we had been right in concluding that the abundance of *A. cruciata* is determined largely by weather, then it should be possible to define a particular climatic zone such that its boundaries are closely related to the boundaries of the distribution of the grasshopper. In seeking to test this hypothesis, we were guided by the information given in section 13.124.

In southern Australia the climate becomes more arid as the distance from the coast increases. Therefore, the limit to the northern (inland) distribution of the grasshopper might be largely determined by the severity and frequency of drought, especially during winter and spring, when rain is needed to insure food for the active stages of the grasshopper. The mean monthly ratio of rainfall to evaporation (P/E) has often been used as an index of climate. The value of $P/E = 0.5$ has been used to estimate the duration of the growing season for crops and pastures in general (Davidson, 1936*c;* Fig. 11.01). Since we were concerned with especially hardy grasses, we chose to use the value $P/E = 0.25$. Since the most dangerous month for the nymphs is October (if the nymphs survive October, the adults are likely to find enough food to lay at least a few eggs), we chose to plot on the map the isocline for $P/E = 0.25$ for October. Fortunately, this isocline had been calculated and mapped (though not published) by Davidson some years earlier. So we had an independent estimate of the climate with which to test our hypothesis. The agreement in Figure 13.07 between the climatic line and the northern (inland) boundary of the distribution of *A. cruciata* is remarkably close.

The southern (coastward) limit to the distribution of *Austoicetes cruciata* is likely to be determined by the risk of high humidity, favoring the spread of disease. Since the active stages are present during spring when the weather is becoming drier, the greatest risk of encountering a spell of dangerously wet weather occurs early rather than later in the lives of the nymphs. The usual month for hatching is September, so we chose this month. We had available from Davidson's working drawings the isocline for $P/E = 1.0$, which, being double the value usually chosen to indicate sufficient moisture to maintain plant growth, seemed a suitable index for our purpose. In Figure 13.07 the isocline for $P/E = 1.0$ for September follows the southern boundary of the distribution of *A. cruciata* quite closely. The hypothesis has therefore been supported, and we considered this good supporting evidence for our general conclusion that the distribution and abundance of *A. cruciata* are determined largely by weather; there is no evidence for "density-dependent factors."

In Figure 13.07 there are quite large areas between the two climatic lines from which *Austroicetes cruciata* was absent or where local populations were found here and there in isolated situations. In some quite extensive areas the soil was unsuitable for oviposition; in others the persistence of the original vegetation has resulted in a lack of suitable plants for *Austroicetes* to eat. This was explained by Andrewartha (1944*b*). There is no need to go into the matter here.

13.13. *The Distribution and Abundance of* Porosagrotis orthogonia *in North America*

The ecology of the pale western cutworm *Porosagrotis orthogonia* in the Great Plains region of western North America was studied by Parker, Strand, and Seamans (1921), Seamans (1923), and Cook (1924, 1926, 1930). This species is common along the eastern edge of the Rockies, and isolated specimens have also been taken in California, Nevada, Arizona, and New Mexico. Cook was able to study the influence of weather on the rate of increase of *P. orthogonia* during a great outbreak which began about 1917, reached a climax about 1920, and declined during 1921 and 1922. He mapped the limits of the area where outbreaks of this species were likely to occur and he estimated the probable frequency of outbreaks.

13.131 THE BIOLOGY OF *Porosagrotis orthogonia*

The adults of *Porosagrotis orthogonia* fly during the autumn, and the females lay, on the average, about 300 eggs each. There is a faint diapause in the egg stage which is nevertheless sufficient to insure that the hatching of the larvae will nearly always be delayed until the spring. The larvae live in the soil and feed on the roots and underground stems of wheat and other plants. They complete their development by about the end of June and emerge as moths during August and September.

In seeking a place to lay eggs, the females show a strong preference for fields of stubble, in which the soil is loose at least in places; eggs were often laid in the hoofprints left in soil by horses. The newly hatched larvae live on the surface of the soil for perhaps 2 weeks. Then they burrow into the soil and remain below the surface for the rest of their lives except when they are driven above ground by excessive wetness. The following quotation from Cook (1930, p. 11) shows how sensitive they are to the moisture content of the soil:

Many examinations of infested fields have led to the conclusion that the depth at which they feed is regulated by the moisture in the soil. In May and June in Montana the rainfall is very irregular. Following a shower the surface soil will dry out and crumble, leaving a very definite depth at which the soil is still moist. Farmers call this depth the "moisture line" and use its position as an indicator of the amount of moisture still available for the wheat. In extremely dry seasons the line may lie as much as 6 inches below the surface, but it is usually found within 3 inches of the surface. As the moisture line is the place of highest evaporation, it is a cool place, definitely cooler than the soil half an inch above it. Careful study has shown that *orthogonia* larvae will usually be found just above the moisture line while they are active. . . .

In wet weather the moisture line comes to the surface, and so do the larvae. Several times larvae have been observed wandering around on the surface in bright sunlight just after a heavy shower. As soon as the excessive moisture drains from the surface layer, the larvae dig into the soil and again follow the moisture line.

Cook did not measure fluctuations in the birth-rates of natural populations. It is likely that fluctuations in the value of r would be determined largely by fluctuations in death-rates and that variation in the birth-rate would be relatively unimportant.

13.132 CAUSES OF DEATHS AMONG NATURAL POPULATIONS OF *Porosagrotis orthogonia*

Cook made certain observations and did several experiments which showed that larvae in soil that was uniformly friable moved at random, except when very close to the stem of a wheat plant or some other food. But wherever the soil was unevenly compact, the larvae moved along the line of least resistance. In a field of wheat which had been sown with a drill, they moved along the lines made by the drill. As a result of this behavior, larvae that are living in a field of wheat find food quite readily. But in uncultivated land, even if the vegetation is equally dense, their chance of finding food may be less. That is, they may suffer from a relative shortage of food in the sense described in section 11.12.

A quotation from Cook (1930, p. 50) emphasized this point:

Before 1911 *P. orthogonia* was a rare insect. There are a few records of the capture of the moths but no records of damage. From a study of its present habits, it seems safe to assume that in former times it was confined to places where the soil was very loose and dry. Such situations are found in the lighter soils of Montana on knolls and along the edges of benches. It is evident that such situations were not large in area, so that the total area suitable for this species must have been small. Only in these places was it able to feed underground, and, if it occupied other places, where it must feed at the surface, it would be subject to the same parasites and enemies as our other common cutworms, and be kept to small numbers by them. With the extension of dry-land wheat-growing the situation was changed. Not only was the area of loose soil enormously increased, but the soil was planted with a tender and succulent plant. This enabled the species to increase in numbers, and gradually to accumulate a normal population so large that it formed a nucleus for a population so large that it would cause severe damage when conditions became favourable. In order to reduce it permanently to its former status of rarity, the cultivation of grain would have to be discontinued.

The presence of extensive areas of wheat certainly permitted *Porosagrotis orthogonia* to achieve relatively enormous numbers during the outbreak of 1917–21. In certain local situations there were too many for the supplies of food, and many larvae died from starvation or were eaten by their fellows. Nevertheless, they did not, even during their period of greatest abundance, destroy more than 45 per cent of the wheat in the area studied by Cook. So the decline, when it came in 1921 and 1922, was not due to any substantial or widespread shortage of food.

Seamans (1923) observed that the numbers of *Porosagrotis orthogonia* were usually depleted during a generation which was forced to the surface frequently by wet weather; and he postulated that *r* would remain negative (or, if positive, it would be low) for any generation which was exposed to more than 10 wet days (0.25 inch or more per day) during May and June. He attributed the high death-rate on these occasions to predation by a variety of insects and several birds, which, he argued, would kill many more caterpillars if they were exposed on the surface than if they were hidden in the soil. But Cook studied certain populations at the time when the outbreak was declining in 1922; the death-rate exceeded 90 per cent, and Cook found that more than half the deaths were due to disease. Having listed the names of 10 insects and 3 birds which preyed on the larvae of *P. orthogonia*, he concluded: "In spite of this list of enemies, it must be said that they were apparently of small value in terminating the outbreak of *orthogonia* in 1922. In that year, when the population of cutworms was already greatly reduced, the combined efforts of parasites and predators accounted for about 50 to 60 per cent of the remaining population, near Havre, but there is no indication that their work did any more than slightly aid climatic factors and disease in terminating the outbreak."

13.133 THE INFLUENCE OF WEATHER ON THE DISTRIBUTION AND ABUNDANCE OF *Porosagrotis orthogonia*

Cook did not measure the absolute numbers of *Porosagrotis orthogonia*. But he collected, for a large number of farms, records which showed, for each generation, whether it was larger or smaller than the one which had preceded it. He did this for the five years 1919–23 for a large number of districts. He then compared the monthly records of rainfall and temperature with these data and demonstrated a close relationship between the rainfall during May, June, and July and the rate of increase of *P. orthogonia*. The rate of increase was determined largely by the survival-rate, which was likely to be high when the monthly rainfall for May, June, and July averaged less than 1.3 inches, and likely to be low if the monthly rainfall during this period exceeded 1.7 inches.

A hypothetical climate, based on the results of this study, was arbitrarily chosen as being optimal for the multiplication of *Porosagrotis orthogonia*. The winter was dry, with about ½ inch of rain per month; the spring was damp, with April, May, and June averaging about 1.5 inches per month; the summer was dry, with about 4 inches falling during July–October. The temperature ranged from about $-5°$ C. in December to about 20° C. in July. The records for monthly rainfall and temperature for several hundred stations in western United States were then analyzed in the same way and compared with this hypothetical climate.

In this way Cook delimited a zone which included nearly the whole of

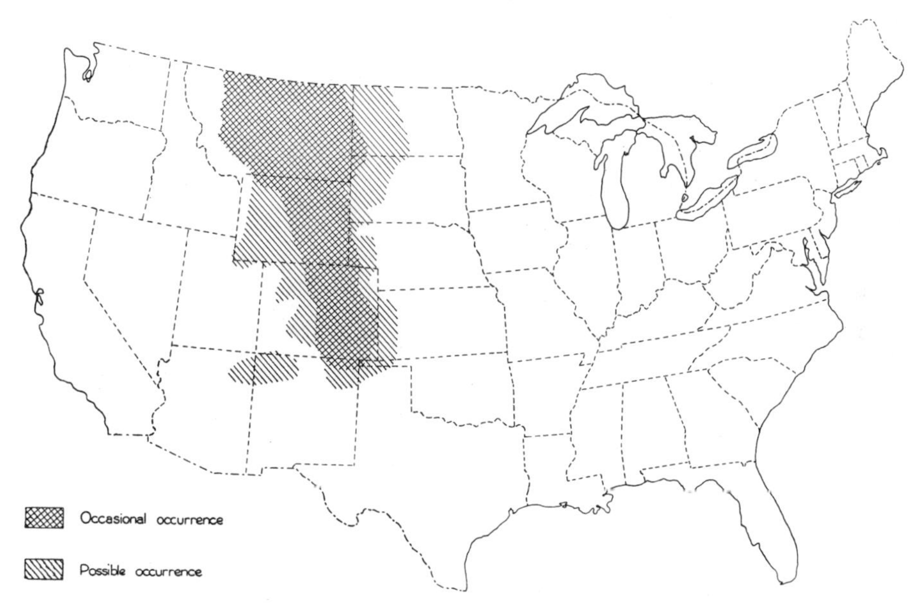

FIG. 13.08.—The area in the United States where *Porosagrotis orthogonia* might be expected to maintain dense populations from time to time. (After Cook, 1930.)

Montana and extended southward through Wyoming and Colorado. He postulated that, inside this zone, *Porosagrotis orthogonia* would be a common species, occasionally multiplying to numbers that constituted a plague. He mapped another narrower zone bordering the first, where, he postulated, *P. orthogonia* would be absent or present in lower numbers but where it might much less frequently multiply to great numbers. These zones are shown in Figure 13.08. He tested his hypothesis by mapping all the records of all known plagues of *P. orthogonia*. Without exception, they fell inside the limits of the zone which he had mapped entirely by reference to meteorological records.

Cook also estimated the frequency of outbreaks of *Porosagrotis orthogonia* in Montana. We give his own account of his calculations:

A study of the Montana outbreaks indicates that one favourable year may increase the number of cutworms sufficiently to cause slight damage and local outbreaks, but two successive favourable years are necessary to produce a severe and widespread outbreak. Curves were fitted to the rainfall distributions for May, June and July for [a number of stations]. . . . It was found that single years with less than 4 inches of rainfall from May 1 to July 31 should occur, on the average, once in 4 years near Helena, once in 5 years near Havre and Crow Agency, once in 6 years near Miles City, once in 8 to 9 years near Poplar and Glendive and less than once in 10 years at Bozeman. These figures indicate the probability of mild outbreaks of a local character [Fig. 13.09].

In order to determine the probability of two successive dry years, it was necessary to form a frequency curve for the differences in rainfall between successive years and compute the probability of a variation of less than one inch in successive years. This probability, multi-

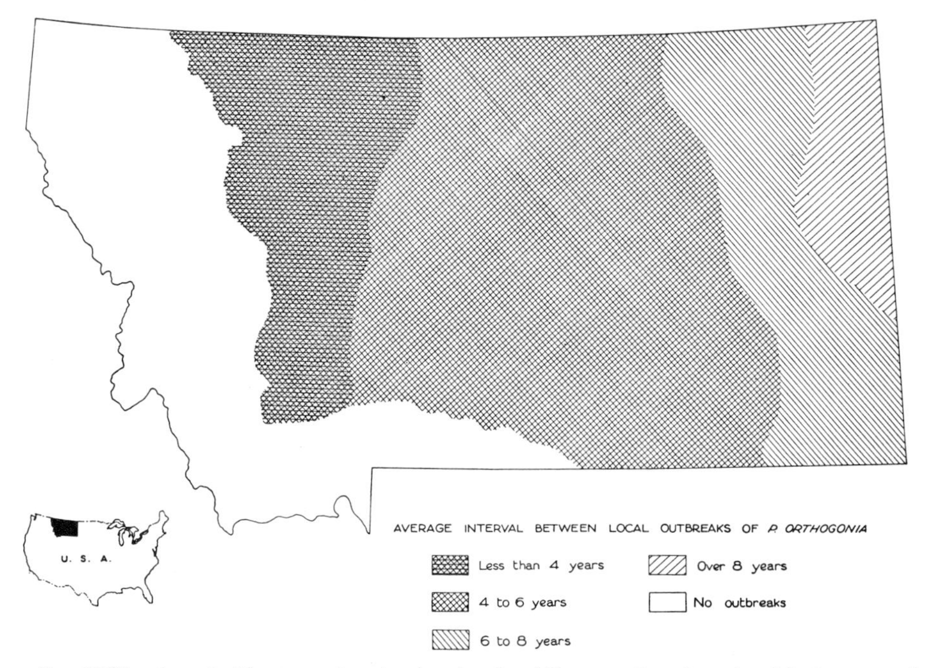

FIG. 13.09.—Areas in Montana where local outbreaks of *Porosagrotis orthogonia* might be expected to develop with the frequencies indicated by the shading. (After Cook, 1930.)

plied by the probability of one favourable year, gave the probability of two successive dry years. Two successive dry years may be expected about once in 16 years near Helena, once in 23 years near Havre and Crow Agency, once in 30 years near Miles City, once in 40 years near Bozeman and once in 60 or 70 years near Poplar and Glendive. These figures indicate that severe and widespread outbreaks may be expected only at long intervals in most parts of Montana [Fig. 13.10].

13.134 GENERAL CONCLUSIONS

The amount of readily available food and the area of friable soil which is suitable as a place where *Porosagrotis orthogonia* may live have been increased enormously as the country was developed for agriculture. The numbers of *P. orthogonia* have increased as a consequence. Not only do they become much more numerous during periods of maximal abundance, but they also maintain a larger population during periods of minimal abundance.

Nevertheless, *Porosagrotis orthogonia* does not fully occupy all the suitable places to live or consume all the available food, even when the caterpillars are most numerous, far less so when they are scarce. The two components of environment which we have called "food" and "place to live" are never in short supply, at least in the absolute sense; there is no suggestion of any "competition" for these resources, except perhaps locally during an outbreak; and there-

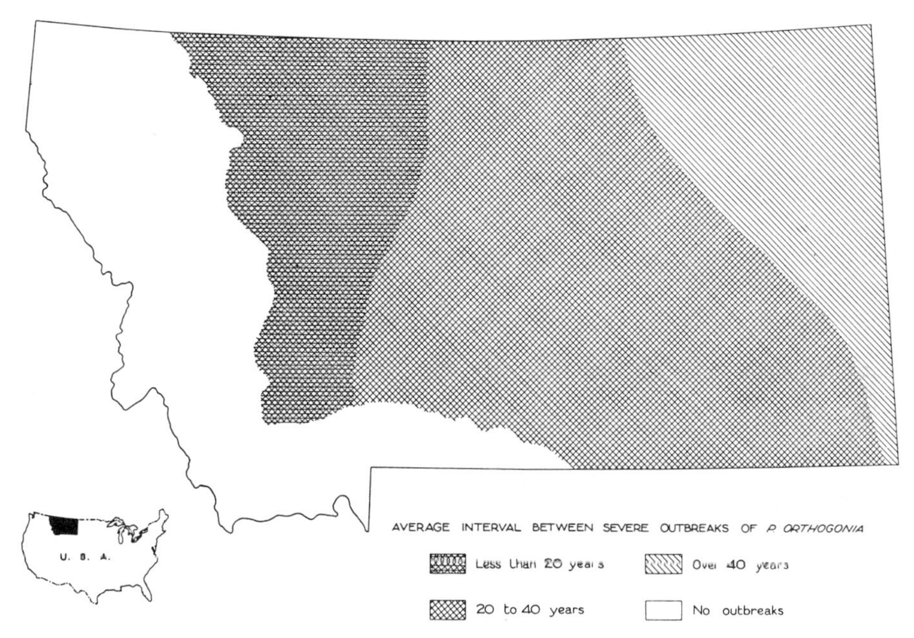

FIG. 13.10.—Areas in Montana where severe and widespread outbreaks of *Porosagrotis orthogonia* might be expected with the frequencies indicated by the shading. (After Cook, 1930.)

fore it is not necessary to consider either "food" or "place to live" in looking for a "density-dependent factor" in the environment of *P. orthogonia.*

Although these two components had changed greatly in the past, they had become relatively stable by the time of this investigation. With "food" and "place to live" not varying much, the numbers attained during periods of relative abundance depended on the weather, especially on the duration of favorable periods. A run of several consecutive favorable years might result in enormous numbers, like those observed during 1920. But the probability of such a run is low. Usually the period of increase would be expected to change to one of decrease before the population had become so dense as this.

Cook showed how catastrophe in the form of excessive wetness at a critical time may overtake a flourishing population and cause it to decline with dramatic suddenness. He indicated the probable frequency of such catastrophes and the depth to which the numbers might decline after an outbreak. He did not investigate how the numbers were determined during periods when the insects were few and decreasing. This is always a most difficult part of any investigation. It is true that *Porosagrotis orthogonia* may be eaten by several species of insects and birds; but in 1922, after the outbreak had been in progress for 5 years, the combined activities of all predators destroyed less than half the population. It is therefore unlikely that predators would have much influence on the numbers of *P. orthogonia* when they were already few. We do not know the answer to the question: Why does not *Porosagrotis orthogonia* become extinct in the Great Plains region? But we can say with assurance that this investigation provided no evidence that the right answer might be: Because its numbers are regulated by "density-dependent factors."

13.14 *Fluctuations in the Numbers of* Choristoneura fumiferana *in Canadian Forests*

The spruce budworm *Choristoneura fumiferana* is indigenous in North America. From time to time, outbreaks have occurred which have endured for perhaps 5 or more years and, in some instances, extended over thousands of square miles of forest. The caterpillars eat the buds, the flowers, and the leaves of spruce, balsam fir, and other coniferous trees. When they are numerous, the budworms may defoliate the trees, doing enormous damage, especially to spruce. After an outbreak has subsided, many years may pass before the budworms become numerous in the same area again. There is evidence, based especially on the study of growth-rings, that the numbers of this indigenous insect have been fluctuating like this for centuries.

Outbreaks may be associated with weather of a particular sort, and this relationship has been studied especially by Wellington *et al.* (1950) and Wellington (1952). The following brief summary is taken chiefly from these two papers. There are numerous other relevant publications dealing especially with the biology of *Choristoneura fumiferana*, and most of these are quoted in the two papers which we have cited. Wellington's studies on the behavior and physiology of *C. fumiferana* have been summarized in sections 7.12, 7.16, 8.31, and 8.32.

A firm obligate diapause occurs in the first larval instar of *Choristoneura fumiferana*. The larvae, hatching during the autumn, do not feed but immediately seek winter quarters, where they spin a "hibernaculum," in which they remain until the following spring. The active stages are present during May, June, and July. There is, because of diapause, only one generation a year, and the same stage in the life-cycle recurs at the same season each year.

The survival-rate among the hibernating larvae is likely to be great when snow falls early and the temperature remains consistently low. Conversely, the death-rate may be high when fluctuations in temperature cause repeated thawing and freezing. The survival-rate among the active stages is greater, the greater the abundance of staminate flowers and the greater the insolation of the places where the larvae are living. Balsam firs that are approaching maturity tend to produce staminate flowers in abundance. Stands of spruce and balsam fir in which the trees are approaching maturity allow the penetration of more light and radiant heat from the sun. So the value of *r* for *Choristoneura fumiferana* is likely to be higher in stands of mature balsam firs or in mixed stands of spruce and balsam firs in which most trees are approaching maturity. No matter how favorable the other circumstances may be, there is little likelihood of *r* being large in stands where the trees are immature (for this is not the most suitable food) or where the conifers are shaded by mature poplars and birches (for the temperature is too low in such places).

Thus, after the destruction of the mature trees of spruce and balsam fir by the budworms or by fire, which may follow severe defoliation by the caterpillars, or by logging, there is a period during which the forest will support only a sparse population of *C. fumiferana*, no matter how favorable the weather may be. Later, when the coniferous trees on which the budworms feed are approaching maturity, the insects may increase in numbers if other circumstances favor them. There is no evidence that this depends to any important extent on predators or any sort of "density-dependent factor." This point was emphasized by Wellington *et al.* (1950, p. 329): "A native insect that has this indicated background and that exhibits violent fluctuations in population density cannot be governed by parasites and diseases during its periods of minimal or initially increasing density so much as by a combination of forest type and climate. The current survey has indicated that endemic populations in the susceptible foci begin to grow to outbreak proportions when the climatic control is relaxed. There has been no indication of control or release of populations by parasitic or pathogenic agents during the long period of minimal density or during the initial population growth in the foci."

The active stages of the budworm are favored by sunny, cloudless weather with little rain during June and July. Figure 13.11 shows how closely the activity of the caterpillars was associated with the rainfall during June. Larvae move away from places that are too wet (secs. 7.12, 7.16). This interrupts their feeding, and many of the larvae die. The hibernating larvae are also favored by keen cloudless weather during the winter. This sort of weather, during both summer and winter, is associated with clear, dry air which is usually found in what the meteorologists call polar-maritime or polar-continental air masses. These names refer to the regions where the air masses originated and acquired their characteristic qualities. The other sort of air mass which commonly occurs in the boreal region of Canada is called "tropical-maritime." This is likely to be associated with warm, humid weather in the winter and rains and cloud during summer; such weather does not suit *Choristoneura fumiferana*.

Centers of low pressure (cyclones) occur relatively infrequently when the weather is being dominated by polar-continental or polar-maritime air masses. So the frequency of cyclones provides an indirect measure of the relative frequencies of the two sorts of air mass and hence of the weather. The meteorological services of Canada maintain records going back many years, from which Wellington was able to calculate the annual frequencies of cyclones in different parts of the country. Figure 13.12 shows the annual number of cyclones for each of 8 districts. The analysis covered 20 years for each district,

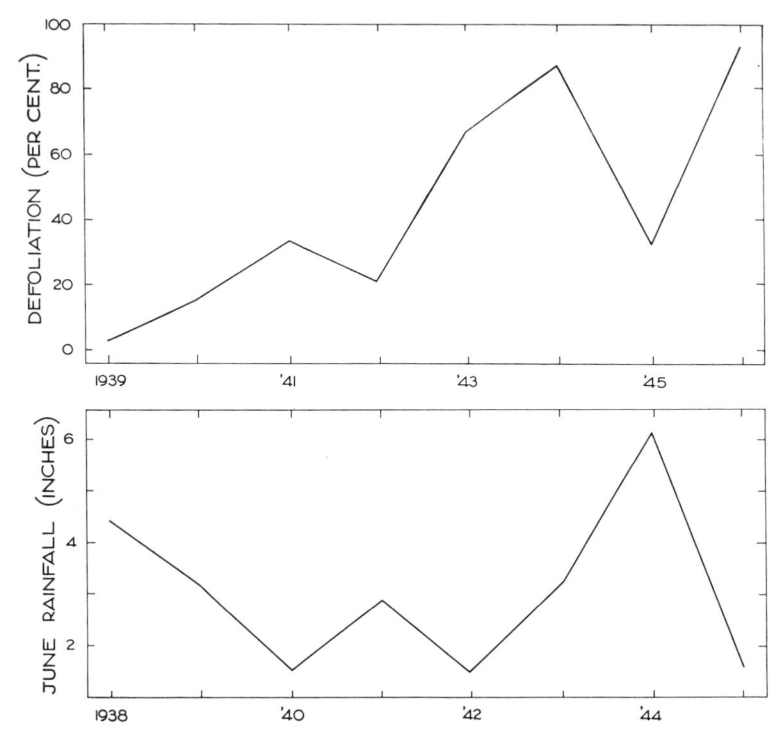

FIG. 13.11.—The upper curve shows the extent to which *Choristoneura fumiferana* defoliated certain trees of balsam fir each year between 1939 and 1946. The lower curve shows the amount of rainfall during June each year between 1938 and 1945. Note the shift of 1 year in the abscissae. (After Wellington *et. al.*, 1950.)

starting in each case 14 years before the beginning of an outbreak of *Choristoneura fumiferana*. Since the outbreaks occurred on different dates in each district, the curves, though they appear in the one diagram, cannot be related to a common date.

The dates of the beginnings of the outbreaks are indicated by arrowheads in Figure 13.12. In some instances the date on which the outbreak "began" was fixed by direct observation of severe defoliation; but in most instances it was inferred from a study of growth-rings. Wellington concluded from these results that outbreaks have usually begun during periods when the weather was being dominated by polar air masses, which would probably result in below-average rain and cloud and above-average hours of sunlight during the summer and consistent cold with a continuous covering of snow during the winter.

After an outbreak has been in progress for several years, predators and diseases become relatively more abundant. But, as the following quotation from Wellington *et al.* (1950, p. 329) shows, it is doubtful whether they are of

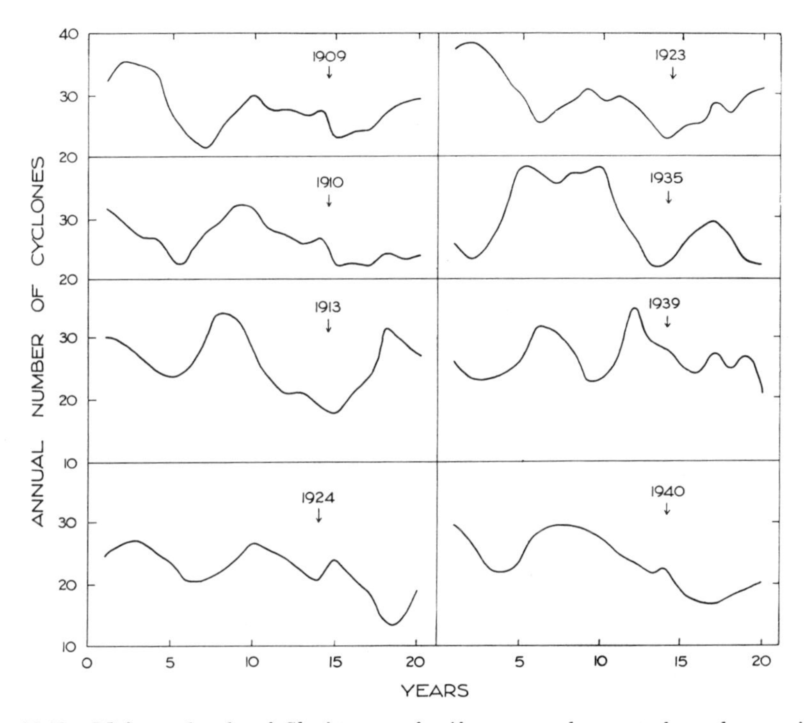

FIG. 13.12.—Major outbreaks of *Choristoneura fumiferana* were known to be under way in Canadian forests, in one place or another, in 1909, 1910, 1913, 1923, 1924, 1935, 1939, and 1940. Each curve shows the number of cyclones recorded in the particular area during the 15 years before and the 5 years after these dates. For further explanation see text. (After Wellington, 1952.)

primary importance in causing the outbreak to decline: "During the first portion of the outbreak period, biological control agents increase in numbers; but observations in the Lake Nipigon infestation indicate that the spruce budworm population begins to experience other troubles, some of which, like starvation, are self-inflicted, shortly before the biological agents assume an important role. The relationships are complex but there are now field indications that starvation and specific reactions to meteorological factors that drive large numbers of larvae from the trees prior to pupation may decrease the spruce budworm population just enough to make it extremely vulnerable to the rapidly increasing populations of biological agents."

13.2 OF FOUR NATURAL POPULATIONS IN WHICH THE NUMBERS ARE DETERMINED LARGELY BY "OTHER ORGANISMS OF DIFFERENT KINDS"

It is well known for many species of invertebrates that their numbers are determined largely by predators (secs. 10.321, 12.23). The circumstances in which the predators are likely to keep the prey scarce are fairly well understood: briefly, the numbers of the prey are likely to be fewer, the more the predator excels the prey in its capacity for multiplication, dispersal, and searching; these qualities are relative and may change with circumstances—for example, weather or the sorts of places available to the prey to live in. This generalization may be made with assurance, at least with respect to insects, because much work has been done by entomologists interested in the "biological control" of pests. The economic aspect has been emphasized. There has been a tendency to classify predators according to whether or not they exercise "economic control" over the pest. The "unsuccessful" ones have received little attention; the "successful" ones have afforded better opportunities for observation, and more has been written about them. Nevertheless, we do not know of a natural population in which the numbers are largely determined by predators that has been the subject of a sustained quantitative investigation comparable with those mentioned in section 13.1.

The importance to be attached to predators of vertebrates is more doubtful. With species like the muskrat and the bobwhite quail, which show a keen awareness of "territory," it seems fairly clear that predators are not so important as the "carrying capacity" of the area where the animals are living (secs. 10.322, 12.31, 13.34). With the larger ungulates and perhaps also with some of the lower orders, predators may be more important. But we do not give any examples of this.

In this section we have, in Lloyd and Reynoldson's studies of the ecology of certain animals which live in sewage-filter beds, two good examples of natural populations in which the numbers are largely determined by nonpredators which also live in the sewage beds. We have an instructive analysis by Ullyett of the causes of deaths in local populations of *Plutella maculipennis;* this reveals an important interaction between predators, diseases, and weather. And we have in *Rhagoletis* a nicely worked-out example of an interaction between predators and place to live in. The activities of a nonpredatory animal (man) result in a large increase in the number of places for *Rhagoletis* to live in. The outcome is that a predator which preys on *Rhagoletis* finds itself with a shortage of food in the relative sense described in section 11.12. But we do not have a well-documented study of a natural population in which the numbers are determined chiefly by predators.

13.21 *The Numbers of* Enchytraeus albidus *in a Sewage-Filter Bed near Huddersfield*

The ecology of the oligochaete worm *Enchytraeus albidus* was discussed by Reynoldson (1947*a*, *b*, 1948) and by Lloyd (1943); the last paper is about the fly *Psychoda alternata*, which is the only other animal of any importance in the environment of *E. albidus*.

The sewage beds at Huddersfield differ from others which were studied (see sec. 13.22), in that the sewage contains an unusually large proportion of "chemical waste" from factories. It is so toxic that the large variety of insects and worms usually found in sewage beds is, at Huddersfield, reduced to a few, two of which, namely, *Enchytraeus albidus* and *Psychoda alternata*, are the only ones that are common. Although *E. albidus* can survive in the sewage bed at Huddersfield, they do not attain the high numbers which are characteristic of the other species which live in other beds where the sewage is not toxic. Experiments in the laboratory showed that the sewage from Huddersfield was toxic in some degree even to *E. albidus*, especially to the young stages, but it was much more toxic to *Lumbricillus lineatus*, which is common in other beds.

The sewage was sprayed onto the surface of the bed, which consists of a layer of "clinkers" 6 feet deep. It percolated down, becoming less toxic as it penetrated more deeply, and eventually the effluent flowed away at the bottom. About 11 million gallons were applied daily to 2 beds, each 200 feet in diameter.

A coating of fungi, bacteria, algae, and inert organic matter formed over the surface of the clinker. This was the chief source of food for both *Enchytraeus albidus* and *Psychoda alternata*. The feeding of the latter aerated the mass; in their absence, anaerobic putrefaction would make the bed quite unsuitable for *E. albidus*, and none could live there. Nearly all the larvae of *P. alternata* were found in the top 12 inches, whereas most of the worms occurred below this depth, so there was no question of the one eating the food that the other required. Nevertheless, in another, much more subtle, way the activities of *P. alternata* were largely the cause of the prevailing low numbers of *E. albidus* in the beds at Huddersfield.

The feeding of a lot of larvae of *Psychoda alternata* caused the coating of fungi, algae, etc., to slough away from the clinker. It was then washed down through the beds and out with the effluent. The worms, partly because they have the habit of clinging to small loose particles of solid, were washed out of the bed in great quantities. Reynoldson sampled the effluent during the spring (when this phenomenon was most pronounced) and estimated that there were about 19,000 worms and 28,000 cocoons containing 100,000 eggs being washed out of the bed daily. And he commented: "Losses much less than these would be a severe drain on such a small population."

If the worms were exposed to this risk continuously, one would expect the population to become extinct. But invariably the washing-out (or "off-loading," as it is called technically) ceased for a period, during which the clinker became covered with a fresh coating of fungi and algae. The worms multiplied during this period. At the same time, the larvae of *Psychoda alternata* had pupated and later emerged as adults. They mated and laid eggs, which in due course gave rise to another batch of feeding larvae. These, in turn, caused another washing-out or "off-loading" to take place. The worms were once again washed out of the bed, and again their numbers were greatly reduced. The rate of increase of *Enchytraeus albidus* between these periodic disasters was slow, perhaps because of the toxins in the sewage or for other reasons that are not understood, so the worms never attained numbers that are comparable with those attained by other species in beds where the sewage is not toxic. On the other hand, they never became extinct, because the phenomenon of off-loading was temporary and ceased long before the last worm had been washed out.

The periodic recurrence of off-loading was determined by the periodic fluctuations in the numbers of *Psychoda alternata.* Lloyd (1943) studied the ecology of this species and showed that the fluctuations were determined partly by the progression of the seasons and partly by fluctuations in the supply of food brought about by the excessive numbers of the larvae themselves. A similar phenomenon was described in greater detail by Nicholson (1950) for an experimental population of *Lucilia cuprina.*

Figure 13.13 illustrates for one year the trends in the numbers of *Enchytraeus albidus* and *Psychoda alternata.* The worms bred most actively during the cooler parts of the year, especially during September–October and again during March–April. The flies developed slowly during winter; they began to be numerous about May and continued so throughout summer. The numbers of adults emerging from the beds fluctuated widely, with a period of about a month between maxima. There was a corresponding period in the fluctuations of the larvae feeding in the bed. The same phenomenon continued during the winter, but the numbers were lower. Off-loading occurred at a reduced rate and less frequently during the winter, and this enabled the worms to multiply at this season. They reached their greatest abundance during late winter and early spring. They were reduced to low numbers by the massive off-loading which occurred during the spring and did not recover until the next winter. There is no need to invoke the concepts of "density-dependent factors" or "competition" to explain how the numbers in this natural population of *E. albidus* are determined.

13.22 *The Numbers of* Metriocnemus hirticollis *and* Spaniotoma minima *in a Sewage-Filter Bed near Leeds*

Lloyd, Graham, and Reynoldson (1940) listed 3 species of Oligochaeta, 2 species of Mollusca, and 7 species of Diptera which are relatively common in the sewage-filter beds at Knostrop near Leeds, and they discussed the ecology of some of them. Other papers have been published on the same subject, and we have referred especially to those by Lloyd (1937, 1941, 1943).

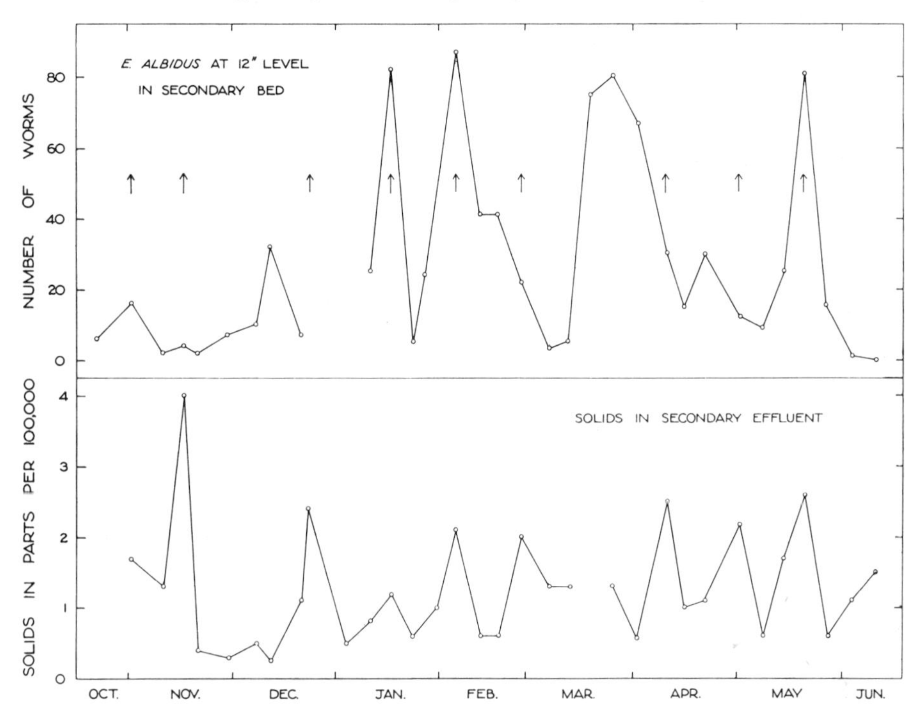

FIG. 13.13.—The number of *Enchytraeus albidus* in a sewage-filter bed near Huddersfield. The lower curve shows the fluctuations in the amount of solids washed out of the bed. The arrows indicate maxima in the process of off-loading. (After Reynoldson, 1948.)

Each bed was essentially a large heap of water-worn gravel, 6 feet high, contained by rectangular walls and provided with a drainage outlet at the bottom. The sewage was sprayed on the surface and allowed to trickle over the stones. In the top foot or so of the bed a dense growth of fungi and algae covered the surface of the stones. This growth, together with solids from the sewage which were trapped in it, supplied a rich and, at certain seasons of the year, abundant source of food for those animals which could live as near the surface as this. The chief food for the animals living in the depths of the bed was a slimy zoögloea which coated the stones in these regions.

Despite the almost continuous application of sewage of constant composition at about the same rate throughout the year, there were marked fluctuations in the amount of food available to the animals living in the bed. This was partly because the growth and senescence of the algae and fungi were influenced by the progression of the seasons and partly because, as we saw in section 13.21, this material had a tendency to slough off the stones and be washed out of the bed when the concentration of animals feeding in it became great.

During the winter the crust of algae and fungi accumulated on the stones, which became heavily incrusted with it. The thicker this covering became, the more likely it was to be loosened by frost and to flake off in big lumps. Such large fragments would sink into the bed and remain there as food. On the other hand, when there were many worms and larvae feeding in this growth, it was more likely to slough off in small fragments, which were mostly washed through the bed and passed out with effluent, carrying many worms and insects with them. This "off-loading" not only caused a periodic shortage of food in the bed but also depleted the populations of certain species by carrying many of them out of the bed. Off-loading went on continuously, but it reached a peak during the spring.

The sewage contained little "chemical waste," and, as a consequence, this bed harbored many more species than did the bed at Huddersfield (sec. 13.21), and their numbers were, on the whole, greater than those of the species which inhabited the bed at Huddersfield. Some species were found only or mostly near the surface; others were found only in the depths; and there were still others which occurred throughout the bed.

The different species were favored by different ranges of temperature, and they also responded differently to other components of the weather. With each one its chance to survive and multiply was influenced by the activities of other species living in the bed. None of these was an obligate predator, but there were some that would destroy and eat certain other species if they happened across them in the course of their feeding. Except for strict predators, all the categories of other animals which we listed in section 10.01 may be recognized in the environments of one or another of the animals living in this sewage-filter bed; and the ecology of these species, as revealed by Lloyd and his co-workers, is both complex and interesting.

The fly *Psychoda alternata*, which comprised over 95 per cent of all the insects in the bed at Huddersfield, was relatively much less abundant at Knostrop. At Huddersfield, *P. alternata* had almost sole use of the rich resources of food in the top foot or so of the bed, because the toxic sewage did not permit other species to live there. But at Knostrop, *P. alternata* was quite rare in the top foot, because this layer was usually dominated by *Metriocnemus longitarsus* and other species which destroyed the eggs and young larvae of *P. alternata*.

Most of the eggs of *P. alternata* were probably laid near the surface, but those that were laid in the depths had a better chance of surviving and probably contributed the greater proportion of adults to the population. It would be interesting to follow the ecology of these and a number of other species more closely, but space permits us to mention only two others, the chironomids *M. hirticollis* and *Spaniotoma minima*. The former was one of the less abundant species; the latter was among the most abundant ones. The annual total number of flies trapped during the 8 years from 1934 to 1941 varied between 76 and 846 for *M. hirticollis* and between 11,000 and 38,000 for *S. minima*. These figures are based on the weekly counts of the numbers of adults emerging from the bed into small traps placed on the surface of the bed.

Generations of *Metriocnemus hirticollis* overlapped, and adults were taken in the traps during every month of the year. They were relatively scarce during autumn and winter and were usually most numerous during April–June. The adults of *M. hirticollis* require to form a swarm in order to mate. The females returned to the bed to lay eggs, but showed little tendency to penetrate deeply; as a consequence, nearly all their eggs were laid in the top foot. The full life-cycle of *M. hirticollis* required about 42 days at 20° C. and 123 days at 10°; none completed its development below 7° C. The temperature of the bed varied from about 8° C. during midwinter to about 17° C. during midsummer. Consequently, the activity of *M. hirticollis* was largely restricted to summer; eggs laid during autumn had little chance of becoming adult before June. The eggs laid during June (when the flights of adults were usually at their greatest) had ample time to become adults before winter, and one might expect the adults of *M. hirticollis* to be abundant during autumn. In fact, they were nearly always scarce at this season. Lloyd (1943) attributed this to the activities of the related chironomid *M. longitarsus*.

This species thrived at much lower temperatures than did *Metriocnemus hirticollis*. A full generation required 26 days at 20° C., 94 days at 10°, 153 days at 6.5°, and 243 days at 2° C. As a consequence, eggs laid early in the autumn emerged as adults during winter; usually circumstances were sufficiently favorable at this season, and another generation was produced before spring. This led to a large flight during May. The females laid their eggs near the surface, and the larvae established themselves in the top foot or so of the bed. Large flights of the first species, *M. hirticollis*, rarely occurred before June. So the females of *M. hirticollis* usually laid their eggs in the very zone that was already densely occupied by partly grown larvae of *M. longitarsus*. The latter are not obligate carnivores, but they readily devoured the eggs and small larvae of the former when they came across them in the course of their feeding. The consequence was that very few of the larvae of *M. hirticollis* survived, and the autumnal flights of this species were usually quite small. This sequence of events happened every year. In some years when *M. longi-*

tarsus were less numerous than usual, the rate of survival among *M. hirticollis* was higher, and the numbers of adults caught in the traps were correspondingly greater; the annual totals varied in the ratio of 1:13. It is therefore pertinent to inquire into the causes of fluctuations in the numbers of *M. longitarsus.*

Lloyd (1943) recognized two important causes for fluctuations in the numbers of *Metriocnemus longitarsus*. The most important was the weather during June, when, as a result of the large flights of adults during May, there were many eggs and young larvae near the surface. Dry weather during June allowed the surface to dry out and caused the deaths of many eggs and young larvae. On the other hand, moist weather during June insured a high rate of survival among this generation, and this, as we have seen before, correspondingly reduced the chances of survival of *M. hirticollis.*

The other cause for fluctuations in the numbers of *Metriocnemus longitarsus* was the weather during the winter: they were usually less numerous after a warm winter. Since this species was able to develop actively at moderately low temperatures, the opposite result might have been expected. Lloyd (1943) explained the anomaly by reference to the activities of the worm *Lumbricillus lineatus*. This species was most active during the winter; a generation was completed in 110 days at 10° C. and in 170 days at 7°. During a warm winter the worms might complete one generation and start another before the spring; but when the winter was colder than usual, the first generation might not be completed until quite late in the spring. Since the feeding of the worms in the crust of fungi and algae which covered the stones was the chief cause of off-loading, and since this depended largely on the concentration of worms in the surface foot or so of the bed, it is clear that off-loading would begin much earlier after a warm winter when the worms had been multiplying rapidly. This was indeed what happened. After a warm winter the stocks of food for the larvae of *Metriocnemus longitarsus* were depleted much earlier than usual, and the larvae themselves faced severer risks of being washed out of the bed. The fluctuations in the numbers of *M. longitarsus* which could be associated with this cause were considerable: the number of adults trapped during May varied from 20 in 1938 to 860 in 1941.

The numbers of *Metriocnemus longitarsus* were not great enough during any of the 8 years between 1934 and 1941 to kill all the *M. hirticollis* in the sewage-filter bed at Knostrop. If, as a result of some extraordinary circumstances, *M. hirticollis* had become extinct from this bed, it is hardly likely to have died out at the same time from all the other beds and other places where it occurs naturally. It is more likely that somewhere in the vicinity some other natural populations would have survived, which would in due course recolonize this particular bed. Nevertheless, the destruction of *M. hirticollis* by *M. longitarsus* was severe enough and consistent enough to prevent *M. hirticollis* from becoming numerous in this particular bed. It is important to notice that the numbers of *M. longitarsus* were not influenced in any way by the numbers of *M. hirticollis;* therefore, *M. longitarsus* was not acting as a "density-dependent factor" in the environment of *M. hirticollis.*

One of the most abundant insects in the sewage-filter bed at Knostrop was the small black chironomid, *Spaniotoma minima.* This species, like *Metriocnemus hirticollis*, emerged from the bed to form mating swarms, and the females returned to lay their eggs near the surface. Also, like *M. hirticollis*, the eggs and young larvae were likely to be destroyed by the older larvae of *M. longitarsus* when the two species occurred together in a densely populated medium. In fact, as with *M. hirticollis*, any egg laid after June had little chance of living long enough to become an adult. Its chance was perhaps a little higher because the larvae of *S. minima* had a tendency to move toward the depths of the bed, and consequently they spent less time in the zone where *M. longitarsus* abounded.

The life-cycle of *Spaniotoma minima* occupied 29 days at 20° C., 80 days at 10°, and 260 days at 2°. This ability to continue developing at low temperatures and the relatively short period required for a generation at moderate temperatures enabled *S. minima* to continue breeding throughout the winter and to produce a relatively large flight in the spring, usually in advance of the major flight of *Metriocnemus longitarsus.* This enabled the next generation of larvae to establish themselves before the chief danger from *M. longitarsus* had developed, and there were usually further large flights of *S. minima* during August. The progeny from these flights were largely destroyed by *M. longitarsus;* and, by autumn, the numbers of *S. minima* had usually dropped well below what they might have been in the absence of *M. longitarsus.* But the numbers of *S. minima* nevertheless remained relatively high, because their short life-cycle, their capacity to develop during winter, and their habit of colonizing the depths of the bed as well as the zone nearer the surface enabled them to recover rapidly from the setbacks imposed by *M. longitarsus.*

Lloyd discussed these results in terms of competition. This usage implies that the egg or small larvae of one species is eaten by its "competitor," while there is still no great shortage of the resource for which they are "competing." This does not seem to come within the limits of any definition of competition that is acceptable. If one must seek to use a single abstract noun to describe these complex phenomena, perhaps "predation" would be nearer the mark. But this is also misleading, because the relationships between *M. longitarsus* and the other species which are eaten are very different from those which characterize the strict predators and their prey, such as were described in section 10.32 and elsewhere. It is better to leave the phenomenon unnamed than to call it by a name that is misleading.

In the summaries given in this section we have considered briefly for both *Metriocnemus hirticollis* and *Spaniotoma minima* how the speed of development and especially the expectation of life may be influenced by the several components of environment. Temperature, moisture, and food were all of some importance, but the component of chief importance was "other animals." As usual, it was necessary to take into account the place where the animal was living in order to assess the influence of the other components on the value of *r*.

13.23 *The Numbers of* Rhagoletis pomonella *in Fields of Blueberries in Maine*

The larva of *Rhagoletis pomonella* lives in the fruits of the blueberry. The following account of the ecology of *R. pomonella* (Diptera, Trypetidae) is drawn from Lathrop and Nickels (1931). In order to appreciate how numbers of *R. pomonella* are determined, it is necessary to know something of how blueberry land is "farmed" in Maine.

Blueberry is the name given to several species of the genus *Vaccinium*. They are low shrubs, growing about 6–12 inches high, and, in certain circumstances, they bear heavy crops of edible berries. The species are indigenous to northeastern North America, where they originally occurred as a sparse undergrowth among tall shrubs or along the margins of forests. When the tall shrubs and the trees were destroyed, the blueberries invaded the areas where the forests had been. It was long ago discovered that if this vegetation were subjected to periodic "burnings," the blueberries would become dominant. Lathrop and Nickels (1931, p. 262) described the procedure:

> On bright, calm days in early spring, after the snow leaves, but before the frost starts out of the ground, and while the blueberry plants are still thoroughly dormant, the surface litter is ignited and under favourable conditions the fire sweeps the land clear of vegetation. . . . The above ground parts of the blueberry plants are removed by the process of burning. The root system is unharmed by the fire, and the plant responds by a greatly accelerated vegetative growth during the summer immediately following the burn, but the plant produces no fruit during this first season. During the second summer a large crop of berries is produced. After this first, abundant crop the production of fruit decreases each season until the yield becomes practically nothing; unless the land is again burnt-over to rejuvenate the blueberry plants. In practice, the commercial blueberry growers have found that two crops of berries between burns are all that the average land will produce profitably, and the three-year cycle usually keeps the weed bushes fairly well under control.

The life-cycle of *Rhagoletis pomonella* includes a firm obligate diapause in the pupal stage. In about 85 per cent of the population, diapause disappeared during the first winter, and most of the remainder emerged after the second winter. The pupal stage is spent in the soil, and most of the pupae occur in the top inch; nevertheless, the burning described above did not cause any deaths. The adults emerge from the soil and lay eggs inside the berries during June–July; the majority of the maggots leave the fruit and enter the soil to pupate during the first half of August.

The maggots of *Rhagoletis pomonella*, living inside the berries, may be sought out and oviposited in by certain predators, of which the braconid *Opius melleus* is the most important. The chance that a maggot will be found by a predator depends partly on the number of predators in the area and partly on the proportion of the blueberries infested by maggots. This is a situation quite analogous to the one we described in sections 11.12 and 10.321 in relation to *Anarhopus sydneyensis* and *Lygocerus* sp.

When the blueberry bushes are growing, as they did originally, as a sparse undergrowth among tall shrubs or along the margins of forests, they bear relatively few berries. In the absence of predators, the numbers of *Rhagoletis pomonella* would probably increase until nearly all the berries were infested. The population would still not be very dense, because there would not be many berries. But the numbers of the berries are not influenced in any way by the numbers of the maggots. However, the population is usually prevented from attaining this density by the activities of the predators.

The sequence of events in areas that are being "farmed" for blueberries is quite different. The account given by Lathrop and Nickels (1931, p. 263) is as follows:

> During the summer immediately following the burn there is a normal emergence of blueberry flies on the burned-over areas, and the absence of a crop deprives the flies of berries in which to oviposit. This may be of little consequence on small patches, but where large areas are burned-over the flies must migrate considerable distances if they are to reach berries in which to oviposit. No maggots are produced on well burned land during the summer immediately following the burn, with the result that the population of the species is greatly reduced. The second summer following the burn is marked by the production of the first crop of berries, usually a heavy yield. The blueberry maggot has been "starved out" during the preceding season, and the beginning of the first crop year finds the population of the species at a low potential. There are two important sources of reinfestation of the new crop of berries: (1) migration of flies from unburned areas and (2) carryover of puparia in the soil from the second preceding season. Migration of flies is undoubtedly important on small burns of a few acres and near the margins of larger burned-over areas. Under usual conditions the migration of flies probably is not a very important factor on solid burns of 10 or 15 acres or more.

The population increases rapidly, because there is an abundance of food and the death-rate from predators is low indeed. This investigation did not reveal where the predators that were found in these populations came from. We do not know whether some of them also had spent two winters dormant in the soil or whether they had all flown in from outside. In either case their numbers were few and their chance of success slight. With such a large crop of berries and such a small proportion of them infested by maggots, the predators have little chance of finding many prey. The numbers of *Rhagoletis pomonella* continue to increase until the third summer after the "burn." Because the vegetation is usually burned again at this stage, the flies again become scarce.

Superimposed on this short-term rhythm in the numbers of *Rhagoletis pomonella*, which is determined directly by the frequency of burning, Lathrop and Nickels discovered a long-term trend which they described as follows:

If the land receives continued care, a period arrives, after several years, when the blueberry maggot population of the area reaches a maximum. At this point the land is relatively free from weed bushes, so an excellent stand of blueberries is supported, and berries are produced in abundance. However, there still remains a sufficient growth of "sprouts" and sweetfern to furnish protection to the adult flies. If the culture is carried beyond this stage, and the sprouts and sweetfern are completely removed from the land, there seems to be a tendency for the maggot population of the area to decrease, probably because of the excessive exposure of the adult flies to rain, wind and sunshine. Certainly, berries from blueberry land at its best development are seldom, if ever, found to have a high percentage of infestation.

If the blueberry land is neglected, the sprouts are not mowed, and the land is not burned-over for a period of years, the vegetation soon reverts to the tall shrub association. After a few years of neglect the yield of blueberries becomes insignificant. The percentage of berries infested by the maggots may be very great, but the total population of blueberry maggots on the area is small.

This study of the ecology of *Rhagoletis pomonella* in Maine shows us that where blueberries grow "naturally" as the sparse undergrowth among shrubs, the maggots are never numerous, because predators eat them. But in the culture of blueberries as practiced in Maine, although the amount of fruit is much greater, the number of maggots is small. This is not because predators eat the maggots but because of man-made catastrophes. In our way of looking at the ecology of an animal, this is an example of the numbers of an animal being determined by the number of suitable places for it, which, in turn, depends upon the activity of a nonpredator—man. The number of blueberries on an acre of land is not in any way determined by the density of the population of *R. pomonella;* nor is the frequency of the burning directly related to the number of flies. Neither can be properly described as a "density-dependent factor."

13.24 *The Causes of Deaths in a Natural Population of* Plutella maculipennis *near Pretoria*

The causes of deaths among a natural population of the tineid *Plutella maculipennis* were studied by Ullyett (1947). Samples were taken at weekly intervals during 1938–39; records were kept of the number of caterpillars on the plants and the causes of deaths occurring during the larval and pupal stages. Deaths among eggs and adults were not recorded, chiefly because of the technical difficulties. It was considered that these omissions did not seriously alter the final conclusions. Ullyett's (1947) paper, from which the following account was taken, dealt chiefly with the quantitative records made during 1938–39, but he stated that these results were confirmed by many less systematic observations made during the 6 years that the investigation lasted.

The life-cycle of *Plutella maculipennis* is not complicated by diapause. In the kindly climate of Pretoria, breeding continued throughout the year. A generation required about 14 days in midsummer and 21 days in midwinter. Generations overlapped, and all stages of the life-cycle were found on the plants at the same time. The females might, in favorable circumstances, lay up to 300 eggs, but in nature this potential was probably not very often attained. Ullyett considered the weather in some detail and concluded that the weather was not likely ever to be directly responsible for the deaths of many *P. maculipennis* in the region of Pretoria.

For food the larvae of *Plutella maculipennis* may eat the leaves of almost any crucifer. On the farm where Ullyett worked, cabbages and other crucifers were usually planted in long narrow plots. Cabbages were growing continuously on the same farm, though not on the same plot, throughout the year, except for 4 months during winter from July to October. However, there was no shortage of food for *P. maculipennis* at this time of the year, because there were plenty of *Nasturtium officinale*, *Brassica pachypoda*, *Lepidium* spp., and other crucifers, either cultivated in gardens or growing as weeds on this farm and elsewhere in its immediate vicinity. At no stage during the investigation did the population of *P. maculipennis* consume more than a small proportion of the food available to it in the area, so we can be reasonably certain that its rate of increase was never seriously reduced by an absolute shortage of food. But the periodic harvesting of each crop of cabbages, the planting of new batches on different plots, and the complete absence of cruciferous vegetables from all the plots during 4 months meant that the distribution of the food was constantly changing. The distances involved may have been small relative to the powers of dispersal of *P. maculipennis*, for Ullyett did not consider this as a serious hindrance to its multiplication (sec. 11.12).

In a kindly climate and with plenty of food and places to live, it was to be expected that the numbers of *Plutella maculipennis* would be determined largely by the other organisms associated with them. Ullyett reared 18 species of Hymenoptera and one species of Diptera from the larvae and pupae of *P. maculipennis*. Nine of them laid their eggs in or on the larva, and four oviposited in or on the pupa; all of them lived inside the body of the prey, that is, they were, in the language of economic entomology, "internal parasites" of *P. maculipennis*. The other six came into our category *e* of section 10.01; they were predators of predators in the environment of *P. maculipennis*. Rather little was known about most of the species in this list. Some were found only rarely, and it is likely that they were casual predators of *P. maculipennis*, preferring some other unknown species. But two species, *Angitia* sp. and *Apanteles halfordi*, were known to prey chiefly, if not exclusively, on *P. maculipennis*. The former was the only one that ever became abundant; in Figure 13.14 the numbers represented by curve *C* were made up chiefly by this species.

The pupae of *Plutella maculipennis*, especially those that were fastened to the outer leaves which touched the ground, were often eaten by beetles of the family Staphylinidae. The staphylinids lived in the soil, and the greater part of their food consisted of a variety of soil-inhabiting insects; hence they were not dependent on *P. maculipennis*. It so happened that they occasionally became quite numerous in the soil under the cabbages, and they sometimes were responsible for a large proportion of the weekly deaths of *P. maculipennis*. The cabbages sometimes harbored large populations of aphids, which attracted

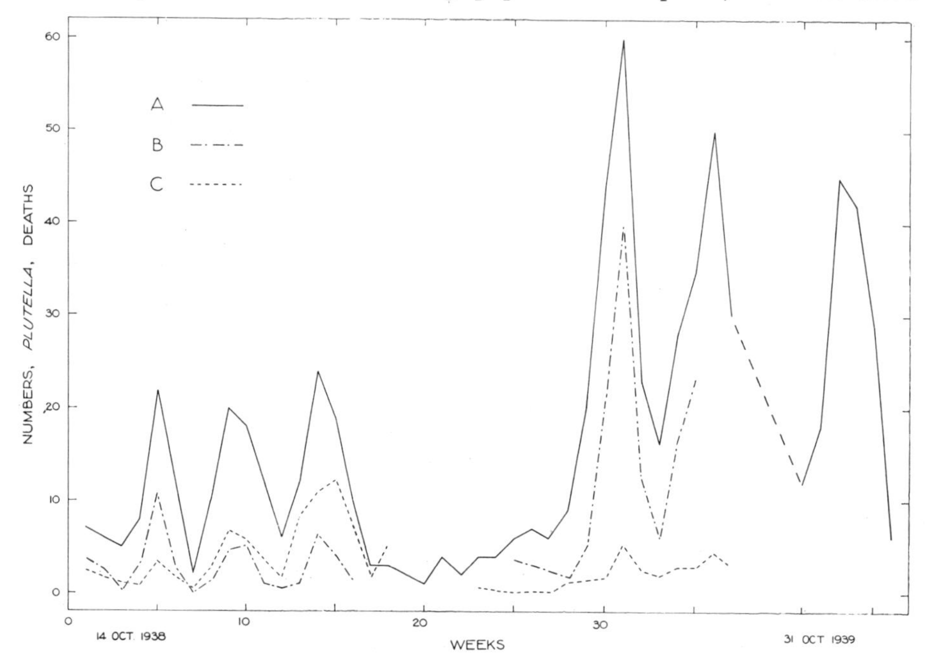

FIG. 13.14.—*A*, the numbers of ***Plutella maculipennis*** in a garden near Pretoria, based on weekly counts of the numbers of larvae and pupae on a sample of cabbage leaves; *B*, the weekly deaths due to facultative predators; *C*, the weekly deaths due to obligate predators. (After Ullyett, 1947.)

their own group of predators. Larvae of certain species of hover flies (Syrphidae) and lacewings (Hemerobiidae or Chrysopidae) and the nymphs of an anthocorid bug were the most important. It so happened that these predators would sometimes consume all or most of the stock of their preferred food (aphids) on a cabbage before they were themselves fully grown. In these circumstances they would turn to the larvae of *P. maculipennis*, and on occasion a high proportion of the weekly deaths was attributed to them. Other, less important, facultative predators include spiders, wasps (*Polistes*), and birds.

These facultative predators had certain qualities in common. Since *Plutella maculipennis* did not constitute a staple part of their diet, their numbers fluctuated independently of those of *P. maculipennis*. Since the caterpillars and pupae of the cabbage moth were not the preferred food of this group of predators, they were not likely to be eaten in numbers unless they were numerous and easy to find. An exception to this occurred when the predators of aphids found themselves deprived of their favorite food. Being wingless, they were unable to leave the cabbage in search of other food, so they ate what they could find. For these reasons the deaths attributed to the activities of this group of facultative predators fluctuated widely and sporadically (Fig. 13.14, curve *B*).

Twice during the period shown in Figure 13.14, widespread outbreaks of a fatal disease caused by the fungus *Entomophthora sphaerosperma* swept through the population of *Plutella maculipennis*, leaving only two or three larvae alive on each cabbage plant. During the first outbreak the weekly counts were maintained, and in Figure 13.14, curve *A* shows that the population remained at this low level for nearly 10 weeks. Precise numbers were not recorded during the second outbreak, but Ullyett (1947, p. 99) wrote: "After the initial peak of the larval infestation had been passed, a period of heavy, continuous rains supervened which prevented field work. This was immediately followed by another epidemic of the fungus disease which reduced the population to a low level comparable with that reached in Period II."

Between outbreaks very few individuals died from this or any other sort of disease. But the fungus must have been present and widespread, because outbreaks of the fungal disease invariably occurred whenever there had been enough rain to keep the surfaces of the leaves of the plants continuously wet for 3 or 4 days running. Light continuous rain was more effective than heavy intermittent rain; even continuous heavy mist would do. This disease was observed to break out with equal severity in a number of different populations of different densities, and Ullyett concluded that the onset of the disease was independent of the density of the population at the beginning but depended only on the weather. While the weather remained favorable to the fungus, the numbers of *Plutella maculipennis* remained at a consistently low level. Ullyett wrote: "The extent to which the population is reduced has been approximately the same on all occasions when the duration of the disease has been comparable. There are indications that this agreement would be even more remarkable if the density of the host population could be correlated with the total surface area of the plant." Once the rain abated and the surface of the plants remained dry for considerable periods between showers, the panzoötic came to an end; and the numbers of *P. maculipennis* increased at a great rate. Figure 13.14, curve *A*, shows the increase quantitatively for the period after the first outbreak of disease, but the curve relating to the period after the second outbreak is qualitative only. Precise records were not kept, but Ullyett commented: "A repetition of the previous events occurred and the end of the following year's observations was marked by an exceedingly heavy infestation of all cruciferous crops which were then on the ground. In this case the resultant damage was so severe that cabbage and cauliflower crops were unmarketable and were fed to stock."

The rate at which the population increased during the intervals between outbreaks of disease depended partly on the fecundity of the moths and the speed of development of the immature stages and partly on the activities of predators. The deaths caused by facultative predators (curve *B* of Fig. 13.14) doubtless retarded the rate of increase somewhat, but they did not prevent the population from becoming so dense that cabbages and cauliflowers were unmarketable and were fed to stock. Since *Plutella maculipennis* were not the favored food of any of these groups of facultative predators and since the numbers of the predators were largely independent of those of *P. maculipennis*, there are good grounds for expecting that these predators would not ordinarily keep *P. maculipennis* scarce. On the other hand, the obligate predator, *Angitia* sp., might be expected to do this, provided that its rate of increase and capacity for increase were adequate (sec. 10.321).

The same disadvantage associated with changing distribution of crucifers on the farm which we mentioned above in relation to *Plutella maculipennis* applied also to *Angitia* sp., which had to find their prey as they spread out to colonize the plants in each new situation. The evidence which we reviewed in section 10.321 would suggest that a small discrepancy in the dispersive powers of prey and predator might make a big difference to the numbers which the prey can maintain in a particular area. Ullyett did not mention this point explicitly, but he implied from his discussion of the results that *Angitia* sp. and the other species represented in curve *C* of Figure 13.14 possessed the necessary qualities for keeping *P. maculipennis* scarce.

The predators of this group were probably chiefly responsible for the low numbers of *Plutella maculipennis* recorded during the first 15 weeks of the period shown in Figure 13.14. But they clearly did little to check the multiplication of their prey during the periods which followed the disappearance of the epizoötic of fungal disease. While the disease was widespread, the prey suffered severely, but the predator suffered even more severe losses: not only were many killed by the disease, but those that survived suffered from a shortage of food. The predators started from a relatively low level after the epizoötic abated. As a consequence, their rate of increase was at first very slow relative to that of the prey; they multiplied steadily but did not become numerous enough to effect any appreciable reduction in their supply of food before catastrophe, in the form of another outbreak of disease, overtook them once again.

Several conclusions of general interest emerged from this investigation: (*a*) At no time did *P. maculipennis* become numerous enough to eat more than a small proportion of the stocks of food available to it. (*b*) The caterpillars were reduced to minimal numbers during epizoötics of fungal disease, the average number of caterpillars per plant being usually 2 or 3 while the epizoötic lasted; but we do not know how this particular density was determined. (*c*) Outbreaks of disease were occasional and temporary; their frequency and duration were determined entirely by the weather. (*d*) The maximal numbers of caterpillars were attained shortly after the epizoötic had abated. (*e*) The numbers prevailing at other times, that is, when there was a prolonged interval between outbreaks of disease, probably depended largely on the activities of predators, especially of the obligate predator *Angitia* sp.

From what has been said, it is clear that in the Pretoria district the relative importance of diseases and predators in determining the prevailing numbers of *Plutella maculipennis* depends on the frequency and duration of spells of wet weather. This matter was not investigated by Ullyett. For the short period covered by Figure 13.14, disease was undoubtedly the more important in relation to both maximal and minimal numbers.

13.3 SOME EXAMPLES WHICH ILLUSTRATE THE IMPORTANCE OF STUDYING THE PLACES WHERE THE ANIMALS MAY LIVE

The first example in this section may seem unduly simple. So far as we can tell, the distribution and abundance of the two species of nematodes *Trichostrongylus vitrinus* and *T. colubriformis* is determined entirely by what we called in sections 12.0 and 12.1 the "qualitative aspect" of the place where an animal may live. By way of contrast, the ecology of *Ixodes ricinus*, which forms the second example, is more complex than any other example which we have discussed. It includes a nice interaction between moisture and "place in which to live"; food is also important in a subtle way, and so is the number of other animals of the same kind. In the third example there is a nice interaction between "food" and "place in which to live"; moisture is also important in its own right. Finally, we discuss the ecology of the bobwhite quail; this is interesting because it is part of the original work on which Errington based his concept of "carrying capacity," which we have had occasion to mention in sections 10.322 and 12.31.

13.31 *The Distribution and Abundance of the Nematodes,* Trichostrongylus vitrinus *and* T. colubriformis *in the Gut of the Sheep*

Several species of nematode worms commonly occur on the surface of the mucous membrane of the gut of the sheep. More often than not, several species occur in the one animal. Some of them are found only in the abomasum, others only in the small intestine. The distribution is even more specific than this, as certain species tend to concentrate in specific sections of the small intestine (Tetley, 1937; see also Fig. 13.15). The three species shown in Figure 13.15 reach their maximal numbers in the first few feet of the small intestine, becoming fewer toward the pylorus and in the lower reaches of the duodenum (Somerville, personal communication). Food is undoubtedly present in abundance, and these striking distributions must reflect a gradient in some quality of the gut; it is almost certainly a chemical gradient, perhaps a gradient of pH.

The information summarized in Tables 13.08 and 13.09 reveal two interesting

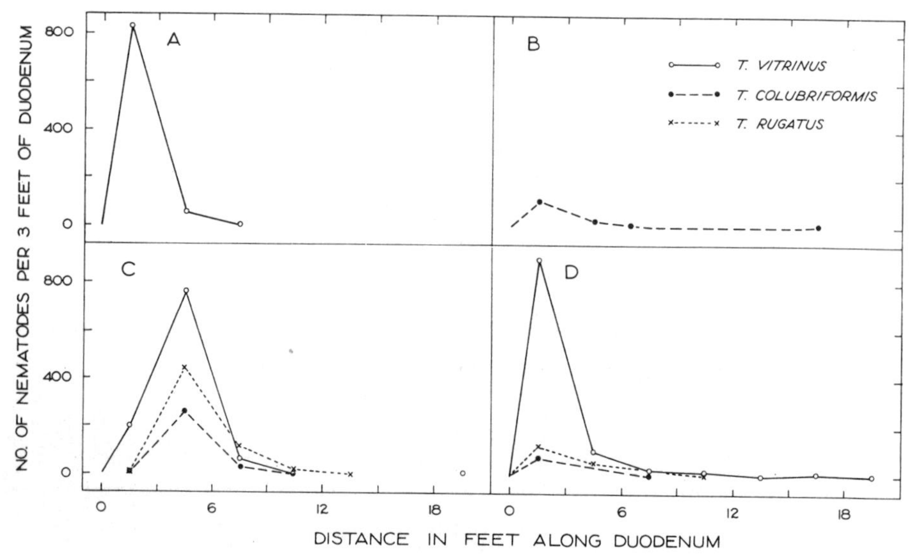

FIG. 13.15.—The distribution and abundance of three species of *Trichostrongylus* along the length of the duodenum of a sheep. Each graph shows the distribution in a single sheep. *A*, *T. vitrinus* by itself; *B*, *T. colubriformis* by itself; *C* and *D*, *T. vitrinus*, *T. colubriformis*, and *T. rugatus* together. (From Somerville, personal communication.)

features about the distributions of *Trichostrongylus vitrinus* and *T. colubriformis:* (*a*) the two species occur in the same part of the gut, and (*b*) the presence of other species in the gut seems not to modify in any way either the

TABLE 13.08*

DISTRIBUTION AND ABUNDANCE OF *Trichostrongylus vitrinus* IN GUT OF 10 SHEEP

OTHER SPECIES PRESENT		DISTRIBUTION IN DUODENUM IN FEET (BEGINNING OF DUODENUM = 0)		MAXIMAL NO. PER 3 FEET OF DUODENUM
No.	Species	Major Distribution	Subsidiary Distribution	
0	....	0–6	....	820
1	*T. colubriformis*	0–6	....	270
		0–12	....	360
2	*T. colubriformis*, *T. rugatus*	0–9	18–31	760
		0–9	15–18	900
		0–12	12–21, 30–36	150
		0–6	....	60
		0–9	36–42	440
2	*T. rugatus*, *T. probolurus*	0–6	....	30
3	*T. colubriformis*, *T. rugatus*, *T. probolurus*	0–9	15–18	40

* From Somerville, personal communication.

TABLE 13.09*

DISTRIBUTION AND ABUNDANCE OF *Trichostrongylus colubriformis* IN GUT OF 10 SHEEP

OTHER SPECIES PRESENT		DISTRIBUTION IN DUODENUM IN FEET (BEGINNING OF DUODENUM = 0)		MAXIMAL NO. PER 3 FEET OF DUODENUM
No	Species	Major Distribution	Subsidiary Distribution	
0	....	0–6	15–18	100
1	*T. vitrinus*	0–6	.....	70
		0–12	.....	60
2	*T. vitrinus*, *T. rugatus*	0–6	9–15	75
		0–9	.....	100
		0–9	.....	260
		0–6	.....	90
		0–6	.....	30
3	*T. rugatus*, *T. probolurus*, *T. vitrinus*	0–12	12–21	3,000
		0–18	21–30	2,250

* From Somerville, personal communication.

distribution or the abundance of either of these two species. If one species interfered with another, it might be expected that the distribution of each species would become restricted in the presence of others (sec. 10.213; and Fig. 10.09). This is clearly not the case.

This example brings out two interesting points: (*a*) a gradient in some attribute of the area where a population is living may be reflected in the abundance of the animals, and (*b*) at least in the special circumstances represented by the gut of the sheep, a number of closely related nonpredatory species can live crowded together in a restricted space.

13.32 *The Distribution and Abundance of* Ixodes ricinus *in Pastures in Northern England*

The tick *Ixodes ricinus* is regarded as a serious parasite of sheep in Britain, especially as it is known to be a vector of disease. It is found on pastures of the uplands but not of the lowlands, where it might have been expected to occur if one considered only climate and the presence of its chief host the sheep. On the upland pastures it is distributed in an irregular way which was for a long time rather puzzling. The ecology of *I. ricinus* was thoroughly investigated by Milne (1943, 1944, 1945*a*, *b*, *c*, 1946, 1947*a*, *b*, 1948, 1949, 1950*a*, *b*, 1951, 1952). There are also several other papers, mention of which may be found in those we have cited. The distribution and abundance of *I. ricinus* are no longer puzzling, and the explanation includes more especially a consideration of (*a*) the ticks' requirements for water and how these may be satisfied in certain sorts of places but not in others and (*b*) the hazards associated with the ticks' "search" for food and mates and how these may be increased by a scarcity of either sheep or ticks.

The life-cycle of *Ixodes ricinus* was described in section 12.22; it occupies 3 years. Except for a brief period of activity during the spring,[2] the tick spends the whole of each year lying concealed at the base of the vegetation near the surface of the soil. One huge meal of blood is taken during the spring, but for the rest of the time the tick neither feeds nor drinks. The unfed larva or nymph loses water by evaporation when it is exposed to air in which the relative humidity is less than 88 per cent; the unfed adult loses water to air in which the relative humidity is less than 92 per cent; but ticks in either stage may absorb water through the cuticle from air in which the relative humidity exceeds these values. So a tick has little chance of remaining alive during the periods which elapse between meals unless it happens to be in a place where the air remains nearly saturated with water vapor continuously throughout the year. The physiology of the tick in relation to moisture is discussed in sections 7.213 and 12.22.

Since the tick can crawl only very feebly, it invariably burrows into the vegetation within an inch or two of where it drops from the host. So the place where the tick spends the interval between meals depends not at all on its own activities (except that it may crawl up or down the grass stems, coming closer to, or going farther away from, the surface of the soil) but is determined entirely by the movements of the host. If a tick is to survive to contribute progeny to the next generation, it must meet with good fortune, with regard to the place where it drops from the host, not once but three times during its life. A tick which falls from a sheep will die from desiccation unless it is lucky enough to fall where there is a dense mat of vegetation and dead organic matter on top of the soil. If it is lucky enough to fall into such a situation, it will burrow down into the thick central part of the mat where the air is most moist. It remains there until it is ready for the next meal of blood (sec. 12.22).

The pastures which grow on uplands, especially on the lower slopes of hills, include coarse grasses and bracken, which form a dense, tangled mat of vegetation and dead organic matter next to the soil. This mat is often several inches thick, and it is likely to remain continuously moist throughout the year. An engorged tick, dropping from its host into this mat, is not likely to die from desiccation before the next meal is due. On the other hand, pastures on the lowlands usually consist of finer grasses, more closely grazed. Except perhaps for a few local situations, they contain no place where the air is likely to remain saturated with water vapor, or nearly so, continuously throughout the year. The chance that a tick dropping from a host into one of these pastures would find a safe place to live is remote; the chance that it would do so three times during its life is so remote that none ever survives in these lowland pastures, despite the fact that sheep infested with ticks are often brought to

2. In some districts the ticks are active during autumn as well as spring; the following account refers especially to those districts where they are active only in the spring.

them. Breeding populations of *Ixodes ricinus* persist only in the rough pastures of the uplands.

Milne sampled the populations of ticks by dragging a blanket over plots of about 100 square yards. It was quite impracticable to look for the ticks while they remained inactive at the base of the vegetation. So the sampling had to be done during the period between April and June, when the ticks were to be found on the tips of the grass stems waiting to attach themselves to a passing host. Although ticks were to be found on the grass stems at any time during this period, the active stage for any individual did not exceed about 12 days. So the numbers on the grass stems and hence the numbers trapped on the blanket depended partly on the numbers in the area and partly on their activity. As in all methods of simple trapping, so it was in this case: there was no easy way of discriminating between these two components. But Milne equalized the component for activity by taking simultaneous samples on the plots that he wished to compare and by spacing the samples uniformly throughout the period when the ticks were active. Usually, because of the labor involved, a limited number of samples was taken from each plot, and these were spaced rather widely in time, so that they were likely to contain an unspecified, but approximately constant, proportion of the total population on the plot. That is to say, his comparisons were based on relative, not absolute, numbers. This is, of course, a characteristic of all methods based on simple trapping.

Milnc compared the numbers of ticks at different altitudes on one farm, where they had been established for many years and where the population was relatively dense. The chief difference in the pastures at different altitudes was in the mat, which was thicker and denser on the lower slopes and shallower near the tops of the hills. The results summarized in Table 13.10 show that the numbers were generally fewer on the higher slopes and virtually absent

TABLE 13.10*

NUMBER OF NYMPHS OF *Ixodes ricinus* AT DIFFERENT ALTITUDES IN SAME "GRAZING"

Location	Date	Altitude (Feet)	Mean Nymphs per "Drag"
Hill A	April 2	500	1.5
		900–1,100	0.0
	April 12	500	9.0
		800	2.0
Hill B	April 25	400–500	19.0
		500–600	19.0
		600–750	0.0
Hill C	June 9	450–500	5.7
		600–800	1.0
		900–1,050	0.3

* Plots were chosen in each case so that the pasture would be typical of the altitude. Data from Milne (1946).

from the tops of the hills. Another experiment showed that this was due to the thinning-out of the mat on the higher slopes.

In some situations, because of minor variations in topography and soil, there were small areas at high altitudes where the mat was thick and dense. In these places the ticks were just as numerous as they were in the thick mat at the lower altitudes (Table 13.11).

TABLE 13.11*
NUMBER OF NYMPHS OF *Ixodes ricinus* AT DIFFERENT ALTITUDES ON SAME "GRAZING" AS IN TABLE 13.10

Date	Height	Vegetation	Mean Nymphs per "Drag"
May 2	450	Thick	1.80
	750	Thick	3.67
May 9	650–900	Thin	1.00
	1,000–1,050	Thin	0.50
	800–1,000	Medium	1.00
	1,100	Medium	2.00
	500–600	Thick	4.50
	1,050	Thick	14.00

* In this case the plots at the higher altitudes were chosen especially from local situations where the mat was unusually thick. Data from Milne (1946).

In another experiment the number of ticks in a "sheep lair" were compared with the number in a representative plot of the pasture where the sheep grazed during the day. The lair was a small area near the top of a hill, where the sheep always went at night to sleep. The pasture on the lair was sparse and the mat thin. Since the sheep spent a greater proportion of their time on the lair than on any other area of comparable size in the meadow, it might have been expected that more engorged ticks would drop from the sheep on the lair than on the pasture elsewhere. Milne found that his samples yielded 0.21 nymphs per "drag" of the blanket on the lair and 5.18 nymphs per "drag" on a representative plot of pasture on the lower hillside. He concluded that the ticks were scarce on the lair because few could survive in the absence of an adequate mat. He added: "Possibly replete ticks do drop off in large numbers on the lairs but most should be unable to breed. Hence the lairing habit of hills sheep probably causes the deaths of many ticks."

Even a heavily infested sheep never carried enough ticks to suggest that food was ever scarce in the absolute sense. Nevertheless, shortage of food in the relative sense in which we discussed it in section 11.12 was probably the chief cause of deaths among ticks situated in good rough pasture where there were plenty of suitable places for them to live. For example, Milne estimated that the population on half of a certain farm included between 70,000 and 116,000 females of *Ixodes ricinus;* there were rather fewer on the other half of the farm, but these were not estimated. During the period when the ticks were actively seeking hosts, this farm was stocked at the rate of about one sheep to the acre. This is the highest rate of stocking likely to be found on upland farms. During this season the entire flock picked up between 15,000 and 45,000 females. With the most conservative estimate possible from these figures, over 60 per cent of the adult ticks on this farm died from lack of food because a tick cannot live through a second summer without food. If the deaths among the larvae and nymphs occurred at the same rate, then the total deaths from this cause during a full generation would be of the order of 94 per cent. It would, of course, be higher on a farm that was more lightly stocked.

If a tick is to become attached to a host, it must first be sitting on the tip of a grass stem at the time when the host brushes past that stem. With few sheep per acre, the frequency with which any one grass stem is brushed by a sheep may be small. Nevertheless, a tick might have a good chance of being picked up if it were able to remain a long time on the tip of a grass stem. But this is just what it cannot do. While a tick is sitting on the tip of a grass stem, it is losing water by evaporation to the air. It is likely to die from desiccation if it remains there for more than 5 days. Before this period has elapsed, if it has not been picked up by a host, it usually returns to the mat near the soil, where it recuperates. On the average, a female tick spends about 9 days exposed on the tip of a grass stem. Usually this is made up of several separate periods of 2 or 3 days each. Including the periods spent in the mat between visits to the tip of a grass stem, the period of activity of any one tick usually extends over no more than 4 weeks (Lees and Milne, 1951). Those which fail to attach themselves to a host during this period die.

In most places ticks become active for a period during autumn and again during spring; in some districts activity is restricted to the spring. Figure 13.16

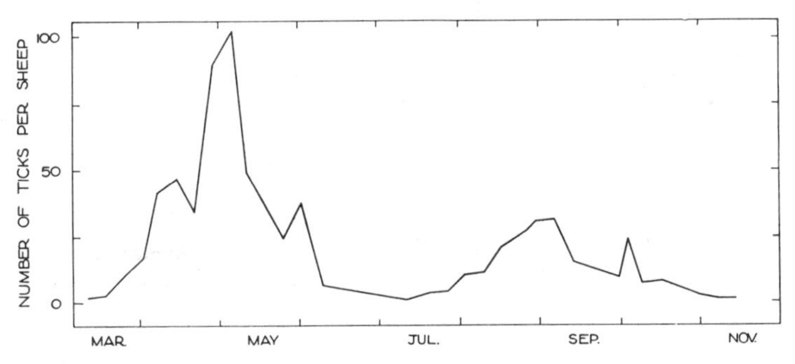

FIG. 13.16.—The average number of *Ixodes ricinus* found on a sheep throughout the year on a farm (Crag) in Cumberland. (After Milne, 1945*a*.)

shows the average number of females found attached to sheep at different dates during 1940 and 1941 in a pasture at Crag. Milne withheld sheep from an experimental plot until 3 weeks after the time for the usual peak of activity in the spring. By the end of the season, the sheep on this plot had picked up only half as many ticks as those on the control plots, showing that about half the population must have ceased activity before the sheep were returned to the pasture. Activity ceased on both the experimental and the control plots at about the same date. This experiment showed that the dates when the ticks become active and cease being active are independent of the presence of the sheep. They are doubtless determined by the progression of the seasons (secs. 8.1, 8.12).

Sheep are not the only hosts of *Ixodes ricinus;* almost any mammal or bird which frequents pastures can probably serve as food for the ticks. Milne recorded as hosts 29 species of mammals, 39 birds, and 1 reptile. The importance of one species of host relative to another depends both on the readiness with which ticks are collected by an individual and on the number of individuals in the area.

In comparing the readiness with which different species of hosts collected ticks, it was necessary to count ticks on different species taken from the same place at the same time, because the number of ticks on the ground varied from place to place and the number of ticks which were active and ready to cling to a passing host varied with the seasons. Milne expressed the results of such counts for a number of species of hosts in terms of "sheep equivalents." The hedgehog, for example, had a "sheep equivalent" of 0.1353. This means that one hedgehog carried as many ticks as 0.1353 of a sheep. The "sheep equivalents" of a number of hosts are shown in Table 13.12.

TABLE 13.12*
"SHEEP EQUIVALENTS" OF NUMBER OF ADULT HOSTS OF FEMALE TICKS

Host	Sheep equivalent	Host	Sheep equivalent
Roe deer	0.1396	Otter	0.0324
Hedgehog	.1353	Fox	.0233
Brown hare	.0771	Pheasant	.0202
Stoat	.0547	Red grouse	.0017
Badger	.0544	Magpie	0.0015
Rabbit	0.0014		

* Mean for five farms in the north of England. After Milne (1949).

Even the "best" wild animal is far behind the sheep as a host. In general, a host which covers more ground will pick up more ticks. The smaller the animal, the less the area of body which is in contact with the vegetation and generally the less the distance traveled in search of food. This means that small animals usually "sweep" a smaller area of ground than large ones do. There are, of course, exceptions, since habits of species vary and some animals touch the vegetation more intimately than others. For example, the mole is larger than the shrew, but it usually harbors fewer ticks than the shrew because the mole spends more time underground and so has less opportunity of meeting ticks.

The figures in Table 13.12 do not indicate the relative importance of these different animals as hosts, for that is also dependent upon the abundance of the hosts. Milne estimated the relative abundance of these various hosts on five farms in England. The mean numbers for these five farms, which had an average acreage of 1,500 acres, is shown in Table 13.13. Some of the animals shown in Table 13.13 spent very little time on the pasture. The otter, for

TABLE 13.13*
ESTIMATED NUMBERS OF HOSTS OF FEMALE TICKS ON 5 FARMS IN NORTH OF ENGLAND WITH AVERAGE ACREAGE OF 1,500 ACRES

Rabbit	692	Hedgehog	12
Sheep	590	Fox	4
Red grouse	322	Roe deer	3
Pheasant	42	Magpie	3
Cattle	30	Badger	2
Brown hare	20	Otter	2
Stoat	16		

* The figures give the mean number of hosts per farm. After Milne (1949).

example, spent most of its time in the water; it visited the farms at infrequent intervals. The badger likewise must be quite unimportant as a host because of its rareness. A pair of badgers in hill and moorland country are said to need at least 500 acres for their upkeep.

The importance of the different species of hosts relative to the sheep can be obtained by multiplying the "sheep equivalent" of each species (Table 13.12) by the number of individuals (Table 13.13). The product is a measure of the relative number of ticks carried by the different hosts and is called by Milne the "population sheep equivalent" of the wild hosts (Table 13.14). The

TABLE 13.14*
"POPULATION SHEEP EQUIVALENTS" OF WILD HOSTS, INDICATING RELATIVE NUMBERS OF FEMALE TICKS CARRIED BY WILD HOSTS

Hedgehog	1.62	Roe deer	0.42
Brown hare	1.54	Badger	.11
Rabbit	0.97	Fox	.09
Stoat	0.87	Otter	.06
Pheasant	0.85	Magpie	0.01
Red grouse	0.55		
		Total	7.09

* After Milne (1949).

total wild hosts on the five farms to which Table 13.14 referred represented an average of 7.09 sheep per farm. This was 1.19 per cent of the total sheep population per farm. In other words, on these farms the sheep provided the food for nearly 99 per cent of the adult females that found hosts of any sort. Corresponding figures for nymphs were worked out on another farm. Sixteen species of wild hosts on this farm were estimated to have a total "population sheep equivalent" of 30.71 sheep. There were 590 sheep on the farm, and they provided the food for 95 per cent of the nymphs that found hosts of any sort. These experiments showed that the ticks were dependent almost entirely on the sheep for food and a place on which to mate. The presence of wild hosts did little to ameliorate the tick's chance of finding food and a mate.

We discussed in section 9.13 how the birth-rate in populations of *Ixodes ricinus* may be reduced by the failure of the ticks to find mates. Since the sexes can meet only on a host, the chance that a female will be fertilized depends on the number of ticks likely to be picked up by one host. This depends largely on the density of the population of ticks; increasing the number of hosts on a particular area may increase the tick's chance of finding food, but not its chance of finding a mate. This is at the root of the explanation of the erstwhile puzzling observation that many farms on which the pastures are undoubtedly well suited to *I. ricinus* have remained for many years free from ticks, despite the fact that ticks must have been repeatedly brought in by wild hosts. But so few are likely to be introduced at any one time in this way that their chance of colonizing the new area must be exceedingly small. The presence of sheep in their usual numbers on these pastures showed that the ticks' failure to become established was due not to the hazards associated with finding food but to the very high risk of not finding a mate. Shortage of mates is associated with a too sparse population of ticks; shortage of food is not associated with a too dense one; the ticks in the same population suffer at one and the same time from the difficulties of finding food and the difficulty of finding a mate. This may seem a paradox to those who always associate shortage of food with "intraspecific competition," but food is not a "density-dependent factor" when it is scarce in the relative sense (sec. 11.12). For another example of this see *Glossina morsitans* (sec. 13.33).

Other hazards which might reduce *r* in natural populations of *Ixodes ricinus* include predators, which are not very prominent, and certain risks which the tick encounters while it is on the host. It may "get lost" in the fleece of the sheep; some parts of the fleece are lethal to the tick; if it happens to become attached in a lethal region, it may die there. The tick has the best chance of surviving on the bare and hairy parts of the body. The tick may be killed by its host after it has become attached. Birds and the smaller mammals "detick" themselves; with birds especially, ticks are found only on the upper regions of the neck and head, where they cannot be reached by the beak; the ones which attach themselves elsewhere on the body are invariably eaten. Larger animals, irritated by the ticks, rub themselves against walls and posts; and some ticks are killed in this way. The skin of a sheep which has been repeatedly infested by ticks may acquire a reaction which may prevent some ticks from engorging or may even kill them. But these risks are less important than the others which we have mentioned.

This study of *Ixodes ricinus* is one of the nicest pieces of ecology that we know. The place where a tick may be living is important because of an interaction with one component of weather, namely, moisture. Food is important: the tick is so lacking in dispersive ability that it suffers severely from shortage of food, notwithstanding the presence of plenty of food in the area if only the tick could find it. The tick also suffers from too few other animals of its own kind in its environment; this is also largely due to its poor dispersive ability. It is noteworthy that Milne was able to offer a thorough and adequate explanation of the distribution and abundance of *I. ricinus* without referring to the dogma of "density-dependent factors."

13.33 *Some Points from the Ecology of Four Species of Tsetse Flies,* Glossina *Spp.*

"Tsetse fly" is the name given to a group of bloodsucking flies of the genus *Glossina* which are found in forests, woodlands, and other shady places in Africa. They have been extensively studied because they carry serious diseases among men and domestic animals. A number of papers dealing with the ecology of *G. morsitans, G. tachinoides, G. palpalis,* and *G. longipalpis* have been published by Jackson (1930, 1933, 1936, 1939, 1949), Nash (1930, 1933*a*, *b*, 1937, 1948), Gaschen (1945), Potts (1937), and Zumpt (1940).

One interesting aspect of this work is that the absolute numbers of flies in certain areas have been estimated by the sampling method known sometimes as "the Lincoln Index" but better called "the method of capture, marking, release, and recapture." The principle of the method is simple. A sample is collected from the population to be measured. The individuals are marked so that they may be recognized again and are then released in such a way that they may be expected to distribute themselves at random with respect to the rest of the (unmarked) individuals in the population. At a later date another sample is taken at random from the population, and the numbers of marked and unmarked individuals in it are recorded. It is important that either the initial marking or the subsequent catching be done evenly over the area selected for study. The second sample will contain some marked and some unmarked individuals. Provided that (*a*) the marked individuals had redistributed themselves at random to the unmarked ones; (*b*) the marked ones are neither more nor less readily caught than the unmarked ones; and (*c*) between the times of release and recapture there have been no gains or losses by births, deaths, or migration, the total population in the area equals

$$\frac{\text{Number marked at first} \times \text{total marked and unmarked recaptured}}{\text{Number of marked ones recaptured}}.$$

In nature, condition *c* is never likely to be fulfilled, but it is the great merit of this method that, by suitably designing the experiment and making the appropriate extensions to the fundamental equation, one can measure not only the total population in the area but also the rate at which it is changing. Moreover, the rate of change may be divided into its four components of births, deaths, immigration, and emigration. A full account of the method as it was used for *Glossina morsitans* may be found in three papers by Jackson (1933, 1936, 1939). Dowdeswell, Fisher, and Ford (1940) discussed the method in some detail in relation to their study of the numbers of a natural population of the butterfly *Polyommatus icarus*. A general account of the method was also given by Ford (1945*a*, p. 270).

Evans (1949) attempted to use the method with a natural population of house mice, *Mus musculus;* but he had to abandon it, because he found that there were some individuals which liked to be trapped and others which were especially shy of the traps. The former had a greater chance of being marked in the first instance and of being recaptured. There was no way of getting over this difficulty. With other species, other sorts of difficulties may arise which make it doubtful whether conditions *a* and *b* have been fulfilled. So the method is not, unfortunately, available for general use in the study of natural populations.

In what follows we shall mention a few interesting points about several species of tsetse flies without going deeply into the ecology of any one.

There is no diapause or quiescence in the life-cycle of *Glossina morsitans* in West Africa, and new individuals are being added to the population throughout the year. For brief periods at certain seasons, births exceed deaths, and r is positive, resulting in a temporary increase in numbers; for the remainder of the year deaths exceed births; r is negative, and the numbers are declining.

The larva of *Glossina morsitans* does not lead an independent existence: it is nourished to maturity inside the body of the female. The fully mature larva is placed by the female on the surface of the soil or among litter, especially under bushes or against logs. The larva crawls into the soil and pupates. The pupal stage may last for several weeks. The pupa may die if the place where it happens to be should become too dry or too wet. The northern limit to the distribution of *G. morsitans* is determined largely by the survival-rate among pupae which have to develop during the dry season; the limit corresponds rather closely to the 30-inch isohyet (Fig. 13.17), which is associated with a dry season of 6 or 7 months.

In the northern parts of their distribution *Glossina morsitans* may increase in numbers during the wet season as well as during the first half of the dry season. But as the dry season advances, r becomes negative. The flies begin to evacuate the drier parts of their range in the open savannah woodland early in the dry season, slowly becoming restricted to the islands of forest or the denser vegetation along the rivers. By March, which is the hottest and driest month in Nigeria, the flies are to be found only in these moister, more shady areas. At this time of the year the flies enter adjacent woodlands to hunt only in the mornings and evenings. As soon as the rains begin again, *G. morsitans* begin to disperse into the savannah woodland, individuals traveling as far as 8 miles from the places where they had spent the dry season. During the dry summer, larvae are deposited only in thickets, but during the wet season they may be placed under logs in open savannah. New generations

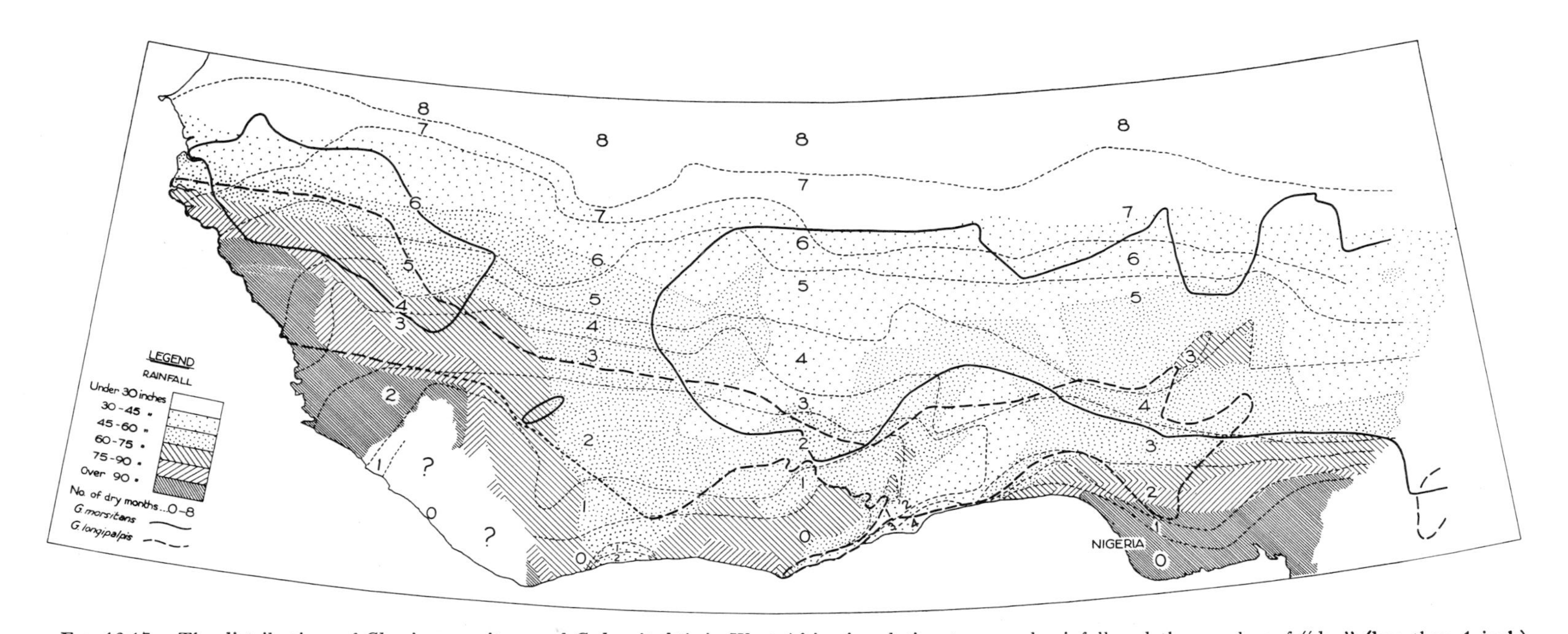

FIG. 13.17.—The distributions of *Glossina morsitans* and *G. longipalpis* in West Africa in relation to annual rainfall and the number of "dry" (less than 1 inch) months in a year. (After Nash, 1948.)

breed in the newly invaded areas, and, in due course, the flies disperse far and wide through the savannah woodland. A period is set to this multiplication and expansion by the return of the dry season.

The southern limit to the distribution of *Glossina morsitans* seems not to correspond to any isohyet. There is evidence that excessive wetness may be harmful to the pupae; in the southern (wetter) parts of their distribution deaths exceed births during the wet season, and r remains negative; the first part of the dry season provides the only period when r remains positive. Furthermore, in East Africa, in Tanganyika, the distribution of the species may expand during a run of dry years into vegetational zones which would normally not be inhabited by them. Nevertheless, the limits of their distribution on the wet side are probably determined less by the direct influence of moisture on the insect itself than by the influence of moisture on the distribution of certain types of vegetation.

In Nigeria the vegetation runs in bands that are roughly parallel to the coast. The coast is fringed by mangrove swamps and swamp forests. The sequence that is found as one proceeds inland is evergreen forest, mixed deciduous forests, heavy savannah woodland with islands of mixed deciduous forest, and, finally, far inland, light savannah woodland. Narrow belts of vegetation typical of wetter zones follow the rivers into the drier areas. Figure 13.17 shows that *Glossina morsitans* is absent from the heavy evergreen forests of the south; the precise explanation for this seems not to have been worked out yet. The influence of vegetation is better understood for some other species, especially for *G. tachinoides* (see below).

The adults of *Glossina morsitans* feed only on the blood of ungulates, especially antelopes, and their distribution may impose a further limit on the distribution of the fly. The numbers of *G. morsitans* are not likely ever to be so great that it suffers from a shortage of food in the absolute sense, but food may sometimes be difficult to find when antelopes are scarce. The following is the gist of a hypothetical example which Jackson (1937, p. 886) made. Suppose that there are 100 tsetse flies in an area which also supports just enough antelopes to enable the flies to maintain their numbers indefinitely without increase or decrease. This does not mean that the total weight of blood is just adequate for the needs of 100 flies but rather that the antelopes are just numerous enough to insure that the average tsetse fly meets with food often enough for it to produce offspring at the rate required to match the death-rate in the population. Suppose that 900 new flies are introduced to the area. The fecundity of each newly arrived fly, depending as it does on the frequency with which the fly meets an antelope, will be just the same as that of the original inhabitants, and the population will continue to maintain itself at a steady level. This is because, even with the larger number of flies, there is still no shortage of food in the absolute sense. Jackson expressed the same point in different words: "It [food] may no more increase its action on a rising population of the tsetse fly than does climate, nor temper its severity towards a diminishing community. [In other words, food is not operating as a 'density-dependent factor.'] It seems that there is no pressure of numbers in the ordinary sense because there is probably no competition for food, and certainly none for shelter, as the writer's experiments indicate that the flies are so sparse that there can be no physical crowding even when the apparent density is comparatively very high." Elsewhere in the same paper Jackson pointed out that there is no evidence that predators or disease influences numbers to any extent, and he concluded: "There remains the possibility that (at moderate densities at least) there are no dependent factors at all acting on tsetse."

The climate in the region where *Glossina morsitans* lives is characterized by a pronounced wet season, during which rainfall may exceed 70 or 80 inches, and a dry season, which may last for 6 or 7 months. A pupa may be likely to drown during the wet season and die from desiccation during the dry; drastic reductions in the numbers of *G. morsitans* may occur during either season. Jackson (1937), after having excluded the possibility of "density-dependent factors" concluded: "Meanwhile the writer is inclined to accept provisionally the view that the density of tsetse about Kakoma is being maintained at a mean level below that which the environment might allow; that the annual passage of a favourable season does not continue sufficiently long to bring the tsetse numbers up to saturation level before the ensuing unfavourable period drives them down again."

The biology of *Glossina tachinoides* is generally similar to that of *G. morsitans*, except that the former can feed on a wider variety of animals, including man. Its distribution, for this reason, is not likely to be restricted by the absence of suitable food. On the other hand, it has a characteristic behavior in hunting for food which, as we shall explain later, severely limits its distribution.

The northern boundary to the distribution of *Glossina tachinoides* tends to follow the isohyet for 30 inches of rain a year, which is associated with a dry season of 6 or 7 months. This species, taking advantage of the moisture and the vegetation associated with rivers, may penetrate into country where the annual rainfall is 16 inches and the dry season lasts for 8 months. The major projection of its distribution into the dry country shown in the northeast part of the map in Figure 13.18 is associated with the great Katagum River.

The southern limit to the distribution of *Glossina tachinoides* is not associated with any isohyet, although moisture is indirectly responsible for the limitation of its distribution in this direction also. This species has the peculiar habit of remaining within a few feet of the ground while hunting for food. They cannot fly through dense undergrowth, and they will not rise above any

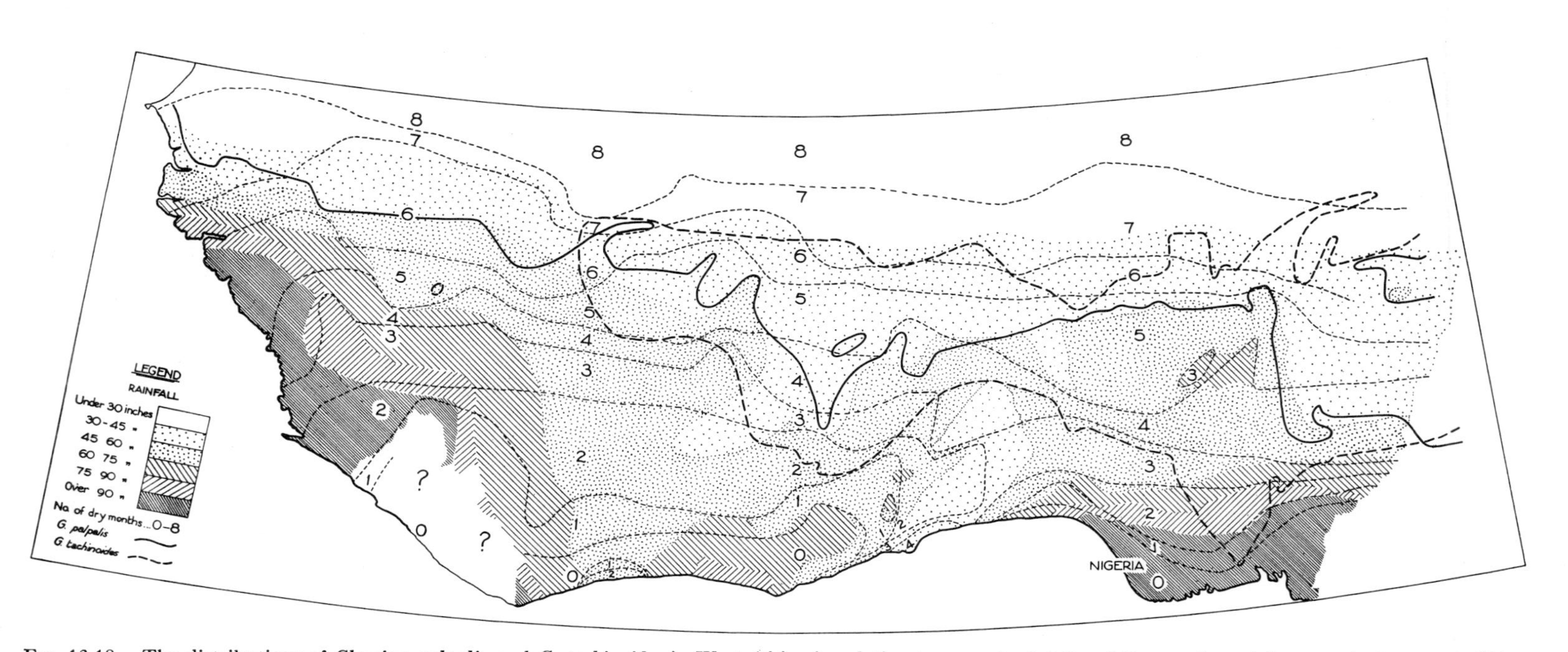

FIG. 13.18.—The distributions of *Glossina palpalis* and *G. tachinoides* in West Africa in relation to annual rainfall and the number of dry months in a year. (After Nash, 1948.)

growth that exceeds about 2 feet in height. Consequently, *G. tachinoides* are not found in areas of forest or woodland where the undergrowth is prominent. As a result, this species is excluded from some of the more humid regions which are inhabited by certain other species of *Glossina*, especially *G. palpalis*.

The northern limit to the distribution of *Glossina palpalis* follows the 45-inch isohyet fairly closely; in this region the dry season lasts for 5 or 6 months. In some places the distribution of this species extends into drier areas, but then it is restricted to the vegetation associated with rivers. In the south the distribution of *G. palpalis* extends to the coast, except that they are absent from the mangrove swamps and the swamp forests which form a fringe along the coast (Fig. 13.18); in areas which support these sorts of vegetation the soil is so wet that there are few places where a pupa is likely to escape drowning. Elsewhere *G. palpalis* is absent from local situations where the undergrowth is dense and continuous, for in these circumstances the flies cannot reach their food. Provided that there are some clearings which permit hunting, *G. palpalis*, in striking contrast with *G. tachinoides*, may colonize forest or woodland where the undergrowth is 6 feet high. Table 13.15 shows the striking

TABLE 13.15*
NUMBERS (AS PER CENT OF TOTALS) OF *Glossina palpalis* AND *G. tachinoides* WHICH SETTLED AT DIFFERENT LEVELS ON MAN STANDING UPRIGHT IN STREAM BED

HEIGHT ABOVE GROUND	LATTER PART OF DRY SEASON		SEASON OF HEAVY RAINS	
	G. palpalis	*G. tachinoides*	*G. palpalis*	*G. tachinoides*
Ground to ankle (0–4 inches)	2	35	2	12
Ankle to knee (4–22 inches)	13	44	23	54
Knee to waist (22–40 inches)	24	18	32	21
Above waist (40–56 inches)	61	3	43	13

* Data from Nash (1948).

differences in the habits of these two species with respect to the height above the ground at which they hunt: a man stood in the bed of a stream and counted the numbers of the two species which settled on him at several different levels The experiment was done twice, once late in the dry season and once during the season of heavy rains. In the latter part of the dry season 15 per cent of *G. palpalis* attacked below the knee, as compared with 79 per cent of *G. tachinoides*. During the wet season *G. tachinoides* fly even closer to the ground, and *G. palpalis* fly higher.

The distributions of four species of *Glossina* are shown broadly in Figure 13.19; they overlap without coinciding. In those areas where the distributions overlap, the resources which the different species need in common are never in short supply relative to the numbers of the flies; the species are not linked with one another through a common predator; nor do they interfere with one another's chance to survive and multiply in any other way that has been discovered Although they are all species of the one genus, they do not enter into one another's environments in any important way. The differences in their distributions are due, as we have seen, to differences in their physiology and behavior. According to current theory of speciation (chap. 16), these differences arose at some time in the past when the populations were isolated from each other.

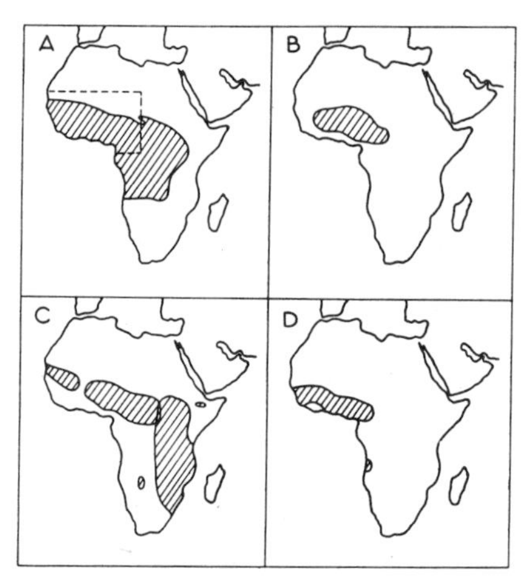

FIG. 13.19.—The distributions of four species of tsetse flies in Africa. *A*, *Glossina palpalis; B, G. tachinoides; C, G. morsitans*, showing eastern and western races; *D, G. longipalpis*. The dotted line in *A* shows the area which was mapped on a larger scale in Figs. 13.17 and 13.18. (After Gaschen, 1945; and Nash, 1948.)

13.34 *Fluctuations in the Numbers of Bobwhite Quail* Colinus virginianus *on 4,500 Acres of Farmland in Wisconsin*

Errington (1945) reported the results of 15 years' continuous study of a population of *Colinus virginianus* on 4,500 acres of farmland in Wisconsin. This is getting near the northern limit of the distribution of this species. Errington said: "To the northward are found only frontier populations of this 'farm-game' bird, usually sparse and discontinuous though sporadically abundant." What Errington called "cover conditions for the bobwhite" had long been deteriorating in this area, and it is likely that the deterioration continued slowly while this investigation was in progress. But what changes occurred were slight and gradual, and it was not considered necessary to take them into account.

The life-cycle and behavior of *Colinus virginianus* were mentioned in sections 10.322 and 12.31. The quail's habit of overwintering in coveys, the nature of the terrain, and the climate made it practicable, at any time during the winter, to count all the quail on the area. The routine varied during the 15 years, so that more information was available for some years than for others; but for every year it was recorded how many quails were present on the area at the beginning of the winter and how many at the end of the winter. These records are given in Figure 13.20, from which one can see at a glance the magnitude of the decreases that occurred during the winter and the increases that were made during the summer. With very few exceptions, the decrease during the winter was due to deaths among the local population. But the increase during the summer was not all due to breeding by the local population. At this time of the year the quail wandered freely over the countryside, and it was not possible to assess the gains and losses due to migration.

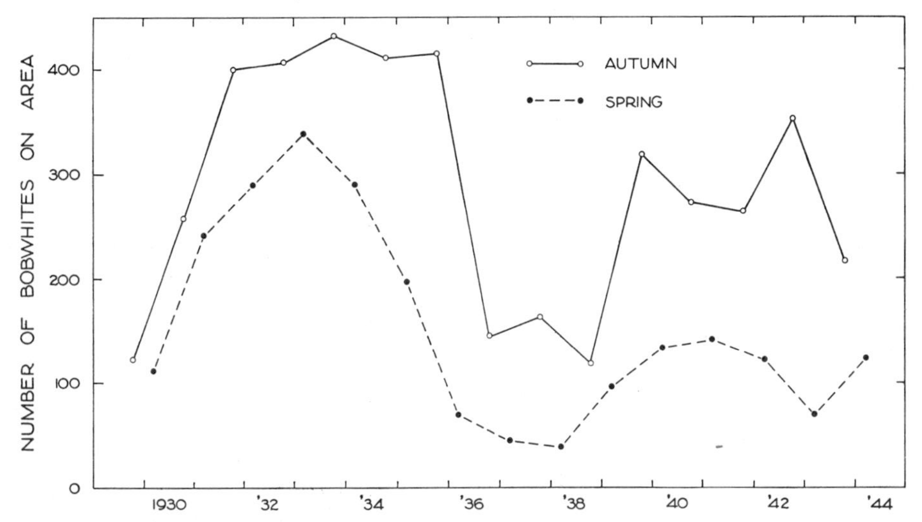

FIG. 13.20.—The numbers of *Colinus virginianus* on 4,500 acres of farmland in Wisconsin. The upper curve shows the numbers attained at the end of each summer; the appropriate comparisons between the upper and lower curves indicate the number of deaths during each winter and the increases which occurred during each summer. (After Errington, 1945.)

Most of the deaths which occurred during the winter could be attributed to starvation, exposure, or the activities of predators. The quail depended on seeds for food. The risk of starvation was greatest during a hard winter, when persistent snow or ice covered the ground and made their food inaccessible. For example, during the winter of 1935–36 there were "several weeks of the severest winter weather shown by Weather Bureau records for the region. Blizzard followed blizzard, with prolonged, intense cold and much snow. One after another of the Prairie du Sac coverts was nearly or quite depopulated of bobwhites. By spring, only 70 of the birds were left alive on the area." There had been 416 at the beginning of winter. But it was not necessary for the weather to be quite so extreme in order to kill many of the birds. The bobwhite starved if their food supply was cut off for a week or two. For example, the winter of 1937–38 "was also mild, but a food shortage aggravated by heavy snows in late January was attended by extreme losses. The area's spring level of 39 bobwhites remaining of an initial 163 was the lowest for which we have accurate measurement."

Deaths from exposure were less frequent and more difficult to assess than deaths from starvation. Errington summarized the position as follows: "In cold weather, the hunger-weakened are the likeliest to succumb. On rare occasions, sound birds in unsheltered places fall victim to cold, and others are imprisoned to starve or suffocate in hard-packed snowdrifts. In 1935/36 . . . bobwhites not only starved in the deep snow or died from the stress of the blizzards but also seemed to be worn down by the sustained cold."

The risks both from starvation and from exposure were accentuated by men who plowed the land during the autumn, thereby burying the seeds that might have served as food for the quail, and who slashed or burned "brush" which might otherwise have provided a place for the quail to seek shelter. Even if these activities did not result in an absolute shortage of food or an absolute shortage of hiding places, they might still increase the risks of starvation because the quail would not use food that was too far from a safe refuge. Or if they did, this would increase the risk of being taken by a predator.

The chief predator of the bobwhite quail was the great horned owl, *Bubo virginianus*. The numbers of these owls on the area varied from 4 to 8, with an average of 6. Other species of predators included a hawk, *Buteo jamaicensis,* gray fox, *Urocyon* sp.; red fox, *Vulpes* sp.; skunk, *Mephites* sp.; mink; and several sorts of weasels, *Mustela* spp., as well as dogs, cats, and men from the farms. During some winters the predators, among them, killed more than half the quail that were overwintering on the area. For example, there were 264 quail on the area at the beginning of the winter of 1941–42, but only 122 were still alive at the end of the winter. The weather was mild, and Errington attributed all the deaths to "non-emergency losses," which is nearly the same as saying that they were due to predators of one sort or another. On the other hand, there were several winters when the deaths attributed to predators were only 7 or 8 per cent. For example, there were 257 quail on the area at the beginning of the winter of 1930–31, and 236 were still alive at the end of the winter; all these deaths were attributed to predators, but they represented only 8 per cent of the numbers that were present at the beginning of the winter.

Since all the predators mentioned by Errington were species which would eat a variety of other foods besides the bobwhite quail and since most of them would, as a rule, tend to take that sort of food which was most plentiful, it was to be expected that the other animals which served as food for the predators might exercise an important influence on a quail's chance of being eaten. Errington called these "buffer species." They included mice, rabbits, and gallinaceous birds. The more likely ones were considered in more or less detail, but no relationship was found between their numbers and a quail's chance of being eaten. After exhausting all other likely possibilities, Errington concluded that the only component in a quail's environment which influenced the risks that it would be eaten by a predator was the number of animals of the same sort that were in the area. This led to the formulation of the "threshold concept," by which was meant that, for any particular area, there was a maximal number of quail which could live through the winter with a high degree of immunity from predators. If the numbers exceeded this "threshold," the supernumeraries would become "insecure" and would probably be eaten by predators. During the 15 years the "threshold" seemed to vary. Errington implied that this might be due to intrinsic changes in the behavior of the quail, but the evidence for this was not convincing (see below).

Figure 13.20 shows that not only did the number of deaths during the winter vary greatly but so did both the absolute and the relative increases during the summer. Scatter diagrams were constructed by plotting the relative increase during the summer against numbers present at the beginning of the summer. Two relationships were discovered in this way: (*a*) the relative increase during the summer was roughly inversely proportional to the number present at the beginning of the summer; (*b*) this relationship became more pronounced when the numbers of bobwhite quail and ring-necked pheasants were taken together. The trend emerged as a flat sigmoid curve. It was not defined precisely, because there was a wide scatter among the points.

We are now in a position to summarize the conclusions which may be drawn from the empirical facts discovered by this investigation: (*a*) The maximal number of quail which overwintered on the area was determined largely by the behavior of the quail themselves in relation to the distribution and abundance of suitable places for them to live (sec. 12.31). (*b*) The minimal numbers were determined largely by the weather; cold weather was directly responsible for some deaths, but more important was the indirect influence of snow and ice in reducing the stock of accessible food. (*c*) The numbers attained by the end of the breeding season depended partly on the numbers present at the beginning of the summer and partly on the relative increase during the breeding season. (*d*) The relative increase during the breeding season depended partly on the numbers of quail and partly on the numbers of a nonpredator (the ring-necked pheasant) which required to share some of the same resources and also probably interfered with the quail in a more subtle way as well. (*e*) There were runs of years in which the quail were plentiful, followed by runs of years in which they were scarce; but the sequence of 15 years' records was too short to permit a statistical test of any hypothesis about cycles or periodicity.

Errington carried the analysis of these data further than this. He analyzed them more especially in terms of the two hypotheses which we mentioned earlier, namely, the "threshold concept" in relation to the rate of survival during winter and the "inverse density law" in relation to the relative increase during summer. He constructed curves to represent these hypotheses and found that whereas most points fell approximately along the trend lines, there were some which deviated widely from the trend defined by the majority. Moreover, there was a relationship between the points which deviated most from the major trend in each case. This is best expressed by saying that a winter in which the number of deaths from predators was unexpectedly low was likely to follow, or be followed by, a summer in which the relative increase was also unexpectedly low. It was suggested that this might be explained by assuming that this population of *Colinus virginianus* experienced "depression phases" analogous to those supposed to be manifested by the populations of certain species of vertebrates (*Lepus*, *Microtus*, *Lynx*, and certain birds and fishes) in which the fluctuations are said to be periodic (sec. 13.4).

These data do not in themselves provide convincing evidence for the reality of "depression phases." Errington preferred this explanation because it also fits in with similar observations on populations of other species, especially the muskrat *Ondatra zibethicus*. It often happens that unusually low birth-rates in populations of muskrats seem to synchronize with low rates of increase not only for the same species in different areas but also for quite unrelated species such as hares or grouse throughout the country. There seems to be no way of explaining these observations in terms of the ordinary components of environment. Nor do they seem to be due to the innate qualities of the animals. In the present state of our knowledge the term "depression phase" may be used to describe the phenomenon without indicating any explanation for it. The idea of "depression phases" has been linked with the idea of "periodicity" in the fluctuations in numbers in certain vertebrates. But one need not necessarily depend on the other.

13.4 THE PHENOMENON OF "PERIODICITY" IN THE NUMBERS OF CERTAIN VERTEBRATES

Elton and Nicholson (1942*b*) consulted the records of the Hudson's Bay Company and compiled tables showing the relative number of furs of the lynx *Lynx canadensis* received by the company each year from 1736 to 1934. They divided the total area into regions. Moran (1953) analyzed the records from the Mackenzie River region for the period 1821–1934. He chose them because this was the only region from which "complete figures are available over a long period of years." Moran also pointed out: "When the total captures for the whole area are considered for each year, a rather blurred picture of the cycle appears, but on splitting up the records into the individual regions, the cyclic behaviour of the population in each region . . . becomes quite clear."

We reproduced these records in Figure 13.21, *B*, because they are probably the best example of the "10-year cycle" that can be found in the literature. It has been said that the numbers of the lynx depend on the numbers of the snowshoe hare, which is its staple food. The numbers for the snowshoe hare

Lepus americanus, which were used in Figure 13.21, *A*, were copied from a small diagram in Dymond (1947), because the original records (Hewitt, 1921) were not available to us. Figure 13.22 shows the number of furs of muskrats *Ondatra zibethica* received by the Hudson's Bay Company (Elton and Nichol-

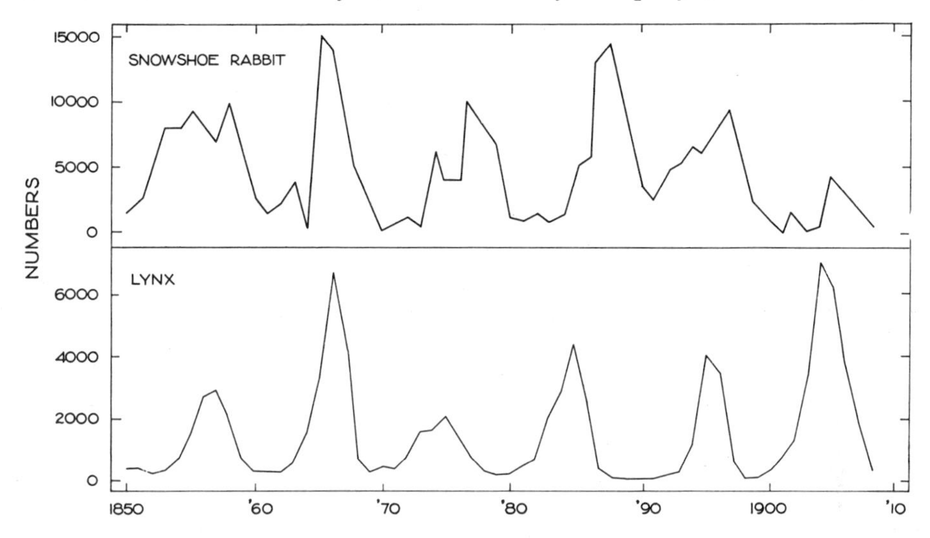

FIG. 13.21.—Fluctuations in the numbers of snowshoe rabbits (total area ?) and the lynx (Mackenzie River region), as indicated by records of pelts kept by the Hudson's Bay Company. (After Hewitt, 1921; and Elton and Nicholson, 1942*b*.)

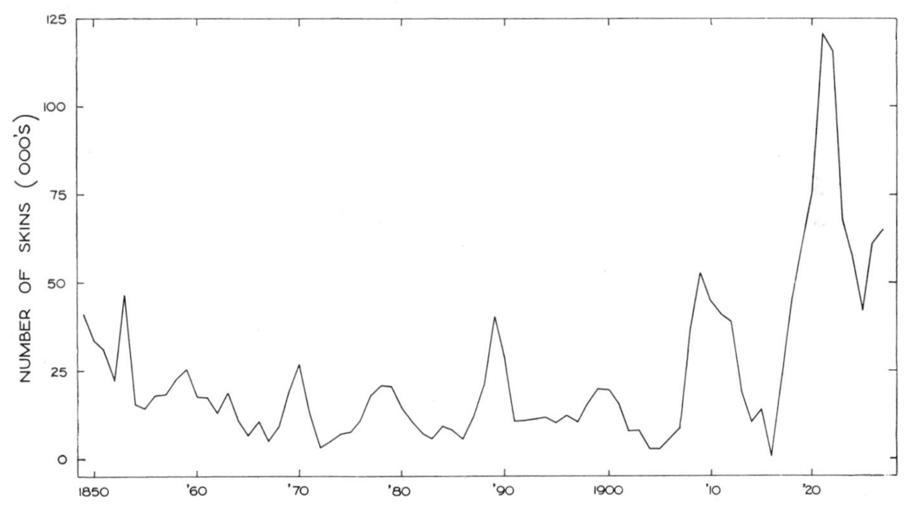

FIG. 13.22.—Fluctuations in the numbers of muskrats *Ondatra zibethica* for the Mackenzie River region, as indicated by the records of the Hudson's Bay Company. (After Elton and Nicholson, 1942*a*.)

son, 1942*a*). Figure 13.23 shows the numbers of red grouse *Lagopus scoticus* shot by sportsmen on three moors in Scotland between 1850 and 1945 (Mackenzie, 1952). The last two series have also been said to provide evidence for "cycles," but they are obviously not so "regular" as that of the lynx, which seems outstanding in this regard. Most writers seem to accept the reality of cycles without question. Nevertheless, the evidence warrants a critical examination.

Two aspects of "cycles" have attracted attention, namely, the amplitude of the fluctuations and their periodicity. For example, Dymond (1947) stated

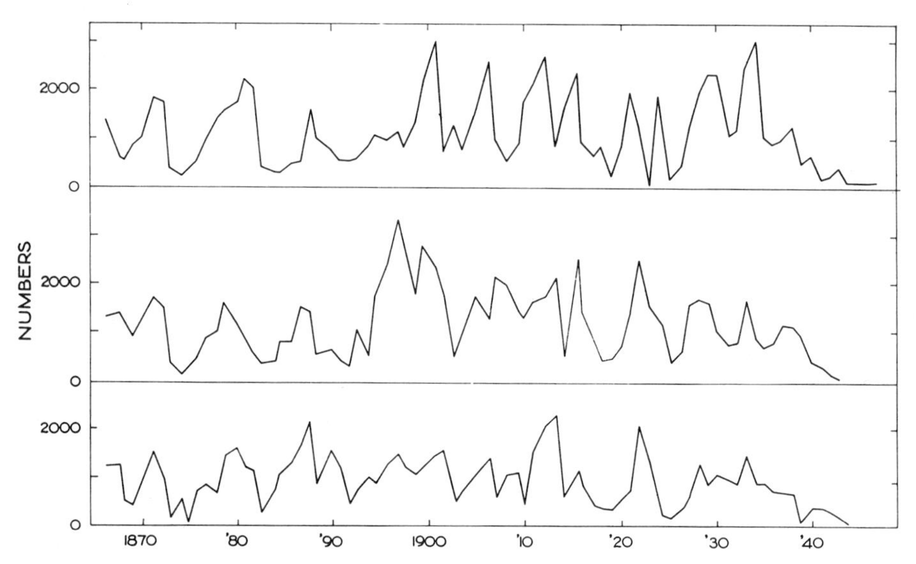

FIG. 13.23.—Fluctuations in the numbers of red grouse shot on three moors in Scotland. (After Mackenzie, 1952.)

that the maximal numbers of *Lynx canadensis* exceeded the minimal numbers in the ratio of 100 to 1 in the northern part of British Columbia, 50 to 1 in the middle part of the province, and 20 to 1 farther south. The number of salmon in the Fraser River, British Columbia, were estimated to have fluctuated between the extremes of 15,000 in 1907 and 240,000 in 1913. These are large fluctuations, and doubtless they have seemed more impressive because of the commercial value of most of the species concerned. But these fluctuations may be small compared with those which occur in natural populations of locusts, grasshoppers, thrips, and many other species of invertebrates. In other words, the muskrat, the lynx, the snowshoe hare, the grouse, and the salmon are not unusual with respect to the amplitude of the fluctuations which are found in their numbers.

A truly cyclic variable, such as the sine or cosine of an angle, waxes and wanes with constant amplitude and phase. A variable may still be properly called "cyclical" if it is determined by two components, one of which is truly cyclical and the other added to it is random. A variable may be properly called "oscillatory" if the amplitude and phase of the fluctuations about the mean (or trend line) are not random with respect to their order along the abscissae. The usual test for randomness in a series is to calculate the serial correlation coefficient (Kendall, 1948, p. 402). The difference between the two sorts of series was summarized by Kendall (1948, p. 398):

> In a cyclical series the maxima and minima, apart from disturbances due to the superposition of a random element, occur at equal intervals of time and are therefore predictable for a long way into the future—for so long, in fact, as the constitution of the system remains unchanged. In oscillatory series, on the other hand, the distances from peak to peak, trough to trough or upcross to upcross, are not equal, but vary very considerably. Similarly, in the oscillatory series the amplitudes of the movements may vary very substantially, whereas in a cyclical series they should be constant (again, except in so far as superposed random elements disturb them) . . . the time-series observed in practice are very rarely cyclical as we have defined the term. . . . The far more usual case is that of varying amplitude and period from peak to peak or upcross to upcross.

The series plotted in Figures 13.21, 13.22, and 13.23 are clearly not cyclical, because the variation in the amplitude of the oscillations is far too great. On the other hand, Moran (1952, 1953) showed by the method of serial correlation that the series represented by the numbers of *Lynx canadensis* and *Lagopus scoticus* were significantly nonrandom with respect to years; and there need be little doubt that the same could be shown for the muskrat (Fig. 13.22). Although they are not cyclical in the strict mathematical meaning of the word (because the amplitude is not constant), these series may be periodic if the turning points recur at regular intervals.

If the minor fluctuations in the curves shown in Figures 13.22 and 13.23 are ignored, the more obvious periods of maximal numbers seem to be repeated at approximately equal intervals. But a dilemma is encountered when one tries to measure the intervals between "peaks" precisely: the regularity largely disappears if the maxima are determined objectively by counting as a maximum every point that is greater than the ones that come immediately before and after it; on the other hand, if the minor fluctuations are ignored and only the big peaks are counted, there is no objective way of measuring the precise intervals between maxima: every maximum and minimum has to be fixed arbitrarily. It is not permissible to smooth the raw data, because an oscillatory series may be generated from a random one by taking moving averages and by certain other devices which may be used for smoothing (Kendall, 1948). The information in Table 13.16 was compiled by Cole (1951) by counting every turning point; the information in Table 13.17 was gathered by Dymond (1947), using arbitrary methods of ascertaining the periods between "big peaks." A number of species occur in both tables, and the discrepancies are striking. Let us follow up Cole's approach first.

If we take many numbers from a table of random numbers and plot them in the order that they come from the tables, they will fluctuate about a horizontal line drawn through their arithmetic mean. Wherever there is a number which

TABLE 13.16

INTERVALS BETWEEN MAXIMA IN NUMBERS OF 8 SPECIES OF VERTEBRATES ESTIMATED OBJECTIVELY BY METHOD OF COLE

Species	Country	No. of Oscillations	Mean Interval (Years)	C.V. (Per Cent)	P
Lynx*	Canada	23	5.82	50.1	< 0.01
Lynx	Finland exc. S.W.	93	3.56	41.5	< .01
Field hare	England	48	3.60	48.1	< .01
Rabbit	England	58	3.53	47.5	< .01
Partridge	England	117	3.21	39.8	< .01
Arctic fox*	Labrador	124	3.50	27.1	Nonsig.
Red fox*	Labrador	124	3.37	31.1	Nonsig.
Red fox	Finland	93	3.66	40.0	< 0.01
Random numbers	....	...	3.0	37.3	.....

* The species marked by an asterisk feature also in Table 13.17. The last column gives the significance of the differences between the coefficients of variation for the empirical series and the random numbers. Data from Cole (1951).

TABLE 13.17

INTERVALS BETWEEN "BIG PEAKS" IN NUMBERS OF 22 SPECIES OF VERTEBRATES IN CANADA

SPECIES		USUAL INTERVAL (YEARS)	EXTREME RANGE (YEARS)
Common Name	Technical Name		
Arctic fox*	*Alopex lagopus*	4	3– 6
Red fox*	*Vulpes fulva*	9–10	8–13
Lynx*	*Lynx canadensis*	9–10	7–12
Snowshoe rabbit	*Lepus americanus*	9–10	8–11
Muskrat	*Ondatra zibethica*	9–10	?8–?12
Field mouse	*Microtus pennsylvanicus*	4	....
Lemming	*Lemmus trimucronatus*	4	....
Marten	*Martes americanus*	9–10	8–11
Fisher	*M. pennanti*	9–10	8–11
Mink	*Mustela vison*	9–10	6–12
Goshawk	*Accipiter gentilis*	9–10	....
Rough-legged hawk	*Buteo lagopus*	4	3–5
Ruffled grouse	*Bonasa umbellus*	9–10	9–12
Willow ptarmigan	*Lagopus lagopus*	9–10	....
Sharp-tailed grouse	*Pediocetes phasianellus*	9–10	....
Horned owl	*Bubo virginianus*	9–11	....
Snowy owl	*Nyctea scandiaca*	4	?3–?6
Great northern shrike	*Lanuis excubitor*	4	?3–?6
Pine grosbeak	*Pinicola enucleator*	5–6	....
Canadian salmon	*Salmo salar*	9–10	....
Sockeye salmon	*Oncorhynchus nerka*	4	....
Pink salmon	*O. gorbuscha*	2	....

* The species marked by an asterisk also occur in Table 13.16. After Dymond (1947).

is larger than the ones on either side of it, there will be a maximum in the curve; wherever there is a number smaller than the ones on either side of it, there will be a minimum in the curve. Half the turning points in the curve will be maxima. Now it is possible, by making use of the algebraic method of permutations and combinations, to calculate the probability that maxima will be separated by 2, 3, 4, . . ., n, places. Cole (1951) did this and obtained the theoretical distribution shown in the second column of Table 13.18; the distribution shown in the third column was calculated by Cole from Schulman's (1948) records for the serial distribution of growth-rings in the Douglas fir;

TABLE 13.18

FREQUENCY-DISTRIBUTIONS OF INTERVALS BETWEEN MAXIMA FOR ANY LONG SERIES OF RANDOM NUMBERS AND EMPIRICAL SERIES BASED ON SIZE OF GROWTH-RINGS IN DOUGLAS FIR

INTERVAL BETWEEN MAXIMA (ORDINAL UNITS FOR RANDOM NUMBERS; YEARS FOR DOUGLAS FIR)	FREQUENCY OF INTERVALS OF SPECIFIED DURATION	
	Random Numbers (Per Cent)	Douglas Fir (Per Cent)
2	39.1	37.2
3	34.4	32.9
4	17.4	19.4
5	6.5	7.7
6	1.9	1.3
7	0.5	0.8
8 and more	0.2	0.6
Mean interval	3.0*	3.07
Coefficient of variation (per cent)	37.3*	37.3

* The numbers marked with an asterisk are repeated in Table 13.16. After Cole (1951).

any ring which was larger than the ones on either side of it was counted as a maximum. The distribution was calculated from 504 intervals. There is close agreement between the two distributions, and Cole concluded that in the Douglas fir growth-rings of varying sizes occurred at random with respect to time.

The mean and the coefficient of variation for the theoretical series are given at the foot of Table 13.16. We have calculated the significance of the differences between the coefficients of variation for the empirical series and the series of random numbers. The means of the empirical series are of the same order as that of the theoretical one, but the coefficients of variation are significantly larger for 6 of the 8 species listed in Table 13.16. This leads to the surprising conclusion that three-quarters of the empirical series are more variable than the theoretical one. None was significantly less variable than the theoretical series, which they would need to be if they were to provide evidence for periodicity. The departures from randomness (which may also be demonstrated by the method of serial correlation) happen, not as has so often been thought, because the maxima recur with a regularity greater than would be expected by random chance, but rather because the maxima recur with less regularity than they would in a random series. In other words, the curves representing these populations are oscillatory but not periodic.

This conclusion is analogous to the one reached in section 13.02. There we showed by the appropriate statistical tests that, for many species of animals, the individuals in natural populations are distributed nonrandomly with respect to space and that the departures from randomness are due to excessive "patchiness" or irregularity in the distributions. Here we have to conclude with Cole (1951, 1954) that (for many species of vertebrates at least) the individuals are distributed nonrandomly with respect to time and that the departures from randomness are due to excessive irregularity (or lack of periodicity) in the phase of the oscillations.

The discrepancies between Table 13.16 and 13.17 for species that are common to both may be due partly to the methods of ascertaining the numbers, as well as to the different methods of measuring the intervals between maxima. Figure 13.21, *B*, refers to the lynx in the Mackenzie River district; we have calculated the mean interval between maxima as 8.1 years, with a coefficient of variability of 31.5 per cent. This differs from the figures given by both Cole and Dymond, which refer to lynx in a wider area. There is always a dilemma to be faced in gathering the raw material for the study of "cycles": in order to have a sufficiently long series of figures, it is necessary to go back a long time into the past; the farther one goes back, the more doubtful are the figures. Possible variations in the area concerned, in the number of trappers, in their skills and objectives, and in a variety of other unknowns combine to throw doubt on the results. The numbers plotted in Figure 13.21, *B*, differ from those published by earlier authors; Elton and Nicholson (1942*b*) went thoroughly into the reasons for departing from the earlier publications.

The series in Figure 13.21, *B*, is definitely nonrandom, because Moran (1953) has shown that the serial correlation coefficient is large and significant. Compared with a random series, the phase and the amplitude of the oscillations are large, but the variability in the phase is only slightly less than that of a random series. This series is therefore markedly oscillatory (nonrandom), but it does not provide strong evidence for periodicity—especially when it is recalled that this series was arbitrarily selected because it was probably the most "cyclical" one to be found in the literature.

It might be argued that Cole's approach, while having the merits of objectivity and precision, lacks reality and that the figures in Table 13.17, though they depend on subjective judgment, are nevertheless more real than those in Table 13.16. This may be so, but there is no way to put the matter to the test. It might also be suggested that since the methods of ascertaining the numbers are so uncertain, it is hardly worth while to apply rigorous methods to their analysis. In this connection it would be well to recall Kendall's (1948, p. 402) warning: "Experience seems to indicate that few things are more likely to mislead in the theory of oscillatory series than attempts to determine the nature of the oscillatory movement by mere contemplation of the series itself."

Cole (1951) considered the arguments in favor of counting every turning point instead of only the big ones. Apart from the obvious statistical reasons for this choice, there are two strong biological arguments against ignoring the minor fluctuations: (*a*) A level of favorableness, operating for a certain period, in the environments of individuals in a small population may result in a small absolute change in numbers, whereas a large population in identical circumstances (except for its own density) may experience a great change in its numbers, measured absolutely. Cole likened this to a small and a large bank balance increasing at compound interest. (*b*) An animal's chance to survive and multiply (and hence the rate of change in the population) is determined by various components of the environment, which are themselves distributed continuously (in the statistical sense). Small and large variations in the environmental components will occur with a frequency which is determined by the means and the variances of the distributions; certainly, these will be reflected in small and large variations in the numbers of the animals. There is just no reason at all for attributing greater reality to the big variations than to the small. We were able to demonstrate this principle quantitatively with respect to *Thrips imaginis* (sec. 13.11). The ecology of *Lepus*, *Lynx*, *Vulpes*, *Lagopus*, etc., may be more complex than that of *T. imaginis;* but the same principles may be expected to hold with them also.

Our success in relating the numbers of *Thrips imaginis* quantitatively to certain environmental components depended largely on three conditions: (*a*) we had adequate knowledge about the physiology and behavior of *T. imaginis;* (*b*) we were working in an area where the geography and especially the climatology had been investigated thoroughly; and (*c*) our records of the relative numbers of *T. imaginis* were adequate with respect to both the number of samples and the precision with which they were taken.

The records on which the discussions in this section have been based are inferior to those for *Thrips imaginis* with respect to all three conditions, especially the first and the last: the number of pelts received by the Hudson's Bay Company or the number of cases of salmon packed along the Fraser River must depend on many things besides the number of animals in the forests or the river. In studying the ecology of these species, as with all others, it is best to set out to explain all the variability in the recorded numbers and to take into account all the environmental components that seem relevant. We reached the same conclusion after discussing the ecology of the bobwhite quail (sec. 13.34). The investigation is more likely to be successful if it is firmly based on a sound knowledge of the biology of the animal and the geography of the area where it is living. The long series of "records" for some of these species may be alluring; but, in the future, progress may depend on getting more precise information about the animals themselves and their environments.

CHAPTER 14

A General Theory of the Numbers of Animals in Natural Populations

From my early youth I have had the strongest desire to understand or explain whatever I observed,—that is, to group all facts under some general laws. These causes combined have given me the patience to reflect or ponder for any number of years over any unexplained problem. As far as I can judge, I am not apt to follow blindly the lead of other men. I have steadily endeavoured to keep my mind free so as to give up any hypothesis, however much beloved (and I cannot resist forming one on every subject), as soon as facts are shown to be opposed to it.

DARWIN, *Autobiography*

14.0 INTRODUCTION

IN THIS chapter we have to build a general theory about the distribution and numbers of animals in nature. This should summarize in general terms as many as possible of the facts which we have discussed in the empirical part of the book and so provide a general answer to the questions we propounded in section 1.1: Why does this animal inhabit so much and no more of the earth? Why is it abundant in some parts of its distribution and rare in others?

Elton (1949, p. 19) wrote: "It is becoming increasingly understood by population ecologists that the control of populations, i.e., ultimate upper and lower limits set to increase, is brought about by density-dependent factors either within the species or between species (see Solomon, 1949). The chief density-dependent factors are intraspecific competition for resources, space or prestige; and interspecific competition, predators or parasites; with other factors affecting the exact intensity and level of these processes." Elton said quite precisely what he meant by "density-dependent factors"; in section 2.12 we discussed some of the other meanings which have been attributed to the same phrase; like many other ecological terms, this one lost most of its strength shortly after it was coined. We hope that Elton overstated the case. We believe that it would be nearer the mark to say that the various assertions about "density-dependent factors" and "competition" which are familiar to ecologists are just about the only generalizations available in this field. The student of ecology may either accept or reject them. Hitherto if he rejected them, he has had nothing to put in their place.

The statement that the ultimate upper and lower limits set to increase in a population can be determined only by "density-dependent factors" may be taken as axiomatic for the highly idealized hypothetical animals of the sort which may be represented by the symbols in a simple mathematical model. In this case the limits which are referred to are theoretical quantities which must be deduced by mathematical argument. They cannot, by their very nature, be related to the empirical quantities which are got by counting the numbers of animals in natural populations. There are two chief reasons for this: (*a*) the idealized hypothetical populations are very different from what is actually found in nature, and (*b*) one would not expect to come across a limiting density in any finite number of observations.

Yet this mistake is commonly made, as Elton has pointed out. The usual generalizations about "density-dependent factors," when they refer to natural populations, have a peculiar logical status. They are not a general theory, because, as we have seen, especially in chapter 13, they do not describe any substantial body of empirical facts. Nor are they usually put forward as a hypothesis to be tested by experiment and discarded if they prove inconsistent with empirical fact. On the contrary, they are usually asserted as if their truth were axiomatic. A good example of this approach is seen in the passage from Smith (1935; quoted in sec. 2.12), where he argued that since weather is known to "regulate" the numbers of animals, it must therefore be a density-dependent factor. These generalizations about "density-dependent factors" and competition in so far as they refer to natural populations are neither theory nor hypothesis but dogma.

We often find the expressions "balance," "steady-density," "control," and "regulate" used in theoretical discussions of populations. Their meanings may be obscure, especially when they are used in relation to natural populations. They stem from the dogma of "density-dependent factors," and they are allegorical. Unless their meanings are made very clear, it is best to be cautious about any passage in which they are used.

Our theory is not concerned with these rather allegorical properties of populations but with numbers that can be counted in nature. In each of the sections which follow we first state the principles of the theory in general terms and with the aid of simple diagrams. Then we refer back to the natural populations which have been discussed in earlier chapters and show how particular empirical observations may be fitted into the general theory; for the theory may be regarded as sound only if it serves to explain all, or most, of the empirical observations that may be brought forward to test it.

14.1 COMMONNESS AND RARENESS

If we say that species A is *rare* or that species B is *common*, we can only mean that individuals of species A are few relative to some other quantity which we can measure, and individuals of species B are numerous relative to the same or some other quantity. We might arbitrarily choose a number relative to a certain area and decide to call a species rare if they are, say, fewer than one (or one million) individuals per square mile. This approach may be necessary for the hunter or fisherman, but it does not, at this stage, help us to think clearly about the theory of ecology. So we shall put it aside for the present.

Another meaning for this sentence might be that in a certain area individuals of species A are few relative to the individuals of species B. Certain mathematical expressions have been developed to describe the relative numbers of different species found in large communities or large samples from communities. The size of the community is measured by the number of individuals of all species that are counted. Williams (1944) gave the name "index of diversity" to the coefficient $\alpha = n_1/x$ derived from the series

$$n_1, \frac{n_1}{2}x, \frac{n_1}{3}x^2, \frac{n_1}{4}x^3, \ldots, \text{etc.},$$

where n_1 is the number of species represented in the sample by one individual and x is a positive number less than 1. Large values of α indicate diversity in the community. Preston (1948) fitted a "truncated" normal curve to the frequencies of species represented in the sample by 0–1, 1–2, 3–4, 5–8, 9–16, . . . , $(2^{n-1} + 1)$–2^n, individuals per species. In this expression the diversity of the community is measured by the mean and the variance of the normal curve and the position of the "truncation."

Both expressions depend on the assumption that in any large community there will be a few species represented by many individuals and many species represented by few individuals. This assumption is confirmed by the goodness of the fit obtained when the theoretical curves are tested against empirical records. In other words, most communities seem to be dominated by a few species. This may be an important fact for students of "community ecology." It is, indeed, on facts of this sort that they base their theory of "dominance" and other important theories about communities and the relationships of the species which constitute them. But this way of looking at commonness and rareness and the studies which stem from it have little to contribute to the general theory of ecology, using "ecology" in the narrow sense defined in section 1.0. So we say no more about them.

A third way of considering commonness and rareness is to relate the number of individuals in the population to the quantities of necessary resources, food, places for nests, etc., in the area that it inhabits. This is the common-sense way for the theoretical ecologist and the practical farmer to look at the matter. The farmer is not interested in whether the caterpillars in his crop of wheat are more or less numerous than, say, the mites which eat the dead grass on the surface of the soil. He wants to know whether the caterpillars are numerous enough to eat much of his crop. The ecologist has the same point of view, but his interest goes deeper. If the caterpillars are so few that they eat only a small proportion of the stock of food available to them, the ecologist has to inquire what other environmental components may be checking their increase.

14.11 *The Conditions of Commonness in Local Populations*

A population of a certain species, living in a certain area, may consistently use all the stocks of a particular resource, say food or places for nesting, that

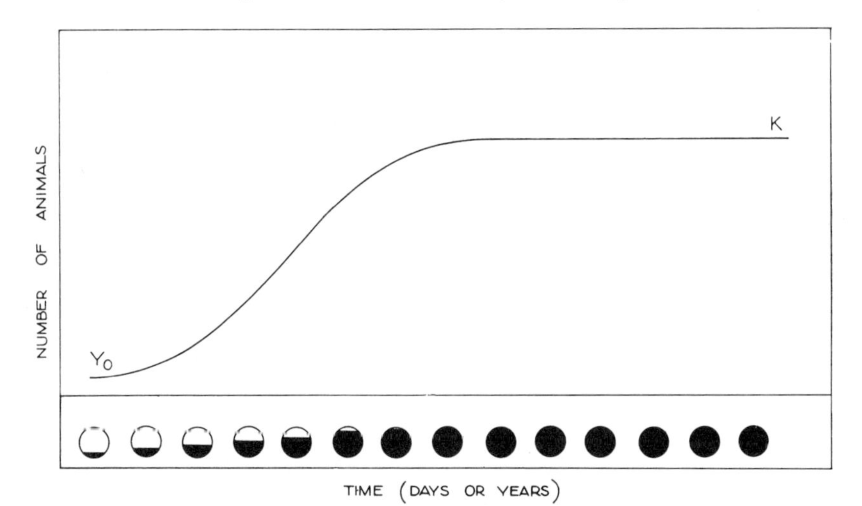

FIG. 14.01.—The growth of a "local population" (i.e., the population in a "locality") whose numbers are limited by the stock of some nonexpendable resource, such as nesting sites. The initial numbers (i.e., the number of immigrants who colonize the locality) are represented by Y_0; the maximal numbers that the resources of the locality will support are represented by K. The circles repeat the information given by the curve. The proportion of the circle shaded represents the number of animals at the specific time as a proportion of the maximum.

occur in the area. The simplest case of this, though perhaps the most unusual in nature, may be found when there is no interaction between the resource and "other animals of the same kind." This happens when the animals can use the resource without destroying it. Such was the case in the imaginary example we gave in section 2.121. The bees in that example did not destroy their nesting places by using them. There was the same number of holes for nest-building generation after generation, irrespective of the numbers of bees in each generation. If nothing else checked the bees, their numbers would be determined entirely by the number of holes for nests. We have illustrated this principle in Figure 14.01. The asymptote for the curve represents the number of bees in a local situation when all the nesting places are used. The shaded parts of the circles represent the population as a proportion of what the total resources of the area will support. A natural population which comes very close to the simplicity of the imaginary one is the population of great tits mentioned by Kluijver (sec. 12.31) as living in a wood where there was a shortage of holes for building nests. Another one, which is nearly as simple, is the one described by Flanders (sec. 12.23) for *Metaphycus helvolus* living on oleanders. In this case living food in the form of migrating scale insects was the resource in short supply. It was fed into the population of *Metaphycus* at a

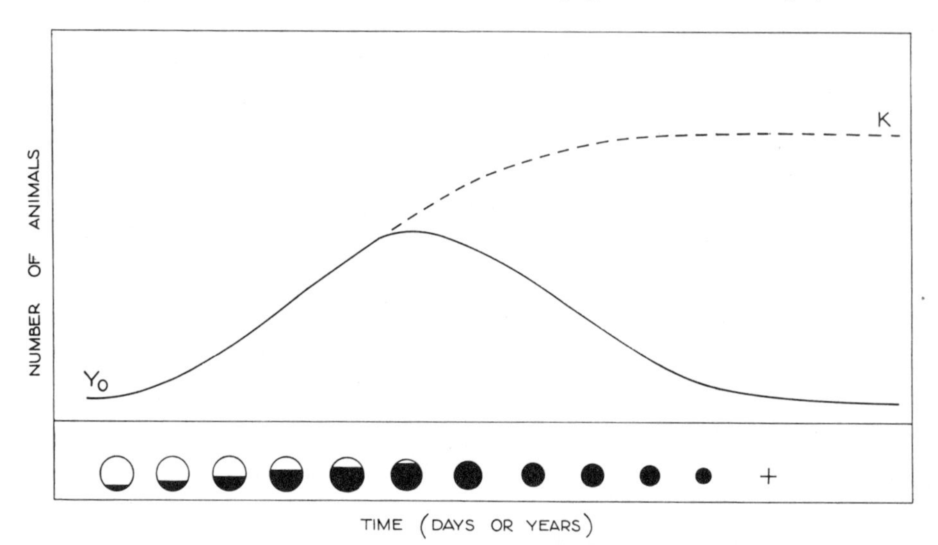

FIG. 14.02.—The solid line represents the growth of a local population whose numbers are limited by a diminishing resource, such as living food, which becomes less, the more animals there are feeding on it. The symbols have the same meanings as in Fig. 14.01; the circles grow smaller because the plant (or population of prey) grows smaller and eventually dies out because there are too many animals eating it. The cross indicates that this is the place where a local population recently became extinct (see Figs. 14.07 and 14.08).

fairly steady rate, which was largely independent of the numbers of animals waiting to eat it. But there were always enough *Metaphycus* to eat all the food that was provided.

When living food is the resource that is fully used up, there may be certain complicated interactions between "food" and "other animals of the same sort"; the result may be a reduction not only in the total stocks of food but also in the proportion of it that is effectively used (sec. 11.22). The sequence of events usually culminates in the complete destruction of localized stocks of food. This accentuates the patchiness of the distribution of food; and the animal's chance of finding food comes increasingly to depend on its powers of dispersal. A characteristic sequence of events was described for *Cactoblastis cactorum* in section 5.0. The general case is illustrated in Figure 14.02. A horizontal line drawn through K would represent the maximal number of animals that could live in this locality if they used up all their resources. The curve represents the rise and fall in the number of animals from the time that they first find the place to the time that they die out from lack of food. The diminution in the amount of food and its ultimate extinction from this place are indicated by the diminishing size of the circles from left to right. The cross indicates a place from which a local population has recently died out. The food becomes less and ultimately disappears, because there are too many animals feeding on it. We mentioned *C. cactorum* feeding on *Opuntia* spp. as an example of this (sec. 5.0); *Ptychomyia remota* is a carnivore for which the same sequence of events was described in section 10.33; other examples were given in section 10.321; we can think of many more examples from carnivorous species than from herbivorous ones. It is unusual to find herbivorous animals eating out their stocks of food, even in local situations. Animals which are living in the circumstances which we have described in this section must be counted as common, no matter how few they may be per square mile. Conversely, animals which are living in the circumstances which we describe in section 14.12 must be counted as rare with respect to their stocks of food, etc., no matter how many of them there may be per square mile.

14.12 *The Conditions of Rareness in Local Populations*

Very few of the natural populations which were described in chapter 13 ever became numerous enough to make use of all their stocks of food, etc. Most natural populations are like this. The numbers fluctuate, perhaps widely; but they do not become numerous enough, even during periods of maximal abundance, to use more than a small proportion of their resources of food, nesting sites, and so on.

The general case for this condition is illustrated in Figure 14.03. The curves and the symbols have the same meanings as in Figures 14.01 and 14.02; note that the size of the circles remains the same from start to finish, because the stocks of food, etc., are not appreciably reduced by the activities of the animals; the circles, of course, never become completely black. The circle with a cross in it provides an alternative to the one above it. These alternatives indicate that the population in a locality runs a risk of being extinguished but also has a chance to survive as a small remnant. In the latter case the population may increase again when circumstances become favorable once more.

The numbers continue to increase while births exceed deaths, that is, while r remains positive; they begin to decrease when r becomes negative. A large part of this book has been devoted to a discussion of the ways that environment may influence the three components of r—fecundity, speed of development, and duration of life. We do not need to reiterate here that any environmental component may have its appropriate influence and that, at any one time, some may be more influential than others. We single out weather and discuss it in section 14.13, partly because its influence is more subtle than that of the other components, and partly because this is a subject that has been much misunderstood.

In section 14.2 we discuss, in relation to the general theory, the way that weather, food, predators, and a place in which to live may influence the value of r and hence determine the numbers in natural populations which occupy substantial areas and which may, or may not, be short of some necessary re-

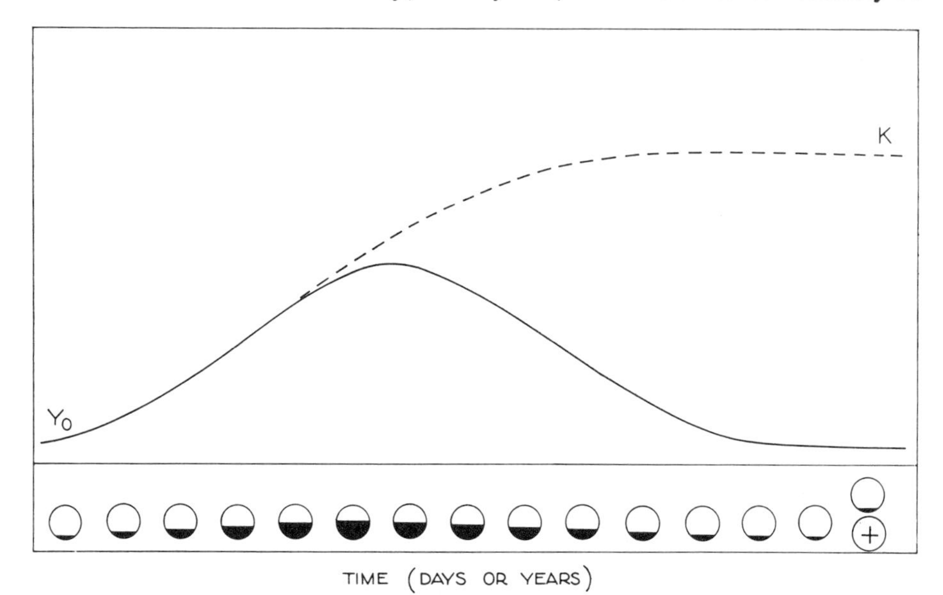

FIG. 14.03.—The solid curve represents the growth of a local population which never uses up all the resources of food, etc., in the locality because its numbers are kept, by weather, predators, food (in the relative sense of sec. 11.12), or some other environmental component, at a level well below the maximum that the resources of the locality could support. The symbols have the same meanings as in Figs. 14.01 and 14.02. The two symbols at the end indicate alternative conclusions to the history of this local population on this occasion: it may be extinguished (*circle with a cross in it*), or a remnant may persist, perhaps to increase again when circumstances change (*circle with a remnant of shading*).

source. We choose these components because they are often important in nature. The reader can readily fit other environmental components into the general theory.

Figure 14.03 refers generally to any population which is not short of any resource, no matter whether r is chiefly influenced by weather, food, predators, or some other environmental component. But it lacks generality in one respect: it does not cover the special case of the population in which a more or less constant proportion is sheltered from a danger which is likely to destroy all those that are not so sheltered. A good example of this was provided by the artificial population of *Ephestia*, which was kept in check by a predator (sec. 10.222). A natural example, which was nearly as good, was described by Flanders for a population of scale insects *Saissetia oleae*, living on the shaded parts of oleander (sec. 12.23). This exception to the general rule is covered by Figure 14.04. The meanings of the curves and symbols are self-evident.

14.13 *The Way in Which Weather May Determine Commonness or Rareness in Local Populations*

We build up our theory about weather from a simple beginning with Figure 14.05. The abscissae are in units of time—days, years, or generations. The

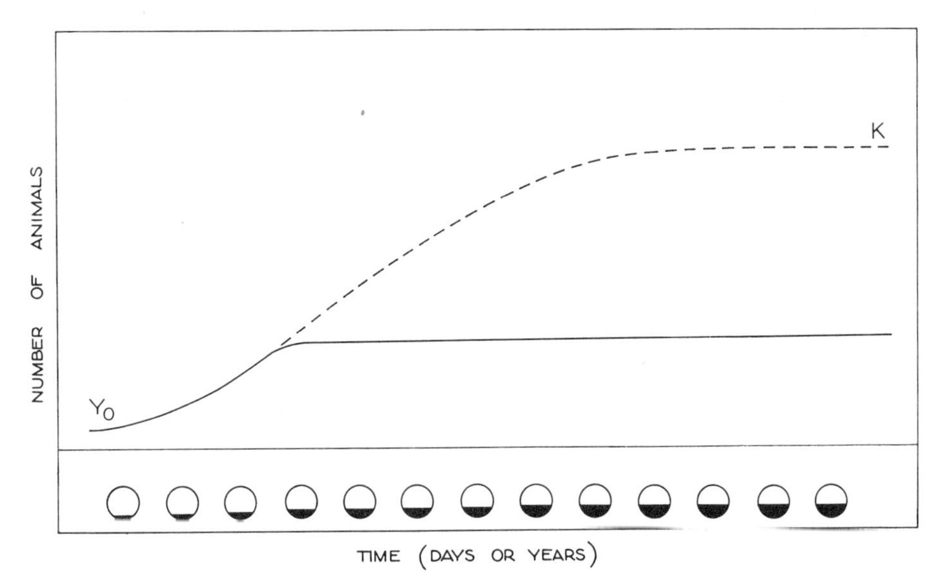

FIG. 14.04.—The solid curve represents the growth of a local population in which a more or less constant number of the individuals are sheltered from some danger which destroys all those which are not so sheltered. This number is not enough to use up more than a small proportion of the total resources of food, etc., in the locality.

ordinates for each curve are numbers of animals; K represents the number of animals which the area could support if the total resources of food were fully used up. The number indicated by Y_0 is the remnant left when the unfavorable period ends and is also the nucleus for multiplication during the next favorable period. The rate of increase which pertains during the favorable period is called r.

Let us suppose that the three pairs of curves represent three ways in which the weather may determine the number of animals in a local population. The curves on the left relate to a place where the weather is favorable, and those on the right relate to a place where the weather is severe. In the top pair the two curves start from the same level at the beginning of the favorable period (Y_0 is constant); and they rise at the same rate because r is constant. But the curve for area A rises farther because the favorable period lasts longer. With the middle pair, the two curves rise at the same rate (r is constant); they continue rising for the same interval of time (favorable period, t, is constant). But the curve for area A rises farther than that for B because it started from a higher level (Y_0 is greater for A). This means that in A the unfavorable period was shorter or else the catastrophe that caused it was less severe, or for some other reasons the population was still relatively large when the weather

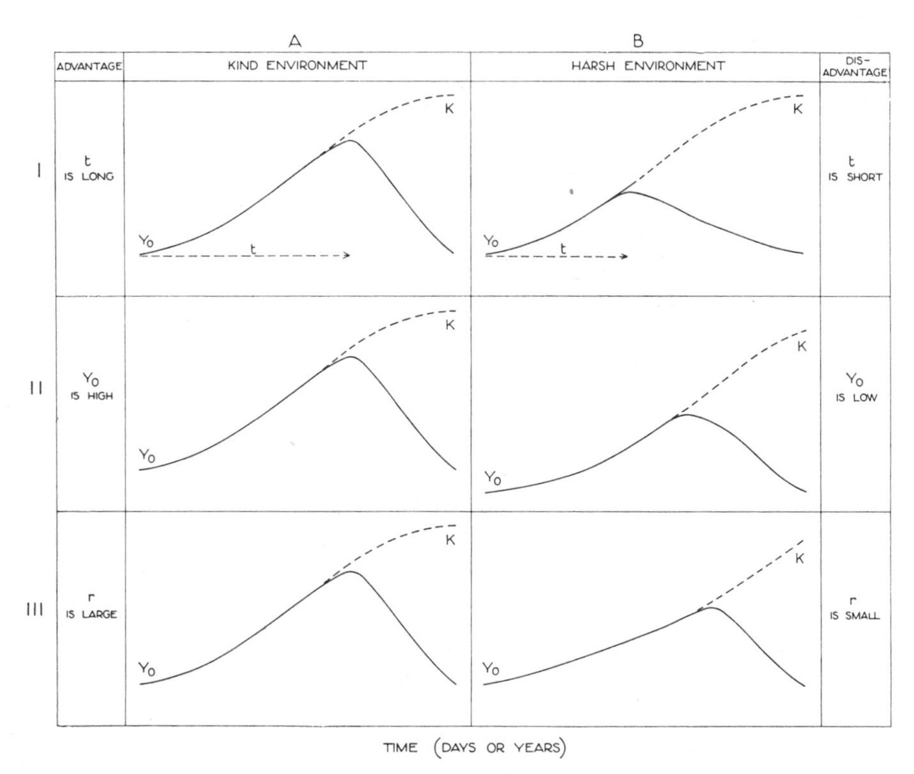

FIG. 14.05.—Three ways in which weather may influence the average numbers of animals in a locality. The numbers in the locality are increasing during spells of favorable weather and decreasing during spells of unfavorable weather. Two areas A and B are compared with respect to three qualities. Quality *I* determines the duration of the favorable period; quality *II* determines the severity of the unfavorable period; and quality *III* determines the rate of increase of the population during the favorable period. The numbers which would be attained if all the resources of food, etc., in the area were made use of are indicated by K; the numbers to which the population declines during the unfavorable period is represented by Y_0. For further explanation see text.

changed and allowed the animals to start increasing again. With the bottom pair, the two curves start from the same level (Y_0 is constant); they continue rising for the same interval of time (favorable period, t, is constant). But the curve in area A rises higher because it is steeper (rate of increase, r, is greater for A). Taking all three curves into account, one can easily see that the animals would be more numerous, on the average, in area A than in area B. This principle has been stated in completely general terms. The model which we describe in section 1.1 provides a particular example of it.

14.2 THE PRINCIPLES GOVERNING THE NUMBERS OF ANIMALS IN NATURAL POPULATIONS

Each of Figures 14.01–14.05 represents the trend in numbers in one locality. A natural population occupying any considerable area will be made up of a number of such local populations or colonies. In different localities the trends may be going in different directions at the same time. It is therefore feasible to represent the condition of the population in a large area by drawing a collection of circles, each one of which represents the condition in a local population. Since we are now considering larger areas, we must take into account, in addition to the principles set out in section 14.1, the dispersive powers of the animals and their food. Also we are now in a position to widen the meaning in which we use "common" or "rare" so that it includes the number of animals in the population relative to the area over which they are distributed (sec. 14.0). We do this in the examples which follow.

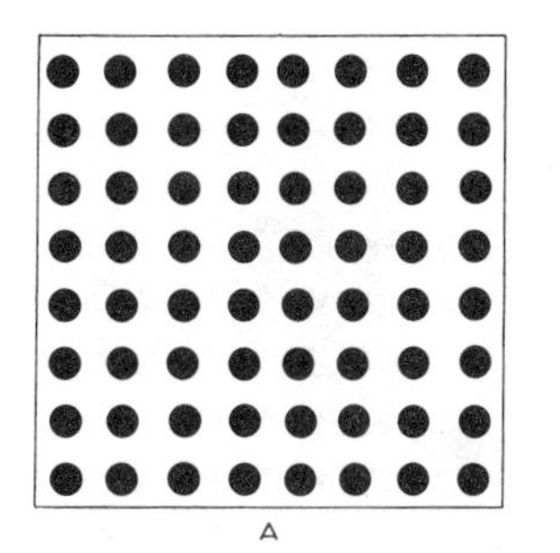

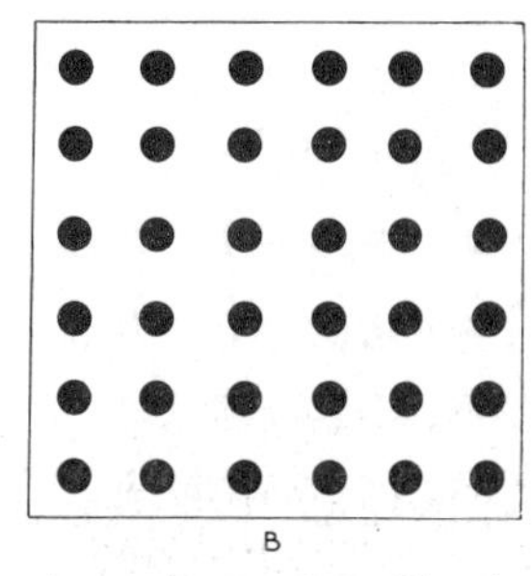

FIG. 14.06.—The populations of large areas are made up of a number of local populations. In the two areas which are compared in this diagram the resources of every locality are fully used up (as in Fig. 14.01), but there are more favorable localities in area *A* than in area *B*.

To start with the simplest case first, we make up an example directly from Figure 14.01. Let us suppose that in Figure 14.06 area *A* includes many localities where fenceposts have many holes suitable for the bees to build their nests in. And suppose that area *B* is like *A* in every respect except that there are few localities where the fenceposts carry many holes that are suitable for nests. Every hole will be used. All the circles are completely blackened in both areas. The two areas are alike, in that the bees are equally common with respect to their stocks of nesting sites. But the areas are different, in that the bees will be more numerous in *A* than in *B;* this is indicated by the numbers of circles in the two areas. This example brings out the two meanings of "common" quite nicely. Which one you emphasize will doubtless depend on whether your interest lies in having empty auger holes or many bees to pollinate your lucerne. The wood where Kluijver put additional nesting boxes for *Parus* was like *A*, except that some were not used; and the other more natural wood, which he described as lacking a sufficiency of tree-holes for nests, was like *B* (sec. 12.31).

The next example is more complex because it deals with living food instead of a lifeless nonexpendable resource like nesting sites. Suppose that, in Figure 14.07, *A* and *B* are two areas where prickly pears grow in colonies; the two areas are alike in climate, soil, distribution of suitable places for *Opuntia* to grow, and every other respect except that *A* supports a population of *Cactoblastis cactorum*, whereas in *B* there is a variant which has very much poorer powers of dispersal. Say we call it *Cactoblastis blastorum*. Area *C* is like the

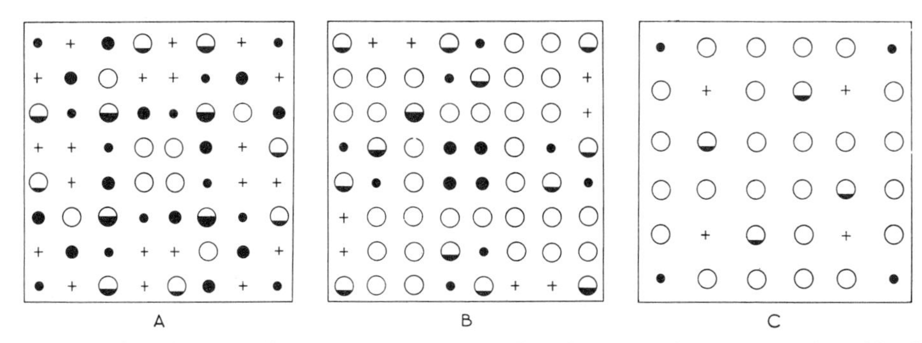

FIG. 14.07.—The populations of the large areas *A*, *B*, and *C* are made up of a number of local populations. Not all the favorable localities support local populations (*hollow circles*). Favorable localities are distributed equally densely in *A* and *B* but more sparsely in *C*. Area *A* is occupied by a species with high powers of dispersal; *B* and *C* are occupied by a species which is similar except that it has inferior powers of dispersal. The symbols are taken from Fig. 14.02. In Area *A* the number of animals is chiefly limited by the stock of food in the absolute sense. In *B* and *C* there is no absolute shortage of food, but the numbers of animals in these areas are limited by a shortage of food in the relative sense (sec. 11.12). This is related to the poor powers of dispersal of the species that lives in these areas. For further explanation see text.

other two except that, for physiographical reasons, colonies of prickly pears must remain very sparsely distributed, irrespective of whether any are destroyed by *Cactoblastis* or not. Area *C* supports a population of *C. blastorum*. The distributions of the circles and the shading in them indicate: in area *A* there are few colonies of prickly pears, but *C. cactorum* is making good use of what are there. Its numbers are clearly being limited chiefly by the absolute amount of food in the area; and the numerous crosses indicate that there is definitely less food in the area as a result of the presence of *C. cactorum*. In area *B* there are more colonies of prickly pears. Relatively few of them harbor local populations of *C. blastorum*. The numbers of *C. blastorum* are not being seriously limited by an absolute shortage of food. Nevertheless, by virtue of their poor powers of dispersal, they are suffering from a relative shortage of food. In area *C* the circumstances are much the same as in *B*, only more so. The food is even harder to find; the death-rate is therefore higher still; and the numbers of *C. blastorum* in the area are few indeed. The presence of a few crosses in both *B* and *C* shows that relative shortage of food (sec. 11.12) is the real cause of the trouble, because, once *C. blastorum* has found a colony of prickly pears, the insects increase rapidly and, in due course, destroy all the food, just as the real *C. cactorum* does in area *A*.

We need not have invented *C. blastorum* for area *B*. Several natural examples have been described, for example, *Chrysomela gemellata* (sec. 11.12) and *Rhizobius ventralis* (sec. 10.321). The fact that the latter feeds on animals instead of plants does not alter the principle. We do not know of an exact parallel for area *C*, but *Thrips imaginis* near Adelaide during summer are kept

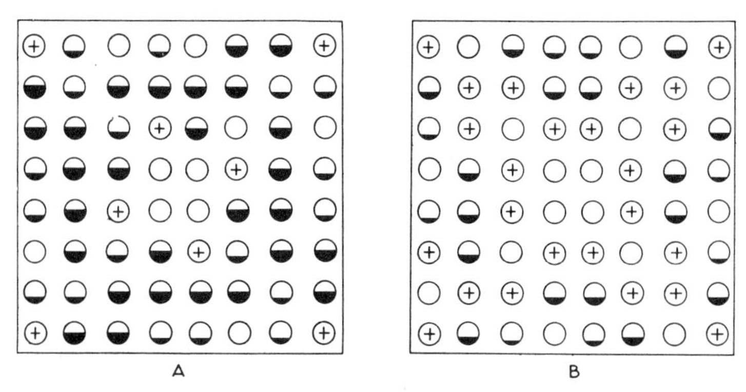

FIG. 14.08.—The populations in the large areas *A* and *B* are made up of a number of local populations. There is an equal number of suitable places where the animals may live in each area. In area *B* there is an active predator whose powers of dispersal and multiplication match those of the prey. This predator is absent from *A*, where its place is taken by one which is more sluggish. Relatively more of the suitable places are occupied by the prey in *A* than in *B*, and the local populations are, on the average, larger. There are more local situations in *B* from which the prey has recently been exterminated. Altogether, the animal is more common in *A* than in *B*. The symbols have the same meaning as in Fig. 14.03.

scarce by a relative shortage of food as *C. blastorum* is in area *C*, but they do not destroy their food so thoroughly as *C. blastorum* is supposed to do.

Let us consider next the case of an animal which has an active predator in its environment. We take the symbols from Figure 14.03 to make Figure 14.08. The two areas, *A* and *B*, have the same number of circles, because the presence of the animals does not influence the stocks of food, etc., in the area. Both *A* and *B* support populations of the same sort of herbivores. But in *B* there is also a species which is an active predator, with powers of dispersal and rate of increase which match those of the prey; in area *A* these predators are replaced by another species with equivalent capacity for increase but inferior dispersive powers. In *B* many localities which are quite favorable are empty; some have become empty only recently (circle with cross in it), and in most of those which are occupied the numbers are low. Either the herbivores have only recently arrived at the place and have not yet had time to become numerous, or else, if they have been there longer, the predators have found them and are in the process of exterminating the local colony. In *A*, relatively more of the favorable places are occupied, and the numbers in the local populations are, on the whole, larger. There are relatively few places from which the prey have been recently exterminated. All this can be explained simply by saying that the prey, being superior to the predators in dispersive powers, enjoy, on the average, a relatively long period of freedom from attack in each new place that they colonize. Several natural examples of this principle were discussed in section 10.321.

A similar diagram may be drawn to illustrate the case in which the numbers

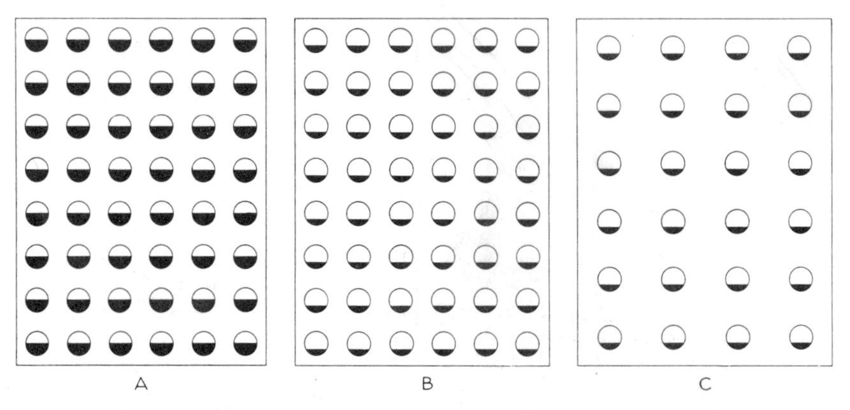

Fig. 14.09.—The populations in the large areas *A*, *B*, and *C* are made up of a number of local populations. The climate in *A* is more favorable than that in *B* or *C*; so the numbers (indicated by the shaded part of the circle) are, on the average, larger in each locality in *A* and in the area as a whole. There are fewer favorable localities in *C* than in *B*; so there are, on the average, fewer animals in area *C* than in *B*, where the climate is the same but the number of places greater.

in the area are determined largely by climate. In Figure 14.05 the two areas *A* and *B* were considered to differ only with respect to climate. The circles in Figure 14.09 may be related to Figure 14.05. In Figure 14.09 there is the same number of suitable places where the animals may live in *A* as in *B*, but the climate is more favorable in *A* than in *B*, so the animals are more common relative to their stocks of food, etc., in *A* than in *B;* there are also more of them in the area. Area *C* is inferior to *B* with respect to the number of places, and it is inferior to *A* with respect to both the number of places and the climate. The animals are uncommon in *C* relative to *A* in both meanings of the word. The natural population of *Porosagrotis orthogonia* (sec. 13.13) is a particular example of this principle, except that the difference between *A* and *B* in that case referred to the difference between the same area before and after it had been developed for agriculture.

The reader may have anticipated the general conclusion that we have been leading up to. The numbers of animals in a natural population may be limited in three ways: (*a*) by shortage of material resources, such as food, places in which to make nests, etc.; (*b*) by inaccessibility of these material resources relative to the animals' capacities for dispersal and searching; and (*c*) by shortage of time when the rate of increase r is positive. Of these three ways, the first is probably the least, and the last is probably the most, important in nature. Concerning *c*, the fluctuations in the value of r may be caused by weather, predators, or any other component of environment which influences the rate of increase. For example, the fluctuations in the value of r which are determined by weather may be rhythmical in response to the progression of the seasons (e.g., *Thrips imaginis*, sec. 13.11) or more erratic in response to "runs" of years with "good" or "bad" weather (e.g., *Austroicetes cruciata*, sec. 13.12). The fluctuations in r which are determined by the activities of predators must be considered in relation to the populations in local situations (Figs. 14.03 and 14.08). How long each newly founded colony may be allowed to multiply free from predators may depend on the dispersive powers of the predators relative to those of the prey.

With respect to the second way in which numbers may be limited, the food or other resource may be inaccessible either because it is sparsely distributed, like the food of *Thrips imaginis* during late summer (sec. 13.11), or because it is concealed, like the food of *Lygocerus* (sec. 10.321). In either case the inaccessibility is relative to the animals' capacities for dispersal and searching. We emphasize, in section 11.12 and elsewhere, that it is the inaccessibility of the resource which is important; the argument is independent of whether the stocks of food, etc., are scarce or plentiful, in the absolute sense of so much per square mile.

Of course, it is not to be supposed that the ecology of many natural populations would be so simple that their numbers would be explained neatly by any one of the principles described in this and the preceding two sections. A large number of systems of varying complexity could be synthesized, in the imagination, from various combinations of the principles set out in these sections. But this is not their purpose: they are intended to help the student to analyze the complex systems which he finds in nature into their simpler components, so that he may understand them better.

14.3 SOME PRACTICAL CONSIDERATIONS ABOUT THE NUMBERS OF ANIMALS IN NATURAL POPULATIONS

Darwin in chapter 3 of *The Origin of Species* wrote: "The causes which check the natural tendency of each species to increase in numbers are most obscure. Look at the most vigorous species; by as much as it swarms in numbers, by so much will its tendency to increase be still further increased. We know not what the checks are in one single instance. Nor will this surprise any one who reflects on how ignorant we are on this head." A little further on he wrote: "It is good thus to try in our imagination to give any form some advantage over another. Probably in no single instance should we know what to do, so as to succeed." In chapter 11 he stated: "Rarity is the attribute of a vast number of species of all classes in all countries. If we ask ourselves why this or that species is rare, we can answer that something is unfavourable in its conditions of life; but what that something is we can hardly ever tell."

During the century that has passed since these passages were written, we have learned enough about "the conditions of life" of at least a few species to be able, at will, to alter their numbers greatly. And we have, unwittingly, altered the "conditions of life" for a great many more species, so that their numbers have increased, or decreased, sometimes to our advantage, sometimes to our disadvantage.

With species that are pests of crops, foodstuffs, clothing, etc., it is to our advantage if we can insure that they never become numerous enough to eat a considerable proportion of their total resources of food. The majority of such species are insects. With respect to most of them, our ignorance is still abysmal, and the only way we know how to alter their "conditions of life" to our advantage is to poison as many of them as we can with insecticides. The chief disadvantage of the method is its costliness; but as the insecticides increase in number and complexity, we run an increasing risk of poisoning ourselves as well as the insects.

Ecologically, the use of insecticides fits into the principle illustrated by Figure 14.03. The difference between a good insecticide and a poor one is illustrated, in Figure 14.05, by the middle pair of curves. The difference between an insecticide applied frequently and one used infrequently is shown by the top pair of curves in Figure 14.05. And the difference between an insecticide applied thoroughly and one applied less well is shown in Figure 14.09. It is incorrect to say that, because an insecticide is not a "density-dependent factor," it cannot "regulate" the numbers in a population. There are millions of farmers who could testify that the proper use of insecticides makes a big difference to the size of the populations of certain insects. Many people all over the world would starve if this were not so.

"Biological control" (that is, the introduction of new sorts of predators into the area) is the next most popular way of altering the "conditions of life" of an insect pest. But other methods may be suggested by the principles set out in Figures 14.05 and 14.09. Modifications in husbandry might be thought out which would shorten the period available to the insect for multiplication or reduce the value of r (Fig. 14.05). Or other methods might be tried which would reduce the number of suitable places for the insects to live in or alter the distribution of such places in a way that would be detrimental to the pest (Fig. 14.09).

With animals that are valued for their flesh or fur or for some other quality, it is necessary to take quite a different point of view. It matters hardly at all whether the populations use up much or little of their total resources of food, etc., so long as the individuals are numerous enough to be hunted or fished with profit. If it so happened that the population conformed to Figure 14.01 or Figure 14.02, then the numbers per square mile might be increased by adding to the stocks of the limiting resource in the area. This method seems to have been greatly successful with the muskrat in North America (sec. 12.31). But if it happened that the population conformed to the principles illustrated by Figures 14.03–14.05, then other methods would be required.

14.4 SOME REFLECTIONS ON THE GREATLY MISUNDERSTOOD SUBJECT OF "EXTINCTION"

We conclude this chapter with an anticlimax. The reader may regard it as an appendix if he likes. However rare they may be with respect to the proportion of their total resources that they use up, most of the species that are studied or observed are common in terms of the number of individuals per square mile. This is easily explained by a number of practical reasons. But the fact that we usually study unusual populations is far too frequently overlooked. The truth is that the vast majority of species are rare, by whatever criterion (sec. 14.1) they are judged. Smith (1935, p. 880) commented on this: "The fact that the number of species which become sufficiently abundant to damage crops is *relatively* small, and that such species form only an insignificant fraction of the total number of phytophagous insects is ignored." Bodenheimer (1930, in Smith, 1935) commented in the same vein: "It is only in rare borderline cases that the food is used up to the possible limit. Any meadow, field, or orchard will prove this sufficiently." And, of course, Darwin was well aware of the fact. In a slightly different context he wrote (chap. 11 of *The Origin of Species*): "To admit that species generally become rare before they become extinct, to feel no surprise at the rarity of the species, and yet to marvel greatly when the species ceases to exist, is much the same as to admit that sickness in the individual is the forerunner of death—to feel no surprise at sickness, but, when the sick man dies, to wonder and to suspect that he died of some deed of violence."

Nevertheless, the misconception prevails that the extinction of a population is a very rare event. This leads our colleagues who hold to the dogma of "density-dependent factors" to propound this riddle. On hearing us expound our views on ecology, they ask: "How is it, if there is no density-dependent factor in the environment, that the population does not become extinct?" This places us in a position like that of the man in the dock who was asked to answer Yes or No to the question: "Do you still beat your wife?" We cannot answer the question until we have cleared up the misconception in the mind of the questioner.

First of all, it is indeed true that the species which we study are less likely than most to become extinct, because invariably we choose for study those in which the populations are large. Elsewhere in this book, and especially in sections 13.11 and 13.12, we explain why certain species of insects did not become extinct during the period that they were studied and infer that they are not likely to become extinct during the immediate future. A quotation from Stevenson-Hamilton (1937, p. 258) shows that the same principles apply to mammals which commonly maintain large populations: "The extermination, under natural conditions, of one indigenous species by another, whether directly through carnivores consuming herbivorous types, or indirectly by one herbivorous type proving too strong for its associates in the same area, except as a final culmination, after Man, or one of the other factors cited above, has first played the principal part, is unknown in natural history so far as our experience extends, and may be safely ruled out in any wild-life reservation *of adequate extent where room exists for seasonal migration* [our italics]."

But the risk of extinction for species in which the populations are small is greater. Darwin in the fourth chapter of *The Origin of Species* wrote: "Any form which is represented by a few individuals will run a good chance of utter extinction during great fluctuations in the nature of the seasons, or from a temporary increase in the numbers of its enemies." A quotation from Ford (1945*a*, p. 143) shows how this risk operated in the particular case of the butterfly known as the "Wood White":

> Any species can survive such periodic fluctuation provided that its numbers are large enough, and that it is somewhere sufficiently well established to tide over the dangerous period when it is reduced to its lowest level. Clearly, a "normal" cycle of this kind may be disastrous to a butterfly which maintains itself precariously, whether in an isolated locality or in the country as a whole. The Wood White survived in Westmorland until about 1905 when it disappeared and has never been seen there again. At that time the species was in general becoming rare, and in the south retracting its range to a few favoured places. From these it could, and did, spread once more; but not in the north, where the one isolated colony was wiped out by a process which had no serious consequences elsewhere. Similarly, it disappeared from the New Forest early this century and has not returned. It was said at the time that this was due to overcollecting, and probably that was true. But I suspect that the butterfly was reduced to dangerously small numbers by natural causes, operating there as elsewhere, and for that reason the activities of the collector were fatal.

Ford quoted several other examples of butterflies which have recently become extinct in Britain.

There is evidence from paleontology to show that extinction of species is commonplace in the time scale of that science. Simpson (1952) estimated that there might now be 2,000,000 species of plants and animals in the world and that the total number of species that have existed since the "dawn of life" may be of the order of 500,000,000. On this estimate, more than 99 per cent of the species that have ever existed are now extinct. Mayr (1942, p. 224) pointed out that many of the species that are known to have become extinct during modern times have lived on small islands, where the terrain would be more uniform and the opportunity for dispersal less than in a larger area.

There is no fundamental distinction to be made between the extinction of a local population and the extinction of a species other than this that the species becomes extinct with the extinction of the last local population. We can witness the extinction of local populations, of even the most abundant species, going on all around us all the time. So if the extinction of a population were proof that the "environment" lacked a "density-dependent factor," we would have ample evidence of the absence of "density-dependent factors" from the "environments" of the animals in local populations of most species.

Species are likely to be "rare" near the margins of their distributions, and outside the distribution they are "extinct." Because distribution and abundance are but two aspects of one phenomenon, the study of abundance in different parts of the distribution is itself a study of the causes of rareness and commonness in species.

PART V

Genetic Aspects of Ecology

CHAPTER 15

Genetic Plasticity of Species

If under changing conditions of life organic beings present individual differences in almost every part of their structure, and this cannot be disputed; if there be, owing to their geometrical rate of increase, a severe struggle for life at some age, season or year, and this certainly cannot be disputed; then, considering the infinite complexity of the relations of all organic beings to each other and to their conditions of life, causing an infinite diversity in structure, constitution, and habits, to be advantageous to them, it would be a most extraordinary fact if no variations had ever occurred useful to each being's own welfare, in the same manner as so many variations have occurred useful to man. But if variations useful to any organic being ever do occur, assuredly individuals thus characterized will have the best chance of being preserved in the struggle for life; and from the strong principle of inheritance, these will tend to produce offspring similarly characterized. This principle of preservation, or the survival of the fittest I have called Natural Selection. It leads to the improvement of each creature in relation to its organic and inorganic conditions of life; consequently in most cases, to what must be regarded as an advance in organisation. . . . As buds give rise by growth to fresh buds, and these if vigorous, branch out and overtop on all sides many a feeble branch, so by generation I believe it has been with the great Tree of Life, which fills with its dead and broken branches the crust of the Earth, and covers the surface with its ever branching and beautiful ramifications.

DARWIN, *The Origin of Species*

15.0 INTRODUCTION

THE continued influence of Darwin's *The Origin of Species* and the recent advances in population genetics have modified the old concept of a species as represented by a "type specimen" in a museum. The ecologist must recognize that the species has a certain plasticity in its genetic composition. By this we mean that individuals differ and that those that are more favored by their environments have a greater chance of leaving progeny behind them. In this sense we might speak loosely of the species being "molded" by the various components of its environment. Thus samples of the same species collected at the same time but at different places may differ significantly with respect to the relative frequencies of the different genes; similarly, differences may occur in samples taken from the same place but at different times. These two phenomena may be regarded as the quantitative and qualitative aspects of distribution and abundance, respectively.

Dobzhansky (1950*b*, p. 405) used the phrase "Mendelian population" to embrace both the species as a whole and its local subdivisions. He defined a Mendelian population as a reproductive community of sexual and cross-fertilizing individuals which share a common pool of genes. The species is the most inclusive Mendelian population; its chief characteristic is that its members do not (no matter how good may be the opportunity) interbreed with members of other Mendelian populations. Populations whose members do not interbreed because they are kept apart by geographic barriers may not be classed as species on this evidence alone; for example, a number of Mendelian populations living on several widely separated oceanic islands may all belong to the one species, even though there is virtually no chance of interbreeding in nature because of the distances separating the islands. On the other hand, if it were found, when they were brought together, that they still did not interbreed, then they would be correctly classed as separate species. Dobzhansky (1951, p. 181) listed "sexual isolation" and "hybrid inviability" among a number of other mechanisms which may prevent species from interbreeding. In chapter 16 we shall consider how one species may become two; that is, the origin of discontinuities (in the distributions of genes) which are irreversible because they occur in Mendelian populations which can no longer interbreed even in the absence of geographic barriers. In this chapter we discuss the changes (largely reversible) occurring within the Mendelian population which retains its unity, that is, remains as one species.

The subordinate Mendelian populations within the species are variously known as subspecies, races, or local populations. Since the relative frequencies of the different genes may differ from one local population to the next and from time to time in the population to be found in the one locality, it is clear that the species, which is no more than the sum of all these local populations, must be changing with time also. From the evolutionary aspect this is important, because, as Wright (1945, p. 415) said: "The elementary evolutionary process is . . . change of gene (or chromosome) frequency." The genetic mechanisms involved have been discussed in relation to interactions between mutation, immigration, natural selection, and the random loss of genes from small populations by Wright (1931, 1948, 1949), Fisher (1930), and Dobzhansky (1951). These strictly genetical discussions are not for us.

As ecologists, we have a practical interest in phenotypical manifestations; we recognize and measure such qualities as innate capacity for increase, dispersal, diapause, and so on. But in so far as these qualities are inherited, it is important for us to appreciate the variability which may arise from this source. If we are interested in, say, the innate capacity for increase in a particular species, we must study this in a sample taken from a natural population. It may not be practicable to sample as widely as desirable; but at least we should know that a sample taken from a certain place at a certain time may differ from one taken from another place at the same time, or from the same place at another time; and we should have some knowledge of the genetic explanations for this.

On the other hand, as ecologists, we may make our own contribution to the subject. The relative frequencies of the different genes in a local population may be influenced by weather, food, and the numbers and kinds of other animals in the area. These are chiefly important in relation to selection but also in relation to Wright's "fourth factor," namely, the random loss of genes from small populations. This chapter is then chiefly concerned with these ecological aspects of "genetic plasticity." In compiling it, we have drawn freely from the work of geneticists, especially from Dobzhansky and his school, emphasizing for our own purposes those aspects of their studies which are important to the study of distribution and abundance.

15.1 GENETIC PLASTICITY IN RELATION TO THE COMPONENTS OF ENVIRONMENT

In chapter 1, taking the species as a unit, we showed that distribution and abundance were merely different aspects of the one phenomenon. This is also true for the smaller units within the species. For example, in the next two sections we shall show how weather influences not only the distribution of geographic races (which are Mendelian populations living in different regions and having different proportions of the various genotypes) but also the seasonal trends in the relative abundance of the different genotypes throughout the year or from year to year in the same place.

15.11 *Weather*

15.111 SEASONAL TRENDS IN THE GENETIC COMPOSITION OF "MENDELIAN POPULATIONS"

A single gene determines whether the beetle *Adalia bipunctata* shall be red or black (Timofeeff-Ressovsky, 1940). Natural populations which contain both color variants are described as "polymorphic" with respect to this character. All Mendelian populations are more or less polymorphic. Some polymorphic variations may be induced by the environment in a way that depends only indirectly on genotype, for example, color in the eggs of *Bombyx* (sec. 8.12); these are of little interest to us here. But in *Adalia* color is a sure indication of genotype. Polymorphism of this sort provides a useful tool for studying genetic plasticity, since it gives a label to particular genes, or groups of genes, which, in addition to causing recognizable morphological variations, may also influence the physiology of the individual in some way which makes it more or less fit to survive and multiply in a certain environment.

Timofeeff-Ressovsky (1940) found that this was so with *Adalia*. In the vicinity of Berlin *Adalia* overwinters in the adult stage. The death-rate during winter is high. Timofeeff-Ressovsky found, in one population which he studied, that 96 per cent of the black forms and 89 per cent of the red ones died during winter. As a consequence, the proportion of black ones in the population decreased as winter progressed. By spring, they represented 37 per cent of the population. However, during the summer they multiplied relatively more rapidly than the red ones, so that, by autumn, they constituted 59 per cent of the population. Another example of a morphological variation which indicates a physiological variation is provided by the banded snail *Cepaea nemoralis*, which has a variety of colors and banding patterns (Cain and Sheppard, 1950, 1952). Schnetter (1951) observed that the yellow variant of this snail increased in relative abundance by 16 per cent during a sequence of drought years (sec. 15.152).

In the moth *Panaxia dominula* there is a color variety known as *bimacula* which is the homozygote of a mutant gene *m;* its constitution may be represented by m/m. The color variety *medionigra* is the heterozygote ($+/m$); and the normals have the constitution $+/+$. Fisher and Ford (1947) determined the frequency of the *medionigra* gene *m* for the years 1928–46 from estimates of the numbers of the different phenotypes in a small isolated population near Oxford. It was less than 1.2 per cent before 1929, rose to 9.2 per cent in 1939, and reached a maximum of 11.1 per cent in 1940. After 1940 the frequency fell (Table 15.01). Fisher and Ford (1947) considered that this must be attributed to selection, but Wright (1948) reanalyzed the data and suggested an alternative explanation in terms of his theory of "genetic drift" in small populations. This explanation depends upon the breeding population having been reduced on several occasions between 1928 and 1940 to numbers of the order of 100. Since the numbers in the breeding population during this period were not known, there is no way of judging the probability of Wright's explanation.

TABLE 15.01*

CHANGE IN FREQUENCY OF GENOTYPES $+/+$, $+/m$, AND m/m IN A LOCAL POPULATION OF THE MOTH *Panaxia dominula*

Years Up To	Estimate of Numbers of Moths	Types of Moths Collected				Frequency of *m* (Per Cent)
		$+/+$	$+/m$	m/m	Total Moths	
1928	....	164	4	..	168	1.2
1939	....	184	37	2	223	9.2
1940	....	92	24	1	117	11.1
1941	2,000–2,500	400	59	2	461	6.8
1942	1,200–2,000	183	22	..	205	5.4
1943	1,000	239	30	..	269	5.6
1944	5,000–6,000	452	43	1	496	4.5
1945	4,000	326	44	2	372	6.5
1946	6,000–8,000	905	78	3	986	4.3

* After Fisher and Ford (1947).

In *Drosophila* the polymorphic variations which have been most used in the study of genetic plasticity are differences in the morphology of chromosomes. For the most part, these are brought about by inversion of short lengths of chromosomes, which result in a rearrangement of certain genes, though not necessarily in any gain or loss of genes. In *Oenothera* a certain rearrangement of genes, by translocation, was, in itself, sufficient to induce a noticeable variation (in the color of the flowers) in the phenotype. When the genes were restored, by crossing over to their original position, normal phenotypes were produced, indicating that the results were due entirely to the position of the genes (Catcheside, 1947). On the other hand, many inversions have been observed which seemed not to be associated with noticeable changes in the phenotype. In *Drosophila*, however, certain of these polymorphic variations in the arrangement of genes within the chromosomes are associated with variations in the physiology of the individual which make it either more or less fit to survive and multiply, in association with other genotypes, in different climates or at other seasons of the year. The first indication of this was given by Dobzhansky's (1943) discovery that in a local population of *Drosophila* the relative abundance of several such polymorphic forms varied with the seasons. This discovery was confirmed by Dubinin and Tiniakov (1945). For the more genetic aspects of these discoveries and the work which arose from them the reader should consult Dobzhansky (1951, chap. 5). We have selected for discussion only those aspects which are of special interest in ecology.

In the third chromosome of *Drosophila pseudoobscura* fifteen different arrangements of genes have been recognized in individuals from wild populations. None ranges over the entire distribution of the species; but the populations of most localities contain more than one and up to eight of them. The different gene arrangements are usually called after the locality where they were first found; thus on Mount San Jacento the three most common ones are Standard (ST), Arrowhead (AR), and Chiricahua (CH). It will be shown below that at least some of these visible morphological variations are as-

TABLE 15.02*

FREQUENCIES (PER CENT) OF CHROMOSOMES OF *Drosophila pseudoobscura* AND *D. willistoni*, WHICH, IN HOMOZYGOTES, ARE ASSOCIATED WITH SPECIFIED PHYSIOLOGICAL VARIATIONS

EFFECTS	*D. willistoni*		*D. pseudoobscura*		
	Second Chromosome	Third Chromosome	Second Chromosome	Third Chromosome	Fourth Chromosome
Lethals and semilethals	41	32	21	14	26
Subvitals	58	49	21	31	41
Sterility	31	28	14	?	8
Retarded development	32	36	54	?	32
Accelerated development	14	3	0.4	?	3
Visible effects	16	16	4	?	2

* After Pavan *et al.* (1951).

sociated with physiological qualities which are important in relation to the animal's "struggle for existence." But there is also a great deal of genetic variation which determines physiological qualities not associated with any visible morphological variations. For example, Pavan *et al.*, (1951) collected some thousands of *D. willistoni* and *D. pseudoobscura* and looked for what they

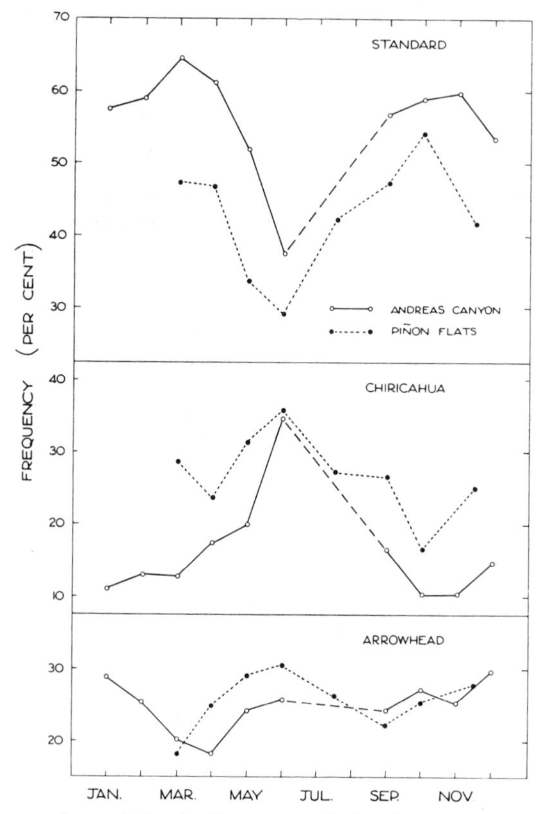

FIG. 15.01.—Changes observed in the frequency of the three gene arrangements: Standard, Chiricahua, and Arrowhead in *Drosophila pseudoobscura* at Piñon Flats and Andreas Canyon in California. The graphs present combined data for 4 years of observations. (After Dobzhansky, 1943.)

called this "concealed genic variability." Some of their results are summarized in Table 15.02.

Dobzhansky (1943) measured the relative frequencies of the three gene arrangements CH, ST, and AR at different seasons of the year for 4 years in local populations of *Drosophila pseudoobscura* living in three localities in California. The relative frequency of ST decreased and that of CH increased from March to June. The trend was in the opposite direction during the hot season, from June to September. The changes in the relative frequencies of AR were less regular, but, on the whole, they seemed to follow a trend somewhat similar to that of CH. There was a remarkable similarity in the seasonal trends in the two localities illustrated in Figure 15.01.

The seasonal trend was repeated each year but with minor variations, which were, no doubt, associated with the vagaries of the weather. The fluctuations in the relative frequency of one gene arrangement (AR) in one locality (Mather, California) are compared month by month for 4 years in Figure 15.02. Between 1945 and 1950 the frequency of AR increased and that of ST decreased. But in 1951 AR became rarer again, and ST became more common. During the same period similar records were kept for *Drosophila persimilis*. In this species

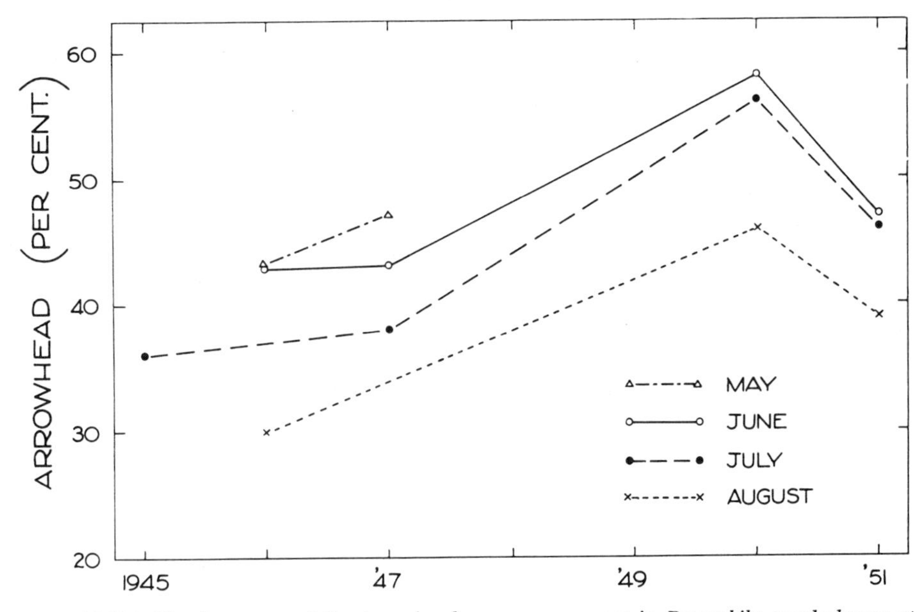

Fig. 15.02.—The frequency of the Arrowhead gene arrangement in *Drosophila pseudoobscura* at Mather in California from 1945 to 1951. (After Dobzhansky, 1948, 1952.)

there was not a distinct seasonal trend, but the changes from year to year were quite marked. The gene arrangement known as "Whitney" (WH) increased in relative abundance from 1945 to 1950 and fell in 1951. The opposite happened to ST and Klamath (KL). It so happened that the years 1946–50 were unusually dry and that the drought ended during the winter of 1951, which had a rainfall above the average. Since the relative frequencies of all the gene arrangements which were studied showed a general trend, either up or down, during 1945–50 which was reversed in 1951, it seems most likely that the genetic changes were related to the drought and its cessation in 1951.

Seasonal trends in the relative frequencies of two genotypes have also been observed in the hamster *Cricetus cricetus*, which has only one generation each year. In Russia, *Cricetus* occurs in two color varieties, black and gray; the difference is thought to be due to a single gene. In the winter in the Poltava region the black ones become fewer relative to the gray ones, but near Chernigov the blacks become relatively more numerous during the winter. In both regions the trends are reversed during the summer (Gershenson, 1945).

There has been a number of experiments which show that temperature is important in determining the seasonal fluctuations in genetic composition observed in natural populations. Dubinin and Tiniakov (1945, 1946*a*, *b*) observed in a natural population of *Drosophila funebris* near Moscow that certain gene arrangements (inversions) increased from spring to autumn and dwindled during the winter. When samples of the different genotypes were exposed to low temperature in the laboratory, it was found that those which normally dwindled during the winter in the field were also the ones with the shortest life-span at low temperatures in the laboratory. The normal form of *Daphnia longi* develops best in laboratory cultures at 20° C. But Banta and Wood (1939) isolated a variant which developed best between 25° and 30° C. Wilkes (1942) placed adults of the chalcid *Microplectron fuscipennis* in a cage in which there was a gradient in the temperature from one end to the other. The usual result would have been a unimodal distribution of insects about a narrow range of temperature (sec. 6.21). But in this case Wilkes observed a trimodal distribution. The highest peak occurred at 25° C., but there were two minor ones at 15° and 8° C. Starting with some of the individuals which preferred 8° C. and by inbreeding and selection, he bred a race of which 40 per cent of the individuals preferred a temperature below 10° C., as compared with 3 per cent in the original stock. The mutant "eversae" of *Drosophila funebris,* which arose in a laboratory stock, had a higher survival-rate at 24°–25° C. than the wild stock; but at 15°–16° C. and also at 28°–30° C. its survival-rate was lower than that of the wild stock (Timofeeff-Ressovsky, 1934; quoted from Dobzhansky, 1951).

Wright and Dobzhansky (1946) showed that trends in the relative frequencies of the different gene arrangements in *Drosophila pseudoobscura,* rather like the seasonal trends which had been observed in nature, could be produced in experimental populations by manipulating the temperature in the cage where the flies were breeding. A number of flies was placed in a "population cage" with an abundance of food, which was replenished at frequent intervals. The cage soon became crowded, and the flies were allowed to continue breeding in these circumstances for a number of generations. Many more eggs were laid than emerged as adults, indicating that "the struggle for existence" was intense. In each experiment the proportion of flies carrying each gene arrangement was known for the initial population, and it was measured at intervals during the experiment by the following method. Some eggs were taken from the cage and reared uncrowded and with plenty of food. The distributions of the gene arrangements in these samples gave reliable estimates of the composition of the population of adults in the cage at the time when the eggs were laid.

Figure 15.03 shows what happened in one such experiment conducted at 25° C. with flies from Piñon Flats, California, of which 11 per cent carried ST, and 89 per cent CH, at the beginning of the experiment. Within about 4 months the proportion of ST chromosomes in the population had quadrupled. It then rose more slowly to about 70 per cent, after which no further change occurred. In other experiments with flies from the same source the same equilibrium was approached, irrespective of whether more or fewer than 70

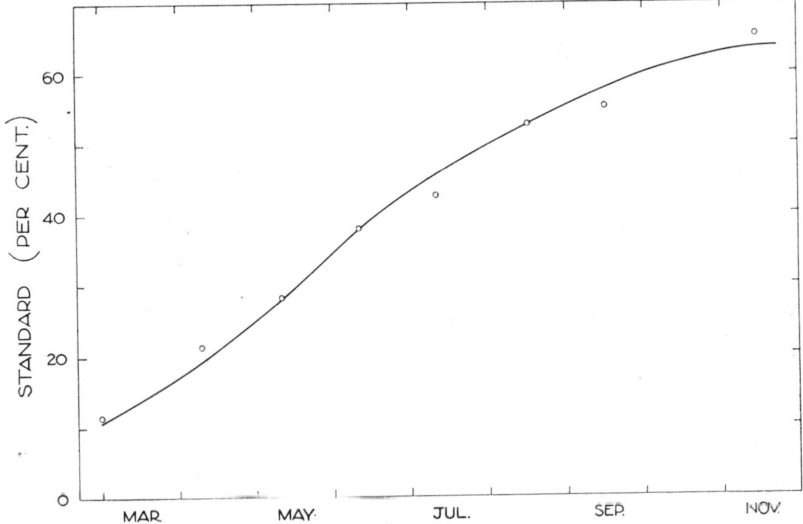

FIG. 15.03.—The frequency of the Standard gene arrangement in *Drosophila pseudoobscura* in different months in a population cage at 25° C. The population was initiated with 10.7 per cent Standard and 89.3 per cent Chiricahua chromosomes from Piñon Flats in California. (After Dobzhansky, 1947.)

per cent of them carried ST at the beginning. The increase of ST from 11 to 70 per cent was due to natural selection. The failure of the selection to proceed beyond the equilibrium of 70 per cent was explained by Wright and Dobzhansky (1946) and Dobzhansky (1951) with the aid of the following hypothesis.

The heterozygotes (individuals having one ST and one CH chromosome) are superior (in crowded population cages at 25° C.) to both homozygotes, CH/CH or ST/ST. In other words, the rate of increase of the heterozygotes in these experiments was greater than the rate of increase of either homozygote. Dobzhansky referred to the relative rates of increase as the "adaptive values" of the different polymorphic forms. From the genetic point of view, the adaptive value is "the relative capacity of carriers of a given genotype to transmit their genes to the gene pool of the following generations" (Dobzhansky, 1951, p. 78). The extent to which the genes are transmitted to the gene pool is related to the rate of increase of the genotypes in the population. Dobzhansky went on to say: "The adaptive value is, then, a statistical concept which epitomizes the reproductive efficiency of a genotype in a certain environment. Now, the adaptive value is obviously influenced by the ability of a type to survive. The adaptive value of a homozygote for a lethal gene is evidently zero. But the individual's somatic vigour, its viability, is only one of the variables which determine the adaptive value. The duration of the reproductive period, the number of eggs produced (fecundity), the intensity of the sexual drive in animals, the efficiency of the mechanisms which conduce to successful pollination in plants, and many other variables are likewise important." Clearly, Dobzhansky's "adaptive value" is like our concept of r, the actual rate of increase (see Chaps. 3 and 9). It has meaning only in relation to the components of environment which help to determine it. A difference is that r measures an absolute rate of increase, whereas "adaptive values" are stated relative to one another and have no meaning in terms of absolute rate of increase. An example will make this clear. In the experiment described above, Dobzhansky arbitrarily gave the heterozygotes an adaptive value of 1 and then worked out the adaptive values of the homozygotes ST/ST and CH/CH as 0.7 and 0.4, respectively. The smooth theoretical curve in Figure 15.03 is derived from these ratios. Wright and Dobzhansky (1946) referred to the differences between the adaptive values of the homozygotes and the heterozygotes as the "selective coefficients" of the homozygotes. The selective coefficients of ST/ST and CH/CH were thus 0.3 and 0.6, respectively. The selective coefficient of the heterozygote was, by definition, zero.

When the same experiment was repeated at 16.5° C., the proportions of ST and CH in the population at the end of the experiment were the same as they had been at the beginning. The advantage which ST had possessed over CH at 25° C. disappeared at the lower temperature. In other experiments of the same sort, Wright and Dobzhansky (1946) and Dobzhansky and Spassky (1944) showed that the relative advantage (as measured by changing proportions in crowded population cages) which one gene arrangement possessed over another depended not only on temperature but also on certain other components of the environment. When genotypes of yet a third gene arrangement (Arrowhead) were introduced into the population cage (at 25° C.) the relative adaptive values of the homozygote ST/ST and heterozygote ST/CH changed; the homozygote became slightly inferior to the heterozygote (Levene *et al.*, 1954). This shows that the adaptive values of the carriers of ST/ST and ST/CH gene arrangements depend upon the other sorts of genotypes in the population cage.

Da Cunha (1951) isolated nine species of yeasts and two species of bacteria from the crops of *Drosophila pseudoobscura* collected in California. These microörganisms, in separate cultures, were fed to populations of flies in which the CH and ST gene arrangements were represented; the experiments were done in the usual "population cages." The adaptive values of the homozygotes

ST/ST and CH/CH and the heterozygotes ST/CH were estimated in the way we have already described, the "adaptive value" of the heterozygote being taken as unity. The results given in Table 15.03 show that the adaptive values

TABLE 15.03*

ESTIMATES OF ADAPTIVE VALUES AND FREQUENCIES AT EQUILIBRIUM OF STANDARD (ST) AND CHIRICAHUA (CH) GENE ARRANGEMENTS IN EXPERIMENTAL POPULATIONS OF *Drosophila pseudoobscura* AT 25° C. WHICH WERE FED WITH DIFFERENT MICROÖRGANISMS

FOOD	"ADAPTIVE VALUE"			FREQUENCY AT EQUILIBRIUM	
	ST/ST	ST/CH	CH/CH	ST	CH
Saccharomyces cerevisiae	0.71	1	0.32	0.70	0.30
Rhodotorula mucilaginosa	0.83	1	0.49	0.75	.25
Candida parapsilosis	0.89	1	0.54	0.80	.20
Zygosaccharomyces dobzhanskii	1.06	1	0.35	1.00	.00
Z. drosophilae	<1	1	<1	±0.73	±.27
Kloeckeraspora apiculatus	1.55	1	1.00	1.00	.00
C. guilliermondii	0.80	1	0.32	0.77	.23
C. krusei	<1	1	<1	±0.75	±.25
Bacteria strain No. 3	1.14	1	0.37	1.00	.00
Bacteria strain No. 9	1.15	1	0.57	1.00	0.00

* After Da Cunha (1951).

of each genotype depended upon the sort of food provided in the cage. The heterozygote was superior to the homozygotes (heterosis) in most of the experiments; but at least two species of yeasts and two species of bacteria caused the adaptive value of the heterozygotes to fall below that of the ST homozygote. El-Tabey *et al.* (1952) stated that most flies collected in nature contained only one species of yeast in the crop at any one locality at any one time. This might be due to the dominance of one yeast in the field or to the selection of food by the flies. Shihata (unpublished data quoted by Da Cunha, 1951) observed that in nature the relative abundance of the different species of yeasts show seasonal fluctuations. Hence it would seem likely that, in nature, food may be one of the more important components of environment determining the fluctuations in relative frequency of the different gene arrangements.

In nature the flies may not be crowded as they were in the cages; also, differential capacity for dispersal would count for nothing in the cages, whereas it might be quite important in nature. So one would not expect to find in the results of these experiments a precise explanation for the phenomena observed in natural populations. But they do indicate that the "adaptive value" of different gene arrangements may depend on environment. Since the adaptive value of a particular gene arrangement in an experimental population may be varied at will by manipulating the temperature or the food in the cage where it is living, there is good reason to expect that the fluctuations in the relative frequencies of different gene arrangements observed in natural populations are due to fluctuations in the weather and other environmental influences. Adaptive value is like the rate of increase, r, in that it may be analyzed into its three components, namely, fecundity, longevity, and speed of development. Any one of these may be influenced independently of the others by the various components of environment.

Levine (1952) reared larvae of the three genotypes ST/CH, CH/CH, and ST/ST separately at 15°, 25°, and 30° C. He found that the speed of development and the survival-rate were the same for all three genotypes. Nor could he find any differences when he reared them at four different levels of crowding. He also measured the survival-rate in pupae and adults at 25° C. and at several different humidities. He found no differences in pupae at relative humidities above 76 per cent; below this, there were slight differences, with the survival-rate among ST/ST greater than those among ST/CH and CH/CH. At relative humidities below 56 per cent, the survival-rate in ST/CH was greater than that in CH/CH. At all humidities used in these experiments the adults with ST/ST gene arrangements lived longest and those with CH/CH the shortest. The conditions of these experiments thus tended to favor ST/ST as compared with CH/CH homozygotes. In a further analysis of "adaptive value" of these homozygotes Moos (personal communication) determined the fecundity, speed of development, and longevity at 25° C. From these data she calculated the innate capacity for increase (see chap. 3) and showed that the innate capacity for increase (expressed as a finite rate of increase per female per week) of ST/ST homozygotes was 36 per cent greater than that of CH/CH homozygotes.

In these experiments the influence of crowding was studied only with reference to crowding by the same genotype. When two or several genotypes within the one Mendelian population are forced to live crowded together, the experiments may be likened to those described in section 10.213, when two species were forced to live crowded together. But they differ in one important respect: the ranks of the less successful homozygotes CH/CH and ST/ST are constantly being replenished by progeny of the more successful heterozygote CH/ST. So long as the heterozygotes remain and breed, the homozygotes cannot be extinguished. This is not so when the units are different species, as described in section 10.213. This calls for special methods in investigating the crowding of one genotype by another in the same Mendelian population. We describe a nice experiment by Dobzhansky which shows how this may be done.

If, in a population of flies living in a cage, the frequencies of the gene arrangements ST and CH be represented by q and $1 - q$, respectively, and if the flies mate at random with respect to gene arrangement and the number of eggs laid by individuals of the different gene arrangements are the same, then the proportions of heterozygotes and homozygotes among the eggs laid will be given by the expansion of the binomial $\{q + (1 - q)\}^2$ which gives:

$$\text{ST/ST} : \text{ST/CH} : \text{CH/CH} = q^2 : 2q(1 - q) : (1 - q)^2. \quad (1)$$

But if the different classes of zygotes survive at different rates, the observed proportions will depart from those predicted by the foregoing expression. If we suppose the survival-rates of ST/ST, ST/CH, and CH/CH to be, respectively, W_1, W_2, and W_3, then we have

$$\text{ST/ST} : \text{ST/CH} : \text{CH/CH} = q^2W_1 : 2q(1-q)W_2 : (1-q)^2W_3. \tag{2}$$

Dobzhansky (1947*a*) determined the relative frequencies of the three genotypes in samples of larvae hatched from eggs which had been deposited in population cages but were reared in culture bottles, where nearly all of them survived. The observed proportions deviated only slightly from those predicted by equation (1). This indicated that the flies were mating at random with respect to genotype and that each genotype was laying about the same number of eggs. Samples of adults were then taken from the same crowded cages and classified. There were too many heterozygotes and too few homozygotes to fit the frequencies predicted by equation (1). This indicated a differential elimination of the homozygotes CH/CH and ST/ST at some stage between egg and adult. Further calculations showed that the change was on a scale that was more than sufficient to account for the change in the ratios of the three genotypes which occurred in the population cages at 25° C. (Fig. 15.03). In wild populations, also, the male homozygotes were less, and the male heterozygotes more, abundant than would have been expected in the absence of differential death-rates at some stage between egg and adult (Dobzhansky, 1948).

Some geneticists have referred to the changes in gene frequency which are caused by components of environment such as weather as "competition" between genotypes. But this is not "competition" as we have defined it in section 2.121, unless there is a shortage of a common resource, such as food, or unless one genotype harms another in its search for the common resource. A gene may, of course, spread rapidly through a population without "competition" (in the strict sense) occurring at all.

The observations and experimental results discussed in this section show that a species may possess a "plasticity" by virtue of its genetic constitution which enables it (or parts of it, as represented by local Mendelian populations) to become adapted, relatively rapidly, to the changing seasons of the year. The other aspect of the same phenomenon, which is discussed in the next section, is the way in which Mendelian populations, living in regions which differ from one another in climate or in some other way, come, by virtue of the same mechanism, to be permanently, though not necessarily irreversibly, different from one another.

15.112 GEOGRAPHIC RACES

Polymorphism in local populations, separated partly or completely either by distance or by some more tangible geographic barrier, is commonplace and widespread among all classes of animals. It has been shown in a number of instances that the morphological variations are associated, through genetic mechanisms, with physiological variations which have a positive selective value in the area where they are common; and it seems a priori likely that most polymorphism is like this. The more ecological aspects of the phenomenon known as "geographic races" are discussed in this section with reference to a few selected examples: chromosomal polymorphism in *Drosophila;* poly-

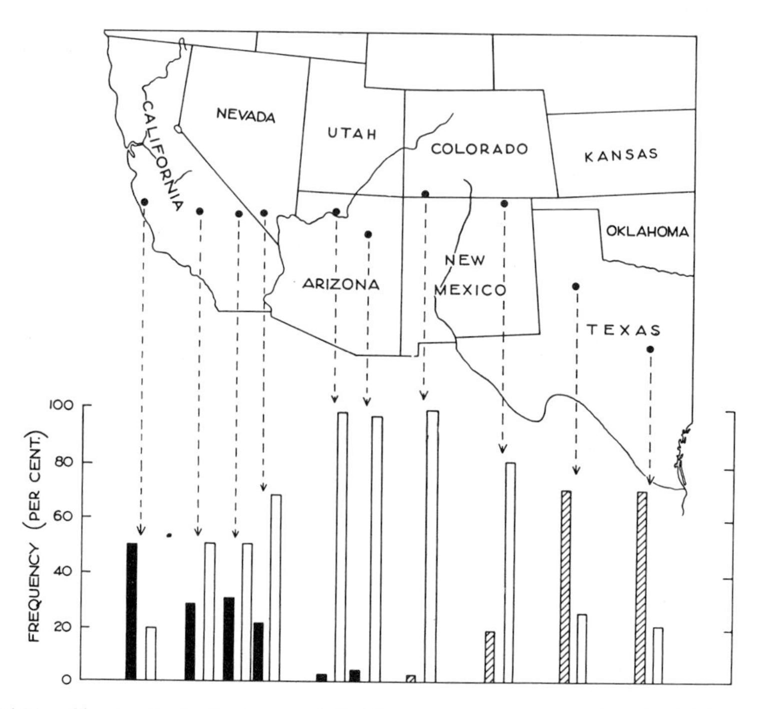

FIG. 15.04.—Changes in the frequency of the three gene arrangements Standard (*black columns*), Arrowhead (*white columns*), and Pikes Peak (*hatched columns*) in *Drosophila pseudoobscura* in a transect across the southwest of the United States. (After Dobzhansky, 1947*b*.)

morphism in the frog *Rana pipiens* associated with variations in the range of temperature that can be tolerated by the eggs; polymorphism in color in the hamster *Cricetus cricetus* associated with type of vegetation in the places where it lives; and, finally, variation in the incidence and intensity of diapause in certain races of insects.

Dobzhansky (1947*b*) collected *Drosophila pseudoobscura* from a number of situations so chosen that they formed an east-west transect of about 1,200 miles across the southwestern part of North America. The flies from each place were examined for the presence of the gene arrangements Standard (ST), Arrowhead (AR), and Pikes Peak (PP). The results are illustrated in Figure 15.04. Note the striking contrast in the distributions of ST and PP. The former is abundant in the extreme west and absent from the central and eastern parts of the transect; the latter is abundant in the extreme east and absent from the west. Another transect was made running north and south through Arizona. A similar gradient was observed. The gene arrangement Cuernavaca (CV) was abundant in the south (New Mexico) and absent from the north (Nevada), whereas ST was present in the north but absent from the south (Dobzhansky, 1951, p. 138). This heterogeneity within the species is, of course, likely to be repeated at each lower level. Dobzhansky found that the individuals carrying, for example, ST in one region might be quite different from those carrying the same gene arrangement in another region. To discuss this would take us too deeply into genetics; the reader is referred to papers by Dobzhansky (1948, 1950*c*).

It is not necessary to look to large distances for changes in the frequencies of chromosomal types in *Drosophila pseudoobscura*. The changes which occur within a few miles at different elevations on a mountainside may be dramatically large. Table 15.04 shows the frequencies of four gene arrangements in local populations living at different elevations in the Sierra Nevada, California (Dobzhansky, 1949*a*). Note how the distributions of ST and AR seem to be complementary, the former decreasing and the latter increasing with altitude. The seasonal trends within these local populations may be summarized by saying that during the summer the upland populations become more like the lowland ones and during the winter the lowland populations become more like the upland ones. So the evidence seems to suggest that the physiological difference between the carriers of ST and AR is closely related to temperature; those carrying ST may be better adapted to higher temperatures, while those carrying AR may have the advantage at lower temperatures. Spiess (1950) described similar gradients in the frequencies of chromosomal types in *D. persimilis* associated with altitude. He also confirmed experimentally that there were differences in the "adaptive values" of the different types.

TABLE 15.04*

FREQUENCY (PER CENT) OF FOUR COMMON GENE ARRANGEMENTS IN THIRD CHROMOSOME OF *Drosophila pseudoobscura* ALONG TRANSECT AT DIFFERENT ELEVATIONS IN SIERRA NEVADA MOUNTAINS OF CALIFORNIA

Locality	Elevation (Feet)	Gene Arrangements			
		Standard	Arrowhead	Chiricahua	Tree Line
Jacksonville	850	46	25	16	8
Lost Claim	3,000	41	35	14	6
Mather	4,600	32	37	19	9
Aspen Valley	6,200	26	44	16	11
Porcupine Flat	8,000	14	45	27	9
Tuolumne Meadows	8,600	11	55	22	9
Timber Line	9,900	10	50	20	10

* After Dobzhansky (1949*a*).

The Brazilian species *Drosophila willistoni* has been found to be especially suitable for studies in geographic races, because it is abundant and widely distributed; also it shows a remarkable degree of chromosomal polymorphism: at least 40 different inversions have been recorded (Da Cunha *et al.*, 1950). The incidence of these inversions differs from place to place throughout the distribution of *D. willistoni* in Brazil. For example, the frequency of heterozygous

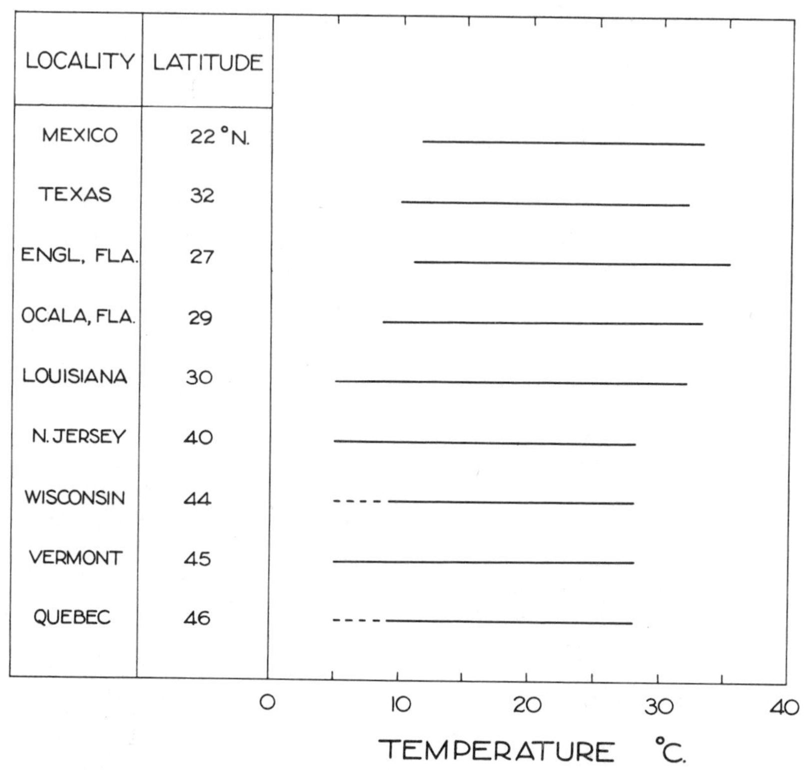

FIG. 15.05.—The temperature ranges for normal development of eggs of the meadow frog *Rana pipiens* from different localities in North America. (After Moore, 1949*a*.)

inversions was 0.8 per female in the deserts of Bahia but as high as 9 in the equatorial rain forests in central and northern Brazil.

The *willistoni* group is known to contain four "sibling" species, namely, *D. willistoni*, *D. paulistorum*, *D. tropicalis*, and *D. equinoxialis*. The first is the most abundant and widespread; the second, which is also abundant, is less widely distributed; and the last two are less numerous and more narrowly distributed. The number of inversions known for these are 40, 34, 4, and 4, respectively (Dobzhansky *et al.*, 1950). So there is a positive correlation between the degree of chromosomal polymorphism and the abundance and versatility of these species with respect to the wideness of their distributions. Dobzhansky *et al.* (1950) postulated that the degree of polymorphism in a species is proportional to the variety in the places where it lives, the causal relationship probably being that a species becomes widely distributed because it is variable rather than vice versa.

In Europe *Drosophila funebris* shows both seasonal and geographical trends in the genetic composition of local populations. The survival-rate from eggs to the emergence of adults (using batches of 150 eggs) in populations from Mediterranean regions was low at low temperatures and high at high temperatures. The races from Russia, on the other hand, have high survival-rates at both high and low temperatures. There would seem to be quite obvious advantages in these adaptations for these two regions (Dobzhansky, 1951, p. 150). Doubtless these geographic races are adapted to the regions in which they live in other ways, just as, for example, the geographic races of the beetle *Carabus nemoralis* have different temperature preferences (Krumbiegel, 1932); certain geographic races of *Daphnia atkinsoni* in England have different thermal death-points (Johnson, 1952).

The distribution in North America of the frog *Rana pipiens* extends from Canada to Panama. In contrast to this, the distributions of other species of *Rana* are restricted to a narrow range of latitude. As might have been expected from a species with such a wide range, *R. pipiens* is made up of a number of geographic races (Moore, 1949*a*, 1950). One difference between races is that the eggs differ in the range of temperature which is favorable for their development (Fig. 15.05). In general, those from the south are favored by a higher range of temperature than those from the north (Table 15.05). In the southern forms the eggs are smaller, and they are laid in relatively small diffuse groups instead of in one large compact mass. There is one exception to all this, in that the most southerly species of all is found high in the mountains in Costa Rica, where the prevailing temperature is low. This race resembles those from the colder northern regions in all these particulars.

TABLE 15.05*

NUMBER OF HOURS REQUIRED BY EGGS OF *Rana pipiens* FROM DIFFERENT REGIONS TO REACH "STAGE 20" IN EMBRYONIC DEVELOPMENT

REGION	TEMPERATURE	
	12° C.	28° C.
Vermont	325	50
Louisiana	348	41
Florida (Ocala)	354	42
Florida (Englewood)	364	38
Texas	429	47
Mexico	396	42

* After Moore (1949*a*).

An indication of the magnitude of the genetic differences between the races is the severity of the defects which develop in the F_1 hybrids. The most severe defects observed by Moore occurred when the parents were from Vermont (cold) and Florida (warm); the F_1 zygotes died at an early stage of embryonic development. Minimal defects occurred when the parents came from regions that were close together, that is, from similar latitudes. And, in general, the severity of the defects were found to be positively correlated with the differences in latitude at the homes of the parents (that is, with the north-south component of the distance separating the homes of the parents). Once again the frogs from the mountains in Costa Rica provided the exception. They were least compatible with the race from Florida, which was closest to them in terms of latitude, and most compatible with the race from Vermont, which was farthest away from them in terms of distance north and south. But the temperature at the high altitude in Costa Rica was rather like the temperature at the high latitude of Vermont. So Moore (1950) concluded that natural selection in *Rana pipiens* in North America has produced two genotypes which differ with respect to the temperature at which they thrive. He attributed the abnormalities in the hybrids to the inharmonious association of the two systems of genes. In certain fishes Hart (1952) has demonstrated the existence of geographic races which differ with respect to the "upper lethal temperature" (sec. 6.321).

We mentioned in the preceding section that the hamster *Cricetus cricetus* has black and gray forms determined by a single gene. Gershenson (1945) found that the frequency of the black form was highest (up to 27 per cent) in the "forest-steppe" vegetation. But farther north in the forest and farther south in the open steppe, few, if any, black ones were found. Gershenson considered that the black race is ecologically better adapted to the forest steppe, where it is most abundant.

If we accept Dobzhansky's (1951) definition of races as "Mendelian populations of a species which differ in the frequencies of one or more genetic variants, gene alleles, or chromosomal structure," then the studies discussed in this section and the preceding one are all good examples of species which consist of a number of races. The number of races in the hamster is clear cut; there are only two known at present. But a difficulty arises in deciding how many races are known for *Drosophila pseudoobscura*. The number is, as Dobzhansky (1951) pointed out, purely a matter of convenience. A race is characterized by its predominating frequency of gene arrangements, and whether or not we would wish to specify a local population by a name will depend upon our intentions.

Many species of insects have a resting stage or diapause in their life-cycle which, broadly speaking, determines whether they have one generation a year (uni-voltinism) or more than one generation in a year (multi-voltinism). Diapause in the life-cycle is partly under genetic control, and its incidence and characteristics frequently show a geographic variation. An insect may be uni-voltine by virtue of an obligate diapause which inhibits its development at some stage of the life-cycle. Others without any obligate diapause in the life-cycle may have a number of generations in a year, but the number may be geared to the environment by a facultative diapause. Uni-voltinism and multi-voltinism are both adaptations insuring that the active stages of the insect are present at the most favorable time of the year. It is not surprising, then, to find that they are characteristics which have a geographic variability in some species which are widely distributed.

In the uni-voltine species with an obligate diapause, the genetic control of diapause may be such that no ordinary variation in the environment is sufficient to prevent diapause from happening. In the multi-voltine species diapause occurs in response to the appropriate stimulus from the environment. In this case the variation which may be observed in the incidence and intensity of diapause may be analyzed into two components, one being directly due to variation in the environment and the other to genetic variation in the population.

Evolutionary changes may be at a minimum in an indigenous species living where its ancestors have lived for generations. But an environment may change as a result of ecological succession or evolutionary changes in the plants or animals in the area. Or a species may bridge a geographic barrier and colonize a new region and so become widespread. When these things happen, they are frequently accompanied by (or preceded by) genetic changes in the nature of diapause. Some of the best examples of this come from records of pest species that have been carried across an ocean or a desert in the course of commerce. In such cases selection may begin to operate in a different direction, and evolutionary changes may be rapid. Several examples may be quoted.

The corn borer *Pyrausta nubilalis* is widely distributed in Europe, where well-defined one-generation and two-generation areas may be recognized. For example, in northern France, Germany, Hungary, and parts of Yugoslavia the species is strictly uni-voltine, but in southern France, Italy, and Greece it is multi-voltine (Babcock, 1927; Babcock and Vance, 1929). The voltinism characteristic of a certain area persists when the insects are reared in a different climate, indicating that there are genetic differences in the population from the different areas. There is, however, a close correlation between climate and voltinism (Babcock, 1927), and it may be inferred that the different races are adapted to the climate of the particular areas which they inhabit, the voltinism manifested by each race being the result of selection operating over a long period.

But this is not the case in North America, where the species is a comparative newcomer. The corn borer was found in pest numbers near Boston, Massachusetts, in 1917 (Vance, 1942); in this region it has two generations each year. Later it was reported from New York, where it is strictly uni-voltine.

The climate in these two places does not seem to be sufficiently different to account for this difference in voltinism. There is experimental evidence that the difference is largely due to differences in the genetic constitutions of the two populations: when insects were taken from New York and reared in Massachusetts, they remained uni-voltine during four successive years (Babcock, 1927). In 1921 the corn borer was found in the Great Lakes states, and it has since spread into the Corn Belt of the Middle West of the United States and into the adjoining provinces of Canada. At first, the populations of these areas were strictly uni-voltine, less than 1 per cent having a second generation. More recently, the proportion of the population completing two generations each year has been increasing. In 1936 it was about 12 per cent in Ohio; but by 1944 it had reached 75 per cent in some places (Bottger and Kent, 1931; Ficht, 1936; Vance, 1942; Neiswander, 1947; Wishart, 1947). In 1944, Arbuthnott showed that the population from Massachusetts was homozygous for multi-voltinism, whereas the population from Ohio (Great Lakes states) was heterozygous. From the latter he was able to select a race that was homozygous for uni-voltinism; but selection in the opposite direction failed to yield a race that was homozygous for multi-voltinism (Arbuthnott, 1944). From these observations we may infer that both uni-voltine and multi-voltine races of *Pyrausta* have been introduced into North America (probably on separate occasions); that a multi-voltine race has never been introduced into the Middle West or, alternatively, has not been able to survive there; and that the present population in this area derives from an original uni-voltine race that is at present evolving toward a greater degree of multi-voltinism.

A similar complex of strains has been demonstrated for the spruce sawfly *Gilpinia polytoma* in Canada and Europe (Prebble, 1941; Smith, 1941). The population in the Canadian Gaspé Peninsula is virtually uni-voltine. Prebble found that a few individuals developed without diapause, but efforts to select a multi-voltine strain from these failed. In every instance, after a few generations all the individuals entered diapause. The population in New Brunswick is heterozygous for voltinism. Prebble found that about 15 per cent of the larvae entered diapause, but from the remainder he selected a number of families which were multi-voltine: one of these he reared for 22 successive generations without diapause.

Both in the Gaspé Peninsula and in New Brunswick, *Gilpinia* reproduces by thelyotokous parthenogenesis: males occur in the ratio of about 1 in 1,200. The females are diploid and the males haploid, as is usual in Hymenoptera; the diploid number of chromosomes is 14 (Smith, 1941). In Europe *Gilpinia* is characteristically multi-voltine and reproduces by facultative arrhenotokous parthenogenesis. In this form the diploid chromosome number is 12. It differs also in slight morphological detail from the Canadian form. In some, though perhaps not all, areas in Europe the populations include individuals identical in morphology and chromosome number with the Canadian forms. They reproduce by thelyotokous parthenogenesis and are mostly uni-voltine (Smith, 1941).

Although *Gilpinia* was first recognized in Canada during a mass outbreak of the species in 1940, there is evidence of its having been present for a longer period, probably since the latter part of the last century (Smith, 1941). The original introduction into Canada may have been a uni-voltine 14-chromosome form, which has persisted relatively unchanged (with respect to voltinism) in the Gaspé Peninsula but has evolved toward a greater degree of multi-voltinism in the New Brunswick area. Alternatively, diverse European forms may have been introduced into Canada. Those having 12 chromosomes and little tendency toward diapause have not survived; those with 14 chromosomes and a stronger tendency toward diapause have. Selection has been toward obligate diapause and uni-voltinism in the Gaspé Peninsula and toward facultative diapause and multi-voltinism in the more southern regions.

The cosmopolitan *Ephestia elutella* is a common moth in granaries in England. With the large international trade in grain, the local population is from time to time reinforced by immigrants from regions with warmer climates. In Florida and also in the Caucasus, *Ephestia* is multi-voltine (Reed and Livingstone, 1937; Ustinov, 1932). In London it is virtually uni-voltine; but a few of the eggs which are laid very early give rise to larvae which pupate during the same summer to produce a small second generation of moths (Richards and Waloff, 1946). From this material Basden has succeeded in selecting a race which, within a certain narrowly limited environment, is multi-voltine and may be reared continuously through successive generations (Waloff, 1949). But in nature the progeny of individuals of the second generation of any one year emerge as moths late the next year, when most of the population of moths consists of individuals which are strictly uni-voltine. Their chance of mating with these is greater than of mating among themselves, and consequently there is little chance of multi-voltinism increasing among the population. In any case, the uni-voltine life-cycle appears to be better adapted to this climate, and doubtless selection toward uni-voltinism is going on continuously.

Some of the pioneering work on races in insects was Goldschmidt's (1934) investigation of the variation in intensity of diapause in the eggs of the gipsy moth *Lymantria dispar*. This species is uni-voltine, and its life-cycle includes a diapause in the egg stage which insures that the larvae emerge with the first flush of foliage on trees in spring. The moth has a tremendously wide distribution, occurring practically all over Europe and northern Asia. The very different climates within this large area have resulted in the differentiation of geographic races which vary, among other characteristics, in the intensity of the diapause in the embryo. In regions with a long winter and short spring, like Korea, the eggs hatch quite quickly at favorable temperatures. This enables the larva to be about at the beginning of spring, when fresh young leaves are available. In regions with a mild winter and a warm summer, like the Mediterranean countries and southwestern Japan, the eggs take much longer to develop at favorable temperatures. Thus diapause prevents them from hatching with the first burst of warmth they happen to experience, when, more probably than not, the vegetation would be unsuitable for the larvae. The intensity of diapause in *Lymantria* is determined by a number of genes.

Another uni-voltine insect in which the timing of the life-cycle is regulated by a diapause, the intensity of which varies with the geographic race, is the Australian plague grasshopper *Austroicetes cruciata*. Two geographic races occur, one in Western Australia and the other in the eastern states. They are separated by the Nullarbor desert, which is about 1,000 miles long. The two races have slight morphological differences, and populations may be classified with certainty, but extremes overlap (Key, personal communication). Both races are strictly uni-voltine by virtue of an intense obligate diapause in the egg. They differ, in that the optimal temperature for diapause-development is about 3° C. higher for the western race. This is correlated with the warmer winter in the area occupied by the western race (Andrewartha, 1944*c*). A comparative study of diapause in eggs of Acrididae shows how plastic this character may be. The grasshopper *Melanoplus mexicanus* has a relatively weak diapause in the egg stage, but it is sufficient in the northern populations in the United States to insure a strictly uni-voltine life-cycle. In the south the species becomes multi-voltine. In most of the true locusts, e.g., *Schistocerca gregaria*, diapause is nonexistent, and multi-voltinism is the rule. At the other extreme, certain plague grasshoppers, e.g., *Austroicetes cruciata* and *Camnula pellucida*, exhibit an intense and universal diapause which insures a strict uni-voltinism. These two extremes are linked by species and races which intergrade almost continuously with respect to the intensity of diapause (Andrewartha, 1945). It is clear that during the evolution of the Acrididae the capacity for diapause has been repeatedly modified.

15.12 *Food*

The literature of economic entomology records that certain species of insects have, quite suddenly, over a period of a few years, extended the range of plants on which they feed, to include a species or a variety previously ignored by them. For example, before 1918 the codlin moth *Cydia pomonella* had not been recorded from walnuts in the United States. A few years later Quayle (1926) reported that walnuts were being consistently infested by this pest under circumstances which suggested that this might be due to the appearance of a new race of *Cydia* with a preference for walnuts. The capsid bug *Lygus pabulinus* is a common insect in East Anglia. Before 1917 it was occasionally found feeding on currants, but never on apple. By 1926 it had become a serious pest of currants and was also quite common on apples (Thorpe, 1930). Similarly, the English willow bug *Plesicornus rugicollis* seems to have turned to apple first about 1918, since when it has become a serious pest (Mayr, 1942). In western United States the larvae of the butterfly *Colias philodice* feed mainly on various clovers and a few other legumes. Before 1948 it was rarely found on lucerne, though there was no lack of opportunity. About 1948 in the upper Arkansas Valley the larvae began to be noticed as a serious pest on lucerne (Hovanitz, 1949*a*). Many other examples might be quoted (Thorpe, 1930, 1931).

The first appearance of this new habit on the part of some individuals in the population may not necessarily be determined genetically, although it may eventually lead to the differentiation of races or even species (see below). The ichneumonid *Nemeritis canescens* in nature is a predator of the larva of *Ephestia kühniella.* Thorpe and Jones (1937) reared some *Nemeritis* on the larvae of the wax moth *Meliphora grisella* and then compared their behavior with that of normal adults reared from *Ephestia.* Both groups were attracted by the smell of *Ephestia.* But the experimental ones were also attracted by the smell of larvae of *Meliphora;* this faculty was quite lacking from the controls. Thorpe (1939) also reared *Drosophila melanogaster* on a medium containing peppermint. The controls were repelled by the smell of peppermint, but those reared from this special medium were attracted by it when they were tested in an olfactometer. The larvae of the ermine moth *Hyponomeuta padella* may be found in England feeding on apple, blackthorn, or hawthorn. Thorpe (1930, 1931) offered leaves of apple to larvae collected from hawthorn. They would not eat the apple while the hawthorn was present, but in the absence of the preferred food they could be reared to maturity on the alternative one. Moths reared in this way on hawthorn were then given a choice of apple or hawthorn for oviposition; 79.3 per cent of the eggs were laid on the hawthorn; the controls which had been reared on apple laid 90.3 per cent of their eggs on apple. A similar experiment done with larvae collected in the field from apple gave corresponding results. Thorpe called this phenomenon "host selection" or "olfactory conditioning." Birch (1954) bred the large "strain" of the weevil *Calandra oryzae* in maize and another lot in wheat for about 15 generations. When given a choice of wheat and maize for oviposition, the maize-bred insects laid 28 per cent of their eggs in maize, whereas the wheat-bred insects laid 15 per cent of their eggs in maize. On the other hand, the small "strain" of *C. oryzae*, which had been reared for a similar period in maize, laid only 11 per cent of its eggs in maize when given the same choice; this was little different from the proportion (7 per cent) laid in maize by the controls which had been bred in wheat. Salt (1941) did similar experiments with *Trichogramma*, but in this case the experimental animals which had been reared for 260 generations on unnatural food (eggs of *Sitotroga* instead of *Ephestia*) were no different at the end from controls which had, during this time, been reared on their natural food. Further examples of this sort are quoted by Dethier (1954). So this phenomenon is not universal; but it does provide a possible explanation for the rapid spread of the new habit—a characteristic which was common to all the examples mentioned above—especially as the evidence sometimes seems not to indicate genetic differences, at least in the early stages.

Quayle (1926) studied the codlin moth from apples and walnuts. He did not find any morphological differences in either larvae or adults which would distinguish those from apples from the ones from walnuts. Nor when he offered moths reared in walnuts a choice of apples or walnuts, could he detect any special tendency to oviposit on the walnut in preference to the apple. Basinger and Smith (1946) collected larvae from apples and walnuts from various districts in California. The collections were made just before winter, and the larvae were kept under observation in a lath house at Riverside. There were recorded for each moth the date on which it emerged, how long it lived, and how many eggs it laid. There was no consistent difference with respect to these qualities between moths from the two sorts of food except for one district, Stockton; of those collected at Stockton, the ones from apple emerged earlier and laid fewer eggs than the ones from walnut. These characters are not likely to be under any simple genetic control, and this experiment must be regarded as suggestive but inconclusive.

Races of insects which differ from their progenitors with respect to their requirements for food have been artificially bred on a number of occasions. For example, Pictet (1912) produced a pine-feeding race of *Lasiocampa quercus* by transferring young larvae (which normally feed on oak) to pine needles. At first, most of them died; later, when they began to feed on the tip of the needle (a habit foreign to them on oak), they were able to establish themselves on pine. However, when once this habit was established, a pine-breeding race was formed. Even the second generation showed a preference for pine and were induced to return to oak only with difficulty. Meyer and Meyer (1946) transferred a race of *Chrysopa vulgaris* to a new prey—the coccid *Pseudococcus comstocki*. Here, too, there were high death-rates in the first three generations. Craighead (1921) confined cerambycid beetles on different timbers and found that, after 4–6 years, races were developed which showed a growing preference for the sort of timber in which they had been confined.

It seems likely that in nature there would usually be more interchange between the old food and the new than in these experiments. Perhaps this is why the differences between the codlin moth from apple and walnut seem rather tenuous. On the other hand, Armstrong (1945) has reported a new race of codlin moth from a pear orchard in Ontario. The moths in this orchard mature 3 weeks in advance of those breeding in apples elsewhere in the region. This coincides with the softening of the ripening pears. Hovanitz (1949*a*) considered that the population of *Colias* on lucerne in the upper Arkansas Valley (see above) should be classed as a race, because he had noticed that for years the few larvae that were found on lucerne grew very slowly and the adults were largely sterile. But after 1948 a pronounced change was noticed in both the numbers and the vigor of the caterpillars on the lucerne. In Japan two morphological varieties of the predatory ladybird beetle *Harmonia axyridis* are known to be genetically determined; they have been called *axyridis* and *conspicua*. Komai and Hosino (1951) measured the relative proportions of these two forms collected from pine and from wheat during the years 1948–50 (Table 15.06). It is not known if these differences are due to preferences in the selection of food by the different genotypes on the different plants. But there is no doubt that these are strictly races, as this term was defined at the end of section 15.112.

TABLE 15.06*
RELATIVE ABUNDANCE (EXPRESSED AS PER CENT OF TOTAL POPULATION) OF TWO FORMS OF *Harmonia axyridis* ON PINE AND ON WHEAT

YEAR	FORM *axyridis*		FORM *conspicua*	
	Pine	Wheat	Pine	Wheat
1948	. .	11	. .	. .
1949	20	7	31	40
1950	33	. .	12	62

* After Komai and Hosino (1951).

The phenomenon seems to have been carried a step further in the sibling species of *Rhagoletis*. It was once considered that *R. pomonella* comprised two races which were morphologically indistinguishable except for size and one of which preferred apple and the other blueberry. But Thorpe (1930) showed that the two "races" could be crossed only with great difficulty in the laboratory and concluded that it was unlikely that they ever interbred in nature. They should therefore be regarded as separate species.

We do not know of any field studies which would indicate just how thoroughly "olfactory conditioning" or some other form of "host-selection" might in particular cases serve as an isolating mechanism. Undoubtedly, with those species in which it occurs, there would be a tendency for individuals to place their progeny on the same sort of food as that on which they had grown up. Selection would move in the direction of retaining genotypes adapted to the new food. Recurrent mutations which had no survival-value on the old food might now be advantageous and so be retained. The original conditioned olfactory response may set the direction of selective processes tending to cause genetic divergence (Thorpe, 1945). Whether this sort of mechanism alone would be sufficient to account for the separation of one Mendelian population into two distinct species, as in the case of *Rhagoletis*, is still being argued. We shall mention this matter again in section 16.2.

15.13 *A Place in Which to Live*

It is not so easy to provide contemporary examples of the dominant influence on the formation of races of that component of the environment which we have called "a place in which to live" (chap. 12), but the following example illustrates the point. In Europe the two butterflies known as the Lulworth Skipper and the Glanville Fritillary are not noted for being confined to the coastal regions; yet in England these two butterflies are restricted to the coast. In commenting on this, Ford (1945*a*, p. 150) wrote:

> It is probable that they both require a greater amount of sunlight than our climate normally affords, but this does not account for their strict limitation to the coast and their extreme localisation there. The truth is that species on the edge of their range, such as these, can only survive by adapting themselves closely to the environment which they find in certain places which chance to suit them particularly well. Thus they form local races whose probable physiological modifications may or may not be reflected in their colour-patterns. British specimens of the Lulworth Skipper and of the Glanville Fritillary are not distinguishable from those caught on the continent, but the Swallow-tail, which has also survived here by adapting itself to peculiar conditions, has come to differ visibly from any other race in the process. This butterfly is purely a fen insect in England, though that is not its normal habitat abroad. It is worth noticing that other organisms, for example plants and birds, may be found in abnormal localities near the edge of their range.

The human louse *Pediculus* has two races, the head louse and the body louse, which are so different morphologically that they have received different names. They have other differences, too—for example, the head louse feeds more frequently and is active at lower temperatures and shows preference for laying its eggs on hair as against cloth. Most of the hybrids are fertile, but some are intersexes, indicating a genetic difference between the two races. Each race is adapted to the sort of place where it usually lives, but the differences are still reversible, for the head louse can be transformed to the body louse if it is kept on the body for four generations (Huxley, 1942, p. 305; Mayr, 1942, p. 210). The Eurasian race of the cormorant *Phalacrocorax carbo* always nests in trees, but the Atlantic race nests only on rocks. Likewise, the flesh-footed race of the herring gull *Larus argentatus* in Europe always nests on the coast, but the yellow-footed race inhabits marshy shores of tundra lakes. These and further examples from among birds are given by Mayr (1949*a*, 1951). The field mouse *Peromyscus maniculatus* in the United States has two main morphological types; the long-tailed, long-eared, large-footed type occurs in forests, and the short-tailed, small-eared, small-footed type occurs in grasslands (Blair, 1950).

It is generally assumed that the component of the environment which we call "a place in which to live" must have been of great importance in the evolution of races and species. The evidence is the very diversity of the sorts of places in which animals are adapted to live. Even very closely related species usually show some difference in the sorts of places they live in (sec. 10.311).

Indeed, the progress of evolution itself adds to the number of different sorts of places in which animals can live. The evolution of the flowering plants made possible the development of various wasps and butterflies. These, in turn, made possible the evolution of certain insectivorous birds, and so on.

15.14 *The Numbers of Other Animals of the Same Species*

We have seen in chapter 13 that natural populations of animals are characterized by fluctuations in numbers. There are two aspects to the influence of numbers of individuals of the species on its genetic composition—the influence of small numbers and the influence of crowding. There is little empirical information about the influence of either "underpopulation" or crowding on the genetic composition of natural populations, but the theoretical implications have been quite extensively analyzed by Wright (1931, 1937) and others.

The possible importance of the random loss of genes in small populations in changing the gene frequency is illustrated by the following two quotations from Wright (1931, p. 106): "The constancy of gene frequencies in the absence of selection, mutation or immigration cannot for example be expected to be absolute in populations of limited size. Merely by chance one or the other of the allelomorphs may be expected to increase its frequency in a given generation and in time the proportions may drift a long way from the original values." On page 158 in the same paper: "Finally in a large population, divided and subdivided into partially isolated local races of small size, there is a continually shifting differentiation among the latter (intensified by local differences in selection but occurring under uniform and static conditions) which inevitably brings about an indefinitely continuing, irreversible, adaptive and much more rapid evolution of the species."

In very small populations the random loss of genes may result in the chance fixation of deleterious genotypes. Huxley (1942, p. 201) suggested that some of the cases in which protection of a remnant of a once abundant species has failed to prevent further decline (as in the heath hen in Massachusetts, sec. 9.12) may be due to this cause. According to Mayr (1942, p. 224), about 97 per cent of all species of birds which are known to have become extinct within the last 200 years lived on small, well-isolated islands, whereas not a single species became extinct on large islands, such as Borneo or New Guinea.

A partly hypothetical example is provided by Elton (1930) in his account of the relative frequency of the white and the blue varieties of the arctic fox *Alopex lagopus*. In most places within the distribution of the fox, both blue animals and white animals occur together, but in some localities either all the foxes are white, or all are blue. The numbers of the arctic fox are characterized by large fluctuations, the peaks which occur every 3 or 4 years are followed by a reduction to quite small numbers. Elton estimated that in

30,000 years there would have been about 8,000 such periods of scarcity. If on any of these occasions the numbers were very small, there would be a greater chance that the survivors might all be of one color in any one local population, just as there is a greater chance of getting all heads in a few tosses of a coin, as compared with many tosses. It is thus possible to imagine that even the rare blue fox may be the sole survivor in some localities in times of scarcity. When the opportunity for multiplication recurred, these small, relatively homogeneous local populations would give rise to local populations consisting exclusively of blue foxes.

Huxley (1942, pp. 194–95 and 202) gave a number of examples of what appears to be nonadaptive variation in local populations of species which are subdivided up into numerous discontinuous local populations. In the absence of any selective differences in the local populations which could account for their variability, it is reasonable to suppose, as he does, that it is due to the chance loss of genes in the small groups. It is Wright's (1940) view that the subdivision in this way of a large number of individuals of the species into small local populations which are partially discontinuous may make for rapid evolutionary change. The partial isolation into small groups forms diversity both by random drift in the gene frequency and by selection. And the incompleteness of the discontinuity between the local populations (through migration) means that the variability provided by the diverse local populations is potentially available to the species as a whole.

In the example of the blue and the white foxes given above, it is supposed that the progeny arising from the small local populations will be uniform with respect to color. Ford (1945*a*) gave an example of a butterfly *Euphydryas aurinia* which was reduced to a very small local population with less than normal variability; but, instead of then giving rise to a homogeneous population, the subsequent increase in numbers was accompanied by a great outburst of variability. In his account of this butterfly Ford said:

> The species was quite common in 1881, and gradually increased until by 1894 it had become exceedingly abundant. After 1897, the numbers began to decline, and from 1906 to 1912 it was quite scarce. From 1913 to 1919 it was very rare, so that a few specimens only could be caught each year as a result of long-continued search, where once they were to be seen in thousands. From 1920 to 1926, a very rapid increase took place, so that by 1925 the butterfly had become excessively common, and so it remained until we ceased our observations in 1935. The amount of variation was small during the first period of abundance and while the species was becoming scarce, and it may be said that a constant form existed at this time, from which departures were infrequent. When the numbers rapidly increased, an extraordinary outburst of variation took place so that hardly two specimens were alike, while extreme departures from the normal form alike in colour, pattern, and shape, were common. A high proportion of these were deformed in various ways, and some could hardly fly. When the rapid increase had ceased, such abnormalities practically disappeared and the colony settled down once more to a comparatively uniform type. This, however, was recognizably different from the one to be found during the first period of abundance.

Ford attributed the increase in variation to the reduced selection as conditions improved, selection becoming stringent again when the population once more became dense. The less stringent selection resulted in survival of genetical recombinations which would never see the light of day when circumstances were not so favorable.

Whereas the random loss of genes in small populations would give rise to nonadaptive variability, the selection associated with crowding must be adaptive. For example, if there is not enough food for the whole population, then those which are more efficient in getting food and in utilizing it will have a greater chance of surviving than individuals without these characteristics. Perhaps this sort of thing happens to artificial populations of *Drosophila* which are crowded in cages for the sort of experiments described in section 15.111. We are inclined to think that crowding as such may not have such an important selective influence in natural populations as has sometimes been supposed. The evidence for this has already been considered in other contexts in chapters 13 and 14, in which analyses of populations of a number of different species indicated that populations were constantly being reduced to low numbers as a result of the onset of unfavorable weather or some other circumstance, rather than because of shortage of food or space in which to live.

15.15 *The Numbers of Other Individuals of Different Species*

A quotation from Darwin strikes the keynote for this section. He was writing about the differences between species on the different islands of the Galapagos Archipelago:

> But how is it that many of the immigrants have been differently modified, though only in small degree, in islands situated within sight of each other, having the same geological nature, the same height, climate, etc.? This long appeared to me a great difficulty; but it arises in chief part from the deeply-seated error of considering the physical conditions of a country as the most important; whereas it cannot be disputed that the nature of the other species with which it has to compete, is at least as important, and generally a far more important element of success . . . when in former times an immigrant first settled on one of the islands, or when it subsequently spread from one to another, it would undoubtedly be exposed to different conditions in the different islands, for it would have to compete with a different set of organisms; a plant for instance, would find the ground best fitted for it occupied by somewhat different species in the different islands, and would be exposed to the attacks of somewhat different enemies. If then it varied, natural selection would probably favour different varieties in the different islands.

15.151 NONPREDATORS

In a population cage containing *Drosophila melanogaster* and *D. simulans* the individuals of each species enter the environment of the other as nonpredators which have similar requirements for food and space. When these two species live together at 25° C. in population cages, it is usual for *D. simulans* to become the less abundant species, until it eventually becomes extinct. Moore (1952*b*) found this to happen in about 100 days in 19 out of 20 cages. But in the twentieth cage a race of *D. simulans* developed which had a higher survival-rate than *D. simulans* in the other cages. The first indication of this was observed on the 74th day, when the proportion of *D. simulans* began to rise instead of fall (as in the other cages). This continued and on the 229th day some *D. simulans* were removed, and their progeny were tested against a sample of normal *D. melanogaster*. It now proved to be much more successful than the initial stock of *D. simulans*, for it continued to persist in the population cages together with *D. melanogaster* for as long as the experiments were continued. There is no doubt that the new race of *D. simulans* was superior to that which had not been selected. It was changed probably as a result of having been crowded with *D. melanogaster*. That it may have been due to the crowding by its own species is not altogether ruled out as a possibility. But the experiments provide no means of testing these alternatives. Assuming that the former alternative is true, these experiments provide an example of a nonpredator which has similar requirements for food and space influencing the genetical composition of another species. This sort of thing might also happen in natural populations, but there is as yet little trustworthy evidence for it. We are not convinced that ecological differences so commonly found among related sympatric species is proof of the efficacy of "competition" in nature, popular belief notwithstanding. Since the evidence for this statement has been given in section 10.311, it will not be repeated here. On the other hand, nonpredators of a different sort, that is to say, those which enter the environment of an animal in a capacity other than to utilize the same food and space, may be of prime importance in providing new opportunities for variation; for example, the evolution of the flowering plants made possible the development of certain wasps and butterflies, which, in turn, made possible the evolution of certain insectivorous birds (sec. 15.13). The general principle of this section may also be applied to the effect of one genotype on the adaptive value of another genotype of the same species (see experiment by Levene *et al.*, 1954, sec. 15.111).

15.152 PREDATORS

Predators may be selective feeders on certain variants within the species which are easier to find because of protective coloration. The American clouded yellow butterfly *Colias philodice* has larvae which are normally green; but there is also a blue-green variant, and this color is inherited as a simple recessive. Gerould (1921) reared the larvae of both the green and the blue variants out-of-doors and found that all the blue-green larvae were eaten by sparrows, while the normal greens escaped, evidently by virtue of their protective coloration. Ford (1945*a*) quoted some experiments of Poulton and Saunders, who exposed the pupae of the small tortoiseshell butterfly *Aglais urticae* in various situations, on banks, on a fence, on walls, and on plants. Table 15.07 shows that fewer survived on fences. Here again the deaths were presumed to be due to predators which found the pupae more readily when they were on fences than in other places. This is an interesting example, as it shows how the presence of different sorts of places where the animals may live may modify the influence exerted by predators on the genetic composition of the population (sec. 10.32).

TABLE 15.07*

SURVIVAL-RATES OF PUPAE OF THE BUTTERFLY *Aglais urticae* IN DIFFERENT PLACES

Place	No. of Pupae	Per Cent of Pupae Which Survived
Banks	219	38
Fences	98	8
Walls	26	12
Nettles	35	15

* After Ford (1945*a*).

A more complex example of the same principle is provided by the snail *Cepaea nemoralis*, which has a number of genetical variants differing in color and banding pattern. Cain and Sheppard (1950) showed that fewer of the yellow variety, relative to the brown and the pink, were taken by thrushes, the greener the background. Sheppard (1951*a*) counted the number of broken shells of the yellow, brown, and pink varieties found around the "anvil stones" of thrushes and compared these with the numbers found in random collections made in the same area that the thrushes hunted over. In the collections made from the anvil stones, the yellow variety dropped from 43 per cent at the beginning of April to 14 per cent at the end of May (Fig. 15.06); but in the random collections the proportion remained nearly constant at 24 and 28 per cent, respectively. So the birds were collecting relatively fewer yellow ones as the season advanced, despite the fact that there was no change in the relative abundance of the different varieties in the area where the birds were hunting. This was attributed to the fact that the background had become greener as the season advanced. And Sheppard (1951*a*) reached the general conclusion: "It is certain that the selective value will alter markedly from season to season, year to year and locality to locality, as the result of slight changes in the background colour in different seasons and years as well as differences in the intensity of predation and other changes in the ecology of the colony."

In some industrial areas in Britain and Europe, almost the whole population of certain moths have become black during the last 80 years. In some cases the black variants have been shown to differ from the normals in a single or a few genes. Ford (1945*b*) listed seven species of geometrids and one thyridid moth which have been studied quite extensively; others have been less well studied. All of them exhibit cryptic coloring in the nonindustrial areas, so that when they rest on trees, walls, and fences they merge with the background. It has been suggested that the black forms ha e been selected out in industrial areas because such insects match their blackened surroundings better than those with normal coloring. But this explanation is too facile. Ford (1945*b*) pointed out that there are few predators of the moth in industrial areas. And, second, this theory fails to account for the additional fact that the black forms

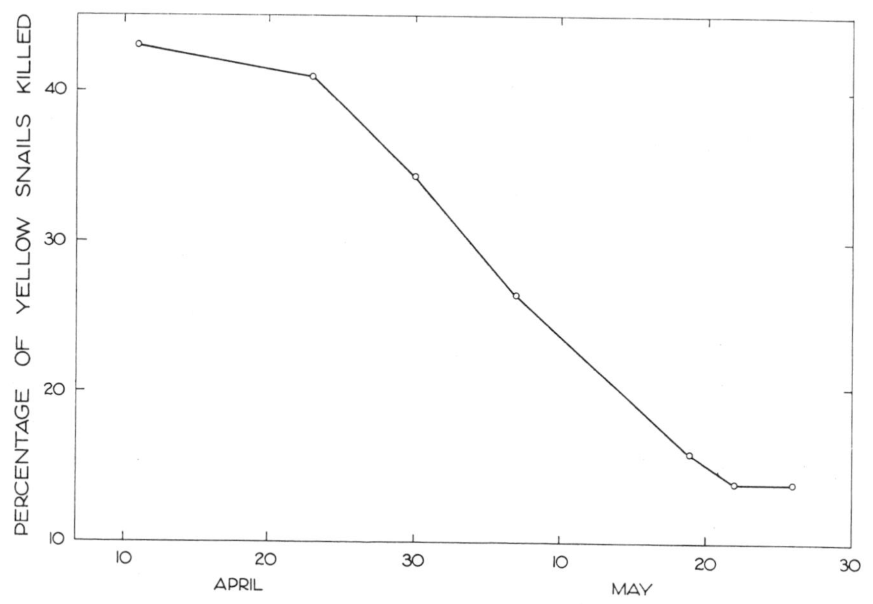

FIG. 15.06.—The seasonal change in the proportion of yellow snails (*Cepaea nemoralis*) killed by thrushes in Marley Wood. (After Sheppard, 1951*a*.)

are hardier than the normals, at least in some species, yet they do not supplant the normals in nonindustrial areas. Ford proposed an alternative hypothesis that the black form in industrial areas is not there because blackness protects it from its enemies but because it is more viable, its capacity for increase is greater. The gene or genes associated with increased viability are evidently also associated with increased melanin production. They doubtless also arise in the nonindustrial areas, but there they would be very conspicuous, and it is presumed they would be ready prey for predators. The lack of cryptic coloring is no disadvantage in industrial areas, where predation is scarcely a factor. The black form has the advantage over the normal form there simply because it is hardier.

Dice (1947) kept different strains of the mouse *Peromyscus maniculatus* which varied in color in cages with soil which also varied in color. Two species of owls were used as predators. The mice which matched the soil background were preyed upon less than those which were more conspicuous. In every experiment in which the predator was evidently using signs to capture the prey, the concealingly colored mice had more than a 20 per cent advantage over the conspicuous mice in escaping capture.

The influence of disease organisms is similar to the influence of predators, for it results in the selection of immune variants. These correspond to the protectively colored prey in the foregoing examples, where the prey were "immune" by virtue of their coloration.

15.16 *Other Components of Environment—Insecticides*

Melander (1914) was the first person to call attention to evidence that an insect of economic importance was becoming increasingly difficult to kill with insecticides. In certain localities in California the San Jose scale *Aspidiotus perniciosus* had become more difficult to control with sprays of lime-sulphur. Melander suggested that the scale had become more resistant to the insecticide. This suggestion met with a good deal of skepticism at the time, though it has never been disproved by any subsequent investigation (Babers, 1949). However, a second example of apparent resistance to an insecticide which followed closely on this one captured the attention of entomologists; careful investigations soon substantiated it fully. This was Quayle's claim in 1916 that the red scale *Aonidiella aurantii* on citrus at Corona, California, had become resistant to cyanide. Smith (1941) gathered together this and other evidence which had accumulated since 1916 and showed quite clearly that a number of insects had developed resistance to insecticides. Besides the red scale, both the black scale *Saissetia oleae* and the citricola scale *Coccus pseudomagnoliarum* had become harder to kill than they once were.

The resistant form of the red scale has persisted at Corona since about 1914 and has since become widespread in California; but there are still areas where the standard fumigation continues to be satisfactory. The results from commercial fumigations have been confirmed by experimental fumigations in the laboratory; the resistant insects from Corona survived greater concentrations and longer exposures than did the nonresistant ones from certain other areas (Quayle, 1938). Dickson (1941) showed that resistance was determined by a single gene on the X chromosome. Although this is a very small genetic difference, the physiological and ecological differences between the two forms are of such importance and so distinct that they are usually referred to as separate races.

The resistant race of red scale is not present in the San Gabriel Valley, although fumigation has been practiced there for a longer period than in any other part of California. Smith (1941) put forward a tentative explanation, which may be summarized as follows: There is experimental evidence that, in the absence of cyanide, the nonresistant race has a higher capacity for increase than the resistant race. If this were not so, it would be difficult to explain why the nonresistant race persists anywhere. It is possible to obtain resistant individuals by collecting in "nonresistant" areas, though clearly they are not abundant there. Whether one race or another becomes predominant in an area must depend upon their relative rates of increase between fumigations and their survival-rates during fumigation. In an area where the resistant race is numerous, fumigation may be practiced annually or even more frequently, while in an area where the nonresistant race is the more numerous, each fumigation may be separated by several years. Evidently, infrequent fumigations enable the nonresistant race to increase to such proportions between fumigations that it becomes the dominant race; but with frequent fumigations, its numbers are kept low relative to the race which breeds more slowly but has a better chance of being alive after a fumigation. This was put forward by Smith as a particular explanation for a specific phenomenon, but the reader will recognize it as an example of the general theory of distribution and abundance which we discuss in chapter 14.

Both genotypes are present in resistant and nonresistant populations, but they occur with different frequencies. Selection accounts for this. Dobzhansky (1951) pointed out that the question as to whether the gene for resistance in the Corona population arose by mutation or was introduced from elsewhere is largely an academic one. The former possibility is, on the whole, more probable, for populations of the red scale are so large that even if the mutation-rate of the gene for resistance were very low, mutant individuals would still be present in the citrus-growing areas at any time.

Resistance to insecticides is known for other groups besides scale insects. The larvae of a particular race of codlin moth *Cydia pomonella* from Colorado can enter apples sprayed with arsenic more successfully than can the "normal" codlin moth larvae. The Colorado race also maintained its resistance when the arsenic was replaced by certain other poisons (Hough, 1928, 1929, 1934). Steiner *et al.* (1944) found a difference in resistance to arsenic in the larvae of codlin moth from two orchards in Ohio which were only 37 miles apart. One of the orchards had been sprayed heavily with arsenic for the preceding 5 years. The larvae from it were much more resistant to arsenic than were those from the other orchard, which had not been sprayed at all during that time. Boyce (1928) developed races of *Drosophila melanogaster* and *Aphis gossypii* in the laboratory which were resistant to cyanide. Races of the house fly *Musca domestica* have been found in the field and produced in the laboratory which are resistant to DDT. Indeed, some races of the house fly which have been developed in the laboratory are resistant to almost every insecticide (Babers, 1949; Babers and Pratt, 1951). These authors reported about fifteen species of insects and one species of mite which have races resistant to insecticides. There seems to be every reason to suppose that this contest between insects and men, the one producing resistant races while the other produces new insecticides, will continue, as it has with bacteria resistant to antibiotics and with fungi which are capable of attacking hitherto resistant plants. These are all striking examples of genetic plasticity leading to evolutionary changes in the species as its circumstances of life change.

15.2 GENETIC PLASTICITY AND ECOLOGICAL ADAPTABILITY

The persistent species are those which retain genetic plasticity, enabling them to take advantage of "ecological opportunities" which are presented from time to time. It may seem that such a species should be less abundant than the more uniform one which is "perfectly adapted" to the circumstances of the moment, but such perfect adaptation may carry the risk of early extinction. The genetically plastic species retains in its population a reserve of genes, many of which are lethal or semilethal in certain circumstances, though they may confer adaptive advantages in other circumstances (Dobzhansky, 1951; Wright, 1949; Wallace and Madden, 1953). Dobzhansky (1947*b*) wrote: "The mutation process furnishes the raw materials without which adaptive changes cannot be constructed, but the same process also unavoidably produces a multitude of poorly adapted variants. Restriction of the supply of heritable material might permit a species to reach a higher level of immediate fitness, but it jeopardizes its adaptability to changing environments."

These principles are illustrated in Dobzhansky's studies on *Drosophila pseudoobscura*. This species has a high degree of genetic plasticity, which enables it to occupy diverse geographic regions and to remain relatively abundant throughout the tremendous seasonal changes which occur in the western part of the United States (sec. 15.111). But this plasticity entails a continuous production of relatively ill-adapted "inversion homozygotes." For example, the constitution CH/CH is a semilethal at temperatures above 20° C. In other circumstances it might confer an advantage on its possessor. There is an advantage in retaining it in the population to meet this exigency. It is, in fact, retained as a result of heterosis or superiority of the heterozygote. Dobzhansky (1951) pointed out that we could imagine a population of *D. pseudoobscura* in which carriers of ST chromosomes would be favored during the hot period of the year, while carriers of AR chromosomes would be superior in the winter. An exceptionally hot summer could result in elimination of all AR chromosomes. A population in which this happened would be placed at a disadvantage during the winter. No such "accidents" can happen if the heterozygote is superior to either homozygote. In that case, natural selection does not eliminate either ST or AR chromosomes, but it moves the frequencies of these types in the population toward a particular equilibrium, depending on the circumstances of the moment. Evidence for the superiority of the heterozygote in other species of *Drosophila* is given by Dobzhansky and Wallace (1953).

Heterosis in *Drosophila* makes it possible for a species to respond rapidly to variations in its circumstances of life. Dobzhansky pointed out that these may vary not only with the seasons and with great distances but also with quite small distances, associated with variation in climate, food, etc., in the places where local populations live in the one geographic region.

In *Drosophila* chromosomal polymorphism has been shown to be an indication of genetic plasticity. This was illustrated in section 15.112 by reference to *D. willistoni* and *D. tropicalis*. The former, having 40 known inversions in its chromosomes, is widely distributed in such a way that its local populations must experience wide extremes of climate, food, etc. The latter, having four known chromosomal inversions, has a more restricted distribution.

An important aspect of adaptability is the speed with which a favored variant may become dominant in the population. This happens with great rapidity in the seasonal changes of the chromosomal variants of *D. pseudoobscura*, indicating that the selective advantage of the favorable variants is great. But it is not necessary for a favored variant to have a large selective advantage if it is to spread in the population. On the contrary, the selective advantage which enables a variant to spread through a population may be so small that it cannot be measured by any experiment. Ford (1945*a*, p. 276) gave a hypothetical example, which we quote:

> Consider two forms of a species, A and B, controlled by a pair of allelomorphs, and let us suppose that by reason of its warning colour and slightly distasteful qualities B has a 1 per cent. advantage over A. By this I mean that for every 100 specimens of B which live to reproduce, only 99 of A do so. On the other hand, this would confer on B what is, from the point of view of evolution, a very heavy advantage. The rate at which it would spread would not be the same when it was common as when it was rare. However B would require about 160 generations to increase from 1 per cent. to 5 per cent. of the population. Evolutionary changes are rarely of so great a magnitude as this. Now consider such a situation from the point of view of experimental study. A difference of this kind might perhaps just be detected by an elaborate and carefully-planned laboratory technique, using very large numbers over a long period of time. To suppose that it could be perceived even by exhaustive observation in the field is quite ridiculous. [See also Allee *et al.*, 1949, pp. 647 and 654.]

Adaptability is also a function of the size of the population and its structure. In section 15.14 we discuss Wright's hypothetical demonstration of the way in which genetic plasticity may be reduced in small populations and how it may be retained in a large population which is subdivided into small local groups.

Essentially, this chapter has dealt with natural selection in local populations. It has been shown that the variations in temperature, moisture, food, predators, etc., from season to season or place to place may be reflected in the frequencies of the different genes in the local populations experiencing these changes. Sometimes, as with the seasons, the variations in temperature, food, etc., may be rhythmical fluctuations about means which remain constant; but sometimes they may constitute secular trends. Then we catch a glimpse of evolution. The species that cannot change with the times become extinct; no species can change quite fast enough to keep up with the times. Wright (1948) described the whole process as one of incessant change; the species struggles to hold its own, while its circumstances of life continue to deteriorate. The species is like an army which must stabilize its position in the face of continuous fire but which can also take the opportunity, when it arises, of advancing into new territory. This carries a warning for the ecologist who might be tempted to take seriously theoretical models of populations depending on oversimplified abstractions of static environments and static animals. And it emphasizes the need for ecological theory to allow, as we have sought to do in the general theory of chapter 14, for the incessant change which is one of nature's most essential features.

CHAPTER 16

Ecology and the Origin of Species

Why, if species have descended from other species by insensibly fine gradations, do we not everywhere see innumerable transition forms? Why is not all nature in confusion, instead of the species being, as we see them, well defined?

DARWIN, *The Origin of Species*

If a species is as plastic as has been depicted, it is perhaps even more puzzling how it can split into several species.

MAYR (1949*b*, p. 517)

Barriers of any kind, or obstacles to free migration, are related in a close and important manner to the differences between the productions of the New and Old Worlds, excepting in the northern parts, where the land almost joins, and where, under a slightly different climate, there might have been free migration for the northern temperate forms, as there now is for the strictly arctic productions. We see the same fact in the great difference between the inhabitants of Australia, Africa, and South America under the same latitude; for these countries are almost as much isolated from each other as is possible. On each continent, also, we see the same fact; for on the opposite sides of lofty and continuous mountain ranges, of great deserts and even of large rivers, we find different productions; . . . .

DARWIN, *The Origin of Species*

16.0 INTRODUCTION

DARWIN remarked in chapter 13 of *The Origin of Species:* "There is a striking parallelism in the laws of life throughout time and space; the laws governing the succession of forms in past times being nearly the same with those governing at the present time the differences in different areas." The laws governing the succession of forms in past times may be discovered by studying fossils, especially of such groups as ammonites, mollusks, echinoderms, and horses. It is quite clear that populations living continuously in the same place changed (probably imperceptibly) from generation to generation, until, after the passage of long periods of time, the descendants came to look very different from the ancestral forms from which they sprang. The morphological differences may be so great that we may be reasonably sure that there would be, associated with them, isolating mechanisms which would prevent interbreeding, if one could, by some necromancy, resurrect the ancestral forms and allow them to mingle with the modern ones. In this sense, then, we refer to those two Mendelian populations, the one derived by direct succession from the other, as two distinct species; and this evolutionary process may be called "the transformation of species in time." This is illustrated in Figure 16.01, *A*. Suppose an island was colonized long ago by a species *a*. Suppose climate (or some other component of environment) manifested a secular trend; this might be reflected in corresponding trends in the frequencies of particular genes in this population. After many generations the accumulated genetic changes make it necessary to recognize a second species *b*, which, in due course, trans-

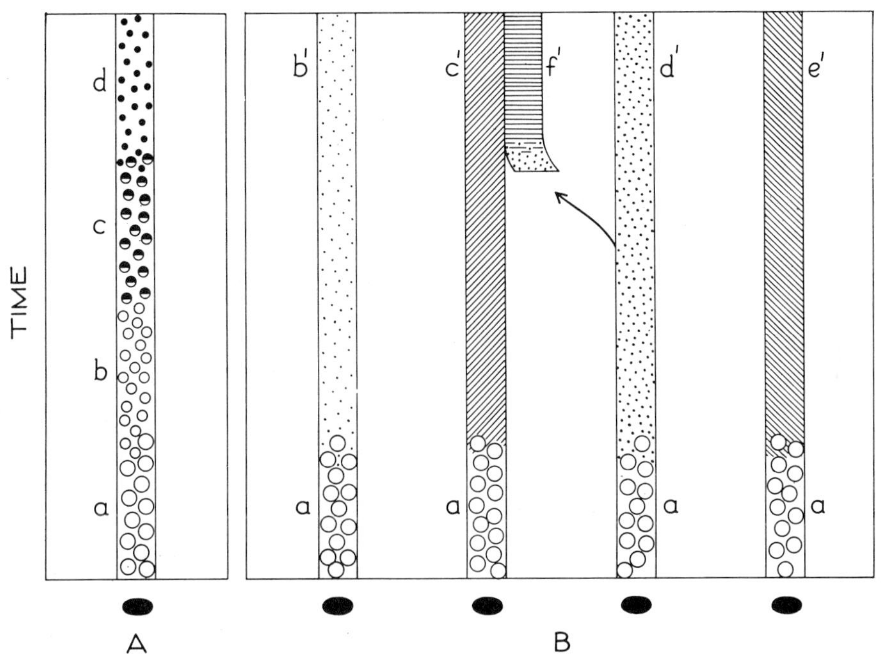

FIG. 16.01.—*A*, transformation of species *a* in time on a single island. *B*, multiplication of species *a* in space on four islands, widely separated in a large ocean, into 5 species *b'*, *c'*, *d'*, *e'*, and *f'*. (Modified from Mayr, 1949*a*.)

forms to *c* and then to *d*. Through the ages one species has given rise to four, though at any one time there has never been more than one species on the island; each was separated from the others only by time, and each was linked to the others by a continuous succession of generations. This aspect of the origin of species does not enter into ecology, and we shall say no more about it. The other aspect of the origin of species may, for contrast, be called "the multiplication of species in space."

This is illustrated in Figure 16.01, *B*. Suppose there are four islands in a large ocean, each one experiencing a different climate and being clothed in a different vegetation. Species *a* colonized each of these islands. Because of the differences in climate and vegetation, the trends in the frequencies of particular genes were different on each island. Eventually the differences in genetic constitution became so great that interbreeding became impossible. Then we would recognize four distinct species, b', c', d' and e', each living on a separate island but at the same time. If, after this stage, one of species were to colonize one of the other islands the two species now on this island would remain distinct but the immigrant species might evolve into still another species f'. It is now generally recognized by geneticists that such complete isolation, sustained for a long period, usually leads to the development of isolating mechanisms, preventing interbreeding, and therefore to the origin of new species. Dobzhansky (1951) mentioned several isolating mechanisms that have been observed. Either the individuals of different species do not mate; or, if they do, fertilization does not occur; or fertilization may occur, but the progeny either fail to develop or are sterile if they do reach maturity.

It is generally agreed that races are incipient species; not all, by any means, will eventually become species; but all have this potentiality. So the problem of the multiplication of species in space is resolved into the problem of the origin of isolating mechanisms which may prevent interbreeding between parts of a Mendelian population which was previously one. We need not repeat the extensive genetic arguments on this subject; the reader is referred to discussions by Mayr (1942, pp. 190 ff., 1949*a*, p. 285), Wright (1949), and Dobzhansky (1949*b*, 1951, chap. 7). The general conclusion is that, except for polyploidy, all isolating mechanisms are controlled by many genes and are of such genetic complexity that they would be most unlikely to develop or be perpetuated unless the two populations had been, for a very long time, isolated from each other so thoroughly that virtually no interbreeding had occurred between them. It would be impertinent for us as ecologists to add to the extensive literature on this subject, were it not that some of the unsolved problems would seem to require an ecological approach. For example, what degree of isolation (or inhibition of the exchange of genes) is required before local populations become distinct species? And in nature what are the most likely causes of such isolation? These two questions are conveniently considered together; quite a lot is known about the latter, but very little about the former.

16.1 GEOGRAPHIC BARRIERS

In the hypothetical example represented by Figure 16.01, *B*, we considered only that the islands were separated by ocean and ignored the complications introduced by the chance that species *a*, having that capacity for dispersal which sufficed to find these four islands in the first place, may continue to disperse back and forth between them. In nature the chance that a particular barrier will be adequate to isolate two local populations from each other must always be considered in relation to the capacity for dispersal shown by the members of each population. Since the habit of dispersal differs so vastly between species, we find that, whereas a mere hedgerow or a plowed field suffices as a barrier for one species, many miles of ocean may be insufficient with another. Dispersal is discussed in section 16.11.

Since a barrier may be constituted by an area which is devoid of food or in which the temperature is harsh; since the chance of a barrier's being surmounted may depend in part on the size of the local populations concerned; since the distribution of local populations with respect to each other may depend on the distribution of food and suitable places in which to live—for these reasons and others like them, it is instructive to consider geographic barriers in relation to the components of environment as these were outlined in chapter 2 and elaborated in the other chapters of Part II. This we do in section 16.12.

16.11 *Dispersal*

Mayr (1942, p. 211) wrote: "It may be said, as a broad generalization, that the more sedentary a species of animal is, the more it will tend to differentiate into geographic races. Conversely, it should be true that the more easily the individuals of a species are dispersed, the less diversification into geographic races takes place." As examples of Mayr's converse proposition, there are many species of Protozoa, Nematoda, and Tardigrada which are world-wide in their distributions, being dispersed as resistant cysts by winds and currents of water. In these groups cosmopolitan species are more numerous than is often recognized. The white cabbage butterfly *Pieris rapae* is migratory, and its populations do not differ in different parts of Great Britain. There are many butterflies which, like this one, have a high capacity for dispersal and whose populations are relatively uniform over wide distributions.

As examples of Mayr's major generalization, there is a number of butterflies with poor capacity for dispersal which tend to split up into local populations, each with its own distinctive characters (Ford, 1945*a*, p. 280). For example, a colony of the heath fritillary *Melitaea attalia*, which was deliberately introduced into Essex within the present century, in now notably smaller and darker than the Kent form, from which it was derived and from which it is now isolated, because of the relatively low innate capacity for dispersal. Some butterflies seem to be quite extraordinarily local in their flights. Sheppard (1951*b*) marked a number of moths in two colonies of *Panaxia dominula* near Oxford. The two colonies were separated by more than a mile of agricultural land; yet during a period of about 4 years in which the moths were marked, there was no exchange of individuals between the two colonies.

On Tean, one of the Scilly Islands, the butterfly *Maniola jurtina* lives in areas where rocky ground is covered with a thick vegetation of bracken, gorse, and long grass. Dowdeswell and Ford (1952) studied three local populations in such areas which were separated by narrow necks of sandy soil carrying a short turf and a few low shrubs. The neck separating one pair of colonies was 150 yards long; the one separating the other pair was 220 yards long. In each colony butterflies were marked with cellulose paint and released; very few left one colony to join another. The island of St. Martins is separated from Tean by 300 yards at its closest point. About 50 butterflies were marked and released on St. Martins, but none of them was recaptured on Tean during the next 10 days, when extensive collections were made (Dowdeswell, Fisher, and Ford, 1949). It seems that local populations of this species may be effectively isolated by a few hundred yards of ocean or of land on which the vegetation is unsuitable.

Isolation in the field mouse *Peromyscus* is enhanced by low capacity for dispersal. Even where *P. maniculatus* ranges most widely in the mesquite association of northern Mexico, the average home range of an individual mouse is less than 5 acres. Most of its dispersal is due to the young animals searching for homes, but the greatest distance they travel is about 3,000 feet (Blair, 1950). The species of land snails which inhabit the valleys of islands of Hawaii and Tahiti are classic examples of extreme local diversity associated with low capacity for dispersal (Dobzhansky, 1951, p. 170). Dispersal is so slight that the snails are evidently unable to surmount the obstacle offered by the ridge between valleys; and there are obvious morphological differences in the snails even from adjacent valleys. Sheppard (1952) counted the frequency of color types of the snail *Cepaea nemoralis* in several localities only a few hundred yards apart. The snails in beech woods had a lower frequency of yellows than did those in downland areas. The same relationship was found in another area, a mile away from the first. Evidently, the breeding colonies in quite small parts of a continuously inhabited area have diverged from one another as a result of natural selection reinforced by low powers of dispersal.

Even closely related species of birds differ in a surprising way in their capacity for dispersal. Mayr (1942, pp. 227, 239) contrasted the Australian silvereye *Zosterops lateralis*, which invaded New Zealand after jumping an ocean gap of 2,000 km., with the closely related *Zosterops rendovae* of the central Solomon Islands. Strikingly distinct races of this species are restricted to islands not more than 2–6 km. apart. A flight of a few minutes would carry the birds from one island to the next, but it is characteristic of the species that such flights do not occur, or, if they do, it is so rarely that they are of little importance in dispersal. Rensch compared the amount of subspeciation in Palearctic birds and found that the number of subspecies per species was about twice as high in sedentary species as in migratory ones (Mayr, 1942). According to Huxley (1942, p. 242), large, strong swimmers and migratory species among fish in American rivers show greater uniformity of characters than do smaller fish with a low capacity for dispersal; these tend to show marked local diversity.

The fauna of the oceans shows what happens when animals with high capacity for dispersal live in the absence of frequent or substantial barriers to dispersal. We quote from Mayr (1942, p. 234): "Of the known species of animals about 93 per cent. live on land and only about 7 per cent. in the sea. The marine fauna of mollusks, crustaceans, or annelids of a given locality is, in most cases, infinitely larger than the corresponding land fauna, but the marine fauna has a very wide distribution (sometimes covering all oceans or at least the entire Pacific), while most of the terrestrial species have a very limited range." The great disparity between the numbers of terrestrial and marine species is probably best explained in terms of the relative lack, in the oceans, of features which would correspond to climatic and topographic barriers on land, the large numbers in marine populations, and the high capacity for dispersal in most marine species.

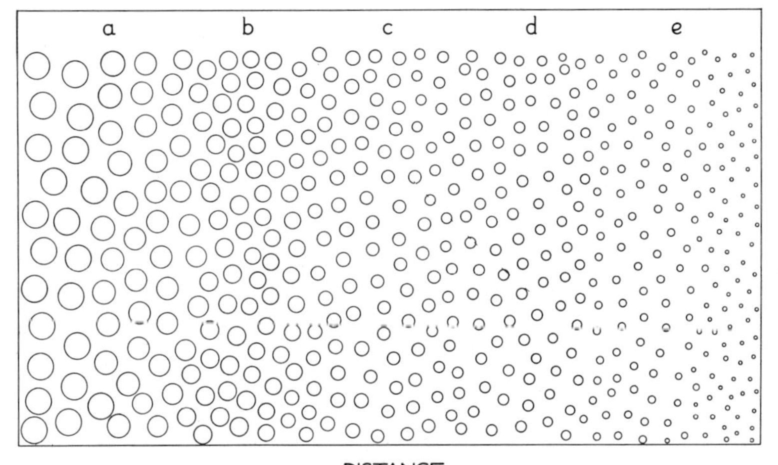

FIG. 16.02.—Diagrammatic representation of a cline within a species. The different stippling represents different gene frequencies in the population across a stretch of country from population *a* to population *e*.

With species of relatively low capacity for dispersal, distance alone may so reduce gene flow as to give rise to a gradient from one end of the distribution of the species to the other. But there is no evidence that distance alone is a sufficient barrier to gene flow to permit of speciation. Rather it seems to give rise to clines within a species. Where the cline extends over a great distance, the local populations at its two extremities may become so diverse that they may be unable to interbreed when brought together. But, in so far as neighboring populations can interbreed, the extreme populations may exchange genes via the intervening populations, as in *Rana pipiens* in North America (sec. 15.112). Figure 16.02 illustrates this diagrammatically. Imagine a large area to be colonized at one end by a single species *a*, which gradually invades the whole area. We may suppose that one end of the area is in a warm climate and the other in a cold climate. There will be a gradient of climate between the two ends of the area, and this will eventually be reflected in a gradient in the frequencies of genes (and the characters associated with them) in local populations. Figure 16.02 represents this by a change in stippling across the area. The populations at each end of the range are quite distinct, though they are connected by a continuous gradation in characters from type *a* through *b* and *c* to *e*. Through these populations, genes may pass from *a* to *e*, though individuals from these two races may not interbreed when they are brought together in the laboratory. The obliteration of populations *b* and *c* would constitute *a* and *e* distinct species, for they would then be completely isolated with respect to the interchange of genes. The division into two species might also be achieved if the gene flow across the distribution were reduced to a certain low rate by an effective barrier such as is illustrated in Figure 16.03. Just how

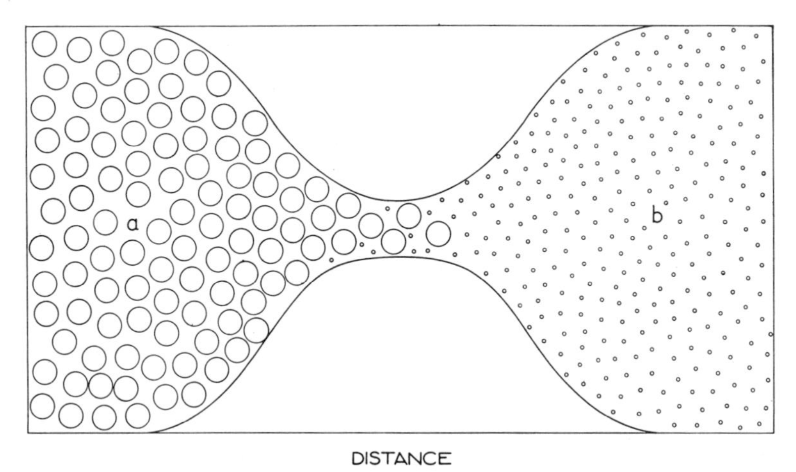

FIG. 16.03.—Diagrammatic representation of incomplete isolation between two species *a* and *b* with small gene flow between them. (Modified from Mayr, 1949*a*.)

complete this barrier has to be, is largely a matter of conjecture in the present state of our knowledge. Perhaps many species could be represented by Figure 16.03; and, as Mayr (1949*a*, p. 286) said, "isolation between populations of a species is rarely complete. Even when there is apparently an absolute barrier, like an ocean strait, some individuals apparently get carried across the gap at irregular intervals."

The question How great must a barrier be in order to insure that a race ultimately becomes a species? clearly cannot be answered in absolute physical units like "so many miles of desert" or "such an expanse of ocean." The answer must take the form, "great enough to insure that less than such a proportion of the population crosses it to breed with those on the other side." Scarcely anything is known about this. We might expect the proportion to vary greatly with the species and the circumstances in which it lives. If, in a certain area, where there lived a local population which had already become a race, pressure of selection against immigrants from another local population was high (i.e., the death-rates among the immigrants and perhaps their hybrids were much higher than among the indigenous forms), then we might expect speciation to proceed in this situation against a higher flow of migration than in another situation where this was not the case (Mayr, 1949*a*, p. 290).

One finds in the literature a tendency to speak of "habitat isolation" or "ecological isolation" when the particular geographic barrier does not include great distances. This usage is confusing and should be abandoned. In a particular species of butterfly a few acres of farmland may serve adequately to keep two local populations apart. In another butterfly it may happen that the barrier takes the form of several hundred miles of desert. In either case they are spatially isolated, the one differing from the other in the distance between the populations and the nature of the intervening terrain. Any attempt to make a clear-cut distinction between "geographic isolation" and "habitat isolation" is, by the nature of the case, quite arbitrary. We agree with Mayr (1947) that to call one sort spatial or "geographic isolation" and the other "habitat isolation" is to suggest that two different processes are involved, which is not so. Mayr (1949*a*, p. 291) pointed out that the usage has arisen from a difference in emphasis: "If Speyer (1938) finds that a valley population of a moth hatches on the average one month earlier than the populations on the neighbouring hills, one may emphasize either the existing difference and the ecological factors that favoured selection for the difference, or one may emphasize the geographical segregation of the two localities which limits gene flow between them. The important fact is that two different populations are involved, each with a locally superior gene combination. Spatial segregation will prevent the destruction of these adaptive gene combinations through gene flow."

It is meaningless to contrast geographic races with ecological races. Mayr (1951) rightly says: "There is no ecological race that is not spatially isolated and no geographic race that does not show certain ecological differences." Nor should we confuse this issue further (as has often been done) with another which is quite distinct, namely, whether a part of a population of a species, by changing its behavior, can effectively isolate itself from the rest of the population. We discuss this problem in section 16.2.

16.12 *Environment*

A glance at Figure 11.01 will show that a southerly extension of the central Australian desert reaches the south coast in the region of the Great Australian Bight. This is the Nullarbor Plain, a vast treeless expanse of semidesert which separates the kinder climates of the southwestern and southeastern parts of the continent. Certain species are found on both sides of the desert, and taxonomists agree that some of them are indigenous in both regions. For example, *Hyla aurea* is found in ponds around Perth in the west and also near Sydney in the east. But Moore (1954) found, when he brought the two together, that they did not interbreed. The two populations must therefore rank as sibling species. Similarly, the grasshopper *Austroicetes cruciata* is found in similar bioclimatic zones on both sides of the desert. The western population differs from the eastern with respect to the temperature range which favors diapause-development (Andrewartha, 1943*b*, 1944*c*). It is not known whether individuals from the two populations would interbreed, but one might expect they would not. There is also evidence from the Calliphorid flies (Tillyard and Seddon, 1933) that speciation has occurred and doubtless is still occurring as a result of the isolation enforced by this great geographic barrier. In the case of *A. cruciata*, distance alone is not important, for the distribution of this species in the east extends over a distance at least as great as that which separates the eastern from the western populations. In this case the important feature of the barrier is the climate of the central region; and the most important aspect of the climate is moisture, best expressed as the ratio of rainfall to evaporation (Davidson, 1936*c;* Fig. 11.01). Of course, it is not practicable to abstract one component like this, because the climate and especially the rainfall influence the character of the soil and vegetation in the area. These also depend on the terrain. Local features of topography influence the effectiveness of the rainfall and the nature of the vegetation (sec. 12.2). In mountainous regions local variations in temperature associated with altitude and aspect may be important.

In the mountainous regions in the western parts of the United States and Mexico oaks are restricted to high altitudes. Each peak and each canyon supports an isolated community of oaks. There are certain gall wasps (Cynipidae) which can breed on oaks, but on no other sort of plant. Each community of oaks supports a population of gall wasps. These are isolated from each other because there is no food for them in the areas separating each peak and each canyon. Kinsey (1942) found that within each species there were as many distinct races as there were isolated populations. In other words, every isolated population had developed into a morphologically distinct race. Taxonomists recognize 347 distinct species of Cynipidae from oak. This is a high number relative to the numbers found on other plants. Kinsey considered that this might be attributed to the very patchy distribution of oaks in most of the area where Cynipidae occur. In this case the most important feature of the geographic barrier was the absence of food from the areas separating the local populations. It is easy to imagine how a shift in climate or perhaps a different terrain might reduce the number of the places where oaks could grow and leave them more widely scattered. The degree of isolation between the races of gall wasps might then have been so great that they ultimately developed into distinct species. Or one might imagine in a continental area changes in climate (say) toward greater aridity, which would change the facies of the vegetation over broad areas while leaving, here and there in the more favored situations, small remnants of the original moisture-loving plants. The local populations of animals living in these places may come to be separated from one another by great areas which contain no food for them. This sort of change is considered to have occurred during past eras in Australia, Africa, and Asia and has been invoked in explanation of some of the speciation which must have occurred in these continental areas. In North America glaciation and the consequent changes in vegetation are considered to have played a similar role (Mayr, 1951).

A well-known example from contemporary times in which the absence of food has constituted an effective geographic barrier concerns the moth *Thera juniperata*, which lives, during its larval stage, entirely on juniper. Owing to the absence of juniper from the midlands in England, populations of *Thera* are restricted to two areas, one in the north and one in the south. Each population is now recognized as a subspecies (Huxley, 1942, p. 185).

In the foregoing examples the presence or absence of food was sufficient to discriminate between the areas where local populations might thrive and those which served as barriers against their dispersal. But in other cases the differences might not be so simply described, because the suitability of a place where the animal may live may depend on interactions between several components of the environment. For example, the butterfly *Coenonympha tullia* inhabits moors and marshes from sea-level up to about 1,800 feet. Once an area has been developed for agriculture, it is no longer suitable for *Coenonympha*. This species is distributed in a long cline from Scotland down to Staffordshire in the south of England. Along the cline there is a gradual increase in spotting and darkening in color from north to south. Before the increase in agriculture, which took place during the eighteenth and nineteenth centuries, local populations of this butterfly doubtless occurred at frequent intervals throughout the main area of its distribution. But now many of the areas which were previously suitable for it have been developed for agriculture, and the local populations are widely separated. The exchange of genes between local populations may not have ceased entirely yet, but it must be much slower than it was two centuries ago (Ford, 1949, p. 311). The wood mouse *Peromyscus leucopus* provides a similar example. The clearing of woodland in North America has left pockets of woods isolated by agricultural lands. This has increased the degree of isolation of local populations of *Peromyscus*, which therefore might be expected to become increasingly diverse. According to Blair (1950), populations in isolated woodlands as little as 3 miles apart differ significantly in color and size.

Even in the absence of any tangible geographic barrier, a certain degree of isolation may occur between the local populations of a species whose numbers are very few. When a species is not abundant, it tends to become subdivided into discontinuous groups relatively isolated from one another. The chance of dispersal between groups is, of course, smaller, the smaller the groups. The smallness of the local populations is essentially the cause of their being isolated, and it provides the possibility of diversity through both selection and the chance loss of genes (sec. 15.14). A possible example given by Huxley (1942, p. 198) is the rare moth *Rhyacia alpicola* which occurs in small restricted areas, in each of which considerable differences have arisen. One subspecies is found in Lapland, another in Ireland and Scotland, a third in the Shetlands, and a fourth in the Carpathians.

16.2 THE POSSIBILITY THAT NONGENETIC VARIATION IN BEHAVIOR MAY GENERATE RACES AND SPECIES

The experiments discussed in section 15.12 showed that certain insects, when reared on unnatural food, would show a preference for this sort of food when they came to seek a place to lay their eggs. Moreover, since this change in behavior could be imprinted on some individuals in the course of one generation, it is not to be explained by selection or any other genetic mechanism. Similar phenomena are well known from other fields. For example, in section 4.43 we mentioned that in the silkworm *Bombyx mori* the tendency toward diapause is transmitted from the mother to her progeny through the cytoplasm of the egg. It is also well known for locusts that the first-instar nymphs which hatch from eggs laid by females in phase *gregaria* will show the behavior and the coloration of the gregarious phase and that these qualities are transmitted through the cytoplasm.

Considerations of this sort have led Thorpe (1945) and others to postulate that new races, and ultimately species, may be generated in the absence of spatial (geographic) barriers in cases where a proportion of the population may, on some occasion, be constrained by some unusual circumstance to grow up in an unnatural place or on unnatural food. The preference for this sort of place or this sort of food is then passed on to their progeny. During the course of some generations, this preference becomes strengthened not only by repetition of the original process but also by selection. In time the preference becomes so strong that it constitutes an adequate isolating mechanism. Thorpe (1945, p. 68) gave a hypothetical illustration of how he considered isolation might develop among birds. "Imagine now an area where two types of habitat 'a' and 'b' (e.g., two different vegetational types) are available in mosaic distribution and a species confined to habitat 'a' within that area. In some exceptional circumstance of crisis, or as a result of some slight germinal change, certain individuals of the species spread into habitat 'b' and the young reared there become imprinted, or otherwise specialized, to the new niche. If this niche provides room for expansion, the birds of the new habitat will rapidly come to fill it, during which time they will be reproductively isolated to a considerable extent from the 'a' habitat birds." This idea is not entirely new, for it was forecast in Baldwin's (1902) theory of organic selection, but it has been crystallized and given empirical backing by Thorpe's experiments—though not, so far, by any convincing examples from nature.

The butterfly *Colias chrysotheme* would seem to provide an opportunity for the investigation of this hypothesis. This butterfly is widely distributed throughout the United States (Hovanitz, 1948). Half the area occupied by the type known as *philodice* is also occupied by the type known as *eurytheme*, and about three-quarters of the area occupied by *eurytheme* is also occupied by *philodice*. Hybrids occur in nature, and these are interfertile with the parental types and among themselves. As a result, a wide range of intergrading forms occurs in the localities where the two types exist together. But Hovanitz (1949*b*) has never found the hybrids to exceed 10 per cent of the total population. Hybridization has been going on for at least 60 years, and yet the two types have remained so distinct that lepidopterists usually refer to them as different species. They have not been replaced by intermediate types, as might have been expected. The two types could be maintained simultaneously if the hybrids were at a selective disadvantage, but Hovanitz (1948) claims that they are as viable as the parental types. If this is so, then the scarcity of hybrids must be brought about in some other way. Throughout most of the present distribution, red clover is the larval food of *philodice* and lucerne is the food of *eurytheme*. Lucerne is not a suitable food for *philodice*, and red clover is unsuitable for *eurytheme;* sterility or abnormal development is common among butterflies reared on the "wrong" food plants, though hybrids develop normally on either of these plants. In other words, a field of lucerne is productive of *philodice* alone, and a field of red clover is productive of *eurytheme* alone. Further, since the butterflies mate soon after emergence (Hovanitz, 1949*b*), the probability of like mating with like must be considerably greater than the mating of unlikes. Food specificity combined with the tendency to mate soon after emergence would tend to keep the populations of the two types separated, especially if dispersal from field to field is low. But the data which Hovanitz provides do not preclude the possibility of other explanations for the small percentage of hybrids in natural populations. This is the sort of phenomenon which is worth exploring further.

In criticizing Thorpe's hypothesis, Mayr (1947, p. 274) pointed out that "it must be demonstrated that these processes [i.e., 'host-selection,' 'olfactory conditioning,' etc.] reduce gene-flow below the point where it interferes with the establishment of genetically-controlled isolating mechanisms." After an examination of the experimental evidence (sec. 15.12), Mayr concluded that none of the experiments had demonstrated that gene flow between the original and the "conditioned" populations was reduced sufficiently to isolate the two populations effectively. For example, in Cushing's (1941) experiments, *Drosophila guttifera*, which in nature lives on fungus, was reared on a medium of corn, molasses, and agar. This colony was then divided into two; one half was left on this medium, and the other was transferred to a medium containing mushroom extract. After many generations, flies from each colony were offered the choice of the two media for oviposition. Those which had been "conditioned" by molasses laid 19.5 per cent of their eggs on this medium and 80.5 per cent on the medium containing the mushroom extract. Those which had been "conditioned" by mushroom laid 92 per cent of their eggs on this medium. Preferences of this low order would not be likely to lead to adequate isolation.

Mayr (1947) also examined a large number of cases in which races of birds were known to differ with respect to their preferences for different foods or places to live and concluded that in every case the best explanation lay in separation in space by some form of geographic barrier. There are, of course, many instances of closely related species which live in the same region but have slightly different requirements or preferences with respect to food or the places in which they live. But there is no satisfactory evidence that the separation into distinct species happened in accordance with Thorpe's hypothesis. For example, the well-known "species flocks" of fishes and Crustacea in old fresh-water lakes are more adequately explained on the hypothesis of initial spatial isolation as a result of great changes in the characters of these lakes, which are considered to have changed from partly dried-up swamps to deep lakes (Mayr, 1947; Brooks, 1950; Greenwood, 1951). It was also pointed out by Mayr (1947, 1951) that the hypothesis of geographic isolation is supported by an abundance of cases of incipient speciation which can be observed in nature. The other hypothesis remains an interesting possibility, especially in the case of a species which is capable of being strongly "conditioned" and, at the same time, has only slight capacity for dispersal. But, so far, the empirical evidence which might confirm it as one of the mechanisms leading to speciation in nature seems not to have been discovered.

16.3 THE SELECTION OF MECHANISMS FAVORING REPRODUCTIVE ISOLATION

Koopman (1950) found that *Drosophila pseudoobscura* and *D. persimilis* could be crossed in the laboratory less readily at higher than at lower temperatures. The male hybrids were sterile, and the back-cross hybrids were poorly viable at all temperatures. The two species were reared together for several generations in a cage kept at a low temperature; the hybrids were removed from the cage as they appeared, in order to simulate selection against hybrids. The amount of hybridization gradually decreased, until eventually there was no more hybridization between the experimental animals at low temperature than between the normal controls at high temperature. Koopman considered that the initial populations contained variants of both species, which differed from one another in their inclination to mate with individuals of the other species. Since the interspecific hybrids were effectively sterile (because they were removed consistently), the variants which mated interspecifically produced progeny which did not contribute to the next generation. There was a selection for types which tended to mate with their own species. In other words, the isolating mechanism was strengthened by selection, which eliminated those which tended to mate with another species. If these results could be extrapolated to apply to races, they might indicate a way in which incipient species might develop into distinct species in the absence of a geographic barrier.

16.4 THE DISTINCTION BETWEEN THE INITIAL MEANS OF ISOLATION OF INCIPIENT SPECIES AND THE MAINTENANCE OF ISOLATION BETWEEN FULL SPECIES

From what has been said in this chapter about the origin of species, one would expect to find in nature a graded series ranging all the way from races which interbreed with complete facility to distinct species which do not interbreed either in nature or in any artificial circumstances which we can contrive. This is, in fact, just what we do find in nature. The intermediates between the two extremes are various. For example, the four toads *Bufo americanus*, *B. fowleri*, *B. terrestris*, and *B. woodhousii* occur in eastern and central America. Their distributions are different, but they overlap. The species remain distinct, but hybrids are found in the areas where the distributions overlap. In the laboratory it is possible to obtain hybrids which are quite viable (Blair, 1941, 1942; Volpe, 1952). The entire range of the frog *Rana palustris* is included in the wider distribution of *R. pipiens*. Hybrids are never found in nature, but they can be produced artificially in the laboratory (Moore, 1949*b*). Similarly, as we mentioned above, *Drosophila pseudoobscura* and *D. persimilis* will produce some hybrids when they are kept together in one cage, but hybrids are never found in nature. With *Rana*, Moore considered that the failure to interbreed in nature may be due to differences in the breeding season reinforced by preferences for different sorts of places in which to live. With *Drosophila* it is known that the hybrid males are sterile and the back-cross hybrids are inviable; but Dobzhansky (1951) and Koopman (1950) considered that something more than this was probably happening in nature to keep the two species from interbreeding. With these intermediates between races and full species, it is sometimes difficult to find out what is serving to keep the species distinct.

When the final step has been taken and races have become species which do not, in any circumstances, interbreed, then they may come together in the same region and still remain distinct. The geneticist usually classifies isolating mechanisms in terms of those which prevent species from exchanging genes. The mechanism may be to prevent mating, as, for example, between remotely related species of *Drosophila;* or, if mating occurs, the species may be isolated if fertilization does not occur, as in the cross *D. mulleri* males × *D. aldrichi* females; if fertilization does take place, the hybrids may not be viable, as in the cross *Rana clamitans* male and *R. pipiens* female; or, if viable, the hybrids may be sterile, as in the case of the cross *D. mulleri* female × *D. aldrichi* male. Mating swarms of the small flies of the family Chironomidae frequently contain several species mixed up together, yet they do not interbreed. In this case the mechanisms which prevent interbreeding are firmly established, needing no reinforcement from whatever means may have served to keep the races apart while they were developing into species.

Bibliography and Indexes

Bibliography and Author Index

The page citations following the items in the Bibliography give the location of reference to the given title in the text and replace the customary author index.

ABBEY, H. 1952. An examination of the Reed-Frost theory of epidemics, Human Biol., **24**: 201–33. P. 483.

ABE, N. 1937. Post larval development of the coral, *Fungia actiniformis* var. *palanensis* Döderlein, Palao Trop. Biol. Sta. Stud., **4**: 73–93. P. 304.

ABELOOS, M. 1935. Diapause larvaire et éclosion chez le Coléoptère *Timarcha tenebricosa*, C.R. Acad. Sci., Paris, **200**: 2112–14. P. 59.

ADOLPH, E. F. 1927. The regulation of volume and concentration in the body fluids of earthworms, J. Exper. Zoöl., **47**: 31–62. P. 226.

———. 1933. Exchanges of water in the frog, Biol. Rev., **8**: 224–40. P. 226.

AHMAD, T. 1936. The influence of ecological factors on the Mediterranean flour moth *Ephestia kühniella* and its parasite *Nemeritus canescens*, J. Anim. Ecol., **5**: 67–93. Pp. 158, 159, 160, 469.

ALLAN, P. F. 1939. Development of ponds for wildlife in the southern high plains, Tr. North Am. Wildlife Conf., **4**: 339–42. P. 543.

ALLEE, W. C. 1931. Animal aggregations: A study in general sociology. Chicago: University of Chicago Press. P. 334.

———. 1938. The social life of animals. New York: W. W. Norton & Co. Pp. 336, 346.

———. 1951. Cooperation among animals. New York: Henry Schuman. P. 346.

ALLEE, W. C.; EMERSON, A. E.; PARK, O.; PARK, T.; and SCHMIDT, K. P. 1949. Principles of animal ecology. Philadelphia: W. B. Saunders Co. Pp. 4, 5, 14, 15, 26, 129, 130, 305, 334, 335, 336, 401, 502, 524, 568, 704.

ALPATOV, W. W. 1929. Growth and variation of *Drosophila melanogaster* larvae, J. Exper. Zoöl., **52**: 407–37. P. 58.

AMADON, D. 1947. Ecology and the evolution of some Hawaiian birds, Evolution, **1**: 63–68. P. 461.

ANDREWARTHA, H. G. 1933. The bionomics of *Otiorrhynchus cribricollis*, Gyll., Bull. Ent. Research, **24**: 373–84. Pp. 59, 259, 272.

———. 1935. Thrips investigation No. 7. On the effect of temperature and food upon egg production and the length of adult life of *Thrips imaginis*, Bagnall., J. Counc. Scient. & Indust. Research Australia, **8**: 281–88, Pp. 172, 505, 507, 569.

———. 1937. Locusts and grasshoppers in South Australia: some records of past outbreaks, J. Dept. Agr. South Australia, **41**: 366–68. Pp. 282, 545.

———. 1939. The small plague grasshopper (*Austroicetes cruciata*, Sauss.), *ibid.*, **43**: 99–106. Pp. 82, 531, 583.

———. 1940. The environment of the Australian plague locust (*Chortoicetes terminifera*, Walk.) in South Australia, Tr. Roy. Soc. South Australia, **64**: 76–94. Pp. 274, 275, 494, 545, 546, 550.

———. 1943*a*. The significance of grasshoppers in some aspects of soil conservation in South

Australia and Western Australia, J. Dept. Agr. South Australia, **46**: 314–22. Pp. 82, 583, 584.

———. 1943*b*. Diapause in the eggs of *Austroicetes cruciata* Sauss. (Acrididae) with particular reference to the influence of temperature on the elimination of diapause, Bull. Ent. Research, **34**: 1–17. Pp. 58, 59, 62, 64, 82, 583, 714.

———. 1944*a*. Air temperature records as a guide to the date of hatching of the nymphs of *Austroicetes cruciata*, Sauss. (Orthoptera), *ibid.*, **35**: 31–41. Pp. 82, 166, 168, 169, 583.

———. 1944*b*. The distribution of plagues of *Austroicetes cruciata*, Sauss. (Acrididae) in Australia in relation to climate, vegetation and soil, Tr. Roy. Soc. South Australia, **68**: 315–26. Pp. 82, 84, 274, 275, 583, 584, 589, 590, 594.

———. 1944*c*. The influence of temperature on the elimination of diapause from the eggs of the race of *Austroicetes cruciata* Sauss. (Acrididae) occurring in Western Australia, Australian J. Exper. Biol. & M. Sc., **22**: 17–20. Pp. 82, 583, 690, 714.

———. 1945. Some differences in the physiology and ecology of locusts and grasshoppers, Bull. Ent. Research, **35**: 379–89. Pp. 60, 122, 690.

———. 1952. Diapause in relation to the ecology of insects, Biol. Rev., **27**: 50–107. Pp. 57, 58, 68, 69, 71, 80, 81.

Andrewartha, H. G. and Birch, L. C. 1948. Measurement of "environmental resistance" in the Australian plague grasshopper, Nature, **161**: 447–48. Pp. 82, 208, 248, 447, 488, 583.

———. 1953. The Lotka-Volterra theory of interspecific competition, Australian J. Zoöl., **1**: 174–77. Pp. 407.

Andrewartha, H. G.; Davidson, J.; and Swan, D. C. 1938. Vegetation types associated with plague grasshoppers in South Australia, Bull. Dept. Agr. South Australia, No. 333. Pp. 494, 583.

Andrewartha, H. V. 1936. The influence of temperature on the rate of development of the immature stages of *Thrips imaginis* Bagnall and *Haplothrips victoriensis* Bagnall, J. Counc. Scient. & Indust. Research Australia, **9**: 57–64. P. 569.

Arbuthnott, K. D. 1944. Strains of the European cornborer in the United States, Tech. Bull. U.S. Dept. Agr., No. 869. Pp. 59, 70, 71, 76, 688.

Armstrong, T. 1945. Differences in the life history of codling moth *Carpocapsa pomonella* (L.) attacking pear and apple, Canad. Ent., **77**: 231–33. P. 692.

Baas Becking, L. G. M. 1946. On the analysis of sigmoid curves, Acta biotheoret., **8**: 42–59. P. 385.

Babcock, K. W. 1927. The European cornborer *Pyrausta nubilalis*, Hübn. II. Discussion of its seasonal history in relation to various climates, Ecology, **8**: 177–93. Pp. 687, 688.

Babcock, K. W., and Vance, A. M. 1929. The cornborer in central Europe, Tech. Bull. U.S. Dept. Agr., No. 135. P. 687.

Babers, F. H. 1949. Development of insect resistance to insecticides. I, Bull. U.S. Bur. Ent., No. E–776, pp. 1–31. Pp. 701, 702.

Babers, F. H., and Pratt, J. J. 1951. Development of insect resistance to insecticides. II, Bull. U.S. Bur. Ent., No. E–818; pp. 1–45. P. 702.

Bach, P. De, 1946. An insecticidal check method for measuring the efficacy of entomophagous insects, J. Econ. Ent., **39**: 695–97. P. 477.

———. 1949. Population studies of the long-tailed mealybug and its natural enemies on citrus trees in southern California, 1946, Ecology, **30**: 14–25. Pp. 471, 492.

Bach, P. De. and Smith, H. S. 1941. Are population oscillations inherent in the host-parasite relation? Ecology, **22**: 363–69. Pp. 19, 441, 442.

Bachmetjew, P. 1907. Experimentelle entomologische Studien vom physikalisch-chemischen Standpunkt aus. Vol. **2**: Einfluss der äussern Factoren auf Insekten. Leipzig: Engelmann. P. 190.

Bacot, A. 1914. A study of the bionomics of the common rat fleas and other species associated with human habitations with special reference to temperature and humidity, J. Hyg. Cambridge, Plague Suppl., **3**: 447–560. P. 58.

Bacot, A. W., and Harden, A. 1922. Vitamin requirements of *Drosophila*. I. Vitamins B and C, Biochem. J., **16**: 148–52. P. 506.

Bacot, A. W., and Martin, C. J. 1924. The respective influences of temperature and moisture upon the survival of the rat flea (*Xenopsylla cheopis*) away from its host, J. Hyg., **23**: 98–105. Pp. 251, 252, 264.

Bagenal, T. B. 1951. A note on the papers of Elton and Williams on the generic relations of species in small ecological communities, J. Anim. Ecol., **20**: 242–45. P. 465.

Baird, A. B. 1918. Some notes on the control of cherry-tree ugly-nest Tortricid, Agr. Gaz. Canada, **5**: 766–71. P. 58.

Baker, F. C. 1935. The effect of photoperiodism on resting tree-hole mosquito larvae (preliminary report), Canad. Ent., **67**: 149–53. Pp. 58, 68, 299.

Baker, J. R. 1938. The evolution of breeding seasons. *In:* Beer, G. R. De (ed.), Evolution, pp. 161–77. Oxford: Clarendon Press. P. 293.

Baker, J. R., and Baker, I. 1936. The seasons in a tropical rain-forest (New Hebrides). 2. Botany, J. Linn. Soc. London (Zoöl.), **39**: 507–19. P. 290.

Baker, J. R., and Ranson, R. M. 1932. Factors affecting the breeding of the field mouse (*Microtus agrestis*). I. Light, Proc. Roy. Soc. London, B, **110**: 313–21. P. 291.

Baker, W. A., and Jones, L. G. 1934. Studies on *Exeristes roborator*, a parasite of the European cornborer in the Lake Erie district, Tech. Bull. U.S. Dept. Agr., No. 460. P. 58.

Baldwin, J. M. 1902. Development and evolution. London: Macmillan & Co., Ltd. P. 717.

Banta, A. M., and Wood, T. R. 1939. Genetical studies in sexual reproduction. *In:* Banta *et al.*, Studies on the physiology genetics and evolution of some Cladocera, sec. viii. Pub. Carnegie Inst. Washington, **513**: 131–81. P. 676.

Barber, G. W. 1939. Injury to sweet corn by *Euxesta stigmatias* Loew in southern Florida, J. Econ. Ent., **32**: 879–80. P. 110.

Barber, G. W., and Dicke, F. F. 1939. Effect of temperature and moisture on overwintering pupae of the corn earworm in the north-eastern states, J. Agr. Research, **59**: 711–23. Pp. 194, 196, 274.

Barnes, H. F. 1943. Studies of fluctuations in insect populations. X. Prolonged larval life and delayed subsequent emergence of the adult gall midge, J. Anim. Ecol., **12**: 137–38. Pp. 58, 59.

Barney, R. L., and Anson, B. J. 1920. Life history and ecology of the pigmy sunfish, *Elassoma zonatum*, Ecology, **1**: 241–56. P. 479.

Basinger, A. J., and Smith, H. S. 1946. Notes on the time of emergence, longevity, and oviposition of codling moth from walnuts, apples and pears, Bull. Dept. Agr. California, **35**: 37–38. P. 692.

Bates, M. 1949. The natural history of mosquitoes. New York: Macmillan Co. P. 330.

Batham, E. J., and Pantin, C. F. A. 1950. Phases of activity in the sea-anemone *Metridium senile* L. and their relation to external stimuli, J. Exper. Biol., **27**: 377–99. P. 321.

Baumberger, J. R. 1914. Studies in longevity of insects. Ann. Ent. Soc. Amer., **7**: 323–53. Pp. 76, 80.

———. 1917. Hibernation: a periodical phenomenon, *ibid.*, **10**: 179–86. P. 80.

Beall, G. 1940. The fit and significance of contagious distributions when applied to observations on larval insects, Ecology, **21**: 460–74. P. 562.

———. 1941*a*. The monarch butterfly, *Danaus archippus* Fab. I. General observations in southern Ontario. Canad. Field Nat., **55**: 123–29. P. 117.

———. 1941*b*. The monarch butterfly, *Danaus archippus* Fab. II. The movement in southern Ontario, *ibid.*, pp. 133–37. P. 117.

———. 1946. Seasonal variation in sex proportion and wing length in the migrant butterfly, *Danaus plexippus* L. (Lep. Danaidae), Tr. Roy. Ent. Soc. London, **97**: 337–53. Pp. 117, 118, 120.

Beall, G., and Williams, C. B. 1945. Geographical variation in the wing length of *Danaus plexippus* (Lep. Rhopalocera), Proc. Roy. Ent. Soc. London, A, **20**: 65–76. P. 118.

Beament, J. W. L. 1946*a*. The waterproofing process in eggs of *Rhodnius prolixus* Stähl, Proc. Roy. Soc. London, B, **133** : 407–18. P. 248.

———. 1946*b*. The formation and structure of the chorion of the egg in an Hemipteran *Rhodnius prolixus*, Quart. J. Micr. Sc., **87** : 393–439. P. 248.

Beauchamp, R. S. A., and Ullyott, P. 1932. Competitive relationships between certain species of fresh-water Triclads, J. Ecol., **20** : 200–208. P. 463.

Bei-Benko, G. Y. 1928. Synopsis of the nymphs of the West Siberian grasshoppers (in Russian with an English summary), Trud. Sibirsk. Inst. S.-Kh. Lesovodstova, **9** : 39 pp. P. 58.

Bělehrádek, J. 1935. Temperature and living matter. Berlin: Borntraeger. Pp. 146, 147, 148.

Bentley, E. W. 1944. The biology and behaviour of *Ptinus tectus* Boie. (Coleoptera, Ptinidae), a pest of stored products. V. Humidity reactions, J. Exper. Biol., **20** : 152–58. Pp. 211, 217.

Bentley, E. W.; Gunn, D. L.; and Ewer, D. W. 1941. The biology and behaviour of *Ptinus tectus* Boie. (Coleop. Ptinidae), a pest of stored products. I. The daily rhythm of locomotory activity especially in relation to light and temperature, J. Exper. Biol., **18** : 182–95. P. 322.

Berezina, V. M. 1940. Effect of hydrothermic soil conditions on the vertical migrations of cockchafer larvae, Bull. Plant. Prot., Leningrad, **5** : 43–56. P. 59.

Berger, B. 1907. Über die Widerstandsfähigkeit der *Tenebrio* larven gegen Austrockung, Arch. f. d. ges. Physiol., **118** : 607–12. P. 232.

Bertani, G. 1947. Artificial "breaking" of the diapause in *Drosophila nitens*, Nature, **159** : 309. Pp. 58, 67.

Birch, L. C. 1942. The influence of temperatures above the developmental zero on the development of the eggs of *Austroicetes cruciata* Sauss. (Orthoptera), Australian J. Exper. Biol. & M. Sc., **20** : 17–25. Pp. 59, 61, 64, 82, 130, 161, 162, 583.

———. 1944*a*. An improved method for determining the influence of temperature on the rate of development of insect eggs using eggs of the small strain of *Calandra oryzae* L. (Coleoptera), *ibid.*, **22** : 277–83. Pp. 151, 152, 286.

———. 1944*b*. The effect of temperature and dryness on the survival of the eggs of *Calandra oryzae* L. (small strain) and *Rhizopertha dominica* Fab. (Coleoptera), *ibid.*, pp. 265–69. Pp. 267, 268.

———. 1945*a*. The mortality of the immature stages of *Calandra oryzae* L. (small strain) and *Rhizopertha dominica* Fab. in wheat of different moisture contents, *ibid.*, **23** : 141–45. P. 44.

———. 1945*b*. Diapause in *Scelio chortoicetes* Frogg. (Scelionidae), a parasite of the eggs of *Austroicetes cruciata* Sauss., J. Australian Inst. Agr. Sc., **11** : 189–90. P. 56.

———. 1945*c*. The influence of temperature on the development of the different stages of *Calandra oryzae* L. and *Rhizopertha dominica* Fab. (Coleoptera), Australian J. Exper. Biol. & M. Sc., **23** : 29–35. Pp. 151, 161, 171, 374.

———. 1945*d*. The influence of temperature, humidity and density on the oviposition of the small strain of *Calandra oryzae* L. and *Rhizopertha dominica* Fab. (Coleoptera), *ibid.*, pp. 197–203. Pp. 171, 172, 279.

———. 1946*a*. The heating of wheat stored in bulk in Australia, J. Australian Inst. Agr. Sc., **12** : 27–31. Pp. 45, 53, 455.

———. 1946*b*. The movements of *Calandra oryzae* L. (small strain) in experimental bulks of wheat, *ibid.*, pp. 21–26. P. 137.

———. 1947. The ability of flour beetles to breed in wheat, Ecology, **28** : 322–24. Pp. 151, 346.

———. 1948. The intrinsic rate of natural increase of an insect population, J. Anim. Ecol., **17** : 15–26. Pp. 34, 36, 37, 43, 44, 45.

———. 1953*a*. Experimental background to the study of the distribution and abundance of insects. I. The influence of temperature, moisture and food on the innate capacity for increase of three grain beetles, Ecology, **34**: 698–711. Pp. 34, 47, 52, 53, 367, 421, 509.

———. 1953*b*. Experimental background to the study of the distribution and abundance of insects. II. The relation between innate capacity for increase in numbers and the abundance of three grain beetles in experimental populations, *ibid.*, **34**: 712–26. Pp. 34, 358, 360, 367, 380, 383.

———. 1953*c*. Experimental background to the study of the distribution and abundance of insects. III. The relation between innate capacity for increase and survival of different species of beetles living together on the same food, Evolution, **7**: 136–44. P. 422.

———. 1954. Experiments on the relative abundance of two sibling species of grain weevils, Australian J. Zoöl., **2** (in press) P. 691.

Birch, L. C., and Andrewartha, H. G. 1941. The influence of weather on grasshopper plagues in South Australia, J. Dept. Agr. South Australia, **45**: 95–100. Pp. 82, 83, 344, 448, 583, 590.

———. 1942. The influence of moisture on the eggs of *Austroicetes cruciata* Sauss. (Orthoptera) with reference to their ability to survive desiccation, Australian J. Exper. Biol. & M. Sc., **20**: 1–8. Pp. 55, 82, 84, 227, 230, 248, 260, 583.

———. 1944. The influence of drought on the survival of eggs of *Austroicetes cruciata* Sauss. (Orthoptera) in South Australia, Bull. Ent. Research, **35**: 243–50. Pp. 82, 583, 588, 589.

Birch, L. C.; Park, T. and Frank, M. B. 1951. The effect of intraspecies and interspecies competition on the fecundity of two species of flour beetles, Evolution, **5**: 116–32. Pp. 372, 428.

Birch, L. C., and Snowball, G. J. 1945. The development of the eggs of *Rhizopertha dominica* Fab. (Coleoptera) at constant temperatures, Australian J. Exper. Biol. & M. Sc., **23**: 37–40. P. 151.

Bishopp, F. C., and Laake, E. W. 1921. Dispersion of flies by flight, J. Agr. Research, **21**: 729–66. Pp. 97, 98.

Bissonette, T. H. 1935. Modifications of mammalian sexual cycles. III. Reversal of the cycle in male ferrets (*Putorius vulgaris*) by increasing periods of exposure to light between October second and March thirtieth, J. Exper. Zoöl., **71**: 341–67. P. 291.

———. 1936. Sexual photoperiodicity, Quart. Rev. Biol., **11**: 371–86. P. 294.

———. 1941. Experimental modification of breeding cycles in goats, Physiol. Zoöl., **14**: 379–83. P. 292.

Blair, A. P. 1941. Variation, isolating mechanisms, and hybridization in certain toads, Genetics, **26**: 398–417. P. 719.

———. 1942. Isolating mechanisms in a complex of four species of toads, Biol. Symp., **6**: 235–49. P. 719.

Blair, W. F. 1950. Ecological factors in speciation of *Peromyscus*, Evolution, **4**: 253–75. Pp. 694, 710, 715.

Blais, J. R. 1952. The relationship of the spruce budworm (*Choristoneura fumiferana*, Clem.) to the flowering condition of balsam fir (*Abies balsamea* L. Mill.) Canad. J. Zoöl., **30**: 1–29. P. 505.

Bliss, C. I. 1926. Temperature characteristics for prepupal development in *Drosophila melanogaster*, J. Gen. Physiol., **9**: 467–95. P. 147.

Bliss, C. I., and Fisher, R. A. 1953. Fitting the negative binomial distribution to biological data and note on the efficient fitting of the negative binomial, Biometrics, **9**: 176–200. P. 568.

Bodenheimer, F. S. 1930. Über die Grundlagen einer allgemeinen Epidemiologie der Insektenkalamitäten, Ztschr. f. Angew. Ent., **16**: 433–50. Pp. 23, 663.

———. 1938. Problems of animal ecology. Oxford: Clarendon Press. Pp. 363, 370, 434.

Bodine, J. H. 1929. Factors influencing the rate of respiratory metabolism of a developing egg (Orthoptera), Physiol. Zoöl., **2**: 459–82. Pp. 59, 230.

———. 1932. Hibernation and diapause in certain Orthoptera. III. Diapause—a theory of its mechanism, *ibid.*, **5**: 549–54. P. 80.

Bodine, J. H., and Evans, A. C. 1932. Hibernation and diapause. Physiological changes during hibernation and diapause in *Sceliphron caementarium*, Biol. Bull. Woods Hole, **63**: 235–45. P. 59.

BOGERT, C. M. 1952. Relative abundance, habitats and normal thermal levels of some Virginian salamanders, Ecology, **33**: 16–30. P. 136.

BOTTGER, G. T., and KENT, V. F. 1931. Seasonal-history studies of the European cornborer in Michigan, J. Econ. Ent., **24**: 372–79. P. 688.

BOUNHIOL, J. J. 1938. Recherches expérimentales sur le déterminisme de la métamorphose chez les Lépidoptères, Bull. Biol. Suppl., **24**: 1–199. P. 81.

BOYCE, A. M. 1928. Studies on the resistance of certain insects to hydrocyanic acid, J. Econ. Ent., **21**: 715–20. P. 702.

———. 1931. The diapause phenomenon in insects with special reference to *Rhagoletis completa*, *ibid.*, **24**: 1018, P. 59.

BOYCE, J. M 1946. The influence of fecundity and egg mortality on the population growth of *Tribolium confusum* Duval, Ecology, **27**: 290–302. P. 369.

BRECHER, G., and WIGGLESWORTH, V. B. 1944. The transmission of *Actinomyces rhodnii* in *Rhodnius prolixus*, Stal. (Hemiptera) and its influence on the growth of the host, Parasitology, **35**: 220–24. P. 57.

BREITENBRECHER, J. K. 1918. The relation of water to the behaviour of the potato beetle in a desert, Pub. Carnegie Inst. Washington, **263**: 343–84. Pp. 58, 231, 259, 271, 272, 280.

BREMER, H. 1926. Über die tageszeitliche Konstanz im Schlüpftermine der Imagines einiger Insekten und ihre experimentelle Beeinflussbarkeit, Ztschr. f. Wissensch. Insekten-Biol., **21**: 209–16. P. 327.

BRETT, J. R. 1944. Some lethal temperature relations of Algonquin Park fishes, Univ. Toronto Stud. Biol., Vol. **52**. P. 201.

BROADBENT, L. 1946. A survey of potato aphids in north-west Derbyshire, 1945, Ann. Appl. Biol. **33**: 360–68. P. 106.

———. 1948. Aphis migration and the efficiency of the trapping method, *ibid.*, **35**: 379–94. Pp. 105, 107.

———. 1949. Factors affecting the activity of alatae of the aphids *Myzus persicae* (Sulzer) and *Brevicoryne brassicae* (L.), *ibid.*, **36**: 40–62. P. 114.

BROCA, P. 1860. Rapport sur la question soumise à la Société de Biologie au sujet de la revivescence des animaux desséchés, Mém. Soc. biol., **2**: 1–140. P. 209.

BROEKHUYSEN, G. J. 1941. A preliminary investigation of the importance of desiccation, temperature and salinity as factors controlling the vertical distribution of certain intertidal marine gastropods in False Bay, South Africa, Tr. Roy. Soc. South Africa, **28**: 255–92. Pp. 259, 261.

BROOKS, J. L. 1950. Speciation in ancient lakes (concluded), Quart. Rev. Biol., **25**: 131–76. P. 718.

BROWMAN, L. G. 1936. Light in its relation to activity and estrus rhythms in the albino rat, Anat. Rec., **67**: 107. P. 322.

BROWN, F. A., JR., and HINES, M. N. 1952. Modifications in the diurnal pigmentary rhythm of *Uca* effected by continuous illumination, Physiol. Zoöl., **25**: 56–70. P. 324.

BROWNING, T. O. 1952*a*. The influence of temperature on the completion of diapause in the eggs of *Gryllulus commodus* Walker, Australian J. Scient. Research, B, **5**: 112–27. Pp. 58, 61.

———. 1952*b*. The influence of temperature on the rate of development of insects, with special reference to the eggs of *Gryllulus commodus* Walker, *ibid.*, pp. 96–111. Pp. 151, 152, 153.

BRUES, C. T. 1939. Studies on the fauna of some thermal springs in the Dutch East Indies, Proc. Am. Acad. Arts & Sc., **73**: 71–95. P. 130.

———. 1946. Insect dietary. Cambridge, Mass.: Harvard University Press. Pp. 505, 507.

BÜNNING, E. 1935. Zur Kenntnis der endonomen Tagesrhythmik bei Insekten und bei Pflanzen, Ber. deutsch. bot. Gesellsch, **53**: 594–623. P. 327.

BULLOUGH, W. S. 1951. Vertebrate sexual cycles. London: Methuen & Co. Pp. 290, 291.

BURDICK, H. C. 1937. The effects of the exposure to low temperatures on the developmental time of the embryos of the grasshopper *Melanoplus differentialis* (Orthoptera), Physiol. Zoöl., **10** : 156–70. Pp. 65, 78.

BURLA, H.; CUNHA, A. B. Da; CAVALCANTI, A. G.; DOBZHANSKY, Th.; and PAVAN, C. 1950. Population density and dispersal in Brazilian *Drosophila willistoni*, Ecology, **31** : 393–404. P. 95.

BURNET, F. M. 1940. Biological aspects of infectious disease. Cambridge: At the University Press. P. 483.

BURNETT, T. 1949. The effect of temperature on an insect host–parasite population, Ecology, **30** : 113–34. P. 468.

BUXTON, P. A. 1923. Animal life in deserts. London: Edward Arnold. P. 224.

———. 1924. Heat, moisture, and animal life in deserts, Proc. Roy. Soc. London, B, **96** : 123–31. P. 204.

———. 1930. Evaporation from the mealworm (*Tenebrio*, Coleoptera) and atmospheric humidity, *ibid.*, **106** : 560–77. P. 232.

———. 1932*a*. Terrestrial insects and the humidity of the environment, Biol. Rev., **7** : 275–320. Pp. 239, 275.

———. 1932*b*. The climate in which the rat-flea lives, Indian J. M. Research, **20** : 281–97. P. 530.

BUXTON, P. A., and LEWIS, D. J. 1934. Climate and tsetse flies: laboratory studies upon *Glossina submorsitans* and *G. tachinoides*, Phil. Trans. Roy. Soc. London, B, **224** : 175–240. P. 275.

CAIN, A. J. and SHEPPARD, P. M. 1950. Selection in the polymorphic land snail *Cepaea nemoralis*, Heredity, **4** : 275–94. Pp. 672, 699.

———. 1952. The effects of natural selection on body colour in the land snail *Cepaea nemoralis*, *ibid.*, **6** : 217–31. P. 672.

CAMERON, J. McB. 1950. The laboratory of insect pathology, Sault Ste. Marie, Ontario, Forest Insects Investigation Bi-monthly Prog. Rep., **6** : No. 4, 1–7. P. 482.

CAROTHERS, E. E. 1923. Notes on taxonomy, development, and life history of certain Acrididae, Tr. Am. Ent. Soc., **49** : 7–24. P. 58.

CATCHESIDE, D. G. 1947. The *p*-locus position effect in *Oenothera*, J. Genetics, **48** : 31–42. P. 673.

CHAPMAN, R. N. 1928. The quantitative analysis of environmental factors, Ecology, **9** : 111–22. Pp. 334, 389.

———. 1931. Animal ecology with especial reference to insects. New York: McGraw-Hill Book Co. Pp. 5, 15, 16, 388.

CHAPMAN, R. N.; MICKEL, C. E.; PARKER, J. R.; MILLER, G. E.; and KELLY, E. G. 1926. Studies in the ecology of sand dune insects, Ecology, **7** : 416–26. P. 204.

CHIANG, H. C., and HODSON, A. C. 1950. An analytical study of population growth in *Drosophila melanogaster*, Ecol. Monogr., **20** : 173–206. P. 380.

CHITTY, D. 1952. Mortality among voles (*Microtus agrestis*) at Lake Vyrnwy, Montgomeryshire in 1936–9, Phil. Trans. Roy. Soc. London, B, **236** : 505–52. Pp. 397, 398.

CHITTY, D., and ELTON, C. 1937. Canadian Arctic wildlife enquiry, 1935–36, J. Anim. Ecol., **6** : 368–85. P. 478.

CHITTY, HELEN. 1950. Canadian Arctic wildlife enquiry, 1943–49, with a summary of results since 1933, J. Anim. Ecol., **19** : 180–93. Pp. 478, 503.

CHRISTENSEN, P. J. H. 1937. Zur Histologie und Embryologie der überwinterten Eier von *Orgyia antiqua* L., Zool. Jahrb., Abt. Anat., **62** : 567–82. P. 59.

CHRISTIAN, J. J. 1950. The adreno-pituitary system and population cycles in mammals, J. Mammal., **31** : 247–59. P. 398.

CLARK, L. B. 1941. Factors in the lunar cycle which may control reproduction in the Atlantic palolo, Biol. Bull., **81** : 278. P. 302.

CLARK, L. B., and HESS, W. N. 1940*a*. Swarming of the Atlantic palolo worm, *Leodice fucata* (Ehlers), Papers Tortugas Lab., **33**: 21–70. P. 300.

———. 1940*b*. The reactions of the Atlantic palolo, *Leodice fucata*, to light, *ibid.*, pp. 71–81. P. 300.

CLARK, L. R. 1947*a*. An ecological study of the Australian plague locust *Chortoicetes terminifera* Walk. in the Bogan-Macquarie outbreak area in N.S.W., Bull. Counc. Scient. & Indust. Research Australia, No. 226. Pp. 134, 135, 546, 547, 552, 558.

———. 1947*b*. Ecological observations on the small plague grasshopper, *Austroicetes cruciata* (Sauss.), in the Trangie District, Central Western New South Wales, *ibid.*, No. 228. Pp. 135, 274, 275.

———. 1949. Behaviour of swarm hoppers of the Australian plague locust, *Chortoicetes terminifera* Walk., *ibid.*, No. 245. P. 135.

———. 1950. On the abundance of the Australian plague locust *Chortoicetes terminifera* (Walker) in relation to the presence of trees, Australian J. Agr. Research, **1**: 64–75. P. 313.

———. 1953. The ecology of *Chrysomela gemellata* Rossi and *C. hyperici* and their effect on St. John's Wort in the Bright district, Victoria, Australian J. Zoöl., **1**: 1–69. Pp. 493, 502.

CLARKE, C. H. D. 1936. Fluctuations in numbers of ruffed grouse, *Bonasa umbellus* (Linné), with special reference to Ontario, Univ. Toronto Stud. Biol., **41**: 1–118. P. 484.

———. 1940. A biological investigation of the Thelon game sanctuary, Bull. Nat. Mus. Canada, **96**: 1–135. P. 481.

CLARKE, G. L. 1932. Quantitative aspects of the change of phototropic signs in *Daphnia*, J. Exper. Biol., **9**: 180–211. P. 317.

———. 1933. Diurnal migration of plankton in the Gulf of Maine and its correlation with changes in submarine irradiation, Biol. Bull., **65**: 402–36. P. 305.

———. 1934. Further observations on the diurnal migration of copepods in the Gulf of Maine, *ibid.*, **67**: 432–55. P. 305.

CLARKE, J. R. 1953. The effect of fighting on the adrenals, thymus and spleen of the vole (*Microtus agrestis*), J. Endocrinol., **9**: 114–26. P. 398.

CLAUSEN, C. P. 1933. Some general considerations in parasite introduction, Proc. 5th Pacific Scient. P. 472.

———. 1940. Entomophagous insects. New York: McGraw-Hill Book Co. Pp. 434, 435.

———. 1951. The time factor in biological control, J. Econ. Ent., **44**: 1–9. Pp. 336, 469, 470, 473.

CLEMENTS, F. E., and SHELFORD, V. E. 1939. Bio-ecology. New York: John W. Wiley & Sons. Pp. 4, 15, 365.

CLEVELAND, L. R. 1928. Further observations and experiments on the symbiosis between termites and their intestinal protozoa, Biol. Bull., **54**: 231–37. P. 504.

CLOUDESLEY-THOMPSON, J. L. 1950. The water relations and cuticle of *Paradesmus gracilis* (Diplopoda, Strongylidae), Quart. J. Micr. Sc., **91**: 453–64. P. 245.

COAD, B. R. 1931. Insects captured by airplane are found at surprising heights, Yearbook U.S. Dept. Agr., 1931, pp. 320–23. P. 107.

COCKERELL, T. D. A. 1934. "Mimicry" among insects, Nature, **133**: 329–30. P. 23.

COLE, L. C. 1946*a*. A study of the Cryptozoa of an Illinois woodland, Ecol. Monogr., **16**: 49–86. P. 562.

———. 1946*b*. A theory for analyzing contagiously distributed populations, Ecology, **27**: 329–41. P. 562.

———. 1951. Population cycles and random oscillations, J. Wildlife Management, **15**: 233–52. Pp. 643, 644, 645, 646.

———. 1954. Some features of random population cycles, *ibid.*, **18**: 2–24. P. 645.

COLLINS, C. W. 1915. Dispersion of gipsy moth larvae by wind, Bull. U.S. Dept. Agr., No. 273. Pp. 104, 107, 109.

COMPERE, H. 1925. New chalcidoid (Hymenopterous) parasites of the black scale, *Saissetia oleae*, Bernard, Univ. Calif. Pub. Ent., **3**: 295–326. P. 475.

Cook, W. C. 1924. The distribution of the pale western cutworm (*Porosagrotis orthogonia* Morr.): a study in physical ecology, Ecology, **5**: 60–69. Pp. 273, 594.

———. 1926. Some weather relations of the pale western cutworm (*Porosagrotis orthogonia* Morr.). A preliminary study, *ibid.*, **7**: 37–47. Pp. 273, 275, 594.

———. 1929. A bioclimatic zonation for studying the economic distribution of injurious insects, *ibid.*, **10**: 282–93. P. 273.

———. 1930. Field studies of the pale western cutworm (*Porosagrotis orthogonia* Morr.), Bull. Montana Agr. Exper. Sta., No. 225. Pp. 273, 594, 595.

Costa Maia, J. D. 1952. Some mathematical developments on the epidemic theory formulated by Reed and Frost, Human Biol., **24**: 167–200. P. 483.

Cotton, R. T. 1930. The effect of light upon the development of the dark mealworm, *Tenebrio obscurus*, Proc. Ent. Soc. Washington, **32**: 58–60. P. 331.

Cousin, G. 1932. Étude expérimentale de la diapause des insectes, Bull. biol. suppl., **15**: 1–341. P. 70.

Cowles, R. B. 1945. Heat-induced sterility and its possible bearing on evolution, Am. Nat., **79**: 160–75. P. 175.

Crabb, W. D. 1948. The ecology and management of the prairie spotted skunk in Iowa, Ecol. Monogr., **18**: 201–32. P. 521.

Craighead, F. G. 1921. Hopkin's host-selection principle as related to certain Cerambycid beetles, J. Agr. Research, **22**: 189–220. P. 692.

Crombie, A. C. 1941. On oviposition, olfactory conditioning and host selection in *Rhizopertha dominica* Fab. (Insecta, Coleoptera), J. Exper. Biol., **18**: 62–79. P. 509.

———. 1942. The effect of crowding upon the oviposition of grain-infesting insects, *ibid.*, **19**: 311–40. P. 373.

———. 1943. The effect of crowding upon the natality of grain-infesting insects, Proc. Zoöl. Soc. London, A, **113**: 77–98. Pp. 374, 381.

———. 1944. On intraspecific and interspecific competition in larvae of graminivorous insects, J. Exper. Biol., **20**: 135–51. Pp. 371, 429.

———. 1945. On competition between different species of graminivorous insects, Proc. Roy. Soc. London, B, **132**: 362–95. Pp. 407, 412, 429, 430, 456, 457.

———. 1946. Further experiments on insect competition, *ibid.*, **133**: 76–109. P. 431.

Crombie, A. C., and Darrah, J. H. 1947. The chemoreceptors of the wireworm (*Agriotes spp.*) and the relation of activity to chemical constitution, J. Exper. Biol., **24**: 95–109. P. 510.

Cunha, A. B. Da. 1951. Modification of the adaptive values of chromosomal types in *Drosophila pseudoobscura* by nutritional variables, Evolution, **5**: 395–404. Pp. 678, 679.

Cunha, A. B. Da; Burla, H.; and Dobzhansky, Th. 1950. Adaptive chromosomal polymorphism in *Drosophila willistoni*, Evolution, **4**: 212–35. P. 684.

Cunha, A. B. Da; Dobzhansky, Th.; and Sokoloff, A. 1951. On food preferences of sympatric species of *Drosophila*, Evolution, **5**: 97–101. P. 461.

Cunliffe, N. 1921. Some observations on the biology and structure of *Ornithodorus moubata*, Murray, Parasitology, **13**: 327–47. P. 283.

———. 1922. Some observations on the biology and structure of *Ornithodorus savignyi*, Audouin, *ibid.*, **14**: 17–26. Pp. 277, 283.

Cushing, J. E., Jr. 1941. An experiment of olfactory conditioning in *Drosophila guttifera*, Proc. Nat. Acad. Sc., Washington, **27**: 496–99. P. 718.

Cutler, D. W. 1923. The action of Protozoa on bacteria when inoculated into sterile soil, Ann. Appl. Biol., **10**: 137–41. P. 433.

D'Ancona, U. 1942. La lotta per l'esistenza. Turin, Italy: Giulio Einaudi, P. 348. (English trans. 1954. The struggle for existence. Bibliotheca Biotheoretica, **6**: 1–274.)

Dane, D. S.; Miles, J. A. R.; and Stoker, M. G. P. 1953. A disease of Manx shearwaters: further observations in the field, J. Anim. Ecol., **22**: 123–33. P. 484.

Darby, H. H., and Kapp, E. M. 1933. Observations on the thermal death points of *Anastrepha ludens* (Loew.), Tech. Bull. U.S. Dept. Agr. No., 400. Pp. 202, 203.

DARLING, F. F. 1938. Bird flocks and the breeding cycle: a contribution to the study of avian sociality. Cambridge: At the University Press. Pp. 337, 343.

DARLINGTON, P. J. 1938. The origin of the fauna of the Greater Antilles, with a discussion of the dispersal of animals over water and through the air, Quart. Rev. Biol., **13**: 274–300. P. 86.

DAVIDSON, J. 1929. On the occurrence of the parthenogenetic and sexual forms in *Aphis rumicis* L. with special reference to the influence of environmental factors, Ann. Appl. Biol., **16**: 104–34. P. 305.

———. 1931. The influence of temperature on the incubation period of the eggs of *Sminthurus viridis* L. (Collembola), Australian J. Exper. Biol. & M. Sc., **9**: 143–52. P. 130.

———. 1932*a*. Factors affecting oviposition of *Sminthurus viridis* L. (Collembola), *ibid.*, **10**: 1–16. Pp. 276, 279.

———. 1932*b*. On the viability of the eggs of *Sminthurus viridis* L. (Collembola) in relation to their environment, *ibid.*, pp. 65–88. P. 272.

———. 1933. The environmental factors affecting the development of the eggs of *Sminthurus viridis*, *ibid.*, pp. 9–23. P. 272.

———. 1936*a*. On the ecology of the black-tipped locust (*Chortoicetes terminifera*, Walk.) in South Australia, Tr. Roy. Soc. South Australia, **60**: 137–52. Pp. 281, 545.

———. 1936*b*. The apple-thrips (*Thrips imaginis* Bagnall) in South Australia, J. Dept. Agr. South Australia, **39**: 930–39. P. 569.

———. 1936*c*. Climate in relation to insect ecology in Australia. 3. Bioclimatic zones in Australia, Tr. Roy. Soc. South Australia, **60**: 88–92. Pp. 82, 494, 570, 577, 593, 714.

———. 1942. On the speed of development of insect eggs at constant temperatures, Australian J. Exper. Biol. & M. Sc., **20**: 233–39. Pp. 149, 151.

———. 1943*a*. The time required for the eggs of the body louse (*Pediculus humanus corporis* de Geer) to develop and hatch at different temperatures, M. J. Australia, June 12, 1943, pp. 533–36. P. 151.

———. 1943*b*. On the speed of development of insect eggs at constant temperatures, Australian J. Exper. Biol. & M. Sc., **20**: 233–39. P. 151.

———. 1944. On the relationship between temperature and rate of development of insects at constant temperatures, J. Anim. Ecol., **13**: 26–38. Pp. 146, 149, 150, 151, 152.

DAVIDSON, J., and ANDREWARTHA, H. G. 1948*a*. Annual trends in a natural population of *Thrips imaginis* (Thysanoptera), J. Anim. Ecol., **17**: 193–99. Pp. 91, 141, 492, 569.

———. 1948*b*. The influence of rainfall, evaporation and atmospheric temperature on fluctuations in the size of a natural population of *Thrips imaginis* (Thysanoptera), *ibid.*, pp. 200–222. Pp. 91, 141, 434, 492, 496, 569, 574, 575, 577.

DAVIES, M. E., and EDNEY, E. B. 1952. The evaporation of water from spiders, J. Exper. Biol., **29**: 571–82. P. 244.

DAVIES, W. M. 1928. The effect of variation in relative humidity on certain species of Collembola, Brit. J. Exper. Biol., **6**: 79–86. P. 240.

———. 1939. Studies on aphides infesting the potato crop. VII. Report on a survey of the aphis population of potatoes in selected districts of Scotland (25 July–6 August, 1936), Ann. Appl. Biol., **26**: 116–34. P. 106.

DAVIS, D. E. 1953. The characteristics of rat populations, Quart. Rev. Biol., **28**: 373–401. P. 522.

DAVIS, D. H. S. 1933. Rhythmic activity in the short-tailed vole, *Microtus*, J. Anim. Ecol., **2**: 232–38. P. 322.

DAWSON, R. W. 1931. The problem of voltinism and dormancy in *Telea polyphemus*. J. Exper. Zoöl., **59**: 87–131. P. 59.

DEAL, J. 1941. The temperature preferendum of certain insects, J. Anim. Ecol., **10**: 323–55. P. 133.

DECKER, G. C. 1931. The biology of the stalkborer, *Papaipema nebris*. Research Bull. Iowa Agr. Exper. Sta., No. 143. P. 59.

DEEVEY, E. S., JR. 1947. Life tables for natural populations of animals, Quart. Rev. Biol., **22** : 283–314. P. 39.

DETHIER, V. G. 1937. Gustation and olfaction in Lepidopterous larvae, Biol. Bull., **72** : 7–23. P. 510.

———. 1947. Chemical insect attractants and repellents. Philadelphia: Blakiston Co. P. 510.

———. 1954. Evolution of feeding preferences in phytophagous insects, Evolution, **8** : 33–54. Pp. 509, 692.

DICE, L. R. 1947. Effectiveness of selection by owls of deer-mice (*Peromyscus maniculatus*) which contrast in color with their background, Contr. Lab. Vertebr. Biol. Univ. Michigan, **34** : 1–20. P. 700.

———. 1952. Natural communities. Ann Arbor: University of Michigan Press. P. 4.

DICK, J. 1937. Oviposition in certain Coleoptera, Ann. Appl. Biol., **24** : 762–96. Pp. 171, 173.

DICKSON, R. C. 1941. Inheritance of resistance to hydrocyanic acid fumigation in the California red scale, Hilgardia, **13** : 515–21. P. 701.

———. 1949. Factors governing the induction of diapause in the oriental fruit moth, Ann. Ent. Soc. Amer., **42** : 511–37. Pp. 58, 76, 259, 297, 300.

DITMAN, L. P.; WEILAND, G. S.; and GUILL, J. H. 1940. The metabolism in the corn earworm. III. Weight, water and diapause. J. Econ. Ent., **33** : 282–95. Pp. 58, 72.

DITMAN, L. P.; VOGT, G. B.; and SMITH, D. R. 1943. The relation of unfreezable water to cold-hardiness of insects, J. Econ. Ent., **36** : 304–11. Pp. 192, 194.

DIVER, C. 1940. The problem of closely related species living in the same area. *In:* HUXLEY, J. (ed.), The new systematics. London: Oxford University Press. P. 462.

DOBZHANSKY, TH. 1943. Genetics of natural populations. IX. Temporal changes in the composition of populations of *Drosophila pseudoobscura*, Genetics, **28** : 162–86. P. 674.

———. 1947*a*. Genetics of natural populations. XIV. A response of certain gene arrangements in the third chromosome of *Drosophila pseudoobscura* to natural selection, *ibid.*, **32** : 142–60. P. 681.

———. 1947*b*. Adaptive changes induced by natural selection in wild populations of *Drosophila*, Evolution, **1** : 1–16. Pp. 682, 703.

———. 1948. Genetics of natural populations. XVI. Altitudinal and seasonal changes produced by natural selection in certain populations of *Drosophila pseudoobscura* and *Drosophila persimilis*, Genetics, **33** : 158–76. Pp. 681, 683.

———. 1949*a*. Observations and experiments on natural selection in *Drosophila*, Proc. 8th Internat. Cong. Genetics (Hereditas, Suppl.), pp. 210–24. P. 683.

———. 1949*b*. On some of the problems of population genetics and evolution, Ric. Sc., Suppl., **19** : 1–9. P. 708.

———. 1950*a*. Heredity, environment and evolution, Science, **111** : 161–66. P. 405.

———. 1950*b*. Mendelian populations and their evolution, Am. Nat., **84** : 401–18. P. 669.

———. 1950*c*. Genetics of natural populations. XIX. Origin of heterosis through natural selection in populations of *Drosophila pseudoobscura*, Genetics, **35** : 288–302. P. 683.

———. 1951. Genetics and the origin of species. 3d ed., rev. New York: Columbia University Press. Pp. 670, 673, 676, 677, 683, 685, 686, 702, 703, 708, 710, 719.

———. 1952. Genetics of natural populations. XX. Changes induced by drought in *Drosophila pseudoobscura* and *Drosophila persimilis*, Evolution, **6** : 234–43. P. 675.

DOBZHANSKY, TH.; BURLA, H.; and CUNHA, A. B. DA. 1950. A comparative study of chromosomal polymorphism in sibling species of the *willistoni* group of *Drosophila*, Am. Nat., **84** : 229–46. P. 684.

DOBZHANSKY, TH., and EPLING, C. 1944. Contributions to the genetics, taxonomy, and ecology of *Drosophila pseudoobscura* and its relatives. I. Taxonomy, geographic distribution, and ecology of *Drosophila pseudoobscura* and its relatives. Pub. Carnegie Inst. Washington, **554** : 1–46. Pp. 324, 325.

DOBZHANSKY, TH., and PAVAN, C. 1950. Local and seasonal variations in relative frequencies of species of *Drosophila* in Brazil, J. Anim. Ecol., **19** : 1–14. Pp. 459, 461, 565.

DOBZHANSKY, TH., and SPASSKY, B. 1944. Genetics of natural populations. XI. Manifestation of genetic variants in *Drosophila pseudoobscura* in different environments, Genetics, **29**: 270–90. P. 678.

DOBZHANSKY, TH., and WALLACE, B. 1953. The genetics of homeostasis in *Drosophila*, Proc. Nat. Acad. Sc. Washington, **39**: 162–71. P. 703.

DOBZHANSKY, TH., and WRIGHT, S. 1943. Genetics of natural populations. X. Dispersion rates in *Drosophila pseudoobscura*, Genetics, **28**: 304–40. Pp. 93, 94, 139, 141.

———. 1947. Genetics of natural populations. XV. Rate of diffusion of a mutant gene through a population of *Drosophila pseudoobscura, ibid.*, **32**: 303–24. Pp. 93, 95.

DODD, A. P. 1936. The control and eradication of prickly pear in Australia, Bull. Ent. Research, **27**: 503–17. P. 501.

———. 1940. The biological campaign against prickly-pear. Brisbane: Comm. Prickly Pear Board. P. 502.

DOHANIAN, S. M. 1942. Variability of diapause in *Melissopus latiferreanus*, J. Econ. Ent., **35**: 406–8. P. 59.

DOLLEY, W. L., JR., and GOLDEN, L. H. 1947. The effect of sex and age on the temperature at which reversal in reaction to light in *Eristalis tenax* occurs, Biol. Bull., **92**: 178–86. P. 315.

DOLLEY, W. L., JR., and HAINES, H. G. 1930. An entomological sheep in wolf's clothing, Scient. Monthly, **31**: 508–16. P. 317.

DOLLEY, W. L., JR., and WHITE, J. D. 1951. The effect of illuminance on the reversal temperature in the drone fly, *Eristalis tenax*, Biol. Bull., **100**: 84–89. P. 315.

DORST, H. E., and DAVIS, E. W. 1937. Tracing long-distance movements of the beet leafhopper in the desert, J. Econ. Ent., **30**: 948–54. P. 110.

DOUDOROFF, P. 1938. Reactions of marine fishes to temperature gradients, Biol. Bull., **75**: 494–509. P. 131.

DOUGLASS, J. R. 1928. Precipitation as a factor in the emergence of *Epilachna corrupta* from hibernation, J. Econ. Ent., **21**: 203–13. P. 58.

DOWDESWELL, W. H.; FISHER, R. A.; and FORD, E. B. 1940. The quantitative study of populations in the Lepidoptera. I. *Polyommatus icarus*, Ann. Eugenics, **10**: 123–36. P. 629.

———. 1949. The quantitative study of populations in the Lepidoptera. II. *Maniola jurtina* L. Heredity, **3**: 67–84. P. 710.

DOWDESWELL, W. H., and FORD, E. B. 1952. The distribution of spot-numbers as an index of geographical variation in the butterfly *Maniola jurtina* L. (Lepidoptera: Satyridae), *ibid.*, **6**: 99–109. P. 709.

DOWNES, J. A. 1948. The history of the speckled wood butterfly (*Pararge aegeria*) in Scotland with a discussion on the recent changes of range of other British butterflies, J. Anim. Ecol., **17**: 131–38. P. 459.

DRESEL, E. I. B., and MOYLE, V. 1950. Nitrogenous excretion of amphipods and isopods, J. Exper. Biol., **27**: 210–25. Pp. 237, 238.

DRIFT, J. VAN DER. 1951. Analysis of the animal community in a beech forest floor, Tijdschr. Ent., **94**: 1–168. P. 511.

DUBININ, N. P. and TINIAKOV, G. G. 1945. Seasonal cycles and the concentration of inversions in populations of *Drosophila funebris*, Am. Nat., **79**: 570–72. Pp. 673, 676.

———. 1946*a*. Structural chromosome variability in urban and rural populations of *Drosophila funebris, ibid.*, **80**: 393–96. P. 676.

———. 1946*b*. Natural selection and chromosomal variability in populations of *Drosophila funebris*, J. Heredity, **37**: 39–44. P. 676.

DUBLIN, L. I., and LOTKA, A. J. 1925. On the true rate of increase as exemplified by the population of the United States, 1920, J. Am. Statist. A., **20**: 305–39. Pp. 41, 43, 44.

DUBLIN, L. I.; LOTKA, A. J.; and SPIEGELMAN, M. 1949. Length of life. Rev. ed. New York: Ronald Press Co. P. 40.

DUCLAUX, E. 1869. De l'influence du froid de l'hiver sur le développement de l'embryon du ver à soie et sur l'éclosion de la graine, C.R. Acad. Sc. Paris, **69**: 1021–22. P. 55.

DUVAL, M., and PORTIER, P. 1922. Limite de résistance au froid des chenilles de *Cossus cossus*, C.R. Soc. biol., Paris, **86**: 2–4. P. 189.

DYMOND, J. R. 1947. Fluctuations in animal populations with special reference to those of Canada, Tr. Roy. Soc. Canada, **41**: 1–34. Pp. 124, 478, 543, 641, 642, 643, 644.

EASTHAM, L. E. S., and MCCULLY, S. B. 1943. The oviposition responses of *Calandra granaria*, Linn., J. Exper. Biol., **20**: 35–42. P. 279.

EDNEY, E. B. 1947. Laboratory studies on the bionomics of the rat fleas *Xenopsylla brasiliensis* Baker and *X. cheopis* Roths. II. Water relations during the cocoon period, Bull. Ent. Research, **38**: 263–80. P. 232.

———. 1951. The evaporation of water from woodlice and the millipede *Glomeris*, J. Exper. Biol., **28**: 91–115. Pp. 232, 245, 260.

ELLIS, P. E. 1951. The marching behaviour of hoppers of the African migratory locust, *Locusta migratoria migratorioides* in the laboratory, Bull. Anti-locust Research Centre, London, Vol. **7**. P. 111.

EL-TABEY, A. M.; SHIHATA, A.; and MRAK, E. M. 1952. Intestinal yeast floras of successive populations of *Drosophila*, Evolution, **6**: 325–32. P. 679.

ELTON, C. 1925. The dispersal of insects to Spitsbergen, Tr. Ent. Soc. London, **73**: 289–99. P. 110.

———. 1927. Animal ecology. London: Sidgwick & Jackson. Pp. 3, 4, 5, 86, 123, 125, 514, 524.

———. 1930. Animal ecology and evolution. Oxford: Clarendon Press. P. 695.

———. 1938. Animal numbers and adaptation. *In:* BEER, G. R. DE (ed.), Evolution: essays on aspects of evolutionary biology, pp. 127–37. Oxford: Clarendon Press. Pp. 489, 557.

———. 1942. Voles, mice and lemmings. Oxford: Clarendon Press. Pp. 477, 478, 484.

———. 1946. Competition and the structure of ecological communities, J. Anim. Ecol., **15**: 54–68. P. 465.

———. 1949. Population interspersion: an essay on animal community patterns, J. Ecol., **37**: 1–23. Pp. 19, 28, 444, 497, 523, 580, 648.

ELTON, C., and NICHOLSON, M. 1942*a*. Fluctuations in numbers of the muskrat (*Ondatra zibethica*) in Canada, J. Anim. Ecol., **11**: 96–125. P. 641.

———. 1942*b*. The ten-year cycle in numbers of the lynx in Canada, *ibid.*, pp. 215–43. Pp. 640, 646.

EMDEN, F. VON. 1933. Ueber die erbliche Bindung von Latenzen an Jahreszeiten, Fifth Cong. Internat. Ent. Paris, 1932, pp. 813–22. P. 58.

ERRINGTON, P. L. 1934. Vulnerability of bobwhite populations to predation, Ecology, **15**: 110–27. Pp. 540, 541.

———. 1939. Reactions of muskrat populations to drought, *ibid.*, **20**: 168–86. P. 538.

———. 1940. Natural restocking of muskrat-vacant habitats, J. Wildlife Management, **4**: 173–85. P. 343.

———. 1943. An analysis of mink predation upon muskrats in north-central United States, Research Bull. Iowa Agr. Exper. Sta., **320**: 797–924. Pp. 88, 343, 479, 481, 491, 503, 538.

———. 1944. Ecology of the muskrat, Rep. Iowa Agr. Exper. Sta., 1944, pp. 187–89. Pp. 491, 542, 543.

———. 1945. Some contributions of a fifteen-year local study of the northern bobwhite to a knowledge of population phenomena, Ecol. Monogr., **15**: 1–34. Pp. 480, 558, 636.

———. 1946. Predation and vertebrate populations, Quart. Rev. Biol., **21**: 145–77 and 221–45. Pp. 364, 478, 481, 482, 503, 540, 543.

———. 1948. Environmental control for increasing muskrat production, Tr. 13th North Am. Wildlife Conf., pp. 596–609. Pp. 542, 543.

ERRINGTON, P. L., and HAMERSTROM, F. N. 1936. The northern bobwhite's winter territory, Research Bull. Iowa Agr. Exper. Sta., Vol. **201**. Pp. 540, 541.

ERRINGTON, P. L., and SCOTT, T. G. 1945. Reduction in productivity of muskrat pelts on an Iowa marsh through depredations of red foxes, J. Agr. Research, **71**: 137–48. P. 538.

EVANS, A. C. 1934. Studies on the influence of the environment of the sheep blowfly *Lucilia sericata* Meig. I. The influence of humidity and temperature on the egg, Parasitology, **26**: 366–77. P. 283.

———. 1938. Physiological relationships between insects and their host plants. I. The effect of the chemical composition of the plant on reproduction and production of winged forms in *Brevicoryne brassicae* L. (Aphididae), Ann. Appl. Biol., **25**: 558–72. Pp. 507, 508.

———. 1944. Observations on the biology and physiology of wireworms of the genus *Agriotes*, *ibid.*, **31**: 235–50. P. 58.

EVANS, F. C. 1949. A population study of house mice (*Mus musculus*) following a period of local abundance, J. Mammal., **30**: 351–63. P. 630.

EVANS, F. C., and SMITH, F. E. 1952. The intrinsic rate of natural increase for the human louse *Pediculus humanis* L., Am. Nat., **86**: 299–310. P. 34.

EWER, D. W., and EWER, R. F. 1941. The biology and behaviour of *Ptinus tectus* Boie (Coleoptera, Ptinidae) a pest of stored products. III. The effect of temperature and humidity on oviposition, feeding and duration of life-cycle, J. Exper. Biol., **18**: 290–305. P. 225.

FAURE, J. C. 1932. The phases of locusts in South Africa, Bull. Ent. Research, **23**: 293–424. P. 68.

FEDETOV, D. M. 1946. On functional changes in the imago of *Eurygaster integriceps*, Zool. Zhur., **25**: 245–50. P. 58.

FELLER, W. 1940. On the logistic law of growth and its empirical verifications in biology, Acta biotheoret., A, **5**: 51–66. P. 385.

FELT, E. P. 1925. The dissemination of insects by air currents, J. Econ. Ent., **18**: 152–56. Pp. 107, 110.

FICHT, G. A. 1936. The European cornborer in Indiana, Bull. Indiana Agr. Exper. Sta., No. 406. P. 688.

FIFE, J. M., and FRAMPTON, V. L. 1936. The pH gradient extending from the phloem into the parenchyma of the sugar beet and its relation to the feeding behaviour of *Eutettix tenellus*, J. Agr. Research, **53**: 581–93. P. 511.

FINNEY, D. J. 1947. Probit analysis: a statistical treatment of the sigmoid response curve. Cambridge: At the University Press. Pp. 178, 179, 180.

———. 1950. Two new uses of the Behrens-Fisher distribution, J. Roy. Statist. Soc., **12**: 296–300. P. 446.

FISHER, J., and WATERSON, G. 1941. The breeding distribution, history and population of the Fulmar (*Fulmarus glacialis*) in the British Isles, J. Anim. Ecol., **10**: 204–72. P. 337.

FISHER, K. 1938. Migrations of the silver-Y moth (*Plusia gamma*) in Great Britain, J. Anim. Ecol., **7**: 230–47. P. 120.

FISHER, R. A. 1930. The genetical theory of natural selection. Oxford: Clarendon Press. P. 670.

———. 1948. Statistical methods for research workers. 10th ed. Edinburgh: Oliver & Boyd. Pp. 94, 578.

FISHER, R. A., and FORD, E. B. 1947. The spread of a gene in natural conditions in a colony of the moth *Panaxia dominula* (L.), Heredity, **1**: 143–74. Pp. 90, 519, 672.

FISHER, R. A., and YATES, F. 1948. Statistical tables for biological, agricultural and medical research. London: Oliver & Boyd. P. 180.

FLANDERS, S. E. 1942. Oösorption and ovulation in relation to oviposition in the parasitic Hymenoptera, Ann. Ent. Soc. Amer., **35**: 251–66. P. 435.

———. 1943. Indirect hyperparasitism and observations on three species of indirect hyperparasites, J. Econ. Ent., **36**: 921–26. Pp. 473, 475.

———. 1944. Diapause in the parasitic Hymenoptera, *ibid.*, **37**: 408–11. P. 56.

———. 1947. Elements of host discovery exemplified by parasitic Hymenoptera, Ecology, **28**: 299–309. Pp. 115, 434, 435, 475, 476.

———. 1948. A host-parasite community to demonstrate balance, *ibid.*, **29**: 123. P. 440.

———. 1949. Black scale, California Agriculture, **3**: 74. P. 539.

FLEMION, F., and HARTZELL, A. 1936. Effect of low temperature in shortening the hibernation period of insects in the egg stage, Contr. Boyce Thompson Inst., **8**: 167–73. P. 67.

FLETCHER, J. J. 1889. Observations on the oviposition and habits of certain Australian batrachians, Proc. Linn. Soc., New South Wales, **4**: 357–87. P. 294.

FORD, E. B. 1945*a*. Butterflies. London: Collins. Pp. 513, 566, 567, 630, 664, 694, 696, 698, 699, 704, 709.

———. 1945*b*. Polymorphism, Biol. Rev., **20**: 73–88. Pp. 699, 700.

———. 1949. Early stages in allopatric speciation. *In:* JEPSON, G. L.; SIMPSON, G. G.; and MAYR, E. (eds.), Genetics, paleontology and evolution. Princeton, N.J.: Princeton University Press. Pp. 336, 715.

FOX, H. M. 1924. Lunar periodicity in reproduction, Proc. Roy. Soc. London, B, **95**: 523–50. P. 303.

———. 1932. Lunar periodicity in reproduction, Nature, **130**: 23. P. 303.

FRAENKEL, G., and BLEWETT, M. 1943*a*. The natural foods and the food requirements of several species of stored products insects, Tr. R. Ent. Soc. London, **93**: 457–88. P. 506.

———. 1943*b*. The basic food requirements of several insects, J. Exper. Biol., **20**: 28–34. P. 505.

———. 1943*c*. The sterol requirements of several insects, Biochem. J., **37**: 692–95. P. 506.

———. 1944. The utilisation of metabolic water in insects, Bull. Ent. Research, **35**: 127–37. Pp. 233, 234, 235, 506.

———. 1946. The dietetics of the caterpillars of three *Ephestia* species, *E. kuehniella*, *E. elutella*, and *E. cautella*, and of a closely related species, *Plodia interpunctella*, J. Exper. Biol., **22**: 162–71. P. 506.

FRAENKEL, G., and GUNN, D. L. 1940. The orientation of animals: kineses, taxes and compass reactions. Oxford: Clarendon Press. Pp. 133, 220, 289, 307, 308, 309, 510.

FRAMPTON, V. L.; LINN, M. B.; and HANSING, E. D. 1942. The spread of virus diseases of the yellows type under field conditions, Phytopathology, **32**: 799–808. P. 102.

FRANK, P. W. 1952. A laboratory study of intraspecies and interspecies competition in *Daphnia pulicaria* (Forbes) and *Simocephalus vetulus* O. F. Müller, Physiol. Zoöl., **25**: 178–204. P. 360.

FREEMAN, J. A. 1945. Studies in the distribution of insects by aerial currents: the insect population of the air from ground level to 300 feet, J. Anim. Ecol., **14**: 128–54. P. 108.

FRISCH, K. VON. 1950. Bees, their vision, chemical senses, and language. New York: Cornell University Press. P. 320.

FRY, F. E. J. 1947. Effects of the environment on animal activity, Univ. Toronto Stud. Biol., **55**: 1–62. Pp. 132, 146, 147, 199, 200.

FRY, F. E. J.; BRETT, J. R.; and CLAUSEN, G. H. 1942. Lethal limits of temperature for young goldfish, Rev. canad. de biol., **1**: 50–56. P. 200.

FRY, F. E. J., and HART, J. S. 1948. Cruising speed of goldfish in relation to water temperature, J. Fish. Research Board Canada, **7**: 169–74. P. 137.

FRY, F. E. J.; HART, J. S.; and WALKER, K. F. 1946. Lethal temperature relations for a sample of young speckled trout *Salvelinus fontinalis*, Pub. Ontario Fish. Research Lab., **66**: 5–35. Pp. 199, 201.

FUKUDA, S. 1940. The determination of voltinism in the silkworm with special reference to the pigment formation in the serosa of the egg (in Japanese; English summary), Zoöl. Mag. Tokyo, **52**: 415–29. P. 79.

———. 1951*a*. Factors determining the production of non-diapause eggs in the silkworm, Proc. Jap. Acad., **27**: 582–86. P. 79.

———. 1951*b*. The production of the diapause eggs by transplanting the suboesophageal ganglion in the silkworm, *ibid.*, pp. 672–77. P. 79.

———. 1952. Function of the pupal brain and suboesophageal ganglion in the production of non-diapause and diapause eggs in the silkworm, Annot. Zoöl. Japan, **25**: 149–55. P. 79.

FULLER, M. E. 1934. The insect inhabitants of carrion: a study in animal ecology, Bull. Counc. Scient. & Indust. Research Australia, No. 82. Pp. 450, 476, 525.

FULTON, R. A., and ROMNEY, V. E. 1940. The chloroform-soluble components of beet leafhoppers as an indication of the distance they move in the spring, J. Agr. Research, **61**: 737–43. P. 102.

GARDNER, T. R. 1938. Influence of feeding habits of *Tiphia vernalis* on the parasitization of the Japanese beetle, J. Econ. Ent., **31**: 204–7. P. 476.

GARLICK, W. G. 1948. A five-year field study of codling moth larval habits and adult emergence, Scient. Agr., **28**: 273–92. Pp. 61, 70.

GASCHEN, H. 1945. Les Glossines de l'Afrique Occidentale Française, Acta trop., Suppl., Vol. **2**. P. 629.

GAST, P. R. 1937. Contrast between the soil profiles developed under pines and hardwoods, J. Forestry, **35**: 11–16. P. 511.

GAUSE, G. F. 1931. The influence of ecological factors on the size of population, Am. Nat., **65**: 70–76. Pp. 356, 366.

———. 1934. The struggle for existence. Baltimore: Williams & Wilkins. Pp. 17, 333, 348, 351, 352, 354, 387, 392, 407, 408, 412, 414, 424, 438, 440, 441, 456.

———. 1935*a*. Vérifications expérimentales de la théorie mathématique de la lutte pour la vie. ("Actualités sc. indust.," Vol. **277**.) Paris: Hermann et Cie. Pp. 402, 438.

———. 1935*b*. Experimental demonstration of Volterra's periodic oscillations in the numbers of animals, J. Exper. Biol., **12**: 44–48. Pp. 424, 431, 439.

———. 1936. The principles of biocoenology, Quart. Rev. Biol., **11**:320–36. Pp. 433, 438.

GAUSE, G. F.; SMARAGDOVA, N. P.; and WITT, A. A. 1936. Further studies of interaction between predator and prey, J. Anim. Ecol., **5**: 1–18. Pp. 439, 440.

GAY, F. J. 1953. Observations on the biology of *Lyctus brunneus* (Steph.), Australian J. Zoöl., **1**: 102–10. P. 523.

GEISTHARDT, G. 1937. Ueber die ökologische Valenz zwerier Wanzenarten mit verschiedenem Verbreitungsgebiet, Ztschr. f. Parasitenk., **9**: 151–202. P. 285.

GEROULD, J. H. 1921. Blue-green caterpillars: the origin and ecology of a mutation in haemolymph color in *Colias* (*Eurymus*) *philodice*, J. Exper. Zoöl., **34**: 385–415. P. 698.

GERSHENSON, S. 1945. Evolutionary studies on the distribution and dynamics of melanism in the hamster (*Cricetus cricetus* L.). 1. Distribution of black hamsters in the Ukrainian and Bashkirian Soviet Socialist Republics (U.S.S.R.), Genetics, **30**: 207–51. Pp. 676, 686.

GILMOUR, D.; WATERHOUSE, D. F.; and MCINTYRE, G. A. 1946. An account of experiments undertaken to determine the natural population density of the sheep blowfly, *Lucilia cuprina* Wied., Bull. Counc. Scient. & Indust. Research Australia, No. 195. Pp. 98, 99, 100, 562, 563, 564.

GIVEN, B. B. 1944. Notes on the physical ecology of *Diadromus* (*Thyraeella*) *collaris*, Grav., New Zealand J. Scient. Tech., **26**: 198–201. P. 187.

GLASER, R. W. 1923. The effect of food on longevity and reproduction in flies, J. Exper. Zoöl., **38**: 383–412. P. 507.

GLENN, P. A. 1922. Relation of temperature to development of the codlin moth, J. Econ. Ent., **15**: 193–98. P. 166.

———. 1931. Use of temperature accumulations as an index to the time of appearance of certain insect pests during the season, Tr. Illinois Acad. Sc., **24**: 167–80. P. 166.

GLICK, P. A. 1939. The distribution of insects, spiders and mites in the air, Tech. Bull. U.S. Dept. Agr., No. 673. P. 108.

———. 1942. Insect population and migration in the air, Pub. Am. A. Adv. Sc., **17**: 88–98. P. 108.

GOLDSCHMIDT, R. 1927. Physiologische Theorie der Vererbung. Berlin: Julius Springer. P. 80.

———. 1934. *Lymantria*, Bibliog. genet., **11**: 1–186. P. 689.

Govaerts, J., and Leclercq, J. 1946. Water exchange between insects and air moisture, Nature, **157**: 483. P. 232.

Graham, S. A., and Orr, L. W. 1940. The spruce budworm in Minnesota, Tech. Bull. Minnesota Agr. Exper. Sta., No. 142. P. 62.

Grange, W. B. 1947. Practical beaver and muskrat farming. Babcock, Wis.: Sandhill Press. P. 543.

Gray, E. 1951. *Stylonichia mytilus* and the lunar periods, Nature, **167**: 38. P. 304.

Gray, J. 1928. The role of water in the evolution of terrestrial vertebrates, J. Exper. Biol., **6**: 26–31. Pp. 226, 251.

Green, R. G., and Evans, C. A. 1940. Studies on a population cycle of snowshoe hares on the Lake Alexander area, J. Wildlife Management, **4**: 220–38, 267–78, 347–58. P. 398.

Green, R. G., and Larson, C. L. 1938. A description of shock disease in the snowshoe hare, Am. J. Hyg., **28**: 190–212. P. 398.

Greenslade, R. M. 1941. The migration of the strawberry aphis *Capitophorus fragariae* Theob., J. Pomol., **19**: 87–106. P. 106.

Greenwood, M.; Hill, A. B.; Topley, W. W. C.; and Wilson, J. 1936. Experimental epidemiology. ("Great Britain M.R.C. Special Rep. Ser.," No. 209.) London: H. M. Stationery Office. P. 483.

Greenwood, P. H. 1951. Evolution of the African cichlid fishes: the *Haplochromis* species-flock in Lake Victoria, Nature, **167**: 19–20. P. 718.

Gregory, F. G. 1948. The control of flowering in plants. Symp. Soc. Exper. Biol. (Growth), **2**: 75–103. P. 292.

Griffin, D. R., and Welsh, J. H. 1937. Activity rhythms in bats under constant external conditions, J. Mammal., **18**: 337–42. Pp. 322, 323.

Grison, P. 1947. Développement sans diapause des chenilles de *Euproctis phaeorrhaea*, L. (Lep. Liparides), Compt. rend. Acad. Sc., Paris, **225**: 1089–90. Pp. 58, 66, 75.

Griswold, G. H. 1944. Studies on the biology of the webbing clothes moth (*Tineola bisselliella* Hum.), Mem. Cornell Agr. Exper. Sta., **262**: 1–59. P. 285.

Gunn, D. L. 1933. The temperature and humidity relations of the cockroach (*Blatta orientalis*). I. Desiccation, J. Exper. Biol., **10**: 271–85. Pp. 240, 255.

———. 1937. The humidity reactions of the wood louse *Porcellio scaber* (Latreille), *ibid.*, **14**: 178–86. Pp. 210, 221.

———. 1940. The daily rhythm of activity of the cockroach *Blatta orientalis* L. I. Aktograph experiments, especially in relation to light, *ibid.*, **17**: 267–77. P. 321.

Gunn, D. L., and Cosway, C. A. 1938. The temperature and humidity relations of the cockroach. V. Humidity preference, J. Exper. Biol., **15**: 555–63. Pp. 217, 221.

Gunn, D. L.; Perry, F. C.; Seymour, W. G.; Telford, T. M.; Wright, E. M.; and Yeo, D. 1948. Behaviour of the desert locust (*Schistocerca gregaria*, Forsk.) in Kenya in relation to aircraft spraying, Bull. Anti-locust Research Centre London, No. 3. P. 138.

Gunn, D. L., and Pielou, D. P. 1940. The humidity behaviour of the mealworm beetle *Tenebrio molitor* L. III. The mechanism of the reaction, J. Exper. Biol., **17**: 307–16. Pp. 220, 221.

Gunn, R. M. C.; Sanders, R. N.; and Granger, W. 1942. Studies in fertility in sheep. 2. Seminal changes affecting fertility in rams, Bull. Counc. Scient. & Indust. Research, Australia, No. 148. P. 175.

Hadley, P. B. 1908. The behaviour of the larval and adolescent stages of the American lobster (*Homarus americanus*), J. Comp. Neurol., **18**: 199–301. P. 317.

Haeckel, E. 1870. Ueber Entwickelungsgang u. Aufgabe der Zoologie, Jenaische Ztschr., **5**: 353–70. P. 13.

Hagan, H. R. 1917. Observations on the embryonic development of the mantid *Paratenodera sinensis*, J. Morphol., **30**: 223–44. P. 230.

Hall, F. G. 1922. The vital limits of exsiccation of certain animals, Biol. Bull., **42**: 31–51. Pp. 209, 260.

HAMILTON, A. G. 1950. Further studies on the relation of humidity and temperature to the development of two species of African locusts—*Locusta migratoria migratorioides*, R. and F., and *Schistocerca gregaria*, Forsk., Tr. Roy. Ent. Soc. London, **101** : 2–56. Pp. 277, 278.

HARDY, A. C., and MILNE, P. S. 1937. Insect drift over the North Sea, Nature, **139** : 510–11. P. 108.

———. 1938*a*. Studies in the distribution of insects by aerial currents. Experiments in aerial tow-netting from kites, J. Anim. Ecol., **7** : 199–229. P. 108.

———. 1938*b*. Aerial drift of insects, Nature, **141** : 602–3. P. 108.

HARRIES, F. H. 1937. Some effects of temperature on the development and oviposition of *Microbracon hebetor* (Say.), Ohio J. Sc., **37** : 165. P. 171.

———. 1939. Some temperature coefficients for insect oviposition, Ann. Ent. Soc. Amer., **32** : 758–76. P. 172.

HARRISON, L. 1922. On the breeding habits of some Australian frogs, Australian Zoöl., **3** : 17–34. P. 294.

HART, D. S. 1951. Photoperiodicity in the female ferret, J. Exper. Biol., **28** : 1–12. P. 292.

HART, J. S. 1952. Geographic variations of some physiological and morphological characters in certain freshwater fish, Pub. Ontario Fish. Research Lab., Vol. **72.** Pp. 202, 686.

HASEGAWA, K. 1952. Studies on the voltinism in the silkworm, *Bombyx mori*, L. with special reference to the organs concerning determination of voltinism. J. Fac. Agr. Tottori Univ., **1** : 83–124. P. 79.

HAZELHOFF, E. H. 1928. Carbon dioxide a chemical accelerating the penetration of respiratory insecticides into the tracheal system by keeping open the tracheal valves, J. Econ. Ent., **21** : 790. P. 239.

HEFLEY, H. M. 1928. Differential effects of constant humidities on *Protoparce quinquemaculata*, Haworth and its parasite *Winthemia quadripustulata*, J. Econ. Ent., **21** : 213–21. P. 469.

HELLER, J. 1926. Chemische Untersuchungen über die Metamorphose der Insekten. III. Ueber die "subitane" und "latente" Entwicklung, Biochem. Ztschr., **169** : 208–34. P. 58.

HENNEGUY, L. F. 1904. Les insectes: morphologie, reproduction, embryogénie. Paris: Masson et Cie. P. 56.

HENRY, C. J. 1939. Response of wildlife to management practices on the Lower Souris Migratory Waterfowl Refuge, Tr. North Am. Wildlife Conf., **4** : 372–77. P. 543.

HENSON, W. R., and SHEPHERD, R. F. 1952. The effects of radiation on the habitat temperature of the lodgepole needle miner *Recurvaria milleri* (Gelechiidae, Lepidoptera), Canad. J. Zoöl., **30** : 144–53. P. 529.

HERRSTROEM, G. 1949. Illumination preferendum and adaptation experiments with *Agonum dorsale* Pont. (Col., Carabidae), Oikos, **1** : 48–55. P. 318.

HEWITT, C. G. 1921. The conservation of the wild life of Canada. New York: Charles Scribner's Sons. P. 641.

HILL, M., and PARKES, A. S. 1930. On the relation between the anterior pituitary body and the gonads. II. The induction of ovulation in the anoestrous ferret, Proc. Roy. Soc. London, B, **107** : 39–49. P. 291.

HODSON, A. C. 1937. Some aspects of the role of water in insect hibernation, Ecol. Monogr., **7** : 271–315. P. 195.

HODSON, A. C., and WEINMAN, C. J. 1945. Factors affecting recovery from diapause and hatching of eggs of the forest tent caterpillar *Malacosoma disstria*, Hubn., Tech. Bull. Univ. Minnesota Agr. Exper. Sta., No. 170. Pp. 59, 62.

HOLDAWAY, F. G. 1932. An experimental study of the growth of populations of the flour beetle *Tribolium confusum* Duval as affected by atmospheric moisture, Ecol. Monogr., **2** : 261–304. Pp. 33, 357, 394.

HOLME, N. A. 1950. Population dispersion in *Tellina tenuis*, da Costa, J. Marine Biol. A. United Kingdom, **29** : 267–80. P. 567.

HOOVER, E. E., and HUBBARD, H. E. 1937. Modification of the sexual cycle in trout by control of light, Copeia, **4**: 206–10. P. 292.

HOUGH, W. S. 1928. Relative resistance to arsenical poisoning of two codling moth strains, J. Econ. Ent., **21**: 325–29. P. 702.

———. 1929. Studies of the relative resistance to arsenical poisoning of different strains of codling moth larvae, J. Agr. Research, **38**: 245–56. P. 702.

———. 1934. Colorado and Virginia strains of codling moth in relation to their ability to enter sprayed and unsprayed apples, *ibid.*, **48**: 533–53. P. 702.

HOVANITZ, W. 1948. Ecological segregation of inter-fertile species of *Colias*, Ecology, **29**: 461–69. P. 717.

———. 1949*a*. Change of host preference in *Colias philodice*, J. Econ. Ent., **41**: 980–81. Pp. 691, 692.

———. 1949*b*. Interspecific matings between *Colias eurytheme* and *Colias philodice* in wild populations, Evolution, **3**: 170–73. P. 717.

HOWARD, L. O., and FISKE, W. F. 1911. The importation into the United States of the parasites of the gipsy moth and the brown-tail moth, Bull. U.S. Bur. Ent., No. 91. P. 16.

HOWE, R. W. 1953*a*. Studies on beetles of the family Ptinidae. VIII. The intrinsic rate of increase of some ptinid beetles, Ann. Appl. Biol., **40**: 121–34. Pp. 34, 43, 45, 50, 52.

———. 1953*b*. The rapid determination of the intrinsic rate of increase of an insect population, *ibid.*, pp. 134–51. P. 34.

HOWE, R. W., and OXLEY, T. A. 1944. The use of carbon dioxide production as a measure of infestation of grain by insects, Bull. Ent. Research, **35**: 11–22. P. 45.

HOWES, N. H., and WELLS, G. P. 1934*a*. The water relations of snails and slugs. I. Weight rhythms in *Helix pomatia*, J. Exper. Biol., **11**: 327–43. Pp. 225, 250, 260.

———. 1934*b*. The water relations of snails and slugs. II. Weight rhythms in *Arion ater*, L. and *Limax flavus*, L., *ibid.*, pp. 344–51. Pp. 225, 250, 260.

HUBER, L. L.; NEISWANDER, G. R.; and SALTER, R. M. 1928. The European cornborer and its environment, Bull. Ohio Agr. Exper. Sta., No. 429. P. 280.

HUECK, H. J. 1951. Influence of light upon the hatching of winter-eggs of the fruit tree red spider, Nature, **167**: 993–94. P. 328.

HUNTSMAN, A. G. 1946. Heat stroke in Canadian maritime stream fishes, J. Fish. Research Board Canada, **6**: 476–82. P. 201.

HUTCHINSON, G. E. 1953. The concept of pattern in ecology, Proc. Acad. Nat. Sc. Philadelphia, **105**: 1–12. P. 568.

HUTCHINSON, G. E., and DEEVEY, E. S., JR. 1949. Ecological studies on populations, Surv. Biol. Prog., **1**: 325–59. Pp. 334, 345, 406.

HUXLEY, J. 1942. Evolution the modern synthesis. London: Allen & Unwin. Pp. 694, 695, 696, 710, 715, 716.

IDE, F. P. 1935. The effect of temperature on the distribution of the mayfly fauna of a stream, Univ. Toronto Stud. Biol., **39**: 1–76. P. 163.

IMMS, A. D. 1906. Anurida, Mem. Liverpool Marine Biol., Vol. **13**. P. 230.

ISELY, D., and ACKERMAN, A. J. 1923. Life history of the codling moth in Arkansas with special reference to factors limiting abundance, Bull. Arkansas Agr. Exper. Sta., No. 189. P. 330.

JACK, R. W., and WILLIAMS, W. L. 1937. The effect of temperature on the reaction of *Glossina morsitans*, Westw., to light, Bull. Ent. Research, **28**: 499–503. P. 315.

JACKSON, C. H. N. 1930. Contributions to the bionomics of *Glossina morsitans*, Bull. Ent. Research, **21**: 491–527. P. 629.

———. 1933. On the true density of tsetse flies, J. Anim. Ecol., **2**: 204–9. P. 629.

———. 1937. Some new methods in the study of *Glossina morsitans*, Proc. Zoöl. Soc. London, 1936, pp. 811–96. Pp. 629, 632, 633.

———. 1939. The analysis of an animal population, J. Anim. Ecol., **8**: 238–46. P. 629.

———. 1949. The biology of tsetse flies, Biol. Rev., **24**: 174–97. P. 629.

JACKSON, C. M. 1926. Storage of water in various parts of the earthworm at different stages of exsiccation, Proc. Soc. Exper. Biol. & Med., **23** : 500–504. P. 260.

JAHN, T. L. 1935. Nature and permeability of grasshopper egg membranes. II. Chemical composition of membranes, Proc. Soc. Exper. Biol. & Med., **33** : 159–63. P. 249.

———. 1936. Nature and permeability of grasshopper egg membranes. III. Changes in electrical properties of the membranes during development, J. Cell. & Comp. Physiol., **8** : 289–300. P. 249.

JAHN, T. L., and CRESCITELLI, F. 1938. The electrical response of the grasshopper eye under conditions of light and dark adaptation, J. Cell. & Comp. Physiol., **12** : 39–55. P. 324.

———. 1940. Diurnal changes in the electrical response of the compound eye, Biol. Bull., **78** : 42–52. P. 324.

JAHN, T. L., and WULFF, V. J. 1943. Electrical aspects of a diurnal rhythm in the eye of *Dytiscus fasciventris*, Physiol. Zoöl., **16** : 101–9. P. 324.

JENKINS, D. W., and CARPENTER, S. J. 1946. Ecology of the tree hole breeding mosquitoes of Nearctic North America, Ecol. Monogr., **16** : 31–47. P. 523.

JENKINS, D. W., and HASSETT, C. C. 1951. Dispersal and flight range of northern mosquitoes marked with radiophosphorus. Canad. J. Zoöl., **29** : 178–87. P. 101.

JOHNSON, C. G. 1934. On the eggs of *Notostira erratica*. L. (Hemiptera, Capsidae). I. Observations on the structure of the egg and the subopercular yolk-plug, swelling of the egg and hatching, Tr. Soc. Brit. Ent., **1** : 1–32. P. 230.

———. 1937. Absorption of water and the associated volume changes in eggs of *Notostira erratica*, L. (Hemiptera, Capsidae) during embryonic development under experimental conditions, J. Exper. Biol., **14** : 413–21. P. 230.

———. 1940. The longevity of the fasting bed bug (*C. lectularius* L.) under experimental conditions and particularly in relation to the saturation deficiency law of water loss, Parasitology, **32** : 239–70. Pp. 179, 180, 184, 256, 264.

———. 1942. Insect survival in relation to the rate of water loss, Biol. Rev., **17** : 151–77. Pp. 264, 266.

———. 1950. Infestation of a bean field by *Aphis fabae* Scop. in relation to wind direction, Ann. Appl. Biol., **37** : 441–50. P. 105.

———. 1951. The study of wind-borne insect populations in relation to terrestrial ecology, flight periodicity and the estimation of aerial populations, Scient. Prog., **39** : 41–62. Pp. 108, 109.

JOHNSON, C. G., and SOUTHWOOD, T. R. E. 1949. Seasonal records in 1947 and 1948 of flying Hemiptera-Heteroptera, particularly *Lygus pratensis* L., caught in nets 50 feet to 3,000 feet above the ground, Proc. Roy. Ent. Soc. London, **24** : 128–30. P. 113.

JOHNSON, D. S. 1952. A thermal race of *Daphnia atkinsoni*, Baird and its distributional significance, J. Anim. Ecol., **21** : 118–19. P. 685.

JOHNSON, M. S. 1939. Effect of continuous light on periodic spontaneous activity of white footed mice (*Peromyscus*), J. Exper. Zoöl., **82** : 315–28. P. 322.

JONES, E. P. 1937. The egg parasites of the cotton boll worm, *Heliothis armigera*, Hubn. (*obsoleta*, Fabr.), in Southern Rhodesia, Pub. Brit. South African Co., **6** : 37–105. P. 435.

JORDAN, E. O., and BURROWS, W. 1945. Textbook of bacteriology. 14th ed. Philadelphia: W. B. Saunders Co. P. 412.

KALMUS, H. 1938. Das Aktogramm des Flusskrebses und seine Beeinflussung durch Organextrakte, Ztschr. f. vergl. Physiol., **25** : 798–802. P. 324.

———. 1940. Diurnal rhythms in the axolotl larva and in *Drosophila*, Nature, **145** : 72–73. P. 327.

KAMENSKII, C. A., and PAIKIN, D. M. 1939. The causes of several years' hibernation of *Cleonus punctiventris*, Bull. Plant Prot., Leningrad, **1** : 49–54. P. 58.

KATÔ, M. 1949. The diurnal activity of a dermestid beetle, *Attagenus japonicus*, Scient. Rep. Tôhoku Univ., No. 4, ser. B, **18** : 195–204. P. 326.

KENDALL, M. G. 1948. Advanced theory of statistics. 4th ed. London: Charles Griffin & Co. Pp. 643, 646.

KENNEDY, J. S. 1937. The humidity reactions of the African migratory locust *Locusta migratoria migratorioides* R. and F., gregarious phase, J. Exper. Biol., **14** : 187–97. P. 221.
———. 1939. The behaviour of the desert locust (*Schistocerca gregaria*, Forsk.) (Orthoptera) in an outbreak centre, Tr. Roy. Ent. Soc. London, **89** : 385–542. Pp. 111, 136, 139.
———. 1947. Personal communication. P. 299.
———. 1950*a*. Host-finding and host-alternation in aphides, Proc. 8th Internat. Cong. Ent. Stockholm, 1948, pp. 423–26. Pp. 105, 106.
———. 1950*b*. Aphid migration and the spread of plant viruses, Nature, **165** : 1024. Pp. 105, 106.
KETTLE, D. S. 1951. The spatial distribution of *Culicoides impunctatus* Goet. under woodland and moorland conditions and its flight range through woodland, Bull. Ent. Research, **42** : 239–91. Pp. 101, 103, 536.
KEVAN, D. K. McE. 1944. The bionomics of the neotropical cornstalk borer, *Diatraea lineolata* Wlk. (Lep. Pyral.) in Trinidad, B.W.I., Bull. Ent. Research, **35** : 23–30. Pp. 58, 75, 76.
KEY, K. H. L. 1938. The regional and seasonal incidence of grasshopper plagues in Australia, Bull. Counc. Scient. & Indust. Research Australia, No. 117. P. 551.
———. 1942. An analysis of the outbreaks of the Australian plague locust (*Chortoicetes terminifera*, Walk.) during the seasons 1937–38 and 1938–39, *ibid.*, No. 146. P. 545.
———. 1943. The outbreak of the Australian plague locust (*Chortoicetes terminifera*, Walk.) in the season 1939–40, with special reference to the influence of climatic factors, *ibid.*, No. 160. P. 545.
———. 1945. The general ecological characteristics of the outbreak areas and outbreak years of the Australian plague locust (*Chortoicetes terminifera*, Walk.), *ibid.*, No. 186. Pp. 545, 546, 549.
———. 1954. The taxonomy, phases, and distribution of the genera *Chortoicetes* Brunn. and *Austroicetes* Uv. (Orthoptera: Acrididae). Canberra: Comm. Scient. & Indust. Research Organization. P. 546.
KINSEY, A. C. 1942. Isolating mechanisms in gall wasps, Biol. Symp., **6** : 251–69. P. 714.
KIRKPATRICK, T. W. 1923. The Egyptian cotton-seed bug (*Oxycarenus hyalinipennis*, Costa.), its bionomics, damage, and suggestions for remedial measures, Ministry Agr. Egypt. Tech. Scient. Serv. Bull., No. 35. P. 264.
KLUIJVER, H. N. 1951. The population ecology of the great tit *Parus m. major*, Ardea, **39** : 1–135. Pp. 124, 398, 490, 544, 545.
KNIGHT, H. H. 1922. Studies on the life history and biology of *Perillus bioculatus* including observations on the nature of the colour pattern, 19th Rep. State Ent. Minnesota, pp. 50–96. P. 198.
KNOCKE, E. 1933. Klima und Nonne. I. Die Entwicklungsruhe des Embryo. Arb. biol. Reichs. Land-w. Forstw., **20** : 193–235. P. 70.
KOGURE, M. 1933. The influence of light and temperature on certain characters of the silkworm—*Bombyx mori*, J. Dept. Agr. Kyushu Imp. Univ., **4** : 1–93. Pp. 58, 77, 294, 295, 296, 332.
KOMAI, T., and HOSINO, Y. 1951. Contributions to the evolutionary genetics of the ladybeetle, *Harmonia*. II. Microgeographic variations, Genetics, **36** : 382–90. P. 693.
KONTKANEN, P. 1948. On the restriction of dominance groups in synecological research on insects, Ann. Ent. Fenn., **14** : 33–40. P. 4.
———. 1949. On the determination of affinity between different species in synecological analysis, *ibid.*, Suppl., pp. 118–25. P. 4.
———. 1950. Quantitative and seasonal studies on the leafhopper fauna of the field stratum on open areas in north Karelia, Ann. Zoöl. Soc. Zool. Bot. Fenn., **13** : 1–91. Pp. 4, 5.
KOOPMAN, K. F. 1950. Natural selection for reproductive isolation between *Drosophila pseudoobscura* and *Drosophila persimilis*, Evolution, **4** : 135–48. Pp. 718, 719.
KORRINGA, P. 1947. Relations between the moon and periodicity in the breeding of marine animals, Ecol. Monogr., **17** : 349–81. Pp. 302, 303.

Koskimies, J. 1947. On movements of the swift *Micropus a. apus* L., during the breeding season, Ornis Fenn., **24**: 106–11. P. 89.

Kostitzin, V. A. 1939. Mathematical biology. London: Harrap. Pp. 348, 354.

Kozhanchikov, I. V. 1938. Physiological conditions of cold-hardiness in insects, Bull. Ent. Research, **29**: 253–62. Pp. 58, 76, 189, 191, 192, 193, 194.

Krause, G. 1938. Die Ausbildung der Körpergrundgestalt im Ei der Gewächschausschrecke *Tachycines asynamorus*, Ztschr. f. Morphol. u. Ökol. d. Tiere, **34**: 499–564. P. 230.

Krogerus, H. 1949. Experimentalla undersökenjav over kropps-temperaturen hos skalbaggav, Ann. Ent. Fenn. Suppl., **14**: 132–33. P. 256.

Krogh, A. 1914. On the influence of the temperature on the rate of embryonic development, Ztschr. f. allg. Physiol., **16**: 163–77. P. 165.

Krogh, A., and Zeuthen, E. 1941. The mechanism of flight preparation in some insects, J. Exper. Biol., **18**: 1–10. P. 256.

Krumbiegel, I. 1932. Untersuchungen über physiologische Rassenbildung, Zool. Jahrb., **63**: 183–280. P. 685.

Künkel, K. 1916. Zur Biologie der Lungenschnecken. Heidelberg: Carl Winter. P. 260.

Lack, D. 1944. Ecological aspects of species-formation in passerine birds, Ibis., **86**: 260–86. Pp. 457, 458, 459, 460, 462.

———. 1945. The ecology of closely related species with special reference to the cormorant (*Phalacrocorax carbo*) and shag (*P. aristotelis*), J. Anim. Ecol., **14**: 12–16. P. 461.

———. 1946. Competition for food by birds of prey, J. Anim. Ecol., **15**: 123–29. Pp. 432, 459, 464, 487.

———. 1947. Darwin's finches. Cambridge: At the University Press. Pp. 459, 460, 461, 462.

Lack, D., and Venables, L. S. V. 1939. The habitat distribution of British woodland birds, J. Anim. Ecol., **8**: 39–70. Pp. 522, 524.

Laing, J. 1938. Host finding by insect parasites. II. The chance of *Trichogramma evanescens* finding its host, J. Exper. Biol., **15**: 281–302. P. 475.

Langenbuck, R. 1941. Zur Biologie des *Grapholita nigricana*. Arb. physiol. f. angew. Ent., **8**: 219–44. P. 58.

Larsen, E. B. 1943. The influence of humidity on life and development of insects. Experiments on flies, Vidensk. Medd. dansk. naturh. Fören. Kbh., **107**: 127–84. Pp. 203, 268, 283.

Lathrop, F. H., and Nickels, C. B. 1931. The blueberry maggot from an ecological viewpoint. Ann. Ent. Soc. Amer., **24**: 260–74. Pp. 612, 613.

Leclercq, J. 1946*a*. Des insectes qui boivent de l'eau. Bull. Ann. Soc. Ent. Belg., **82**: 71–75. P. 225.

———. 1946*b*. Influence de l'humidité sur les cocons du ver à soie (*Bombyx mori*), Experientia, **2**: 1–4. P. 285.

———. 1946*c*. Influence de l'humidité atmosphérique sur les chrysalides d'*Araschnia levana*, L., Lambillionea, **2**: 27–8. P. 285.

Leeper, G. W. 1941. The scientist's English, Australian J. Sc., **3**: 121–23. P. 11.

———. 1952. An index of ease of reading, *ibid.*, **15**: 31–32. P. 11.

Lees, A. D. 1943*a*. On the behaviour of wireworms of the genus *Agriotes* Esch. (Coleoptera, Elateridae). I. Reactions to humidity, J. Exper. Biol., **20**: 43–53. Pp. 210, 219, 221.

———. 1943*b*. On the behaviour of wireworms of the genus *Agriotes* Esch. (Coleoptera, Elateridae). II. Reactions to moisture, *ibid.*, pp. 54–60. P. 210.

———. 1946*a*. The water balance in *Ixodes ricinus* L. and certain other species of ticks, Parasitology, **37**: 1–20. Pp. 231, 232, 233, 271.

———. 1946*b*. Chloride regulation and the function of the coxal glands in ticks, *ibid.*, pp. 172–84. P. 236.

———. 1947. Transpiration and the structure of the epicuticle in ticks, J. Exper. Biol., **23**: 379–410. Pp. 227, 243.

———. 1952. Entomology: Diapause. Sc. Prog., **40**: 306–12. P. 299.

Lees, A. D., and Beament, J. W. L. 1948. An egg-waxing organ in ticks, Quart. J. Mic. Sc., **89**: 291–332. P. 246.

LEES, A. D., and MILNE, A. 1951. The seasonal and diurnal activities of individual sheep ticks (*Ixodes ricinus* L.), Parasitology, **41**: 189–207. P. 625.

LEGENDRE, J. 1934. La longévité chez les larves d'un moustique arboricole. Compt. rend. Acad. Sc. Paris, **198**: 1263–65. P. 299.

LEIGHLY, J. 1937. A note on evaporation, Ecology, **18**: 180–98. Pp. 253, 254.

LEOPOLD, A. 1933. Game management. New York: Charles Scribner's Sons. Pp. 345, 522, 540.

———. 1943. Deer irruptions, Wisconsin Conserv. Bull., August, 1943. P. 502.

LESLIE, P. H. 1948. Some further notes on the use of matrices in population mathematics, Biometrika, **35**: 213–45. Pp. 362, 363, 364, 383.

LESLIE, P. H., and PARK, T. 1949. The intrinsic rate of natural increase of *Tribolium castaneum* Herbst, Ecology, **30**: 469–77. Pp. 34, 384.

LESLIE, P. H., and RANSON, R. M. 1940. The mortality, fertility and rate of natural increase of the vole (*Microtus agrestis*) as observed in the laboratory, J. Anim. Ecol., **9**: 27–52. Pp. 34, 36, 37, 43, 44, 46.

LEVENE, H.; PAVLOVSKY, O.; and DOBZHANSKY, TH. 1954. Interaction of the adaptive values in polymorphic experimental populations of *Drosophila pseudoobscura*, Evolution, Vol. **8**: (in press). Pp. 678, 698.

LEVINE, R. P. 1952. Adaptive responses of some third chromosome types of *Drosophila pseudoobscura*, Evolution, **6**: 216–33. P. 680.

LEWIS, C. B., and BLETCHLEY, J. D. 1943. The emergence rhythm of the dungfly *Scopeuma* (*Scatophaga*) *stercoraria*, L., J. Anim. Ecol., **12**: 11–19. P. 327.

LINDQUIST, A. W. 1952. Radioactive materials in entomological research, J. Econ. Ent., **45**: 264–70. P. 101.

LINDQUIST, B. 1942. Experimentelle Untersuchungen über die Bedeutung einiger Landmollusken für die Zersetzung der Waldstren, K. fysiogrp. Sällsk. Lund. Förh., **11**: 144–56. Pp. 101, 511.

LLOYD, D. C. 1940. Host selection by hymenopterous parasites of the moth, *Plutella maculipennis* Curtis, Proc. Roy. Soc. London, B, **128**: 451–84. P. 435.

LLOYD, L. 1937. Observations on sewage flies: their seasonal incidence and abundance, J. Inst. Sewage Purification (1937), pp. 1–16. P. 607.

———. 1941. The seasonal rhythm of a fly (*Spaniotoma minima*) and some theoretical considerations, Tr. Roy. Soc. Trop. Med. & Hyg., **35**: 93–104. P. 607.

———. 1943. Materials for a study in animal competition. II. The fauna of the sewage beds. III. The seasonal rhythm of *Psychoda alternata*, Say and an effect of intraspecific competition, Ann. Appl. Biol., **30**: 47–60, 358–64. Pp. 452, 605, 606, 607, 609, 610.

LLOYD, L.; GRAHAM, J. F.; and REYNOLDSON, T. B. 1940. Materials for a study in animal competition. The fauna of the sewage bacteria beds, Ann. Appl. Biol., **27**: 122–50. Pp. 452, 607.

LOEB, J. 1918. Forced movements, tropisms and animal conduct. Philadelphia: J. B. Lippincott Co. Pp. 309, 316.

LORENZ, K. Z. 1952. King Solomon's ring. London: Methuen Co. P. 513.

LOTKA, A. J. 1925. Elements of physical biology. Baltimore: Williams & Wilkins. Pp. 34, 42, 347, 406, 407, 412, 413, 414.

———. 1932. The growth of mixed populations: two species competing for a common food supply, J. Washington Acad. Sc., **22**: 461–69. Pp. 348, 406, 407.

———. 1939. Théorie analytique des associations biologiques. Deuxième partie. Analyse démographique avec application particulière à l'espèce humaine, "Actualités sc. indust.," **780**: 1–149 Paris: Hermann & Cie. P. 43.

LUDWIG, D. 1928. The effect of temperature on the development of an insect (*Popillia japonica*), Physiol. Zoöl., **1**: 358–98. P. 187.

———. 1932. The effect of temperature on the growth curves of Japanese beetle (*Popillia japonica*, Newman), *ibid.*, **5**: 431–47. Pp. 59, 62, 71, 72.

———. 1937. The effect of different relative humidities on respiratory metabolism and survival of the grasshopper *Chortophaga viridifasciata* De Geer, *ibid.*, **10**: 342–51. P. 231.

———. 1942. The effect of different relative humidities, during the pupal stage, on the reproductive capacity of the Luna moth, *Tropaea luna* L., *ibid.*, **15**: 48–60. P. 280.

———. 1943. The effect of different relative humidities, during the pupal stage, on the reproductive capacity of the Cynthia moth, *Samia walkeri*, Felder and Felder, *ibid.*, **16**: 381–88. P. 280.

LUDWIG, D., and ANDERSON, J. M. 1942. Effects of different humidities, at various temperatures, on the early development of four saturniid moths (*Platysamia cecropia* Linnaeus, *Telea polyphemus* Cramer, *Samia walkeri* Felder and Felder, and *Callosamia promethea* Drury) and on the weights and water contents of their larvae, Ecology, **23**: 259–74. P. 284.

LUDWIG, D., and CABLE, R. M. 1933. The effect of alternating temperatures on the pupal development of *Drosophila melanogaster*, Meigen, Physiol. Zoöl., **6**: 493–508. Pp. 160, 161, 162.

LUDWIG, D., and LANDSMAN, H. M. 1937. The effect of different relative humidities on survival and metamorphosis of the Japanese beetle (*Popillia japonica*, Newman), Physiol. Zoöl., **10**: 171–79. P. 255.

LUTZ, F. E. 1932. Experiments with Orthoptera concerning diurnal rhythm, Am. Mus. Novit., No. 550. P. 322.

LUYET, B. J., and GEHENIO, M. P. 1940. Life and death at low temperatures. Normandy, Mo.: Biodynamica. ("Monographs on General Physiology," No. 1.) Pp. 77, 190.

MCCABE, T. T., and BLANCHARD, B. D. 1950. Three species of *Peromyscus*. Santa Barbara, Calif.: Rood. P. 519.

MACAN, T. T., and WORTHINGTON, E. B., 1951. Life in lakes and rivers. London: Collins. P. 522.

MCCLURE, H. E. 1943. Ecology and management of the mourning dove *Zenaidura macroura*, Research Bull. Iowa Agr. Exper. Sta., No. 310. P. 544.

MCCRACKEN, I. 1909. Heredity of the race characters univoltinism and bivoltinism in the silkworm (*Bombyx mori*), J. Exper. Zoöl., **7**: 747–64. P. 78.

MCDONOGH, R. S. 1939. The habitat distribution and dispersal of the psychid moth *Luffia ferchaultella* in England and Wales, J. Anim. Ecol., **8**: 10–28. Pp. 515, 516, 517.

MACFADYEN, A. 1949. Population ecology, Scient. Prog., **147**: 532–43. P. 444.

MACKENZIE, J. M. D. 1952. Fluctuations in the numbers of British tetraonids, J. Anim. Ecol., **21**: 128–53. P. 641.

MACKERRAS, I. M., and MACKERRAS, M. J. 1944. Sheep blowfly investigations. The attractiveness of sheep for *Lucilia cuprina*, Bull. Counc. Scient. & Indust. Research Australia, No. 181. P. 455.

MACKERRAS, M. J. 1933. Observations on the life-histories, nutritional requirements and fecundity of blowflies, Bull. Ent. Research, **24**: 353–62. P. 507.

MACLAGAN, D. S. 1932*a*. An ecological study of the "lucerne flea" (*Smynthurus viridis*, Linn.). I, Bull. Ent. Research, **23**: 101–45. P. 276.

———. 1932*b*. The effect of population density upon rate of reproduction, with special reference to insects, Proc. Roy. Soc. London, B, **111**: 437–54. Pp. 372, 381.

MACLEOD, J. 1934. *Ixodes ricinus* in relation to its physical environment: the influence of climate on development, Parasitology, **26**: 282–305. Pp. 282, 534.

———. 1935. *Ixodes ricinus* in relation to its physical environment. II. The factors governing survival and activity, *ibid.*, **27**: 123–44. Pp. 277, 280.

MCLEOD, J. H. 1937. Further notes on the parasites of aphids, Ann. Rep. Ent. Soc. Ontario, **67**: 63–64. P. 476.

MACLULICH, D. A. 1937. Fluctuations in the numbers of the varying hare (*Lepus americanus*), Univ. Toronto Stud. Biol., No. 43. Pp. 123, 484.

MAERCKS, H. 1933. Der Einfluss von Temperatur und Luftfeuchtigkeit auf die Embryonalentwicklung der Mehlmottenschlupfwespe *Habrobracon juglandis* Ashmead, Arb. biol. Abt. (Anst.—Reichsanst.), Berlin, **20**: 347–90. P. 265.

MAIL, G. A. 1930. Winter soil temperatures and their relation to subterranean insect survival, J. Agr. Research, **41**: 571–92. P. 526.

———. 1932. Winter temperature gradients as a factor in insect survival, J. Econ. Ent., **25**: 1049–53. P. 526.

MAIL, G. A., and SALT, R. W. 1933. Temperature as a possible limiting factor in the northern spread of the Colorado potato beetle, J. Econ. Ent., **26**: 1068–75. Pp. 526, 527.

MANTON, S. M., and HEATLEY, N. G. 1937. VI. Studies on the Onychophora. II. The feeding, digestion, excretion, and food storage of *Peripatopsis*, Phil. Trans. Roy. Soc. London, B, **227**: 411–64. P. 238.

MANTON, S. M., and RAMSAY, J. A. 1937. Studies on the Onychophora. III. The control of water loss in *Peripatopsis*, J. Exper. Biol., **14**: 470–72. P. 240.

MARCHAL, P. 1936. Recherches sur la biologie et le développement des Hyménoptères parasites: les Trichogrammes, Ann. epiphyt. phytogenet., **2**: 447–550. Pp. 58, 59, 73.

MARCOVITCH, S. 1924. The migration of the Aphidae and the appearance of the sexual forms as affected by the relative length of daily light exposure, J. Agr. Research, **27**: 513–22. P. 305.

MARSHALL, F. H. A. 1942. Exteroceptive factors in sexual periodicity, Biol. Rev., **17**: 68–89. Pp. 290, 291.

MARSHALL, S. M., and STEPHENSON, T. A. 1933. The breeding of reef animals. I. The corals, Scient. Rep. Great. Barrier Reef Exped., **3**: 219–45. P. 304.

MAST, S. O. 1912. Behaviour of fireflies (*Photinus pyralis* ?) with special reference to the problem of orientation, J. Anim. Behavior, **2**: 256–72. P. 307.

MATTHÉE, J. J. 1951. The structure and physiology of the egg of *Locusta pardalina* (Walk.), Scient. Bull. Dept. Agr. South Africa, No. 316. Pp. 68, 227, 228, 229, 230, 233, 248, 249.

MAYNE, B., and YOUNG, M. D. 1938. Antagonism between species of malaria parasites in induced mixed infections, Pub. Health Rep., Washington, **53**: 1289–91. P. 457.

MAYR, E. 1942. Systematics and the origin of species. New York: Columbia University Press. Pp. 665, 691, 694, 695, 708, 709, 710, 711.

———. 1947. Ecological factors in speciation, Evolution, **1**: 263–88. Pp. 90, 713, 717, 718.

———. 1948. The bearing of the new systematics on genetical problems. The nature of species, Adv. Genetics, **2**: 205–37. Pp. 407, 465.

———. 1949*a*. Speciation and systematics. *In:* JEPSEN, G. L.; SIMPSON, G. G.; and MAYR, E. (eds.), Genetics, paleontology and evolution, pp. 281–98. Princeton: Princeton University Press. Pp. 694, 708, 712, 713.

———. 1949*b*. Speciation and selection, Proc. Am. Phil. Soc., **93**: 514–19. P. 706.

———. 1951. Speciation in birds: Progress report on the years 1938–1950, Proc. 10th Internat. Ornithol. Cong., 1950, pp. 91–130. Pp. 694, 713, 715, 718.

———. 1952. Notes on evolutionary literature, Evolution, **6**: 139–44. P. 124.

MELANDER, A. L. 1914. Can insects become resistant to sprays? J. Econ. Ent., **7**: 167–72. P. 701.

MELLANBY, K. 1932*a*. The effect of atmospheric humidity on the metabolism of the fasting mealworm (*Tenebrio molitor* L., Coleoptera), Proc. Roy. Soc. London, B, **111**: 376–90. Pp. 231, 233.

———. 1932*b*. Effects of temperature and humidity on the metabolism of the fasting bed bug (*Cimex lectularius*) Hemiptera, Parasitology, **24**: 419–28. Pp. 233, 256.

———. 1934*a*. The site of loss of water from insects, Proc. Roy. Soc. London, B, **116**: 139–49. P. 239.

———. 1934*b*. Effects of temperature and humidity on the clothes moth larva, *Tineola biselliella* Hum. (Lepidoptera), Ann. Appl. Biol., **21**: 476–82. P. 255.

———. 1935. The evaporation of water from insects, Biol. Rev., **10**: 317–33. Pp. 252, 266.

———. 1936. Humidity and insect metabolism, Nature, **138**: 124–25. P. 233.

———. 1938. Diapause and metamorphosis of the blowfly *Lucilia sericata*, Parasitology, **30**: 392–402. P. 58.

———. 1939. Low temperature and insect activity, Proc. Roy. Soc. London, B, **127**: 473–87. P. 177.

———. 1942. Metabolic water and desiccation, Nature, **150**: 21. P. 233.

MENDES, L. O. T. 1949. Determinação do potencial biótico da "broca do café"—*Hypothenemus hampei,* Ferr.—considerações população (with English summary), Ann. Acad. bras. cienc., **21** : 275–90. P. 385.

MENUSAN, H., JR. 1934. Effects of temperature and humidity on the life processes of *Bruchus obtectus,* Say, Ann. Ent. Soc. Amer., **27** : 515–26. P. 285.

———. 1935. Effect of constant light, temperature and humidity on the rate and total amount of oviposition of the bean weevil, *Bruchus obtectus,* Say, J. Econ. Ent., **28** : 448–53. Pp. 317, 331.

MERRELL, D. J. 1951. Interspecific competition between *Drosophila funebris* and *Drosophila melanogaster,* Am. Nat., **85** : 159–69. P. 433.

MEYER, N. F., and MEYER, Z. A. 1946. The formation of biological forms in *Chrysopa vulgaris* Schr. (Neuroptera, Chrysopidae), Zool. Jahrb., **25** : 115–20. P. 692.

MILLER, A. 1940. Embryonic membranes, yolk cells, and morphogenesis of the stonefly *Pteronarcys proteus* Newman. (Plecoptera: Pteronarcidae), Ann. Ent. Soc. Amer., **33** : 437–77. P. 230.

MILLER, J. M. 1931. High and low lethal temperatures for the western pine beetle, J. Agr. Research, **43** : 303–21. Pp. 197, 529.

MILNE, A. 1943. The comparison of sheep tick populations (*Ixodes ricinus* L.), Ann. Appl. Biol., **30** : 240–50. P. 621.

———. 1944. The ecology of the sheep tick, *Ixodes ricinus* L. Distribution of the tick in relation to soil and vegetation in northern England, Parasitology, **35** : 186–96. Pp. 535, 621.

———. 1945*a*. The ecology of the sheep tick, *Ixodes ricinus* L. The seasonal activity in Britain with particular reference to northern England, *ibid.*, **36** : 142–52. P. 621.

———. 1945*b*. The ecology of the sheep tick, *Ixodes ricinus* L. Host availability and seasonal activity, *ibid.*, pp. 153–57. P. 621.

———. 1945*c*. The control of the sheep tick (*Ixodes ricinus* L.) by treatment of farm stock, Ann. Appl. Biol., **32** : 128–42. P. 621.

———. 1946. The ecology of the sheep tick, *Ixodes ricinus* L. Distribution of the tick on hill pasture, Parasitology, **37** : 75–81. Pp. 535, 621, 623, 624.

———. 1947*a*. The ecology of the sheep tick, *Ixodes ricinus* L. Some further aspects of activity, seasonal and diurnal, *ibid.*, **38** : 27–33. P. 621.

———. 1947*b*. The ecology of the sheep tick, *Ixodes ricinus* L. The infestations of hill sheep, *ibid.*, pp. 34–50. P. 621.

———. 1948. Pasture improvement and the control of sheep tick (*Ixodes ricinus* L.), Ann. Appl. Biol., **35** : 369–78. P. 621.

———. 1949. The ecology of the sheep tick, *Ixodes ricinus* L. Host relationships of the tick. 2. Observations on hill and moorland grazings in northern England, Parasitology, **39** : 173–97. Pp. 621, 626, 627.

———. 1950*a*. The ecology of the sheep tick, *Ixodes ricinus* L. Microhabitat economy of the adult tick, *ibid.*, **40** : 14–34. Pp. 232, 244, 534, 535, 536, 621.

———. 1950*b*. The ecology of the sheep tick, *Ixodes ricinus* (L.). Spatial distribution, *ibid.*, pp. 35–45. Pp. 338, 341, 534, 535, 621.

———. 1951. The seasonal and diurnal activities of individual sheep ticks (*Ixodes ricinus,* L.), *ibid.*, **41** : 189–208. P. 621.

———. 1952. Features of the ecology and control of the sheep tick, *Ixodes ricinus* L., in Britain, Ann. Appl. Biol., **39** : 144–46. Pp. 534, 621.

MITCHELL, D. F., and EPLING, C. 1951. The diurnal periodicity of *Drosophila pseudoobscura* in southern California, Ecology, **32** : 696–708. Pp. 324, 325.

MOHR, C. O. 1943. Cattle droppings as ecological units, Ecol. Monogr., **13** : 275–98. P. 524.

MOORE, C. R., and QUICK, W. J. 1924. The scrotum as a temperature regulator for the testes, Am. J. Physiol., **68** : 70–79. P. 175.

MOORE, H. W. 1948. Variations in fall embryological development in three grasshopper species, Canad. Ent., **80** : 83–88. Pp. 58, 62.

MOORE, J. A. 1939. Temperature tolerance and rates of development in the eggs of Amphibia, Ecology, **20**: 459–78. P. 164.

———. 1940*a*. Adaptive differences in the egg-membranes of frogs, Am. Nat., **74**: 89–93. P. 164.

———. 1940*b*. Stenothermy and eurothermy of animals in relation to habitat, *ibid.*, pp. 188–92. P. 131.

———. 1942. The rôle of temperature in the speciation of frogs, Biol. Symp., **6**: 189–213. P. 164.

———. 1949*a*. Geographic variation of adaptive characters in *Rana pipiens* Schreber, Evolution, **3**: 1–24. P. 685.

———. 1949*b*. Patterns of evolution in the genus *Rana*. *In:* JEPSEN, G. L.; SIMPSON, G. G.; and MAYR, E. (eds.), Genetics, paleontology and evolution, pp. 315–38. Princeton: Princeton University Press. P. 719.

———. 1950. Further studies on *Rana pipiens:* racial hybrids, Am. Nat., **84**: 247–54. Pp. 685, 686.

———. 1952*a*. Competition between *Drosophila melanogaster* and *Drosophila simulans*. I. Population cage experiments, Evolution, **6**: 407–20. P. 423.

———. 1952*b*. Competition between *Drosophila melanogaster* and *Drosophila simulans*. II. The improvement of competitive ability through selection, Proc. Nat. Acad. Sc., Washington, **38**: 813–17. P. 698.

———. 1954. Geographic and genetic isolation in Australian Amphibia, Am. Nat., **88**: 65–74. P. 714.

MORAN, P. A. P. 1949. The statistical analysis of the sunspot and lynx cycles, J. Anim. Ecol., **18**: 115–16. P. 123.

———. 1950. Some remarks on animal population dynamics, Biometrics, **6**: 250–58. P. 123.

———. 1952. The statistical analysis of game-bird records, J. Anim. Ecol., **21**: 154–58. P. 643.

———. 1953. The statistical analysis of the Canadian lynx cycle. I. Structure and prediction, Australian J. Zoöl., **1**: 163 73. Pp. 640, 643, 646.

MOREAU, R. E., and MOREAU, W. M. 1938. The comparative breeding ecology of two species of *Euplectes* (bishop birds) in Usambara, J. Anim. Ecol., **7**: 314–27. P. 481.

MOYLE, V. 1949. Nitrogenous excretion in Chelonian reptiles, Biochem. J., **44**: 581–84. P. 238.

MUESEBECK, C. F. W., and PARKER, D. L. 1933. *Hyposoter disparis*, Viereck, an introduced ichneumonid parasite of the gipsy moth, J. Agr. Research, **46**: 335–47. P. 476.

MUNGER, F. 1948. Reproduction and mortality of California red scales, resistant and non-resistant to hydrocyanic gas as affected by temperature, J. Agr. Research, **76**: 153–63. P. 187.

MURIE, A. 1940. Ecology of the coyote in the Yellowstone, Bull. Fauna Nat. Parks U.S., No. 4. P. 481.

———. 1944. The wolves of Mt. McKinley, *ibid.*, No. 5. Pp. 481, 482.

MYERS, K. 1952. Oviposition and mating behaviour of the Queensland fruit-fly, *Dacus* (*Strumeta*) *tryoni* (Frogg.) and the *Solanum* fruit-fly, *Dacus* (*Strumeta*) *cacuminatus* (Hering), Australian J. Scient. Research, B, **5**: 264–81. P. 329.

NAGEL, R. H., and SHEPARD, H. H. 1934. The lethal effect of low temperatures on the various stages of the confused flour beetle, J. Agr. Research, **48**: 1009–16. Pp. 184, 186.

NAKAMURA, N. 1941. The effect of light on the regeneration in *Syncoryne nipponica*, Jap. J. Zoöl., **9**: 185–90. P. 332.

NASH, T. A. M. 1930. A contribution to our knowledge of the bionomics of *Glossina morsitans*, Bull. Ent. Research, **21**: 201–56. P. 629.

———. 1933*a*. The ecology of *Glossina morsitans*, Westw., and two possible methods for its destruction. Part I, *ibid.*, **24**: 107–57. P. 629.

———. 1933*b*. The ecology of *Glossina morsitans*, Westw., and two possible methods for its destruction. Part II, *ibid.*, pp. 163–95. P. 629.

———. 1937. Climate, the vital factor in the ecology of *Glossina*, *ibid.*, **28**: 75–127. P. 629.

———. 1948. Tsetse flies in British West Africa. London: H. M. Stationery Office. Pp. 629, 635.

NEEDHAM, J. 1935. Problems of nitrogen catabolism in invertebrates. II. Correlation between uricotelic metabolism and habitat in the phylum Mollusca, Biochem. J., **29**: 238–51. Pp. 236, 237.

NEISWANDER, C. R. 1947. Variations in the seasonal history of the European cornborer in Ohio, J. Econ. Ent., **40**: 407–12. P. 688.

NEL, R. G. 1936. The utilisation of low temperatures in the sterilisation of deciduous fruit infested with the immature stages of the Mediterranean fruit fly *Ceratitis capitata*, Scient. Bull. Dept. Agr. South Africa, No. 155. P. 186.

NICHOLSON, A. J. 1933. The balance of animal populations, J. Anim. Ecol., **2**: 132–78. Pp. 19, 20, 406, 418, 443, 444.

———. 1947. Fluctuation of animal populations, Rep. 26th Meet. A.N.Z.A.A.S. Perth, 1947. Pp. 9, 15, 19, 20, 87.

———. 1950. Population oscillations caused by competition for food, Nature, **165**: 476–77. Pp. 360, 450, 497, 498, 606.

NICHOLSON, A. J., and BAILEY, V. A. 1935. The balance of animal populations. Part I, Proc. Zoöl. Soc. London, 1935, pp. 551–98. Pp. 19, 413, 415, 416, 417, 418, 442, 443.

NIELSEN, E. T., and GREVE, H. 1950. Studies on the swarming habits of mosquitoes and other Nematocera, Bull. Ent. Research, **41**: 227–58. P. 330.

NORRIS, M. J. 1933. Contributions towards the study of insect fertility. II. Experiments on the factors influencing fertility in *Ephestia kühniella* Z. (Lepidoptera, Phycitidae), Proc. Zoöl. Soc. London, 1933, pp. 903–34. P. 173.

ODUM, E. P. 1953. Fundamentals of ecology. Philadelphia, W. B. Saunders Co. P. 525.

OOSTHUIZEN, M. J. 1939. The body temperature of *Samia cecropia* Linn. (Lepidoptera, Saturniidae) as influenced by muscular activity, J. Ent. Soc. South Africa, **2**: 63–73. P. 256.

ORTON, J. H. 1926. On lunar periodicity in spawning of normally grown Falmouth oysters (*O. edulis*) in 1925, with a comparison of the spawning capacity of normally grown and dumpy oysters, J. Marine Biol. A., United Kingdom (N.S.), **14**: 199–225. P. 303.

ORWELL, G. 1946. The English language, J. &. Proc. Australian Chem. Inst., September, 1946. P. 11.

PARK, O. 1935. Studies in nocturnal ecology. III. Recording apparatus and further analysis of activity rhythm, Ecology, **16**: 152–63. P. 322.

———. 1940. Nocturnalism—the development of a problem, Ecol. Monogr., **10**: 485–536. Pp. 321, 322.

PARK, O., and KELLER, J. G. 1932. Studies in nocturnal ecology. II. Preliminary analysis of activity rhythm in nocturnal forest insects, Ecology, **13**: 335–46. Pp. 322, 323.

PARK, O., and SEJBA, O. 1935. Studies in nocturnal ecology. IV. *Megalodacne heros*, Ecology, **16**: 164–72. P. 322.

PARK, T. 1933. Studies in population physiology. II. Factors regulating initial growth of *Tribolium confusum* populations, J. Exper. Zoöl., **65**: 17–42. Pp. 334, 368, 372.

———. 1938. Studies in population physiology. VIII. The effect of larval population density on the post-embryonic development of the flour beetle *Tribolium confusum* Duval, *ibid.*, **79**: 51–70. Pp. 368, 374, 378.

———. 1948. Experimental studies of interspecies competition. I. Competition between populations of the flour beetles *Tribolium confusum* Duval and *Tribolium castaneum* Herbst, Ecol. Monogr., **18**: 265–308. Pp. 357, 358, 423, 426.

———. 1954. Competition: an experimental and statistical study. *In:* Statistics and mathematics in biology. Ames: Iowa State College Press. Pp. 394, 395, 423, 427.

PARK, T.; GREGG, E. V.; and LUTHERMAN, C. Z. 1941. Studies in population physiology. X. Interspecific competition in populations of granary beetles, Physiol. Zoöl., **14**: 395–430. Pp. 360, 423.

PARK, T.; MILLER, E. V.; and LUTHERMAN, C. Z. 1939. Studies in population physiology.

———. 1935. The influence of external factors on the spawning date and migration of the common frog, *Rana temporaria temporaria* Linn., *ibid.*, 1935, pp. 49–98. P. 523.

SAVELY, H. E. 1939. Ecological relations of certain animals in dead pine and oak logs, Ecol. Monogr., **9**: 321–85. P. 524.

SAVORY, T. H. 1930. Environmental differences of spiders of the genus *Zilla*, J. Ecol., **18**: 384–85. P. 212.

SCHALLEK, W. 1942*a*. The vertical migration of the copepod *Acartia tonsa* under controlled illumination, Biol. Bull., **82**: 112–26. Pp. 305, 312.

———. 1942*b*. Some mechanisms controlling locomotor activity in the crayfish, J. Exper. Zoöl., **91**: 155–66. P. 322.

SCHMIDT, P. 1918. Anabiosis of the earthworm, J. Exper. Zoöl., **27**: 57–72. Pp. 209, 259, 260.

SCHMIEDER, R. G. 1933. The polymorphic forms of *Melittobia chalybii*, Ashm. and the determining factors involved in their production (Hymenoptera: Eulophidae), Biol. Bull. Woods Hole, **65**: 338–54. Pp. 59, 74.

———. 1939. The significance of two types of larvae in *Sphecophaga burra* (Cresson) and the factors conditioning them (Hymenoptera: Ichneumonidae). Ent. News, **50**: 125–31. Pp. 59, 74, 75.

SCHNEIDER, F. 1939. Ein Vergleich von Urwald und Monokulture in Bezug auf ihre Gefährdung durch phytophage Insekten, auf Grund einiger Beobachtungen an der Ostküste von Sumatra, Schweiz. Ztschr. f. Forstwesen., 1939, Nos. 2–3. P. 475.

SCHNETTER, M. 1951. Veränderungen der genetischen Konstitution in natürlichen Populationen der polymorphen Bänderschnecken, Verh. deutsch. zool. Gesellsch., **50**: 192–206. P. 672.

SCHULMAN, E. 1948. Dendrochronology in north-eastern Utah, Tree-ring Bull., **15**: 2–14. P. 644.

SCHULZ, F. N. 1930. Zur Biologie des Mehlwurms (*Tenebrio molitor*). I. Der Wasserhaushalt, Biochem. Ztschr., **227**: 340–53. P. 234.

SCHWERDTFEGER, F. 1941. Über die Ursachen des Massenwechsels der Insekten, Ztschr. f. angew. Ent., **28**: 254–303. Pp. 448, 459.

———. 1944. Weitere Beobachtungen zur Lebensweise der Kiefernschonungsgespintstblattewespe, *Acantholyda erythrocephala*, *ibid.*, **30**: 364–71. P. 58.

SCOTT, T. G. 1947. Comparative analysis of red fox feeding trends on two central Iowa areas, Research Bull. Iowa Agr. Exper. Sta., No. 353. P. 538.

SCOTT, W. N. 1936. An experimental analysis of the factors governing the hour of emergence of adult insects from their pupae, Tr. Roy. Ent. Soc. London, **85**: 303–29. P. 328.

SEAMANS, H. L. 1923. Forecasting outbreaks of the pale western cutworm in Alberta, Canad. Ent., **55**: 51–53. Pp. 273, 594, 596.

SERFLING, R. E. 1952. Historical review of epidemic theory, Human Biol., **24**: 145–66. P. 483.

SERVENTY, D. L., and WHITTELL, H. M. 1951. Handbook of the birds of Western Australia (with the exception of the Kimberley division). 2d ed. Perth, West Australia: Paterson Brokensha. P. 491.

SHELFORD, V. E. 1907. Preliminary note on the distribution of the tiger beetles (*Cicindela*) and its relation to plant succession, Biol. Bull., **14**: 9–14. P. 461.

———. 1918. A comparison of the responses of animals in gradients of environmental factors with particular reference to the method of reaction of representatives of the various groups from Protozoa to mammals, Science, **48**: 225–30. Pp. 211, 218.

———. 1927. An experimental study of the relations of the codling moth to weather and climate, Bull. Illinois Nat. Hist. Surv., **16**: 311–440. Pp. 162, 166, 170, 192.

———. 1929. Laboratory and field ecology. The responses of animals as indicators of correct working methods. Baltimore: Williams & Wilkins. P. 56.

SHELFORD, V. E., and MARTIN, L. 1946. Reactions of young birds to atmospheric humidity, J. Wildlife Management, **10**: 66–68. P. 211.

SHEPPARD, P. M. 1951*a*. Fluctuations in the selective value of certain phenotypes in the polymorphic land snail, *Cepaea nemoralis* (L.), Heredity, **5**: 125–34. Pp. 699, 700.

———. 1951*b*. A quantitative study of two populations of the moth *Panaxia dominula* (L.), *ibid.*, pp. 349–78. Pp. 90, 91, 709.

———. 1952. Natural selection in two colonies of the polymorphic land snail *Cepaea nemoralis, ibid.*, **6**: 233–38. P. 710.

Simmonds, F. J. 1948. The influence of maternal physiology on the incidence of diapause, Phil. Trans. Roy. Soc. London, B, **233**: 385–414. Pp. 58, 59.

Simpson, C. B. 1903. The codling moth, Bull. U.S. Div. Ent., No. 41. P. 165.

Simpson, G. G. 1949. The meaning of evolution: a study of the history of life and of its significance for man. New Haven, Conn.; Yale University Press. P. 404.

———. 1952. How many species? Evolution, **6**: 342. P. 664.

Skoblo, I. S. 1941. Ripening and fertility of females of *Habrobracon brevicornis* as dependent on development conditions during pre-imaginal phases, Compt. rend. Acad. Sci. (N. S.), U.S.S.R., **33**: 424–26. Pp. 58, 72.

Slifer, E. H. 1932. Insect development. IV. External morphology of grasshopper embryos of known age and with a known temperature history, J. Morphol., **53**: 1–22. Pp. 62, 65.

———. 1938. The formation and structure of a special water-absorbing area in the membranes covering the grasshopper egg, Quart. J. Micr. Sc., **80**: 437–57. P. 227.

———. 1946. The effects of xylol and other solvents on diapause in the grasshopper egg, together with a possible explanation for the action of these agents, J. Exper. Zoöl., **102**: 333–56. Pp. 80, 230.

Slobodkin, L. B. 1954. Population dynamics in *Daphnia obtusa* Kurz., Ecol. Monogr., **24**: 69–88. P. 360.

Smith, C. N., and Cole, M. M. 1941. Effect of length of day on the activity and hibernation of the American dog-tick *Dermacentor variabilis* (Acarina), Ann. Ent. Soc. Amer., **34**: 426–31. Pp. 58, 299.

Smith, H. S. 1935. The rôle of biotic factors in the determination of population densities, J. Econ. Ent., **28**: 873–98. Pp. 15, 17, 18, 19, 23, 649, 663.

———. 1939. Insect populations in relation to biological control, Ecol. Monogr., **9**: 311–20. Pp. 116, 473, 474, 475.

———. 1941. Racial segregation in insect populations and its significance in applied entomology, J. Econ. Ent., **34**: 1–13. Pp. 58, 688, 689, 701.

Smith, H. S., and Armitage, H. M. 1931. The biological control of mealybugs attacking citrus, Bull. Calif. Agr. Exper. Sta., No. 509. Pp. 472, 474.

Snyman, A., 1949. The influence of population densities on the development and oviposition of *Plodia interpunctella* Hubn. (Lepidoptera), J. Ent. Soc. South Africa, **12**: 137–71. Pp. 335, 371, 380.

Solomon, M. E. 1949. The natural control of animal populations, J. Anim. Ecol., **18**: 1–35. Pp. 13, 15, 648.

Spencer, W. 1928. Wanderings in wild Australia. 2 vols. London: Macmillan & Co., Ltd. P. 224.

Speyer, W. 1938. Ueber das Vorkommen von Lokalrassen des kleinen Frostspanners (*Cheimatobia brumata* L.). Ein Beitrag zur Verständnis der verschiedenen Flugzeiten, Arb. physiol. u. angew. Ent. Berlin-Dahlem, **5**: 50–76. P. 713.

Spiess, E. B. 1950. Experimental populations of *Drosophila persimilis* from an altitudinal transect of the Sierra Nevada, Evolution, **4**: 14–33. P. 683.

Spieth, H. T., and Hsu, T. C. 1950. The influence of light on the mating behaviour of seven species of the *Drosophila melanogaster* species group, Evolution, **4**: 316–25. P. 329.

Spooner, G. M. 1933. Observations on the reactions of marine plankton to light, J. Marine Biol. A., United Kingdom, **19**: 385–438. P. 309.

Squire, F. A. 1937. A theory of diapause in *Platyedra gossypiella*, Saund., Trop. Agr., **14**: 299–301. P. 75.

———. 1939. Observations on the larval diapause of the pink bollworm, *Platyedra gossypiella*, Saund., Bull. Ent. Research, **30**: 475–81. Pp. 75, 76.

———. 1940. On the nature and origin of the diapause in *Platyedra gossypiella*, Saund., *ibid.*, **31**: 1–6. Pp. 59, 75, 80.

STABLER, H. P. 1913. Red spider spread by wind, Monthly Bull. Calif. Comm. Hort., **2**: 777–80. P. 104.

STANLEY, J. 1932. A mathematical theory of the growth of populations of the flour beetle, *Tribolium confusum*, Duv., Canad. J. Research, **6**: 632–71. P. 385.

STEELE, H. V. 1941. Some observations on the embryonic development of *Austroicetes cruciata* Sauss. (Acrididae) in the field, Tr. Roy. Soc. South Australia, **63**: 329–32. Pp. 60, 64, 82.

STEENBURGH, W. E. VAN, and BOYCE, H. R. 1937. The simultaneous propagation of *Macrocentrus ancylivorus* and *Ascogaster carpocapsae* on the peach moth *Laspeyresia molesta*, Busck.: a study in multiple parasitism, Ann. Rep. Ent. Soc. Ontario, 1937. P. 476.

STEGGERDA, F. R. 1937. Comparative study of water metabolism in amphibians injected with pituitrin, Proc. Soc. Exper. Biol. & Med., **36**: 103–6. P. 226.

STEINER, L. F. 1940. Codling moth flight habits and their influence on results of experiments, J. Econ. Ent., **33**: 436–40. P. 101.

STEINER, L. F.; ARNOLD, C. H.; and SUMMERLAND, S. A. 1944. The development of large differences in the ability of local codling moths to enter sprayed apples, J. Econ. Ent., **37**: 29–33. P. 702.

STEINHAUS, A. 1946. Insect microbiology: an account of the microbes associated with insects and ticks, with special reference to the biologic relationships involved. New York: Comstock Pub. Co. P. 275.

STEINHAUS, E. A. 1949. Principles of insect pathology. New York: McGraw-Hill Book Co. Pp. 482, 483, 484, 485.

STEPHENS, G. C.; FINGERMAN, M.; and BROWN, F. A., JR. 1952. A non-birefringent mechanism for orientation to polarized light in arthropods, Anat. Rec., **113**: 559–60. P. 321.

STEVENSON-HAMILTON, J. 1937. South African Eden: from Sabi game reserve to Kruger national park. London: Cassell & Co., Ltd. Pp. 502, 664.

STIRRETT, G. M. 1931. Preliminary observations on the winter mortality of the larvae of the European cornborer in Ontario in 1930, 61st Ann. Rep. Ent. Soc. Ontario, pp. 48–52. P. 197.

———. 1938. A study of the flight, oviposition, and establishment periods in the life cycle of the European cornborer *Pyrausta nubilalis* Hbn., and the physical factors affecting them. II. The flight of the European cornborer. Annual cycle of flight. Flight to light trap, Scient. Agr., **18**: 462–84. Pp. 174, 197.

SVÄRDSON, G. 1949. Competition and habitat selection in birds, Oikos, **1**: 159–74. P. 458.

SWAN, D. C., and BROWNING, T. O. 1949. The black field-cricket (*Gryllulus servillei* Saussure) in South Australia, J. Dept. Agr. South Australia, **52**: 323–27. P. 61.

SWEETMAN, H. L. 1929. Precipitation and irrigation as factors in the distribution of the Mexican bean beetle *Epilachna corrupta* Muls., Ecology, **10**: 228–44. P. 195.

———. 1931. Preliminary report on the physical ecology of certain *Phyllophaga* (Scarabaeidae), *ibid.*, **12**: 401–22. P. 233.

———. 1935. Successful examples of biological control of pest insects and plants, Bull. Ent. Research, **26**: 373–77. P. 467.

———. 1936. The biological control of insects. Ithaca: Comstock Pub. Co. P. 336.

———. 1938. Physical ecology of the firebrat, *Thermobia domestica*, Packard, Ecol. Monogr., **8**: 285–311. P. 204, 285.

———. 1939. Responses of the silverfish *Lepisma saccharina*, L. to its physical environment, J. Econ. Ent., **32**: 698–700. P. 204.

TANSLEY, A. G. 1935. The use and abuse of vegetational concepts and terms, Ecology, **16**: 284–307. P. 14.

TATE, P. 1932. The larval instars of *Orthopodomyia pulchripalpis*, Parasitology, **24**: 111–20. Pp. 59, 299.

TAYLOR, C. B. 1936. Short-period fluctuations in the numbers of bacterial cells in soil, Proc. Roy. Soc. London, B, **119**: 269–95. P. 434.

TAYLOR, T. H. C. 1937. The biological control of an insect in Fiji. An account of the coconut leaf mining beetle and its parasite complex. London: Imperial Institute of Entomology. Pp. 336, 487.

TETLEY, J. H. 1937. The distribution of nematodes in the small intestine of the sheep, New Zealand J. Scient. Tech., **18**: 805–17. P. 619.

THEODOR, O. 1934. Observations on the hibernation of *Phlebotomus papatasii*, Bull. Ent. Research, **25**: 459–72. P. 59.

THERON, P. P. A. 1943. Experiments on terminating the diapause in larvae of codlin moth, J. Ent. Soc. South Africa, **6**: 114–23. P. 58.

THOMAS, E. L., and SHEPARD, H. H. 1940. The influence of temperature, moisture, and food on the development and survival of the saw-toothed grain beetle, J. Agr. Research, **60**: 605–15. P. 186.

THOMPSON, D. H. 1941. The fish production of inland streams and lakes. *In:* A symposium on hydrobiology. Madison, Wis.: University of Wisconsin Press. P. 479.

THOMPSON, V., and BODINE, J. H. 1936. Oxygen consumption and rates of dehydration of grasshopper eggs (Orthoptera), Physiol. Zoöl., **9**: 455–70. P. 249.

THOMPSON, W. R. 1928. Host selection in *Pyrausta nubilalis*, Hubn., Bull. Ent. Research, **18**: 359–64. P. 16.

———. 1939. Biological control and the theories of the interactions of populations, Parasitology, **31**: 299–388. P. 405.

———. 1943. A catalogue of the parasites and predators of insect pests. London: Imperial Parasite Service. P. 466.

THOMSON, R. C. M. 1938. The reactions of mosquitoes to temperature and humidity, Bull. Ent. Research, **29**: 125–40. Pp. 219, 221.

THORPE, W. H. 1930. Biological races in insects and allied groups, Biol. Rev., **5**: 177–212. Pp. 691, 693.

———. 1931. Biological races in insects and their significance in evolution, Ann. Appl. Biol., **18**: 406–14. P. 691.

———. 1939. Further studies on pre-imaginal olfactory conditioning in insects, Proc. Roy. Soc. London, B, **127**: 424–33. P. 691.

———. 1945. The evolutionary significance of habitat selection, J. Anim. Ecol., **14**: 67–70. Pp. 693, 716.

THORPE, W. H., and CANDLE, H. B. 1938. A study of the olfactory responses of insect parasites to the food plants of their host, Parasitology, **30**: 523–28. P. 511.

THORPE, W. H.; CROMBIE, A. C.; HILL, R.; and DARRAH, J. H. 1947. The behaviour of wireworms in response to chemical stimulation, J. Exper. Biol., **23**: 234–66. P. 510.

THORPE, W. H., and JONES, F. G. W. 1937. Olfactory conditioning in a parasitic insect and its relation to the problem of host selection, Proc. Roy. Soc. London, B, **124**: 56–81. P. 691.

THORSON, T., and SVIHLA, A. 1943. Correlation of the habitats of amphibians with their ability to survive the loss of body water, Ecology, **24**: 374–81. Pp. 251, 260, 261, 262.

TILLYARD, R. J., and SEDDON, H. R. 1933. The sheep blowfly problem in Australia. Report No. 1 by the Joint Blowfly Committee. ("Pamphlets of the Council for Scientific and Industrial Research, Australia," No. 37.) P. 714.

TIMOFEEFF-RESSOVSKY, N. W. 1934. Über die Vitalität einiger Genmutationen und ihrer Kombinationen bei *Drosophila funebris* und ihre Abhängigkeit vom "genotypischen" und vom äusseren Mileau. Z. i. A. V., **66**: 319–44. P. 676.

———. 1940. Zur Analyse des Polymorphismus bei *Adalia bipunctata*, L., Biol. Zentralbl., **60**: 130–37. P. 671.

TITSCHAK, E. 1926. Untersuchung über das Wachstum den Nahrungsverbrauch und Eierzeugung. II. *Tineola biselliella*, Hum. Gleichzeitig ein Beitrag zur Klärung der Insektenhäutung, Ztschr. f. wissensch. Zool., **128**: 509–69. P. 81.

TOPLEY, W. W. C. 1926. The Milroy lectures on "experimental epidemiology," Lancet, **210**: 477, 531, 645. P. 483.

TOTHILL, J. D.; TAYLOR, T. H. C.; and PAYNE, R. W. 1930. The coconut moth in Fiji. London: Imperial Bureau of Entomology. P. 486.

TOYAMA, K. 1913. Maternal inheritance and Mendelism, J. Genetics, **2**: 351–405. P. 77.

TRUMBLE, H. C. 1937. The climatic control of agriculture in South Australia, Tr. Roy. Soc. South Australia, **61**: 41–62. P. 577.

TULESCHKOV, K. 1935. Ueber Ursachen der Ueberwinterung der *Lymantria dispar*, L., *L. monacha*, und anderen Lymantriiden im Eistadium, Ztschr. f. angew. Ent., **22**: 97–117. Pp. 58, 59, 62, 70.

TURČEK, F. J. 1949. The bird populations in some deciduous forests during a gipsy moth outbreak, Bull. Inst. Forestry Research Czechoslovakia, 1949, pp. 108–31. P. 537.

UDA, H. 1923. On "maternal inheritance," Genetics, **8**: 322–35. P. 77.

UDVARDY, M. D. F. 1951. The significance of interspecific competition in bird life, Oikos, **3**: 98–123. Pp. 452, 458, 537.

ULLYETT, G. C. 1945. Oviposition by *Ephestia kühniella* Zell., J. Ent. Soc. South Africa, **8**: 53–59. Pp. 335, 435.

———. 1947. Mortality factors in populations of *Plutella maculipennis*, Curtiss (Tineidae Lepidoptera) and their relation to the problem of control, Mem. Dept. Agr. South Africa, **2**: 77–202. Pp. 503, 614, 617.

———. 1949*a*. Distribution of progeny by *Chelonus texanus*, Cress. (Hymenoptera: Braconidae), Canad. Ent., **81**: 25–44. P. 435.

———. 1949*b*. Distribution of progeny by *Cryptus inornatus* Pratt (Hymenoptera: Ichneumonidae), *ibid.*, pp. 285–99; **82**: 1–11. Pp. 435, 436.

———. 1950. Competition for food and allied phenomena in sheep-blowfly populations, Phil. Trans. Roy. Soc. London, B, **234**: 77–174. Pp. 371, 374, 379, 405, 419, 435, 450, 497.

UMEYA, Y. 1926. Experiments on ovarian transplantation and blood transfusion in silkworms (*Bombyx mori*, L.) with special reference to alternation of voltinism, Bull. Sericult. Exper. Sta. Chosen, Japan, **1**: 1–27. P. 78.

USTINOV, A. A. 1932. A review of pests of tobacco in Abkhazia observed in 1931, Sukhum. Abkhazk. tabachn. Zonal. Sta., 1931. P. 689.

UTIDA, S. 1941*a*. Studies on experimental population of the Azuki bean weevil, *Callosobruchus chinensis* (L.). I. The effect of population density on the progeny populations, Mem. Coll. Agr. Kyoto, **48**: 1–30. Pp. 374, 376.

———. 1941*b*. Studies on experimental population of the Azuki bean weevil, *Callosobruchus chinensis* (L.). II. The effect of population density on progeny populations under different conditions of atmospheric moisture, *ibid.*, **49**: 1–20. P. 376.

———. 1941*c*. Studies on experimental population of the Azuki bean weevil, *Callosobruchus chinensis* (L.). III. The effect of population density upon the mortalities of different stages in the life cycle, *ibid.*, pp. 21–42. P. 371.

———. 1941*d*. Studies on experimental population of the Azuki bean weevil, *Callosobruchus chinensis* (L.). IV. Analysis of density effect with respect to fecundity and fertility of eggs, *ibid.*, **51**: 1–26. P. 374.

———. 1941*e*. Studies on experimental population of the Azuki bean weevil, *Callosobruchus chinensis* (L.). V. Trend of population density at the equilibrium position, *ibid.*, pp. 27–34. Pp. 360, 384.

———. 1942. Studies on experimental population of the Azuki bean weevil, *Callosobruchus chinensis* (L.). VII. Analysis of the density effect in the preimaginal stage, *ibid.*, **53**: 19–31. Pp. 376, 384.

UVAROV, B. P. 1928. Insect nutrition and metabolism, Tr. Ent. Soc. London, **76**: 255–343. Pp. 111, 123.

———. 1931. Insects and climate, *ibid.*, **79**: 1–247. Pp. 177, 530.

VANCE, A. M. 1942. Studies on the prevalence of the European cornborer in the east north central states, Circ. U.S. Bur. Ent., No. 649. Pp. 687, 688.

———. 1949. Some physiological relationships of the female European cornborer moth in controlled environments, J. Econ. Ent., **42** : 474–84. P. 172.

Varley, G. C. 1941. On the search for hosts and the egg distribution of some chalcid parasites of the knapweed gall-fly, Parasitology, **33** : 47–66. P. 435.

———. 1947. The natural control of population balance in the knapweed gall-fly (*Urophora jaceana*), J. Anim. Ecol., **16** : 139–87. Pp. 19, 20, 434, 443, 444.

———. 1949. Special review: population changes in German forest pests, *ibid.*, **18** : 117–22. Pp. 448, 449.

Varley, G. C., and Butler, C. G. 1933. The acceleration of development of insects by parasitism, Parasitology, **25** : 263–68. P. 58.

Vaughan, T. W. 1919. Corals and the formation of coral reefs, Ann. Rep. Smithsonian Inst., 1917, pp. 189–238. P. 232.

Verwey, J. 1930. Depth of coral reefs and penetration of light, with notes on oxygen consumption of corals, Proc. 4th Pacific Scient. Cong. Java, **2A** : 277–99. P. 332.

Volpe, E. P. 1952. Physiological evidence for natural hybridization of *Bufo americanus* and *Bufo fowleri*, Evolution, **6** : 393–406. P. 719.

Volterra, V. 1926. Variazioni e fluttuazioni del numero d'individui in specie animali conviventi, Mem. Acad. Lincei Roma, **2** : 31–113. Pp. 17, 406, 407, 413, 414.

———. 1931. Leçons sur la théorie mathématique de la lutte pour la vie. ("Cah. scient.," Vol. 7.) Paris: Gauthiers-Villars. Pp. 347, 348, 406, 407, 408, 413, 416.

Volterra, V., and D'Ancona, U. 1935. Les associations biologiques au point de vue mathématique. ("Actualités scient. et indust.," Vol. **243**.) Paris: Hermann. & Cie P. 441.

Vowles, D. M. 1950. Sensitivity of ants to polarised light, Nature, **165** : 282–83. P. 321.

Wadley, F. M. 1931. Ecology of *Toxoptera graminium*, Ann. Ent. Soc. Amer., **24** : 325–95. P. 172.

Wadley, F. M., and Wolfenbarger, D. O. 1944. Regression of insect density on distance from centre of dispersion as shown by a study of the smaller European elm bark beetle, J. Agr. Research, **69** : 299–308. P. 102.

Wagner, R. P. 1944. The nutrition of *Drosophila mulleri* and *D. aldrichi*. Growth of the larvae on a cactus extract and the microorganisms found in cactus, Pub. Univ. Texas, **4445** : 104–28. P. 403.

Waitzinger, L. A. 1933. Effect of various illuminations upon the silkworm during its growth, Lingnan Scient. J., Canton, **12** : 349–65, 507–40, Suppl., pp. 165–72. P. 331.

Wakeland, C. 1934. The influence of forested areas on pea field populations of *Bruchus pisorum*, L. (Coleoptera, Bruchidae), J. Econ. Ent., **27** : 981–86. P. 102.

Walker, M. G. 1940. Notes on the distribution of *Cephus pygmaeus* Linn. and of its parasite *Collyria calcitrator*, Bull. Ent. Research, **30** : 551–73. P. 476.

Wallace, B., and Dobzhansky, Th. 1946. Experiments on sexual isolation in *Drosophila*. VIII. Influence of light on the mating behaviour of *Drosophila subobscura*, *Drosophila persimilis* and *Drosophila pseudoobscura*, Proc. Nat. Acad. Sc., Washington, **32** : 226–34. P. 329.

Wallace, B., and Madden, C. 1953. The frequencies of sub- and supervitals in experimental populations of *Drosophila melanogaster*, Genetics, **38** : 456–70. P. 703.

Waloff, N. 1941. The mechanism of humidity reactions of terrestrial isopods, J. Exper. Biol., **18** : 115–35. Pp. 211, 213, 221.

———. 1948. Development of *Ephestia elutella*, Hb. (Lep. Phycitidae) on some natural foods, Bull. Ent. Research, **39** : 117–30. Pp. 76, 77, 506.

———. 1949. Observations on larvae of *Ephestia elutella*, Hubner (Lep. Phycitidae) during diapause, Tr. Roy. Ent. Soc. London, **100** : 147–59. Pp. 58, 689.

Waloff, Z. 1946*a*. A long-range migration of the desert locust from southern Morocco to Portugal, with an analysis of concurrent weather conditions, Proc. Roy. Ent. Soc. London, A, **21** : 81–84. P. 121.

———. 1946*b*. Seasonal breeding and migrations of the desert locust (*Schistocerca gregaria* Forsk.) in eastern Africa, Mem. Anti-locust Research Centre London, Vol. **1**. Pp. 121, 122.

WALOFF, Z., and RAINEY, R. C. 1951. Field studies on factors affecting the displacements of desert locust swarms in eastern Africa, Bull. Anti-locust Research Centre London, **9**: 1–50. P. 121.

WATERHOUSE, D. F. 1947. The relative importance of live sheep and of carrion as breeding grounds for the Australian sheep blowfly *Lucilia cuprina*, Bull. Counc. Scient. & Indust. Research Australia, No. 217. Pp. 450, 454, 476, 525.

WATERMAN, T. H. 1950. A light polarization analyzer in the compound eye of *Limulus*, Science, **111**: 252–54. P. 321.

WAY, M. J.; HOPKINS, B.; and SMITH, P. M. 1949. Photoperiodism and diapause in insects, Nature, **164**; 615. Pp. 58, 299.

WEISS, H. B. 1943. Color perception in insects, J. Econ. Ent., **36**: 1–17. P. 320.

WEISS, H. B.; SORACI, F. A.; and MCCOY, E. E., JR., 1941. Notes on the behavior of certain insects to different wave-lengths of light, J. New York Ent. Soc., **49**: 1–20, 149–59. P. 320.

———. 1942. The behavior of certain insects to various wave-lengths of light, *ibid.*, **50**: 1–34. P. 320.

WELLINGTON, W. G. 1948. The light reactions of the spruce budworm, *Choristoneura fumiferana* Clemens (Lepidoptera: Tortricidae), Canad. Ent., **80**: 56–82. Pp. 289, 308, 309, 312, 313, 314, 318.

———. 1949*a*. Temperature measurements in ecological entomology, Nature, **163**: 614–15. Pp. 133, 134.

———. 1949*b*. The effects of temperature and moisture upon the behaviour of the spruce budworm, *Choristoneura fumiferana* Clemens (Lepidoptera: Tortricidae). I. The relative importance of graded temperatures and rates of evaporation in producing aggregations of larvae, Scient. Agr., **29**: 201–15. Pp. 132, 133, 212.

———. 1949*c*. The effects of temperature and moisture upon the behaviour of the spruce budworm, *Choristoneura fumiferana* Clemens (Lepidoptera: Tortricidae). II. The responses of larvae to gradients of evaporation, *ibid.*, pp. 216–29. Pp. 133, 212, 213.

———. 1950. Variations in the silk-spinning and locomotor activity of larvae of the spruce budworm *Choristoneura fumiferana* (Clem.), at different rates of evaporation, Tr. Roy. Soc. Canada, **44**: 89–101. Pp. 113, 221, 222, 529.

———. 1952. Air-mass climatology of Ontario north of Lake Huron and Lake Superior before outbreaks of the spruce budworm, *Choristoneura fumiferana* (Clem.), and the forest tent caterpillar, *Malacosoma disstria* Hbn. (Lepidoptera: Tortricidae; Lasiocampidae), Canad. J. Zoöl., **30**: 114–27. P. 600.

WELLINGTON, W. G.; FETTES, J. J.; TURNER, K. B.; and BELYEA, R. M. 1950. Physical and biological indicators of the development of outbreaks of the spruce budworm, *Choristoneura fumiferana* (Clem.) (Lepidoptera: Tortricidae), Canad. J. Research, D, **28**: 308–31. Pp. 274, 275, 600, 601, 603.

WELLINGTON, W. G., and HENSON, W. R. 1947. Notes on the effects of physical factors on the spruce budworm, *Choristoneura fumiferana* (Clem.), Canad. Ent., **79**: 168–70, 195. P. 113.

WELLINGTON, W. G.; SULLIVAN, C. R.; and GREEN, G. W. 1951. Polarized light and body temperature level as orientation factors in the light reactions of some Hymenopterous and Lepidopterous larvae, Canad. J. Zoöl., **29**: 339–51. Pp. 315, 321.

WELSH, J. H. 1938. Diurnal rhythms, Quart. Rev. Biol., **13**: 123–39. P. 324.

WEST, A. S. 1947 The California flatheaded borer (*Melanophila californica*, van Dyke) in ponderosa pine stands of north eastern California, Canad. J. Research, D, **25**: 97–118. P. 508.

WHEELER, J. F. G. 1937. Further observations on lunar periodicity, J. Linn. Soc. London (Zoöl.), **40**: 325–45. P. 303.

WHEELER, W. M. 1893. A contribution to insect embryology, J. Morphol., **8**: 1–160. Pp. 56, 230.

WHITE, H. C. 1937. Local feeding of kingfishers and mergansers, J. Biol. Board Canada, **3**: 323–38. P. 479.

———. 1939. Bird control to increase the Margaree River salmon, Bull. Fish. Research Board, Canada, No. 58. P. 479.

WHITEHEAD, A. N. 1926. Science and the modern world. Cambridge: at the University Press. P. 13.

———. 1947. Essays in science and philosophy. New York: Philosophical Library. P. 12.

WIGGLESWORTH, V. B. 1931*a*. A curious effect of desiccation on the bed bug (*Cimex lectularius*); Proc. Roy. Ent. Soc. London, **6**: 25–26. P. 231.

———. 1931*b*. The physiology of excretion in a blood-sucking insect, *Rhodnius prolixus* (Hemiptera, Reduviidae), J. Exper. Biol., **8**: 411–51. P. 236.

———. 1934. The physiology of ecdysis in *Rhodnius prolixus*. II. Factors controlling moulting and "metamorphosis," Quart. J. Micr. Sc., **77**: 191–221. Pp. 57, 81.

———. 1936. The funetion of the corpus allatum in the growth and reproduction of *Rhodnius prolixus* (Hemiptera), *ibid.*, **79**: 91–121. P. 80.

———. 1937. Wound healing in an insect (*Rhodnius prolixus*, Hemiptera), J. Exper. Biol., **14**: 364–81. P. 242.

———. 1939. The principles of insect physiology. London: Methuen & Co., Ltd. P. 177.

———. 1941. The sensory physiology of the human louse, *Pediculus humanus corporis*, de Geer (Anoplura), Parasitology, **33**: 67–109. Pp. 211, 218, 221.

———. 1945. Transpiration through the cuticle of insects, J. Exper. Biol., **21**: 97–114. Pp. 242, 243.

———. 1947. The epicuticle in an insect, *Rhodnius prolixus* (Hemiptera), Proc. Roy. Soc. London, B, **134**: 163–81. P. 242.

———. 1948*a*. The structure and deposition of the cuticle in the adult mealworm, *Tenebrio molitor* L. (Coleoptera), Quart. J. Micr. Sc., **89**: 197–217. P. 242.

———. 1948*b*. The insect cuticle, Biol. Rev., **23**; 408–51. P. 242.

———. 1948*c*. The insect as a medium for the study of physiology, Proc. Roy. Soc. London, B, **135**: 430–46. Pp. 57, 80.

———. 1950. The principles of insect physiology. London: Methuen & Co., Ltd. P. 320.

WILKES, A. 1942. The influence of selection on the preferendum of a chalcid (*Microplectron fuscipennis* Zett.) and its significance in the biological control of an insect pest, Proc. Roy. Soc. London, B, **130**: 400–415. Pp. 132, 676.

———. 1946. The introduction of insect parasites of the spruce budworm into eastern Canada, Canad. Ent., **78**: 82–86. P. 59.

WILLIAMS, C. B. 1930. The migration of butterflies. Edinburgh: Oliver & Boyd. P. 117.

———. 1939. An analysis of four years' captures of insects in the light trap. I. General survey; sex proportion; phenology and time of flight, Tr. Roy. Ent. Soc. London, **89**: 79–132. P. 144.

———. 1940. An analysis of four years' captures of insects in a light trap. II. The effect of weather conditions on insect activity; and the estimation and forecasting of changes in the insect population, *ibid.*, **90**: 227–306. P. 144.

———. 1944. Some applications of the logarithmic series and the index of diversity to ecological problems, J. Ecology, **32**: 1–44. P. 650.

———. 1947. The generic relations of species in small ecological communities, J. Anim. Ecol., **16**: 11–18. P. 465.

WILLIAMS, C. B.; COCKBILL, G. F.; and GIBBS, M. A. 1942. Studies in the migration of Lepidoptera, Tr. Roy. Ent. Soc. London, **92**: 101–283. Pp. 91, 117.

WILLIAMS, C. M. 1946. Physiology of insect diapause: the rôle of the brain in the production and termination of pupal dormancy in the giant silkworm *Platysamia cecropia*, Biol. Bull., **90**: 234–43. Pp. 59, 63, 66, 80.

———. 1947. The endocrinology of diapause, Bull. Biol. Franç. Belg., **33**: 52–56. P. 80.

———. 1948. Extrinsic control of morphogenesis as illustrated in the metamorphosis of insects, Growth Symp., **12**: 61–74. P. 80.

WILLIS, E. R., and ROTH, L. M. 1950. Humidity reactions of *Tribolium castaneum* (Herbst.), J. Exper. Zoöl., **115**: 561–87. P. 211.

Wilson, F. 1946. Interaction of insect infestation, temperature, and moisture content in bulk-depot wheat, Bull. Counc. Scient. & Indust. Research Australia, No. 209. P. 455.

Wiltshire, E. P. 1946. Studies in the geography of Lepidoptera. III. Some Middle East migrants, their phenology and ecology, Tr. Roy. Ent. Soc. London, **96**: 162–82. P. 123.

Winsor, C. P. 1932. The Gompertz curve as a growth curve, Proc. Nat. Acad. Sc., Washington, **18**: 1–8. P. 385.

Wishart, G. 1947. Further observations on the changes taking place in the cornborer population in western Ontario, Canad. Ent., **79**: 81–83. P. 688.

Wolf, E. 1930. Die Aktivität der japanischen Tanzmaus und ihre rhythmische Verteilung, Ztschr. f. vergl. Physiol., **11**: 321–44. P. 322.

Wolfenbarger, D. O. 1946. Dispersion of small organisms. Distance dispersion rates of bacteria, spores, seed, pollen and insects: incidence rates of diseases and injuries, Am. Midland Nat., **35**: 1–152. P. 102.

Wood, J. G. 1937. The vegetation of South Australia. Adelaide: British Scientific Guild (South Australia Branch). P. 494.

Woodroffe, G. E., and Southgate, B. J. 1950. Birds' nests as a source of domestic pests, Proc. Zoöl. Soc. London, **121**: 55–62. P. 525.

Wright, S. 1931. Evolution in Mendelian populations, Genetics, **16**: 97–159. Pp. 670, 695.

———. 1937. The distribution of gene frequencies in populations, Proc. Nat. Acad. Sc., Washington, **23**: 307–20. P. 695.

———. 1940. Breeding structure of populations in relation to speciation, Am. Nat., **74**: 232–48. Pp. 88, 696.

———. 1945. Tempo and mode in evolution: a critical review, Ecology, **26**: 415–19. P. 670.

———. 1948. On the rôles of directed and random changes in gene frequency in the genetics of populations, Evolution, **2**: 279–94. Pp. 670, 672, 705.

———. 1949. Adaptation and selection. *In:* Jepsen, G. L.; Simpson, G. G.; and Mayr, E. (eds.), Genetics, paleontology and evolution, pp. 365–89. Princeton: Princeton University Press. Pp. 670, 703, 708.

Wright, S., and Dobzhansky, Th. 1946. Genetics of natural populations. XII. Experimental reproduction of some of the changes caused by natural selection in certain populations of *Drosophila pseudoobscura*, Genetics, **31**: 125–56. Pp. 676, 677, 678.

Yeates, N. T. M. 1949. The breeding season of the sheep with particular reference to its modification by artificial means using light, J. Agr. Sc., **39**: 1–43. Pp. 291, 292.

Zolotarev, E. Kh. 1938. Summer and autumn rearing of *Antherea pernyi* as influencing the diapause of the pupa (in Russian with an English summary), Zool. Zhur., **17**: 622–33. P. 58.

Zumpt, F. 1940. Die Verbreitung der *Glossina palpalis*—Subspezies in Belgisch—Kongogebiet, Rev. Zoöl. & Bot. Africa, **33**: 136–49. P. 629.

Zwolfer, W. 1931. Studien zur Oekologie und Epidemiologie der Insekten. I. Die Kieferneule, *Panolis flammea*, Schiff., Ztschr. f. angew. Ent., **17**: 475–562. P. 279.

Index

[Printed in U.S.A.]